Carnivorous Plants

Carnivorous Plants

Physiology, ecology, and evolution

EDITED BY

Aaron M. Ellison

Harvard University, USA

Lubomír Adamec

Institute of Botany of the Czech Academy of Sciences, Czech Republic

Great Clarendon Street, Oxford, OX2 6DP,
United Kingdom

Oxford University Press is a department of the University of Oxford.
It furthers the University's objective of excellence in research, scholarship,
and education by publishing worldwide. Oxford is a registered trade mark of
Oxford University Press in the UK and in certain other countries

First published 2018
Reprinted with corrections 2018
First published in paperback 2019

Impression: 2

Published in the United States of America by Oxford University Press
198 Madison Avenue, New York, NY 10016, United States of America

British Library Cataloguing in Publication Data
Data available

Library of Congress Cataloging in Publication Data
Data available

ISBN 978–0–19–877984–1 (Hbk.)
ISBN 978–0–19–883372–7 (Pbk.)

DOI 10.1093/oso/9780198779841.001.0001

Printed in Great Britain by
Bell & Bain Ltd., Glasgow

To Flossie—fierce carnivore

and

to all those past, present, and future who study and conserve carnivorous plants

Contents

Preface

Carnivorous plants have been, and continue to be, an endless source of fascination for scientists and nonscientists alike. Scientists learn much about general theoretical principles from exceptions to the rules, and the idea of "killer" plants is a staple trope of science-fiction novels, horror films, and Broadway shows. The two editors of, and the dozens of contributors to, this volume, came to our studies of carnivorous plants through a range of different routes, but all of us share an abiding curiosity in the form, function, and evolution of what Charles Darwin called "the most wonderful plants in the world."

Although a walk through any bookstore anywhere in the world will turn up a seemingly endless array of "coffee-table" books replete with lavish—and occasionally lurid—photographs of carnivorous plants, there are surprisingly few synthetic, scientific monographs about them. Charles Darwin (1875) wrote the first one; Francis Lloyd (1942) wrote the second; and Barrie Juniper, Richard Robins, and Daniel Joel (1989) wrote the third. All of these authors focused attention on the unique traits of carnivorous plants, provided detailed descriptions of carnivorous organs, and unequivocal demonstrations that some plants indeed eat animals.

Both of us editors have, at intervals, been approached by colleagues and publishers to synthesize the current state-of-the-science of carnivorous plants, but in the three decades since Juniper et al. was published, the range of studies and the number of papers published about carnivorous plants have grown far too large for any single person or small group of people to handle. This new synthesis of research on carnivorous plants, therefore, is a joint effort of many individuals and research teams working around the world, and we hope that it is as inspiring to future students of carnivorous plants as Darwin (1875), Lloyd (1942), and especially Juniper et al. (1989) have been to all of us.

In contrast to Darwin, Lloyd, and Juniper et al., we focus attention on the evolution and systematics (Chapters 3–11), physiology (Chapters 12–19), and ecology (Chapters 2, 21–26) of carnivorous plants, and use the existing data to explore how knowledge about carnivorous plants can inform our understanding of noncarnivorous ones—and *vice versa*—and provide opportunities for drug development and biomimetic technologies (Chapter 20). The application of molecular systematics in the last two decades has firmly laid to rest the hypothesis that carnivorous plants form a single evolutionary lineage (Croizat 1960). New species within established carnivorous genera continue to be described (Chapters 4–9), and carnivorous species in two other genera—*Paepalanthus* and *Philcoxia*—have been discovered (Chapter 10). The set of traits that defines the "carnivorous syndrome" has narrowed (Chapter 1) as it has become clear that carnivory not only can be gained in evolutionary lineages but lost in them, too (Chapter 3).

New details about how traps work have been revealed by a combination of technological advances (e.g., high-speed cinematography) and elegant experiments (Chapters 12–15). Biochemical analysis of digestive physiology (Chapter 16) in combination with genomic analysis (Chapter 11) have given rise to new insights into how carnivorous traits evolved from mechanisms that defend plants from herbivores and pathogens. The cost-benefit model for the evolution of carnivory in plants (Givnish et al. 1984), which has inspired decades of research into carnivorous plant ecophysiology, has been refined further (Chapter 18). Syntheses of experimental results clearly illustrate the importance of nutrient budgets for understanding the evolution of carnivory by plants while highlighting key differences between terrestrial and aquatic carnivorous plants (Ellison and Adamec 2011; Chapters 17, 19), differences that can be surprisingly large.

Studies of carnivorous plants also have moved out of the laboratory and into the field. Observational data collected in the last thirty years have amplified dramatically findings of previous generations of naturalists, while manipulative experiments have led to new discoveries about the myriad direct and indirect interactions that contribute to successful carnivorous plant lineages (Chapters 21–26). Carnivorous plants and their symbionts have been developed as model ecological systems for studies of food-web dynamics (Chapters 23–25), ecological tipping points (Sirota et al. 2013), and evolutionary responses to climatic change (Chapter 24).

Finally, Juniper et al. devoted but a single, brief paragraph at the end of their Preface to the conservation and preservation of carnivorous plants. In the last three decades, overcollection and poaching of carnivorous plants from the field, loss of unique and irreplaceable habitats, and climatic change have become existential threats to many carnivorous plant species (Chapters 4–10, 27, 28). It is no longer enough only to provide data to authorities and agencies in the hopes that they will be convinced by the data to conserve and protect carnivorous plants. As the scientists who know them best, we have an obligation to get involved in actions to ensure the integrity of their habitats and their continued persistence in them (Ellison 2016).

Aaron M. Ellison & Lubomír Adamec
Petersham, Massachusetts,
USA & Třeboň, Czech Republic
May 2017

Editors and contributors

***Aaron M. Ellison* (lead editor; lead author: Chapters 1, 29; co-author: Chapter 28)** is the Senior Research Fellow in Ecology at Harvard University's Harvard Forest. He received his Ph.D. (1986) in evolutionary ecology from Brown University. He has been studying the physiology, ecology, and evolution of carnivorous plants since the mid-1990s. He is the author of >170 peer-reviewed papers (including dozens on carnivorous plants); co-author of the books *A Primer of Ecological Statistics* (with N.J. Gotelli; Sinauer Associates, 2004/2012 [2nd edition]); *A Field Guide to the Ants of New England* (with N.J. Gotelli, E.J. Farnsworth, and G.D. Alpert; Yale University Press, 2012); and *Vanishing Point* (BookBaby, 2017); and co-editor of the book *Stepping in the Same River Twice: Replication in Biological Research* (with A. Shavit; Yale University Press, 2017).

***Lubomír Adamec* (co-editor; lead author: Chapters 17, 19; co-author: Chapters 1, 29)** is the Senior Research Scientist in the Section of Plant Ecology of the Institute of Botany, Czech Academy of Sciences at Třeboň, Czech Republic. He received his M.Sc. (1982) in plant physiology from the Faculty of Sciences, Charles University, Prague, Czechoslovakia, and has been working in the Institute of Botany since 1986. Since 1990, his research has focused on the ecophysiology of carnivorous plants, especially nutrition dynamics and aquatic carnivorous species. He is the author of ≈100 peer-reviewed papers, including ≈80 on carnivorous plants. He is the curator of the world's largest professional collection of aquatic carnivorous plants.

***Victor Albert* (co-lead author: Chapter 11)** is Empire Innovation Professor of Biological Sciences at the University at Buffalo. He is the author of >125 peer-reviewed papers and chapters, including 11 on carnivorous plants. He also has co-edited two books on systematics, phylogenetics, and genomics. He has researched the molecular evolutionary biology of carnivorous plants since the 1990s, particularly on the bladderwort family (Lentibulariaceae), for which his most recent work was presentation and analysis of the genome sequence of *Utricularia gibba*.

***Bruce Anderson* (co-author: Chapters 10, 26)** is a Professor at the University of Stellenbosch, South Africa. He received his Ph.D. (2003) on digestive mutualisms between arthropods and *Roridula* from the University of Cape Town, South Africa. His primary interest is in the evolutionary ecology of plant–animal interactions, especially the pollination of plants. He has published >40 peer-reviewed papers, many of which have focused on arthropod interactions with carnivorous plants.

***Paulo Cesar Baleeiro* (co-author: Chapter 8)** is a Ph.D. student at the University of São Paulo (USP), Brazil. He received his M.Sc. (2011) in biological science (botany) from the Federal University of Rio de Janeiro, Brazil, for research on the diversity of *Utricularia* in the Brazilian Cerrado. His current research interests include systematic studies of *Utricularia* using morphometric and molecular data.

***Jiří Bárta* (co-author: Chapter 25)** is a Research Scientist at the University of South Bohemia in České Budějovice, Czech Republic. Throughout his career, he has been interested in the diversity and functioning of complex microbial communities in terrestrial and aquatic ecosystems. His specific focus is on interaction of bacterial and fungal communities. He is interested in molecular-based methods (metagenomics, metatranscriptomics) and he is author or co-author of more than 20 peer-reviewed papers.

***Ulrike Bauer* (lead author: Chapter 15; co-author: Chapters 12, 14, 21)** is a Royal Society University Research Fellow at the University of Bristol, UK. She became interested in the biomechanics and ecology of prey capture by carnivorous pitcher plants during her undergraduate studies at the University of Würzburg, Germany, in 2004, and has continued studying the interactions of pitcher plants and animals ever since. She received her Ph.D. (2010) from the Department of Zoology, University of Cambridge, UK, and has held two independent Early Career Fellowships prior to her current position. She has published work on the evolution of trapping strategies and the environmental and developmental influences on trap function, and discovered previously unknown trapping mechanisms.

***Leonora S. Bittleston* (lead author: Chapter 23)** is a James S. McDonnell Foundation Postdoctoral Fellow in Complex Systems at the Massachusetts Institute of Technology in Cambridge, Massachusetts, USA. She studies the ecology and evolution of multispecies interactions and community dynamics. She received her Ph.D. in Organismic and Evolutionary Biology from Harvard University, where her research focused on the insect and microbial communities of pitcher plants. Before beginning her Ph.D., Leonora studied symbiont-host interactions at the University of California, Berkeley, and researched the effects of fungal endophytes on the behavior of leaf cutter ants at the Smithsonian Tropical Research Institute in Panama. She has collected hundreds of microcosm samples from *Nepenthes* pitchers in locations ranging from Singapore to the remote Maliau Basin of Sabah's Lost World.

***Jakub Borovec* (co-author: Chapter 25)** is a Research Scientist at the Biology Centre of the Czech Academy of Sciences and a lecturer at the University of South Bohemia, České Budějovice, Czech Republic. His research focuses on biogeochemical cycling within aquatic ecosystems, especially at the sediment-water interface, and includes studies of the interactions between phosphorus, metals, and microorganisms in lentic systems. He is the author or co-author of >20 peer-reviewed limnological papers.

***William Bradshaw* (co-author: Chapter 24)** is Professor of Biology in the Institute of Ecology and Evolution at the University of Oregon's Laboratory of Evolutionary Genetics in Eugene, Oregon, USA. He studied animal physiology and ecology at the University of Michigan and the Bio Labs at Harvard University, working mainly on photoperiodism and seasonal development of midges and mosquitoes. Since 1971, he has been working on the genetics, evolution, physiology, population biology, community ecology, systematics, and biogeography of container-breeding mosquitoes, especially the pitcher-plant mosquito, *Wyeomyia smithii*, and the tree-hole mosquitoes of eastern and western North America. Most recently, he has been studying genetic response to recent rapid climate change and the genomics of blood feeding and seasonal adaptations in *W. smithii*.

***J. Stephen Brewer* (lead author: Chapter 2)** is a Professor of Biology at the University of Mississippi, USA. He has been studying the ecology of carnivorous plants, including competitive interactions with non-carnivorous plants, since 1998. He has published over 50 peer-reviewed scientific articles, including 14 related to carnivorous plants. His current research on carnivorous plants focuses on carnivory as an adaptation to hypoxic soils and substrates and implications for the distribution of carnivorous plants. He also studies how disturbances such as fire mediate competitive interactions between carnivorous and non-carnivorous plants and promote species coexistence in hyper-diverse wet pine savanna of the southeastern USA.

***Lijin Chin* (co-author: Chapter 26)** is currently pursuing her Master's degree in the economics of natural resource and environmental management at Cranfield University, UK. She was awarded first-class honors for her Bachelor's Degree in environmental management from Monash University (Malaysian campus) in 2010, with specific research work done on plant–mammal interactions of *Nepenthes rajah* and *Nepenthes macrophylla*, and resource partitioning by Bornean pitcher plant species. Both studies have been published in international peer-reviewed journals. Lijin's current research interests have broadened to consider the interrelations between ecosystem services, economics, and development.

Charles Clarke **(lead author: Chapters 5, 27; co-author: Chapter 26)** is an Adjunct Senior Research Associate at the Australian Tropical Herbarium. He has been studying the ecology and diversity of *Nepenthes* pitcher plants since the late 1980s. His current research interests include interactions between *Nepenthes* and vertebrates, and the conservation of carnivorous plant species.

Adam Cross **(lead author: Chapters 10, 22; co-author: Chapters 4, 27)** is a Research Fellow at the Department of Environment and Agriculture, Curtin University, and Kings Park and Botanic Garden, Perth, Western Australia. He received his Ph.D. in botany from the University of Western Australia (2014), and is the author of a number of peer-reviewed papers on seed biology and community ecology in aquatic and carnivorous plants, and the book *Aldrovanda: The Waterwheel Plant* (Redfern Natural History Publications, 2012). He has strong research interests in seed biology, community ecology and phytosociology, freshwater aquatic ecosystems, carnivorous plants, and conservation biology.

Douglas W. Darnowski **(lead author: Chapter 21; co-author: Chapter 20)** received his Ph.D. from Cornell and currently is an Associate Professor of Biology at Indiana University Southeast, a regional campus of Indiana University. His research for ≈20 years has dealt with several genera of carnivorous and protocarnivorous plants, most commonly *Drosera, Utricularia, Genlisea,* and *Stylidium*. He is the author or co-author of ≈12 papers and one book on these plants.

Art Davis **(co-author: Chapter 22)** is Professor in the Department of Biology at the University of Saskatchewan, Saskatoon, Canada. His research career has led to >50 peer-reviewed articles devoted to plant–insect interactions at the cytological, developmental and ecological levels, especially in pollination biology. His interest in the potential pollinator–prey conflict of carnivorous plants began while supervising the study of *Drosera* spp. by Gillian Murza.

Bruno Di Giusto **(co-author: Chapter 12)** is an Assistant Professor in the Journalism and Mass Communication Department of the International College, Ming Chuan University at Taipei, Taiwan. He received his Ph.D. (2001) in evolutionary biology and ecology from the Faculty of Sciences and Techniques in Montpellier, France. His studies on the ecology and evolution of plant–insect interactions include herbivory, community ecology, chemical ecology, and coevolution. Since 2004, his research has focused on the ecology of *Nepenthes*, especially the use of volatiles commonly found in generalist flowers to lure potential pollinators and prey in the vicinity of the pitchers.

Kingsley Dixon **(co-author: Chapter 4)** is a Professor and Research Scientist specializing in conservation biology, ecology, and restoration ecology at Curtin University in Western Australia. He also holds a personal Chair at The University of Western Australia in the School of Plant Biology, and established the major research facility at Kings Park and Botanic Garden where he also holds a Visiting Professorship. He started his research career working on the ecology, biology, and nutrition of *Drosera*, and made the first demonstration of nitrogen uptake from invertebrate capture by the Western Australian endemic *Drosera erythrorhiza*.

Kimberly Farr **(co-author: Chapter 11)** is a Ph.D. student in the Department of Biological Sciences at the University at Buffalo, working in the laboratory of Professor Victor A. Albert. She focuses her research on the rapid evolution of angiosperms. Her current research focuses on the convergent evolution of carnivory in plants by analyzing genomic structure using comparative genomic approaches.

Matthew Fitzpatrick **(lead author: Chapter 28)** is an Associate Professor at the University of Maryland Center for Environmental Science, Appalachian Laboratory, and the author of >40 scientific publications. His research focuses on the development and application of spatial modeling methods for mapping patterns of biodiversity and predicting the dynamics of range expansion. He has long been fascinated with carnivorous plants and has photographed them in their natural habitats across North America and Australia. He maintains a small personal collection of rare and unusual plants, including many carnivorous species.

***Andreas Fleischmann* (lead author: Chapters 3, 4, 6, 7; co-author: Chapters 10, 22)** is Curator of Vascular Plants and Research Scientist in Systematic Botany at the Munich Herbarium, Germany. He has been studying and growing carnivorous plants since his childhood, and has studied them in the field in Africa, South America, and Australia. He received his Ph.D. in systematic botany from the University of Munich in 2011, and has published numerous peer-reviewed papers, notably on the carnivorous plant genera *Drosera*, *Genlisea*, *Utricularia*, and *Heliamphora*.

***Kenji Fukushima* (co-author: Chapter 11)** is a Research Scholar (funded by Japan Society for the Promotion of Science) at the University of Colorado, USA. He received his Ph.D. from SOKENDAI, Japan's Graduate University for Advanced Studies, for research on the evolution of leaf morphology and digestive enzymes in carnivorous plants. His current research focuses on genomic convergence of independently derived carnivorous plant lineages.

***Robert Gibson* (co-author: Chapter 4)** is an Ecologist in the New South Wales Office of Environment and Heritage, Australia. He received his Ph.D. (2008) in *Drosera* taxonomy from the School of Environmental & Rural Science, University of New England, Armidale. He is the author of 5 peer-reviewed papers, and more than 50 general journal articles, primarily on the genus *Drosera*, and continues to study this carnivorous plant genus.

***Thomas J. Givnish* (lead author: Chapter 18; co-author: Chapters 3, 10)** is the Henry Allan Gleason Professor of Botany and Professor in the Nelson Institute for Environmental Studies at the University of Wisconsin-Madison, USA. He received his Ph.D. (1976) in ecology from Princeton University. He has studied the ecology and evolution of carnivorous plants since the mid-1980s, discovered carnivory in bromeliads, traced its evolutionary roots in the genus *Brocchinia*, and developed a cost-benefit model for the evolution of botanical carnivory. He and one of his students, Will Sparks, are currently investigating apparent paradoxes in the ecology of carnivory. He is the author of >120 peer-reviewed papers, and editor of *On the Economy of Plant Form and Function* (Cambridge University Press, 1985) and *Molecular Evolution and Adaptive Radiation* (Cambridge University Press, 1997).

***Paulo M. Gonella* (co-author: Chapter 4)** is a Ph.D. student at the University of São Paulo (USP), Brazil. He received his M.Sc. (2012) in botany from USP, with a dissertation on taxonomy of Brazilian *Drosera* species. His current research interests are in *Drosera* taxonomy, phylogeny and evolution, especially in the species-rich *campos rupestres* of central Brazil. His research recently resulted in the discovery, via Facebook, of *Drosera magnifica*, the largest known New World sundew.

***Melinda Greenwood* (co-author: Chapter 26)** is a freelance ecologist living in Northern New South Wales, Australia. She received her honors degree (2011) in Environmental Science from Monash University, Melbourne, Australia. Her findings on the interactions between *Nepenthes* and a small mammal community in the highlands of Borneo have been published internationally, and her research footage was featured in the David Attenborough documentary "Kingdom of Plants" (2011).

***Cástor Guisande* (co-author: Chapter 8)** is a Professor of Ecology at the University of Vigo, Spain. He received his Ph.D. (1989) from the University of Seville, Spain, and did postdoctoral research at the Max Planck Institute for Evolutionary Biology in Plön, Germany and at Royal Holloway College and Bedford New College, UK. He is the author of >100 peer-reviewed papers, nearly all of them in the field of aquatic ecology, with several papers on aquatic carnivorous plants and a review about bladderworts.

***Mitsuyasu Hasebe* (co-author: Chapter 11)** is a Professor of Evolutionary Biology at Japan's National Institute for Basic Biology. He received his Ph.D. from the University of Tokyo on molecular phylogeny of land plants. He has published papers on the phylogeny of Droseraceae and the mechanisms of *Sarracenia* pitcher development. He was a core member of the moss *Physcomitrella* and the lycopod *Selaginella* genome consortia and he is currently working on the *Cephalotus follicularis* and *Drosera spatulata* genome projects. He has been awarded the Society Prize from the Botanical Society of Japan and the Pelton Award from the Botanical Society of America.

***Luis Herrera-Estrella* (co-author: Chapter 11)** received a Ph.D. in plant molecular biology from the State University of Ghent, Belgium. He is currently a Professor of Plant Biology and the Director of Langebio. He has made important contributions to the field of plant molecular biology, especially in the study of gene regulation and in the development of gene transfer methods. His current research is focused on the development of transgenic plants better adapted to marginal soils and on genomics analysis of endemic plants from México.

***Christine Holzapfel* (co-author: Chapter 24)** is a Senior Research Associate in Biology in the Institute of Ecology and Evolution at the University of Oregon's Laboratory of Evolutionary Genetics. She studied systematic biology and evolution at the University of Michigan and the Gray Herbarium at Harvard University, working mainly on the evolution of plants and grasshoppers in the Canary Islands, and co-authored the first flora of that archipelago. Since 1971, she has been working on the genetics, evolution, physiology, population biology, community ecology, systematics, and biogeography of container-breeding mosquitoes, especially the pitcher-plant mosquito, *Wyeomyia smithii*, and the tree-hole mosquitoes of eastern and western North America. Most recently, she has been studying genetic response to recent rapid climate change and the genomics of blood feeding and seasonal adaptations in *W. smithii*.

***John Horner* (lead author: Chapter 12; co-author: Chapters 21, 22)** is a Professor in the Department of Biology at Texas Christian University. He received his Ph.D. from the University of New México, Albuquerque, USA. His research has focused primarily on the ecology and evolution of plant–insect interactions. He is the author or co-author of >30 peer-reviewed articles in ecology and evolution, including several on carnivorous plant ecology.

***Steven J. Hunter* (co-author: Chapter 18)** is a Ph.D. student at the University of Wisconsin-Madison, USA. He holds B.Sc. degrees in mathematics (Massachusetts Institute of Technology, Cambridge, Massachusetts, USA) and plant sciences (Australian National University, Canberra, Australia). His current research focuses on systematic studies of the Hawaiian lobeliads.

***Enrique Ibarra-Laclette* (co-author: Chapter 11)** obtained a Ph.D. in Sciences from the Research and Advanced Studies Center of the IPN (CINVESTAV). He is currently a Research Scientist at the Instituto de Ecología A. C. (INECOL). His specialty is the genomics and transcriptomics of non-model plant species, and recently, of native species of México with importance at the pharmacological level. He has published >26 international publications in the area of cell, molecular, and genetic biology, and genomics and transcriptomics of plants.

***Reinhard Jetter* (co-author: 15)** is Professor in the Departments of Botany and Chemistry at the University of British Columbia. He obtained his Ph.D. (1993) in plant biology from the University of Kaiserslautern, Germany, and has worked at the Institute of Biological Chemistry in Pullman, Washington, and the Julius-von-Sachs-Institute at the University of Würzburg, Germany. His research on the chemical composition, morphology, and biosynthesis of plant surface waxes has led to over 80 peer-reviewed publications and four book chapters. He has special interests in the ecological and physiological functions of plant surface compounds, including prey capture in carnivorous plants.

***Richard Jobson* (lead author: Chapter 8)** is a Research Scientist at the National Herbarium of New South Wales, Royal Botanic Gardens & Domain Trust, Australia. He received his Ph.D. (2003) in plant systematics and evolution from the School of Botany at The University of Queensland, Australia. He did postdoctoral research at several universities, including Cornell (Ithaca, New York, USA), Michigan (Ann Arbor, Michigan, USA), Montpellier (France), and Oxford (UK). At the National Herbarium of New South Wales, he is the curator of the Droseraceae and Lentibulariaceae. He is the author of >35 peer-reviewed papers, including 18 on carnivorous plants. His research involves the ecology and evolution of grasses and aquatic plants, especially the carnivorous plant family Lentibulariaceae.

***Daniel M. Joel* (Foreword)** is an Emeritus Senior Researcher and former Head of the Department of Weed Research of the Agricultural Research Organization at the Volcani Institute, Israel. He first specialized in plant anatomy, ultrastructure, and physiology, and since 1976 has been studying interactions between plants and other organisms, with a focus on carnivorous and parasitic plants. His studies on carnivorous plants started in 1979 at the Botany School, Oxford University (UK). He has authored >200 scientific publications, co-authored the book *The Carnivorous Plants* (with B.E. Juniper and R.J. Robins; Academic Press, 1989), and was the senior editor of the book *Parasitic Orobanchaceae —Parasitic Mechanisms and Control Strategies* (with J. Gressel and L.J. Musselman; Springer, 2012).

***Andreas Jürgens* (co-author: Chapter 22)** is a Professor of Plant Chemical Ecology at the Technical University of Darmstadt, Germany and Honorary Associate Professor in the School of Life Sciences at the University of KwaZulu-Natal, South Africa. He received his Ph.D. (1998) from the University of Ulm, Germany. His research has focused primarily on the chemical ecology and evolution of pollination systems. As a postdoctoral scholar, he worked in a joint project between Landcare Research and Plant & Food Research on the pollination biology of different *Drosera* species in the Southern Alps in New Zealand. Since then, he has published several articles on pollinator–prey conflict and chemical ecology of carnivorous plants.

***Nick Kalfas* (co-author: Chapter 10)** is a Ph.D. candidate in the Department of Genetics and Evolution at the University of Adelaide, South Australia. His research focuses on the co-evolution and collective ecology of *Cephalotus follicularis* and its associated wingless fly *Badisis ambulans*, involving large-scale genetic studies and investigating the nature of the relationship, fire ecology, and life histories in the system. In his undergraduate studies, he studied the phylogeography of *Cephalotus*, beginning his keen interest in carnivorous plants, population genetics, and the diversity of flora in the southwest of Australia.

***Tianying Lan* (co-author: Chapter 11)** is a Research Scientist in the Department of Biological Sciences at University at Buffalo. She received her Ph.D. from Nankai University, Tianjin, China, where she studied plant sex chromosomal evolution patterns and epigenetic signatures. Her research currently focuses on genome architectures and evolutionary history of various plant and animal species.

***Laurent Legendre* (lead author: Chapter 20)** is a Professor at the University of Lyon (France). He received his Ph.D. (1993) from the Chemistry Department at Purdue University for research into physiological and biochemical aspects of plant–pathogen interactions. Since then, his research has focused on molecular aspects of plant defenses, their elicitors, and cellular aspects of plant intracellular signaling. Recently, he has been studying plant metabolomic changes during interactions of plants with plant growth-promoting rhizobacteria. At the same time, he has maintained a research program on carnivorous plants, publishing on hydroponic and *in vitro* medium formulations, and a phylogenetic study of the genus *Pinguicula*.

***Ildikó Matušíková* (lead author: Chapter 16)** is the Senior Research Scientist in the Department of Ecochemistry and Radiology at the University of Ss. Cyril and Methodius in Trnava, Slovak

Republic. She graduated in biochemistry (1996) and received her Ph.D. (2000) in genetics, both from Comenius University in Bratislava, Slovakia. She studies the biochemistry of, and transcriptomic changes in, stressed plants, has published more than 45 peer-reviewed papers and contributed to the *Handbook of Plant and Crop Stress* (CRC Press, Taylor & Francis Group, 2011). She also does research on the molecular biology of *Drosera* and has published several research papers on the role of hydrolytic enzymes in prey digestion by carnivorous plants. She is an executive editor of *Nova Biotechnologica et Chimica*.

***Marcos Méndez* (co-author: Chapter 21)** is a Senior Lecturer in Biodiversity and Conservation at Rey Juan Carlos University, Spain. He has broad interests in plant evolutionary ecology, from resource allocation and life histories to plant–animal interactions and sexual expression. He is the author of >40 peer-reviewed papers, including some on photosynthesis of carnivorous plants and nutrient stoichiometry in relation to reproduction of carnivorous plants.

***David Merritt* (co-author: Chapter 22)** is a Senior Research Scientist at Kings Park and Botanic Garden, Perth, Western Australia. He received his Ph.D. (2002) in seed science from the University of Western Australia; his research interests include seed storage physiology and longevity, seed dormancy and germination, and promoting the use of seeds for the conservation of biodiversity and the restoration of degraded habitat.

***Thomas E. Miller* (lead author: Chapter 24)** is a Professor in the Department of Biological Science at Florida State University. He studied for his Ph.D. at Kellogg Biological Station and Michigan State University, where he was first introduced to *Sarracenia purpurea*. He began research on inquiline communities in pitcher plants in the mid-1990s after moving to North Florida. His research has progressed from studies of top-down vs. bottom-up control, to studies of succession, then metacommunity studies, to now studying evolution among protozoa in individual pitcher plant leaves. He is also interested in models of evolution in multispecies systems and the dynamics of vegetation in coastal dune systems.

***Jonathan Moran* (lead author: Chapter 26; co-author: Chapter 5)** is an Associate Professor in the School of Environment and Sustainability at Royal Roads University in Victoria, Canada. He received his Ph.D. (1992) from the University of Aberdeen, Scotland, and his research since then has centered on plant–animal interactions. Some of this research has focused on the biology of *Nepenthes* pitcher plants, on which he has authored >20 peer-reviewed papers. His work on *Nepenthes* includes investigations into prey-trapping strategies, mutualism with both vertebrate and invertebrate partners, the nutritional benefits of carnivory, the physiology of nutrient uptake, and the influence of climate on the mechanics of prey capture.

***Ljudmila E. Muravnik* (co-author: Chapter 13)** is the Senior Research Scientist in the Laboratory of Plant Anatomy and Morphology at the Komarov Botanical Institute, Russian Academy of Sciences in Saint Petersburg, Russia. She has been working at the Komarov Botanical Institute since 1974, and she received her Ph.D. (1986) from there, too. Since 1979, her research has focused on the ultrastructural features of carnivorous plant glands relative to their functions. She is the author of ≈50 peer-reviewed papers, including 30 on carnivorous plants.

***Gillian L. Murza* (co-author: Chapter 22)** coordinates undergraduate biology labs at the University of Saskatchewan, Canada. Her M.Sc. (2002) research on the pollinator–prey conflict in *Drosera anglica* was completed at the University of Saskatchewan, Canada. She has published work on the pollinator–prey conflict of *D. anglica* and on the flower structure of related species of sundews as it relates to their reproductive biology.

***Robert F.C. Naczi* (lead author: Chapter 9)** is the Arthur J. Cronquist Curator of North American Botany at The New York Botanical Garden, Bronx, New York, USA. His research focuses on floristics of the eastern United States, systematics of sedges (Cyperaceae), and systematics and evolution of the western hemisphere pitcher plants (Sarraceniaceae) and their arthropod symbionts, particularly histiostomatid mites and sarcophagid flies. He is the lead editor of *Sedges:*

Uses, Diversity, and Systematics of the Cyperaceae (Missouri Botanical Garden Press, 2008), and author of >75 peer-reviewed papers. He received his Ph.D. (1992) in Botany from the University of Michigan, Ann Arbor, Michigan, USA.

Maria Paniw **(co-author: Chapter 10)** is a Ph.D. candidate at the Departamento de Biología, University of Cadiz, Spain. Her dissertation is on the population dynamics of the carnivorous *Drosophyllum lusitanicum*. She has strong research interests in population ecology, quantitative ecology related to the stochastic modeling of successional systems, and the eco-evolutionary dynamics of carnivorous plants.

Andrej Pavlovič **(co-author: Chapters 16, 17, 18)** is an Associate Professor in the Department of Biophysics at the Palacký University in Olomouc, Czech Republic, and the Department of Plant Physiology, Comenius University in Bratislava, Slovakia. He received his Ph.D. (2009) in plant physiology from the Faculty of Natural Sciences of Comenius University. In 2015, he defended his habilitation thesis entitled *Cost/benefit Model for the Evolution of Botanical Carnivory: From Molecules to Ecology*. His research is focused on the photosynthesis and enzymology of carnivorous plants. He is the author of ≈30 peer-reviewed papers, including ≈15 on carnivorous plants. He is an associate editor of *Photosynthetica* and is a national representative of Slovakia to The Federation of European Societies for Plant Biology (FESPB).

Bartosz J. Płachno **(lead author: Chapter 13; co-author: Chapters 12, 21)** is an Adjunct Professor at Jagiellonian University in Kraków (Poland), where he received his Ph.D. for a study of the embryology and trap evolution of Lentibulariaceae. He has received stipends from both the Foundation for Polish Sciences and the Ministry of Science and Higher Education (Poland). He is author of >59 peer-reviewed papers, mainly on carnivorous plants (especially on the embryology, trap anatomy, and algae in the traps of *Genlisea, Utricularia,* and *Heliamphora*). Currently, he is conducting embryological and pollination biology studies (floral morphology) of carnivorous plants.

Simon Poppinga **(lead author: Chapter 14; co-author: Chapter 15)** is a post-doctoral scholar and research group leader for plant movements, biomimetics, and elastic architecture in the Plant Biomechanics Group at the University of Freiburg, Germany. He also received his Ph.D. from Freiburg University. His research focuses on biomechanics and functional morphology of moving plant structures, especially of carnivorous plants traps. His other research interests include functional plant surfaces and the transfer of functional principles found in nature into technical applications (bionics/biomimetics).

Tanya Renner **(co-lead author: Chapter 11; co-author: Chapter 16)** is an Assistant Professor in the Department of Entomology at Pennsylvania State University. She studied for her Ph.D. at the University of California, Berkeley, where she began to examine evolutionary patterns and processes that drive functional diversification among the carnivorous plants of the Caryophyllales. Since then, she has been particularly interested in how multi-species interactions shape biodiversity at the micro-evolutionary scale and influence form and function. Her research focuses on the underlying genetics both linking and separating carnivorous plant defense and digestion, and morphological adaptations for insect capture.

Barry Rice **(co-author: Chapter 27)** is a Professor at Sierra College (Rocklin, California, USA) and an Associate Scientist at the Center for Plant Diversity at the University of California, Davis, California, USA. He has been working with conservation of carnivorous plants since the 1990s. He worked for 11 years for The Nature Conservancy, and he was the founding Director of Conservation Programs for the International Carnivorous Plant Society. He has written numerous popular interest articles, refereed articles, books, and book chapters on carnivorous plants. His current work is focused on the distribution of carnivorous plants in the western United States.

Alastair Robinson **(co-author: Chapter 5)** is a field botanist who has studied *Nepenthes* across virtually its entire range. Parallel interests include terrestrial orchids, particularly those in the subtribe Acianthinae. He originally has a background in plant pathology and physiology, culminating in a Ph.D. from the University of Cambridge (UK), for research into the molecular control mechanisms of cell division in plants.

Aymeric Roccia **(co-author: Chapter 6)** is a plant biologist, geneticist and chemist, who studies scented, aromatic, and medicinal plants. He received his Ph.D. (2013) in plant biology from Jean Monnet University, Saint Etienne, France. He recently described a new natural butterwort hybrid, contributed the *Pinguicula* chapter to *Flora Gallica*, and is the lead author of *Pinguicula* of the Temperate North, one of the two volumes of the most recent revision of the genus *Pinguicula*.

André V. Scatigna **(co-author: Chapter 10)** is a Ph.D. student at the University of Campinas (UNICAMP), Brazil. He received his M.Sc. degree (2014) in plant biology from UNICAMP for a dissertation on phylogenetics and conservation genetics of *Philcoxia*. His current research interests are in the systematics and evolution of the tribe Gratioleae (Plantaginaceae). His research recently resulted in the discovery of three new species and other new populations of *Philcoxia*.

Jan Schlauer **(co-author: Chapters 2, 3, 5)** is a biochemist by education who has been interested in carnivorous plants for over 35 years. He has been the scientific co-editor of the *Carnivorous Plant Newsletter* (the official journal of the International Carnivorous Plant Society) since 1996, and has written or co-authored >40 papers dealing with enzymology, secondary metabolism, chemotaxonomy, carnivorous plant classification, and chorology. Currently he is conducting floristic and taxonomic studies in Europe and abroad with special emphasis on carnivorous and other heterotrophic plant species.

Stephan C. Schuster **(co-author: Chapter 11)** received a Ph.D. in organic chemistry from the Max-Planck-Institute for Biochemistry. He currently is a Professor working at the Center for Comparative Genomics and Bioinformatics (PSU), and is a founding member of the Singapore Centre on Environmental Life Sciences Engineering at Nanyang Technical University, Singapore. His current research interests include metagenomics, evolutionary biology, and ancient DNA.

Dagmara Sirová **(lead author: Chapter 25)** is a postdoctoral researcher at the University of South Bohemia in České Budějovice and research scientist at the Institute of Hydrobiology, Biology Centre of the Czech Academy of Sciences, Czech Republic. Throughout her career, she has been interested in the ecology of complex microbial communities and in the interactions between microorganisms and their eukaryotic hosts, using the *Utricularia*-associated microbiome as a model. She is the author of several peer-reviewed papers on the topic.

Stephen A. Smith **(co-author: Chapter 3)** is an Assistant Professor of Ecology and Evolutionary Biology at the University of Michigan, Ann Arbor, Michigan, USA. He studies evolutionary and phylogenetic questions regarding plants and their rates of evolution, biogeography, and molecular evolution. He has recently initiated an analysis of the Caryophyllales using transcriptomes and genomes to better understand the intersection of molecular evolution and shifts in ecology and phenotype. This has included more directed analyses of the carnivorous Caryophyllales, in which he is examining gene and genome duplications related to the evolution of carnivory.

K. William Sparks **(co-author: Chapter 18)** is a Ph.D. candidate at the University of Wisconsin-Madison. He received his B.A. in biology from Swarthmore College, Swarthmore, Pennsylvania, USA. His dissertation research incorporates studies on the physiological ecology of carnivorous plants and computational methods in ecological and evolutionary research.

Thomas Speck **(co-author: Chapter 14)** is Professor of Botany, Functional Morphology, and Biomimetics, and Director of the Botanic Garden of the University of Freiburg, Germany. He also acts as Deputy Managing Director of the Freiburg Center for Interactive Materials and Bio-Inspired Technologies (FIT) and is member of the Materials Research Centre Freiburg. He has been studying biology at the University of Freiburg, where he received his Ph.D. (1990) and his habilitation (1996) in botany and biophysics. He is the author of >200 peer-reviewed papers and book articles. His main interest in carnivorous plants comes from functional morphology and biomechanics with the aim of deriving inspiration from them for biomimetic materials and structures.

***Shane Turner* (co-author: Chapter 22)** is a Research Fellow at the School of Plant Biology, University of Western Australia, and Kings Park and Botanic Garden, Perth, Western Australia. He received his Ph.D. in plant conservation biotechnology from Curtin University in 2002. He has ongoing research interests in seed ecology, plant biotechnology, conservation biology, and carnivorous plants.

***Alexander Volkov* (co-author: Chapter 14)** is a Professor at Oakwood University, Huntsville, Alabama, USA. He is the co-author or editor of eight books: *Liquid-Liquid Interfaces: Theory and Methods* (CRC Press, 1996), *Liquid Interfaces in Chemistry and Biology* (Wiley, 1998), *Plant Energetics* (Academic Press, 1998), *Liquid Interfaces in Chemical, Biological, and Pharmaceutical Applications* (M. Dekker, 2001), *Interfacial Catalysis* (M. Dekker, 2003), *Plant Electrophysiology* (Springer, 2006), *Plant Electrophysiology—Methods and Cell Electrophysiology* (Springer, 2012), *and Plant Electrophysiology—Signaling and Responses* (Springer, 2012). He is the author of >190 peer-reviewed papers, two patents, and 49 reviews and book chapters, including 23 publications on Venus' flytrap electrophysiology and biomechanics. Dr. Volkov received his M.S. degree (1973) from Moscow State University (Russia), and his Ph.D. degree (1982) from the Frumkin Institute of Electrochemistry (Moscow, Russia).

***Jaroslav Vrba* (co-author: Chapter 25)** is the Professor of Hydrobiology at the University of South Bohemia in České Budějovice and the Senior Research Scientist in the Institute of Hydrobiology, Biology Centre of the Czech Academy of Sciences, Czech Republic. He has been studying the role of hydrolytic enzymes and microbial commensals in nutrition of aquatic carnivorous plants since the late 1990s. He is the author of ≈100 peer-reviewed papers on aquatic microbial ecology and limnology (including a dozen on carnivorous plants), and Editor-in-Chief (since 1998) of *Silva Gabreta*.

Foreword

Daniel M. Joel

. . . scientific literature is not a compendium of certified verities, but the record of a continuing conversation about how best to make sense of reality. **—Franklin M. Harold (2016: 90)**

. . . we see how little has been made out in comparison with what remains unexplained and unknown. **—Charles Darwin (1875: 223)**

The possibility that plants could "eat" animals inspires legends and science-fiction movies. Indeed, carnivorous plants may be among the most interesting plants on earth. Nevertheless, for many centuries scientists could not believe that carnivorous plants could exist, because they were bound to the paradigm that only animals eat. Though some plants were known to capture insects, they were not thought to exploit them as a source of nutrition. Thus, the potential existence of botanical carnivory was almost completely ignored and research conducted on the plants now known to us as carnivorous was concerned mainly with morphological characterization of their bizarre traps and taxonomic affiliation.

It was only during the late nineteenth century that scientists began to explore botanical carnivory; these efforts culminated in the desire of Charles Darwin to understand its nature and the activity of carnivorous plants. He employed available scientific instruments, including a primitive optical microscope, and was the first to demonstrate the power of digestion within traps, their excitability and movement, and to suggest the existence of a system analogous to the nerves in some of them. His 1875 book, *Insectivorous Plants*, described carnivorous mechanisms in a variety of plants belonging to several different plant families and provided a detailed account of the reality of plant carnivory, which was then recognized and accepted by the broader scientific community. One year later, Burdon-Sanderson and Page (1876) recorded electrical action potentials in the leaf of *Dionaea muscipula* (the Venus' fly-trap) and linked them to the triggering and closing of the trap (Williams and Pickard 1979). The first demonstration that plants can gain a growth advantage from carnivory was provided but three years later by Charles Darwin's son Francis (Darwin 1878), and by Kellermann and von Raumer (1878).

As microscopes grew more powerful and plant physiology became a leading research discipline, the structure and behavior of carnivorous plants was investigated further. The knowledge accumulating during the sixty years following Darwin was summarized by Francis E. Lloyd in *The Carnivorous Plants* (1942). Despite having what seems to us today to be limited research facilities and equipment, scientists working in the late nineteenth and early twentieth century contributed much basic knowledge about carnivorous plants that is still relevant today.

A major step forward in the understanding of botanical carnivory came with the shift in twentieth-century biology from descriptions to experiments on the physiological and biochemical mechanisms that facilitate carnivory. Use of the newly developed electron microscope enabled exploration of subcellular activities during digestion and uptake of animal matter. *The Carnivorous Plants*, by Barrie E. Juniper, Richard Robins, and Daniel M. Joel (1989), provided a comparative treatment of the various carnivorous plants grouped by their mechanisms of attraction, trapping, digestion, and absorption rather than by their taxonomic affiliation. That book also detailed the dynamic ultrastructure of digestive glands, described the ecological requirements and specifications of carnivorous plants, and discussed their mutualistic and exploitive interactions with insects.

In the last thirty years, several factors have led to accelerations in investigations of carnivorous plants. The development of new research disciplines, including whole genome analysis, transcriptomics, and proteomics has facilitated the analysis of biological systems at the molecular level, thus enabling a thorough investigation of evolutionary trends in carnivorous plants and leading to a better understanding of biochemical processes, and physiological and genetic aspects of carnivorous plant activity. The mass propagation by tissue culture of carnivorous plants has ensured a regular supply of cloned plant material for research. At the same time, the public media has given carnivorous plants a popular image, leading to the establishment of carnivorous plant societies and the emergence of carnivorous plant nurseries that promoted growing carnivorous plants as a hobby. More people have become keen to visit and protect the plants in their natural habitats and to explore their multifaceted interaction with insects.

Considering the intense fascination and interest that carnivorous plants hold nowadays, a need has developed for an updated source of scientific information about these plants. The present book fills this need and provides the widest possible account of the state-of-the-art scientific information on botanical carnivory, the various carnivorous plants that are currently known, and their basic physiological mechanisms, ecology, evolution, and genetics. It also discusses the use of carnivorous plants as models for the study of ecological systems and as biomimetic models for the development of novel technological products. The book is particularly important now because many species of carnivorous plants are on the verge of extinction as the result of loss of habitats, climatic change, and no less significant, the frequent collection from the field of rare carnivorous plants for commercialization. A special chapter on the conservation of carnivorous plants is therefore rightly included in the book.

Aaron M. Ellison and Lubomír Adamec have done an outstanding job in editing this comprehensive review of carnivorous plants, which will be an important source of knowledge on the bookshelf of anyone, whether professional, student, or layperson who is interested in carnivorous plants and wishes to deepen their knowledge of these wonderful plants.

PART I

Overview

CHAPTER 1

Introduction: what is a carnivorous plant?

Aaron M. Ellison and Lubomír Adamec

We perceive the relationship between plants and animals asymmetrically. As a rule, plants are producers and animals are consumers—of both plants and other animals. Most animals actively move freely through their environment, whereas plants are rooted in place and comparatively immobile. And while animals are on the offensive, searching for and eating plants, the plants are on the defensive, warding off animals physically and chemically. This asymmetry has resulted in a coevolutionary arms race between plants and animals that has, at least in part, spurred their diversification (Ehrlich and Raven 1964).

There are, however, at least 800 species of plants (Appendix) that have turned the tables on animals and become predators of animals. These carnivorous plants, which literally swallow (Latin: *vorare*) flesh (*carnis*), have evolved independently in at least ten evolutionarily separate lineages (Chapter 3). Despite their relative rarity among the > 350,000 species of flowering plants, carnivorous plants have attracted the attention of many scientists in a wide range of disciplines, from botany to zoology, anatomy and physiology to ecology and evolution, and biophysics to bioengineering.

As with much of contemporary biology, serious study of carnivorous plants began with Darwin (1875, Chase et al. 2009). In *Insectivorous Plants*, Darwin (1875) raised several fundamental questions: How do carnivorous plants capture and kill insects, do they obtain nutrients from the captured prey that increase the fitness of carnivorous plants, and what is the evolutionary origin (i.e., homology) of the specialized organs (e.g., glands, hairs, "tentacles" of the sundews [*Drosera* spp.] and other taxa then placed in the Droseraceae, and active traps of *Dionaea, Aldrovanda*, and *Utricularia*). These questions have been revisited in synthetic reviews at roughly 50-year intervals (Lloyd 1942, Juniper et al. 1989, and this present volume).

1.1 The carnivorous syndrome

Answering any of Darwin's questions, and others that have arisen in the last 140 years, depends on accurate identification of carnivorous plants and delimitation of them from noncarnivorous ones. Identifying and delimiting carnivorous plants requires understanding their evolutionary histories (Chapters 3–10); identification of their essential morphological and physiological traits (Chapters 12–19); and parsing of the details of their interactions with their prey, pollinators, and symbionts (Chapters 21–26). Like many plants, carnivorous ones are sparsely distributed and restricted to particular habitats (Chapter 2), many of which are continually altered by human activities and are vulnerable to climatic change (Chapters 27–28). As the knowledge of carnivorous plants has expanded, so too has their relevance for understanding noncarnivorous ones (and *vice versa*), and, after more than a century of detailed study, carnivorous plants no longer are seen simply as botanical monstrosities.

The 19 genera of carnivorous plants represent at least ten evolutionarily different lineages that are united by a group of shared, functional traits that enable the plants to obtain most of their nutrients

Ellison, A. M., and Adamec, L., *Introduction: what is a carnivorous plant?* In: *Carnivorous Plants: Physiology, ecology, and evolution*. Edited by Aaron M. Ellison and Lubomír Adamec:
Oxford University Press (2018). © Oxford University Press. DOI: 10.1093/oso/9780198779841.003.0001

from animal prey: rapid movements of traps that are regulated electrophysiologically; secretion of hydrolytic enzymes that digest proteins, chitins, and other organic compounds; foliar uptake of nutrients; stimulation by foliar nutrient uptake of uptake of nutrients by roots; and stimulation of plant growth by foliar nutrient uptake (Adamec 2011c). Although each of these functional traits or physiological processes also can occur—usually singly—in noncarnivorous plants, they normally co-occur in carnivorous plants where they are coupled in series firmly to another. Together, they form a coordinated cluster of characters referred to as the "carnivorous syndrome" (Juniper et al. 1989; Adamec 1997a).

The expression of each of these traits varies among carnivorous plants (Chapters 12–16). The main benefit of carnivory is the uptake of growth-limiting mineral nutrients from prey (Adamec 1997a; Chapters 16–19), and so we consider there to be five essential traits of the carnivorous syndrome (cf. Lloyd 1942, Givnish 1989, Juniper et al. 1989, Adamec 1997a, 2011c, Rice 2011c, Pavlovič and Saganová 2015):

1. capturing or trapping prey in specialized, usually attractive, traps;
2. killing the captured prey;
3. digesting the prey;
4. absorption of metabolites (nutrients) from the killed and digested prey;
5. use of these metabolites for plant growth and development.

Only plants that possess all five of these traits should be considered as "carnivorous" plants.

Many researchers also have considered attraction and retention of prey to be two other essential traits within the carnivorous syndrome (Juniper et al. 1989; Chapters 12, 13, 16, 21). Although these traits may improve the efficiency of carnivory, they do not appear to be required. For example, attraction has been studied and confirmed only for some carnivorous plants and does not occur in others (Givnish 1989, Guisande et al. 2007; Chapter 12). Prey retention close to the trap entrance following attraction (i.e., transiently before prey capture itself) occurs only in *Dionaea* and pitcher plants (Juniper et al. 1989), but most prey attracted to the latter leave unscathed (Newell and Nastase 1998, Dixon et al. 2005).

As all plants can absorb organic substances from soil by roots (e.g., from dead animals), the criterion of actively killing prey captured in traps (Chapters 13–15) separates carnivorous from mycoparasitic (a.k.a. saprophytic) plants. Digestion of prey can occur by a variety of means. Most carnivorous plants secrete hydrolytic enzymes (Chapter 16), but some rely on trap commensals (digestive mutualists: Chapters 23, 24, 26) to digest prey and mineralize the organic nutrients (e.g., Givnish 1989, Jaffe et al. 1992, Butler et al. 2008). The role of autolysis in digestion of prey carcasses has been underestimated and understudied (Chapter 16), but may play a significant role in carnivorous plants that do not secrete digestive enzymes. Many carnivorous plants can also acquire organic and mineral nutrients from detritus, pollen, algae, or microorganisms (Chapters 17, 19, 25).

1.2 Subsets of carnivorous plants

By analogy with parasitic plants, in which holoparasitic plants are differentiated from hemiparasitic ones, Joel (2002) distinguished holocarnivorous plants (e.g., *Aldrovanda, Dionaea, Drosera, Drosophyllum, Pinguicula, Utricularia, Nepenthes*), that secrete their own digestive enzymes, from hemicarnivorous ones (e.g., *Darlingtonia, Heliamphora, Brocchinia, Roridula*), that do not. Rice (2011c) referred to the latter as paracarnivorous plants.

Alternatively, carnivorous plants can be subdivided based on how they obtain nutrients from prey, whether or not they secrete digestive enzymes. All genera of carnivorous plants except for *Roridula* obtain nutrients directly from their prey (Adamec 2011c). The two *Roridula* species capture prolific prey but they do not digest it. Rather, their captured prey is consumed by hemipteran bugs (*Pameridea* spp.) that live only on *Roridula* and which defecate on its leaves; the plants absorb the bug-processed nutrients through specialized cuticular gaps (Ellis and Midgley 1996, Anderson 2005; Chapters 10, 26). In *Roridula*, as in *Sarracenia purpurea* (Chapter 24) mineral nutrients from prey are gained indirectly, through excrement of digestive mutualists. A handful of species in a few other lineages of carnivorous plants also appear to have lost the carnivorous habit (Chapter 3), and at least four species of *Nepenthes*—*N. lowii*, *N. macrophylla*, *N. rajah*, and *N. hemsleyana*—obtain some or all of their nutrients from excrement of tree shrews or bats that, respectively, are attracted and feed on exudates from the highly modified pitchers that feed these mammals or roost within the pitchers (Chapters 5, 13, 26). Other species are partly detritivorous (Chapter 17).

1.3 Other plants that share some carnivorous characteristics

The ability to trap passively small insects, digest them, and absorb at least some nutrients from their carcasses has been observed in many other plant species. Spomer (1999) detected surface proteinase activity in 15 species of North American plants (including the common potato, *Solanum tuberosum*) with glandular leaves, stems, or flowers; he called these plants protocarnivorous, but we think this term should be avoided because it implies that plants with surface proteinases are ancestral to carnivorous plants or that evolution of plants is progressing toward carnivory. Two of the 19 species he studied, *Geranium viscosissimum* and *Potentilla arguta*, also absorbed carbon from ^{14}C-labeled algal protein into their leaves, a phenomenon that has been observed infrequently in carnivorous plants (Chapter 16). Similarly, some *Stylidium* species entrap insects in their inflorescences and have some protease activity associated with their sticky glands (Darnowski et al. 2006).

Although thousands of vascular plants have sticky, glandular organs that can ensnare insects, and most plants can absorb nutrients through their stems and leaves, most of these plants grow in relatively fertile soils, whereas carnivorous plants normally grow only in nutrient-poor soils or dystrophic, barren waters (Chapters 2, 17–19). Sticky glandular organs with proteinase activity apparently have evolved as a defense against small arthropod herbivores or microbial pathogens (Chapter 16) but the episodic capture of very small prey appears to contribute little to the mineral nutrient budget in plants such as *Geranium viscosissimum*, *Potentilla arguta*, or *Stylidium* spp. In contrast, prey capture and digestion account for most nutrients obtained by carnivorous ones (Płachno et al. 2009a; Chapter 17–19).

1.4 The benefits and costs of carnivory

The quantity (biomass) of captured prey per plant and the efficiency of prey use are the principal factors determining the ecological benefits and evolutionary fitness of carnivory under natural conditions (e.g., Givnish et al. 1984, Adamec 1997a, 2011c, Farnsworth and Ellison 2008; Chapters 2, 17, 18). There are, however, substantial structural, physiological, and ecological costs to attracting, capturing, and digesting prey (e.g., Givnish et al. 1984, Ellison and Gotelli 2002; Chapter 18). For carnivory to evolve, the marginal benefits of either direct or indirect carnivory (§1.2) must exceed their marginal costs (Givnish et al. 1984; Chapter 18). These benefits most often outweigh the costs in nutrient-poor, well lit, and wet environments (Givnish et al. 1984, Benzing 1987, 2000; Chapters 2, 18).

Even after the plants have digested and mineralized nutrients from their prey, the dismembered carcasses still may contain relatively high amounts of mineral nutrients, especially nitrogen in chitin (Adamec 2002; Chapter 17). After traps senesce and decay, the carcasses and dead traps that contact wet topsoil decompose in proximity to the plant's (usually weakly developed) roots. Because of this additional decomposition, spent prey carcasses can be said to fertilize the soil close to the roots, which can thus take up additional, prey-derived nutrients.

1.5 The future: learning from carnivorous plants

In *Insectivorous Plants*, Darwin wrote that "as it cannot be doubted that this process [carnivory] would be of high service to plants growing in very poor soil, it would tend to be perfected through natural selection" (Darwin 1875: 362–363). Many noncarnivorous plants also grow in nutrient-poor soils, and an obvious question is, if carnivory is so useful in such environments, why aren't all plants that grow in them carnivorous?

In the last century and a half, as researchers have defined and delimited plant carnivory, they have also sought to understand its evolution and explain its rarity. The results have shed new light on the evolution and workings of physical and chemical defenses across the plant kingdom (Chapter 16), identified key genetic changes responsible for the shift from defensive chemistry to offensive carnivory (Fukushima et al. 2017), and highlighted the myriad ways that plants can extract scarce nutrients from nutrient-poor substrates (Benzing 1987, Givnish 1989; Chapter 18). Carnivorous plants illustrate general principles of coevolution and convergent evolution; in turn, a better understanding of both processes in carnivorous plants has led to the concept of evolutionarily convergent interactions that are revealing new symbiotic relationships among animals, plants, fungi, and microbes (Bittleston et al. 2016b). As more and more genomes of carnivorous plants are sequenced and annotated (Chapter 11), the data are being combined with detailed natural history observations and experimental results to test and develop new theories about plant form, function, ecology, and evolution. It is reasonable to suggest Darwin's "most wonderful plants in the world" increasingly will become model biological systems for field research (Chapters 24, 25) and classroom study (Ellison 2014) in the same way that *Escherichia coli*, *Arabidopsis thaliana*, and *Drosophila melanogaster* have been used in the laboratory.

This vision will be realized only if viable populations of carnivorous plants persist in the field. The widespread fascination with carnivorous plants, their rarity, and their occurrence in unique, often patchy habitats makes them very vulnerable to over-collection and poaching, land-use changes, and ongoing climatic change. While scientists, land-use managers, and conservation professionals are working to preserve populations of threatened and endangered carnivorous plant species and their habitats (Chapter 27), researchers are using their unique distributional characteristics to test assumptions and reveal shortcomings in existing models used to forecast how species respond to climatic change (Chapter 28). The refined models, based on ideas gleaned from studying carnivorous plants, will benefit all species, rare or common, carnivorous or not.

CHAPTER 2

Biogeography and habitats of carnivorous plants

J. Stephen Brewer and Jan Schlauer

2.1 Introduction

Although carnivorous plants are considered to be physiologically specialized, they grow in many different environments and in unevenly distributed locations around the world. Two underlying mechanisms most frequently explain global scale distribution patterns: migration (differentiation during or after dispersal to new locations), and vicariance (differentiation during or after fragmentation of a larger, contiguous original area caused by geological or climatological changes). Even after differentiation, hybridization may cause introgression or polyphyletic speciation wherever two taxa contact one another. At regional and local scales, most investigators (beginning with Darwin 1875) have hypothesized that botanical carnivory is most beneficial in habitats with nutrient-poor soils. Others, including Juniper et al. (1989), recognized the potential importance of additional factors, including calcium concentration of soils, soil moisture, shade, and competition with noncarnivorous plants.

In this chapter, we discuss global distribution patterns of carnivorous plants with respect to biogeography and climate. We then present data on occurrence of terrestrial carnivorous and noncarnivorous plants in relation to soil fertility, light, and soil moisture within a region (Mississippi, USA) as a case study for testing the hypotheses that carnivory puts carnivorous plants at a competitive disadvantage in shady habitats with dry, nutrient-rich soils. We further consider different hypotheses to explain the high species diversity of wet savannas vis-à-vis mechanisms of coexistence of carnivorous and noncarnivorous herbs. We conclude that additional comparative ecological studies of carnivorous and noncarnivorous plants will be crucial not only for understanding distribution of carnivorous plants but also for understanding general controls of plant diversity and distribution.

2.2 Global biogeography

Carnivory in plants has evolved independently numerous times (Chapter 3), and their evolutionary history has led to a number of geographic "hotspots" with large numbers of species (e.g., *Nepenthes* in Southeast Asia [Chapter 5]; *Drosera* in Australia and South Africa [Chapter 4]; and *Sarracenia* in the Southeast United States [Chapter 9]). Only in comparatively few cases is there reasonable evidence for long-distance dispersal in carnivorous plants, and more or less continuous spread and diversification is clearly the dominant pattern. Biogeographical analogies between unrelated taxa reflect only very general common trends, not synchronous or congruent events (Schlauer 2010). Global-scale distribution patterns, summarized in Table 2.1, can be characterized by floristic composition (Takhtajan 1986) and climatic conditions (Troll and Paffen 1964, Kottek et al. 2006).

Carnivorous plants are most abundant and diverse in tropical to subtropical regions of all continents and have a preference for warm to hot, humid to wet climates. In contrast, they tend to be rare or absent in polar ice deserts, in arid to dry climates, or on remote oceanic islands (Table 2.1). The speciose genera of carnivorous plants occupy contrasting

Brewer, J. S., and Schlauer, J., *Biogeography and habitats of carnivorous plants*. In: *Carnivorous Plants: Physiology, ecology, and evolution*. Edited by Aaron M. Ellison and Lubomír Adamec:
Oxford University Press (2018). © Oxford University Press. DOI: 10.1093/oso/9780198779841.003.0002

Table 2.1 Geographic and climatological distribution of major taxonomic carnivorous plant groups.

Genus-Subgenus-Section	# of spp.	Floristic Region/Province of Takhtajan (1986)*	Köppen-Geiger Climate Classification**	Troll-Paffen Climate Classification***
Lentibulariaceae				
Utricularia-Bivalvaria-Aranella	10	Caribbean, Guayana Highland, Amazonian, Brazilian, Guineo-Congolian & Sudano-Zambezian rr. (65–67, 70, 117, 118, 120, 121, 123, 124, 127, 129)	Af, Am, Aw, Bsh	V_1–V_3
Utricularia-Utricularia-Candollea	1	Sudano-Zambezian r.; Zambezian p. (70)	Aw	V_2
Utricularia-Utricularia-Martinia	1	Guayana Highland & Amazonian rr. (120, 121)	Af, Am	V_1
Utricularia-Bivalvaria-Calpidisca	9	mainly Guineo-Congolian r., Sudano-Zambezian r.; Zambezi & Sudan pp., occasionally Karoo–Namib r.; Namaland p. & Madagascan r.; Eastern Madagascar p. (64–67, 69, 70, 72, 73, 80, 81, 84, 92, 117, ?121, 130, 133)	Af, Am, Aw	V_1, V_2
Utricularia-Bivalvaria-Nigrescentes	3	Sudano-Zambezian r.; Zambezi p. & Madagascan r.; Eastern Madagascar p., Eastern Asiatic, Indian, Indochinese, Malesian & Northeast Australian rr. (16, 18, 20, 21, 23–28, 60, 70, 84, 90–95, 97–105, 108, 109, 113, 131–133)	Af, Aw, Cfa	V_1, V_2, IV_7
Utricularia-Bivalvaria-Meionula	5	Eastern Asiatic r.; Japan–Korea & Southeastern China pp., Indian, Indochinese, Malesian & Northeast Australian rr. (18, 24, 90–92, 94–96, 98–105, 108, 109, 131–133)	Af, Aw, Cfa	V_1, V_2, IV_7
Utricularia-Polypompholyx-Australes	3	Northeast Australian, Southwest Australian & Neozeylandic rr. (133–135, 148, 151)	Cfb, Csb	III_2, IV_1, IV_7
Utricularia-Bivalvaria-Lloydia	1	Caribbean r.; Central American p., Guayana Highland, Amazonian & Brazilian rr., Guineo-Congolian r., Sudano-Zambezian r.; Zambezi p., Eritreo–Arabian r.; Somali–Ethiopian p., Eastern Asiatic r.; Khasi–Manipur p., Indian r.; Upper Gangetic Plain p. (28, 65–67, 70, 73, 93, 117, 120, 121, 124, 126)	Am, Aw, Bsh, Cwa	V_1–V_3
Utricularia-Polypompholyx-Polypompholyx	2	Southwest Australian r. (133–135)	Cfb, Csb	IV_1
Utricularia-Polypompholyx-Tridentaria	1	Southwest Australian r. (135)	Csb	IV_1
Utricularia-Polypompholyx-Pleiochasia	25	Northeast Australian & Southwest Australian rr. (116, 131–135, 148–150)	Aw, Cfb, Csb	IV_1, IV_7, V_3
Utricularia-Utricularia-Orchidioides	10	Caribbean, Guayana Highland, Amazonian & Andean rr. (117, 118, 120–123, 128, 129)	Af, Am, Aw	V_1–V_3
Utricularia-Utricularia-Iperua	6	Guayana Highland & Brazilian rr. (120, 124, 126, 127)	Af, Am, Aw	V_2
Utricularia-Bivalvaria-Stomoisia	2	North American Atlantic r. and adjacent Canada, Caribbean, Amazonian, Brazilian rr., Guineo-Congolian r.; Upper Guinean p. (15, 29–31, 65, 117, 118, 121, 124)	Af, Am, Aw, Cfb, Dfb	III_8, IV_7, V_1–V_3

(*continued*)

Table 2.1 (*Continued*)

Genus-Subgenus-Section	# of spp.	Floristic Region/Province of Takhtajan (1986)*	Köppen-Geiger Climate Classification**	Troll-Paffen Climate Classification***
Utricularia-Utricularia-Stylotheca	1	Caribbean, Guayana Highland, Amazonian & Brazilian rr. (117, 118, 120, 121, 124)	Af, Am, Aw	$V_1–V_3$
Utricularia-Utricularia-Choristothecae	2	Guayana Highland r. (120)	Af	V_2
Utricularia-Bivalvaria-Benjaminia	1	Guayana Highland, Amazonian & Brazilian rr. (120, 121, 124, 127)	Af, Am, Aw	$V_1–V_3$
Utricularia-Bivalvaria-Oligocista	43	Pantropical (18, 21, 23–28, 60, 65–67, 69, 70, 72, 73, 84, 89–109, 113, 116, 117, 120, 121, 124, 126–133, 140)	Af, Am, Aw, Bsh, Cfb, Cwa, Cwb	$V_1–V_3$
Utricularia-Bivalvaria-Avesicarioides	2	Guineo-Congolian r.; Upper Guinean & Nigeria–Cameroon pp. (65, 66)	Af, Am, Aw	$V_1–V_3$
Utricularia-Utricularia-Chelidon	1	Guineo-Congolian r.; Nigeria–Cameroon p. (66)	Af, Am, Aw	$V_1–V_3$
Utricularia-Bivalvaria-Enskide	3	Northeastern Australian r., Malesian r.; Papuan p. (109, 131, 132)	Af, Am, Aw, BSh, Cfa	$V_1–V_3$
Utricularia-Utricularia-Kamienskia	2	Eastern Asiatic r.; Central & Southeastern Chinese pp. (23, 24)	Cfa	V_2
Utricularia-Bivalvaria-Phyllaria	16	Majority restricted to East Asiatic, Indochinese & Malesian rr., only one species extending to Guineo-Congolian, Usambara–Zululand, Sudano-Zambezian & Indian rr. (21, 23, 25–28, 60, 65–67, 70, 73, 90, 92, 93, 95–98, 100–109)	Af, Am, Aw, Cfa, Cwa	IV_7, $V_1–V_3$
Utricularia-Utricularia-Oliveria	1	Guineo-Congolian, Usambara–Zululand, Sudano-Zambezian & Madagascan rr. (66, 67, 70, 84)	Af, Am, Aw, Cwa	$V_1–V_3$
Utricularia-Utricularia-Psyllosperma	10	Madrean r.; Mexican Highland p., Caribbean r.; Central American p., Guayana Highland, Amazonian, Brazilian & Andean rr. (64, 117, 120, 121, 124–127, 129, 140)	Af, Am, Aw	$V_1–V_3$
Utricularia-Utricularia-Foliosa	13	Caribbean r.; Central American p., Guayana Highland, Amazonian, Brazilian & Andean rr. (117, 118, 120–122, 124–129, 140)	Af, Am, Aw	$V_1–V_3$
Utricularia-Utricularia-Sprucea	1	Caribbean, Amazonian, Brazilian, Andean & Chile–Patagonian rr. (117, 120, 121, 123)	Af, Am, Aw, Cfa, Cfb	$V_1–V_3$
Utricularia-Utricularia-Avesicaria	2	Guayana Highland, Amazonian & Brazilian rr. (120, 121, 124)	Af, Am, Aw	$V_1–V_3$
Utricularia-Utricularia-Mirabiles	2	Guayana Highland r. (120)	Af	V_2
Utricularia-Utricularia-Steyermarkia	3	Guayana Highland r. (120)	Af	V_2
Utricularia-Utricularia-Lecticula	2	North American Atlantic, Caribbean & Guayana Highland rr. (15, 29, 30, 117, 118, 120)	Af, Am, Aw, Cfb, Dfb	III_8, IV_7, $V_1–V_3$
Utricularia-Utricularia Setiscapella	9	Pantropical & subtropical, but predominantly American (15, 28–30, 39, 65–67, 69, 70, 72, 80, 84, 98, 101–103, 117, 118, 120–129, 131–133, 140)	Af, Am, Aw, Cfa, Cfb, Dfb	$V_1–V_3$
Utricularia-Utricularia-Nelipus	3	Indochinese, Malesian & Northeast Australian rr. (96, 97, 99, 101, 102, 108, 131–133)	Af, Am, Aw, Cfa	$V_1–V_3$

(*continued*)

Table 2.1 (*Continued*)

Genus-Subgenus-Section	**# of spp.**	**Floristic Region/Province of Takhtajan (1986)***	**Köppen-Geiger Climate Classification****	**Troll-Paffen Climate Classification*****
Utricularia-Utricularia-Utricularia	38	Almost cosmopolitan (1–18, 20–33, 36, 39–42, 44, 45, 49, 51, 54, 55, ?57, 60, 62–74, 78, 80, 81, 84, 87, 88, 90–109, 113, 115–135, 140, ?148, ?149)	azonal (aquatic)	azonal (aquatic)
Utricularia-Utricularia-Vesiculina	3	North American Atlantic, Caribbean, Guayana Highland, Amazonian, Brazilian, Andean & Chile–Patagonian rr. (15, 29, 30, 117, 118, 120, 121, 124, 127, 129, 140)	Af, Am, Aw, Cfa, Cfb, Dfb	III_8, IV_6–IV_7, V_1–V_3
Utricularia-Utricularia-Biovularia	3	North American Atlantic, Caribbean, Guayana Highland, Amazonian, Brazilian & Andean rr. (29, 30, 117, 118, 120, 121, 124, 129)	Af, Am, Aw, Cfb, Dfb	III_8, IV_7, V_1–V_3
Pinguicula-Isoloba-Isoloba	6	North American Atlantic r.; Atlantic & Gulf Coastal Plain pp. (30, 118)	Cfa	IV_7
Pinguicula-Isoloba-Cardiophyllum	3	Circumboreal r; Atlantic European p. & Mediterranean r; Southwestern Mediterranean, Ligurian–Tyrrhenian, Adriatic & Eastern Mediterranean pp. (2, 39, 43–45)	Csa, Csb, Cfb	III_2, IV_1
Pinguicula-Isoloba-Ampullipalatum	5	Andean r.; Northern Andes, Central Andes, Chile–Patagonian r.; Patagonia & Magellania pp. (128, 129, 139, 141, 142)	Cfb, Cfc, ET	III_1, III_2, V_1, V_2
Pinguicula-Pinguicula-Pinguicula (without P. vulgaris)	21	Circumboreal r.; Pontus Euxinus p., Mediterranean r. (5, 39, 41, 43–45)	Csa	IV_1
P. vulgaris (P.-P.-Pinguicula)	1	Circumboreal, Eastern Asiatic, Rocky Mountain & Mediterranean rr. (1–4, 7–11, 14, 15, 17, 18, 32, 39, 41, 43, 44)	Cfb, Csa, Csb, Cwa, Cwb, Dfb, Dfc, Dfd, Dwb, Dwc	I_4, II_2–II_3, III_2–III_8, III_{11}, IV_1
Pinguicula-Temnoceras-Temnoceras	5	Madrean r.; Mexican Highlands p., Caribbean r.; Central America p. (64, 117)	Bsh, Cwa, Cwb	V_1–V_3
Pinguicula-Temnoceras-Crassifolia	4	Madrean r.; Mexican Highlands p. (64)	Bsh, Cwa, Cwb	V_1–V_3
Pinguicula-Temnoceras-Longitubus	5	Madrean r.; Mexican Highlands p., Caribbean r.; Central America p. (64, 117)	Bsh, Cwa, Cwb	V_1–V_3
Pinguicula-Temnoceras-Microphyllum	7	Madrean r.; Mexican Highlands p. (64)	Bsh, Cwa, Cwb	V_1–V_3
Pinguicula-Temnoceras-Orcheosanthus	13	Madrean r.; Mexican Highlands p., Caribbean r.; Central America p. (64, 117)	Bsh, Cwa, Cwb	V_1–V_3
Pinguicula-Temnoceras-Agnata	4	Madrean r.; Mexican Highlands p. (64)	Bsh, Cwa, Cwb	V_1–V_3
Pinguicula-Temnoceras-Heterophyllum	5	Madrean r.; Mexican Highlands p. (64)	Bsh, Cwa, Cwb	V_1–V_3
Pinguicula-Temnoceras-Homophyllum	18	Madrean r.; Mexican Highlands p., Caribbean r.; Central America & West Indies pp. (64, 117, 118)	Bsh, Cwa, Cwb	V_1–V_3
Pinguicula elongata (P.-Temnoceras-Heterophylliformis)	1	Andean r.; Northern Andes p. (128)	Af	V_1
Pinguicula-Temnoceras-Micranthus	1	Circumboreal & Eastern Asiatic rr., Central Asiatic r.; Tibetan p. (1–3, 8–12, 22, 23, 25, 27, 60)	Cfb, Cwa, Cwb, Dfb, Dfc, Dfd, Dwb, Dwc	I_3, II_2–II_3, III_2–III_8, III_{11}

(*continued*)

Table 2.1 (*Continued*)

Genus-Subgenus-Section	# of spp.	Floristic Region/Province of Takhtajan (1986)*	Köppen-Geiger Climate Classification**	Troll-Paffen Climate Classification***
Pinguicula-Temnoceras-Nana	4	Circumboreal, Eastern Asiatic & Rocky Mountain rr. (1, 8–18, 32)	Dfc, Dfd, Dwb, Dwc, ET	I_3, II_1–II_3, III_8
Genlisea-Tayloria-Tayloria	8	Brazilian r.; Central Brazilian Upland p. (124, 126)	Bsh	V_2
Genlisea-Genlisea-Genlisea	13	Mainly Guayana Highland, Amazonian & Brazilian rr. (117, 118, 120, 121, 123, 124, 126, 127)	Bsh, Aw	V_2
Genlisea-Genlisea-Africanae	9	Guineo-Congolian r., Sudano-Zambezian r.; Zambezian p. (65–67, 70, 72)	Af, Am, Aw	V_1–V_3
Genlisea-Genlisea-Recurvatae	3	Sudano-Zambezian r.; Zambezian p., Madagascan r.; Eastern Madagascan p. (70, 84)	Af, Am, Aw	V_1–V_3
Droseraceae				
Drosera-Regiae-Regiae	1	Cape r. (130)	Csb	IV_1
Drosera-Arcturia-Arcturia	2	Southeast Australian & Tasmanian pp., Neozeylandic r. (133, 134, 148–150)	Cfb	III_2, IV_7
Drosera-Ergaleium-Phycopsis	1	Northeast Australian, Southwest Australian & Neozeylandic rr. (133–135, 148–151)	Cfb, Csb	IV_1, IV_7
Drosera-Ergaleium-Bryastrum	49	Northeast Australian, Southwest Australian (most) & Neozeylandic rr. (133–135, 148–150)	Cfb, Csb	IV_1, IV_7
Drosera-Ergaleium-Meristocaulis	1	Guayana Highland r.; Guyana p. (120)	Af	V_1
Drosera-Ergaleium-Ergaleium	31	Mainly Southwest Australian r., few in Eastern Asiatic, Indochinese, Malesian, Neocaledonian, Northeast Australian & Neozeylandic rr. (18, 21, 23, 24, 26–28, 60, 90–94, 97–99, 104, 106, 108–109, 131–135)	Am, Aw, Cfa, Cfb, Csb, Cwa	IV_1, IV_7, V_2–V_3
Drosera-Ergaleium-Erythrorhiza	12	Northeast Australian (few) & Southwest Australian (most) rr. (133, 135)	Csb	IV_1
Drosera-Ergaleium-Stoloniferae	4	Southwest Australian r. (135)	Csb	IV_1
Drosera-Ergaleium-Coelophylla	1	Northeast Australian & Southwest Australian rr. (133–135)	Cfb, Csb	IV_1, IV_7
Drosera-Ergaleium-Lasiocephala	15	Northeast Australian r. (108, 109, 131, 132)	Aw	V_2–V_4
Drosera-Drosera-Arachnopus	11	Guineo-Congolian, Sudano-Zambezian, Madagascan, Indian, Eastern Asiatic, Indochinese, Malesian & Northeast Australian rr. (18, 21, 24, 65–67, 70, 84, 90–92, 96–100, 103–107, 109, 131–133)	Af, Aw, Cfa	V_1, V_2, IV_7
Drosera-Drosera-Prolifera	3	Northeast Australian r.; Queensland p. (132)	Am	V_2
Drosera-Drosera-Thelocalyx	2	Trop. Asia, Northeast Australian, Amazonian & Brazilian rr., ?Guineo-Congolian r. (18, 21, 24, 28, ?65, 90–92, 94–98, ?99, ?100, ?101, 102–109, 113, 121, 123, 124, 126, 131–133, 136)	Af, Am, Aw	V_1–V_3
Drosera-Drosera-Stelogyne	1	Southwest Australian r. (135)	Csb	IV_1

(*continued*)

Table 2.1 (*Continued*)

Genus-Subgenus-Section	# of spp.	Floristic Region/Province of Takhtajan (1986)*	Köppen-Geiger Climate Classification**	Troll-Paffen Climate Classification***
Drosera-Drosera-Psychophila	2	Chile–Patagonian & Neozeylandic rr. (141, 142, 148–150, 152)	Cfb, Cfc, ET	I_4, III_1, III_2, III_{10}
Drosera-Drosera-Drosera	11	Trop. Africa, especially Sudano-Zambezian r.; Zambesi p. (65–70, 72, 84)	Aw	V_3
	14	Cape r. (130)	Csb	IV_1
	40	Central & South America, mainly Guayana Highland, Amazonian & Brazilian rr. (119–129, 140–142)	Af, Am, Cwb	V_1, V_2
	6	North American Atlantic r.; mainly Atlantic & Gulf Coastal Plain pp. (29–33, 117, 118)	Cfa	IV_7
	3	Circumboreal & Eastern Asiatic rr. (1–18, 21–24, 41, 43, 45, 46, 61, 115)	Cfb, Dfb, Dfc	II, III
	4	Eastern Asiatic, Indochinese, Malesian, Neocaledonian, Northeast Australian & Neozeylandic rr. (18, 21, 24, 97, 103–105, 107–109, 113, 116, 132–134, 148–150)	Af, Cfb, Cwa	III_2, V_1, V_2
Dionaea	1	North American Atlantic r.; Atlantic & Gulf Coastal Plain pp. (30)	Cfa	IV_7
Aldrovanda	1	Widespread in Old World (3, 5, 6, 7, 16, 18, 43, 44, 53, 65, 67, 70–72, 94, 106, 131, 133)	azonal (aquatic)	azonal (aquatic)
Nepenthaceae				
Nepenthes-Nepenthes-Nepenthes	15	Eastern Asiatic r.; Khasi–Manipur p., Madagascan, Indian, Malesian & Neocaledonian rr. (28, 84, 89, 90, 107–109, 116)	Af (most), Am, Aw	V_1 (most)–V_3
Nepenthes-Nepenthes-Urceolatae	5	Indochinese, Malesian, Polynesian & Northeast Australian rr. (96–98, 100–105, 107–109, 113, 132)	Af (most), Am, Aw	V_1 (most)–V_2
Nepenthes-Nepenthes-Tentaculatae	9	Malesian r.; Borneo & Celebes pp. (103, 107)	Af	V_1
Nepenthes-Nepenthes-Insignes	12	Malesian r.; Borneo, Philippines, Celebes & Moluccas pp. (103,104,107,108)	Af	V_1
Nepenthes-Nepenthes-Villosae	19	Malesian r.; Borneo, Philippines (103, 104)	Af	V_1–V_2
Nepenthes-Nepenthes-Regiae	48	Malesian r.; Borneo, Philippines, Celebes, Moluccas & Papua pp. (103, 104, 107–109)	Af	V_1–V_2
Nepenthes-Nepenthes-Pyrophytae	23	Indochinese r., Malesian r.; Malaya, Borneo, Sumatra (few) pp. (96, 98, 100–103, 105)	Af (most), Am, Aw	V_1–V_3
Nepenthes-Nepenthes-Montanae	29	Malesian r.; Borneo (few), Sumatra (most) & South Malesia (few) pp. (103, 105, 106)	Af (most), Am, Aw	V_1–V_2
Sarraceniaceae				
Heliamphora	23	Guayana Highland r.; Guyana p. (120)	Af	V_1
Darlingtonia	1	Rocky Mountain r.; Vancouver p. (32, ?62)	Csb	III_1
Sarracenia	11	North American Atlantic r.; predominantly in Atlantic & Gulf Coastal Plain pp. (15, 29, 30)	Cfa	IV_7

(*continued*)

Table 2.1 (*Continued*)

Genus-Subgenus-Section	# of spp.	Floristic Region/Province of Takhtajan (1986)*	Köppen-Geiger Climate Classification**	Troll-Paffen Climate Classification***
Byblidaceae				
Byblis-Byblis-Anisandra	2	Southwest Australian r. (135)	Csb	IV_1
Byblis-Byblis-Byblis	6	Northeast Australian r.; North Australia & Queensland pp., just reaching nearby New Guinea (108, 131, 132)	Aw	V_2–V_4
Roridulaceae				
Roridula	2	Cape r. (130)	Csb	IV_1
Cephalotaceae				
Cephalotus	1	Southwest Australian r. (135)	Csb	IV_1
Drosophyllaceae				
Drosophyllum	1	westernmost Mediterranean r. (39)	Csb	IV_1
Dioncophyllaceae				
Triphyophyllum	1	Guineo-Congolian r.; Upper Guinean p. (65)	Af, Am	V_1

*r./rr.: floristic region/s; p./pp.: floristic province/s.

** *Main climates* (first letter): A–equatorial; B–arid; C–warm temperate; D–snow; E–polar; *Precipitation* (second letter): W–desert; S–steppe; f–fully humid; s–summer dry; w–winter dry; m–monsoonal; *Temperature* (last letter): h–hot arid; k–cold arid; a–hot summer; b–warm summer; c–cool summer; d–extremely continental; F–polar frost; T–polar tundra.

***I (Polar and Subpolar Zone): I_1–High-polar ice-cap climates; I_2–Polar climates; I_3–Subarctic tundra climates; I_4–Highly oceanic subpolar climates. II (Cold-temperate Boreal Zone): II_1–Oceanic boreal climates; II_2–Continental boreal climates; II_3–Highly continental boreal climates. III (Cool-temperate Zone, Woodland Climates): III_1–Highly oceanic climates; III_2–Oceanic climates; III_3–Suboceanic climates; III_4–Subcontinental climates. III (Cool-temperate Zone, Continental climates): III_5–Continental climates with cold, slightly dry winters; III_6–Highly continental climates; III_7–Humid- and warm-summer climates; III_8–Permanently humid, warm-summer climates. III (Cool-temperate Zone, Steppe Climates): III_9–Humid steppe climates with cold winters; III_{10}–Dry steppe climates with cold winters; III_{11}–Humid-summer steppe climates with cold winters; III_{12}–Semi-desert and desert climates with cold winters. IV (Warm-temperate Subtropical Zone): IV_1–Dry-summer Mediterranean climates with humid winters; IV_2–Dry-summer steppe climates with humid winters; IV_3–Steppe climates with short summer humidity; IV_4–Dry-winter climates with long summer humidity; IV_5–Semi-desert and desert climates; IV_6–Permanently humid grassland climates; IV_7–Permanently humid climates with hot summers. V (Tropical Zone): V_1–Tropical rainy climates; V_2–Tropical humid-summer climates; V_3–Wet-and-dry tropical climates; V_4–Tropical dry climates; V_5–Tropical semi-desert and desert climates.

centers of high diversity: *Nepenthes* in tropical Asia (frequently at elevated altitudes), *Drosera* in Australia, *Pinguicula* in Central America and the Mediterranean (predominantly at elevated altitudes), *Genlisea* in tropical South America and Africa, and *Utricularia* in tropical America, Africa, and Australia. More details on the biogeography and distribution of each family and genus are provided in the chapters devoted to each taxon (Chapters 4–10).

2.3 Habitat specificity defines regional distributions

Carnivorous plants do not grow in all habitats, either because they have not yet reached some of them (§2.2) or because they have particular adaptations that allow them to survive in nutrient-poor soils (Darwin 1875). Darwin based his hypothesis that botanical carnivory is an adaptation to nutrient-poor soils on the observation that *Drosera* grows on nutrient-poor soils and the related inference that this habitat specificity is related to poorly developed roots. Thus, the evolution of modified leaves to capture nutrients from insects and other small animals is seen as an alternative to additional investment in roots in nutrient-poor soils (Tilman 1988, Juniper et al. 1989).

2.3.1 Hypotheses concerning co-occurrence of carnivorous and noncarnivorous plants

In general, carnivorous plants have weak root systems (e.g., Juniper et al. 1989, Adamec 2005a, Adlassnig et al. 2005b, Brewer et al. 2011), and understanding the benefits of carnivory in part centers around understanding the relative advantages of plants investing in additional root absorptive area versus investing in carnivory. If increased root allocation and carnivory are simply different ways to accomplish the same goal in nutrient-poor habitats, then carnivorous plants should co-occur with noncarnivorous plants that have either relatively large root systems or have other adaptations for surviving in nutrient-poor soils (e.g., mycorrhizae or *Rhizobia*). Carnivorous plants would not be expected to be at a competitive disadvantage in such habitats, but rather nutrient niche complementarity (Brewer 2003), dispersal limitation, or other stochastic processes (Myers and Harms 2009) would result in their coexistence with noncarnivorous plants. Alternatively if carnivorous and noncarnivorous plants with equally effective adaptations for nutrient-poor soils do not co-occur, then environmental factor(s) other than soil nutrient availability (Givnish et al. 1984, Juniper et al. 1989; Brewer et al. 2011) or interspecific competition may limit the distribution of carnivorous plants.

Juniper et al. (1989) emphasized the role of calcium concentration in the soil, suggesting that most carnivorous plants are "calcifuges" (unable to tolerate conditions related directly or indirectly to high soil or water calcium concentrations) restricted to acidic soils with low nutrient availability. Although calcium toxicity has been demonstrated in a couple of species (Adlassnig et al. 2006), others are associated with high calcium or ultrabasic soils (Adlassnig et al. 2005b, van der Ent et al. 2015). Givnish et al. (1984) and Juniper et al. (1989) also identified that carnivorous plants grow mainly in nutrient-poor habitats that also are wet and unshaded.

Givnish et al. (1984; Chapter 18) explained this pattern using an evolutionary cost/benefit model based on the photosynthetic benefit of carnivory. Light-saturated photosynthesis of leaves of a noncarnivorous shrub is strongly and positively correlated with leaf nitrogen, which in turn requires greater investment of carbon to roots when soil nitrogen is in low supply (Gulmon and Chu 1981). By extension, the increased marginal photosynthetic benefit resulting from increased nutrient capture associated with additional marginal investment in carnivory should decline with increasing availability of substrate nutrients and with decreasing availability of water and light (Givnish et al. 1984; Chapters 17–19). In a review of studies purporting to test the cost/benefit model, Ellison (2006) concluded there were inadequate data to assess the prediction that carnivorous plants should be competitively disadvantaged in dry, shady, or nutrient-rich habitats.

2.3.2 Regional patterns of co-occurrence

To compare regional distributions of carnivorous and noncarnivorous plants, we evaluate the available evidence to test the prediction that terrestrial carnivorous plants are more likely to be associated with wet, unshaded, nutrient-poor habitats. We also address the related prediction that noncarnivorous plants are under-represented in habitats in which carnivorous plants occur. Support for the second prediction would support the cost/benefit model and could imply that carnivorous plants have a competitive advantage in wet, unshaded, nutrient-poor habitats, which would also support the cost/benefit model (Ellison 2006).

Carnivorous plants are more strongly associated with wet habitats than with nutrient-poor habitat in wet pine savannas of southern Mississippi, USA (Brewer et al. 2011). Carnivorous plants and noncarnivorous plants with deep, aerenchymatous roots were more strongly associated with the wettest microsites (Brewer et al. 2011). Variation in soil nutrient availability within the wet savannas was negligible and explained virtually no variation in abundance or occurrence of carnivorous plants (Brewer et al. 2011). Finally, carnivorous plants are disproportionately associated with wet habitats and wetter microsites within wet pine savannas, but these habitats also have some of the highest total plant species richness in the region (Peet and Allard 1993, Palmquist et al. 2014, Noss et al. 2015). This result suggests that the same factors responsible for the general restriction of carnivorous plants to wet, unshaded, nutrient-poor habitats simultaneously are responsible for extraordinarily high plant diversity overall (Peet et al. 2014).

These results suggest that it is important not only to address why carnivorous plants have high habitat specificity but also what allows them to coexist with noncarnivorous plants in the same habitats. Explanations for the absence of carnivorous plants in other habitats include competitive exclusion or lack of environmental tolerance.

Nutrient availability and competitive exclusion. Both Darwin's hypothesis and the cost/benefit model predict that carnivorous plants will be at a competitive disadvantage in nutrient-rich soils. Numerous studies have examined variation in carnivory investment or variation in performance in relation to substrate nutrient availability (reviews in Juniper et al. 1989, Adamec 1997a, Ellison and Gotelli 2009; Chapters 17–19). Most studies have focused on one or a few carnivorous species grown without competition from noncarnivorous plants, and have found that investment in carnivory decreases with substrate nutrient addition or increased nutrient availability (Adamec 1997a, Ellison 2006: Chapter 17). In some cases, decreased investment in carnivory with increasing substrate nutrient availability is adaptive (Thorén and Karlsson 1998, Zamora et al. 1998, Thorén et al. 2003), but in others increased nutrient deposition reduces both trap production and individual plant fitness (Ellison and Gotelli 2002, Gotelli and Ellison 2002).

These studies do not address the question of competitive disadvantage in nutrient-rich soils (Ellison 2006). A shift in allocation by a carnivorous plant from trap production to increased chlorophyll concentrations in leaves or to leaves with greater surface area is functionally similar to a noncarnivorous plant increasing its leaf-to-shoot ratio as a function of increased substrate nutrient supply (Olff et al. 1990); both can prevent or delay competitive displacement (Dybzinski and Tilman 2007). Demonstration that carnivorous plants are so disadvantaged in nutrient-rich substrates that their distribution would be affected requires data showing that either carnivorous plants are affected more negatively by increased nutrient availability than are noncarnivorous plants; or noncarnivorous plants competitively suppress or exclude carnivorous ones in nutrient-rich substrates.

Although a few investigators have examined competition between carnivorous and noncarnivorous plants (Wilson 1985, Svensson 1995, Brewer 1998a, 1999b, 1999c, 2003), only one has examined soil fertility-related differences in competitive effects (Abbott and Brewer 2016). That study examined the competitive effects of noncarnivorous plants on *Sarracenia alata* transplanted from its native bog into a productive, nutrient-rich marsh within its dispersal range, but where no carnivorous plants grew. Although the growth rate and survival rate of transplants of pitcher plants were indeed significantly lower in the marsh and in marsh soil than in the bog or bog soil, there was no evidence of a competitive effect in either the bog or the marsh (Figure 2.1). The poor performance of transplants in savanna soil with marsh neighbors intact indicated the potential for competitive exclusion in the marsh, a potential not realized, however, because of the greater importance of the harsh edaphic conditions (i.e., lower redox potential) in marsh soil (Figure 2.1). With but scant data, there is little support for the hypothesis that carnivorous plants are competitively excluded from nutrient-rich habitats. Abiotic stress factors also need to be evaluated to determine why carnivorous plants do not grow in many open and unshaded but nutrient-poor sites.

Light. The cost/benefit model predicts that carnivorous plants are intolerant of shade or are poor competitors for light, and most species are associated with open, herbaceous habitats (Juniper et al. 1989, Givnish 1989). With the possible exception of *Drosophyllum lusitanicum*, which could be considered a dwarf tree, there are no carnivorous trees, and even *D. lusitanicum* grows in sunny sites. Although *Triphyophyllum peltatum* is somewhat shade-tolerant (Juniper et al. 1989), it is carnivorous only during part of its life (Chapter 10). Like *Triphyophyllum*, other carnivorous plants that grow in forested habitats (e.g., *Nepenthes*; Chapter 5), tend to be climbers that rely on trees to get above the tree canopy or stolons to move into gaps or edges to take advantage of increased light availability. *Triphyophyllum* does not require production of carnivorous leaves to make the transition from rosette to liana; nutrient deficiency, not light, appears to trigger the production of carnivorous traps in the field, but the mechanism for the ontogenetic shift remains unknown (Bringmann et al. 2002). In Sierra Leone, *Triphyophyllum* often is found at forest edges and in secondary forests with reduced overstory canopies (Jonathan 1992), a finding that is consistent with a hypothesis of light-triggered investment carnivory.

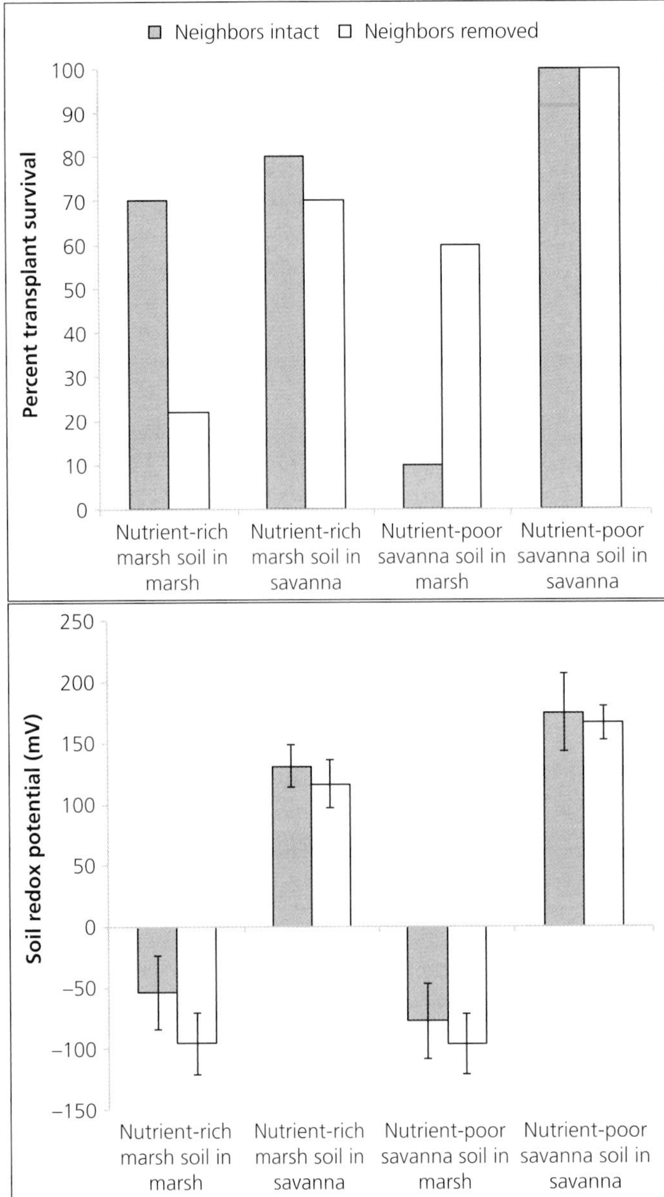

Figure 2.1 Percent survival of transplanted *Sarracenia alata* and soil redox potential (means ± 1 standard error of the mean) as a function of site (marsh, wet savanna), soil (nutrient-rich marsh soil, nutrient-poor savanna soil), and neighbor removal (neighbors intact, neighbors removed). Data from Abbott and Brewer (2016).

Field studies demonstrating competitiveness as a function of light availability or differential shade tolerance are rare. Zamora et al. (1998) reported reduced performance of *Pinguicula vallisneriifolia* when shaded in the field but they did not examine competition for light. A field competition experiment examining above- versus below-ground competitive effects of neighboring vegetation on *Drosera capillaris* failed to show that shade was the primary factor excluding it from dense woody thickets within a wet pine savanna (Brewer 1999b). However, *D. capillaris* showed increased emergence from a bank of dormant seeds following fire, and this was a response to shade reduction, not fire-mediated changes in the soil (Brewer 1999b; Chapter 22). Nevertheless, reduction of above-ground biomass in woody thickets where the abundance of *D. capillaris* was low had no greater positive effect on seedling emergence than did such reduction in open sedge meadows, where *D. capillaris* was abundant (Brewer 1999b). Growth of adult transplants of *D. capillaris* was lower in woody thickets than in open sedge meadows, but it did not benefit more from a reduction in above-ground biomass in the former than in the latter, suggesting that below-ground competition with woody neighbors was more important than above-ground competition. In sum, support for the hypothesis that carnivorous plants are poor competitors for light remains weak. Although carnivorous plants are rare in shady forests, it seems likely that factors other than light availability play a role in their absence.

Substrate moisture. Among habitats with similarly nutrient-poor soils, the cost/benefit model predicts that carnivorous plants will be at a disadvantage in drier soils. Noncarnivorous plants can adapt to poor, dry soils (via natural selection or phenotypic plasticity) by increasing allocation to roots. In contrast, if investment in carnivory conflicts with investment in roots, then the reduced benefit-to-cost ratio of investment in carnivory as soil moisture declines puts carnivorous plants at a disadvantage. Demonstrating that this disadvantage is large enough to affect their distributions requires evidence either that carnivorous plants are less tolerant of dry soils or that they cannot compete with noncarnivorous plants for water in drier soils; such evidence is lacking.

As with data on performance in different light environments, support for the soil-moisture hypothesis is inferred from studies of phenotypic plasticity in carnivorous investment and growth responses by individual species in response to substrate moisture supply. Zamora et al. (1998) found that growth of *Pinguicula vallisneriifolia* responded more positively to increased light and prey when water was not limiting, which is consistent with the prediction that carnivory is less beneficial in drier substrates. Some carnivorous plants do grow in seasonally dry habitats (e.g., *Drosera* in Western Australia) and rely on seed or bud dormancy or water storage to persist through dry periods (Dixon and Pate 1978, Juniper et al. 1989, Luken 2007). How *Drysophyllum lusitanicum* thrives in consistently dry soils along the Mediterranean coast of Iberia and North Africa is unclear but may be related to foliar absorption of sea mist or the production of deep roots (Juniper et al. 1989, Müller and Deil 2001, Correia and Freitas 2002, Adlassnig et al. 2005a, Paniw et al. 2015).

Reduced performance and efficiency of carnivory when water availability is low could indicate greater sensitivity to drought, but a proper test requires comparative studies. Brewer et al. (2011) examined the responses of carnivorous and noncarnivorous plants to experimentally manipulated soil moisture availability in the field (via substrate elevation in a wet savanna) and found that *Drosera* and *Sarracenia* spp. responded more negatively to elevating the substrate (reduced water availability) than did co-occurring noncarnivorous plants. In the same study, structural equation modeling and nonparametric multiplicative regression were used to quantify the potential role that competition played in determining distributions of these species along a moisture gradient from a mesic savanna ecotone to wet savanna. Consistent with the results of the substrate elevation experiment, the statistical models revealed that carnivorous plants were associated more strongly with wetter microsites. Abundances of carnivorous plants were low in the drier, mesic soils, irrespective of the abundance of mesophytic noncarnivorous plants with deep roots. Hence, the greater relative frequency of carnivorous plants in the wetter microsites largely resulted from a lower tolerance by carnivorous plants of drier soils. Hydrophytic noncarnivorous plants with well-developed root aerenchyma, however, also appeared to be intolerant of dry soils. Hence, the lack of tolerance of carnivorous plants to dry soils is comparable to that of obligate wetland noncarnivorous ones, and competition for water did not appear to determine species distributions along a moisture gradient in the savanna.

Although it is generally the case that carnivorous plants are less tolerant of low soil moisture, the cost/benefit model is not the only explanation for why they are more strongly associated with wet habitats than noncarnivorous plants. Many carnivorous plants are obligate wetland species and carnivory could be an alternative to the production of deep roots with well-developed aerenchyma as a means of maximizing nutrient uptake in hypoxic substrates (Brewer et al. 2011). Instead of producing well-developed root aerenchyma (Seago et al. 2005), carnivorous plants may avoid substrate hypoxia by producing shallow roots (Brewer et al. 2011). To compensate for the reduced access to soil nutrients, carnivorous plants invest in traps to capture nutrients above-ground. One consequence of producing shallow roots (or no roots), however, is extreme sensitivity to dry conditions. Other functional plant types with specialized adaptations for nutrient capture but which do not generally produce shallow roots (e.g., legumes) rarely are restricted to wet habitats (Depuy and Dreyfus 1992, Cornelissen et al. 2003) and in some cases cannot tolerate wet substrates (Saur et al. 2000).

2.4 Mechanisms of coexistence in wet, unshaded, nutrient-poor soils

In the absence of experimental data, Juniper et al. (1989) asserted that carnivorous plants were, generally speaking, poor competitors with noncarnivorous ones, a conclusion at odds with that of Lloyd (1942). Furthermore, contrary to their claim that carnivorous plants typically grow only in close association with other carnivorous plants, *Sphagnum*, and a few orchids (Juniper et al. 1989: 24), some plant communities known to have among the highest local plant species diversity in temperate North America have numerous species of carnivorous and noncarnivorous plants growing together (Peet and Allard 1993, Palmquist et al. 2014, Noss et al. 2015). We suggest that a better understanding of two aspects of coexistence of carnivorous and noncarnivorous plants—niche complementarity and fire-mediated stochasticity—will improve our understanding of general mechanisms of plant species coexistence.

2.4.1 Niche complementarity

Brewer (2003) asked why pitcher plants do not compete with noncarnivorous plants for nutrients. An obvious answer is that the growth and reproduction of the different species are limited by difference nutrients. If the hypothesis that carnivory is an alternative to producing deep roots with well-developed aerenchyma is correct, then these alternative adaptations could represent different nutrient-capture niches. If so, as predicted by the competitive exclusion principle, a lack of niche overlap could permit coexistence. Testing the competitive exclusion principle requires experimental manipulation of the niche of one of the competitors so that it is more similar to that of the other competitors. Although such experimental niche manipulation had been done before by manipulating root co-occurrence (e.g., Berendse 1982), it was easier to manipulate the nutrient-capture niche of *Sarracenia alata*. Because *S. alata* produced shallower and less extensive roots than most of its noncarnivorous neighbors (Brewer 2003), Brewer predicted that it might be a relatively poor competitor for soil resources (including soil nutrients). To test this idea, Brewer (2003) measured how the growth of juvenile *S. alata* individuals responded to being denied prey and to the reduction of neighbors (most of which were noncarnivorous). Prey was excluded from *S. alata* by filling the pitchers with fabric batting; noncarnivorous neighbors were reduced by clipping them to the ground repeatedly over the course of a growing season and by uprooting plants in the immediate vicinity of *S. alata*. Hence, he attempted to reduce both above- and below-ground competition.

The growth of *S. alata* (but not survival) increased dramatically in response to the sustained reduction of the neighbors (Brewer 2003); there was clear evidence of competition. However, the competitive release was not consistent with different nutrient niches (Figure 2.2). *Sarracenia alata* did not show significantly reduced growth when denied prey but when neighbors were left intact. Rather, the negative effect of denying prey was significant only when the neighbors were reduced (Figure 2.2). Brewer (2003) interpreted these results as evidence that *S. alata* increased its demand for prey upon receiving more light, which occurred when he reduced its neighbors. This result is likely an example of adaptive phenotypic plasticity in response to varying light levels, whereby the demand for nutrients (and thus the negative effect of prey exclusion) increases with reduced competition for light (Brewer 2006). Alternatively, competition for prey and light among conspecifics limited prey capture and growth. Prey capture was indeed greater when neighbors were removed (although not on a per pitcher volume basis Brewer [2003]). Gibson (1991a) found reduced prey capture in *Drosera tracyi* when conspecifics were

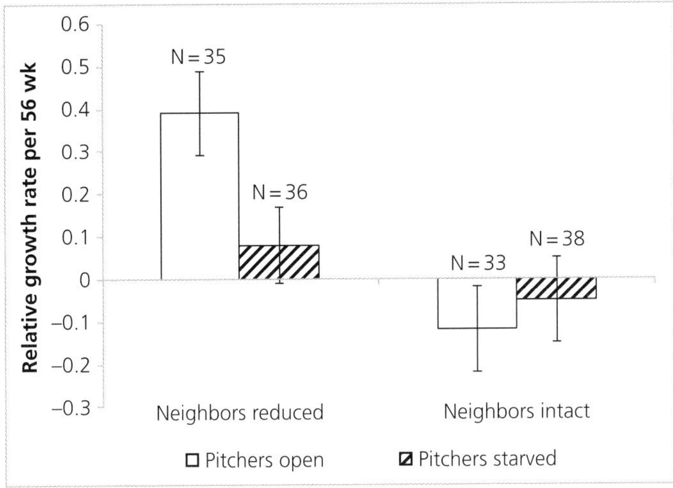

Figure 2.2 Effects of neighbor removal and prey exclusion on relative growth rate over a 56-week period. Least-squares means (± 1 SE of the mean) of relative growth rate are adjusted for the initial diameter of the largest pitcher on each target plant. Figure modified from Brewer (2003) and reproduced with permission from the Ecological Society of America.

added. Greater negative effects of pitcher-plant neighbors relative to noncarnivorous ones could promote species coexistence, provided the latter also exhibit density dependence (Chesson 2000).

2.4.2 Fire-mediated stochasticity

Absent niche complementarity, another possible explanation for coexistence of carnivorous and noncarnivorous herbs is that they compete for resources equally well (or poorly) in wet pine savannas, and repeated fires maintain nonequilibrium populations of each. The reduction of competition by repeated fires benefits both carnivorous plants (Roberts and Oostings 1958, Folkerts 1982, Barker and Williamson 1988, Schnell 2002, Correia and Freitas 2002, Garrido et al. 2003, Kesler et al. 2008, Paniw et al. 2015), and many co-occurring herbaceous noncarnivorous ones (Glitzenstein et al. 2003, 2012, Hinman and Brewer 2007, Palmquist et al. 2014).

To test whether losses of herbaceous species associated with prolonged fire exclusion in wet pine savannas in southern Mississippi were stochastic or deterministic, Brewer analyzed effects of natural additions of competitively dominant trees and associated shrubs on assemblages of carnivorous and noncarnivorous herbs (Brewer 2017). Losses of herb species associated with shrub thickets that develop underneath mature slash pines (*Pinus elliottii*) that established during a period of fire exclusion provides a long-term (up to 65-yr) "natural" competition experiment. The establishment of trees in wet pine savannas reduces dispersal limitation of shrubs but not herbs, nucleating the addition of dominant woody competitors (Hinman et al. 2008). If herb species losses associated with tree and shrub establishment resulted from deterministic processes such that certain herb species consistently were more vulnerable to competitive displacement than were others, then herbaceous beta diversity of areas near trees (i.e., spatial variation in species composition *sensu* Anderson et al. 2011) should be lower than in open areas. Thus, resulting assemblages near trees would be compositionally more similar to one another than would be assemblages in open areas away from trees. One way in which beta diversity could decline near trees is if certain functional groups (e.g., carnivorous herbs) consistently were excluded from the neighborhoods of the dominant woody plants (Figure 2.3). Alternatively, if losses of herb species associated with tree and shrub establishment resulted from local stochastic processes, then herbaceous beta diversity of areas near trees should be similar to that in open areas. Assuming no dispersal or resource niche differences between carnivorous and noncarnivorous herbs (Myers and Harms 2009), competitive displacement of either group would be equally likely (Figure 2.3).

Brewer (2017) contrasted alpha and beta diversity of neighborhoods associated with slash pine/shrub thickets to those in open areas away from thickets in each of three wet pine savanna sites in southern Mississippi (Wolf Branch, Sandy Creek, and Little Red Creek) using a permutation-based dispersion

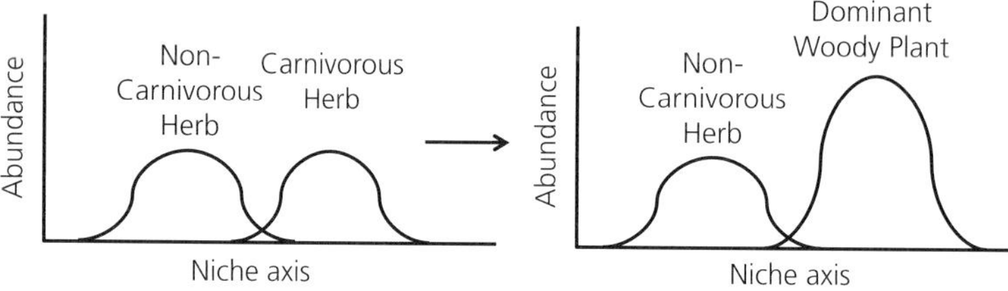

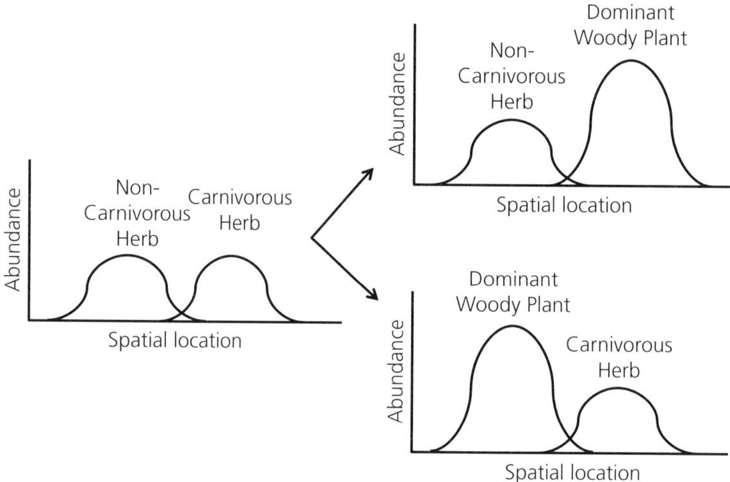

Figure 2.3 Predicted patterns of competitive displacement of herbs by dominant woody plants in wet pine savannas when coexistence of carnivorous and noncarnivorous herbs results from either niche differences (**a**) or neutrality and chance (**b**).

test (Permdisp: M.J. Anderson 2006). To ensure that differences in multivariate dispersion (and thus beta diversity) were not the result simply of differences in alpha diversity (which was significantly lower near slash pine trees than away from them; Brewer 1998b), Brewer calculated Raup-Crick distances using the raupcrick function in the R vegan package (Chase et al. 2011).

In all three savannas, beta diversity did not differ significantly between tree/shrub thickets and open areas away from thickets (Wolf Branch $F_{1,28} = 0.019$; $p = 0.91$; Sandy Creek $F_{1,30} = 0.43$; $p = 0.49$; Little Red Creek $F_{1,30} = 0.72$; $p = 0.41$), and Brewer concluded that trees or associated shrubs stochastically eliminated herb species from their neighborhoods. If this conclusion is true, herbaceous species that coexist in open areas do not differ functionally from one another in ways that affect their ability to compete with the dominant woody species. Hence, we suggest that increases in dominant woody competitors following reductions in fire frequency (Hinman et al. 2008) result in random losses of herbaceous species and that carnivorous herbs are no more or less vulnerable to competitive exclusion than are noncarnivorous herbs. These results do not mean that niche differences were unimportant entirely. Negative feedbacks unrelated to how herbs compete with woody plants could contribute to long-term species coexistence (Adler et al. 2010).

2.5 Future research

A recurring theme throughout much of this chapter is that our understanding of the biogeography and ecology of carnivorous plants requires comparisons with noncarnivorous plants. For whatever reasons, such comparisons are rarely done (Ellison 2006, Brewer et al. 2011), and research in these areas has

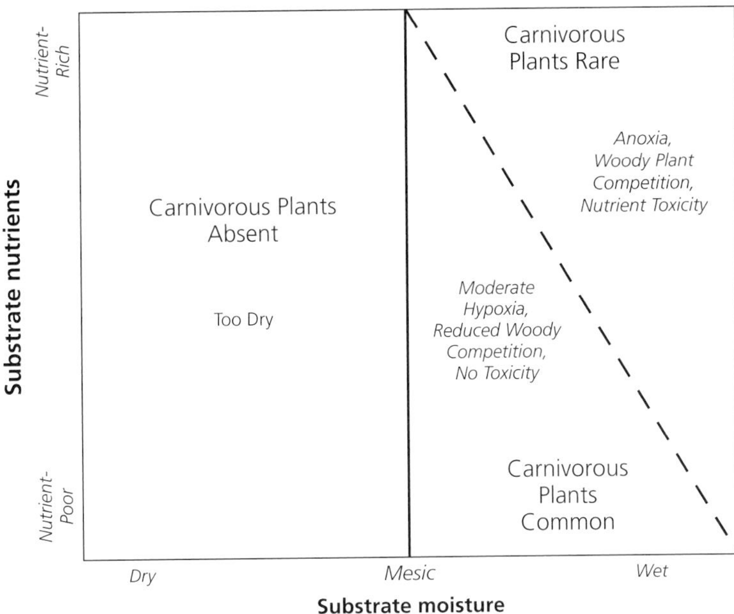

Figure 2.4 Predicted habitat distributions of terrestrial, wetland carnivorous plants and hypothesized environmental factors determining their distributions.

focused on phenotypic plasticity of carnivorous effort or have compared differences among carnivorous plant species. Contrasting the drivers of distributions of carnivorous and noncarnivorous plants will improve our understanding of both limits to carnivorous plant distribution and to mechanisms of plant coexistence in diverse assemblages (Figure 2.4).

Data available to date suggest that terrestrial carnivorous plants generally are less tolerant of dry soils than are most noncarnivorous ones (Juniper et al. 1989). Exceptions arise in seasonally dry habitats, where carnivorous plants reduce or suspend investment in carnivory during the dry season, persisting as tubers, rhizomes, leafless stipule clusters, or dormant seeds (Chapter 22). *Drosophyllum lusitanicum* is an exception to this general rule, and additional research is needed to determine if it relies on absorption of sea mist through its leaves or absorption of water through deep roots.

In the southeastern United States, the positive association of carnivorous plants with nutrient-poor soils is not as strong as their association with wet soils. This difference between carnivorous plants and noncarnivorous plants may result from carnivorous plants being less tolerant of dry soils than of nutrient-rich soils. Many carnivorous plants are absent from some habitats with nutrient-rich soils, but the reasons why remain unclear. To date, there is no evidence to support the hypothesis that carnivorous plants simply cannot compete with noncarnivorous plants in nutrient-rich soils. More competition experiments are needed (including those that manipulate competition for water with woody plants), and alternative explanations (nutrient toxicity, anoxia intolerance) need to be explored more fully.

Carnivorous herbs are outcompeted by woody plants (and in some cases, large grasses) but do not appear to be less competitive than co-occurring noncarnivorous herbs. Traits that carnivorous and noncarnivorous herbs share that make them vulnerable to competition from woody plants (e.g., small size, slow growth) deserve greater attention.

Finally, explanations for coexistence of carnivorous and noncarnivorous herbs when woody plants are reduced by fire or other disturbances remain elusive. There is some limited evidence for neutral coexistence of carnivorous and noncarnivorous herbs in wet pine savannas and no evidence yet supports partitioning of nutrient niches. The roles played by intraspecific competition for prey by carnivorous plants and soil-mediated negative feedbacks on noncarnivorous neighbors in promoting coexistence need to be examined.

CHAPTER 3

Evolution of carnivory in angiosperms

Andreas Fleischmann, Jan Schlauer, Stephen A. Smith, and Thomas J. Givnish

3.1 Introduction

Carnivorous plants are an ecologically defined group of organisms (Darwin 1875), not a single lineage marked by common descent. To date, > 800 species of angiosperms—in five orders, 12 families, and 19 genera—are broadly recognized as carnivorous plants (Table 3.1, Figure 3.1; Appendix).

3.1.1 Evolution of carnivory

Charles Darwin (1875) provided the first conclusive evidence that some plants could trap and digest animals. Ever since, evolutionary biologists have asked how and when carnivory evolved among plants. Darwin (1875) himself was convinced that there had been a number of different, independent origins of plant carnivory, but establishing the relationships between various carnivorous genera was, until the advent of molecular systematics, often hampered by convergent, parallel, or divergent evolution of morphological traits associated with carnivory (Ellison and Gotelli 2009).

By definition, carnivorous plants must have some unequivocal adaptation(s) to attract, trap, or digest prey, be capable of absorbing nutrients from killed animals next to their surfaces, and obtain some benefit thereby in terms of increased growth, survival, or reproduction (Givnish et al. 1984, Givnish 1989, Juniper et al. 1989, Adamec 1997a; Chapter 1). The "carnivorous syndrome" (Chapter 1) manifests itself through numerous morphological and physiological adaptations, from gross morphology and leaf form (Chapters 5–10), through glandular structure and function, to gene expression and evolution (Chapters 11–16). Many species have specific ecological associations (digestive mutualism, myrmecophily, coprophagy) with other organisms that contribute to prey capture and digestion or plant protection against herbivores (Givnish 1989, Ellis and Midgley 1996, Anderson and Midgley 2003, Alcaraz et al. 2016; Chapters 23–26). Several of the morphological adaptations of carnivorous plants (such as secretory glands in Caryophyllales and Lamiales), and many of the genes activated in the trap leaves (e.g., Bemm et al. 2016) appear to have evolved or been repurposed from herbivore defense mechanisms (Darwin 1875, Kerner von Marilaun 1878, Müller et al. 2004, Alcalá et al. 2010; Chapters 13, 16).

Five general types of traps have evolved in carnivorous plants: adhesive ("flypaper") traps, consisting of sticky glandular leaves; pitcher ("pitfall") traps, formed by tubular leaves or, in the case of tank-forming monocots, rosettes of leaves; "snap-traps", formed by rapidly closing laminar lobes; specialized eel (or "lobster-pot") traps, formed by narrow, tubular leaves that are internally lined with retrorse hairs; and suction ("bladder") traps, which are highly modified, sac-like leaves. Some of these trapping methods have evolved only once—e.g., suction traps in *Utricularia*—whereas others have evolved convergently in several different clades—e.g., epiascidiate pitchers in Cephalotacaceae, Nepenthaceae, and Sarraceniaceae, in which the adaxial (upper) side of the leaf forms the pitcher interior (Arber 1941, Lloyd 1942, Franck

Fleischmann, A., Schlauer, J., Smith, S. A., and Givnish, T. J., *Evolution of carnivory in angiosperms.*
In: *Carnivorous Plants: Physiology, ecology, and evolution.* Edited by Aaron M. Ellison and Lubomír Adamec:
Oxford University Press (2018). © Oxford University Press. DOI: 10.1093/oso/9780198779841.003.0003

Table 3.1 Carnivorous plant taxa. For each family, the number of total genera and the number of carnivorous genera (if different) are given in parentheses. Estimated phylogenetic (stem) age is in millions of years before present (Mya). For each genus, the total number of species and the number of carnivorous species (if different) are given in parentheses. Modified from Fleischmann (2010), species numbers based on and updated from McPherson (2011), McPherson et al. (2011), Fleischmann (2012a, 2012b, 2015a), Lowrie (2013), Givnish et al. (2014a), Gonella et al. (2016), Scatigna et al. (2017), and APG IV (2016). Age estimations based on Figure 3.1 (S. Smith and T. Givnish *unpublished data*), except for carnivorous Ericales (Ellison et al. 2012) and for *Genlisea and Utricularia* (Ibarra-Laclette *et al.* 2013).

Order	Family	Phylogenetic age (estimate)	Genus	Phylogenetic age (estimate)	Trap type	Distribution
Poales	Bromeliaceae (58; 2)	21.2	*Brocchinia* (20; 2)	1.9 (*B. reducta*)	Pitfall	Guayana Highlands
			Catopsis (≈20; 1)	2.6 (*C. berteroniana*)	Pitfall	Neotropics
	Eriocaulaceae (6; 1)	89.5	*Paepalanthus* (≈450; 1)	2.7 (*P. bromelioides*)	Pitfall	Brazil
Caryophyllales	Droseraceae (3)	84.8	*Drosera* (≈250)	53.4	Adhesive	Cosmopolitan
			Dionaea (1)	48.0	Snap	Eastern USA
			Aldrovanda (1)	48.0	Snap	Old World
	Nepenthaceae (1)	84.8	*Nepenthes* (≈130–160)	84.8	Pitfall	Southeast Asia, India, Australia, Madagascar, Seychelles
	Drosophyllaceae (1)	70.4	*Drosophyllum* (1)	70.4	Adhesive	Western Mediterranean
	Dioncophyllaceae (3; 1)	54.2	*Triphyophyllum* (1)	6.9	Adhesive	Tropical western Africa
Ericales	Sarraceniaceae (3)	48.6	*Sarracenia* (11)	22.8	Pitfall	Eastern USA + Canada
			Darlingtonia (1)	35.0	Pitfall	Western USA
			Heliamphora (23)	22.8	Pitfall	Guayana Highlands
	Roridulaceae (1)	38.1	*Roridula* (2)	38.1	Adhesive	Cape of South Africa
Oxalidales	Cephalotaceae (1)	32.2	*Cephalotus* (1)	32.2	Pitfall	Southwest Western Australia
Lamiales	Byblidaceae (1)	44.5	*Byblis* (8)	44.5	Adhesive	Australia
	Lentibulariaceae (3)	43.4	*Pinguicula* (≈96)	33.5	Adhesive	Cosmopolitan, excluding Australia
			Genlisea (30)	31.0	Eel	Tropical Africa, Neotropics
			Utricularia (≈240)	31.0	Suction	Cosmopolitan
	Plantaginaceae (≈90/1)	44.2	*Philcoxia* (7)	19.3	Adhesive	Brazil

1976, Froebe and Baur 1988, Juniper et al. 1989, Fukushima et al. 2015; Figure 3.2). In Droseraceae and Lentibulariaceae, divergent trap types have evolved in closely related genera. Adhesive traps can be active (with mobile glands or leaves: *Drosera, Pinguicula*; Chapter 14) or passive (not capable of movement upon prey capture: *Byblis, Drosophyllum, Philcoxia, Roridula, Triphyophyllum*; Figure 3.3; Chapter 15). Some species of *Drosera* (e.g., *D. glanduligera*) have traps that combine functional properties of adhesive traps and snap-traps (Poppinga et al. 2012; Chapter 14).

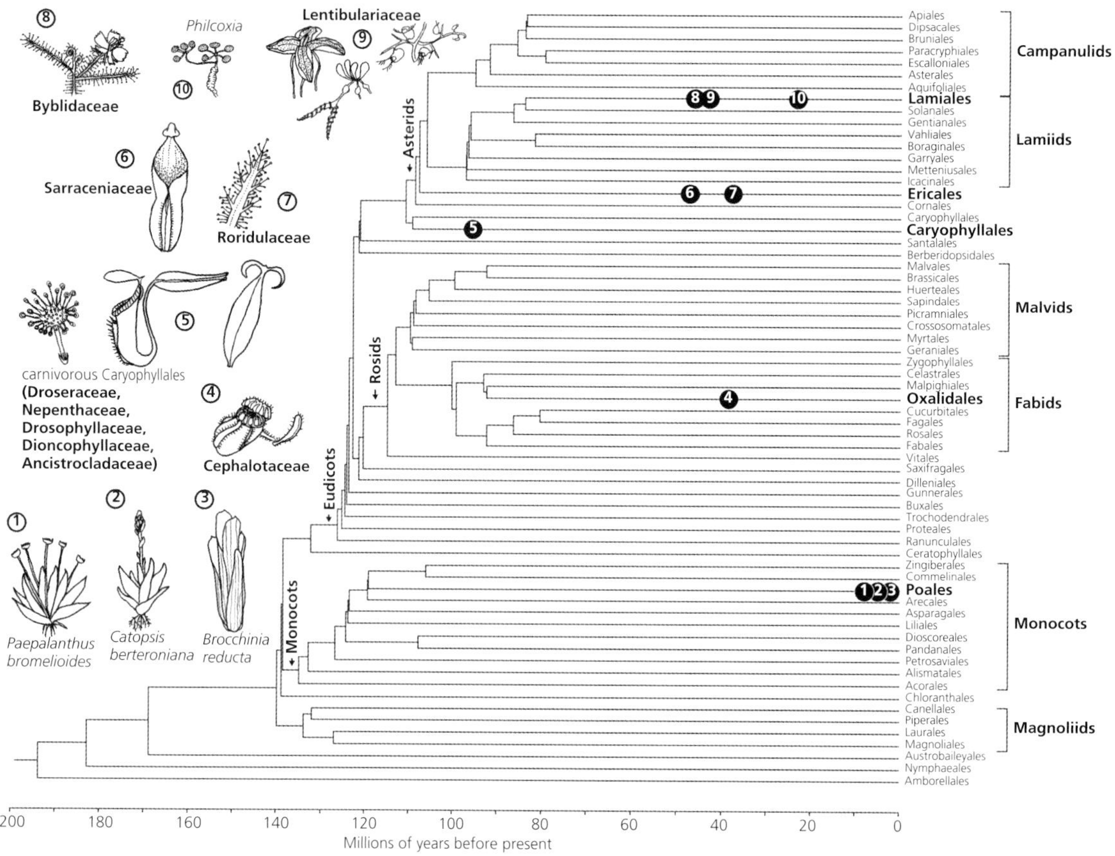

Figure 3.1 Distribution and dates of origin of the ten carnivorous clades of flowering plants. Ordinal classification follows APG (2016). Ages of nodes (including stem ages of orders) were estimated from a maximum-likelihood analysis of multiple plastid loci, constrained by the topology of the tree presented by APG (2016). For noncarnivorous taxa, all families were represented by a single placeholder; for carnivorous clades, all species with appropriate sequences in GenBank were included. Twenty-three primary calibration points were obtained from Magallón et al. (2015); four secondary calibration points for the crown ages of monocots, Asparagales, Poales, and Bromeliaceae were obtained from Givnish et al. (2015) and S. Smith and T.J. Givnish (*unpublished data*). Dots indicate stem ages of carnivorous clades: Byblidaceae, Lentibulariaceae, and *Philcoxia* in Lamiales; Roridulaceae and Sarraceniaceae in Ericales; the Droseraceae-Nepenthaceae-Drosophyllaceae-Dioncophyllaceae-Ancistrocladaceae clade in Caryophyllales; Cephalotaceae in Oxalidales; and *Paepalanthus, Brocchinia reducta*, and *Catopsis berteroniana* in Poales. Age of *Paepalanthus bromelioides* calculated from data of Trovó et al. (2013) run on RAxML; ultrametric tree formed using chronoPL assuming stem age of *Paepalanthus* = 48.5 Mya. The age of *Catopsis berteroniana* is based on the stem age of *Catopsis* estimated here and branch lengths within the genus given by Gonsiska (2010). Illustration by Andreas Fleischmann.

3.1.2 Origins of carnivory

Although Croizat (1960) proposed that carnivorous plants had a single origin, contemporary researchers have used molecular systematics to demonstrate that carnivory evolved independently among flowering plants at least ten times (Albert et al. 1992, Chase et al. 1993, Givnish et al. 1997, 2011, 2014a, Cameron et al. 2002, Müller et al. 2004, 2006, Heubl et al. 2006, Ellison and Gotelli 2009, Fleischmann 2010, Schäferhoff et al. 2010, Ellison et al. 2012, Pereira et al. 2012, Givnish 2015; Schwallier et al. 2016, Stephens et al. 2015b; Figure 3.1, Table 3.1). Carnivory likely arose once in the Caryophyllales ("non-core Caryophyllales" *sensu* APG IV 2016; §3.2) and Oxalidales, twice in the Ericales, and three times each in the Lamiales and the Poales (Figure 3.1). Nearly 98% of all carnivorous plant species are found in just

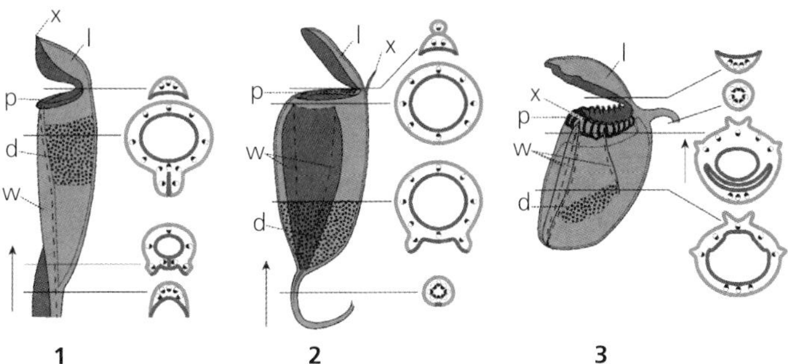

Figure 3.2 (Plate 1 on page P1) Convergent evolution of foliar pitchers in unrelated taxa. External pitcher appearance, surfaces, and schematic cross sections of **1**. *Sarracenia*; **2**. *Nepenthes*; and **3**. *Cephalotus*. All have pitcher leaves of epiascidiate ontogeny but of fundamentally different morphology and anatomy. Arrows: position and growth direction of shoot axis, axis in cross sections located below; *d*: digestive (glandular) zone inside pitcher; *w*: wing (ala) of pitcher outer surface; *p*: peristome; *l*: lid; *x*: true leaf apex; light gray (green in color plate): abaxial surface; dark gray (red in color plate): adaxial surface; partially black-and-white filled ellipses: main vascular bundles in cross section, black: xylem, white: phloem. Pitchers not shown in correct size relations. Pitfall traps of the carnivorous monocots (Chapter 10) are made up of the entire rosette, while in the taxa illustrated here, each pitcher is derived from a single leaf. Illustration by Jan Schlauer.

Figure 3.3 (Plate 2 on page P1) Convergent and homologous evolution of adhesive traps in carnivorous plants. The peculiar outwardly circinate vernation of (**a**) *Drosophyllum* also is observed in the related *Triphyophyllum* (**b**), both in the emerging carnivorous (top) and noncarnivorous leaves (bottom; both Caryophyllales). (**c**) This is paralleled in the unrelated *Byblis* (illustrated by *B. aquatica*; Lamiales). *Drosera* (Droseraceae, Caryophyllales; illustrated by (**d**) *Drosera tracyi* and (**e**) *D. capensis*) and (**f**) *Pinguicula* (early-branching Lentibulariaceae, illustrated by *Pinguicula heterophylla*), have similar active flypaper traps, but with inward circination. In both, the leaves of several species are motile upon stimulation by prey. Photographs by Andreas Fleischmann.

the Caryophyllales, Lentibulariaceae, and Ericales, 93% in the first two clades alone (Ellison and Gotelli 2009; Table 3.1). None of these five angiosperm orders are exclusively carnivorous, but nine of the 12 families in which carnivorous plants occur are exclusively carnivorous or very nearly so.

Four families—Bromeliaceae, Eriocaulaceae, Dioncophyllaceae, Plantaginaceae—comprise predominately noncarnivorous genera and species (Table 3.1) and apparently involve quite recent transitions to or losses of the carnivorous habit. Carnivory also has been lost a few times in the Caryophyllales—in the Ancistrocladaceae and Dioncophyllaceae (*Habropetalum* and *Dioncophyllum*; §3.7). Carnivorous lineages are especially numerous in regions with open, nutrient-poor sites and abundant rainfall relative to evaporation, where the economics of nutrient capture and plant growth are likely to favor carnivory as an ecological adaptation (Givnish 1989; Chapter 18).

More than one trap type has evolved in two unrelated clades—Lentibulariaceae and Caryophyllales (Figure 3.1)—but in all cases, they appear to have originated from adhesive traps (Müller et al. 2004, Heubl et al. 2006, Fleischmann 2010). This implies a minimum of one or two transitions between trap types within these lineages, and raises questions about the homology among diverse trap types (e.g., sticky leaves in *Pinguicula*, below-ground eel traps in *Genlisea*, and suction traps in *Utricularia* in Lentibulariaceae; adhesive traps in *Drosera*, *Drosophyllum*, and *Triphyophyllum*, snap-traps in *Aldrovanda* and *Dionaea*, and pitcher traps in *Nepenthes* within carnivorous Caryophyllales).

Carnivorous Ericales also have two different trap types (pitcher traps in Sarraceniaceae, sticky traps in Roridulaceae), but in this lineage, independent origins of the traps is more likely (§3.4.1). Resin-secreting glands are widespread among Ericales (e.g., several species of *Rhododendron*, Ericaceae), and many species are associated with capsid bugs that feed opportunistically on casually caught arthropods (e.g., Sugiura and Yamazaki 2006). Carnivory could have evolved easily from such ancestors and the digestive mutualism of *Roridula* (Chapters 10, 26) could represent simply a continuation and intensification (via obligate mutualism of plant and animal partner) of an Ericalean exaptation of sticky plants and associated, scavenging plant bugs. In contrast, the fundamentally different trap type of Sarraceniaceae, an epiascidiate pitcher, appears to have evolved de novo as a carnivorous trap from noncarnivorous foliar leaves (e.g., Arber 1941, Franck 1976, Fleischmann 2010, Fukushima et al. 2015; Chapter 18).

Carnivory appears to have evolved at least three times in the Lamiales: in Byblidaceae, Lentibulariaceae, and *Philcoxia* of Plantaginaceae (Schäferhoff et al. 2010, Pereira et al. 2012). Many members of Lamiales are strongly glandular with high secretory potential, which appears to constitute a certain exaptation for carnivory (Müller et al. 2004), and several other glandular members of Lamiales repeatedly have been suspected as being carnivorous or "proto-carnivorous" (*sensu* Givnish et al. 1984). These include *Ibicella* and *Proboscidea* (Martyniaceae) (e.g., Beal 1875, Mameli 1916), and *Lathraea*, *Tozzia*, and *Bartsia* (Orobanchaceae) (e.g., Kerner von Marilaun and Wettstein 1886, Groom 1897, Heslop-Harrison 1976). However, nutrient uptake from casually or intentionally caught animals has not been detected in any of these genera (Schmidt and Weber 1983, Juniper et al. 1989, Rice 1999, Płachno et al. 2009a).

Carnivory also appears to have arisen at least three times in the monocot order Poales, twice in Bromeliaceae (two species of *Brocchinia* [Givnish et al. 1984, 1997] and *Catopsis berteroniana* [Fish 1976]), and once in the otherwise noncarnivorous Eriocaulaceae (*Paepalanthus bromelioides*; Nishi et al. 2013). In all three lineages, a pitfall trap evolved from a rosette of leaves with tightly overlapping bases that impound rainwater. This trap is very different in design from those in the three eudicot pitcher-plant families, and is quite long-lived; the plant's single rosette persists as individual leaves are borne and die. Digestive glands are unknown in these monocot carnivores, but absorptive hairs on the bases of individual leaves take up nutrients.

3.1.3 Phylogeography and timing of origin

Dating the origins of carnivorous plants is challenging. Zanne et al. (2014) provide a recent but controversial (Edwards et al. 2015) fossil-calibrated phylogeny for all angiosperms, and Givnish et al. (2011, 2014a) for bromeliads and monocots. Maximum ages for the origins of carnivory can, at least in principle, be estimated from the stem (root) ages of carnivorous lineages on those trees, assuming that the last common ancestor of the lineage already had been carnivorous. However, carnivory may have evolved at any point between the stem and crown age of a carnivorous lineage; the latter is the first date at which extant species or genera within a lineage began diverging from each other. Further, several studies of deep-node angiosperm phylogenies (e.g., Soltis et al. 2011, Zanne et al. 2014, Tank et al. 2015) either do not resolve the closest relatives of some clades of carnivorous plants, or under-represent carnivorous lineages and their relatives in the sampling, so that phylogenetic relationships reconstructed often are not very meaningful (e.g., the phylogenetic position of Lentibulariaceae as sister to Schlegeliaceae in Refulio-Rodriguez and Olmstead 2014 and APG IV 2016). Thus, the ages of certain groups, especially in Lamiales, and their precise closest relatives remain conjectural.

With these provisos, the maximum (stem) age of carnivorous lineages appears to range from 1.9 Mya for *Brocchinia reducta* (Bromeliaceae) to 95.1 Mya for the carnivorous clade of Caryophyllales (Table 3.1, Figure 3.1). Within the Caryophyllales, the species-rich genus *Drosera* is nearly cosmopolitan, and Nepenthes is widespread in the Austral-Asian tropics. Among the remaining, monotypic genera, *Aldrovanda* is widespread but sparsely distributed in the Old World, whereas *Dionaea, Drosophyllum, Triphyophyllum* have more restricted ranges (Table 3.1). Within the Ericales, *Darlingtonia* and *Sarracenia* (Sarraceniaceae) are endemic to western and eastern North America, respectively, whereas *Heliamphora* is restricted to the Guayana Shield of northern South America. *Roridula* (Roridulaceae) today is found only in the Cape region of South Africa but also is known from ≈35–47 Mya fossils from the margin of the Baltic Sea in northwestern Europe (Sadowski et al. 2015), supporting age estimations of ≈38 Mya for Roridulaceae (Ellison et al. 2012).

Both Caryophyllales (stem age of the order ≈109 Mya, and of the carnivorous lineage, ≈95 Mya) and the Sarraceniaceae + Roridulaceae clade (stem age ≈65 Mya; ≈51 Mya according to Ellison et al. 2012) are phylogenetically old enough to have rafted via continental drift to several of the southern continents and subcontinental fragments as Gondwana broke up. The apparently younger (≈43 Mya) Lentibulariaceae also have a nearly cosmopolitan distribution, that principally reflects the range of widespread species that have evolved most recently (hibernacula-forming temperate *Pinguicula*, the aquatic *Utricularia* subg. *Utricularia*) and that have in many cases been involved in postglacial range extensions. In contrast, the similarly aged *Cephalotus* (32.2 Mya) and Byblidaceae (44.5 Mya) have two of the narrowest distributions. *Cephalotus* is restricted to a small portion of southwest Australia, whereas *Byblis* grows in southwest Australia, northern Australia, and southern New Guinea (Chapter 10). The four occurrences of carnivory in genera or families that are not themselves wholly carnivorous involve the most recent origins of carnivory: *Philcoxia* (19.3 Mya in our reconstructions; 3.4–8.4 Mya in Zanne et al. 2014); *Brocchinia* (1.9 Mya) and *Catopsis* (2.6 Mya); and *Paepalanthus* (2.7 Mya) (Table 3.1, Figure 3.1).

Phylogenetic ages of carnivorous clades do not correlate with diversification. The similarly species-rich genera *Drosera* and *Utricularia* are among the oldest and youngest clades, respectively, of carnivorous plants. Extant species diversity is always a result of speciation and extinction events, and especially in the phylogenetically old carnivorous plant lineages, we find many species-poor or even monotypic genera or families, indicating extinction events (Table 3.1; Figure 3.4). Most of the monotypic genera—*Dionaea, Aldrovanda, Drosophyllum, Cephalotus*—are considered paleoendemics; their extant species likely are the sole survivors of formerly more diverse and species-rich lineages. The evolution of carnivory is not associated with subsequent diversification in any lineage, and does not appear to be a "key innovation" that boosted speciation. Not even the evolution of a novel trap type can be linked to rapid speciation in most cases. For example, one might expect that the evolution of pitcher traps in *Nepenthes* from sticky ancestors could have driven speciation. This is not the case. All early-branching lineages of *Nepenthes* are species-poor, whereas the majority of extant species diversity occurs in derived lineages with limited molecular divergence or short branch lengths (Meimberg et al. 2001, Merckx et al. 2015, Schwallier et al. 2016). This pattern implies geographic radiation (*sensu* Simões et al. 2016), including relatively recent speciation events and reticulate evolution following colonization of and adaptation to highland habitats on Malesian islands.

Early-branching lineages of *Drosera* also are species-poor and geographically isolated, whereas most extant diversity in this genus is found in derived clades with very short branch lengths (Rivadavia et al. 2003, 2012, Fleischmann et al. *unpublished data*), reflecting recent and likely sympatric speciation and reticulation. A similar pattern occurs in *Utricularia*—most species diversity is found in the derived lineages (Jobson et al. 2003, Müller and Borsch 2005)—but in this case transition

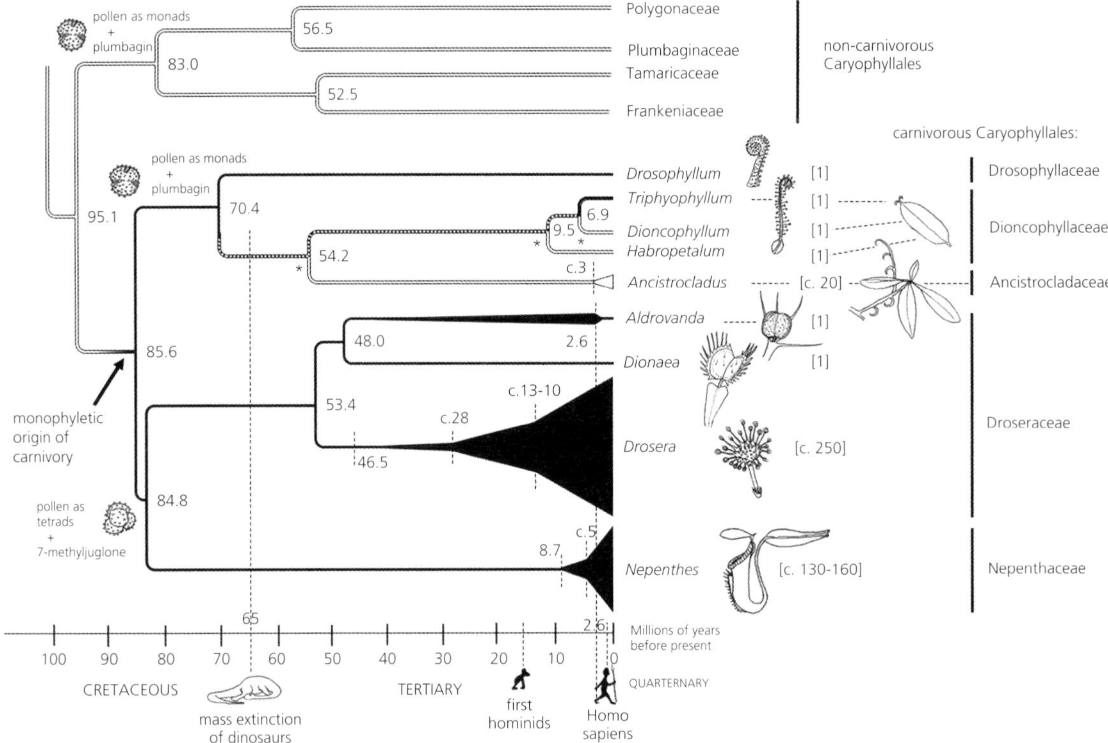

Figure 3.4 Dated phylogenetic tree of Caryophyllales; topology based on APG (2016) and underlying data, divergence times based on relaxed clock estimations (S. Smith and T. Givnish *unpublished data*). Carnivorous lineages shown in black, noncarnivorous lineages in white; dotted lines = both character states possible. Divergence times given for each branch. Asterisks mark nodes with loss of carnivory. Diversification of *Ancistrocladus, Drosera*, and *Nepenthes* are based on divergence times of species-rich clades. Possible extinction events were not considered except for *Aldrovanda*; data on fossil *Aldrovanda* taxa from Degreef (1997). Numbers of extant species for each genus given in square brackets. Illustration by Andreas Fleischmann.

from apparently passive traps of the first-branching *U.* sect. *Polypompholyx* (three species) to active suction traps in all subsequently branching lineages (the other ≈240 species; Chapter 8) could have been a key innovation promoting diversification (Westermeier et al. 2017).

3.2 Caryophyllales

In Caryophyllales, carnivory appears to have evolved only once (Albert et al. 1992, Meimberg et al. 2000, Rivadavia et al. 2003, Heubl et al. 2006; Figure 3.4). The carnivorous lineage ("carnivorous Caryophyllales:" Droseraceae, Nepenthaceae, Drosophyllaceae, Dioncophyllaceae, Ancistrocladaceae) is sister to an entirely noncarnivorous clade. Carnivorous Caryophyllales were treated by APG IV (2016) as "non-core" Caryophyllales. The members of this monophyletic group differ from their sister core-Caryophyllales (APG IV 2016) in lacking betalains and frequently containing acetogenic naptho- and anthraquinones (Hegnauer 1990, Schlauer 1997b). Relaxed molecular clock estimations (S. Smith and T. Givnish *unpublished*; Figures 3.1, 3.4) date the stem age of carnivorous Caryophyllales to 95.1 Mya (Magallón et al. 2015: 83 Mya); doubtlessly this is the phylogenetically oldest lineage of carnivorous plants, of Late Cretaceous (Cenomanian) and putatively Gondwanan origin.

Sister to the carnivorous Caryophyllales is a clade composed of Frankeniaceae, Tamaricaceae, Plumbaginaceae, and Polygonaceae (Meimberg et al. 2000, Heubl et al. 2006; Figure 3.4), most of which also have glandular hairs or active secretory tissues, strongly suggesting their presence in the common ancestor of both nepenthalean groups. Special multicellular, vascularized glands that excrete chalk or salts occur in Frankeniaceae, Plumbaginaceae, and Tamaricaceae (Wilson 1890, Cuénoud et al. 2002, Heubl et al. 2006) and have excretory functions; members of these families often occur in extreme, saline, sulfur-rich, or calcareous soils in which N or P are likely to be limiting. Darwin (1875) felt that such sticky glandular hairs—presumably serving as defensive or excretory functions—provided a natural first step (exaptation) in the evolution of carnivory.

In carnivorous Caryophyllales, leaf glands secrete digestive fluids; those species with adhesive traps also secrete aqueous mucilage for trapping. The glands themselves respond to various tactile and chemical stimuli in all of the species, but only in Droseraceae do the glands and leaf laminae respond with nastic and tropic movements upon stimulation by prey. These movements can be very fast, as in the active snap-traps of *Aldrovanda* and *Dionaea*, and certain *Drosera* species (Chapters 4, 14).

The immobile traps found in the other carnivorous Caryophyllales conventionally are called "passive," reflecting only their lack of mobility, not their physiological activity (Chapters 12, 15). For example, the traps of some *Nepenthes* species that grow in areas of heavy rainfall use a wettable peristome and viscoelastic pitcher fluid to capture prey, whereas some of those in drier areas use epicuticular waxes to precipitate prey into the pitcher (Moran et al. 2013; Chapters 5, 12, 15). Several species of *Nepenthes* are specialized on capturing ants or termites (Chapters 15, 21) and some acquire nutrients from vertebrate excreta (Clarke et al. 2009, Chin et al. 2010; Chapters 5, 15, 26).

Molecular data indicate unequivocally that *Dionaea* and *Aldrovanda* are sister to each other, and jointly sister to *Drosera*; these three genera of Droseraceae (Chapter 4), in turn, are sister to *Nepenthes* (Nepenthaceae; Chapter 5). Sister to this clade are *Drosophyllum* (Drosophyllaceae; Chapter 10) and Dioncophyllaceae + Ancistrocladaceae (Schäferhoff et al. 2009, Brockington et al. 2009, Zanne et al. 2014, Magallón et al. 2015, APG IV 2016; Figure 3.4).

The position of *Nepenthes* in these recent reconstructions differs from that in earlier phylogenies (Meimberg et al. 2000, 2001, Cameron et al. 2002, Heubl et al. 2006, Renner and Specht 2011), that did not place it sister to Droseraceae, but in a grade Drosophyllaceae [Dioncophyllaceae + Ancistrocladaceae]. However, the sister relationship Droseraceae + Nepenthaceae is well-supported by morphological and phytochemical synapomorphies, including echinate pollen tetrads (Takahashi and Sohma 1982), the presence of 7-methyljuglone and its presumed precursor shinanolone, and the respective isomers plumbagin and isoshinanolone (Schlauer et al. 2005). In contrast, all members of the Drosophyllaceae-Dioncophyllaceae-Ancistrocladaceae clade share pollen monads and exclusively plumbagin or isoshinanolone. Both features also are found in Plumbaginaceae and thus can be considered plesiomorphic in Caryophyllales (Figure 3.4).

Within Dioncophyllaceae, noncarnivorous *Habropetalum* and *Dioncophyllum* are consecutive sisters to the part-time carnivorous *Triphyophyllum*. Dioncophyllaceae in turn is sister to noncarnivorous Ancistrocladaceae, a monogeneric family of ≈20 species of paleotropical lianas (Figure 3.4). Heubl et al. (2006) assume that carnivory was lost in early branching Dioncophyllaceae and subsequently regained in *Triphyophyllum*. It seems more plausible, however, that continuous loss of carnivory happened in all derived members of Dioncophyllaceae + Ancistrocladaceae that live in wet tropical rainforest habitats where sticky traps with water-based glue are ineffective (Fleischmann 2010, 2011b). *Triphyophyllum* only produces carnivorous leaves during the less rainy part of the year (Green et al. 1979) and might be in an evolutionary transition away from carnivory. For the same ecological reason, pitcher traps might have evolved in *Nepenthes* from sticky trap ancestors to cope with rainforest habitats in tropical latitudes.

Heubl et al. (2006) and Renner and Specht (2011) outline different evolutionary pathways to the diversity of traps seen in carnivorous Caryophyllales from closely related outgroup taxa (Plumbaginaceae, Polygonaceae, Frankeniaceae). Renner and Specht (2011) envision one of two initial paths—leaf pinnation or emargination—leading from a leaf with sessile glands to one with stalked glands on the leaf perimeter and sessile glands studding the surface of the leaf interior. It is not clear why these routes should be necessary, given that a stalked gland simply could evolve from a sessile one through elongation of the basal cells and vascularization (Heubl et al. 2006). In the noncarnivorous, sister Plumbaginaceae, stalked, vascularized glands are present e.g., on the calyx lobes, where they are not confined to the margins.

Once a gland becomes stalked, selection should favor vascularization if secretion of substantial amounts of fluids must be maintained far from the veins of the lamina. In Lamiales, however, higher secretory activity is a result of polyploid tissues and special reservoir cells at the base of the stalked gland (§3.4.2). Heubl et al. (2006) and Gibson and Waller (2009) propose that the stalked glands evolved into trigger hairs or marginal teeth in both *Dionaea* and *Aldrovanda*. *Nepenthes* apparently lost stalked glands, and evolved some pitted glands, presumably from ancestral sessile glands or as a sunken version of stalked glands (Heubl et al. 2006, Fleischmann 2010). Renner and Specht (2011) assume an independent origin of stalked glands in Droseraceae, *Drosophyllum*, and *Triphyophyllum*, whereas Heubl et al. (2006) and Fleischmann (2010) propose a common origin from noncarnivorous ancestors and subsequent loss in several taxa (or reversal into pitted glands in Nepenthaceae).

3.2.1 Drosophyllaceae

Drosophyllum is isolated both geographically and systematically. It was described originally as a species of *Drosera*, but based on morphology (Chrtek et al. 1989) and molecular sequence data (Meimberg et al. 2000), this genus is placed in its own, monogeneric family. Almost every character beyond carnivory (i.e. woody habit, glandular trichomes on the abaxial leaf surface, reverse circinate vernation, pantoporate pollen in monads, axial placentation) contradicts inclusion of *Drosophyllum* in Droseraceae. Many different lines of evidence place it in a clade containing Dioncophyllaceae and Ancistrocladaceae. With the only carnivorous representative of the former family, *Triphyophyllum*, it shares passive adhesive traps and reversely circinate leaf vernation, which is paralleled in the angiosperms only in the Lamialean *Byblis* (Figure 3.2). Phylogenetic age estimations date the lineage to 70.4 Mya (Magallón et al. 2015: 57.9 Mya), and the single extant species *Drosophyllum lusitanicum* most likely is a paleoendemic of a once more diverse and widespread lineage.

3.2.2 Dioncophyllaceae

This family contains two noncarnivorous genera (*Dioncophyllum* and *Habropetalum*) and carnivorous *Triphyophyllum*. All are monotypic and endemic to tropical West Africa. *Triphyophyllum* only produces carnivorous leaves during a short part of its juvenile phase (Green et al. 1979), probably to acquire extra nutrients to reach maturity and flowering. Carnivory is expressed during the rainy season but before its peak, when heavy downpours would likely wash away secretions (Green et al. 1979). However, a carnivorous stage is not essential to complete its life cycle (Bringmann et al. 2002), and for the largest part of its life the species is noncarnivorous (Fleischmann 2011a). *Habropetalum* and *Triphyophyllum* are sympatric, but the latter apparently is more closely related to the disjunct *Dioncophyllum* (Meimberg et al. 2000). Its fruits open before the seeds mature, and, uniquely among vascular plants, the seeds surpass the ovary in size.

The taxonomic affinity of Dioncophyllaceae long has been discussed, and a position near carnivorous Nepenthaceae (among other noncarnivorous families erroneously assigned) was discussed first by Airy-Shaw (1951). Molecular phylogenetic evidence (Cameron et al. 1995, Meimberg et al. 2000, 2001), convincing similarities in anatomy and pollen morphology, and the presence of acetogenic naphthylisoquinoline alkaloids provide strong support for Ancistrocladaceae being the most closely related sister group of Dioncophyllaceae (Dahlgren 1980). These characters also suggest a close relationship of Dioncophyllaceae to *Drosophyllum*.

Fossils from the Eocene of Raychikha in the Amur district have been interpreted as seeds of Dioncophyllaceae (Fedotov 1982), but these fossils, which are larger than the seeds of extant Dioncophyllaceae, also could belong to quite different families, and could represent, for example, fossil fruits near *Paliurus* (Rhamnaceae). If these fossils were, however, assigned accurately to Dioncophyllaceae, a West African diversification of a formerly much more widespread family must have taken place relatively recently. Divergence time of Dioncophyllaceae is estimated at ≈54 Mya (S. Smith and T. Givnish *unpublished data*; Magallón et al. 2015: 36.2 Mya) but the extant species are much younger (9.5–6.9 Mya; Figure 3.4). The same holds true for the entirely noncarnivorous sister Ancistrocladaceae, where diversification of extant species is dated to ≈3 Mya (i.e., to a major split of the African and Asian lineages of *Ancistrocladus*; Figure 3.4).

3.2.3 Nepenthaceae

Nepenthaceae contains only the extant genus *Nepenthes*. Although the family is phylogenetically old, dated to the Late Cretaceous (84.8 Mya based on divergence time from Droseraceae; Table 3.1; Magallón et al. 2015 estimated it to be 76.8 Mya), diversification of the extant species of *Nepenthes* is much more recent (≈8.7 Mya for the earliest-branching lineages and ≈6–4 Mya for the species-rich Malesian clades; Figure 3.4). This likely represents an adaptive radiation in newly formed montane habitats of the Malay Archipelago (Merckx et al. 2015, Schwallier et al. 2016; Chapter 5). Pollen originally assigned to Droseraceae (*Droseridites*) from the Kerguelen Islands tentatively has been transferred to *Nepenthes* (Krutzsch 1985). In this context *Droseridites parvus* from the Mid-Palaeocene of Assam (Sah and Dutta 1974) should be considered as possible *Nepenthes* pollen (*Nepenthidites*; Kumar 1995). Fossil pollen assigned to *Nepenthes* also has been discovered in the mid-Miocene of north Borneo (Anderson and Müller 1975), and its presence here in a center of recent diversity (Merckx et al. 2015, Schlauer 2000, Schwallier et al. 2016) is unsurprising. But the assignment of European Tertiary pollen to the same genus (Krutzsch 1985) is at least as puzzling as the dubious *Triphyophyllum* seed from Siberia (Fedotov 1982; §3.2.2).

The combination of apparently plesiomorphic and apomorphic characters within Nepenthaceae—dioecious, four petaloid perianth segments, stamina fused in a column, pollen in tetrads (a synapomorphy with Droseraceae), and axial placentation—isolates the family far from any hypothetical ally. These isolated characters, besides chorology, could also suggest an older phylogenetic age for the family as a whole. The affinity of Nepenthaceae to Dioncophyllaceae (e.g., Dahlgren 1980) is weakly supported, and the divergence between them must have taken place at an early stage in the evolution of Caryophyllales.

In all *Nepenthes*, the leaves of mature plants are composed of a basal blade (expanded, photosynthetic part, not necessarily the lamina) that terminates in a tendril that supports a pitcher with two ventral wings and a rim at its orifice that is formed by a slippery, radially ribbed peristome, a dorsal spur, and a covering lid. Tendrils, wings, or peristome ribs may be reduced in some species but their position and frequently some rudiments usually are apparent. Despite this morphological uniformity, there are at least eight hypotheses for the ontogeny of the pitchers and their various appendages (Franck 1976, Fukushima and Hasebe 2014; Chapter 18).

As none of the close relatives of *Nepenthes* have structures even nearly resembling pitchers, the evolution of the genus remains somewhat conjectural, but all metamorphoses that have led to its general pitcher morphology must have occurred before its phylogenetic divergence. That all of its carnivorous relatives derive their traps exclusively from the leaf lamina similarly implies that the *Nepenthes* pitchers likewise are derived from the lamina. As their most recent carnivorous relatives have a sticky adaxial leaf surface, and various forms of invagination or even peltation can be observed in *Drosera* leaves, it is reasonable to assume that this adaxial surface evolved to become the interior, glandular surface of the pitcher. The boundary between adaxial and abaxial surfaces at the orifice of the *Nepenthes* pitcher corresponds to the leaf margin; it is located just below the inner border of the peristome where glands terminate vascular bundles. The peristome itself is formed by asymmetric divisions of epidermal cells (Owen and Lennon 1999). The pitcher lid is a unique structure without a close parallel in any related genus. It is formed by fusion of two lateral lobes just beneath the spur that is the original leaf apex (Schmid-Hollinger 1970), and the pair of subapical hooks of the climbing leaves in Dioncophyllaceae may be an ontogenetic homologue. Similarly, some species in *Drosera* subg. *Ergaleium* have a pair of auricles or crescentic outgrowths of the lamina (e.g., *D. peltata*) or dichotomously divided leaves (*D. binata*).

3.2.4 Droseraceae

Considerable differences in floral morphology (stamina many vs. equal in number to petals and sepals; placentation basal or axial vs. parietal) separate *Dionaea* from *Aldrovanda* and *Drosera*, but the striking similarity and monophyletic origin of the traps of *Dionaea* and *Aldrovanda* supports the inclusion of *Dionaea* into Droseraceae. This placement is fully corroborated by pollen morphology (Takahashi and Sohma 1982) and molecular phylogenetic reconstructions (Williams et al. 1994, Meimberg et al. 2000, Rivadavia et al. 2003, Heubl et al. 2006, Renner and Specht 2011; Figure 3.4). The subdivision of Droseraceae into several families (Chrtek et al. 1989) seems unjustified beyond the exclusion of *Drosophyllum*. Phylogenetic reconstructions (Cameron et al. 2002, Heubl et al. 2006, Renner and Specht 2011) show a sister relationship of the snap-trap genera *Aldrovanda* and *Dionaea*, with *Drosera* being sister to both (Chapter 4).

Senonian fossils that were described initially as seeds under the name *Palaeoaldrovanda splendens* (Knobloch and Mai 1984) have been re-identified as insect eggs (Heřmanová and Kvaček 2010) and thus cannot contribute to fossils constraining the origin of Droseraceae. Another series of Eocene fossil seeds and possibly traps of *Aldrovanda* (Degreef 1997, Schlauer 1997a), including several different and now extinct species and genera (*Saxonipollis*) occurs through large parts of temperate Eurasia. These fossils are congruent with the widely scattered Palearctic distribution of *Aldrovanda vesiculosa* (Chapter 4).

Pollen of Droseraceae are highly diagnostic: their endoaperturate and echinate pollen tetrads are unique among extant angiosperms, albeit with superficial similarities to Annonaceae pollen. Nepenthaceae have very similar, but much smaller, echinate, and inaperturate tetrads (Takahashi and Sohma 1982, Fleischmann et al. *unpublished data*).

Droseraceae has been widespread since the Early Tertiary; its pollen is represented in Eocene strata of central Australia (*Fischeripollis halensis*; Truswell and Marchant 1986) and in Eocene to earliest Oligocene strata of Antarctica (*Fischeripollis*; Macphail and Truswell 2004). *Drosera* pollen also has been recorded from the Lower Miocene from New Zealand (Mildenhall 1980), and Miocene pollen (as *Droserapollis* and *Droserapites*) of uncertain affinity within the Droseraceae has been found in Taiwan (Huang 1978).

It is unlikely that recent species of *Drosera* existed in Europe before the Pliocene, although several finds of mid-Miocene pollen from Europe have been assigned either to *Drosera* (*Droserapollis*) or *Nepenthes* (*Droseridites*; Krutzsch 1985). A single record from central European mid-Miocene (*Fischeripollis*) has been assigned to *Dionaea* (Krutzsch 1970). The earlier European fossils may be attributed to now extinct lines of Droseraceae or even to other families. Regardless, the fossil record of Droseraceae is the richest of any carnivorous plant lineage, and it testifies a wide distribution of the family since the Early Tertiary.

Molecular clock estimates date the stem age of Droseraceae to the Late Cretaceous (84.8 Mya; 76.8 Mya in Magallón et al. 2015); the split of the snap-trap *Dionaea* + *Aldrovanda* clade from *Drosera* at 53.4 Mya; and the divergence of the earliest-branching lineage of *Drosera* to 46.5 Mya (Table 3.1; Figure 3.4). The species-rich lineages of *Drosera* are comparatively older (≈10–13 Mya) than those of sister genus *Nepenthes*, implying a much earlier diversification of the former (Figure 3.4).

3.3. Oxalidales

3.3.1 Cephalotaceae

The monotypic family Cephalotaceae is endemic to a small area in southwest Australia. The narrow-endemic Albany pitcher plant, *Cephalotus follicularis*, is its only species. No fossils have been attributed to Cephalotaceae, and morphology provides no convincing clues for a related and recent plant genus (based on apparent floral similarities, it has been thought to be close to Crassulaceae and Saxifragaceae). Its unique floral morphology (Chapter 10) and pitcher development isolate the Cephalotaceae from all other carnivorous plant families; it has frequently been considered a paleoendemic.

The first molecular phylogenetic reconstructions of carnivorous angiosperms placed *Cephalotus* close to *Oxalis* (Albert et al. 1992). Although these two genera have only a few morphological synapomorphies, subsequent DNA sequence analyses confirmed Cephalotaceae as being deeply nested within Oxalidales (Figure 3.5). More recent phylogenetic reconstructions identified the monotypic Brunelliaceae (the Neotropical *Brunellia*, ≈60 species of evergreen trees in the Andes, Mexican Highlands, and the Caribbean) as the closest relative of *Cephalotus* (e.g., Bradford and Barnes 2001, Soltis et al. 2011, Sun et al. 2016). This grouping had been suggested earlier by Engler (1897) on morphological grounds.

Heibl and Renner (2012) retrieved a slightly different topology, with Elaeocarpaceae (≈600 species of (sub)tropical trees and shrubs) as immediate sister to Cephalotaceae, and Brunnelliaceae as common sister to both. This result could be explained by limited taxon sampling, as this part of the Oxalidales tree was not focus of their study (C. Heibl *personal*

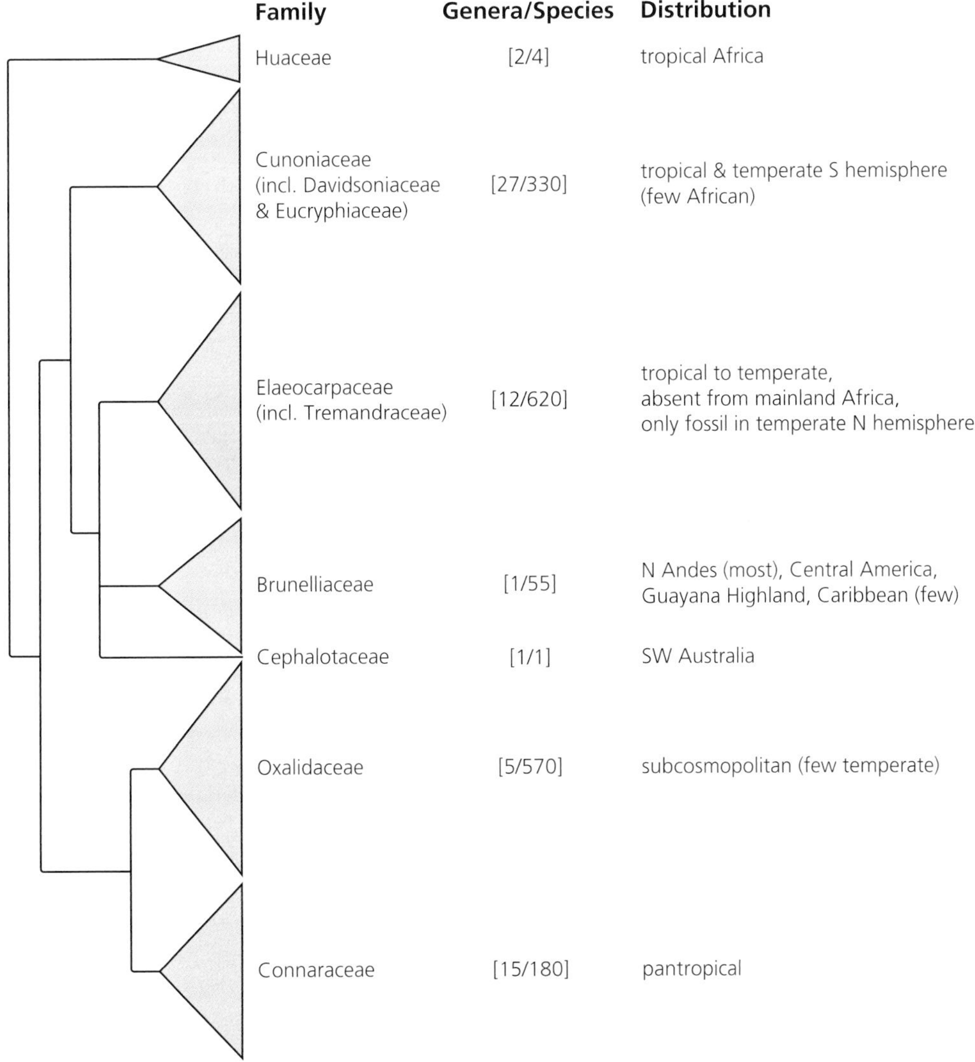

Figure 3.5 Phylogenetic tree of Oxalidales and Cephalotaceae. Illustration by Jan Schlauer.

communication). Nevertheless, Cephalotaceae can be expected to have a position in crown group Oxalidales (these are ≈110 Mya; Tank et al. 2015) within an early branch of the rosid eudicots (APG IV 2016) originating from an Early Tertiary or Late Cretaceous lineage.

The development of the *Cephalotus* pitcher is unique among the epiascidiate carnivorous pitcher plants. In *Cephalotus*, the lid is an excrescence of the lamina base (or the transversal zone of a hypothetical ancestral peltate or pinnate leaf, according to Froebe and Baur 1988) and the rim of the peristome is homologous to the apical leaf margin (Figure 3.3). That is, the lid of the *Cephalotus* pitcher constitutes what is the pitcher body in *Nepenthes* and Sarraceniaceae. A *Cephalotus* pitcher hence may be regarded as an "upside-down pitcher."

The pitchers of the three pitcher-plant families, although of similar shape and identical function, are analogous, not homologous (Figure 3.3). Pitchers of *Cephalotus* and Sarraceniaceae consist of the entire leaf blade, whereas the leaf blade of *Nepenthes* contributes the foliar part, tendril, and pitcher. In Sarraceniaceae and *Nepenthes*, the pitcher lid is the terminal part of the leaf (it still increases size from growth after the pitcher has opened), whereas in *Cephalotus*, the pitcher body is the terminal leaf part, and the lid is the basal part. The *Cephalotus* pitcher grows from the lid to the base; the lid reaches its final size and position first, while the pitcher bottom continues to inflate and increase in size.

Froebe and Baur (1988) were not convinced that the three alae of *Cephalotus* pitchers were simple epidermal outgrowths of the outer pitcher wall (assuming an epiascidiate, peltate leaf). Instead, they hypothesized a "rhachis pitcher" formed from a modified pinnate leaf (twice paired, tetramerous): the basal pair of pinnae would fuse congenitally to become the lid, the subsequent pair of pinnae would equal the two lateral alae, and the median wing would be formed by a terminal pair of pinnae. Their theory largely has been neglected or even rejected (except by Conran 2004a) but it merits reconsideration in the light of phylogeny. Most members of crown group Oxalidales—the majority of species of the immediate sister Brunelliaceae, all Oxalidaceae, all Connaraceae, and most Cunoniaceae—possess compound or pinnate laminae.

3.4 Asteridae: Ericales

3.4.1 Roridulaceae

Roridula has long been considered a paleoendemic of the Cape Flora (Warren and Hawkins 2006), but findings of 35–47 Mya old Eocene amber inclusions from the Baltic (Sadowski et al. 2015) demonstrate that the genus once was more widespread. The *Roridula* lineage has established a symbiotic relationship—digestive mutualism—with carnivorous capsid bugs (Miridae: Hemiptera) of the genus *Pameridea* (Ellis and Midgley 1996, Anderson and Midgley 2003, Anderson 2005; Chapter 26) to overcome the lack of its own digestive enzymes in the resinous glands.

Resinous glands are quite common in various genera of Ericaceae (e.g., *Rhododendron, Erica*) many of which casually trap insects (Darwin 1875, Sugiura and Yamazaki 2006). Hydrolytic enzymes cannot operate in hydrophobic resin in which water is unavailable, so active digestive enzymes are not found in the sticky resin droplets secreted by the glands of Ericales, including *Roridula* (Lloyd 1934, Ellis and Midgley 1996, Płachno et al. 2006, 2009a). This seems to be an evolutionary dead end for a sticky trap. However, through a digestive mutualism, the *Roridula* lineage established an alternative carnivorous pathway of gaining nutrients from captured prey (Anderson and Midgley 2003).

Capsid bugs and other arthropods frequently are found on numerous glandular, albeit noncarnivorous, plants, feeding on the adhering insects and depositing their feces directly onto the plant surface. Most plants can take up dissolved nutrients applied directly to their leaves, and any nutrients in the feces similarly could be absorbed. Unlike glandular noncarnivorous plants that trap insects only haphazardly, the carnivorous *Roridula* appears to attract prey to its sticky, scented leaves (Fleischmann 2010) and obtain a substantial fraction of its nitrogen budget (≈70%) indirectly from its insect prey (Anderson and Midgley 2002). In contrast, a careful study of *Rhododendron macrosepalum*—an example of a plant in Ericales with sticky leaves, buds, and sepals that entrap large numbers of insects, which in turn are consumed by associated mirid bugs—showed no uptake of nitrogen via those bugs (Anderson et al. 2012).

The leaves of *Roridula* are highly absorptive of ultraviolet light (Midgley and Stock 1998), a feature *Roridula* shares convergently with sticky traps of *Drosophyllum* and *Drosera* (Joel et al. 1985, Juniper et al. 1989). The cuticle of the leaf surface of *Roridula* contains cuticular gaps, and the epidermal layer underneath consists of highly absorptive cells (Anderson 2005), very similar to those observed in the glands and digestive surfaces of other carnivorous plants, including the digestive epithelium of the related Sarraceniaceae (Joel and Juniper 1982, Juniper et al. 1989). The pores and absorptive cells of *Roridula* rapidly take up nutrients from the bug feces (Anderson 2005), and to a lesser degree directly from caught prey that contacts the leaf surface (Płachno et al. 2009a).

3.4.2 Sarraceniaceae

No reliable fossils of Sarraceniaceae have been collected. The compression/impression fossil *Archaeamphora longicerva* from the Early Cretaceous of China, was assigned erroneously to Sarraceniaceae by Li (2005). These fossils do not represent early pitcher leaves, but are leaf galls of gymnosperm leaves (Wong et al. 2015).

Unlike the pitchers of *Nepenthes* or the bladders of *Utricularia*, the pitchers of the Sarraceniaceae do not seem to be derived from sticky ancestors (Figure 3.6). All extant members of Sarraceniaceae have pitcher leaves, thus the most reasonable assumption is their common ancestor also had ascidiate pitcher leaves (§3.2.3, §3.3.1; Chapter 9). Sarraceniaceae may have evolved from plants with leaves that formed natural (water-filled) ascidiate phytotelmata (Fleischmann 2010).

3.5 Asteridae: Lamiales

Lamiales includes the majority of extant carnivorous plant species (Ellison and Gotelli 2009, Fleischmann 2010), and represent the only angiosperm order in which entirely carnivorous lineages evolved several times in distantly related families (the entirely carnivorous Byblidaceae and Lentibulariaceae, and *Philcoxia* of Plantaginaceae; Fritsch et al. 2007, Schäferhoff et al. 2010, Pereira et al. 2012). The multiple origins of carnivory in Lamiales may be related to the widespread occurrence of glandular hairs in this order; the exaptation of such hairs may have been important in the evolution of carnivory (§3.2). The presence of such hairs would also make the evolution of carnivory more likely from a cost/benefit viewpoint (Chapter 18).

3.5.1 Byblidaceae

Byblis had been considered to be closely related to Lentibulariaceae (Albert et al. 1992, Jobson et al. 2003), but morphological characters including floral symmetry, corolla morphology, and gland anatomy imply an independent origin of Byblidaceae. More recent phylogenies (Müller et al. 2004, 2006, Schäferhoff et al. 2010) that are based on more comprehensive sampling of Lamiales taxa clearly support *Byblis* as a distinct lineage in the basal group of this asterid order, not closely related to Lentibulariaceae of crown Lamiales.

3.5.2 Plantaginaceae

Carnivory appears only in a single genus in Plantaginaceae. *Philcoxia* is a morphologically isolated lineage in the family (e.g., peltate leaves), and its phylogenetic position within Plantaginaceae remains somewhat unclear because of limited taxon sampling of New World Plantaginaceae in most published phylogenetic reconstructions (A. Scatigna *unpublished data*), although Fritsch et al. (2007) had a comparatively good taxon sampling and revealed it to be phylogenetically close to *Gratiola* and *Bacopa*. Any theories about evolution of carnivory in this genus without knowing its closest relatives remain purely speculative, but carnivory of *Philcoxia* is linked to its nutrient-poor quarzitic sand habitats (Fritsch et al. 2007, Pereira et al. 2012; Chapter 10).

3.5.3 Lentibulariaceae

The phylogenetic affinities of Lentibulariaceae are still unknown, and the family has repeatedly fallen into an unresolved clade within the crown-group of Lamiales (Müller et al. 2006, Schäferhoff et al. 2010). A questionable (and only weakly supported) sister relationship with Schlegeliaceae

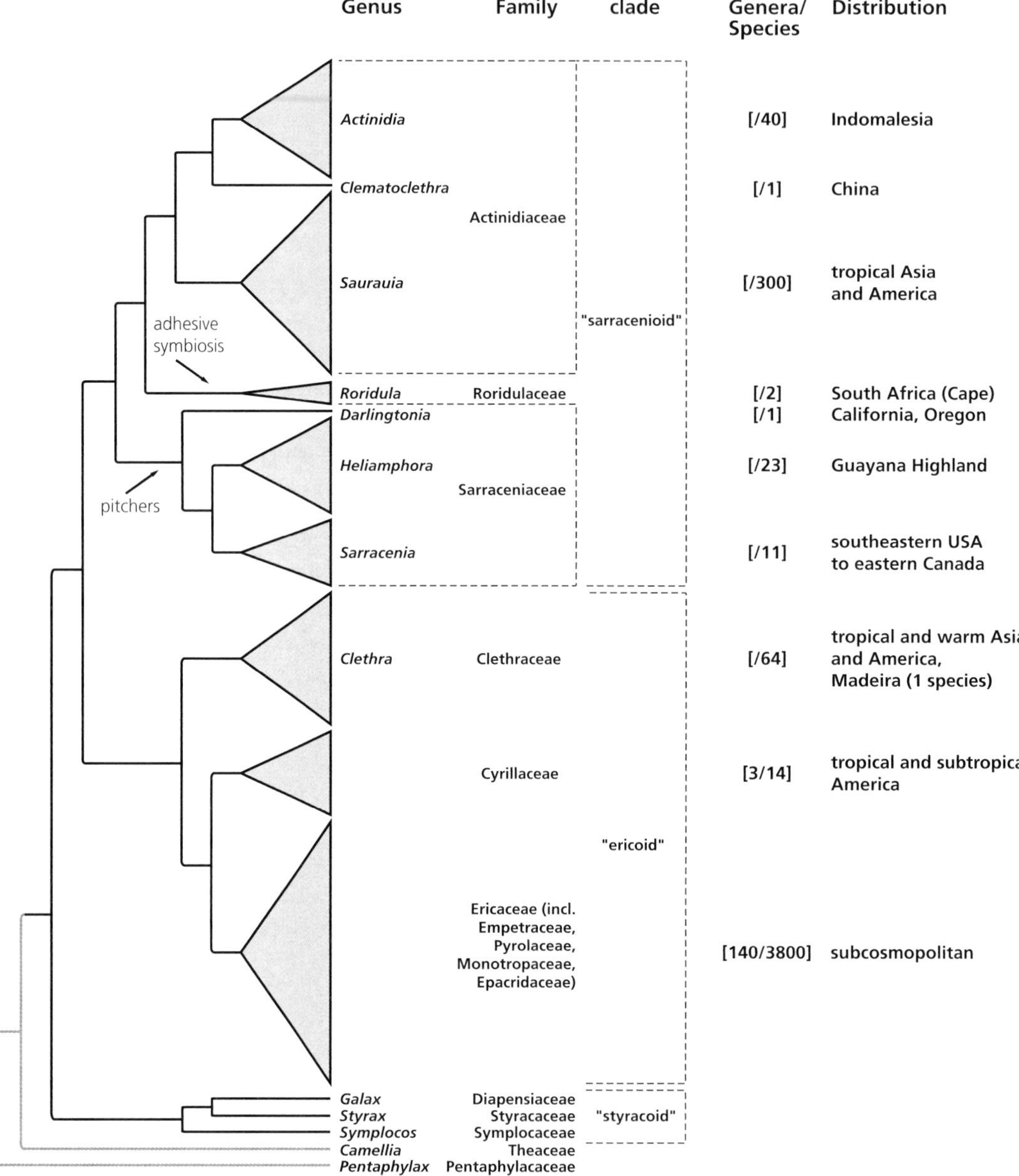

Figure 3.6 Phylogenetic tree of carnivorous Ericales. Illustration by Jan Schlauer.

was obtained by Refulio-Rodriguez and Olmstead (2014) and adopted by APG IV (2016), but this was almost surely the result of limited taxon sampling. Plausible sister groups are those with glandular hairs, which are widespread among the Lamiales. Small arthropods frequently have been reported as adhering to such glands, especially on the flowers. The main purpose of such "defensive killing" (Juniper et al. 1989) is probably to exclude non-pollinating insects from the flowers, and to protect the generative organs from herbivores (Kerner von Marilaun 1878, Fleischmann 2010). There is some experimental evidence that the carnivorous glands of *Pinguicula* also play a defensive role against herbivores (Alcalá et al. 2010), and the sticky foliage may have evolved first for defensive purposes, only later becoming modified into a successful flypaper trap (Darwin 1875).

One candidate relative is the Martyniaceae, which includes many strongly glandular genera. Two of these, *Ibicella* and *Proboscidea*, have been hypothesized to be carnivorous (Beal 1875, Mameli 1916) or "proto-carnivorous" (Rice 1999). Both genera have very sticky, glandular leaves of the flypaper trap type, and catch numerous arthropods (Rice 1999), but they are unable to absorb any nutrients from their putative prey (Płachno et al. 2009a). The glandular hairs of Martyniaceae also are not specialized, but have a generalized anatomy found ubiquitously in glandular Lamiales genera (Müller et al. 2004).

The dense glandular hairs in several other genera of the Lamiales often excrete water (Groom 1897). These secretory glands show a remarkable similarity in design and function to the digestive glands of carnivorous Lamiales and could represent an exaptation for carnivory in Lamiales (Müller et al. 2004). In both Lentibulariaceae and Byblidaceae, we find a gland specialization to stalked secretory glands and sessile digestive glands. This parallels the gland dimorphism observed in the carnivorous Caryophyllales (Juniper et al. 1989, Heubl et al. 2006, Renner and Specht 2011), and represents a further specialization toward carnivory.

Unlike the glands of the sticky-trap genera in carnivorous Caryophyllales (*Drosophyllum*, *Triphyophyllum*, and *Drosera*), which are vascularized to exchange digestive fluid and nutrients (Heslop-Harrison 1975, Juniper et al. 1989, Heubl et al. 2006), the unicellular or pluricellular stalks of the glands of carnivorous and noncarnivorous Lamiales are not lined with vascular bundles. Therefore, the attachment of the digestive glands to vascular tracheid elements must have been a key innovation for the evolution of carnivory in the Lentibulariaceae (Müller et al. 2004). This anatomical change is accomplished by a single, large, basal cell that is embedded in the epidermis of the leaf (Chapter 13). This prominent "reservoir cell" of lentibulariacean glands is physiologically connected to the subjacent tracheid cells by plasmodesmata and has storage functions related to prey digestion and nutrient uptake (Heslop-Harrison 1975, 1976). The cuticle of the gland head cells also has become modified in the Lentibulariaceae: it bears several cuticular gaps that secrete mucus, release enzymes, and take up nutrients from dissolved prey (Juniper et al. 1989).

Although the shape of the glandular hairs found in the three genera of Lentibulariaceae is very different, even differing among members of the same genus in the case of *Utricularia* (Taylor 1989), their functional anatomy is identical (Heslop-Harrison 1975, 1976, Juniper et al. 1989, Płachno et al. 2007a). Their different structures are adaptations to different trap types and ecosystems. The quadrifid glands in aquatic traps of *Utricularia* also are convergent with those in snap-traps of *Aldrovanda* (Caryophyllales).

Genlisea and *Utricularia* are immediate sister genera, and *Pinguicula* is sister to both (Jobson et al. 2003, Müller et al. 2004, 2006; Chapter 6). The rhizophylls of *Genlisea* and the bladder traps of *Utricularia* are homologous to one another, and to the adhesive foliar leaves of *Pinguicula* (Müller et al. 2004, Fleischmann 2012a; Figure 3.7; Chapter 7). The traps of *Genlisea* and *Utricularia* are epiascidiate in ontogeny (Juniper et al. 1989), and so a likely scenario for their evolution is a continued inward folding and final fusion of the lateral margins of adhesive leaves of the presumed common ancestor (Fleischmann 2012a).

Such involute folding happens temporarily in the motile leaves of many extant *Pinguicula* species after they capture prey, mainly to prevent loss of prey by rain or kleptoparasites (Chapters 6, 14). In two small butterwort species, one that may grow partially submerged under water in some habitats (*Pinguicula lusitanica*) and one whose leaves are buried in *Sphagnum* (*P. villosa*), the leaf margins always are highly involute. The margins of these leaves nearly touch and create almost tubular leaves whose glandular adhesive upper laminae are protected inside the tube. Such an ecological scenario may have occurred in the (possibly aquatic) environment where the trapping organs of the common ancestor of *Genlisea* and *Utricularia* evolved (Fleischmann 2012a).

Whereas it is unlikely that all the rosette leaves were transformed through involution, heterophylly—one type for photosynthesis, one for prey capture—appears to have taken shape early

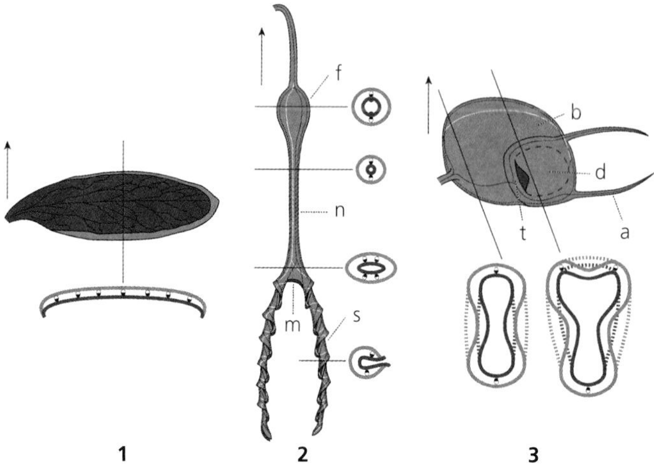

Figure 3.7 (Plate 3 on page P2) Homology of tissues among traps in Lentibulariaceae. External appearance, surfaces and schematic cross sections of **1**. *Pinguicula*; **2**. *Genlisea*; and **3**. *Utricularia* (before suction and after door opening, dashed lines in cross sections indicate shape after suction). Arrows: position and growth direction of shoot axis, axis in cross sections located below; *f*: flask; *n*: neck; *m*: mouth; *s*: spiral arm; *b*: bladder; *t*: threshold; *d*: door; *a*: antenna; light gray (green in color plate): abaxial surface; dark gray (red in color plate): adaxial surface; partially black-and-white filled ellipses: vascular bundles in cross section, black: xylem, white: phloem. Traps not shown with true size relations. Illustration by Jan Schlauer.

in the evolution of the *Genlisea–Utricularia* lineage; all species of *Genlisea* and all of the phylogenetically early branching species of *Utricularia* are heterophyllous, rosette-forming plants (Fleischmann 2012a). Heterophylly could have resulted in ancestral plants that formed some rolled, tubular leaves from a rosette of conventional foliar leaves. These tubular leaves can be envisaged as having had an apical opening and interior surfaces covered with (carnivorous) glands. Any small prey entering these tubular leaves would have become stuck to the glands and subsequently digested. Such a trap system is comparatively resistant against loss of prey to rain or kleptoparasitism and the narrow tubular traps also would have worked under water. The tubular, sticky leaves of the hypothesized ancestor of the *Genlisea–Utricularia* lineage would have had a small cross-sectional diameter, limiting the size of prey to very small organisms. To increase the numbers of trapped small prey, or to exploit different substrates, these tubular leaves might have formed below the soil or water surface.

3.6 Poales

Carnivory has been demonstrated for three tank-forming species of Bromeliaceae: two species of *Brocchinia* (Givnish et al. 1984, 1997, 2014a, Benzing 1987, Płachno et al. 2006) and one species of *Catopsis* (Fish 1976, Frank and O'Meara 1984, Benzing 1987). Carnivory also has been proposed for one member of the Eriocaulaceae, *Paepalanthus bromelioides* (Nishi et al. 2013; Chapter 10). There is no monocot family comprising entirely carnivorous members, and the four species of carnivorous monocots known today each are isolated within genera of noncarnivorous species (Table 3.1), indicating a very recent transition to carnivory.

Carnivory has evolved more recently in the Poales (within the last ≈3 Mya) than in any eudicot lineage. Their trap architecture is also completely different: simple pitfall traps made from the entire plant body, which does not differ morphologically from that of their noncarnivorous relatives.

3.6.1 Bromeliaceae

Within the Bromeliaceae, the documented carnivorous plants are *Brocchinia hechtioides* and *B. reducta* (Brocchinioideae) (Givnish et al. 1984, 1997, 2014a, Benzing 1987, Płachno et al. 2006) and *Catopsis berteroniana* (Tillandsioideae) (Fish 1976, Frank and O'Meara 1984, Givnish et al. 2014a; Chapter 10). In all three species, the evolution of the tank habit and of absorptive leaf trichomes are key innovations for the evolution of carnivory. In other species in these genera, these innovations are applied to other specialized nutrient acquisition strategies, including myrmecophily, nitrogen fixation, and epiphytism with inputs from fallen vegetable detritus or frog excrement; epiphytism is the only alternative strategy in *Catopsis* (Givnish et al. 1997, 2011, 2014a,

Givnish 2017). Across the family, the tank habit appears to have evolved three times, always in tropical montane conditions and often among epiphytes, (Givnish et al. 2011, 2014a).

In *Brocchinia*, the earliest diverging lineages—the so-called "Prismatica clade" and "Melanacra clade"—lack tanks, have a very small total surface area of trichomes, and often grow in sandy lowland areas or well-drained upland sites in the Guayana Shield. The two late-diverging lineages—the "Micrantha clade" and the "Reducta clade"—are sister to one another, have tanks and relatively large areas of leaf trichomes, and are found at moderate to high elevations on *tepuis* of the Guayana Shield (Givnish et al. 1997). *Brocchinia* species appear to have begun diverging from each other ≈13 Mya; impounding and non-impounding lineages diverged ≈9 Mya and carnivorous taxa evolved ≤ 5 Mya (Givnish et al. 2011; cf. ≈12 Mya for *B. reducta* based on Zanne et al. 2014 vs. 1.2 Mya for one of the carnivorous species based on our data presented here; Figure 3.1). Carnivory in *Brocchinia* evolved in association with the wet, extremely infertile habitats on the slopes and summits of the Guayanan *tepuis*.

Both carnivorous species of *Brocchinia* have a nearly cylindrical rosette of steeply inclined, bright yellow-green leaves, with a fine waxy dust on the inner leaf surface, and substantial areas of relatively large, live trichomes on the leaf bases (Givnish et al. 1984, 1997, Gaume et al. 2004). The tank fluid emits a sweet nectar-like odor, is highly acidic, and collects ants and other insects that do not otherwise live in pools of water.

Brocchinia reducta is the shorter of the two carnivorous species, and specializes on ants, whereas *B. hechtioides* has a rosette roughly twice as tall, and appears to be specialized on bees and wasps (Chapter 10). Nutrient inputs via carnivory are so substantial that both species can grow on bare sandstone. On Chimantá *tepui*, the nocturnal treefrog *Tepuihyla obscura* (Hylidae) takes shelter in the tanks of *B. hechtioides* and *B. reducta* by day (Kok et al. 2015), suggesting that these species also obtain nutrients from frog excrement. The tillandsioid bromeliad *Vriesea bituminosa* obtains roughly 25% of its nitrogen budget from frog excrement (Romero et al. 2010). Spiders living above bromelioid bromeliads (both tank- and non-tank-forming species) contributed to their growth and nitrogen budget (Romero et al. 2006, Goncalves et al. 2011).

Catopsis berteroniana is an epiphyte with bright yellow leaves that collects large numbers of dead flying insects in its central tank. A number of other species of *Catopsis* have a similar growth form and may also be carnivorous or approaching carnivory. The initial argument for carnivory in *C. berteroniana* (Fish 1976) was relatively weak. Subsequent experiments by Frank and O'Meara (1984) showed that *C. berteroniana* trapped prey 12 times faster than several control tank-bromeliads, and isotope data revealed enrichment of ^{15}N-nitrogen in plant tissue, as expected if a substantial amount of its nitrogen budget is of animal origin (Gonsiska 2010).

3.6.2 Eriocaulaceae

Paepalanthus (≈450 species) is one of the largest genera in the Eriocaulaceae. The leaf rosette of *Paepalanthus bromelioides* is massive, several cm wide, and analogous to the tanks of many bromeliads (Chapter 10). Jolivet and Vasconcellos-Neto (1993) and Figueira et al. (1994) proposed that this species is carnivorous. Its unusually large rosette impounds rainwater, its leaves are covered with a slippery wax, and its leaf bases bear absorptive trichomes. Many micro-predators live in or above the rosettes and help capture prey and deliver nutrients to the plants via excrement or carcasses. This carnivorous pathway accounts for 27% of N inputs, whereas 67% comes from the termite nests that envelop its roots (Nishi et al. 2013).

Paepalanthus bromelioides grows only in the Serra do Cipó highlands of the Serra do Espinhaço mountain range in Minas Gerais, Brazil, where it grows in open, fire-swept *campos rupestres* ("rocky fields") vegetation over nutrient-poor sandstone. Its growth form is somewhat similar to that of *Bonnetia maguireorum* (Bonnetiaceae) from the tallest of the *tepuis*, the Serra de la Neblina (T.J. Givnish *personal observation*). Givnish et al. (1986) argue that the massive rosette and largely unbranched habit of *B. maguireorum* arose as an adaptation to fire that occurs on rocky, highly unproductive surfaces. One of us (T.J. Givnish) therefore hypothesizes that this growth form evolved in *Paepalanthus* subg.

Platycaulon, which is endemic to the fire-swept Serra do Espinhaço. This evolution proceeded from relatively small individuals (the earliest diverging *P. macropodus*) to much more massive, derived forms including *P. bromelioides*, *P. vellozioides*, and *P. planifolia* (phylogenetic relationships established by Trovó et al. 2013). These large rosettes form phytotelmata that likely would serve as pre-adaptation for carnivory.

3.7 Loss of carnivory

There are at least a few evolutionary reversals from carnivorous to noncarnivorous lineages (Fleischmann 2010, 2011a) that may be caused by adaptation to new habitats where carnivory is selected against (e.g., adhesive traps in very wet environments). In tropical wet forests, part-time and facultative (*Triphyophyllum*; Figure 3.8) or complete loss of carnivorous traits (in *Dioncophyllum*, *Habropetalum*, and *Ancistrocladus*; §3.2.2) has occurred in the otherwise carnivorous clade of Caryophyllales. Similarly, *Drosera schizandra* grows in relatively rich rainforest soil in understory habitats in rainy montane forests of Queensland (A. Fleischmann *personal observation*). Its large leaves bear comparatively few, scattered glands that, unlike most other *Drosera* spp., do not regenerate mucilage after it has been washed away by rain (Bourke 2009, Fleischmann 2011a). Occasional spontaneous mutants of *Drosera* species bear no tentacles (e.g., *D. erythrorhiza* with fully eglandular laminae from Western Australia; Dixon et al. 1980, K. Dixon and A. Cross *personal observations*; Figure 3.8). *Drosera caduca* from tropical northern Australia grows on the nutrient-poor sandy soils (Lowrie 2013). Unlike

Figure 3.8 (Plate 4 on page P3) Loss of carnivory in carnivorous Caryophyllales (**a–e**) and Lamiales (**f**). (**a**) Normally developed *Drosera caduca* (Droseraceae) leaves consist of an enlarged petiole, the lamina greatly reduced to often a single, apical tentacle, or fully absent. This *Drosera* produces carnivorous foliage only in juvenile plants and after dormancy, but for the largest part of its life, it is a noncarnivorous sundew. (**b**) An almost entirely eglandular, noncarnivorous, naturally occurring mutant of *Drosera erythrorhiza*. (**c**) The predominant habit of the part-time carnivorous liana *Triphyophyllum peltatum* is noncarnivorous (shoot with the double-hooked climbing leaves shown). (**d**) The post-carnivorous Ancistrocladaceae (illustrated by *Ancistrocladus abbreviatus* from Sierra Leone); the inset shows the typical Mettenian glands that link it to Nepenthaceae. (**e**) *Nepenthes lowii*, a coprophagous rather than carnivorous pitcher plant (Chapter 26). (**f**) Shoots of the aquatic rheophyte *Utricularia neottioides* from Brazil usually lack traps almost entirely. Photograph (**b**) by Kingsley Dixon, all others by Andreas Fleischmann.

its fully carnivorous congeners, *D. caduca* has leaves with carnivorous laminae only as a juvenile plant or when freshly emerging from dry dormancy. For the rest of its life cycle (and during anthesis), the leaves consist only of a greatly enlarged, noncarnivorous petiole (Fleischmann 2011a, Lowrie 2013; Figure 3.8). The causes of this seasonal heterophylly is not known.

Exploitation of alternative nutrient sources is another mechanism by which carnivory may be lost. The rainwater-impounding pitchers of *Nepenthes* have allowed a few species to evolve mutualisms with tree shrews, bats, or rodents that use them as latrines or roosts and provide abundant nutrients in excrement (Clarke et al. 2009, Chin et al. 2010, Grafe et al. 2011, Greenwood et al. 2011, Schöner et al. 2013; Chapters 5, 26; Figure 3.8).

Some *Utricularia* species, especially in soft-water lakes, appear to depend less on carnivory and more on consumption of algae or pollen (Richards 2001, Peroutka et al. 2008, Koller-Peroutka et al. 2015). At least two aquatic rheophyte species of *Utricularia* are largely noncarnivorous. The Brazilian *U. neottioides* and the African *U. rigida* rarely develop any traps along their foliar shoots (Taylor 1989, Fleischmann 2011a, Adamec et al. 2015a; Figure 3.8). In the swiftly floating water of their natural habitats, suction traps likely would not work, and these species may be in the process of evolving away from carnivory. Populations of *U. neottioides* from red soil streams of coastal Brazil do produce traps (V. Miranda *personal communication*), whereas populations from blackwater streams in the *campos rupestres* of central Brazil do not (Adamec et al. 2015a, A. Fleischmann *personal observation*). These might represent two different evolutionary lineages employing different carnivorous strategies. Last, some species of tank-forming *Brocchinia* obtain nutrients from myrmecophily, leaf-fed epiphytism, N_2-fixation via cyanobacterial plugs, or possibly mutualisms with frogs (Givnish et al. 1997, Givnish 2017). However, these examples represent transitions within the tank-forming lineage itself, not losses of carnivory.

3.8 Future research

Over the past two decades, molecular phylogenetics have identified the relationships among different groups of carnivorous plants and their noncarnivorous ancestors. As a result, we are now in a far better position to understand the evolution of carnivorous plants, based on our knowledge of time and location of divergence from noncarnivorous ancestors; inferences about exaptations; and identification of evolutionary drivers of carnivory that acted on ecological differences between carnivorous plants and their noncarnivorous relatives.

Genomics and evolutionary developmental biology provide new opportunities and methods to explore the evolution of carnivorous plants (e.g., Bemm et al. 2016, Fukushima et al. 2017). We can gather a more detailed understanding of genetic shifts that led to carnivory by investigating evolutionary relationships of the genes involved in carnivorous structures and functions among close relatives of carnivorous plants and common ancestors of carnivorous and noncarnivorous sister groups. We could explore developmental, physiological, and ecological consequences of modified genes or highly amplified families of genes (e.g., using CRISPR) in noncarnivorous relatives.

Genomics may provide the most compelling data for deciding whether carnivory arose once or twice in Ericales by examining whether the same genes or the same orthologous copies of those genes are involved in Sarraceniaceae and *Roridula*, and possibly also in the sister-group, Actinidiaceae.

Last, *Brocchinia* offers opportunities for tracing the evolution of carnivory using sequence- and genomic-level data together with accurate time-calibrated phylogenetic trees. This genus includes both carnivorous and noncarnivorous species; species with a series of other highly specialized means of nutrient capture; and the evolution of carnivory is comparatively recent, which might allow for the use of comparative genomics, development, and morphology to study the first stages of carnivory and other specialized means of nutrient capture.

PART II

Systematics and Evolution of Carnivorous Plants

CHAPTER 4

Systematics and evolution of Droseraceae

Andreas Fleischmann, Adam T. Cross, Robert Gibson, Paulo M. Gonella, and Kingsley W. Dixon

4.1 Introduction

Droseraceae is a family of carnivorous herbs in the "non-core Carophyllales" (*sensu* APG IV 2016) of almost cosmopolitan distribution, comprising three genera: *Drosera*, *Aldrovanda*, and *Dionaea* (Figure 4.1). Both *Aldrovanda* and *Dionaea* are monotypic, but *Drosera* is the largest carnivorous plant genus (slightly exceeding *Utricularia*; Chapter 8) with ≈250 species currently recognized, depending on species concept (Lowrie 2013, Gonella et al. *unpublished data*, Fleischmann et al. *unpublished data*). *Drosophyllum*, which historically was included in Droseraceae, is now placed in a family of its own, the nepenthalean Drosophyllaceae (Chapter 10).

The monophyly of Droseraceae supported by molecular phylogenetic reconstructions (Williams et al. 1994, Rivadavia et al. 2003; Figure 4.2) is substantiated by morphological synapomorphies, such as endoaperturate, echinate pollen tetrads that are unique among angiosperms (Takahashi and Sohma 1982); phytochemistry (Culham and Gornall 1994); and genetic evidence (including the unique loss of the *trnK* intron in all three genera; P. Nevill et al. *unpublished data*). Despite these synapomorphies, the three Droseraceae genera are highly divergent in overall appearance and trap function.

Two trapping mechanisms occur in the Droseraceae: the "snap-traps" of *Aldrovanda* and *Dionaea*, and the "flypaper traps" of *Drosera* (Figure 4.1). Snap-traps consist of leaves with bilobed lamina that close rapidly when triggered to form a chamber in which prey is trapped and digested (Figure 4.1b, 4.1c; Chapter 14). In contrast, the flypaper traps of *Drosera* are made up by a lamina that is adaxially covered with numerous long-stalked, capitate glandular emergences ("tentacles") on the adaxial surface (Figure 4.1a; Chapter 13). These stalked glands secrete a clear, highly viscous, water-based muco-polysaccharide to trap prey (Chapter 16). The lamina and tentacles of many species of *Drosera* are capable of significant movement to maximize contact with captured prey, whereas trap closure in *Dionaea* and, most notably, *Aldrovanda* represents one of the fastest movements in the plant kingdom (Lloyd 1942, Cross 2012a; Chapter 14). Once prey has been captured, proteolytic enzymes are secreted by stalked (*Drosera*), sessile (*Dionaea*), or quadrifid (*Aldrovanda*) glands to break down prey and assimilate nutrients (Chapter 16).

Some authors regard snap-traps as an adaptation to catch larger prey items more effectively (Gibson and Waller 2009, Poppinga *et al.* 2013a). However, as each trap of *Dionaea* and *Aldrovanda* can capture only one prey item at a time, and as large *Drosera* species can trap similarly-sized prey in even larger numbers, this explanation seems unsatisfactory. Fast-moving traps comparable to snap-traps are only paralleled in the bladder traps of *Utricularia* (Chapter 8). In both aquatic genera—*Aldrovanda* and *Utricularia*—the fast trap movement is needed to capture prey together with the surrounding water. Therefore, it seems likely that snaps traps evolved from adhesive traps as an adaptation to

Fleischmann, A., Cross, A. T., Gibson, R., Gonella, P. M., and Dixon, K. W., *Systematics and evolution of Droseraceae.* In: *Carnivorous Plants: Physiology, ecology, and evolution.* Edited by Aaron M. Ellison and Lubomír Adamec: Oxford University Press (2018). © Oxford University Press. DOI: 10.1093/oso/9780198779841.003.0004

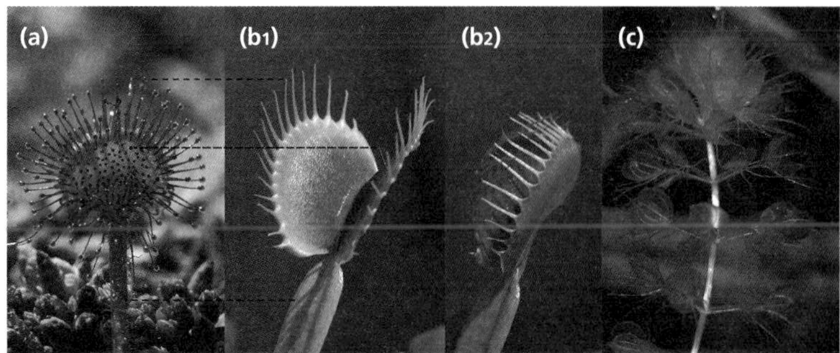

Figure 4.1 Basic trap types in the Droseraceae. (**a**) "flypaper trap" of *Drosera*, in this case the generic type species, *D. rotundifolia*. (**b**) "snap-traps" of *Dionaea muscipula* and (**c**) *Aldrovanda vesiculosa*. Photographs by A. Fleischmann.

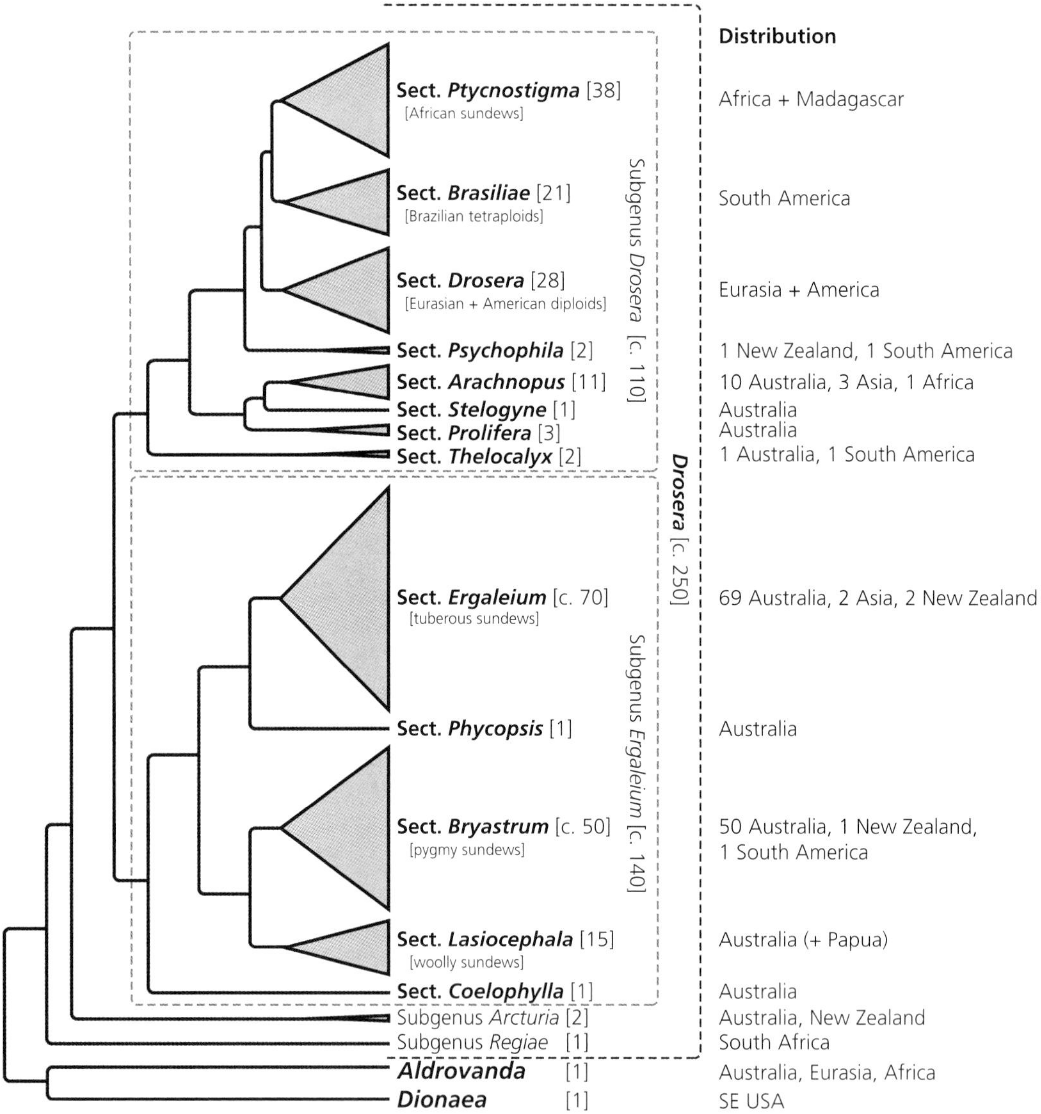

	Distribution
Sect. *Ptycnostigma* [38] [African sundews]	Africa + Madagascar
Sect. *Brasiliae* [21] [Brazilian tetraploids]	South America
Sect. *Drosera* [28] [Eurasian + American diploids]	Eurasia + America
Sect. *Psychophila* [2]	1 New Zealand, 1 South America
Sect. *Arachnopus* [11]	10 Australia, 3 Asia, 1 Africa
Sect. *Stelogyne* [1]	Australia
Sect. *Prolifera* [3]	Australia
Sect. *Thelocalyx* [2]	1 Australia, 1 South America
Sect. *Ergaleium* [c. 70] [tuberous sundews]	69 Australia, 2 Asia, 2 New Zealand
Sect. *Phycopsis* [1]	Australia
Sect. *Bryastrum* [c. 50] [pygmy sundews]	50 Australia, 1 New Zealand, 1 South America
Sect. *Lasiocephala* [15] [woolly sundews]	Australia (+ Papua)
Sect. *Coelophylla* [1]	Australia
Subgenus *Arcturia* [2]	Australia, New Zealand
Subgenus *Regiae* [1]	South Africa
Aldrovanda [1]	Australia, Eurasia, Africa
Dionaea [1]	SE USA

Subgenus *Drosera* [c. 110]

Subgenus *Ergaleium* [c. 140]

Drosera [c. 250]

Figure 4.2 Simplified cladogram of Droseraceae and the genus *Drosera* based on molecular phylogenetic reconstructions; topology based on Rivadavia et al. (2003, 2012) and Fleischmann et al. (*unpublished data*). Triangle size corresponds to number of species in each clade; species numbers for each monophyletic lineage are indicated in square brackets. Species numbers largely following Lowrie (2013) and Gonella et al. (*unpublished data*). Illustration by A. Fleischmann.

living in an aquatic environment. In this evolutionary scenario, *Dionaea* would be interpreted as a terrestrial descendant of an aquatic plant lineage. This scenario is supported because traps of submerged *Dionaea* work perfectly well under water (Roberts and Oosting 1958, Poppinga *et al.* 2016a).

Dionaea and *Aldrovanda* are sister genera with high molecular and morphological support (including similarities in trichome and gland morphology). A single evolutionary origin of the fast-moving snap-traps therefore seems reasonable (Williams et al. 1994, Cameron et al. 2002, Rivadavia et al. 2003, Heubl et al. 2006, P. Nevill et al. *unpublished data*; Chapters 3, 14), an hypothesis that has additional support from fossil biogeography. The modern native distribution of *Aldrovanda* and *Dionaea* displays extreme geographical disjunction (Figure 4.3), but fossil evidence indicates that *A. vesiculosa* also occurred in North America during the Late Tertiary (Matthews and Ovenden 1990) and a common paleo-biogeographic origin of a "snap-trap lineage" leading to the two extant genera appears to be likely. The sister pair *Aldrovanda* and *Dionaea* is consecutive sister to a monophyletic genus *Drosera* (Figure 4.2).

4.2 *Dionaea*

4.2.1 Morphology and systematics

Dionaea is a monotypic genus, comprising the sole extant species *Dionaea muscipula*, which is endemic to the coastal plain of North and South Carolina on the eastern seaboard of the United States (Roberts and Oosting 1958, Schnell 1976; Figure 4.3). It is an herbaceous, rosette-forming, perennial (hemi-)cryptophyte that grows from a rhizome with fibrous roots whose upper part is covered in the persistent fleshy remains of old leaf bases. The ≤ 7-cm-long leaves have a prominent mid-vein and comprise an expanded, obovate leaf-like petiole, a short stalk made up by the mid-vein continuing beyond the expanded section of petiole, and the lamina that comprises two lobes to form the carnivorous snap-traps (Figure 4.1b). The semicircular lamina lobes are ≤ 24-mm long, ≤ 14-mm wide, and are united along the midrib (Bailey and McPherson 2012). Both lobes face each other at nearly right angles and have a thickened margin that supports up to about 20 stiff, bristle-like teeth, each ≤ 18-mm long. The center of each trap lobe typically bears three to five trigger hairs,

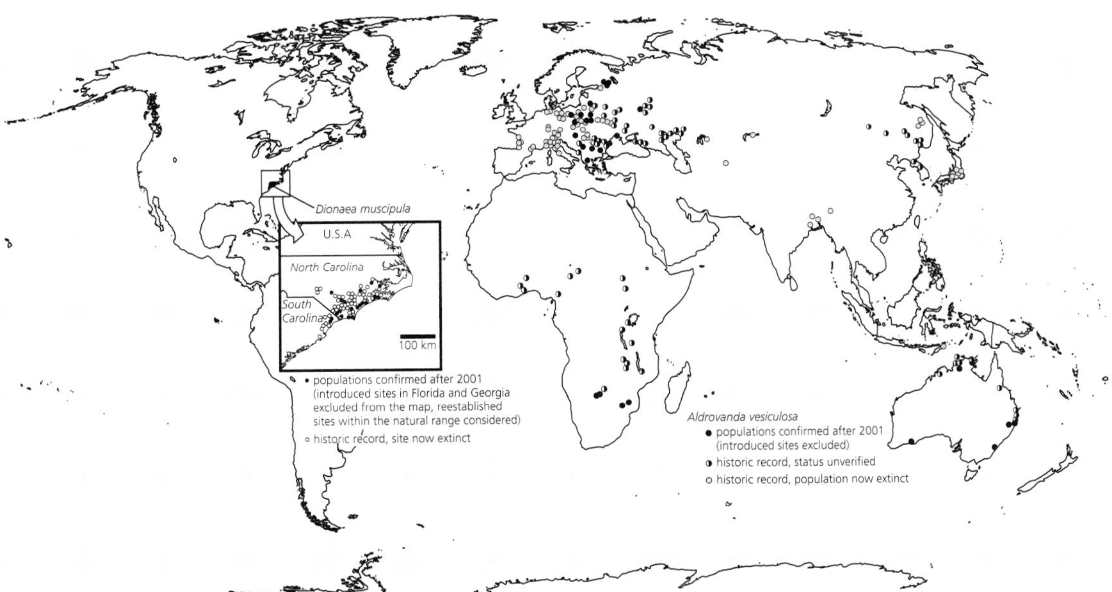

Figure 4.3 Distribution of *Aldrovanda* (based on Cross 2012b) and *Dionaea* (updated from Roberts & Oosting 1958, Schnell 1976, and with information kindly supplied by B. Rice). Map drawn by A. Fleischmann.

each ≤ 6-mm long (Lloyd 1942; Chapter 14). Both the trigger hairs and the marginal teeth are homologous to the glandular emergences of *Drosera* (Heubl et al. 2006; Figure 4.1). The leaves are held flat against the ground but in some plants may be semi-erect during anthesis.

Flowers are produced on a cymose inflorescence ≤ 40-cm tall that arises from the center of mature rosettes. The flowers are pentamerous and actinomorphic, ≤ 2-cm diameter, with five ovate sepals ≤ 12-mm long, five white obovate petals ≤ 14-mm long; both calyx and corolla are persistent in fruit. The superior, unilocular, spherical ovary, made up by five fused carpels, is nectariferous and surmounted by a single central pistil (made up by the fused styles) with a terminal tuft of multiple-branched stigmas, and which is surrounded by a ring of 15 erect stamens (resulting from secondary polyandry: deduplication of the five staminal primordia that otherwise are found in the perfectly tetracyclic flowers of the two other Droseraceae genera, which have five stamens; Diels 1906). The flowers are protandric, and each flower is open for about three days. The pollen tetrads are tetrahedral and the pollen grains are operculate (Takahashi and Sohma 1982). The fruits ripen about two months after pollination and produce ≤ 40 shiny, black, ovoid seeds ≤ 1.4 mm long.

4.2.2 Carnivory

Carnivory in *Dionaea* was proposed by John Ellis (1770) and experimentally tested by Darwin (1875). Prey of *Dionaea* largely consists of terrestrial arthropods (Gibson and Waller 2009, Bailey and McPherson 2012). The snap-trap mechanism is assumed to have evolved from an ancestor with adhesive traps (Chapter 3) to capture larger prey items (Cameron et al. 2002, Gibson and Waller 2009; Chapter 21).

4.2.3 Ecology

Dionaea grows in nutrient-poor, acidic, quartz sand, *Sphagnum* moss, peat, or loam soils in seepage zones, along drainage lines, and in shrublands of open *Pinus* woodlands. It is endemic to the Atlantic coastal plain of North and South Carolina in the United States (Bailey and McPherson 2012), although it has been introduced to, and established in, the New Jersey Pine Barrens, ≈800 km to the north (Chapter 28) and to a few sites in Florida. The sites where *Dionaea* grows characteristically are located on small slopes between freshwater wetlands and drier longleaf pine (*Pinus palustris*) woodlands (Roberts and Oosting 1958, Schnell 1976, Luken 2005, Bailey and McPherson 2012; Chapter 2).

Because *Dionaea* grows in a region that experiences cool winters with frequent frosts, hot humid summers, and infrequent fires, it has a strongly seasonal growth pattern that includes a winter-dormant period as hemi-cryptophytes or cryptophytes (bulbous geophytes). New growth typically begins in late winter or early spring. Scapes emerge in late spring, flowering occurs in early summer, and seeds are set by mid-summer.

4.3 *Aldrovanda*

4.3.1 Morphology and systematics

Aldrovanda vesiculosa is a perennial free-floating rootless macrophyte, comprising a simple or sparsely-branched stem ≤ 20 cm long that possesses successive whorls of bristled, prey-catching leaves. Each whorl consists of four to nine basally adnate leaves with a swollen, dorsally flattened petiole, terminating in a bi-lobed trap of two semicircular convex lobes joined along one side by a continuation of the petiole midrib (Figure 4.1c). Older whorls senesce as new growth is produced, with secondary axillary branches fragmenting from the parent plant to form vegetatively produced clonal progeny (Chapter 19). The solitary, actinomorphic flowers are small, pentamerous, and tetracyclic, with five free sepals, five free, white petals (both calyx and corolla are persistent in fruit), five bithecate stamens, and a single, spherical ovary consisting of five fused carpels (Diels 1906). Flowers open for only two to three hours, and are held above the water surface on a short peduncle that reflexes upon pollination, so that the indehiscent fruits become submerged (Cross 2012a). The tetrahedral pollen tetrads are operculate (Takahashi and Sohma 1982), and the seeds are small (1–2 mm), black, and spherical, with an intricate exotesta structure (Cross et al. 2016; Chapter 22).

4.3.2 Distribution

Aldrovanda is at least of Tertiary origin (65–55 Mya) and has experienced numerous pronounced cycles of extinction and speciation in response to paleoclimatic factors. Nineteen fossil species have been described, and *A. vesiculosa* is the only species that survived beyond the Pleistocene (Cross 2012a).

The historical distribution of *A. vesiculosa* was vast; paleontological specimens or herbarium records have been collected on every continent except South America and Antarctica (Figure 4.3). The species has been found throughout equatorial Africa from South Africa to Chad and from Ghana to Tanzania; from coastal and high rainfall areas of southwestern, northern, and eastern Australia; along major Asian avian flyways from East Timor to Japan, India, China, and the Koreas; and throughout the entire European continent from Kazakhstan to western France (Cross 2012a). Fossil specimens exist from the Porcupine River region of Alaska in northwestern North America (Matthews and Ovenden 1990). Few angiosperms possess a similarly pan-continental distribution, and the range of even fewer encompasses such a wide variety of climates and habitats: *A. vesiculosa* occurs in all of the major Köppen-Geiger climate types except polar areas (Peel et al. 2007; Table 2.1).

4.3.3 Carnivory

Numerous studies have confirmed that the traps of *Aldrovanda* are capable of the capture and digestion of prey (Cross 2012a). The margins of the two lobes are bent inwards and fringed with small, inward-pointing, membranous teeth that may prevent trapped prey from escaping. Traps are oriented to open away from the stem, presumably in an optimal position for prey capture (Cross 2012a). The inner surfaces of the trap lobes are studded with stalked, quadrifid digestive glands and long, double-articulated hairs that trigger trap closure when they are stimulated (Ashida 1934, 1935, Poppinga and Joyeux 2011). Trap closure is smooth and continuous, and is described as kinematic amplification of the bending deformation of the midrib (Poppinga and Joyeux 2011; Chapter 14). Closure occurs in as little as 0.01 seconds (Ashida 1935, 1934).

Glands within the traps produce proteases and phosphatases (Płachno et al. 2006; Chapter 16), and the uptake of nutrients into plant tissues has been traced using radioisotope-labelled *Daphnia* (Fabian-Galan and Salageanu 1968). It is unknown whether *A. vesiculosa* produces volatiles to attract prey, but it has been suggested that the filamentous trigger hairs mimic the appearance of algae and act as prey attractants (Schell 2003, Cross 2012a). Prey most frequently include ostracods, amphipods, daphnids, tardigrades, and the larvae of mosquitoes and non-biting midges. Larger invertebrates or small vertebrates occasionally are captured (Cross 2012a).

Although *A. vesiculosa* is not reliant upon carnivory for growth and development in cultivation (Cross 2012a), field observations and prey exclusion studies show that prey capture contributes positively to growth rate, fecundity, investment in carnivory, and branching (Kamiński 1987a, Adamec 2000, Adamec et al. 2010, Cross 2012a; Chapter 19). Carbon availability is considered a primary factor in apical branching (Adamec 2000, 2011e), and if free CO_2 is limiting, carbon also may be assimilated from captured prey (Adamec 1999).

4.3.4 Ecology and conservation

Aldrovanda vesiculosa is restricted to oligo-mesotrophic and dystrophic (humic) freshwater habitats such as fens, billabongs, lakes, lagoons, and slow-flowing river deltas (Adamec 1995, Cross 2012a). The species competes poorly with other macrophytes and is generally limited to shallow microhabitats harboring loose and species-poor plant communities (Kamiński 1987a, Adamec 1995, 1999, 2005b, Cross et al. 2015). Its growth and development is responsive to water depth, irradiance, pH-value, temperature, and associated vegetation (Kamiński 1987b, Adamec 1999, Cross 2012a, Cross et al. 2015), and is affected detrimentally by high levels of nitrates and phosphates in the water column (L. Skates and A. Cross *unpublished data*). Wetland eutrophication is considered to be the primary cause of local population declines (Adamec 1995, 1999, Jennings and Rohr 2011, Cross 2012a).

The growth rate of *A. vesiculosa* is rapid under optimal conditions, with biomass doubling in as little as eight days, and new axillary branches produced every one to two weeks (Adamec 1999, Adamec and Lev 1999, Adamec and Kovářová 2006, Cross 2012a). A single individual can produce vegetatively up to 40 progeny in the course of a single growing season (Adamec 1999, Adamec and Kovářová 2006, Cross 2012a); shoot fragments transported by avian vectors are believed to be its major dispersal unit (Cross 2012a, Cross et al. 2016). Because of its highly clonal nature, the species displays remarkably low levels of genetic variation throughout its global range (Adamec and Tichý 1997, Maldonado San Martin et al. 2003, Hoshi et al. 2006, Elansary et al. 2010, A. Cross et al. *unpublished data*).

Sexual reproduction is highly variable and contributes few offspring (Cross et al. 2016). The small white flowers are produced in late summer when conditions are favorable, but generally fail to yield many viable seed and these do not appear to persist for long periods in the sediment seed bank (Cross et al. 2016). Fecundity is linked to habitat factors such as competition, prey capture, and water quality (Cross et al. 2015). Condensed, hibernating shoot apices (turions) are produced in late autumn in temperate populations, and these organs persist through winter on the bottom of water bodies before rising to the surface and continuing growth in spring (Adamec 1999, Cross 2012a). Plants from populations in tropical and subtropical regions also produce turions (Cross 2013). In these environments turions may not exhibit the same innate dormancy as those from temperate regions (Adamec 2003a).

The number of *A. vesiculosa* populations worldwide has declined by nearly 90% over the last century, and it is believed that as few as 50 populations remain (Cross 2012a; Figure 4.3). Two thirds of these persist in the Pripyat River basin in Belarus, northern Ukraine, and Poland, and the remainder show a remarkable disjunction across four continents. The species has been confirmed extinct in Austria, Slovakia, Czech Republic, France, Germany, Italy, Bangladesh, India, Japan, Uzbekistan, and East Timor, remains unverified in another 20 countries, and has been listed as Endangered B2ab(iii,v) by the IUCN (Cross 2012b; Figure 4.3; Chapter 27). However, large and stable populations have established after successful introductions to suitable sites in Switzerland, Czech Republic, and Poland (Adamec 1995, 2005b, Adamec and Lev 1999).

4.4 *Drosera*

4.4.1 Life history and morphology

Life history. *Drosera* comprises about 250 species of herbaceous carnivorous plants. These exhibit a great diversity of growth habits and life strategies (Figure 4.5), partially associated with phylogenetic groups (Figure 4.2). Sixteen species (≈7%) are annual therophytes (all species of *D.* sects. *Coelophylla, Thelocalyx*, and *Arachnopus*, and two members of *D.* sect. *Lasiocephala*); all others are perennials. One third of the genus are geophytes, producing either shoot-derived tubers (≈70 species of "tuberous sundews" comprising *D.* sect. *Ergaleium*), subterraneous bulb-like perennating organs formed by the fleshy petiole bases (about half of the species of the northern Australian "woolly sundews" of *D.* sect. *Lasiocephala*), or succulent, cylindrical tap roots (20 species from the South African and American *D.* sects. *Ptycnostigma, Brasiliae*, and *Drosera*). The remaining species are above-ground perennials, hemicryptophytes, or chamaephytes. These species usually display a more or less well-pronounced winter or dry-season dormancy by producing an apical resting bud or size-reduced, dormant leaves (hibernacula, stipule buds, or dormant rosettes, in *D.* sects. *Drosera, Brasiliae, Ptycnostigma, Bryastrum*, and *Lasiocephala*, respectively).

Leaves. Independent of life-history strategy, *Drosera* are either rosette- or stem-forming herbs, with much reduced or well-pronounced internodes between the alternately (spirally) arranged leaves (Figures 4.5h, 4.5i). The leaves are petiolate (in some species, the petiole is greatly reduced), with the lamina covered by glandular emergences that constitute the carnivorous trap (Figure 4.1a). The laminae range from narrowly linear or thread-like to cuneate, spathulate, or oval, reniform, or crescentic with a peltate petiole attachment; lamina shape usually reflects adaptations to habitat, trapping strategy, or prey type, and does not correspond to phylogenetic relationships (e.g., Rivadavia et al. 2003, 2012). Leaves are entire in all species with the exception of *D. binata* (in the monotypic *D.* sect. *Phycopsis*), which produces a (multiple) dichotomously forked lamina (Figure 4.5f). The majority of *Drosera* spp. bear a persistent, variously fringed, papery stipule at the petiole base. This stipule is sometimes much reduced to fimbriate setae, e.g., in *D.* sect. *Psychophila* and some members of *D.* sect. *Ptycnostigma*, or can be greatly expanded to conspicuous organs protecting the stem or apical growing bud in the Australian pygmy sundews of *D.* sect. *Bryastrum* (Figure 4.5c) and some members of *D.* sect. *Brasiliae*.

The two phylogenetically early branching lineages of *D.* subg. *Regiae* (comprising only the paleoendemic South African species *D. regia*) and *D.* subg. *Arcturia* (comprising the two closely related species *D. arcturi* and *D. murfetii* from montane-to-alpine southeastern Australia and New Zealand) probably show the most ancestral morphological features. Their linear leaves are comparatively poorly differentiated, without a very distinct petiole, lacking stipules, and with a single, uniform type of carnivorous gland. All species of the *Drosera* core group (i.e., *D.* subg. *Ergaleium* and *Drosera*) have stipules and show gland dimorphism. They bear long-stalked unifacial marginal glands (or spoon glands, or "snap-tentacles"; Chapter 14) that are confined to the lamina apical margin, and stalked terete glands on the entire lamina surface (at least on juvenile leaves; Leavitt 1903, Conran et al. 2007). However, there is a secondary loss of stipules and unifacial marginal glands in all tuberous *Drosera* (*D.* sect. *Ergaleium*), *D.* sect. *Arachnopus*, some members of *D.* sect. *Ptycnostigma*, and a single member of *D.* sect. *Lasiocephala* (Conran et al. 2007, Fleischmann et al. *unpublished data*).

Inflorescences and flowers. *Drosera* inflorescences are simple scorpioid cymes (bracteose, but the floral bracts are inconspicuous and even caducous in several taxa). Some species produce double or multiple cymes (*D.* sect. *Phycopsis*, some species of *D.* sects. *Ergaleium* and *Brasiliae*), or reduced to single-flowered scapes (*D.* subg. *Arcturia*, *D.* sect. *Psychophila*, some species of *D.* sects. *Ergaleium*, *Bryastrum*, and *Drosera*). The hermaphrodite flowers are actinomorphic, tetracyclic, and generally pentamerous; one species in each of *D.* sect. *Lasiocephala* and *Bryastrum* has tetramerous flowers, and a single species of *D.* sect. *Ergaleium* has 8-12-merous flowers. The five basally adnate sepals are persistent in fruit. The five petals are free, thin, and ephemeral. The five stamens bear bithecate anthers; the gynoecium is made up of five to (more frequently) three fused carpels.

Stylar characters have been used as the major key character for infrageneric classification (Planchon 1848, Diels 1906), and they also support the phylogenetic clades. The styles vary from (2)3–5, depending on species and affinity, and can be variably fused, ranging from five free styles (e.g., in *D.* sect. *Thelocalyx* and some members of *D.* sect. *Bryastrum*) to a single pistil made up of three entirely fused styles (the monotypic *D.* sect. *Stelogyne*). The styles further can be entire and undivided (e.g., *D.* sect. *Bryastrum*), or dichotomously forked to various degrees: divided once into two style arms, as found in the majority of derived species (e.g., *D.* sects. *Drosera, Brasiliae*, and *Ptycnostigma*) or divided multiple times resulting in many fimbriate stylar branches (e.g., most species of *D.* sect. *Ergaleium*).

Pollen. The specific pollen tetrad types (following Takahashi and Sohma 1982, Rivadavia et al. 2012, Fleischmann et al. *unpublished data*) support phylogeny and infrageneric classifications; the early-branching *D.* subg. *Regiae* and *Arcturia* have operculate pollen (which links to the sister genera *Dionaea* and *Aldrovanda*), the more derived *D.* subg. *Ergaleium* and *Drosera* have inoperculate pollen and within each clade show an evolutionary trend in reduction of the number and size of apertures (Fleischmann et al. *unpublished data*).

Fruit and seeds. The fruit is a dehiscent dry capsule, which opens loculicidally by three to five longitudinal slits (majority of species), or rarely valves (in a few species that use ombrochory for seed dispersal; Fleischmann et al. 2007). The numerous seeds are generally small (wind- or water-dispersed) with a thin (majority of species) or multi-layered testa (in the comparatively larger seeds of *D.* sect. *Ergaleium*; Diels 1906) with a reticulate, foveolate, or papillose surface, that in some species has bipolar outgrowths (Chapter 22).

Clonal reproduction. In addition to sexual reproduction, many species reproduce clonally. All members of *D.* sect. *Bryastrum* except *D. meristocaulis* produce gemmae in annual cycles (leaf-derived bulbils; Goebel 1908, Rivadavia et al. 2012; Figure 4.5d). Many species of *D.* sect. *Ergaleium* readily produce daughter tubers (Dixon and Pate 1978, Chen et al. 1997). Several of the perennial species from the other sections have the ability to grow adventitious plantlets from roots or leaves.

4.4.2 Phylogeny and taxonomy

The genus *Drosera* was retrieved as monophyletic in all phylogenetic reconstructions (Rivadavia et al. 2003, 2012), with four major clades evident which are here referred to as subgenera (Figure 4.2). The early branching *D.* subg. *Regiae* and *Arcturia*, which are consecutive sisters to the remainder of the genus, constitute the two ancestral-looking, paleoendemic, monotypic/species-poor lineages with species that produce monopodial, woody rhizomes (Diels 1906, Gibson 1999), no indumentum (except for a single uniform type of carnivorous glands and the glandular calyx of *D. regia*, the plants in these two subgenera are glabrous), no stipules, and operculate pollen (Fleischmann et al. *unpublished data*).

Phylogeography. *Drosera* subg. *Ergaleium* includes more than half of the *Drosera* spp. currently described (≈140 species; Figure 4.2). It comprises exclusively Australian lineages, although all sections except *D.* sect. *Coelophylla* have single members that spread outside Australia (Gibson et al. 2012, Rivadavia et al. 2012). This subgenus has the greatest diversity in life form, habit, and morphology, and the largest palynological diversification (in terms of number of pollen types and exine ornamentation; Takahashi and Sohma 1982, Fleischmann et al. *unpublished data*), and the most diverse generative traits (floral characters and seed shape; Lowrie 2013). This clade likely diversified during the Middle Miocene disruption with adaptation to the drying of the Australian continent and establishment of a Mediterranean climate with seasonal aridity (Yesson and Culham 2006).

Sister to *D.* subg. *Ergaleium* is *D.* subg. *Drosera*, a clade that mirrors an "out-of-Australia" movement and diversification. The two first-branching lineages (comprising four sections) contain almost exclusively Australian species, although long-distance dispersal may have caused the disjunction between Australia and South America of the two sister species of *D.* sect. *Thelocalyx* (Rivadavia et al. 2012). Ten of the 11 species, and the greatest morphological diversity of consecutive sister clade, *D.* sect. *Arachnopus*, also are found in northern Australia (Lowrie 2013). However, two species of this section also occur in tropical Asia, and one, *D. indica*, is found in Asia and (sub)tropical Africa, but not on the Australian continent. This implies an Australian origin of this section, followed by a westward expansion that probably occurred comparatively recently. The consecutive sister to the remainder of the species, *D.* sect. *Psychophila* has a similar disjunct range in the Southern Hemisphere. It comprises two sub-Antarctic species, one endemic to New Zealand and the other to the southern tip of South America.

The remainder of the genus (≈80 species) is placed in three clades supported by ploidy level (Rivadavia et al. 2003, Fleischmann et al. *unpublished data*). These occur in tropical to subtropical Africa (*D.* sect. *Ptycnostigma*), South America (*D.* sects. *Drosera* and *Brasiliae*), and Asia (*D.* sect. *Drosera*), as well as in the temperate Northern Hemisphere (*D.* sect. *Drosera*). The sole exception is *D. spatulata* (*D.* sect. *Drosera*), which also occurs in Australia and New Zealand. As this species belongs to an early-branching group of species within the *D.* sect. *Drosera* lineage, an Australian origin of the subgenus is likely, with recent diversification outside the Australian continent.

Only three species—*D. rotundifolia*, *D. anglica*, and *D. intermedia*—occur in Eurasia, although *D. intermedia* does not occur anywhere in Asia itself (Figure 4.4). All three species are also present and widespread in North America. The allopolyploid *D. anglica* is most likely of North American origin, as one of its parent species, *D. linearis*, is not recorded from the Old World. The other two species also are likely to be of New World origin. Long-distance dispersal is well known in *Drosera* (Rivadavia et al. 2012), and some tropical Pacific outliers of the otherwise boreal *D. anglica* (on Hawai'i) and *D. rotundifolia* (in New Guinea and the Philippines; Coritico & Fleischmann 2016) might represent traces of continuous westward diaspore dispersal from the American continent. They also may have colonized Eurasia via Beringia. However, the extant distribution of *D. intermedia* in western to central Europe, but not in Asia, suggests, at least for this species, a second, independent colonization from America on eastward dispersal. Temperate Eurasia most likely was colonized by *Drosera* in geologically recent times, probably during or after the Pleistocene. In the Eurasian flora, *Drosera* is restricted to very specialized, typically postglacial habitats (Schlauer 2010).

A Gondwanan origin of the genus *Drosera* has been proposed by some authors based on

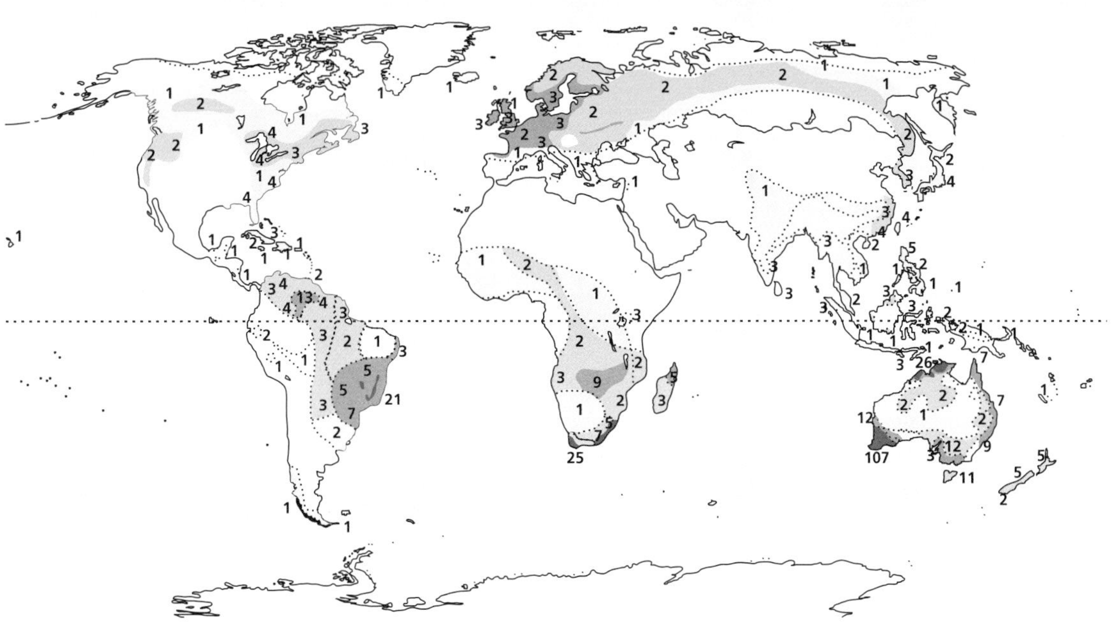

Figure 4.4 Global distribution of *Drosera*. Ranges and species numbers based on, and updated from, Diels (1906), Meusel et al. (1965), Gibson et al. (2012), Lowrie (2013), Coritico and Fleischmann (2016), Gonella et al. (*unpublished data*), and Fleischmann et al. (*unpublished data*). Map drawn by A. Fleischmann.

biogeography (Meimberg et al. 2000, Heubl et al. 2006) and fossil dating of the family (Cameron et al. 2002). However, the Late Cretaceous *Palaeoaldrovanda* "seeds" used as fossil constraint for Droseraceae by Cameron et al. (2002) were misidentified arthropod eggs (Heřmanová and Kvaček 2010), and there currently is no dated molecular phylogenetic evidence for a Gondwana/Late Cretaceous occurrence of the *Drosera* lineage. Fossil findings of 40 Mya Eocene pollen tetrads of likely affinity to Droseraceae (i.e., *Fischeripollis*) from the shores of Antarctica (Macphail and Truswell 2004) could further corroborate a Gondwanan origin. Current estimates of Caryophyllales divergence times place the age of the Droseraceae in the Early Tertiary (67–55 Mya) and that of *Drosera* (i.e., the split from the *Dionaea–Aldrovanda* lineage) at ≈42 Mya (Wikström et al. 2001, Yesson and Culham 2006).

4.4.3 Distribution

The genus *Drosera* occurs on all continents except Antarctica, although Eocene pollen fossils of affinity to *Drosera* are recorded from Antarctic soil (Macphail and Truswell 2004; Chapter 3). However, species diversity is not evenly distributed around the world: the centers of species diversity are found in the Southern Hemisphere (Figure 4.4). The greatest diversity is found in Australia (165 species [Lowrie 2013], 153 of them [93%] endemic), especially in southwest Australia (113 species [Lowrie 2013], 108 [96%] endemic, often narrowly endemic with very small ranges). Additional hotspots are South America (42 species [Gonella et al. *unpublished data*], 39 [93%] endemic), especially Brazil (32 species, 18 [56%] endemics [Gonella et al. *unpublished data*]) and the Guayana Shield (16 species, 10 [63%] endemics); and the wider Cape Region of South Africa (29 species [Fleischmann et al. *unpublished data*], 24 (83%) endemic). In contrast, areas comparatively poor in *Drosera* include the temperate Northern Hemisphere (10 species, three endemics [30%]), New Zealand (7 species, 1 endemic [14%]), the Asian tropics (12 species, 3 endemics [25%]), and the African tropics including Madagascar (12 species, 7 endemics [58%]) (Fleischmann et al. *unpublished data*).

Figure 4.5 (Plate 5 on page P4) Morphological diversity of *Drosera*. (**a**) The annual therophyte *D. sessilifolia* (*D.* sect. *Thelocalyx*). (**b–d**) Pygmy sundews of *D.* sect. *Bryastrum*: (**b**) in most species, the flower rivals the vegetative part in size, illustrated here by *D. barbigera*; (**c**) dormant stipule bud, illustrated here by *D. androsacea*; (**d**) vegetative propagation by gemmae, illustrated here by *D. pedicellaris*. (**e**) *Drosera ordensis* of the woolly sundews, *D.* sect. *Lasiocephala*. (**f**) The dichotomously branched lamina of *D. binata* (*D.* sect. *Phycopsis*). (**g**) Filiform leaves of *D. spiralis* (*D.* sect. *Brasiliae*). (**h, i**) A remarkably similar geophyte habit evolved in parallel in some Australian tuberous, illustrated here by (**h**) *D. zigzagia* (*D.* sect. *Ergaleium*), and members of the South African *D.* sect. *Ptycnostigma*, illustrated here by (**i**) *D. cistiflora*. Photographs (**a–e, g, i**) by Andreas Fleischmann; (**f**) by Robert Gibson, and (**h**) by Laura Skates.

4.4.4 Carnivory

Drosera was the first plant genus in which carnivory was confirmed (Darwin 1875; Chapters 12, 13, 14, 16, 17, 21). Darwin (1875) did many experiments with *Drosera* and other Droseraceae, and his concept of "insectivorous plants" was largely based on this experimental work.

4.4.5 Ecology and habitats

The highest species diversity and endemism in *Drosera* is found in Mediterranean or seasonally arid climates (Yesson and Culham 2006), where the plants occur in a variety of exposed, nutrient-poor, oligotrophic habitats. *Drosera* are generally limited to oligotrophic substrates, and the vast majority of species are strict calcifuges which occur only on acidic soils. Only three species are known to occur on oligotrophic, alkaline limestone soils: *D. linearis* from northeastern America; its amphiploid progeny *D. anglica* in America and Europe; and the tuberous *D. schmutzii* from Kangaroo Island in southern Australia (Schnell 1976, Lowrie 2013). Two other species, *D. neocaledonica* and *D. ultramafica* (both sect. *Drosera*), are confined to serpentine soils, whereas

D. arcturi can tolerate soils that are rich in heavy metals (Gibson et al. 1992, Fleischmann et al. 2011b).

The southwest of Australia is a biodiversity hotspot for *Drosera*, and most species in this region are found in Kwongan shrubland (Lambers 2014), swamplands, and granite outcrops. Members of *D.* sects. *Bryastrum* and *Ergaleium* occur in different parts of the landscape, including exposed ridges and rises on variably dissected laterite weathering profiles such as skeletal soils flanking ironstone (e.g., *D. barbigera, D. prophylla*) or granite outcrops (e.g., *D. andersoniana, D. lasiantha, D. moorei*); hillslopes and rises with ironstone gravel (pisolith) soils (e.g., *D. barbigera, D. silvicola, D. stolonifera*); permanent to ephemeral wetlands and drainage lines with heavy clay loams and peaty-sand soils (e.g., *D. gigantea, D. nitidula, D. pulchella*); sandplains with free-draining sandy soils (e.g., *D. micrantha, D. erythrorhiza, D. zonaria*); and along inland paleodrainage lines including the shores of salt lakes (e.g., *D. bicolor, D. salina, D. zigzagia*) (Lowrie 2013). Many of these habitats also support the annual *D. glanduligera* (sect. *Coelophylla*), which also occurs widely across southern Australia in areas with winter-maximum rainfall (Lowrie 2013). *Drosera hamiltonii* (*D.* sect. *Stelogyne*) and *D. binata* (*D.* sect. *Phycopsis*) are confined to peaty wetlands and creekbanks along the south coast; the latter also is found more widely in permanently wet habitats of southeastern Australia and New Zealand where it often co-occurs with *D. spatulata* (*D.* sect. *Drosera*). *Drosera pygmaea* (*D.* sect. *Bryastrum*) and several tuberous species (*D.* sect. *Ergaleium*) also occur in seasonally to permanently wet soils in southern Australia (Lowrie 2013, Salmon 2001).

The other center of diversity for *Drosera* in a Mediterranean climate is in the Cape Region of South Africa. Here, many species of *D.* sect. *Ptycnostigma* grow in fynbos shrubland and sedgeland on ranges overlying Table Mountain sandstone (Cowling and Richardson 1996). Several species grow in well-drained, sandy loam soil that is seasonally moistened by frequent coastal mists (e.g., *D. cistiflora, D. hilaris*, and *D. ramentacea*). Many species occur in perennially wet seepage zones on sandstone and in wet peaty soils along and beside drainage lines (e.g., *D. aliciae, D. capensis*, and *D slackii*). Species such as *D. coccipetala, D. alba*, and *D. trinervia* grow in seasonally sodden clay-loam soils flanking sandstone outcrops in ranges of the northern part of the Western Cape that are baked dry in summer. In the Western Cape, several species also grow on the coastal plains in seasonally moist clay-loam soils over meta-sediments or on the flanks of low granite outcrops (e.g., *D. pauciflora*). *Drosera regia* (*D.* subg. *Regiae*) is confined to a single valley in the Western Cape, where it occurs in and beside often poorly defined drainage lines. Evergreen to winter-dormant sundews in *D.* sect. *Ptycnostigma*, including *D. collinsiae, D. dielsiana*, and *D. natalensis*, occur in permanent wetlands—seeps and the banks of perennial streams—and in the coastal plains and along ranges in eastern South Africa. The distribution of most of these species also extends north into eastern tropical Africa (Obermeyer 1970a). The annual *D. indica* (*D.* sect. *Arachnopus*) occurs widely across tropical Africa, where it grows in seasonal wetlands and in ephemeral flush vegetation on ferricrites and inselbergs.

Tropical northern Australia also represents a significant center of *Drosera* diversity, and this monsoon tropical region harbors the "woolly sundews" (*D.* sect. *Lasiocephala*). Members of this complex occur in seasonally to permanently wet soils that either are part of the laterite weathering profile or are associated with sandstone. Species with comparatively hairless petioles, including *D. dilatatopetiolaris, D. petiolaris*, and *D. fulva*, often grow in the wetter parts of the landscape, whereas those with the most conspicuous hair cover on their petioles (e.g., *D. broomensis, D. ordensis, D. lanata*) tend to grow in free-draining sandy soils. Some species, (e.g., *D. paradoxa, D. subtilis*) frequently grow in sections of creeks in hard meta-sediment ranges or highly ephemeral seepage habitats. Congeners with broad petioles and large ovoid lamina (e.g., *D. falconeri* and *D. kenneallyi*) grow in heavy clay soils on the floodplains which may be flooded in summer and baked dry during the dry season. This region is also the center of diversity of the *Drosera indica* complex (*D.* sect. *Arachnopus*), species of which often grow sympatrically with members of *D.* sect. *Lasiocephala* and *D. burmannii* (*D.* sect. *Thelocalyx*). The wet tropics of northeastern Australia supports the three "rainforest sundews" of *D.* sect. *Prolifera* (*D. adelae, D. prolifera*, and *D. schizandra*). These closely related species occur in variably shaded and humid locations associated with drainage lines, waterfalls, and in clay-loam rainforest soil on the coastal ranges near Cairns (Lowrie 2013).

In the Neotropics, *Drosera* is most diverse on the ancient quartzitic formations of the *campos rupestres* (tropical montane grasslands and shrublands) in central-eastern Brazil and the Guayana Shield in northern South America. Twenty-two of the 32 Brazilian species occur in *campos rupestres* (13 are endemic to this vegetation type), whereas only two species (*D. cayennensis* and *D. communis*) occur in both regions (Gonella et al. *unpublished data*). Both of these regions are dominated by open-field upland and highland areas with nutrient-poor, sandy soils on sandstone and quartzite outcrops, although some *Drosera* spp. spread toward the surrounding lowlands and may even be restricted to it.

Representatives of *D.* sects. *Drosera* and *Brasiliae* are found in the *campos rupestres*, usually in seasonal or perennial springs and seepages, or along the banks of streams, but may also occur on hilltops with free-draining sandy soil with quartz gravel, relying on the morning mist as the main water source during the dry season (e.g., *D. quartzicola, D. schwackei*). During the dry winter, the growth of most species will slow and leaf size may be reduced, but the species do not go dormant; a few geophyte species die back to the roots and re-sprout in the wet season (e.g., *D. cayennensis, D. hirtella, D. montana*).

Fires are frequent in *campos rupestres*, but *Drosera* spp. quickly recover by new growth sprouting from persistent stems and roots. *Drosera spiralis* is hypothesized to protect its new leaves and meristem from fires with large, enclosing stipules (Gonella et al. 2012). A few other species are restricted to high-montane (> 1500 m a.s.l.) areas of *campos rupestres*, where they experience a less pronounced dry season and less frequent fires, and attain larger sizes (e.g., *D. graminifolia* and *D. magnifica*). Only a few Neotropical species also occur on clayish soils (e.g., *D. sessilifolia, D. communis, D. montana, D. latifolia*) or in shaded areas along banks of rivers under riparian forests (*D. riparia, D. tomentosa*; Gonella et al. *unpublished data*).

On the Guayana Shield, the main diversity center of *D.* sect. *Drosera* in America, most species grow on *tepuis* (sandstone massifs) and other Amazonian mountains, on outcrops of bare sandstone, or on pockets of soil over sandstone or granitic rock, where they experience ≈constant rainfall throughout the year, growing and flowering continuously. In the surrounding lowlands, *Drosera* spp. grow in white quartz sand savannas, and may experience a short dry season, surviving as root-geophytes (*D. cayennensis*) or annuals (*D. sessilifolia*). Some species are found in seasonally flooded areas along river banks, may be submerged for part of the year, and flower during the drier months (*D. amazonica, D. biflora*; Rivadavia et al. 2009).

Except for four species from *D.* sect. *Brasiliae* (*D. montana, D. cendeensis, D. condor*, and *D. peruensis*), *Drosera* is largely absent from the nutrient-rich soils of the Andes. *Drosera condor* and *D. peruensis* grow in the sandstone sub-Andean cordilleras in Ecuador and Peru, in vegetation reminiscent of that of the Guayana Highlands (Gonella et al. 2016). *Drosera uniflora* (*D.* sect. *Psychophila*) also grows in the temperate foothills of the southern Andes, in peaty wetlands among cushion plants. This species forms a hibernacula during the winter, similar to its New Zealand sister-species, *D. stenopetala*.

Several species of *Drosera* grow in seasonally to permanently wet sandy loams on the coastal plains of southeast North America (*D.* sect. *Drosera*), including *D. brevifolia, D. capillaris, D. filiformis*, and *D. tracyi* (Schnell 1976). They may co-occur with *D. intermedia* or *D. rotundifolia*, which together with *D. anglica*, are circumboreal species that grow more commonly in *Sphagnum* bogs and peaty wetlands (Diels 1906). *Drosera anglica* also tolerates the much more alkaline marl bogs and fens.

4.4.6 Conservation

Although a number of *Drosera* are widely distributed (e.g., *D. anglica, D. burmanii, D. indica, D. intermedia, D lunata, D. rotundifolia*, and *D. spatulata*), many species have narrow ranges and are endemic to small areas or specific landforms. The majority of these occur in Western Australia (e.g., *D. bicolor, D. graniticola, D. leioblastus, D. nivea, D. monticola*), whereas exceptions include *D. schizandra* and *D. prolifera* from northern Queensland (Lowrie 2013); *D. ericgreenii* and *D. regia* from South Africa

(Fleischmann et al. 2008, Stephens 1926); *D. humbertii* from northern Madagascar (Exell and Laundon 1956); *D. chimaera, D. graminifolia, D magnifica, D. meristocaulis, D. quartzicola, D. schwackei, D. solaris, D. spirocalyx, D. villosa*, and *D. yutajensis* from South America (Maguire and Wurdack 1957, Fleischmann et al. 2007, Gonella et al. 2012, 2015); and *D. linearis* in North America (Schnell, 1976).

Primary threats to all these species include habitat loss, invasive species, and altered hydrology (Lowrie 2013, Gonella et al. 2012, 2015; Chapter 27). The full distribution and population status of most *Drosera* spp. are known inadequately, which limits their representation on lists of priority species in relevant conservation legislation. However, some species with very small ranges grow within well-managed conservation reserves (e.g., *D. gibsonii, D. prostratoscaposa*), and in the absence of tangible documented threats are not currently considered at risk (A. Cross et al. *unpublished data*). A concerted international effort to assess empirically the distribution, population dynamics, threats to, and status of, all *Drosera* spp., is urgently needed (Chapter 27).

4.5 Future research

Suggested areas of future research into taxonomy and systematics of Droseraceae include refining taxonomic limits and the application of taxonomic rank in the case of *Drosera*, so that the classification of members of this family more accurately reflects its phylogeny. Several species complexes within *Drosera* will need to be studied with molecular, morphological, and ecological approaches. Further examination of the fossil record, particularly pollen and seed, is essential to gain more insight into evolution and diversification of Droseraceae. Inter- and intrapopulation genetics, especially in case of *Drosera*, are almost unknown. The latter could help focus conservation efforts on genetically novel taxa or populations.

CHAPTER 5

Systematics and evolution of *Nepenthes*

Charles Clarke, Jan Schlauer, Jonathan Moran, and Alastair Robinson

5.1 Introduction

Nepenthes is one of the largest genera of carnivorous plants, comprising at least 160 species. These species produce highly modified pitchers that have evolved to target a greater variety of nutritional sources, including specific arthropod taxa, leaf litter, and mammal feces, than any other group of carnivorous plants (Clarke and Moran 2016). The physical environment and plant–animal interactions are key drivers of diversification within and among species, which challenges taxonomists who continue to seek effective means of applying species concepts and accurately classifying the members of the genus.

In recent years, ecologists and evolutionary biologists have made significant progress in detecting drivers of diversification in *Nepenthes*. Advances in genetic sequencing methods are yielding new insights into recent evolutionary events within the genus. In this chapter, we review current trends and practices in the descriptive taxonomy of *Nepenthes* and provide recommendations to improve future work in this field. We also present an overview of evolution within the genus, focusing on the contribution of genetic analyses to our current understanding of the major groups of species within the genus.

5.2 Taxonomy and systematics

After the initial description of *Nepenthes* in 1753, the cumulative number of described species increased steadily from the mid-1880s until 1928, leveled off until the mid-1980s, and has steeply increased since then (Figure 5.1). The proliferation prior to 1928 of newly described taxa was driven by botanical exploration of Southeast Asia focused on the islands of the Sunda Shelf that were under colonial rule. In some cases, descriptions of new species were grossly deficient in detail and rigor (e.g., Masters 1890), whereas others were based on unstable or taxonomically uninformative traits.

These deficiencies were addressed by Danser's (1928) landmark revision of *Nepenthes*. He described 17 new species, and recognized a total of 67 species—only nine more than the previous revision by Macfarlane (1908). Unlike prior students of *Nepenthes*, Danser (1928) identified a suite of taxonomically informative morphological traits, dismissed uninformative and highly variable ones, applied a rigorous species concept, and did not recognize any rank below species. He was scathing in his criticism of earlier efforts, stating that:

> *the polymorphy of the species in general is too little to make such a distinction in the same weight in each case. This seems to be the right way to avoid complication of the nomenclature of a genus so little known. Moreover, the forms which have been distinguished as varieties up to the present, usually are not more than extreme variations of polymorphous species, having no value from a taxonomical point of view, since they give no idea of the polymorphy of the species in general. Often authors have gone so far as to describe specimens with extremely large or broad leaves, or with very long internodes, and different stages of growth as varieties or even species. Continuing in this way can only discredit systematics, especially that of the forms within species.* (Danser 1928: 250)

Clarke, C., Schlauer, J., Moran, J., and Robinson, A., *Systematics and evolution of* Nepenthes.
In: *Carnivorous Plants: Physiology, ecology, and evolution*. Edited by Aaron M. Ellison and Lubomír Adamec:
Oxford University Press (2018). © Oxford University Press. DOI: 10.1093/oso/9780198779841.003.0005

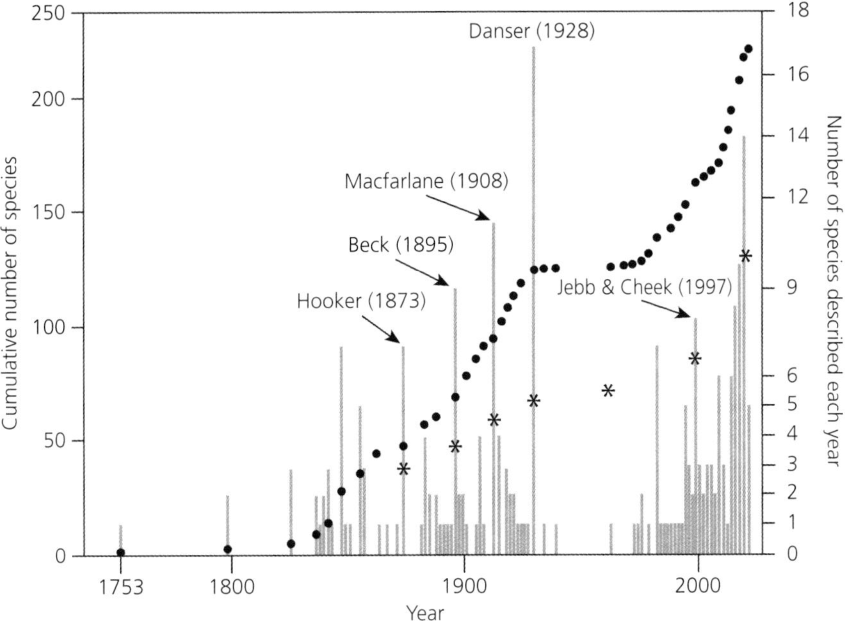

Figure 5.1 *Nepenthes* taxa described between 1753 and 2016. Solid circles indicate the cumulative number of species described in particular years (scale on the left *y*-axis), and vertical bars are the numbers of species described in individual years (scale on the right *y*-axis). Labeled bars identify authors of significant revisions and asterisks denote the number of species recognized by each of those authors (value for 2016 is for species recognized by J. Schlauer that were used to generate Figure 5.4).

After the revision by Danser (1928), there was an extended period of taxonomic inactivity around *Nepenthes*. In part, this resulted from reduced rates of exploration of Southeast Asia coincident with the Great Depression and World War II, both of which curtailed sharply the market for exotic plants in Europe. To a greater degree, however, the taxonomic stasis between 1928 and the mid-1980s reflects the high quality of Danser's research and his application of a conservative interpretation of the species concept in *Nepenthes*. Although his work is not free of errors, the majority of deficiencies were revealed only by detailed field observations that occurred after the global interest in exotic plants was revived in the 1980s.

A surge in publications on the systematics of *Nepenthes* followed from this new fieldwork. Jebb and Cheek (1997) revised the entire genus, and then the Malesian taxa (Cheek and Jebb 2001). Clarke (1997, 2001) provided accounts based largely on field observations of the species from Borneo, Sumatra, and Peninsular Malaysia. Changes to Danser's (1928) interpretations include the reinstatement of *N. eustachya* (Jebb and Cheek 1997), a Sumatran endemic that Danser (1928) treated as a synonym of *N. alata*, and the reinstatement of *N. ramispina* (Jebb and Cheek 1997) and *N. alba* (McPherson 2009), both of which Danser (1928) considered synonymous with *N. gracillima*.

These new interpretations highlight a significant shift in the application of the species concept by the recent authors. The reinstated species are morphologically very similar either to *N. alata* or *N. gracillima*, and the taxonomic importance and stability of distinguishing characteristics remains contentious. A near-exclusive reliance upon subjective interpretations of the taxonomic value of minor variations in pitcher characteristics justified restoration of *N. eustachya*, *N. ramispina*, and *N. alba*, implying a less conservative species concept and a return to the practices that frustrated Danser nearly 90 years earlier.

Taxonomic inflation also has contributed to the rapid recent increase in the number of

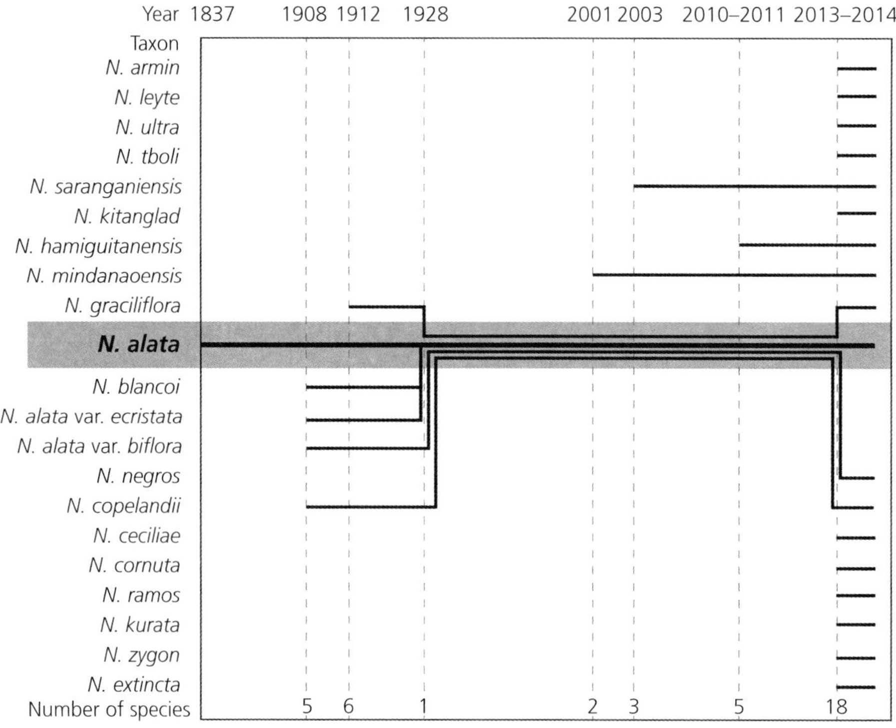

Figure 5.2 Revision of taxa in the *N. alata* group between 1837 and 2014 (Cheek and Jebb 2014). *Nepenthes alata* is represented by a horizontal solid black line embedded within a broader gray bar. All other taxa are represented by discrete, solid black lines. Lines for taxa that overlap the gray bar at various times are or were considered synonyms of *N. alata* for those periods.

Nepenthes species. *Nepenthes alata*, a species endemic to the Philippines, provides an illustrative example (Figure 5.2). This species was described in 1837, followed by *N. blancoi* in 1852, and three additional taxa in 1908: *N. copelandii, N. alata* var. *biflora*, and *N. alata* var. *ecristata* (Macfarlane 1908). Danser (1928) synonymized all of these, and the Sumatran *N. eustachya*, within *N. alata* (Figure 5.2). Several collections from Peninsular Malaysia were also included in *N. alata* by Danser, but Jebb & Cheek (1997) placed these in *N. gracillima*, where they remained until Clarke (1999) demonstrated that they belong to a previously undescribed species, *N. benstonei*. Jebb and Cheek (1997) also reinstated *N. eustachya* as a Sumatran endemic and treated *N. alata* as a widespread, polymorphic species that is endemic to the Philippines. Kurata (2001, 2003) described two new species (*N. mindanaoensis* and *N. saranganiensis*) that may be closely related to *N. alata*, McPherson (2009) reinstated *N. copelandii*, and Gronemeyer et al. (2010, 2011) described *N. ceciliae* and *N. hamiguitanensis*. Cheek and Jebb (2013a, 2013b, 2013c, 2014) described several new species and defined an 18-species "*N. alata* group" (Figure 5.2), circumscribed by several more-or-less shared pitcher and leaf-blade characters.

Since 2014, the *N. alata* group has undergone further revision. Cheek and Jebb (2015) raised it to section status, as *N.* sect. *Alatae*. This section includes the 18 species of the *N. alata* group (Figure 5.2) along with *N. pantaronensis, N. petiolata, N. pulchra, N. robcantleyi, N. sumagaya, N. talaandig*, and *N. truncata*. Gronemeyer et al. (2016) added *N. justinae* to *N.* sect. *Alatae*, but synonymized *N. kurata* into *N. ramos*. Genetic analyses (§5.3.3) indicate that *N.* sect. *Alatae* is not supported because it is polyphyletic, comprising members that appear to lie within both *N.* sect. *Regiae* and *N.* sect. *Villosae*. As a consequence, we do not support the establishment of *N.* sect. *Alatae sensu* Cheek and Jebb (2015).

5.2.1 Determinants of change in *Nepenthes* taxonomy

Nepenthes taxonomists have ceased to use Danser's (1928) conservative interpretation of the species concept for several reasons. First, there has been a substantial increase in knowledge about geographical ranges and morphological variation within and between taxa. For example, analysis of gene sequence similarity (Alamsyah and Ito 2013, Bunawan et al. 2017; §5.3.3) places *N. benstonei* and *N. eustachya*, which are geographically isolated from the *N. alata* group, not in *N.* sect. *Alatae* but rather in *N.* sect. *Pyrophytae* and *Montanae*, respectively; *N. alata sensu stricto* is placed in *N.* sect. *Regiae*. The genetic analysis implies that the observed morphological similarity is convergent, not homologous.

Second, the rapid expansion of the *N. alata* group reflects a change in how *Nepenthes* taxonomists interpret the taxonomic significance of different kinds of morphological variation in the genus. For example, many montane species have highly restricted geographical ranges and distinctive morphological features, which readily distinguish them from other species. In contrast, widespread lowland species, or species that occur on several low mountain ranges (or archipelagos), tend to be less distinctive and demonstrate greater levels of morphological variation, especially among geographically disjunct subpopulations (Clarke and Moran 2016).

Danser (1928) revised the genus at a time when many widespread lowland species (including several that are now assigned to the *N. alata* group) were poorly known but many highly distinctive montane species recently had been discovered. This enabled him to adopt a conservative species concept and treat several lowland taxa as widespread, polymorphous species. This approach was superseded by detailed studies of several polymorphous species that revealed the extent of morphological variation within widespread taxa. This encouraged a revisionist approach to defining species in the genus, and a proliferation of newly named species (Jebb and Cheek 1997, Cheek and Jebb 2001, 2016b, Clarke 2001).

The *Nepenthes alata* group is one example of this trend; the species that comprise the recently established *N.* sect. *Pyrophytae* provide another (Catalano 2010). *Nepenthes copelandii*, synonymized with *N. alata* by Danser (1928) and then reinstated and placed in the *N. alata* group by Cheek and Jebb (2015), has been recorded from three disjunct mountain ranges on the island of Mindanao. Genetic data indicate that *N. alata* and *N. copelandii* are related only remotely (Alamsyah and Ito 2013) and the latter species is now regarded as distinct by most taxonomists. Gronemeyer et al. (2011, 2014) went further, distinguishing two of the three subpopulations of *N. copelandii* as separate species: *N. ceciliae* and *N. cornuta*.

These three taxa are distinguished from one another based on geographical location, growth habit (i.e., terrestrial versus epiphytic), and minor variations in pitcher morphology. Whether *N. ceciliae* and *N. cornuta* are considered separate species is largely subjective and depends on two factors: how individuals choose to apply the species concept, and the quality of the research that underpins the authors' decision to recognize a species. The first of these factors will always be a challenge. Even the four authors of this chapter cannot agree upon the status of approximately 10% of *Nepenthes* species.

The descriptions of *N. ceciliae* and *N. cornuta* provided by Gronemeyer et al. (2011, 2014) were deficient in a different way. The latter species was distinguished on the basis of field observations of macroscopic morphological variation only (the type [*T. Gronemeyer and F. Coritico CMUH00008547*, CUMH (Gronemeyer et al. (2014)], is the only specimen cited). This approach is useful for determining variation within and among (sub-)populations and detecting introgression with local congeners, but vital microscopic variation is ignored, as is all information held by botanical institutions.

Better research could have been applied, but it has not been. Of the 18 species that comprised the *N. alata* group in 2014 (Figure 5.2), nine were described by Cheek and Jebb (2013a, 2013b, 2013c, 2014) using only herbarium collections and records. Many of the taxa they studied were represented by only one or, at best, very small numbers of collections that were unaccompanied by detailed information about habitat, ecology, or geography. Thus, the authors were unable to make accurate judgments about distribution, population characteristics, habitat preferences, introgression with local congeners, or intraspecific variation in key morphological traits for each of their taxa.

5.2.2 Toward an improved taxonomy of *Nepenthes*

Contrasting practical and philosophical approaches to describing and distinguishing species are entirely legitimate and result in valid publication of species, but they are not necessarily useful to other researchers (e.g., physiologists, ecologists) who are not engaged in descriptive taxonomy. Because modern, quantitative methods that can enhance the quality of descriptive taxonomy have yet to be applied to *Nepenthes*, taxonomists need to consider alternative approaches to investigating the degrees and drivers of diversity within the genus.

Taxonomy as an hypothesis. Deciding between competing interpretations of *Nepenthes* taxa is more problematic if published species are treated as statements of established fact. However, the description of a new species is a potentially falsifiable hypothesis (Gaston and Mound 1993). Thus, the status of any published taxon depends on available evidence and different individuals may weight such evidence differently.

For example, by describing *N. extincta*, Cheek and Jebb (2013b) established the hypothesis that this taxon is a distinct species and described its characteristics. They also presented some qualitative evidence to support their hypothesis, but relied entirely upon the type specimen (which lacks lower and upper pitchers, and inflorescences), which they compared with two other putative close relatives (*N. alata* and *N. mindanaoensis*) while ignoring other co-occurring species. In our opinion, this is a critical omission and the evidence provided by Cheek and Jebb (2013b) is insufficient to support the hypothesis of the unique species, *N. extincta*. We assert that prior to compiling and publishing a description one should determine whether the taxon exists as a stable, reproductively isolated, population in the field (i.e., it is not an isolated natural hybrid or unestablished hybrid swarm) and whether this population demonstrates significant, stable morphological differences from populations of co-occurring *Nepenthes* species. Although the lack of such information does not preclude the valid description of a species, some effort should at least be made to confirm that this information does not exist (or could not be reasonably obtained) before proceeding to (incompletely) describe the species.

Use Danser's sections. We are in broad agreement about the status of ≈90% of *Nepenthes* species. Our remaining disagreements concern minor variations in our application of the species concept, degree of reliance on field vs. institutional resources, and how we evaluate the evidence provided by the authorities of a number of species. As a result, the current list of *Nepenthes* species (Appendix) does not reflect entirely any of our personal views.

In the context of our research into the evolution, diversification, biogeography, and ecology of *Nepenthes*, our disagreements are relatively unimportant, because the role of each of these factors can be examined effectively by focusing on the sections established by Danser (1928). For many years after this treatment, his sections received little attention. Now, because of the uncertainty surrounding the status of many recently described species, Danser's sections are proving increasingly useful. They provide a framework for grouping taxa that share similar morphologies without requiring researchers assign a particular hierarchical rank. This approach avoids the need to address controversial species concepts, while allowing the inclusion of all relevant taxa in analyses of diversity and variation in the genus (§5.3).

Adopt quantitative methods in descriptive taxonomy. Institutional sources of information (i.e., herbarium collections, historical records) and field observations used to publish a new species should be complemented by quantitative morphological studies, molecular data, and ecological data wherever possible.

Traits vary within populations, and taxonomists distinguish species from one another on the basis of apparent disjunctions in variation in characteristics. However, patterns of variation can overlap, and intraspecific variation in a particular trait can be considerable. When distinguishing two or more species, all apparent variation (and its sources) should be considered when detecting disjunctions. Such consideration necessarily will involve assessment of intra- and interspecific variation using sufficient sample sizes to provide reasonable levels of statistical power.

The examination and measurement of morphological traits in multiple individuals of the same taxon can generate large amounts of quantitative data. Such data can be analyzed using morphometric analysis ("morphometrics"), which is used widely in zoological taxonomy but less so by plant systematics (Viscosi and Cardini 2011). Classical morphometrics compares simple measurements of traits, such as leaf length, leaf width, or pitcher height (Stace 1982). In contrast, modern geometric morphometrics ("GMM") analyzes the location and arrangement of anatomical "landmarks" at many points on the surface or margins of an object. This approach permits the comparative analysis of the geometries of objects by testing for significant intra- and inter-specific variation in the locations of the landmarks. Many aspects of structures of organs (for *Nepenthes*, e.g., overall shape of the leaf blade, structure of the apex and base, degree of development of petiolar wings, structure of the margins, arrangement of major veins, and the insertion of the tendril) can be compared simultaneously, both within and among two or more taxa (Viscosi and Cardini 2011).

The simultaneous comparison of both state and geometric arrangement of multiple characters represents a significant increase in the sophistication of morphometrics. GMM also can be applied to both two- and three-dimensional objects, facilitating comparisons of natural pitcher geometry as well. If it is used together with field studies and examinations of herbarium material, GMM could serve as a powerful tool for testing hypotheses about the stability of leaf and pitcher traits in *Nepenthes*, and whether they are taxonomically informative.

Consider molecular data. The use of genetic analysis has revolutionized taxonomy, but the application of these methods to questions about the status of particular *Nepenthes* species has not yet been successful (§5.3.3). Many of the technical impediments to progress soon will be removed, and this will lead to the establishment of a substantial sequence database. The status of new taxa can then be tested by comparing their genome data against this database. More powerful comparative studies that incorporate morphological and genomic data depend on sequence and morphometric data to be obtained from the same individual specimen for each taxon involved. Going forward, it is essential that multiple specimens of proposed new taxa be obtained from wild populations, sequenced, and accurately catalogued upon deposition in herbaria.

Take ecological information into account. To an observer of pressed specimens in a herbarium, variations in morphological characteristics may constitute merely points along an intraspecific continuum. However, in some cases, such variation may be of functional significance and provide justification for splitting a single species.

For example, *Nepenthes macrophylla*, which is endemic to Mount Trusmadi in northern Borneo, has a pitcher with a distinctive peristome: its surface ridges develop into pronounced ribs. This trait is shared with *Nepenthes villosa* and *Nepenthes edwardsiana*, which occur approximately 60 km to the northwest, on Mounts Kinabalu and Tambuyukon (Clarke 1997). Because of this similarity, Marabini (1987) described *Nepenthes macrophylla* as a subspecies of *N. edwardsiana*. Small differences between *N. macrophylla* and *N. edwardsiana* in the size and shape of the peristome ridges, pitcher lid, and lamina led Jebb and Cheek (1997) to elevate the former to species rank. More than a decade later, this taxonomic hypothesis was supported by Chin et al. (2010), who demonstrated that *N. macrophylla* pitchers (along with those of *N. rajah* and *Nepenthes lowii*) collect the feces of montane tree shrews (*Tupaia montana*). Because of changes in pitcher size and geometry, *N. macrophylla* has evolved to access a nutritional resource that is unavailable to *N. edwardsiana*, even though the latter also occurs sympatrically with *T. montana* (Chapter 26).

Another example is the taxonomy of *Nepenthes hemsleyana*, which is endemic to northwest Borneo. This species first was described by Macfarlane (1908), although Burbidge (1880) appears to have collected the first specimens, which he described as a variant of *N. rafflesiana*. Subsequently, Danser (1928) synonymized *N. hemsleyana* within *N. rafflesiana*.

It long has been apparent to field workers that *N. hemsleyana* differs morphologically from the nominate form of *N. rafflesiana* in a number of ways. The most obvious difference is the former's elongated pitchers, which led to it being given the informal name "*N. rafflesiana* var. *elongata*" (Clarke

1997). In addition, Moran (1991) and Bauer et al. (2011) noted that the pitchers of *N. hemsleyana* seldom caught any invertebrate prey, whereas pitchers of *N. rafflesiana s.s.* often contained hundreds of prey items. Aerial pitchers of *N. hemsleyana* do not emit the typical fragrance of *N. rafflesiana*, and addition of fragrant pitcher fluid from the latter to the former increases prey capture significantly (Moran 1991, 1996). Finally, the peristome of *N. hemsleyana* produces very little nectar and is not ultraviolet-absorbing, in contrast with *N. rafflesiana* (Clarke and Moran 2011).

Grafe et al. (2011) then demonstrated that aerial pitchers of *N. hemsleyana* serve as roosting sites for Hardwicke's woolly bats (*Kerivoula hardwickii*), which in turn provide the plant with nitrogen via their excreta (Chapter 26). Clarke et al. (2011) used the morphological and ecological data to redescribe this plant as *Nepenthes baramensis*, but because Macfarlane (1908) already had described it as *N. hemsleyana*, the latter name has priority (Scharmann and Grafe 2013).

Taxonomy clearly can benefit from ecological information, and Clarke and Moran (2011) proposed an improved protocol for the collection and preservation of *Nepenthes* material to improve subsequent descriptions. The principal recommendations are as follows.

Stem sections should be collected, bearing both aerial and terrestrial pitchers, and with pitcher contents collected, dried, and attached to the specimen in paper envelopes. One aerial and one terrestrial pitcher should be dissected longitudinally and mounted so that both inner and outer surfaces are accessible for inspection. As *Nepenthes* are dioecious, collect both male and female inflorescences whenever possible, along with the supporting stem.

Pitcher morphology—especially that of the lid—is a crucial trait in mutualistic associations with mammals (Chapter 26). Therefore, pitchers should be pressed so as to preserve the original angle of reflexion and degree of concavity or convexity of the pitcher lid. If this is not possible, photographs can be taken to illustrate these traits (Chin et al. 2010) and the (approximate) lid angle should be measured.

Finally, collectors should take detailed notes describing the habitat in which the specimen was located (vegetation, altitude, substrate), together with observations about any animal visitors (vertebrates and invertebrates) to the pitchers and inflorescences. Features that are unlikely to be preserved, such as fragrance, viscosity of pitcher fluid, etc., should be noted at the time of collection.

5.2.3 Best practices for describing new taxa in *Nepenthes*

Taxonomists describing new species of *Nepenthes* should use economical and informative molecular data and quantitative GMM instead of relying solely on qualitative morphological descriptions. However, the successful use of these methods depends upon solid data and there is little evidence that taxonomists are following the recommendations proposed by Clarke and Moran (2011) and outlined in §5.2.2 to collect such data. Although the requirements of valid publication do not compel authors to collect sufficient data to support their arguments for distinguishing a new species, researchers need to complement consultations of herbarium collections with field observations; couple examinations of macroscopic features with detailed microscopy; and quantitatively analyze all taxonomically informative data.

To be clear, we suggest that any testable description of a new species of *Nepenthes* should include documentation of all attempts to locate the taxon at as many localities as possible, with particular emphasis on searching disjunct localities close to the type locality that could reasonably be expected to support additional populations. Fieldwork should: thoroughly describe the habitat in which the specimen was encountered (e.g., forest type, substrate, soil, degree of disturbance); eliminate the possibility that the taxon is a natural hybrid that does not form a reproductively isolated, stabilized population; determine the stability of traits that appear to be taxonomically informative (e.g., position of the pitcher lip, estimates of leaf and pitcher size) and quantify their covariation with location and local environmental conditions; generate comprehensive herbarium collections that are representative of the taxon and incorporate all key morphological characteristics (i.e., lower and upper pitchers and their contents, male and female inflorescences, fruits and seeds, immature rosettes, and fragments of climbing stems); and document any observed interactions with animals, and pitcher characteristics that are unlikely to be evident in preserved specimens (e.g., fluid viscosity, fragrance).

Consultation of herbarium collections should, at a minimum: identify all collections of the taxon to be described; compare the new taxon to collections of all hypothesized close relatives and co-occurring congeners; and highlight well-preserved, taxonomically informative characters that are difficult to observe in the field, such as characteristics of the indumentum and glands.

The description itself should include either a verbal description or, whenever possible, quantitative measurements of: overall plant architecture; salient details of the rosette and the immature plant; dimensions of leaves, lower pitchers, upper pitchers, and male and female inflorescences; other key traits of leaves, lower and upper pitchers, and male and female inflorescences; the indumentum; pitcher contents; and any observed interactions with invertebrates or vertebrates.

The publication of a description of a new or reinstated species that includes all of these data will provide a sound hypothesis for the concept of a species that is amenable to subsequent observational and experimental testing.

5.3 Evolution in *Nepenthes*

5.3.1 Phylogeography

Evolution within the genus *Nepenthes* is still poorly understood. The current global geographical distribution of the genus includes much of Southeast Asia, parts of South Asia, and many isolated islands in the Indian and western Pacific Oceans (Figure 5.3). The region with the highest diversity includes the islands of Borneo, Sumatra, and the southern Philippines (Jebb and Cheek 1997). The existence of outlying species in India and on the islands of Madagascar, the Seychelles, and Sri Lanka prompted Danser (1928) to propose that the genus was very old, originating before the breakup of Gondwana in the late Cretaceous.

Several geographical outliers (e.g., *N. madagascariensis*, *N. pervillei*, and *N. distillatoria*) are basal (§5.3.3). These lend support to the hypothesis that the genus dispersed to Southeast Asia where it subsequently underwent rapid, relatively recent diversification (Merckx et al. 2015) as the Indian

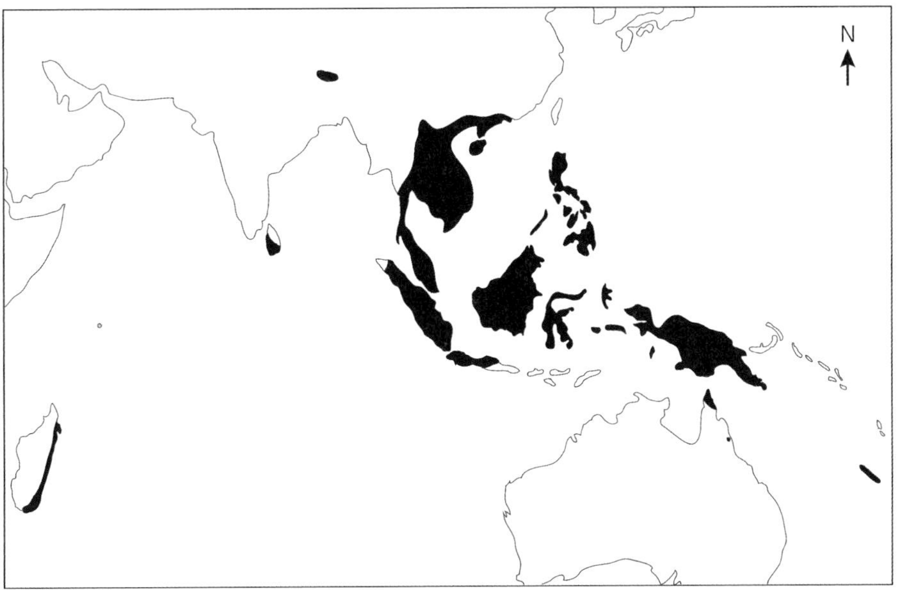

Figure 5.3 Map of Asia and the Indian Ocean showing the geographical distribution of *Nepenthes*.

subcontinent drifted northwards after the breakup of Gondwana (Danser 1928, Clarke and Moran 2016). However, analysis of molecular data indicates that several of the main sections within *Nepenthes* evolved within 10 Mya, and many species evolved within 5–1 Mya, implying that the genus is much younger than previously thought. An alternative hypothesis, therefore, is that dispersal to outlying locations such as Madagascar, Sri Lanka, and the Seychelles took place long after the breakup of Gondwana, but prior to rapid diversification within Malesia.

There is no clear mechanism for the long-distance dispersal events required by the post-Gondwana diversification hypothesis. Although the seeds of *Nepenthes* are wind dispersed, they rarely travel more than a few km. The seeds could be transported by vectors such as wetland birds, which have been shown to disperse seeds of other types of plants over considerable distances (Nogales et al. 2012), but there are no obvious bird species or migratory routes that are reasonable candidates to explain the current distribution of the geographical outliers.

5.3.2 Drivers of diversification

Recent research has identified several actual or potential drivers of diversification in *Nepenthes*. These include biogeographical and environmental processes, and animal–plant interactions that are facilitated mostly by characteristics of pitchers and flowers (Moran 1996).

Local biogeography. The majority of *Nepenthes* species occur on islands in the Malay Archipelago, which is among the most complex biogeographic regions on Earth. Dispersal of seeds from one island to another depends on the strength and direction of prevailing winds and the proximity of other islands that support suitable habitats for *Nepenthes* (Clarke and Moran 2016). Thus, populations of many *Nepenthes* species either are fragmented, if they occur on more than one landmass or mountain range, or are located in specialized habitat patches that are spatially or temporally isolated from others of the same type. This type of isolation promotes allopatric speciation through genetic drift or changes to local environmental selection pressures (e.g., climatic variables).

Environmental specialization for niche habitats or nutrient sources is a prominent trait of *Nepenthes* species. For example, *N. campanulata* from Borneo grows only on a particular type of limestone formation. Although it is not necessarily reproductively isolated from congeners in different habitat types nearby, its specialized growth habit and pitcher geometry appears to be successful only in this type of habitat (Clarke et al. 2014, Clarke and Moran 2016). *Nepenthes* species colonize a variety of nutrient-deficient habitats throughout their range (Chapter 3) and, like the archipelagos on which they occur, these tend to be variable and patchy in both space and time (van der Ent et al. 2015, Clarke and Moran 2016). Thus, over relatively short periods of evolutionary time, tectonic processes (particularly vicariance), combined with variations in local and regional climate regimes, result in habitats whose availability, size, and connectivity vary substantially and that potentially can exert strong selective pressure on morphological traits of individual species.

Climate. The critical role of climate as a driver of pitcher morphology has been demonstrated repeatedly (Bauer et al. 2008, 2012a, Moran et al. 2013). Species whose pitchers have extensive waxy zones on the inner surface and narrow peristomes are favored by more seasonal climates than those with reduced waxy zones and expanded peristomes. Local climatic regimes vary significantly throughout the Malay Archipelago, with the species bearing the most highly modified and specialized pitcher geometries being confined largely to the perhumid regions of the Sunda Shelf (Clarke 1997, 2001, Clarke and Moran 2016).

Local fauna. An additional driver of diversification in *Nepenthes* is the composition of the local animal community that can be exploited for supplementary nutrition. In species-rich regions such as northwestern Borneo, several *Nepenthes* species typically colonize the same habitat patch. Chin et al. (2014) investigated prey capture patterns among mixed populations of *Nepenthes* species at several lowland and highland sites in Borneo, and found that there were significant interspecific differences in invertebrate prey spectra. The pitcher geometries of the species present also were highly variable (Clarke 1997). Chin et al. (2014) hypothesized that these species were partitioning resources, and that different trap morphologies had evolved to capture different combinations of invertebrate prey and reduce the effects of interspecies competition. However, Chin et al. (2014) also found that although the species they examined occurred within a very small geographical region (less than 1 km^2), they typically occupied slightly different environmental niches within the site. This could mean that the pitcher traits of each species had evolved to exploit the arthropod fauna within a given environmental niche, rather than to avoid interspecific competition at larger spatial scales.

The importance of the composition and diversity of the local arthropod community as a driver of diversification in *Nepenthes* has yet to be determined, but there are several well-publicized examples of specialization by *Nepenthes* for alternative sources of nutrients: *N. albomarginata* exploits termites as a primary source of nutrients (Moran et al. 2001); *N. ampullaria* derives a significant proportion of foliar N from leaf litter (Moran et al. 2003); *N. bicalcarata* has a complex mutualistic association with the ant, *Camponotus schmitzi*; and four additional species deploy modified pitchers that specialize in "trapping" mammal feces (Chapter 26).

These examples provide suggestive examples of evolutionary processes within *Nepenthes*, and how environmental factors can select for specialized pitcher traits. However, the majority of *Nepenthes* species produce pitchers that appear to be less specialized and it is necessary to use genetic analyses to trace the relationships among specialized and apparently unspecialized species.

5.3.3 Molecular evolution in *Nepenthes*

Genetic analyses have revolutionized systematics, and promise to yield significant insights into our understanding of evolution in *Nepenthes* in the near future. Mullins (2000, *unpublished data*) used chloroplast (*trnL–trnF*) and nuclear (5S-NTS) sequences together with some morphological data to generate a cladogram for the 74 species studied. This cladogram exhibited widespread incongruence among the datasets for some taxa. Meimberg and Heubl (2006) used *trnK* and *matK* sequences from 85 *Nepenthes* taxa and reported that many of the early-branching species from outlying western locations (e.g., *N. madagascariensis*, *N. pervillei*, and *N. khasiana*) were easily distinguished from the others, but that the majority of species from the center of species diversity in Malesia showed considerable incongruence. Meimberg and Heubl (2006) concluded that these sequences were insufficient for an informative phylogenetic reconstruction of *Nepenthes*.

A likely cause for the observed incongruence among Malesian species is that introgression represents a significant mechanism for speciation in *Nepenthes* and, as the genus is dioecious and thus obligately outbreeding, plastid genomes are inherited maternally. It also appears that much of the diversification in Malesian *Nepenthes* has been very recent (Lowrey 1991, Heubl and Wistuba 1995, Meimberg et al. 2001, Heubl et al. 2006, Merckx et al. 2015), so highly conserved nuclear sequences cannot resolve relationships between most species. Reticulate evolution is thought to be widespread in the genus, and chloroplast markers like *trnL-F* and *rbcL* evolve too slowly to show good results in rapidly evolving groups, giving low resolution for species complexes (such as the closely related species from Indochina). In contrast, more rapidly evolving chloroplast markers like *matK* and *rps16* will show only the matrilineal line, thereby obscuring reticulate evolution and potentially portraying very different species as sister lineages because of more ancestral origins.

Alamsyah and Ito (2013) presented their own phylogeny, but the results did not improve substantially on Mullins (2000) or Meimberg and Heubl (2006). They also may have used ITS markers without sufficient cloning, or it may be present in multiple copies due to the hybridogenic origin of many species in the genus. Schwallier et al. (2016) analysed ITS and *trnK* sequences in 45 species and provided improved resolution among groups of species from Peninsular Malaysia, Sumatra, and Borneo. Merckx et al. (2015) and Bunawan et al. (2017) have contributed ITS and *trnK* or, respectively, *trnL* sequence data of several species, but problems persist surrounding the selection of suitable markers and appropriate modes of analysis.

5.3.4 Infrageneric classification

Branching patterns vary considerably between phylogenetic reconstructions based on chloroplast or

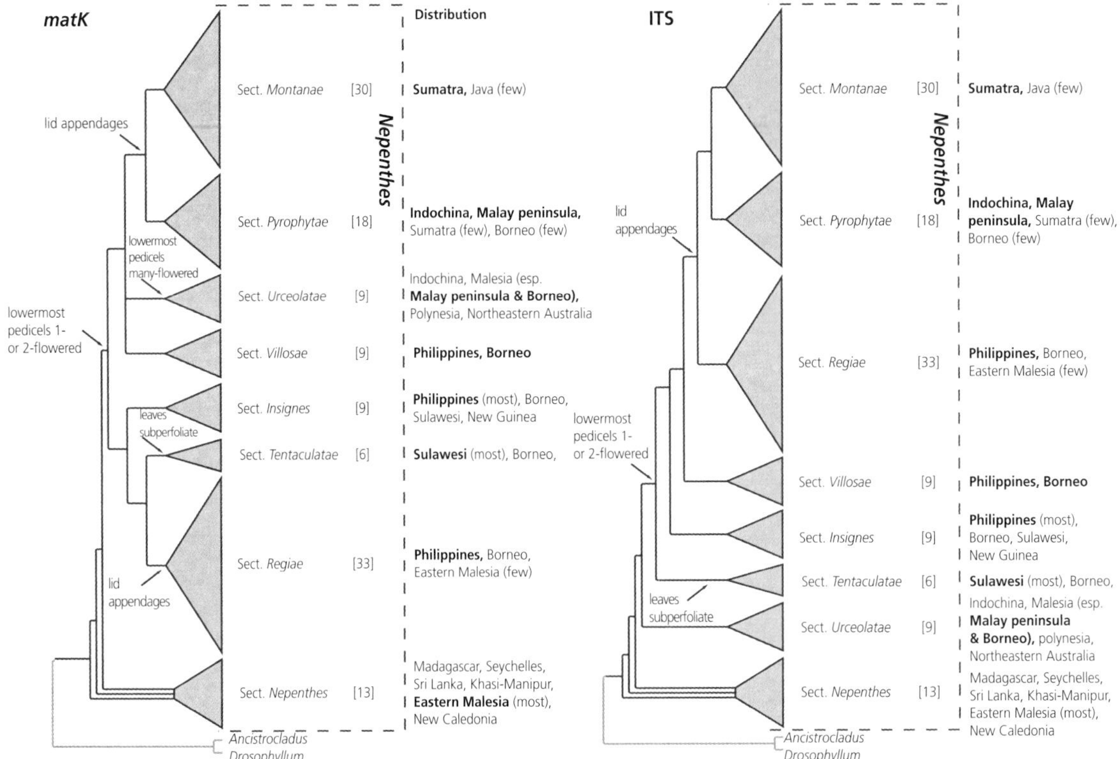

Figure 5.4 Phylogenetic trees of Nepenthaceae.

nuclear sequence data (Figure 5.4), but both tend to confirm mutual affinities among species clustering in groups (clades or grades) that to a certain degree also share common morphological or biogeographical features. Although Danser (1928) used informal "Groups" in his subdivision of *Nepenthes*, he suggested that they should be classified as sections.

Both nuclear and chloroplast-derived data support a basal position of the western species (from Madagascar to Sri Lanka and the Khasi Hills in northeastern India). These species and several further lines (most from New Guinea and one from New Caledonia; "Clade III" in Meimberg et al. 2001; part of "Group I" in Alamsyah and Ito 2013) branch successively sister to the remainder of the genus, rather than commonly in a monophyletic group. Thus, *N.* sect. *Nepenthes* (an autonym implicitly created by subdivision that represents the core of *N.* sect. *Vulgatae* of Danser (1928), rendering the latter a superfluous synonym because both inseparably contain the type species of the genus, *N. distillatoria*) is an early-branching but not necessarily ancestral basal grade, not a monophyletic clade.

Other former members of *N.* sect. "*Vulgatae*" (Danser 1928) apparently belong to several different sections that are possibly monophyletic: *N. papuana*, *N. gracilis*, and *N. mirabilis* share common ancestry with the two species (*N. ampullaria* and *N. bicalcarata*) originally placed in *N.* sect. *Urceolatae* (part of "Clade I"; part of "Group I"). *Nepenthes tentaculata* is the type of the more derived *N.* sect. *Tentaculatae* (part of "Clade II"; "Group II") that was defined by Cheek and Jebb (2016a) and which has its highest local (alpha) diversity in Sulawesi. The Philippine species *N. alata* and *N. philippinensis* are apparently related to the predominantly Bornean *N.* sect. *Regiae* (part of "Clade II" *sensu* Meimberg et al. 2001; "Group V").

The western Malesian and Indochinese species *N. reinwardtiana*, *N. albomarginata*, *N. smilesii*, *N. thorelii* (originally *N.* sect. *Vulgatae*), *N. gracillima*, *N. macfarlanei*, and *N. sanguinea* (originally *N.* sect.

Montanae) belong to an expanded *N.* sect. *Pyrophytae* (Cheek and Jebb 2016c; part of "Clade I"; "Groups VI and VII"), the sister group to *N.* sect. *Montanae* (part of "Clade I"; "Group VII") that is additionally augmented by *N. tobaica* (originally *N.* sect. *Vulgatae*), *N. neglecta*, and *N. spectabilis* (originally assigned to *N.* sect. *Nobiles*) from Sumatra. *Nepenthes gymnamphora* Reinw. ex Nees is here designated as the lectotype (*lectotypus, hic designatus*) of *N.* sect. *Montanae* (Danser 1928: 405).

The remaining species originally placed in *N.* sect. *Nobiles* likewise are absorbed elsewhere: *N. hirsuta* (*N. leptochila*) probably in *N.* sect. *Urceolatae s. l.* (although ITS sequence data support a placement near *N.* sect. *Insignes*; Alamsyah and Ito 2013); and *N. deaniana* in *N.* sect. *Regiae*.

Nepenthes sect. *Insignes* (part of "Clade II"; "Group IV") with a center of diversity in the Philippines loses *N. rafflesiana* (from the Malay Peninsula to Borneo) to *N.* sect. *Pyrophytae* (according to matK data in Meimberg et al. 2001); *N. villosa* from Borneo to *N.* sect. *Villosae* (Cheek and Jebb 2015); and *N. treubiana* from New Guinea to *N.* sect. *Nepenthes*. The position of the Bornean *N. northiana* is somewhat uncertain. With the exception of the notable additions mentioned, the predominantly Bornean and Philippine *N.* sect. *Regiae* remains essentially as defined by Danser (1928).

The aforementioned basal position of *N.* sect. *Nepenthes*, and a derived one for *N.* sect. *Montanae* is recovered frequently irrespective of the gene used for phylogenetic reconstruction. In contrast, *N.* sect. *Urceolatae* is found sister to *N.* sect. *Montanae* in plastid DNA-derived phylogenies (Mullins 2000, Meimberg et al. 2001, Meimberg and Heubl 2006), but it is the second basal branch (next to the *N.* sect. *Nepenthes* grade) in the nuclear DNA-derived ones (Alamsyah and Ito 2013, Merckx et al. 2015, Schwallier et al. 2016). The latter show more parallels in patterns of morphology and distribution and are thus somewhat more plausible, but the plastid marker phylogenies do not account for reticulate evolution and may overestimate the importance of biogeography because of chloroplast capture. The branching sequence among the derived sections is more volatile in the nuclear marker phylogenies.

Future molecular studies will require much longer sequences and methods that can distinguish species whose recent evolution is thought to have occurred through reticulate evolution and extensive introgression. Until recently, such methods have been either too expensive or too challenging to implement for genera such as *Nepenthes*, but recent advances in the field have removed most of these hurdles. The advent of Next Generation Sequencing (NGS) techniques (Glen 2012, Peterson et al. 2012) and other methods such as restriction site associated DNA markers that can resolve groups of closely related taxa offer considerable promise of improved resolution, both among populations of closely related species and at the level of the entire genus. Several research projects that use these methods already are underway. The next few years will see significant improvements in our understanding of evolution and diversification in *Nepenthes*.

5.4 Future research

The adoption of modern quantitative methods, particularly advances in genetic sequencing, promises to revolutionize our understanding of the systematics and evolution of *Nepenthes*. This is a revolution that cannot come too soon, as current trends in traditional descriptive taxonomy are outdated and have led to a proliferation of new species whose utility to scientists working in other fields is limited by inadequate supporting data. Danser's (1928) observation that continuing in this way can only discredit systematics remains as true today as it did nearly 100 years ago.

We advocate a re-appraisal of the way descriptive taxonomy of *Nepenthes* is done and urge the adoption of a more conservative application of the species concept. We are optimistic that the adoption of ecological, morphometric, and sequencing methods will lead to substantial improvements in our understanding of the processes that drive diversification in *Nepenthes*. Climate, vicariance, and plant–animal interactions appear to be key drivers of the evolution of trap morphology and geometry and, by extension, speciation through modification of pitcher components and methods of supplementary nutrient sequestration. The diversity of trap characteristics in *Nepenthes* will continue to provide scientists with a model system for testing hypotheses relating to the evolution and diversification of specialized plant organs.

CHAPTER 6

Systematics and evolution of Lentibulariaceae: I. *Pinguicula*

Andreas Fleischmann and Aymeric Roccia

6.1 Introduction

Lentibulariaceae is a monophyletic family within the Lamiales in the asterid crown-group of eudicots (APG IV 2016), that includes ≈360 species (Appendix). It is a derived and comparatively young family that originated ≈42–28 Mya (upper and lower limits of divergence time from molecular clock estimations of Wikström et al. 2001 and Bell et al. 2010, respectively).

The family comprises three carnivorous genera, *Pinguicula, Genlisea*, and *Utricularia*. All are hygrophilous herbs; several species of the rootless *Genlisea* (Chapter 7) and *Utricularia* (Chapter 8) are hydrophytes that grow as submerged, affixed water plants ("rhizophytic aquatics") or as aquatics freely floating underneath the water surface ("mesopleustophytes").

6.2 Life history and morphology

Pinguicula comprises ≈96 currently recognized species (Roccia et al. 2016, Rivadavia et al. 2017; Appendix). All are rosette-forming, herbaceous carnivorous plants, with typical lamialean, bilabiate, tubular flowers (Figure 6.1). Although the general morphology is fundamentally similar in all species, the species display a wide range of life-history strategies, leaf shapes, and corolla morphologies that have evolved in adaptation to different habitats and perhaps also to varying types of prey.

6.2.1 Life-history strategies

Most species are perennials; <10% (10 species) are either facultative or obligate annuals. Two of these are in the phylogenetically early-branching *Pinguicula* subg. *Isoloba* (*P. pumila* in the North American *P.* sect. *Isoloba* and *P. lusitanica*, the sole member of the Mediterranean-Atlantic *P.* sect. *Pumiliformis*; Figure 6.2 and §6.3). The others are Neotropical species in the derived *P.* sect. *Temnoceras* (Casper 1966, Domínguez et al. 2014, Lampard et al. 2016).

The perennials include "homophyllous" plants that produce only carnivorous leaves throughout the year and "heterophyllous" (*sensu* Roccia et al. 2016) hemicryptophytes and cryptophytes that produce both carnivorous and noncarnivorous leaves during a single growth cycle (Figure 6.2). The heterophyllous species produce carnivorous leaves only when growing actively and survive a dry or cold period either as hibernacula (condensed resting buds formed from scale-like leaves: "temperate heterophyllous growth type" in all members of *P.* sects. *Pinguicula, Nana, Micranthus*, and *Heterophylliformis*; Figures 6.1, 6.2), or as open, compact rosettes or subterraneous bulbs formed of more or less succulent, noncarnivorous leaves, usually with greatly reduced lamina ("tropical heterophyllous growth type" in the Central American and Caribbean *P.* sect. *Temnoceras*). Within that section, intermediate forms between the noncarnivorous "winter rosettes" and bulb-like rosettes, and several reversals to homophyllous growth

Fleischmann, A., and Roccia, A., *Systematics and evolution of Lentibulariaceae: I.* Pinguicula.
In: *Carnivorous Plants: Physiology, ecology, and evolution*. Edited by Aaron M. Ellison and Lubomír Adamec:
Oxford University Press (2018). © Oxford University Press. DOI: 10.1093/oso/9780198779841.003.0006

Figure 6.1 (Plate 6 on page P5) Morphological diversity and growth types of *Pinguicula*. A lithophytic habit with strap-shaped leaves has evolved in parallel in all three subgenera: (**a**) *P. megaspilaea* (*P.* subg. *Isoloba*); (**b**) *P. mundi* (*P.* subg. *Pinguicula*); (**c**) *P. calderoniae* (*P.* subg. *Temnoceras*). Those three species flower from the carnivorous rosettes. In contrast, (**d**) flowering from the noncarnivorous winter rosettes in *P. rotundiflora* (frequent in Mexican members of *P.* subg. *Temnoceras*); (**e**) sprouting hibernaculum of *P. grandiflora* (*P.* subg. *Pinguicula*). (**f**) *P. alpina* with hibernaculum formed in the center of a rosette of carnivorous leaves. (**g**) The (facultative) therophyte *P. lusitanica* has small leaves with strongly involute margins. (**h**) In most species, such as this *P. jarmilae*, the leaf margins are motile, enrolling over caught prey. (**i**) Species with thread-like leaves and revolute margins, such as this *P. heterophylla*, cannot move. Photographs (**b, e**) by Aymeric Roccia, (**c**) by Fernando Rivadavia, and remaining photos by Andreas Fleischmann.

also have occurred (Cieslak et al. 2005, Kondo and Shimai 2006; Figure 6.2).

6.2.2 Leaves

The leaves of *Pinguicula* are entire, more or less succulent to membranous, usually inconspicuously petiolate to sessile; the lamina is of various shape, most frequently broadly oblanceolate to ovate (usually when the leaves are appressed to the ground), or oblong, narrowly oblanceolate to linear or filiform, in which case they usually are held upright or hang down when the plant is growing on vertical habitats. The upper surface of the lamina of all species is lined densely with two types of non-vascularized, carnivorous glands (stalked and sessile; Chapters 13, 14); in few species (e.g., *P. gigantea*, *P. longifolia*) the lower leaf surface also is glandular. The leaf margins generally are slightly or conspicuously involute (the latter usually in species with leaves flat on the ground), but a few species have either flattened or revolute leaf margins (the latter being those species with upright or hanging leaves). In most species, the leaf margins enfold over captured prey (Figure 6.1; Chapter 14), but in species with filiform, upright leaves or species with flattened leaf margins (e.g., *P. agnata*, *P. gigantea*), the leaves are immobile.

The carnivorous leaves of both homophyllous and heterophyllous species can be either uniform in shape throughout ("isophyllous") or vary distinctively in size and shape during the growing season ("anisophyllous"). Examples of the latter include the flat rosetted and long strap-shaped leaves produced by homophyllous *P. megaspilaea*

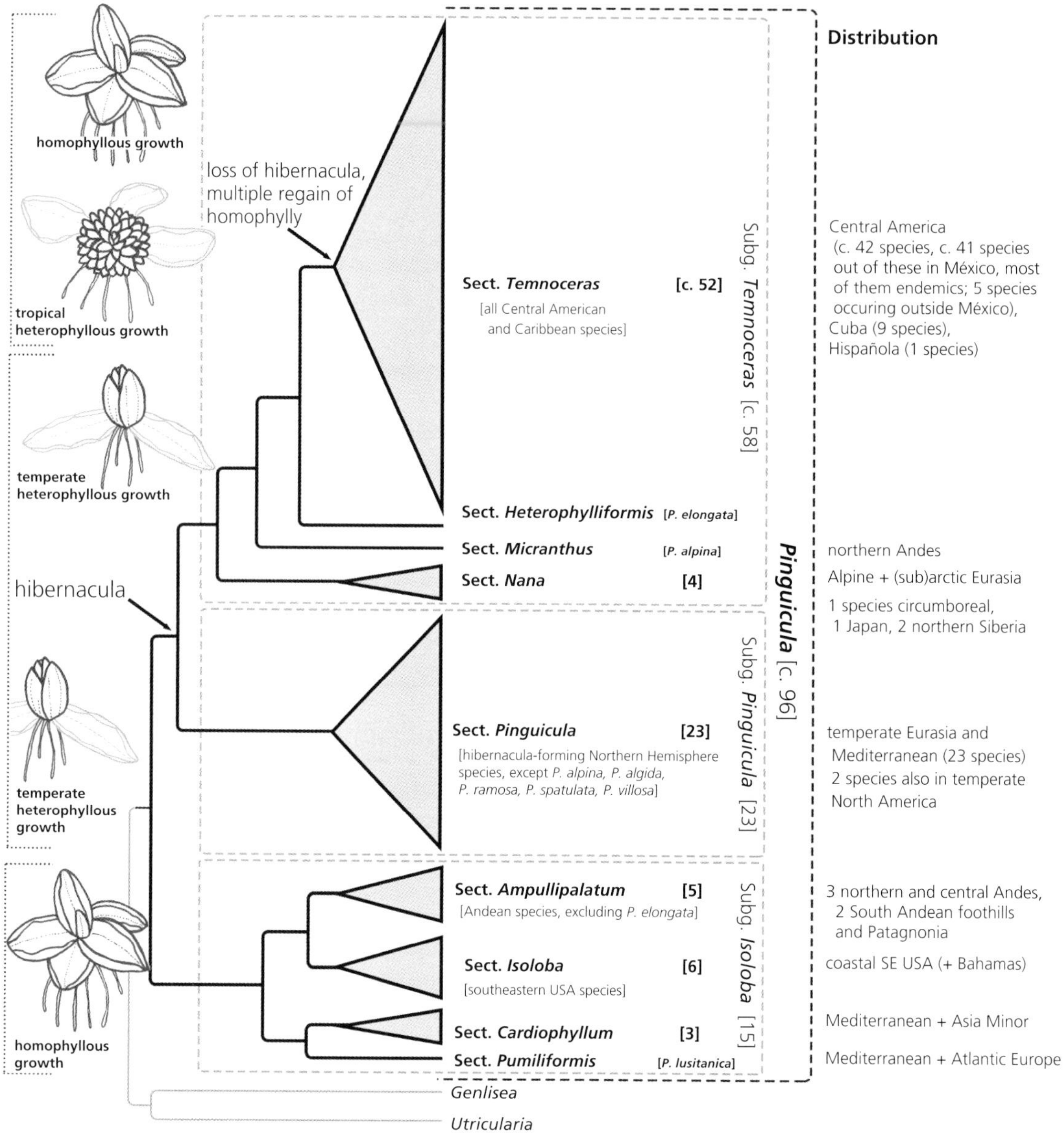

Figure 6.2 Simplified phylogeny of the genus *Pinguicula*, based on plastid DNA data (Beck et al. 2008, Fleischmann 2011b). Illustration by Andreas Fleischmann.

and by heterophyllous *P. vallisneriifolia*. Casper (1966) used the term "temperate-heterophyllous" for the former growth type, although it is not a case of true heterophylly. Rather, it is a heterophyllous species in terms of carnivorous and noncarnivorous foliage. It also produces anisophyllous carnivorous leaves.

6.2.3 Inflorescences and flowers

A solitary scape arises from the rosette among either the carnivorous leaves during active growth or the noncarnivorous leaves during dormancy. In *P. alpina* and the three members of *P.* sect. *Nana*, the flower buds develop in the dormant hibernacula at

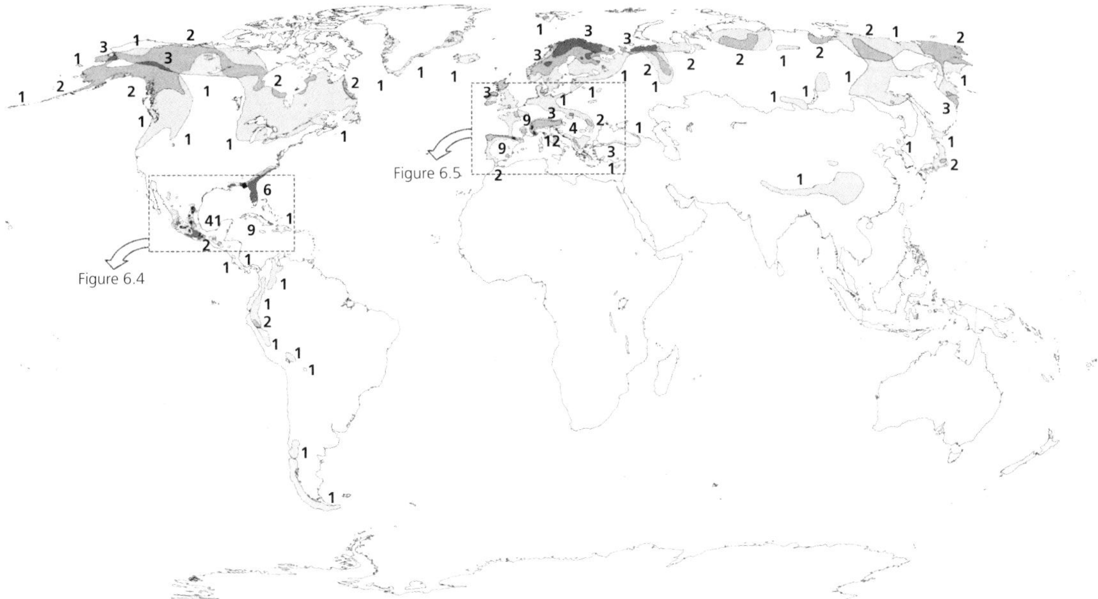

Figure 6.3 Global distribution of *Pinguicula*, with species numbers indicated for a country or region (data from Meusel et al. 1965, Casper 1966, Zamudio 2001, Heslop-Harrison 2004, Fleischmann 2011b, 2015b, and additional information retrieved from herbarium records). Map drawn by Andreas Fleischmann.

the end of the previous growing season (the autumn in advance of flowering); the inflorescence emerges early in the season together with the newly formed leaves. In contrast, flower bud initiation in *P.* sect. *Pinguicula* takes place in the same year of vegetative growth (Degtjareva and Sokoloff 2012).

The underlying basic inflorescence type of *Pinguicula* is a few-flowered, sessile, frondulose terminal umbel (sometimes reduced to a solitary flower; Degtjareva and Sokoloff 2012) with inconspicuous, small scale-like, flower-subtending bracts hidden at the base of the foliar leaves (technically, therefore, the scapes of the flowers, often referred to as "peduncles" are actually pedicels). Shoot growth always terminates with an inflorescence, and new vegetative shoot growth resumes from a dormant lateral bud in the axil of the ultimate foliar leaf below the umbel (Degtjareva and Sokoloff 2012). Because of the compact rosette habit of *Pinguicula* (densely condensed shoots lacking any obvious internodes), the terminal inflorescence nature with resumed lateral growth is rarely obvious.

Flowers in all but two species are solitary on their scapes; the closely-related *P. algida* and *P. ramosa* (*P.* sect. *Nana*) normally have apically bifurcate scapes supporting two (or rarely more) flowers (Casper 1966). The flowers themselves are hermaphroditic, pentamerous, and tetracyclic. They are zygomorphic, with a sympetalous, tubular corolla typical of the Lamiales, including a nectariferous spur. The five sepals are basally adnate and form either a bilabiate (three sepals forming the upper, and two the lower, lip) or spreading calyx, which is persistent in fruit. The five petals are fused to a bilabiate, spurred, throat-like corolla: two petal lobes form the upper lip and three form the lower lip. The corolla shape is either distinctly zygomorphic with the lower lip spreading widely from the upper lip (flowers typically held vertically) or nearly isolobous and radial (flowers frequently facing sky). Corolla color and overall size, size and shape of its tube, shape and hair cover of the palate, patterns of nectar guides, and shape and length of the nectar spur vary widely between species, and appear to reflect adaptation to different pollinators (Fleischmann 2016b) rather than taxonomic affinity. Early classifications relied on corolla morphology (e.g., Casper 1996), but it has proven rather unreliable for infrageneric classification because of the large overlap of floral characters and parallel evolution of corolla design in distantly related groups (Cieslak et al. 2005, Beck et al. 2008).

In all three genera of Lentibulariaceae, the androecium is reduced to two anterior stamens, each with bithecate anthers and curved, dilatated filaments that clasp the superior ovary. The thecae are located on the infertile rear surface of the bilabiate stigma, and thus cannot touch with its receptive surface ("herkogamy," which mechanically avoids self-pollination). However, some annual and small-flowered perennial *Pinguicula* species facultatively self-pollinate at the end of anthesis by growth of the filaments or style, so that the thecae touch the stigma (Chapter 22). The subglobose ovary has central placentation and its short bilabiate style is persistent in fruit. The fruit is a dry capsule with bivalvate dehiscence (Casper 1966). The small seed (0.4–1.0 mm long) is ellipsoidal to fusiform with a micropylar appendage and a reticulate testa in most species (Degtjareva et al. 2004); false vivipary can be observed in the two epiphytic Caribbean species, *P. lignicola* and *P. casabitoana* (Lampard et al. 2016). Cotyledon number of the embryo and seedling varies between species from two to one (by reduction of the second cotyledon), but this does not correlate either with taxonomic affinity or habitat (Haccius and Hartle-Baude 1957, Degtjareva et al. 2004).

6.2.4 Chromosome numbers

Chromosome numbers range from $2n = 16$ to $2n = 128$; species are diploids ($2n$), tetraploids ($4n$), octoploids ($8n$), and hexadecaploids ($16n$), based on estimated chromosome base numbers of $x = 6, 8, 9, 11$, and 14 (Casper and Stimper 2006, 2009). Chromosome numbers do not correspond to phylogenetic clades and they are rather unusable characters for infrageneric classification (§6.3). Different ploidy levels and "base numbers" *sensu* Casper and Stimper (2009) occur within each major clade, possibly indicating reticulate evolution. In contrast to the ultra-small genomes (<100 Mbp) reported for some *Genlisea* and *Utricularia* species (Chapters 7, 8), the genomes of their common sister *Pinguicula* are of medium-size (Greilhuber et al. 2006, Veleba et al. 2014).

6.2.5 Clonal growth

Vegetative propagation (clonal growth) occurs in some species. For example, *P. primuliflora* (*P.* sect. *Isoloba*) frequently forms adventitious plantlets from the tips of its leaves. Many species produce new plantlets from the base of the leaf petiole (especially in hibernacula of *P.* subg. *Pinguicula*; Heslop-Harrison 1962) or the lamina surface of dissected leaves (e.g. the succulent noncarnivorous leaves of most Mexican species, some of which easily detach from the mother rosette). Other species multiply by division from the center of the growing rosette. Adventitious plantlets formed at the tips of stolon runners have been reported for the European *P. vallisneriifolia* and *P. longifolia* (*P.* sect. *Pinguicula*; Casper 1966, Roccia et al. 2016), the Andean *P. jarmilae* and *P. calyptrata* (*P.* sect. *Ampullipalatum*; Fleischmann 2011b, Lampard et al. 2016), and the Mexican *P. stolonifera* and *P. gigantea* (*P.* sect. *Temnoceras*; Fleischmann 2011b, Lampard et al. 2016). These species all are matt-forming lithophytes that grow on vertical cliffs or exposed rock. However, many other lithophytic species do not form runners; this mode of clonal growth has evolved several times in the genus and appears to be unrelated to habitat preferences (Fleischmann 2011b).

6.3 Phylogeny and taxonomy

6.3.1 Phylogeography

The genus is assumed to have originated in temperate Eurasia, followed by an Early Tertiary migration to North America (the "boreotropics" hypothesis; Jobson et al. 2003). During the evolutionary history of the genus, at least five radiations in distinct geographical regions have happened, leading to the present-day diversity of *Pinguicula* (Cieslak et al. 2005). Incongruences have been found between plastid and nuclear marker datasets. However, both show the Eurasian Alpine *P. alpina* as sister to a Central American clade (Cieslak et al. 2005, Degtjareva et al. 2006, Kondo and Shimai 2006). Both datasets further revealed the boreal-subarctic *P. villosa* as sister to *P. ramosa* + *P. spathulata* (as *P. variegata*), but not assigned closely to *P.* sect. *Pinguicula*, where it was placed by Casper (1966). All show *P.* sects. *Cardiophyllum* and *Isoloba* as monophyletic, early-branching, and closely related to *P. lusitanica*, whose phylogenetic position differs between nuclear and plastid datasets.

It is reasonable to assume that early-branching *P.* subg. *Isoloba* is phylogenetically old among *Pinguicula*, judging from its long branch lengths (substitutions) and odd biogeography (Cieslak et al. 2005, Beck et al. 2008). This subgenus comprises a Mediterranean clade (*P. lusitanica* and *P.* sect. *Cardiophyllum*) as immediate sister to a New World clade (*P.* sects. *Isoloba* and *Ampullipalatum*).

Hibernacula are found in all members of the grade comprising *P.* sects. *Pinguicula, Nana, Micranthus*, and *Heterophylliformis* (Figure 6.2), implying that hibernacula evolved only once in the genus. They apparently were lost (or evolved into bulb-like organs or succulent winter rosettes) in *P.* sect. *Temnoceras* as it adapted to warmer, seasonally dry climates in Central America. In contrast, homophyllous growth is not a monophyletic trait in the genus, but rather is an apomorphy of *P.* subg. *Isoloba* (hence could be considered a plesiomorphic state), and evolved again in *P.* subg. *Temnoceras* sect. *Temnoceras* (Figure 6.2). In *P.* sect. *Temnoceras*, homophylly most likely evolved several times, either as an adaptation to stable environmental conditions—e.g., in *P. emarginata* from México and several Cuban species that grow in permanently wet habitats—or connected to annual life strategy.

The hibernacula-forming species are basal to the most diverse and species-rich clade of the genus, *P.* sect. *Temnoceras* (equivalent to the "Mexican-Central American-Caribbean clade" of Cieslak et al. 2005; Figure 6.2). This lineage seems to have diversified in concert with the biogeography of the region: the rich mosaic of closely co-occurring, heterogeneous topographies and climatic conditions in the Mexican Highlands have been considered the major factors driving speciation there (Zamudio 2001, 2005, Cieslak et al. 2005). This would represent a case of geographic/climatic radiation *sensu* Simões et al. (2016), and phylogeographic analyses of the Trans-Mexican Volcanic Belt showed that the climatic and geological changes associated with the volcanic transformation of the Mexican Highlands during the Pleistocene led to the present topography and climate. These were followed by rich allopatric and parapatric speciation of montane plants that are significantly associated with the rich plant biodiversity in that region (Myers et al. 2000, Mastretta-Yanes et al. 2015). Unfortunately, phylogenetic relationships within the species-rich, monophyletic *P.* sect. *Temnoceras* are not yet fully resolved, but it is clear that the taxonomic subgroups of earlier authors (e.g., Casper 1966) are largely para- or polyphyletic (Cieslak et al. 2005, Shimai and Kondo 2007). Reticulate evolution and hybrid speciation seem to account for part of the species richness in Europe and México (Cieslak et al. 2005, Degtjareva et al. 2006, Kondo and Shimai 2006), and most of the European populations are likely to be of post-glacial origin, reflected in polyploid complexes, hybrid swarms, and ongoing speciation (Casper 1966, Casper and Stimper 2006, 2009, De Castro et al. 2016, Roccia et al. 2016).

6.3.2 Infrageneric classification

The following changes to the infrageneric classification of Casper (1966) are proposed to achieve monophyly of all groups. We consider the three major clades evident from phylogenetic reconstructions (Cieslak et al. 2005, Degtjareva et al. 2006, Beck et al. 2008) to be subgenera; species concepts (Appendix) follow Roccia et al. (2016).

Pinguicula subg. *Isoloba* Barnhart, Mem. N.Y. Bot. Gard. 6: 47 (1916), emend. Casper, Bot. Jb. 82: 329 (1963). Type: *Pinguicula pumila* Michx.

It is redefined here to include:

(1) *P.* sect. *Isoloba* Casper, Bot. Jb. 82: 330 (1963). Casper's (1966) *P.* subsection *Pumiliformis* is excluded to become a distinct section.

(2) *P.* sect. *Cardiophyllum* Casper, Feddes Repert. spec. nov. 66: 34 (1962).

(3) *P.* sect. *Pumiliformis* (Casper) Roccia & A. Fleischm. stat. nov. Basionym: *P.* subsect. *Pumiliformis* Casper, Bibliotheca Botanica 127/128: 71 (1966). Type: *P. lusitanica* L.

(4) *P.* sect. *Ampullipalatum* Casper, Bot. Jb. 82: 334 (1963).

This section is moved from *P.* subg. *Temnoceras sensu* Casper (1966). *Pinguicula elongata* now is excluded from this section, which was polyphyletic in the circumscription of Casper (1966). This species was revealed in phylogenetic reconstructions (Beck et al. 2008) as sister to the Central American species here classified as *P.* sect. *Temnoceras* in *P.* subg. *Temnoceras*.

Pinguicula sects. *Heterophyllum, Agnata*, and *Discoradix* are excluded from *P.* subg. *Isoloba* because they fall within a single major clade in phylogenetic reconstructions (Cieslak et al. 2005, Shimai and Kondo 2007, Shimai et al. 2007) that does not include the subgeneric type *P. pumila* nor any other *Isoloba*-members *sensu* Rafinesque (1836). This clade, here circumscribed as *P.* section *Temnoceras*, comprises all Central American and Caribbean taxa formerly assigned to *P.* subg. *Isoloba*.

Pinguicula subg. *Pinguicula* L. Type: *P. vulgaris* L.

In its new circumscription as a monophyletic entity it comprises only a single section:

(5) *P.* section *Pinguicula* as circumscribed by Casper (1966), but excluding his *P.* sect. *Nana*.

Pinguicula sects. *Crassifolia, Homophyllum, Longitubus, Orcheosanthus, Orchidioides*, previously assigned to *P.* subg. *Pinguicula* (Casper 1966), now are moved to *P.* subg. *Temnoceras*.

Pinguicula subg. *Temnoceras* Barnhart, Mem. N.Y. Bot. Gard. 6: 47 (1916), emend. Casper, Bibliotheca Botanica 127/128: 109 (1966). Type: *P. crenatiloba* DC.

Ironically, Barnhart's (1916) *P.* subg. *Temnoceras* initially comprised only *P. crenatiloba*, which displays a somewhat unusual corolla morphology, but now comprises the majority of species in the genus. De Candolle (1844) was the first to group all Central American species together (in *P.* sect. *Orcheosanthus*), but Barnhart's *P.* subg. *Temnoceras* has nomenclatural priority on subgenus rank. This subgenus includes:

(6) *P.* sect. *Temnoceras* Casper, Bot. Jb. 82: 333 (1963). (incl. *P.* sects. *Heterophyllum, Agnata, Discoradix, Homophyllum, Orcheosanthus, Crassifolia, Longitubus*, and *Orchidioides*).
(7) *P.* sect. *Micranthus* Casper, Feddes Repert. 66: 45 (1962).
(8) *P.* sect. *Nana* Casper, Feddes Repert. 66: 41 (1962).
(9) *P.* sect. *Heterophylliformis* (Casper) A. Fleischm. & Roccia stat. nov. Basionym: *P.* subsection *Heterophylliformis* Casper, Bibliotheca Botanica 127/128: 113 (1966). Type: *P. elongata* Benj.

6.4 Distribution

6.4.1 Global patterns of diversity

Pinguicula species grow on all continents except Australia and sub-Saharan Africa (Figure 6.3), although its occurrence in Africa is limited to only two species in a few scattered locations on the Mediterranean northern coast and in the Rif Mountains. The global range in Eurasia and the Americas is not evenly distributed (Figure 6.3), and the hygrophilous, monticolous genus *Pinguicula* is widely absent from large lowland areas (such as rainforests, savannas, grassland, and deserts). *Pinguicula* species have colonized a large variety of habitats, ranging from subarctic Greenland, Scandinavia, and northern Siberia in the Northern Hemisphere, the southernmost tip of the South American continent and Tierra del Fuego in South America, and in subtropical Central America and tropical Cuba. They grow from sea-level to mountainous regions up to 4200 m a.s.l. (*P. alpina* in the Himalayas and *P. calyptrata* in the Andes; Casper 1966). Some species grow in bogs, fens, marls, and swamps; others in grassy seepages; and still others on shallow soils, on wet, dripping walls or in crevices of bare or moss-covered rocks; and last (two species in the Caribbean), epiphytically on tree trunks or branches. The only common feature among these habitats is the presence of moist to wet soils (or wet air in the case of epiphytic species or the Japanese *P. ramosa*) during active growth phases. Soil and water acidity levels, and mineral contents, vary enormously among these habitats (Heslop-Harrison 2004). Adaptation to new habitats (and therefore to new geographic zones) appears to be one of the major forces driving evolution and diversification of *Pinguicula* (Cieslak et al. 2005, Zamudio 2005, Degtjareva et al. 2006, Shimai et al. 2007).

Unlike most other carnivorous plant genera, the majority of *Pinguicula* species are calcicoles, and usually grow on alkaline substrates. Most European *Pinguicula* species grow on wet limestone outcrops, tufa dripping walls, humid pockets of soil on limestone cliffs, calcareous meadows, and along rivulets in alpine alkaline wetlands. *Pinguicula grandiflora, P. alpina*, and *P. vulgaris* display wider edaphic tolerances, also growing on decalcified substrate in limestone mountain ranges, on parapeaty soils under pine trees, on basaltic cliffs, along rivulets in heathlands, on bare peat, and among *Sphagnum* mosses in peat bogs. Only a handful of species are calcifuges that grow strictly on acidic substrates: *P. lusitanica, P. macroceras, P. nevadensis, P. ramosa, P. villosa, P. spathulata*, the six southeast USA species, several of the Cuban species from quarzitic sands or serpentine soils, and five of the six Andean species (the notable exception is *P. involuta*, which also can grow on limestone-based soils; Casper 1966). Some species are found occasionally or almost exclusively on serpentine rock, e.g., *P. balcanica, P. crystallina, P. christinae, P. cubensis, P. lusitanica, P. hirtiflora, P. macroceras*, and *P. megaspilaea*.

6.4.2 México: the center of diversity

The main center of *Pinguicula* biodiversity is in México, which has ≈40 species (equaling 38% of the total generic diversity), 37 (90%) of which are endemics, some of which are microendemics confined only to a single valley or mountain top (Zamudio 2001, 2005, Lampard et al. 2016, Rivadavia et al. 2017). Most Mexican species are confined to seasonally semiarid to arid climate of colline to montane altitudes, where they occur among xerophytic vegetation on exposed rock or in open, dry deciduous forests (Zamudio 2005). *Pinguicula* grow in all the Mexican mountain ranges (Figure 6.4), with the highest diversity being in the Sierra Madre Oriental (25 species, 22 endemics; Zamudio 2005, Lampard et al. 2016) and the Sierra Madre del Sur (15 species, ten endemics). Both mountain ranges are least seasonally arid, have bedrock ranging from limestone and gypsum to shale, granites, and basalts and a well-structured geological relief, all of which create many distinct habitats and abiotic factors favorable for *Pinguicula* (Zamudio 2001, 2005, Rivadavia et al. 2017).

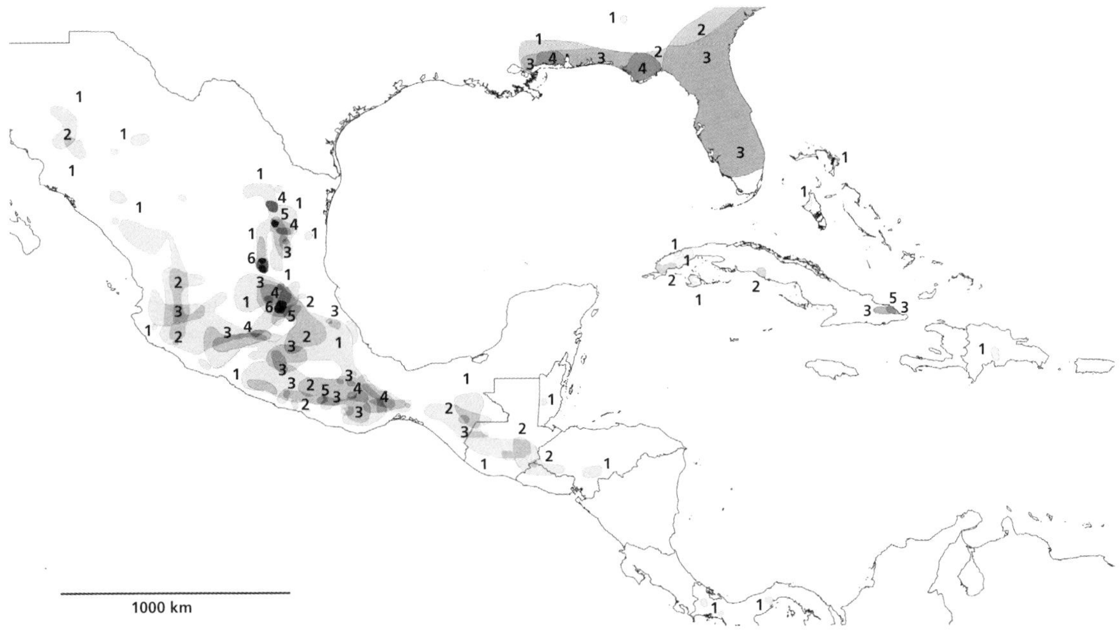

Figure 6.4 Distribution and number of species of *Pinguicula* in Central America, the Caribbean, and southeast USA (data from Casper 1966, Zamudio 2001, 2005, and additional information retrieved from herbarium records and *unpublished data* of Fernando Rivadavia). Map drawn by Andreas Fleischmann.

The widespread and polymorphic *P. moranensis* tolerates almost any kind of substrate, including basalt (Zamudio 2001, 2005). Several Mexican *Pinguicula* species are found growing on decalcified soils in oak and pine forests, including the micro-endemics *P. moctezumae, P. elizabethiae, P. emarginata*, and *P. gigantea*. Seasonally wet ravines formed in the otherwise dry gypsum hills host the microendemics *P. gypsicola, P. takakii, P. colimensis, P. rotundiflora, P. pygmaea, P. nivalis*, and *P. immaculata*. The hemiepiphytes *P. hemiepiphytica* and *P. mesophytica* usually grow on moss-covered tree trunks.

6.4.3 Diversity of other regions

At least nine species grow on Cuba (Domínguez et al. 2014; Lampard et al. 2016), with substantial variation between its eastern and western sides. On the west side, the endemic annual *P. filifolia* grows in white quartz sand in damp open areas at very low altitude, whereas the narrow endemic *P. cubensis* is restricted to serpentine soil near a waterfall at intermediate altitude (Domínguez et al. 2014). The annual *P. albida* has a broader ecological tolerance and grows on both soil types, and on peaty sand in savannas. The two central Cuban species are found in montane tropical forests growing on seasonally wet limestone rock, whereas the eastern Cuban species usually grow in laterite or clay soils (Lampard et al. 2016). Finally, Cuba and the nearby Dominican Republic host two true epiphytic species that grow on bare or moss-covered tree stems and branches: *P. lignicola* and *P. casabitoana*.

Six widely scattered and mostly geographically disjunct species grow in the Andes (Figure 6.3). *Pinguicula elongata, P. calyptrata*, and *P. involuta* grow primarily in the high Andean *paramos*—low grasslands dominated by scattered *Espeletia* (Asteraceae) and that are ecologically similar to alpine heathlands—and *yungas*—wet tropical montane and cloud forests. *Pinguicula jarmilae* is known only from a single dripping wall of sandstone on a roadside at 2100–2500 m altitude (Beck et al. 2008). *Pinguicula chilensis* occurs in wet meadows and in wet places on volcanic gravels. At the very south of the continent, *P. antarctica* grows in peat bogs among *Sphagnum* and the cushion plant *Donatia fascicularis* (Stylidiaceae).

Other centers of high species richness are the European Alps, including the Apennines (12 species, nine [75%] endemics), the Baetic Mountains of the Iberian Peninsula (five species, four endemics; Roccia et al. 2016; Figure 6.5), and the southeast coast of the Unites States (six endemic species; Schnell 1976, Lampard et al. 2016, Roccia et al. 2016).

6.5 Carnivory and other plant–insect interactions

6.5.1 Prey

Carnivory in *Pinguicula* was first postulated and experimentally demonstrated by Darwin (1875), and enzymatic activity was studied by Heslop-Harrison (1975). Prey usually consists of small-to-very small arthropods, predominantly midges and mites (Heslop-Harrison 2004, Fleischmann 2016b). Prey probably are attracted by the combination of volatile musty, fungus-like scent emitted by *Pinguicula* leaves (Lloyd 1942, Heslop-Harrison 2004, Fleischmann 2016b), and the visual attraction of the glistening, wet mucilaginous lamina (Juniper et al. 1989). The open, exposed adhesive leaf rosettes also casually catch air-borne debris; >50% of the "prey" of *P. vulgaris* can be made up of pollen, largely from wind-pollinated plants (Karlsson et al. 1994).

6.5.2 Associated arthropods

A mutualistic plant–arthropod interaction has been documented between *P. longifolia* and the small symbiotic mite, *Oribatula tibialis* (Antor and García 1995). The butterwort offers food and protection to *O. tibialis* while *P. longifolia* benefits from scavenging activities of the mites, which eat fungal hyphae and insect remains, and prevent molds from damaging the leaves (Antor and García 1995).

Hemipteran kleptoparasites of the capsid bug family Miridae have been observed on *P. vallisneriifolia*, where they feed on trapped insects (Zamora 1995). Similar bugs have been observed on *P. dertosensis*, freely moving on the viscous adhesive leaves without being trapped (A. Roccia *personal observation*). Other species of Miridae are known to live on a large variety of sticky plants (including carnivorous *Drosera, Byblis*, and *Roridula*; Chapter 10). A

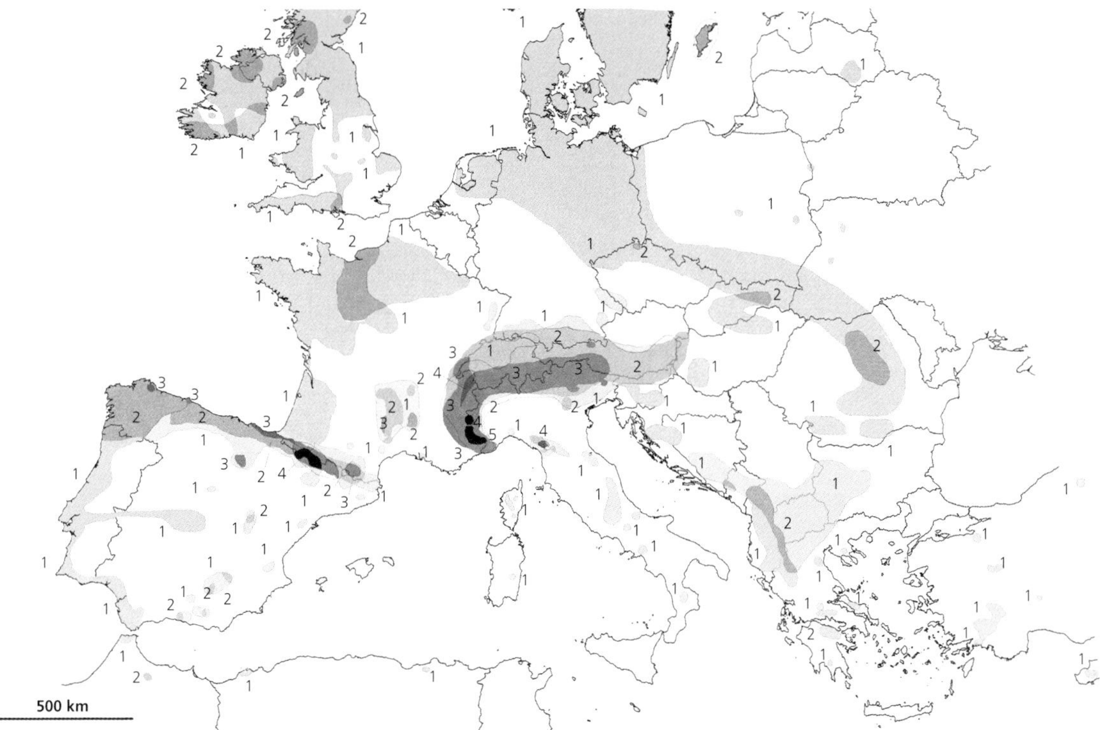

Figure 6.5 Distribution and number of species of Pinguicula in Europe and Asia Minor (data from Meusel et al. 1965, Casper 1966, Heslop-Harrison 2004, Fleischmann 2015b, and additional information retrieved from herbarium records). Map drawn by Andreas Fleischmann.

second kleptoparasite has been documented from *P. vallisneriifolia*: the slug *Deroceras nitidum* can consume a large portion of the prey caught by the plant (Zamora and Gómez 1996). Ants may steal prey from a number of other *Pinguicula* species (Zamora 1990a).

6.6 Conservation

The most severe threats to *Pinguicula* populations are from human activities, especially when the hydrology of their often fragile, oligotrophic seepage habitats are disturbed by source channeling, roadside constructions, or eutrophication. Other threats include over-collection and poaching by plant collectors. Neither the majority of *Pinguicula* species nor their habitats are under any special protection, and no *Pinguicula* species is currently listed on CITES. The IUCN Red List only includes three species under the category "threatened:" *P. fontiqueriana* (vulnerable B1ab(iii) + 2ab(iii)), *P. mundi* (vulnerable D2), and *P. nevadensis* (Endangered B2ab(iii,v)). Two of these are native to Spain where they are listed on the Spanish Red List along with *P. dertosensis* and *P. vallisneriifolia*. But this listing does not offer any protection, only a recognition of their threatened status. However, most, if not all, populations of *P. nevadensis* occur within the Parque Nacional de la Sierra Nevada, many *P. vallisneriifolia* and *P. dertosensis* populations are found in various Natural Parks, and one of the two known *P. mundi* locations is included in the Parque Natural de los Calares del Río Mundo y de la Sima. Although not listed on the national red list, *P. longifolia* is also found in many places in the Parque Nacional de Ordesa y Monte Perdido and in its counterpart in France, Parc National des Pyrénées.

Similar situations apply elsewhere in the world. *Pinguicula crystallina* is considered endangered on Cyprus, the endemic lowland species *P. bohemica* is critically endangered in Czech Republic, the majority of the southeastern US species are threatened by habitat loss (the endemic and endangered *P. ionantha* is protected by law), and many populations of widespread species are threatened on country level or local scale.

In several European countries (e.g., Great Britain, France, Germany, Switzerland), all naturally occurring species of *Pinguicula* are fully protected by law: it is forbidden to collect any plant parts, including seeds or roots. *Pinguicula arvetii, P. caussensis, P. grandiflora, P. longifolia, P. lusitanica, P. reichenbachiana*, and *P. vulgaris* benefit from this protection in France. Although this protection is effective only on a regional scale, for some taxa it does include their entire range (e.g., *P. grandiflora* subsp. *rosea* and *P. reichenbachiana*). Even *P. hirtiflora* benefits from some kind of protection from European and international directives, although it is considered as an alien invasive species in France.

Three species have not been relocated since their original discovery: *P. greenwoodii, P. imitatrix*, and *P. utricularioides*. All three are from México, and despite various attempts, they have not been observed again. Most of the Central American species grow in inaccessible habitats, although habitat loss by human activities sometimes happens on a very large scale. For example, the endemic *P. moctezumae* was thought to have gone extinct after much of its habitat in the Moctezuma valley was flooded after the construction of the Zimapán Hydroelectric Dam. The species was later found in a small side branch of the valley (F. Rivadavia *unpublished data*).

6.7 Future research

A well-resolved, comprehensive phylogenetic reconstruction based on large taxon sampling is still lacking for the genus. Incongruences between nuclear and plastid datasets, and low resolution of, or weak support for, some species groups in published phylogenies do not allow for subsectional evolutionary inferences or classifications. This is especially true for the species-rich but apparently young *P.* sect. *Temnoceras* from México and the Caribbean, which is morphologically and ecologically diverse, but for which existing classifications based on flower morphology and growth type seem to be rather artificial, and do not consider parallel evolution of morphological traits. The same holds true for the apparently young, polyploid species complexes of *P.* subg. *Pinguicula* from the Alps and Apennines (several of which might be the result of hybridogenesis). To reliably reconstruct the evolutionary history of these species-rich clades, and to overcome incongruences between plastid and nuclear DNA data, phylogenetic reconstructions considering reticulate evolution are needed, and will require sampling of many taxa, and large samples of geographically widespread and morphologically variable ones (e.g., *P. moranensis, P. ehlersiae, P. reichenbachiana*, and *P. vulgaris*).

Much less is known about the species interactions with other organisms, including their pollinators and the range of their prey. Field observations could fill this gap, and might shed light on whether different growth types and leaf shapes have evolved only as adaptations to certain habitats, or if they also mirror prey specialization. The diversity of corolla shapes and colors observed in the Central American species of *P.* sect. *Temnoceras* likely is related to different pollinator groups or pollination strategies, but the floral biology of Mexican and Caribbean species is unstudied. Successful conservation of these species demands a complete understanding of their biology and life history.

CHAPTER 7

Systematics and evolution of Lentibulariaceae: II. *Genlisea*

Andreas Fleischmann

7.1 Life history and morphology

Genlisea comprises 30 species (Fleischmann 2012a, Fleischmann et al. 2017; Appendix) of entirely rootless, rosette-forming, hygrophilous carnivorous herbs. Ten species are (facultative or obligate annual) therophytes, 19 are evergreen perennials, and one species, *Genlisea tuberosa*, is a tuberous geophyte (Fleischmann 2012a, Rivadavia et al. 2013). *Genlisea* grows in habitats similar to the majority of carnivorous plants (Chapter 2): open, exposed, nutrient-poor, and at least seasonally moist to very wet habitats, usually on soils flushed by seeping to swiftly moving water. They grow in and around springs, seeps, and ephemeral flushes of granitic inselbergs, sandstone plateaus (*tepuis*), ferricretes, in clearings of wet savanna vegetation, and on wet quarzitic sand fields (Fleischmann 2012a). All species are strict calcifuges that occur only on acidic to neutral soils.

7.1.1 Leaves

All species of *Genlisea* are strictly heterophyllous, producing two contrastingly different types of leaves. Both types of leaves are arranged alternately and with alternate (spiral) phyllotaxis along short, condensed stems (29 species), or on long, prostrate, stolon-like, horizontally creeping subterraneous stems with prolonged internodes (*Genlisea repens*; Figure 7.1). The first type of leaf is photosynthetic, bifacial, flat, foliar, petiolate with spatulate or oblong to lanceolate lamina, arranged in dense or lax rosettes that are appressed to the ground. The second type of leaf is the so-called rhizophyll ("root-leaf"): a subterranean, tubular, hollow organ of inverted Y-shape, which functionally replaces the absent roots by anchoring the plants to the soil and taking up nutrients (Adamec 2008c).

The rhizophylls are the complex, carnivorous eel traps unique to the genus (Figure 7.2; Chapters 12, 13, 15). They are of epiascidiate ontogeny (as are the pitcher leaves of pitcher plants; Chapters 3, 9) and exhibit positive geotropic growth (Kilian 1951, Juniper et al. 1989, Fleischmann 2012a). Each rhizophyll ("trap") consists of a solid stalk that widens at its distal end into a globose to ovoid or spindle-shaped, hollow chamber (the trap vesicle or digestive chamber, often compared to a "stomach"). At the opposite end, the vesicle opens into a terete, hollow "tubular neck." The tubular neck widens and bifurcates at its end into two hollow, helically twisted trap arms (of opposite rotation; Fleischmann 2012a). The interior of the helical arms and tubular neck is lined with serial rows of stiff, retrorse bristles (the numerous "bows" of the eel trap), the vesicle interior surface is covered with quadrifid digestive glands (Figure 7.2), and the rhizophyll external surface is sparsely covered with sessile glands. Traps are ≈1–20 cm in overall length (including the stalk), with the tubular neck part ≈0.1–0.5 mm in diameter (Fleischmann 2012a).

Species from the phylogenetically early-branching *G.* subg. *Tayloria* and *G.* sect. *Africanae* (§7.3) develop just a single monomorphic type of rhizophyll, whereas plants of *G.* sects. *Recurvatae* and *Genlisea* have dimorphic traps: comparatively short, thick, and wide (inner diameter) "surface

Fleischmann, A., *Systematics and evolution of Lentibulariaceae: II.* Genlisea. In: *Carnivorous Plants: Physiology, ecology, and evolution*. Edited by Aaron M. Ellison and Lubomír Adamec:
Oxford University Press (2018). © Oxford University Press. DOI: 10.1093/oso/9780198779841.003.0007

Figure 7.1 (Plate 7 on page P6) Vegetative and generative morphology of *Genlisea*. (**a, d**) *Genlisea flexuosa* (*G.* subg. *Tayloria*), with monomorphic traps. (**b, e**) *G. margaretae* (*G.* sect. *Recurvatae*) with dimorphic traps (note the short, thick surface traps and the longer, filiform deep-soil traps). (**c, f**) *G. repens* (*G.* sect. *Genlisea*), the sole species with a prostrate stoloniferous habit. (**g**) the unique multiple-circumscissile capsule dehiscence pattern of *G.* subg. *Genlisea*, illustrated here by *G. hispidula*. Photographs by Andreas Fleischmann.

traps" just beneath the soil surface; and comparatively long-stalked, thin and filiform "deep-soil traps" that grow vertically downward, reaching into deeper soil layers (Fleischmann 2012a; Figure 7.1). This trap dimorphism might serve to exploit different soil organisms from different soil strata.

7.1.2 Inflorescences and flowers

The inflorescences of *Genlisea* are bracteose racemes; the scape can be glabrous, covered with eglandular or glandular hairs, or a mixture of both types of hairs. The species-specific indumentum often is a reliable taxonomic character.

Each flower is subtended by a basifixed bract and two lateral bracteoles. The flowers of *Genlisea* follow the general design of Lentibulariaceae: they are zygomorphic, hermaphroditic, pentamerous, and tetracyclic, with a personnate, spurred, bilabiate corolla, two bithecate stamens and a superior, subglobose ovary with persistent bilabiate style and free central placentation. The corolla throat of all but one species (*G. exhibitionista*) is covered by a gibbose, upwardly arching swelling of the corolla lower lip that creates a "masked flower." The calyx consists of five spreading, basally adnate sepals and is persistent in fruit. The corolla is bilabiate: the upper lip is made up of two petals which can be fused only to the base, with two free lobes spreading (*G.* subg. *Tayloria*), or entirely fused (*G.* subg. *Genlisea*). The lower lip of the corolla consists of three fused petals, which also make up the corolla throat and spur. The spur either parallels the pedicel (spreading from the corolla lobes; *G.* subg. *Tayloria*) or the corolla lower lip (*G.* subg. *Genlisea*; Fleischmann et al. 2010, Fleischmann 2012a). This results in two different flower types. Flowers of *G.* subg. *Tayloria* (most notably in *G. violacea*, less expressed in some other species) are

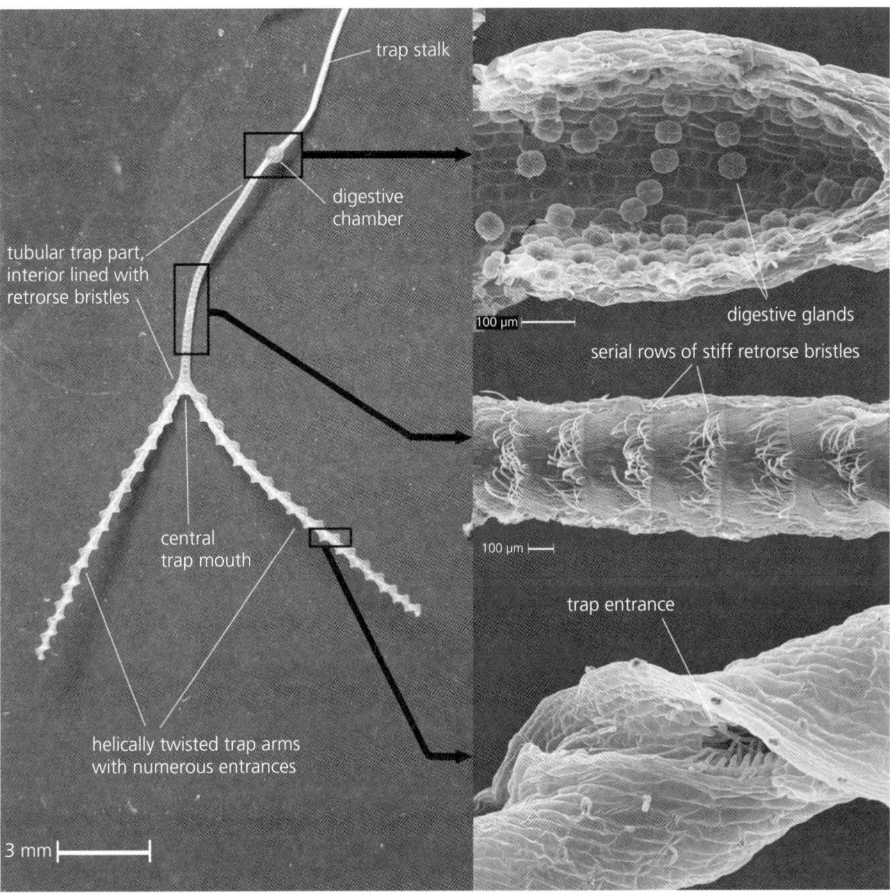

Figure 7.2 The *Genlisea* rhizophyll. Overview of trap on the left (*G. repens* shown); SEM details of digestive chamber (*G. violacea*); tubular neck (*G. repens*); and trap entrance (*G. flexuosa*). Photographs by Andreas Fleischmann (reproduced in part from Fleischmann 2012a).

salverform: the spreading corolla lobes create a landing platform, and only the pollinators' proboscis or mouthparts can be inserted into the abruptly narrowed corolla tube and spur. Members of *G.* subg. *Genlisea* display masked flowers of the snapdragon type, with an upwardly arching gibbose palate that seals the entrance to the corolla throat: to enter the corolla tube to reach the nectar secreted inside the spur, the pollinator has to push down the lower lip of the corolla (Fleischmann 2012a).

7.1.3 Fruits and seeds

The capsules are (sub)globose, often lined with species-specific indumentum, and either held upright in fruit (*G.* sects. *Africanae* and *Genlisea*) or on a downward-curving pedicel (*G.* subg. *Tayloria* and *G.* sect. *Recurvatae*). Ripe *Genlisea* capsules display two contrasting dehiscence types, which correspond to the two subgenera: in *G.* subg. *Tayloria* the capsules are longitudinally bivalvate, whereas in *G.* subg. *Genlisea* they are circumscissile with a single or multiple ring-like dehiscence lines, or display a spiral dehiscence unique among angiosperms (Figure 7.1).

The numerously produced seeds are globose to prismatic. Their size, shape, and testa ornamentation differ among the different sections: *G.* subg. *Tayloria*: prismatic and dorsiventrally compressed, with reticulate or papillate testa; *G.* sects. *Africanae* and *Recurvatae*: both ovoid to globose with very regular, isodiametric reticulation, but of different size classes; *G.* sect. *Genlisea*: pyramidal to angulate-ovoid, with smooth testa (Fleischmann 2012a).

7.2 Carnivory

Warming (1874) first studied and illustrated the morphology and anatomy of the rhizophylls, and Darwin (1875) first considered *Genlisea* to be a carnivorous plant. He was also the first to understand the trapping principle of the peculiar rhizophyll, and compared it with an eel trap or bow net. Heslop-Harrison (1975) documented that proteolytic digestive enzymes are secreted from the trap interior glands, and Barthlott et al. (1998) used radioisotope traces to demonstrate that *Genlisea* takes up nutrients from prey. Captured animals are probably killed inside the traps by anoxia (Adamec 2007b) and subsequently digested by a pool of continuously secreted proteolytic enzymes that are released from the quadrifid glands that line the vesicle and upper neck region of the rhizophyll interior (Heslop-Harrison 1975, Barthlott et al. 1998, Płachno et al. 2006, 2007a).

Genlisea apparently unselectively catches a diversity of soil organisms that are small enough to fit the trap entrance. Barthlott et al. (1998) hypothesized that *Genlisea* specializes on protozoa, but all other data suggest that its prey spectrum is broader than is known for any other carnivorous plant (Darwin 1875, Goebel 1891b, 1893, Kuhlmann 1938, Lloyd 1942, Heslop-Harrison 1975, Studnička 1996, 2003a, Płachno et al. 2005a, 2008, Płachno & Wołowski 2008, Fleischmann 2012a). Trap contents include cyanobacteria and non-photosynthetic eubacteria, protozoa (ciliates, flagellates, and thecate amoebae), and large numbers of nematodes, small soil crustaceans (especially copepods), collembola, mites, and various unicellular "algae" (diatoms, desmids, chrysophytes, euglenophytes, and single-celled green algae). Some of the trapped microorganisms apparently are able to survive within the traps (Płachno and Wołowski 2008), and some non-trophic interactions between *Genlisea* and trap-inhabiting microbes may occur (Caravieri et al. 2014, Cao et al. 2015).

7.3 Phylogeny and evolution

Genlisea and *Utricularia* (Chapter 8) are sister genera (Figure 7.3; Jobson et al. 2003, Müller et al. 2004, Fleischmann et al. 2010); the rhizophylls of *Genlisea* and the bladder traps of *Utricularia* are homologous to one another, as both are to the sticky traps of their common sister genus *Pinguicula* (Fleischmann 2012a; Chapter 3), and to the common foliar leaves of plants. The traps of *Genlisea* are epiascidiate (§7.2.1), and a likely scenario for their evolution is a continued inward folding and final fusion of the lateral margins of adhesive leaves of the presumed common ancestor (Chapter 3).

7.3.1 Infrageneric classification

Two subgenera (Fromm-Trinta 1977, Fischer et al. 2000) and four sections (Fleischmann et al. 2010) have been proposed. The two subgenera, *G.* subg. *Tayloria* and *G.* subg. *Genlisea*, were circumscribed based on capsule dehiscence (Fromm-Trinta 1977, Fischer et al. 2000), and these are supported further by flower morphology and seed ultrastructure (Fleischmann 2012a), and by phylogeny and cytology (Fleischmann et al. 2010, 2014).

7.3.2 Phylogeography

Genlisea is likely to have originated in the Neotropics, like its sister genus *Utricularia* (Jobson et al. 2003), and the highest extant species diversity of *Genlisea* occurs in the southeastern Brazilian highlands (Fleischmann et al. 2011a, Fleischmann 2012a; Figure 7.4). Phylogenetic reconstructions of *Genlisea* (Jobson et al. 2003, Müller et al. 2004, Fleischmann et al. 2010, 2014) have revealed two major sister clades within the genus, which correspond to the two subgenera *Tayloria* and *Genlisea*.

Within the Brazilian *G.* subg. *Tayloria*, the large, perennial species (*G. uncinata, G. oligophylla, G. metallica, G. flexuosa*) represent early-branching lineages, whereas the more derived species are annuals or short-lived polycarpic species (*G. violacea, G. lobata, G. nebulicola, G. exhibitionista*; Fleischmann et al. 2010, 2011a). Subgenus *Genlisea* comprises three clades, two of them exclusively African—*G.* sects. *Africanae* and *Recurvatae*—and

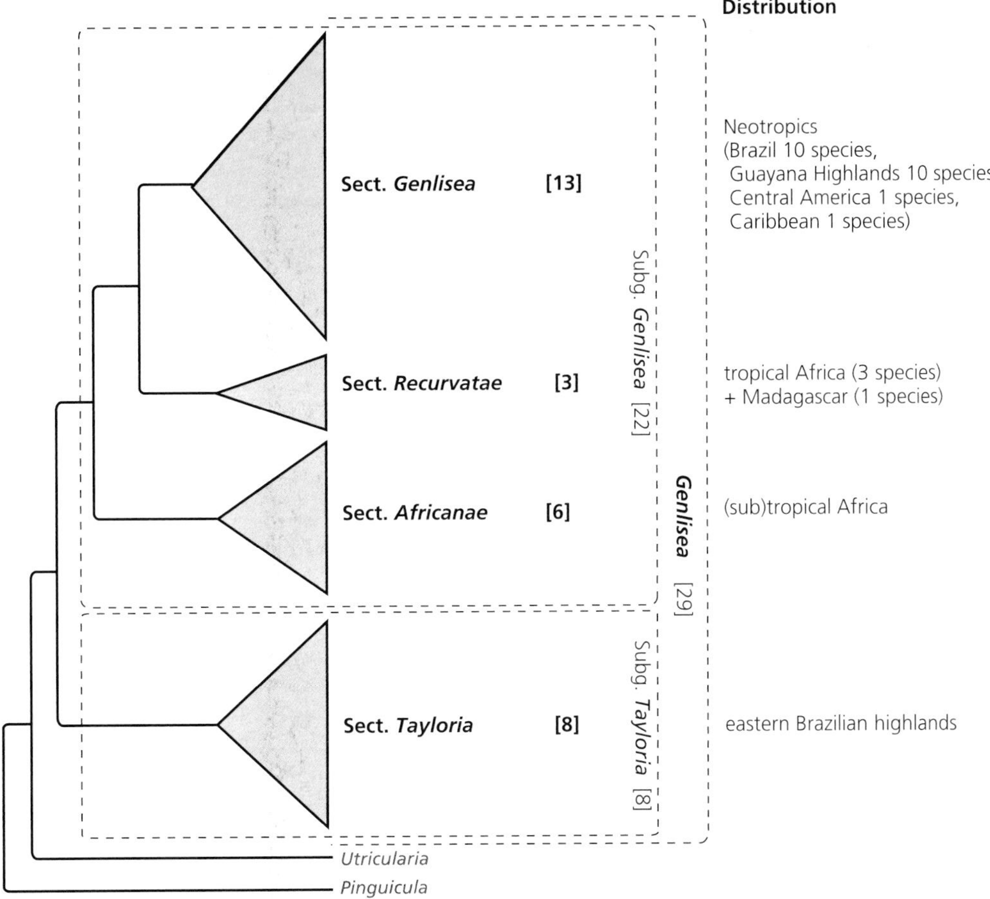

Figure 7.3 Simplified phylogeny of *Genlisea*, topology based on Fleischmann et al. (2010).

one exclusively Neotropical—*G.* sect. *Genlisea* (Fleischmann et al. 2010; Figure 7.3).

All extant Neotropical species of *G.* subg. *Genlisea* share a common ancestor with the African *G.* sect. *Recurvatae*. In contrast to the paraphyletic African species, the derived Neotropical species in *G.* sect. *Genlisea* form a monophyletic group, implying a single colonization event of *G.* subg. *Genlisea* in South America via trans-Atlantic long-distance dispersal (Fleischmann et al. 2010; Fleischmann 2012a). *Genlisea* thus has colonized South America twice: first by a radiation of *G.* subg. *Tayloria*, and later following recolonization from Africa by derived members of *G.* subg. *Genlisea* (Fleischmann et al. 2010).

The two subgenera are immediate sister groups in phylogenetic reconstructions (Fleischmann et al. 2010), and there is as yet no evidence supporting either clade as the last common ancestor. However, species within *G.* subg. *Tayloria* share several floral characters with *Pinguicula*, and these can thus be considered plesiomorphic in *Genlisea*: bivalvate capsule dehiscence, a corolla with bilobed upper lip, and a spur that spreads from the corolla lower lip (paralleling the pedicel). In contrast, species of *G.* subg. *Genlisea*, share certain floral characters with the common sister genus *Utricularia*: an entire upper corolla lip and the spur paralleling the lower corolla lobe. In *Utricularia* (including *U.* sects. *Calpidisca, Setiscapella*, and *Utricularia*; Chapter 8) and in *G.* subg. *Genlisea*, there is a switch from lilac-blue to yellow flower color in derived lineages. This switch certainly evolved in parallel in both genera and in all sections, given their continentally disjunct ranges, different habitats, but it could have a common genetic basis and might be an adaptation to similar pollinator guilds.

7.3.3 Chromosome numbers

Species within *G.* subg. *Tayloria* have a common chromosome number of $2n = 16$ and comparatively large chromosome size (Fleischmann et al. 2014). Within *G.* subg. *Genlisea*, a large reduction in chromosome size and an increase of chromosome number has occurred. In *G.* sect. *Africanae*, chromosome numbers of $2n = 32$ and 40 have been reported for *G. hispidula* (Fleischmann et al. 2014, Vu et al. 2015), whereas *G. margaretae*, in the consecutive sister group, *G.* sect. *Recurvatae*, has $2n = 38$ (Fleischmann et al. 2014, Tran et al. 2016). Species in *G.* sect. *Genlisea* have large numbers of minute chromosomes (Greilhuber et al. 2006; Fleischmann et al. 2014, Vu et al. 2015). This pattern is accompanied by an evolutionary genome size reduction.

7.3.4 Genome size

Genome size among *Genlisea* species varies 25-fold (Greilhuber et al. 2006, Fleischmann et al. 2014, Vu et al. 2015, Tran et al. 2015b); this huge range is exceeded by very few other plant genera (e.g., a 68-fold variation between the largest and smallest known genome sizes has been found within the parasitic plant genus *Cuscuta*; I. Leitch *unpublished data*). Some members of *Genlisea* have the smallest genome sizes currently known among angiosperms, with holoploid genomes of only 61 Mbp (*Genlisea tuberosa*) or 64 Mbp (*Genlisea aurea*), respectively (Greilhuber at al. 2006, Fleischmann et al. 2014). At the same time, the largest genomes known in Lentibulariaceae also occur in *Genlisea*. Genome sizes of the immediate sister genus *Utricularia* and the common sister of both, *Pinguicula*, all fall in between the largest and smallest *Genlisea* genomes (Greilhuber et al. 2006, Fleischmann et al. 2014, Veleba et al. 2014).

In *Genlisea*, genome size appears to reflect chromosome size (and to a lesser degree karyotype) and phylogenetic affinity (Fleischmann et al. 2014). Fleischmann et al. (2014) therefore postulated an evolutionary trend in genome reduction from an ancestor with large genomes (still present in extant members of *G.* subg. *Tayloria* and *G.* sect. *Africanae*) to the ultra-small genomes observed in extant members of the derived *G.* sect. *Genlisea*. However, Vu et al. (2015) found the genome size in *Genlisea* to be of no evolutionary significance, and hypothesized that the contemporary variation in genome size evolved from a common ancestor with medium-sized genomes, leading to the comparatively large genomes of *G.* subg. *Tayloria* and *G.* sect. *Africanae* on the one hand, and very small to ultra-small genomes in *G.* sects. *Recurvatae* and *Genlisea* on the other side.

Genlisea species with large and ultra-small genomes can have similar or identical chromosome numbers, but differ greatly in chromosome structure and cell size (Tran et al. 2015b), and in DNA composition of telomere and centromere regions (Tran et al. 2015a). Genome size appears to have little effect on morphology or physiology in *Genlisea*, and no association has been found between genome size and morphology, habitat, or life history (annual vs. perennial) (Fleischmann et al. 2014, Vu et al. 2015). On the other hand, genome size does appear to be related to rate of evolutionary diversification: the 16-species clade with ultra-small genomes (*G.* sects. *Genlisea* and *Recurvatae*; Figure 7.3) is more species rich than its immediate sister, *G.* sect. *Africanae*, a clade of six species with large genomes. In a few other plant lineages, including *Veronica* (Lamiales), a similar scenario of significant genome downsizing preceding increased diversification has been observed (Meudt et al. 2015). However, a causal relationship between genome size and diversification is unknown.

7.4 Distribution

7.4.1 Global patterns of diversity

Genlisea shows an interesting amphi-Atlantic distribution pattern, with 21 species in the Neotropics (South to Central America, and Cuba in the Caribbean) and nine in tropical Africa including Madagascar (Figure 7.4). The genus is almost exclusively tropical; only *G. hispidula* extends into subtropical latitudes in southern Africa, whereas *G. aurea* and *G. repens* cross the tropic of Capricorn in South America (Fleischmann 2012a). *Genlisea*

Figure 7.4 Global distribution of *Genlisea*, with species numbers indicated (from data in Fleischmann et al. 2010, Fleischmann 2012a).

species grow from sea level to ≈2800 m a.s.l., but the majority are found between 600 and 1500 m a.s.l. (Fleischmann 2012a).

Not a single species of *Genlisea* occurs on both South America and Africa (Fleischmann et al. 2010, 2017, Fleischmann 2012a). The areas with greatest species numbers and diversity in South America are the central Brazilian highlands and the Guayana Shield, and just a single, widespread, annual species (*G. filiformis*) extends the range of the genus to the north, with isolated populations in Cuba, Belize, Nicaragua, and southern México (Fleischmann 2012a; Figure 7.4).

7.4.2 Brazil: the center of diversity

All but three of the 18 Neotropical species (and 60% of all *Genlisea* species) occur in Brazil; the three others, *G. glabra*, *G. sanariapoana*, and *G. pulchella*, are endemic to the Guayana Highlands. At least ten of the Brazilian species are endemic to the country (Fleischmann 2012a, Fleischmann et al. 2017). The eight species of *G.* subg. *Tayloria* all are Brazilian endemics, occupying a small area in the highlands of eastern Brazil, mainly in the Serra do Espinhaço (Fleischmann et al. 2011a). Of the three Guayana Highland endemics, *G. glabra* and *G. roraimensis* are confined to the summits of the *tepuis*, whereas *G. sanariapoana* is a lowland species endemic to the *llanos* plains along the Upper and Middle Orinoco River (Fleischmann 2012a).

7.4.3 African species

The nine African species grow in tropical West and East Africa; *G. margaretae* extends this overall range into central Madagascar, where it occurs in a few outlying populations (Figure 7.4). The centers of diversity in Africa are on the Angolan and Zambian Plateaus, from where *G. hispidula* has spread to form several disjunct populations in Afromontane regions. In parallel with the widespread Neotropical annual, *G. filiformis*, the small annual *G. stapfii* has the widest range of distribution among the African species, extending from Senegal southwest (absent in the dry areas of the Ghana Dry Zone and the Dahomey Gap) to the Central African Republic in the west and the Republic of Congo (Congo-Brazzaville) in the south (Fischer et al. 2000; Fleischmann 2012a).

7.5 Future research

Although the small genus *Genlisea* is taxonomically, morphologically, anatomically, and phylogenetically relatively well-studied, the exact functioning of its rhizophyll traps, especially regarding prey attraction, is still not fully understood. The basic functional mechanism and anatomy of these eel traps long has been known (Warming 1874, Darwin 1875), but it is still unclear whether the rhizophylls are purely passive traps that lure prey to their interior by some kind of attractant or if they actively create a water current that sucks small soil organisms into the traps. There is published evidence to support both theories. Studies with detached traps found no evidence for a water current in *Genlisea* rhizophylls (Adamec 2003b, Płachno et al. 2005a, 2008). On the other hand, immobile prey items and soil particles frequently are present in the traps, bifid glands similar to those which work as water pumps in *Utricularia* traps occur on the rhizophylls, and ink tracers *in vivo* together suggest an active trap (Juniper et al. 1989, Meyers-Rice 1994, Studnička 1996, 2003a, Fleischmann 2012a).

Although Barthlott et al. (1998) hypothesized chemotactic attraction of prey by *Genlisea*, no chemical attractant yet has been detected from *Genlisea* traps (Płachno et al. 2008). Other hypothesized attractants include secreted mucilage as bait (Goebel 1891b, Lloyd 1942) and rhizophyll entrances acting as deceptive soil shelters (Studnička 2003a). Both Studnička (2003a) and Adamec (2007b) hypothesized that oxygen released from the traps could attract prey in otherwise anaerobic soils. This hypothesis seems more likely for the deep-soil traps in species with rhizophyll dimorphism than in most species that form traps just beneath the soil surface, where aerobic conditions prevail (Fleischmann 2012a).

Last, the basic ecology of *Genlisea*, including population dynamics and interspecific interactions (prey spectrum, associated biota, and pollination biology) barely has been explored, although the first pollinator observations were reported by Fleischmann (2012a) and Aranguren Díaz (2016).

CHAPTER 8

Systematics and evolution of Lentibulariaceae: III. *Utricularia*

Richard W. Jobson, Paulo C. Baleeiro, and Cástor Guisande

8.1 Introduction

Utricularia (bladderworts) currently comprises >230 species (Taylor 1989, Rutishauser 2016), divided into three subgenera (*Polypompholyx, Utricularia*, and *Bivalvaria*) and 35 sections (Taylor 1989, Müller and Borsch 2005, Lowrie et al. 2008, Jobson et al. 2017; Appendix). The genus is distributed worldwide with the exception of both poles and most oceanic islands (Taylor 1989). The Neotropics has the highest species diversity, followed by Australia (Taylor 1989, Guisande et al. 2007). *Utricularia* has diversified more rapidly than its sister genera *Pinguicula* (Chapter 6) and *Genlisea* (Chapter 7), a pattern that has been linked to its extreme morphological flexibility (Jobson and Albert 2002, Rutishauser 2016) and that is apparent in its many forms and habitats (Jobson et al. 2003, Müller and Borsch 2005). The evolution of its highly modified Bauplan, including its suction bladder trap, has long attracted researchers (e.g., Darwin 1875, Lloyd 1942).

8.2 Phylogeny and taxonomy

8.2.1 Early classification and delimitation

Linneaus (1753) collated the seven known species of *Utricularia* and typified the suspended aquatic *U. vulgaris*. Each of these seven species were described as having two stamens, a single pistil, a two-parted calyx, a gamopetalous corolla, a one-celled ovary and capsule, and a free-central placenta. By 1895, at least 13 genera of bladderworts had been recognized; Kamieński (1895) placed all, except for some species in *Biovularia* and *Polypompholyx*, into *Utricularia*. Barnhart (1916) recognized and reassigned the aforementioned 13 bladderwort genera based almost entirely on the position and shape of inflorescence bracts and bracteoles.

Taylor (1989) united all previously recognized bladderwort genera into *Utricularia*, separating them into two subgenera—*Polypompholyx* (three species) and *Utricularia* (211 species)—that were delimited based entirely on the presence of a four- versus two-parted calyx respectively. Taylor (1989) further synonymized many of the then ≈900 recognized species, and proposed and described novel taxa based on biogeography and morphology. The end result was 35 sections—*U.* sects. *Polypompholyx* and *Tridentaria* in *U.* subg. *Polypompholyx* and 33 others in *U.* subg. *Utricularia*—representing 214 species (Taylor 1989; Figure 8.1).

8.2.2 Contemporary phylogenies

Relationships among subgenera. The majority of Taylor's (1989) sections were well supported as monophyletic groupings by molecular phylogenies using the plastid markers *rps16* and *trnL-F* (Jobson et al. 2003) and the plastid *trnK* intron (Müller et al. 2004). However, these two studies were inconsistent in their support for the relationship between *U.* subg. *Polypompholyx* and *U.* subg. *Utricularia*. The *rps16*/*trnL-F* results of Jobson et al. (2003) showed strong branch support for a sister relationship between *U.* subg. *Polypompholyx* and *Utricularia* sensu Taylor (1989), whereas the *trnK*

Jobson, R. W., Baleeiro, P. C., and Guisande, C., *Systematics and Evolution of Lentibulariaceae: III.* Utricularia.
In: *Carnivorous Plants: Physiology, ecology, and evolution*. Edited by Aaron M. Ellison and Lubomír Adamec:
Oxford University Press (2018). © Oxford University Press. DOI: 10.1093/oso/9780198779841.003.0008

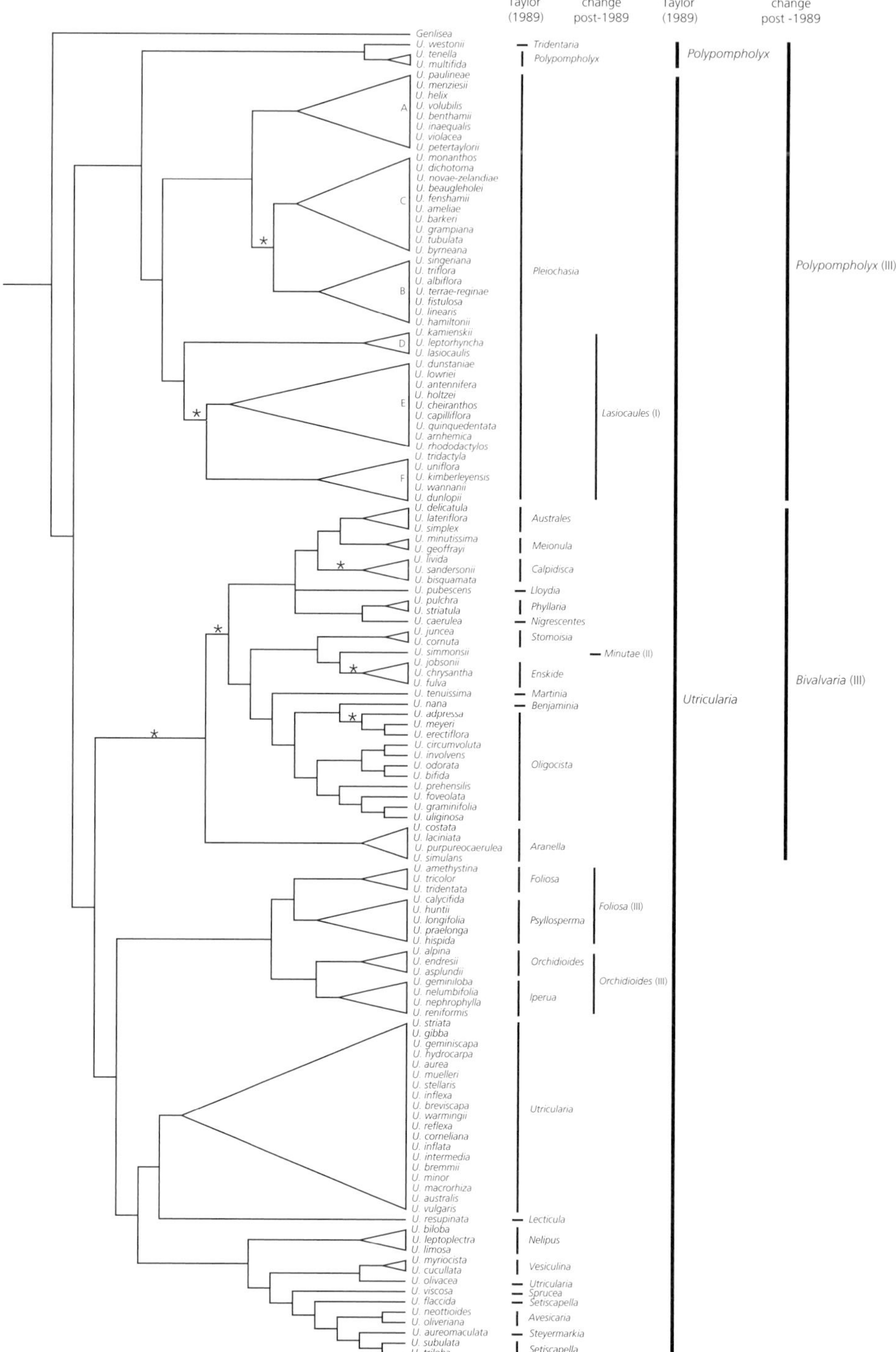

Figure 8.1 Molecular phylogeny of *Utricularia* representing 80% of all sections and 44% of all described species; modified from Jobson et al. (2003) and Jobson and Baleeiro (*unpublished data*) with monophyletic sections collapsed. This maximum credibility tree is derived from Bayesian inference analysis based on sequences of *rps16* and *trnLF*. Sections assigned by Taylor (1989) and those assigned since 1989 are shown with thin bars; subgenera assigned by Taylor (1989) and subsequent changes by Müller and Borsch (2005) are shown with thick bars. Clades discussed in the text (*U.* sect. *Pleiochasia sensu* Taylor) are labeled A–F. Asterisks above branches indicate weak jackknife support (60–75%). All other branches are well supported (76–100%).

results placed *U.* subg. *Polypompholyx* within *U.* subg. *Utricularia* (Müller et al. 2004). Jobson's finding of a sister relationship between the two subgenera was supported further by Müller et al. (2006), who used a supertree approach in which branch support values from each phylogenetic hypothesis are statistically weighted, with branching order assembled according to overall support.

In both Jobson et al. (2003) and Müller et al. (2004, 2006), *U.* subg. *Utricularia* was paraphyletic: *U.* sect. *Pleiochasia* was revealed as sister to *U. multifida* of *U.* sect. *Polypompholyx* (*U.* subg. *Polypompholyx*). Jobson et al. (2003) recommended, and Müller and Borsch (2005) formally proposed, including *U.* sect. *Pleiochasia* within *U.* subg. *Polypompholyx* (Figure 8.1). Reut and Jobson (2010) supported this sister relationship in a study of the *rps16* plastid gene sampled from 23 species of *U.* subg. *Polypompholyx sensu* Müller and Borsch (2005). Jobson et al. (2017) used *rps16, trnL-F,* and *trnDT* sampled from all recognized species to further support the sister relationship between *U.* subg. *Polypompholyx sensu* Müller and Borsch (2005) and *U.* subg. *Utricularia sensu* Taylor (1989), and also supported Taylor's (1989) placement of *U.* sect. *Tridentaria* within *U.* subg. *Polypompholyx*. Müller and Borsch (2005) also proposed dividing *U.* subg. *Utricularia* into two subgenera—*Utricularia* and *Bivalvaria*—based on clearly recognizable monophyletic groups. *Bivalvaria* resurrected a previously synonymized subgenus (Taylor 1989) and now includes all sections contained within clade U2–U4 of Jobson et al. (2003) (Figure 8.1).

Resolving sections. Molecular studies also revealed a handful of Taylor's (1989) sections within *U.* subg. *Utricularia* to be polyphyletic or paraphyletic (Jobson et al. 2003, Müller et al. 2004). Both groups found *U.* sect. *Psyllosperma* to be paraphyletic. Jobson et al. (2003) included in *U.* sect. *Psyllosperma* the species *U. amethystina, U. tricolor* (both in *U.* sect. *Foliosa*), and an additional sequence attributed to a specimen identified as *U. huntii* (*U.* sect. *Psyllosperma*). However, this *U. huntii* accession subsequently was determined to have been a misidentified member of the *U.* sect. *Foliosa* complex (Baleeiro et al. 2016). Thus, the results of Jobson et al. (2003) should be interpreted as *U.* sect. *Foliosa* forming a clade sister to a clade containing members of *U.* sect. *Psyllosperma*. Müller et al. (2004) included in *Psyllosperma* a single species of *U.* sect. *Foliosa* (*U. tridentata*) and four from *U.* sect. *Psyllosperma*, but *U. calycifida* (*U.* sect. *Psyllosperma*) was sister with *U. tridentata*. Work is currently underway (Baleeiro et al. *unpublished data*) to determine whether these sections are monophyletic (Figure 8.1).

Baleeiro et al. (2016, *unpublished data*) combined morphometric and molecular data in an attempt to disentangle the high levels of morphological variation observed within the Neotropical *U.* sect. *Foliosa sensu* Taylor (1989). The work focused on the diverse species complex *U. amethystina*, into which Taylor (1989) had synonymized 31 previously recognized species. Baleeiro et al. (2016) used inflorescence characters to identify eight distinct entities within this species complex. Using nuclear ITS and three plastid genes (*rps16, trnL-F,* and *trnDT*), Baleeiro et al. (*unpublished data*) resolved the *U. amethystina* type as sister to *U. tricolor*, while the eight entities identified by Baleeiro et al. (2016) formed independent clades outside this type clade. From this work Baleeiro et al. (*unpublished data*) propose the resurrection of four previously synonymized taxa, and identify two new species.

Jobson et al. (2003) resolved *U.* sects. *Iperua* and *Orchidioides* as monophyletic, whereas Müller et al. (2004) included *U. humboldtii* (*U.* sect. *Iperua*) in their sequencing and found that it was a member of *U.* sect. *Orchidioides*; Müller and Borsch (2005) then proposed that *U.* sect. *Iperua* should be lumped within *U.* sect. *Orchidioides*. Gomes Rodriques et al. (2017) provide further support for this proposal using both chloroplast and nuclear markers (Figure 8.1). Jobson et al. (2003) found *U. nana*, the only species within the South American *U.* sect. *Benjaminia*, to be derived within *U.* sect. *Oligocista*, and sister to the other South American taxa included in their study. Müller et al. (2004) included an African and Asian species from *U.* sect. *Oligocista* and found that *U. nana* was a sister to *U.* sects. *Oligocista* + *Avesicarioides* (Figure 8.1). Although this result may be a result of a lack of South American *U.* sect. *Oligocista* species in the study, further work is required to clarify the sister relationships within this section.

Last, Jobson et al. (2003) revealed *U.* sect. *Utricularia* to be polyphyletic because their South American accession of *U. olivacea* (*U.* sect. *Utricularia*) is sister to section *Vesiculina*. In contrast, Müller and Borsch (2005) placed their United States accession of *U. olivacea* within *U.* sect. *Utricularia*. Recent re-sampling and sequencing of an additional *U. olivacea* accession confirmed the position sister to *U.* sect. *Vesiculina* (Silva et al. 2016; Figure 8.1). A future study that includes related species *U. naviculata* and *U. biovularioides* will help resolve relationships within this unusual group, possibly requiring formation of a new section.

New sections proposed for *Utricularia*. Since Taylor (1989) there have been two proposals for new sectional classification. The first is *U.* sect. *Minutae* to include *U. simmonsii* (Lowrie et al. 2008), which has the smallest known flowers in the genus and bladder traps that resemble those found in *U.* sect. *Enskide*. Reut and Jobson (2010) included *U. simmonsii* in a re-analysis of the sequence matrix of Jobson et al. (2003) and found it to be sister to *U. chrysantha* (*U.* sect. *Enskide*). More recently the two other members of *U.* sect. *Enskide*, *U. fulva* and *U. jobsonii*, have been included in an analysis demonstrating that *U.* sect. *Enskide* forms a clade sister to *U.* sect. *Minutae* (Jobson and Baleeiro *unpublished data*; Figure 8.1).

The second involves *U.* sect. *Pleiochasia*, in which Reut and Jobson (2010) identified two major clades (1 and 2). The first includes species distributed in both tropical northern and temperate southern Australia, whereas the second includes mostly tropical northern species. Jobson et al. (2017), including all recognized species of *U.* subg. *Polypompholyx*, resolved the same two strongly supported clades as A–C and D–F respectively (Figure 8.1). With the type species of *U.* sect. *Pleiochasia* (*U. dichotoma*) contained within the former clade, Jobson et al. (2017) propose that clades D–F should represent a new section: *Lasiocaules* (Figure 8.1).

8.3 Evolution of life histories and morphology

8.3.1 Habitats and life history

The mapping of habit across the genus indicates that the ancestral state of *Utricularia* was terrestrial (Jobson et al. 2003, Müller and Borsch 2005), with subsequent evolution of forms possessing affixed aquatic, affixed subaquatic, suspended aquatic, epiphytic, lithophytic, and rheophytic habits (Figure 8.2). Shifts to the lithophytic habit have occurred independently, within *U.* sects. *Iperua* (*U.* subg. *Utricularia*), *Phyllaria*, and *Lloydia* (*U.* subg. *Bivalvaria*) (Jobson et al. 2003), and in *U. wannanii* from northern Australia (*U.* subg. *Polypompholyx*) (Jobson and Baleeiro 2015). Jobson et al. (2003) found that the epiphytic habit evolved from the terrestrial habit independently at least twice in the Afrotropical-Indomalayan-Australasia section *Phyllaria* (*U.* subg. *Bivalvaria*) and within the Neotropical sections *Iperua* and *Orchidioides* (*U.* subg. *Utricularia*) (Figures 8.1, 8.2). These species are specialized for the epiphytic habit by having water-storage tubers, with two members of section *Iperua* specialized for growth in the tanks of epiphytic bromeliads (Taylor 1989).

Shifts from terrestrial *Utricularia* into aquatic habitats have occurred at least twice with the rheophytic habit, as well as with suspension in the water column (Jobson et al. 2003, Müller et al. 2004) (Figure 8.2). Rheophytic *Utricularia* are specialized for growth in fast-flowing waters (Taylor 1989), with ventrally papillate, claw-like rhizoids for adherence to rock surfaces and reduced or pinnate leaves (Taylor 1989). The African *U.* sect. *Avesicarioides* is sister to *U.* sect. *Oligocista* in *U.* subg. *Bivalvaria* (Müller et al. 2004), whereas the rheophytic section *Avesicaria* (*U. neottioides*, *U. oliveriana*) is derived within the mainly terrestrial *U.* sect. *Setiscapella* (*U.* subg. *Utricularia*; Figure 8.1). The affixed aquatic habit has arisen multiple times across all three subgenera (Jobson et al. 2003, 2017, Reut and Jobson 2010). Overall, the diversification in habitat and life history by *Utricularia* has led to a great deal of morphological evolution.

8.3.2 Stolons, rhizoids, and leaves

All species of *U.* sects. *Polypompholyx*, *Tridentaria*, and several species within clade A (Figure 8.1) of *U.* sect. *Pleiochasia* lack stolons and form stemmed rosettes similar to those of the sister genus *Genlisea* (Taylor 1989, Fleischmann et al. 2010). Most species

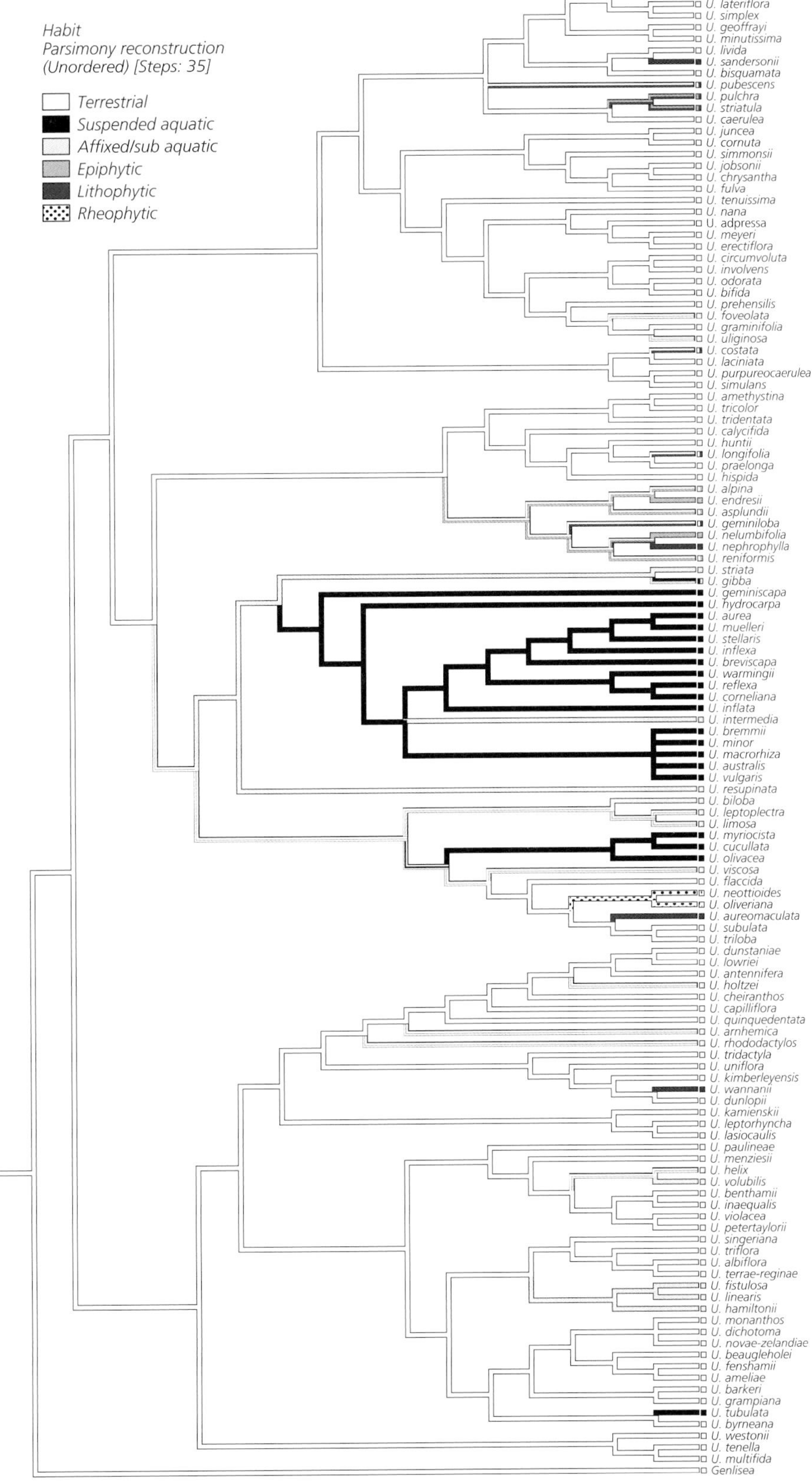

Figure 8.2 Cladogram of the strict consensus tree showing parsimoniously optimized analyses based on habitat and life form. Equivocal branches indicate parsimoniously optimized plesiomorphic characters (modified from Jobson et al. 2003)

of *Pleiochasia* possess stolons, and Taylor (1989) postulated that presence of stolons in most species is a derived feature, and that the rosette-forming species are ancestral. However, Jobson et al. (2017) have shown that the lack of stolons in members of *U.* subg. *Polypompholyx* is just as likely to be a reversal to a more ancestral state.

Rhizoids resemble true roots but they lack typical root structure and appear to function only to anchor the plants in wet substrates. Although true functional roots are present in all *Pinguicula* species, they are aborted during early embryogenesis in all *Genlisea* and *Utricularia* species (Lloyd 1942, Juniper et al. 1989). Rhizoids are produced from the base of the peduncle or nodes of the stolons, but they are reduced in *U.* sects. *Vesiculina* and *Utricularia*. They are usually capillary, simple or branched, and, in members of *U.* sect. *Pleiochasia*, helically twisted. In a few lithophytic and rheophytic species, the undersurface of the rhizoid possesses trichomes modified for adhering to surfaces (Taylor 1989).

Leaves range from small (≈0.5 mm), linear, elliptic, obovate, and cordate, to the reniform, rounded blades of *U.* sect. *Orchidioides/Iperua* that can reach 65 cm in length (*U. reniformis*) or 10 cm in diameter (*U. nelumbifolia*). In contrast, the bladeless aquatic and rheophytic species possess dichotomous ramifications that resemble the nerves of normal leaves, whereas the suspended aquatic *U. olivacea* and *U. biovularioides* possess leafless stolons. Further, water storage tubers that develop from stolons have evolved multiple times across the genus (Taylor 1989). Gomes Rodriques et al. (2017) found that these important adaptations for water storage have evolved at least twice within the epiphytic/terrestrial *U.* sect. *Orchidioides/Iperua*.

The general lack of distinction between leaves, leaf-like organs, and stolons in *Utricularia* has been discussed in two ways. Taylor (1989) considered leaf structures, stolons, and rhizoids to be independent homologous organs to lend clarity to his taxonomic revision of the genus based on morphology. However, he clearly recognized that this suggestion was inadequate to provide a developmental explanation for observed morphological patterns in the genus. Alternatively, Sattler and Rutishauser (1990), Rutishauser and Isler (2001), and Rutishauser (2016) proposed a continuum ("fuzzy") morphology model, in which organs are considered morphological transformations derived from developmental program amalgamation, rhizoids and stolons are considered stem homologs (Troll and Dietz 1954), and leaf-like organs are considered leaf homologs (Goebel 1891b, Kumazawa 1967). Genomic analysis of *Utricularia* (Chapter 11) will enable future investigation into the evolution of genetic programs involved in the developmental pathways that have led to its unusual morphologies (Ibarra-Laclette et al. 2013).

8.3.3 Bladder-trap morphology

All *Utricularia* species have modified leaves that form bladder traps that must be submerged in or surrounded by water to capture prey (Lloyd 1942, Juniper et al. 1989; Chapters 14, 19). The traps actively pump internal fluid to the exterior when the trap is reset (Juniper et al. 1989, Płachno et al. 2015a; Chapters 13, 14). The traps themselves are hollow, ovoid, or globose modified leaves that are 0.2–12 mm in length and develop in a diverse range of positions on, and attached by a stalk to, the plant (Figure 8.3). The trap entrance ("mouth") is described as basal when positioned adjacent to the stalk, terminal when the mouth is opposite the stalk, and lateral when intermediate on the ventral side of the trap (Taylor 1989).

The homologous traps of both *Utricularia* and *Genlisea* develop from an inward-rolling of the adaxial leaf surface with subsequent marginal fusion ("epiascidiate" leaves). In *Genlisea*, traps form from cylindrical primordia with an invaginated tip, whereas in *Utricularia* a spherical invagination forms the primordial trap (Lloyd 1942, Juniper et al. 1989, Albert et al. 1992, Reut 1993). The large variation found in its external morphology (Meierhofer 1902, Lloyd 1942, Taylor 1989) is generally monophyletic among sections in all three subgenera (Jobson and Albert 2002, Albert et al. 2010; Figure 8.3).

In *U.* subg. *Polypompholyx*, traps arise only from the base of the peduncle in all species of *U.* sects. *Polypompholyx*, *Tridentaria*, and some members of *Pleiochasia*. In other members of *U.* sect. *Pleiochasia* they also arise from stolon nodes, occasionally

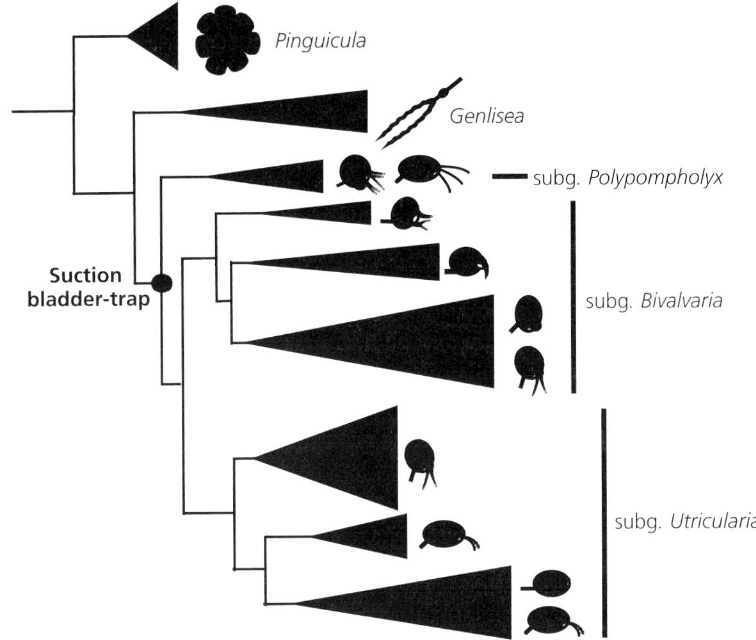

Figure 8.3 Diagram representing the Lentibulariaceae phylogeny (modified from Albert et al. 2010), with schematics of trap types per clade showing rosettes of sticky leaves for *Pinguicula*, passive corkscrew traps of *Genlisea*, and active suction bladder traps of *Utricularia*. Trap forms across *Utricularia* generally are lineage-specific. Branch lengths inferred from nucleotide substitutions are ± relative to one another; collapsed clade sizes represent the taxonomic sampling of Jobson and Albert (2002).

at internodes, and only from whorl nodes in suspended aquatic *U. tubulata*. Traps are positioned only terminally on leaves in affixed aquatic sister species *U. volubilis* and *U. helix*, both of which have dimorphic trap appendage forms specialized either for substrate or water-column positions (Taylor 1989, Jobson et al. 2017).

Trap structures in *U.* subg. *Polypompholyx* are diverse, but in general consist of a usually simple, bifid, or trifid dorsal appendage (rostrum), two lateral setiform or wing-like appendages on either side of the mouth, and a pair of ventral lamella-like wings. Any or all of these may be polymorphic, reduced, or absent, depending on species (Taylor 1989; Figure 8.3). Taylor (1989) differentiated *U.* sect. *Polypompholyx* from *U.* sect. *Tridentaria* based on the presence of a bifid or trifid dorsal rostrum, respectively. Appendage types have arisen multiple times across *U.* subg. *Polypompholyx* and generally correspond with aquatic or terrestrial habits. Reut and Jobson (2010) found that the evolution of filiform antennae appendages around the mouth was associated with the shift from the terrestrial to affixed aquatic habit.

In the epiphytic sections of *U.* subg. *Utricularia* (*U.* sects. *Orchidioides, Iperua*) and *U.* subg. *Bivalvaria* (*U.* sect. *Phyllaria*) with well-developed aerial leaves, traps develop on stolons, rhizoids, and the base of the peduncle, but not on leaves, probably because leaves are rarely submerged in water. In the remaining terrestrial, lithophytic, and epiphytic species, traps can develop not only on stolons, rhizoids, and at the base of the peduncle, but also laterally on the leaf lamina (Taylor 1989). Appendage types include a single dorsal rostrum in *U.* sect. *Calpidisca, Nigrescentes, Meionula, Phyllaria*, and some species in *U.* sect. *Aranella*. In *U.* sects. *Candollea, Martinia*, and several species of *Calpidisca*, an additional appendage is a bifid or gland-covered ventral rostrum (Taylor 1989; Figure 8.3).

In several species of *U.* sect. *Oligocista*, the ventral rostrum is reduced to a bulge-like protrusion. In the closely related *U.* sects. *Enskide, Minutae*, and *Stomoisia*, and in the monotypic *U.* sects. *Benjaminia* and *Sprucea*, traps are mostly naked with a slight dorsal bulge. Members of *U.* sects. *Oligocista, Orchidioides, Iperua, Foliosa, Psyllosperma, Chelidon, Kamienskia*, and *Oliveria* bear two, often simple, sometimes gland-covered dorsal appendages (Taylor 1989)—a character found to be polyphyletic (Jobson and Albert 2002). In the mostly affixed aquatic *U.* sects. *Nelipus, Lecticula, Setiscapella*, and *Steyermarkia*, the two dorsal appendages are beset with setae, while in the suspended aquatic *U.* sect. *Utricularia* these appendages are often branched setiform antennae. In the suspended aquatic section *Vesiculina*, traps are borne terminally on whorled leaves, with a single ventral appendage or completely naked (Taylor 1989, Jobson and Albert 2002; Figure 8.3). In three species placed within *U.* sect. *Utricularia* (Taylor 1989), traps resemble those of other members of the section but they are attached to leafless stolons (e.g., *U. olivacea*; Figure 8.3).

The internal surfaces of the bladder trap are adorned with two main internal gland types; two-armed trichomes usually located near the threshold at the entranceway, and four-armed quadrifid trichomes scattered on most of the inner wall surface. Both gland types are highly variable between sections and also vary in shape and length of arms between species. Although Taylor (1989) does not use these structures as strong differentiating characters, he does provide a description for most species in the genus. The structures rarely have been used for taxonomy, with the exception of Thor (1988 and previous work cited therein) and Płachno and Adamec (2007), who have used the length and shape of quadrifid arms as a character in combination with others to differentiate between two or more species. Future work is needed to assess the utility of these trichomes for taxonomic characterization across the genus.

8.3.4 Bladder-trap evolution

Most species have well developed trigger hairs on the outer surface of the trapdoor that act to release the door during firing (Juniper et al. 1989, Reifenrath et al. 2006, Płachno et al. 2015a; Chapter 14). The traps of species within *U.* sect. *Polypompholyx* often are considered ancestral because of their inflated stalk and the structure of the shortly bifid dorsal appendage that folds over the trap mouth forming a funnel-like entrance-way (Taylor 1989, Reifenrath et al. 2006). Their traps also are thickened with four-layered walls, and the trigger hairs either are highly reduced (Taylor 1989) or not present (Reifenrath et al. 2006). This suggests that there should be a corresponding absence of "active" bladder-trap function in these species and that their traps should function more like the passive eel traps of species in the sister genus *Genlisea* (Chapters 7, 15). However, Lloyd (1942) found that these bladders functioned actively, like others in the genus. The "passive" hypothesis is further belied by the lack of a funnel-like entrance-way and presence of highly developed trigger hairs in the sister species *U. westonii* (*U.* sect. *Tridentaria*). This scenario suggests that seemingly ancestral traps of *U.* sect. *Polypompholyx* may be derived within the lineage (Jobson et al. 2017). Traps within *U.* sect. *Pleiochasia* also are active; studies of *U. monanthos* report typical suction action (Sydenham and Findlay 1973, Juniper et al. 1989). Finally, Płachno et al. (2015a) found that the number of cell-wall layers, which varies from two to five across *U.* subg. *Polypompholyx*, and from two to four layers within the same species complex, has no obvious relationship with efficiency of trap firing.

Jobson et al. (2004) hypothesized that the metabolically expensive bladders led to adaptive innovations during radiation of the genus. For example, hypothesized adaptive advantages of different trap appendage forms include: exclusion of soil particles from around the mouth by overhanging wings (mainly in terrestrial species; Lloyd 1942, Taylor 1989); support for functionally vital surface films of water at the trap mouth by thickly set dorsal wings in some epiphytic species (Taylor 1989); and formation of funnels that may guide potential prey organisms to the trap door (in suspended aquatic taxa; Darwin 1875, Lloyd 1942, Meyers and Strickler 1979, Juniper et al. 1989, Taylor 1989). Appendage variation across the genus may have permitted specialization on particular prey by different *Utricularia* species in both aquatic and terrestrial habitats (Harms 1999, Jobson and Morris 2001, Guisande et al. 2004; Chapter 21). Finally, positive selection for molecular changes in the mitochondrial subunit 1 of cytochrome *c* oxidase have been implicated in the altered energetics of trap resetting (Jobson et al. 2004, Laakkonen et al. 2006, Albert et al. 2010; Chapters 11, 14),

8.3.5 Inflorescences, flowers, and pollen

Lentibulariaceae have zygomorphic flowers with persistent calyces that form an upper and lower lip, a sympetalous corolla tube, a basally spurred or saccate lower lip, two anthers, four sporangia and one or two confluent thecae, two fused carpels, and a unilocular ovary (Casper 1966, Fromm-Trinta 1981, Taylor 1989; Chapters 6, 7). *Pinguicula* have single-flowered scapes, but *Genlisea* and *Utricularia* have synapomorphic racemes of few to many flowers (Casper 1966, Fromm-Trinta 1981, Taylor 1989, Jobson et al. 2003). Within *Utricularia U.* subg. *Polypompholyx*, shifts in habitat and life-form tend to correspond with shifts in inflorescence size and corolla color: five independent shifts from terrestrial to affixed and suspended aquatic life-forms that correspond with shifts from small to medium/large inflorescences. Two of these shifts also correspond with a shift from purple to white corolla color, while for the only suspended aquatic species in the subgenus (*U. tubulata*) the shift is to a pale pink corolla.

Inflorescences. Along with racemes has come an array of different arrangements of bracts, bracteoles, and scales (sterile bracts) upon the peduncle (Taylor 1989). Bracts are present in all species of *Utricularia*, and adjacent bracteoles are present in all sections excepting *U.* sects. *Sprucea, Avesicaria, Mirabiles, Steyermarkia, Lecticula, Setiscapella, Nelipus, Utricularia,* and *Vesiculina* (Taylor 1989). Scales are present in all species except those circumscribed to *U.* subg. *Polypompholyx*, to *U.* sects. *Minutae, Vesiculina* (including *U. olivacea*), and to several species of *U.* sects. *Utricularia* and *Lecticula* (*U. resupinata, U. aurea, U. geminiscapa,* and *U. inflata*) (Taylor 1989). Several suspended aquatic species within *U.* sect. *Utricularia* possess whorled flotation organs at various positions on the peduncle that maintain the aerial position of the inflorescence (Taylor 1989). The floats seem to have evolved only once, with reversal to non-float-bearing peduncles in at least two species (Jobson and Baleeiro, *unpublished data*).

Flower structure. The calyx of *U.* sect. *Polypompholyx* and *Tridentaria* has four sepals, whereas that of *U.* subg. *Utricularia* has two sepals. Taylor (1989) hypothesized that the four-parted calyx of *U.* subg./*U.* sect. *Polypompholyx* may be evolutionarily intermediate between *Genlisea* and *U.* subg. *Utricularia*. However, an optimization showing the earliest node as having two sepals (Jobson et al. 2003), and the lateral sepals forming an epicalyx in *U.* sect. *Polypompholyx* and thus not being homologous to those of *Genlisea*/*Pinguicula*, suggest that the four-parted calyx is an independent apomorphy. During the initial developmental stages of the floral bud of *U.* subg. *Utricularia* and *Bivalvaria*, five sepals are evident; the anterior part has two sepals and the posterior has three. These sets each fuse, forming two sepals later in development (Lang 1901).

Floral color in subgenus *Polypompholyx*. Jobson et al. (2017) used a parsimony-based ancestral state reconstruction to examine shifts in habit that correspond with the evolution of corolla lower lip color and inflorescence size. The earliest node shows a split between pink flowers in *U.* sects. *Polypompholyx* + *Tridentaria*, and a combined character state containing various shades of purple as ancestral for *U.* sect. *Pleiochasia*. Within *U.* sect. *Pleiochasia*, Jobson et al. (2017) found that a red corolla has arisen once (in *U. menziesii*) from a violet ancestor, a trait that is accompanied by a corolla spur 2–3× longer that the lower lip (Taylor 1989), whereas a pink corolla has arisen twice independently within *U.* sect. *Pleiochasia* and *Lasiocaules* from purple ancestors. A white corolla has arisen six times mainly from purple ancestral nodes, whereas the cream-colored corolla in *U. holtzei* and the pink of its sister species *U. cheiranthos* evolved from an apricot ancestor. Also from a purple ancestor, a white corolla evolved in the lithophytic *U. wannanii* and an apricot-colored corolla evolved in its terrestrial sister species, *U. dunlopii* (Reut and Jobson 2010, Jobson et al. 2017). Along with a shift to an apricot-colored corolla and from very small to minute inflorescences, Reut and Jobson (2010) also showed that filiform corolla appendages evolved independently three times: twice from the upper lip lobe and once from the lower lip lobe.

Pollen. Lobreau-Callen et al. (1999) described nine pollen types for *Utricularia* mainly based on aperture number and features of the exine and that, with few exceptions, is associated more with the phylogeny than habitat specialization. Morphological variation of pollen within *Utricularia* is complex, although the exines of most species within the genus are generally smooth or rugose. An exception is the rheophytic *U.* sect. *Avesicaria* (*U. neottioides, U. oliveriana*), that have an exine covered with micro-spines (Sohma 1975, Taylor 1989, Lobreau-Callen et al. 1999). A general trend in lower aperture number, between three and eight, is seen in the mostly terrestrial *U.* subg. *Polypompholyx* and *Bivalvaria*. In contrast, aperture number ranges from 8 to 23 within *U.* subg. *Utricularia*, except for *U.* sect. *Orchidioides* with 3–4 (Thanikaimoni 1966, Huynh 1968, Sohma 1975, Taylor 1989, Lobreau-Callen et al. 1999). Aperture number also appears to vary widely within a species (e.g., 8–15 in the *U. amethystina* complex), but this may reflect cryptic taxonomic diversity rather than intraspecific variation (Baleeiro et al. *unpublished data*).

8.3.6 Cytology

Chromosome numbers have been determined for 28 species. Values do not follow any clear pattern apart from the prevalence of low, seemingly aneuploid counts of n = 5, 7, 9, 10, 11, or 14 within *U.* subg. *Polypompholyx* and *Bivalvaria*, excepting the terrestrial *U. livida* (n = 18) and *U. caerulea* (n = 18/20). Counts for the suspended aquatic *U.* sect. *Utricularia* are usually greater (e.g., n = 9 to 22; Taylor 1989), and although it is evident that polyploidy has occurred during the evolution of the genus it has been hypothesized that extant species rarely hybridize (Kondo 1972a, Jérémie and Jeune 1985, Taylor 1989, Raynal-Roques and Jérémie 2005). However, at least some temperate species in *U.* sect. *Utricularia* (*U. australis, U. bremii, U. ochroleuca, U. stygia*) are evidently (Kameyama and Ohara 2006) or probably (Taylor 1989) derived from hybrid origins.

8.3.7 Fruits and seeds: structure and dispersal

The range of capsule dehiscence and seed types appear to correspond with both habit and phylogeny (Jobson et al. 2003). Fruits from the earliest node of the *Utricularia* clade dehisce from a single ventral longitudinal suture that varies in length and thickness (Jobson et al. 2003). Dehiscence types involving both dorsal and ventral slits occur in the terrestrial *U.* sects. *Tridentaria* and *Australes*, and several species in *U.* sect. *Oligocista* (Taylor 1989). A deviation from the dorsi-ventral orientation of the suture occurs in *U.* sect. *Foliosa sensu* Taylor (1989) where the entire capsule is dorsi-ventrally bivalvate (Taylor 1989, Jobson et al. 2003).

Most species with this ancestral dehiscence type are terrestrial (Figure 8.2), and it may be that all other *Utricularia* dehiscence types are derived from it (Jobson et al. 2003). Considering that fruits of both *Pinguicula* and *Genlisea* have more complex, laterally bivalvate or multi-sutured dehiscence (Casper 1966, Taylor 1989), respectively, it seems that dehiscence in *Utricularia* has become simplified. The earliest node of the *Utricularia* clade has seeds that are globose/ovoid in shape, mostly with reticulated testa and raised anticlinal walls (Jobson et al. 2003). Such a surface is similar to those found in *Genlisea* (Robins and Subramanyam 1980, Taylor 1989) and may allow seeds to remain afloat. This flotation may aid dispersal via water birds (Taylor 1989).

Although seed morphology varies across the genus, in most sections they are small or very small (0.2–1 mm long), ovoid to cylindrical, and subglobose with reticulated testa (Taylor 1989). Members of *U.* sects. *Phyllaria* and *Iperua* have seed coats with multicellular outgrowths that may be a specialization for host attachment, whereas those of *U.* sect. *Orchidioides* are cylindrical or fusiform in shape with smooth coats that may be associated with wind dispersal (Robins and Subramanyam 1980, Taylor 1989). Dehiscence types, seed shapes, and seed surfaces are similar in *U.* sects. *Avesicaria* and *Avesicarioides*, which have seeds covered in mucilage that presumably aids dispersal in fast-flowing streams (Robins and Subramanyam 1980, Taylor 1989). Similar seeds are found among other Neotropical rheophytes in *U.* sects. *Choristothecae* and *Mirabiles*. The affinity of these two sections with *U.* sects. *Avesicaria* and *Setiscapella* currently is unknown, so it is uncertain whether or not the habit and seed type has evolved twice or more.

In the mostly suspended aquatic clades of *U.* sect. *Utricularia* (Figure 8.1), equatorially circumscissile dehiscence predominates (Jobson et al. 2003); the few exceptions (e.g., *U. gibba*) have laterally bivalvate dehiscence (Taylor 1989, Jobson et al. 2003). In circumscissile dehiscence, the equatorial plane of the capsule is separated, presumably giving greater space for dispersal of the large and usually winged prismatic, lenticular seeds; seeds of several species also have surface appendages (Taylor 1989). The seeds of *U.* sect. *Utricularia* have a diverse array of shapes ranging from globose/ovoid, prismatic/conical, to discoid/lenticular, with many species having multicellular outgrowths such as wings (e.g. *U. gibba*) (Robins and Subramanyam 1980, Taylor 1989).

In contrast, the sister of *U.* sect. *Utricularia*, the subaquatic *U.* sect. *Lecticula*, dehisces via a longitudinal slit that is similar to that of *U.* sects. *Nelipus* and *Setiscapella*; the seeds in these sections range in shape from prismatic and globose to ellipsoid (Taylor 1989, Jobson et al. 2003). In the terrestrial/subaquatic sections *Nelipus* and *Setiscapella* and suspended aquatic section *Vesiculina* (Figures 8.1, 8.2), dehiscence is mostly from a single ventral longitudinal slit (Taylor 1989), and seeds are generally ovoid to globose in shape with varying forms of testa ornamentation, including multicellular outgrowths (Taylor 1989).

Indehiscence, a situation in which the seeds are released upon breakdown of the capsule walls, is found in suspended aquatic species circumscribed to *U.* sect. *Utricularia* (*U. biovularioides*, *U. olivacea*, and *U. naviculata*) that produce a single or few ovoid and smooth seeds that are adnate to the capsule wall (Robins and Subramanyam 1980, Taylor 1989). Indehiscence is also found in *U. tubulata* (*U.* sect. *Pleiochasia*) and involves release of a few seeds with elongated papilla-like testa cells (Taylor 1989).

8.4 Population dynamics

8.4.1 Population genetics

Population genetic studies are scant, and are so far restricted to suspended/affixed aquatic species. In a Japanese form of *U. australis* pollinators have not been observed, and the stigmatic surface and anthers are adjacent to each other at dehiscence, possibly promoting self-fertilization (Yamamoto and Kadono 1990, Khosla et al. 1998, Araki and Kadono 2003). However, the virtually complete absence of outcrossing, and to a lesser extent inbreeding, is a barrier to gene flow (Yamamoto and Kadono 1990). In different populations of the same species, Araki and Kadono (2003) found that clonal dominance predominated and seed contributed little to population genetic structure. Using isozymes, Kameyama and Ohara (2006) found no within-population genetic variability and they hypothesized that sterility may have resulted from a prior hybridization event.

8.4.2 Pollination

Lineages within derived clades have evolved highly specialized floral structures such as functional nectar discs, probable vector-specific spurs, and floral sizes and shapes that tend to correlate with vector-specific color and scent (Taylor 1989, Hobbhahn et al. 2006). Taylor (1989) noted visits by various insects and hummingbirds to *Utricularia* flowers. Visitation by insects was observed for three Indian species of *U.* sect. *Oligocista* (Hobbhahn et al. 2006) and the Neotropical terrestrial/epiphytic species *U. reniformis* (*U.* sect. *Iperua*; Clivati et al. 2014). Highly modified corolla lobes have evolved multiple times in *U.* sect. *Pleiochasia* that likely were driven by pollinator specialization (Reut and Jobson 2010, Płachno et al. 2015b). In addition, Gloßner (1992) described ultraviolet patterns presumed to guide vectors in the flowers of at least one species of *Utricularia*.

In a study of the sympatric terrestrial sister species *U. cornuta* and *U. juncea* (*U.* sect. *Stomoisia*), Kondo (1972a) observed a reduction in floral size in *U. juncea* that may have led to a high incidence of cleistogamy. This was hypothesized to have resulted from a lack of selective pressure on floral form because of the absence of pollinators and predominance of autogamy among populations (Kondo 1972a). Although Clivati et al. (2014) did show evidence of pollinator interaction and possession of a sensitive stigma to avoid self-pollination, frequency of visitation was low, which may affect gene flow and lead to a loss of genetic diversity. This result supports the observations of Jérémie and Jeune (1985) who used morphometric analyses and observations on pollination vectors within populations of the Neotropical terrestrial/epiphytic *U. alpina* (*U.* sect. *Orchidioides*). They found little evidence of insect pollination and strong variation between plants occupying terrestrial versus epiphytic microhabitats and they hypothesized that sympatric isolation was occurring between these subpopulations. Given the highly specialized floral structures across the lineage, autogamy within populations of most studied *Utricularia* species is probably a relatively recent evolutionary development (Jobson and Albert 2002). It also seems likely that the wide variation in corolla color, size, and shape (§8.3.5), the majority of which is phylogenetically independent, indicates strong pollinator selection pressure across the lineage. Further work is required to understand more fully pollination systems across the genus (Guisande et al. 2007; Chapter 22).

8.4.3 Clonal growth

Many of the aquatic *Utricularia* species readily reproduce asexually via excised vegetative segments and turions that could become highly mobile in the varied aquatic habitats, possibly transported to new sites by water birds (Taylor 1989). From the scant data on gene flow it appears that *Utricularia* fits an "island" model in which little migration occurs. When it does, it consists mostly of migration of asexual (vegetative) clones that likely have little effect on gene frequencies in already established populations (Ellison et al. 2003).

8.5 Contemporary biogeography and phylogeography

8.5.1 Global patterns of diversity

The genus *Utricularia* has a circumglobal distribution (Figure 8.4), although no species grow at the poles (Taylor 1989), on most oceanic islands, or in all arid regions with exception of the mound spring habitats of Australia's Great Artesian Basin (Jobson 2013). Species richness is higher in the southern hemisphere, with maximum species and sectional diversity occurring in the Neotropical realm (Figure 8.4).

The number of neighboring realms sharing species is very low, indicating that many are endemic (range size <10,000 km^2; Figure 8.4). The percentage of endemic species is highest in the Indomalayan realm (52.8%), followed by the Afrotropics (33.3%) and Neotropics (26.8%). Only a few species have wide distributions: *U. australis* (Palaearctic, Indomalaya, Australasia, Afrotropics), and *U. minor* (Palaearctic, Indomalaya, Australasia, and Nearctic realms) are present in four realms, whereas

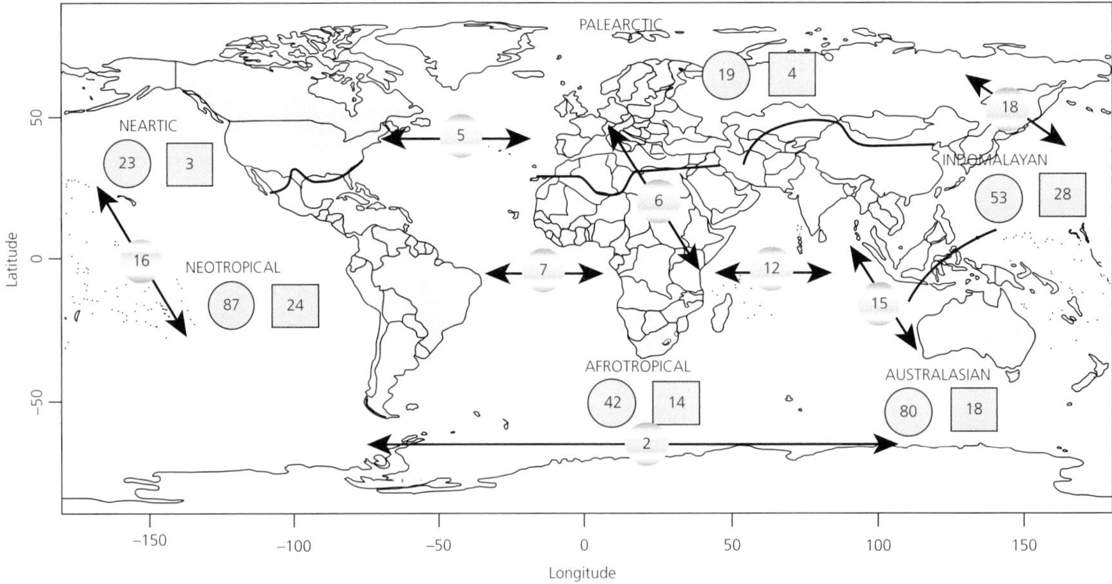

Figure 8.4 Global distribution of *Utricularia*. Black lines represent the border of each biogeographic realm. Number of species for each biogeographic realm are inside circles, number of endemic species (extent restricted < 10,000 km^2) are inside squares, and number of species shared between neighboring areas are inside shaded circles with arrows. The grey pattern shows the distribution around the world estimated by applying kernel distribution to a dataset of 28,773 records obtained from different sources (Guisande et al. 2004, ALA 2016, GBIF 2016).

U. subulata and *U. gibba* are found in at least five realms. In all non-neighboring realms, the number of shared species is low. For example, the Neotropical and Australasian realms share only the two subcosmopolitan species (Figure 8.4).

Terrestrial species are the dominant group in the southern hemisphere and in the Indomalayan realm (Guisande et al. 2007). However, in the northern hemisphere the richness of affixed and suspended aquatic species is very similar to that of the terrestrial species, and in the western Palaearctic there are fewer terrestrial than aquatic species (Guisande et al. 2007). In the Afrotropics and Neotropics, the proportion of the different groups is similar except for the Neotropical endemic bromeliad-tank epiphytes (Guisande et al. 2007).

8.5.2 Phylogeography

The geographical origin of *Utricularia* is unknown, but morphological and phylogenetic studies indicate a probable Neotropical origin of the genus (Taylor 1989, Jobson et al. 2003, Müller and Borsch 2005), with further dispersion, initially to the Afrotropics and Australasia (Figures 8.4, 8.5). The terrestrial habit has been inferred to be plesiomorphic (§8.3.1), and the proportion of terrestrial species is clearly higher in the Neotropics (Guisande et al. 2007), supporting this hypothesized place of origin. In the northern hemisphere, there is a higher proportion of aquatic species, which corresponds with the probable evolutionarily derived condition (Guisande et al. 2007).

In the Neotropics, sequences of nuclear rDNA (ITS region) revealed that haplotypes of five populations of the pantropical affixed aquatic species *U. gibba* were shared between Northern Brazil and Cuba, suggesting a recent dispersal from South to Central America (Marulanda et al. *unpublished data*). Jobson et al. (*unpublished data*) used nuclear and plastid markers across 270 populations of the morphologically variable *U. dichotoma* complex to study its phylogeographic patterns across Australia (all states except Northern Territory) and the New Zealand distribution. Independent clades representing five species (Gassin 1993, Jobson 2013) were resolved for what previously had been thought to be habitat-specific variants of *U. dichotoma* occupying deep coastal wallum swamps, shallow ephemeral pans, arid discharge and recharge mound springs, montane hanging swamps, alpine meadows, wet heathlands, and high elevation soakages (Reut and Fineran 2000).

Future population genetic and phylogeographic studies are now made more tractable by the recent development of broad spectrum microsatellite primers applicable to *Utricularia* (Clivati et al. 2012), and the sequencing of the full chloroplast genomes of *U. foliosa* and *U. reniformis* (Silva et al. 2016, 2017). In addition, the nuclear genome of *U. gibba* could provide a framework for high-throughput genome-wide genotyping in other species (Chapter 11).

8.5.3 Diversification and molecular rate acceleration

The *Utricularia/Genlisea* clade is substantially more species-rich and morphologically divergent than its sister group, *Pinguicula* (Chapter 6). Sampling across major clades using genes from all three genomic compartments, Jobson and Albert (2002) found that *Utricularia/Genlisea* genomes evolve significantly faster than those of *Pinguicula* (Figure 8.3). In an attempt to explain this heterogeneity in rates of genome evolution, Jobson and Albert (2002) first tested a generation time effect, hypothesizing that diversification rates were associated with life-history differences (annual vs. perennial). They found no significant association between life history and diversification rate, and instead suggested that the data could be explained better by the "speciation rate hypothesis," which postulates a relationship between increased nucleotide substitution rate and increased cladogenesis (Jobson and Albert 2002). They further suggested that flexible vegetative development in *Utricularia* could have been a key innovation that reduced selection pressure, enabling the invasion of underutilized nutrient-poor niches.

Using a broader taxonomic comparison across 292 angiosperm genera, including representative sequences from most other carnivorous plant lineages, Müller et al. (2004) found molecular rates in the chloroplast *matK/trnK* gene to be highest for the *Utricularia/Genlisea* lineage. Müller et al. (2004) postulated that this was caused by a positive feedback between more abundantly available prey-derived

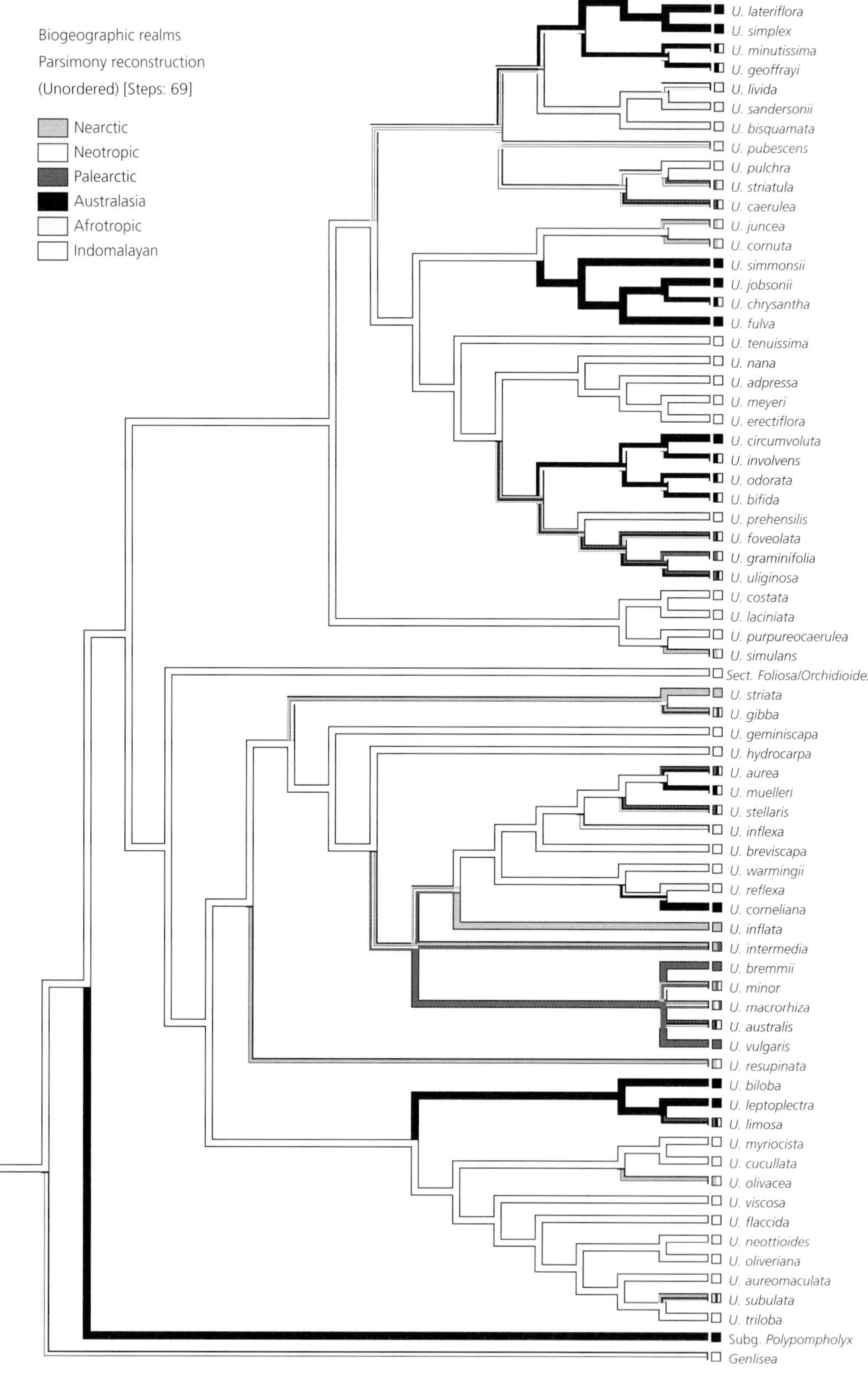

Figure 8.5 Cladograms of the strict consensus tree showing parsimoniously optimized analyses of biogeographic area. Equivocal branches indicate parsimoniously optimized plesiomorphic areas (modified from Jobson et al. 2003).

biosynthetic building blocks and morphological evolution, recently termed the "predictable prey capture hypothesis" (Ellison and Gotelli 2009; Chapter 11).

8.5.4 Diversification time and biogeographic shift in subgenus *Polypompholyx*

Although there is a lack of reliable fossil data for Lentibulariaceae, recent molecular divergence dating by Bell et al. (2010), based on fossil calibration points from across angiosperms, provides an estimated divergence time for most families. Using these data, Jobson et al. (2017) estimated that the split between *U.* subg. *Polypompholyx* and *Bivalvaria* + *Utricularia* occurred ≈31 million years ago (Mya), with the subsequent split between *U.* sect. *Polypompholyx* and *U.* sects. *Pleiochasia* + *Lasiocaules* occurring ≈15 Mya. This suggests that the *U.* subg. *Polypompholyx* ancestral lineage was present in Australia for ≈15 Mya prior to the mid-Miocene when its warm and wet inland regions began a process of aridification (Byrne et al. 2011).

The molecular dating and vicariance-dispersal analyses of Jobson et al. (2017) reveal that *U.* subg. *Polypompholyx* underwent a major vicariance event ≈15 Mya between southwest Western Australia (SW) and northwest Western Australia/Northern Territory (NW) coincident with aridification of drainages across the intermediate linkage Pilbara-Gascoyne (GP) and North Western Plateau (NWP) regions (Figure 8.6). At around this same time, a coastal incursion of the far western edge of GP + NWP also occurred, with subsequent drying of the surrounding region (Byrne et al. 2011).

The incursion was a likely dispersal barrier until it receded to the current coastline ≈10 Mya, after which time it may have provided suitable oligotrophic swampy habitats until the GP + NWP drainages dried up ≈6 Mya (Byrne et al. 2011). This process of loss (15–11 Mya), gain (11–6 Mya), and

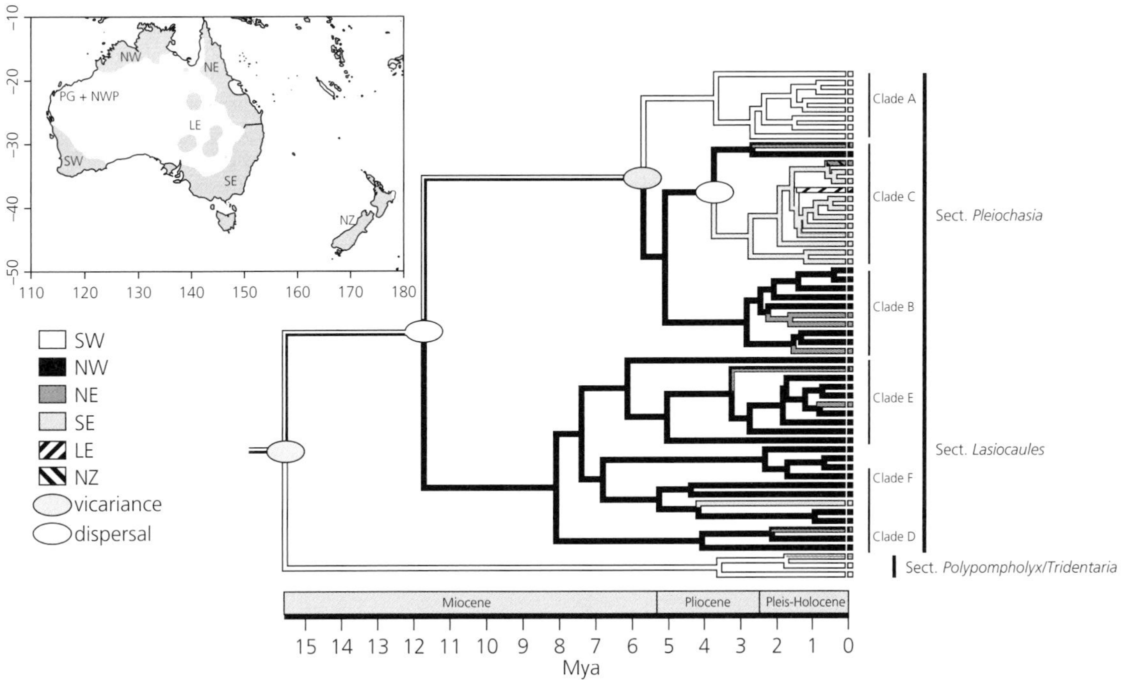

Figure 8.6 Ancestral state reconstruction for six biogeographic areas of the subgenus *Polypompholyx* distribution. Regions optimized to nodes shown on Australian/New Zealand map (southwest = SW, northwest = NW, northeast = NE, southeast = SE, Lake Eyre basin = LE, New Zealand = NZ). Clades A–F are shown behind thin lines and sections are behind thick lines. Ovals at key nodes indicate vicariance (gray) or dispersal (white) events as predicted from S-DIVA analysis, with geological epoch at base of tree. Modified from Jobson et al. (2017) with permission.

loss (6 Mya–present) of linkage habitats between SW and NW corresponds with the first vicariance event at the establishment of the two major clades (*U.* sects. *Polypompholyx* + *Tridentaria* and *Pleiochasia* + *Lasiocaules*), a subsequent dispersal event NW to SW at ≈11 Mya, and a second major vicariance event (*U.* sects. *Pleiochasia* and *Lasiocaules*) between SW and NW at ≈6 Mya that led to the diversification of a second SW clade within *U.* sect. *Pleiochasia* (clade A of Figures 8.1, 8.6).

Jobson et al. (2017) also found that between six and one Mya, several subsequent dispersal events occurred from either the NW or northeast (NE) regions to the southeast (SE) region. They show that dispersal of species from the SW region to the SE, and SE to the mound spring habitats of the Lake Eyre Basin (LE) and New Zealand occurred 4–1 Mya. Each of the two SW clades contain a species with disjunct distributions in SE (*U. violacea* and *U. tenella*), predicted to have dispersed across the Nullarbor Plain (NP) from SW to SE ≈2 Mya (Figure 8.6). This pattern corresponds with the geological history of the NP with a reversion to wet climates during the mid-Pliocene ≈3.5 Mya. At this time there possibly were suitable oligotrophic swampy habitats that persisted until a period of extreme cooling and drying during the Pleistocene 2–1 Mya (Byrne et al. 2011), resulting in the isolation of populations either side of the region (Figure 8.6).

8.6 Conservation issues

From the estimated > 230 species of *Utricularia*, there is information only for 29 species in the IUCN Red List (IUCN 2016); almost all of them are categorized as "unknown." Despite the restrictive criterion used for considering endemic species—only those species restricted to a particular region and whose range size <10,000 km^2—there are 69 species (≈30%) of *Utricularia* that are considered endemic. This percentage of endemic species is very high relative to other plant groups (Joppa et al. 2013). Endemic species could tend to have lower reproductive effort and dispersal capacity and be more vulnerable to disturbances or habitat changes than more widespread species (Brook et al. 2008, Pelayo-Villamil et al. 2015). As small geographical range sizes also increase extinction probabilities (Gaston 1994), the high proportion of endemic species in *Utricularia* potentially means a high proportion of vulnerable species in this genus.

8.7 Future research

Although much progress has been made in resolving phylogenetic relationships within *Utricularia*, the inclusion and phylogenetic resolution of species representing sections that are currently missing or poorly sampled are needed. These mostly involve taxa from the Neotropics, Afrotropics, and Indomalaya. Ideally, these and other new examinations of phylogenetic relationships and phylogeographic patterns should focus on fast evolving genes and take advantage of next-generation sequencing techniques. As full genomes and transcriptomes become available, future investigations into the genetic programs involved in the development, function, and energetics of bladder traps; nutrient uptake and use; and distinctions between leaves and shoots should become possible.

Studies of macroecology, conservation, and biogeography will require a detailed analysis of the taxonomic limits and of the quality of specimen records. These data could help identify the main factors affecting the diversity and distribution of *Utricularia* and the abiotic and biotic factors contributing to its speciation rate. Finally, population-level studies, including population genetics, breeding systems, pollination, and basic population dynamics should focus on narrow endemics and threatened species with an eye toward developing conservation measures for rare *Utricularia*.

CHAPTER 9

Systematics and evolution of Sarraceniaceae

Robert F. C. Naczi

9.1 Introduction

The Western Hemisphere pitcher plants (Sarraceniaceae) include ≈35 species of flowering plants native to portions of North America and northern South America (Appendix). These perennial, herbaceous plants produce tubular, pitcher-shaped leaves ("pitchers"). A suite of adaptations enables pitchers to attract, trap, digest, and absorb nutrients from a variety of animal prey, primarily arthropods. The Sarraceniaceae are morphologically and phylogenetically quite different from the other two pitcher-plant families (Cephalotaceae [Chapter 10] and Nepenthaceae [Chapter 5]).

Since the last global review of Sarraceniaceae evolution (Juniper et al. 1989), field, herbarium, and laboratory research have contributed much new information about diversity, classification, and relationships of the family. This chapter synthesizes the current state of knowledge of Sarraceniaceae systematics and evolution, identifies significant open questions, and recommends future research needed to advance our understanding of the evolution of the Sarraceniaceae.

9.2 Taxonomy

9.2.1 *Darlingtonia*

Darlingtonia includes only *Darlingtonia californica*, native to western Oregon and northern California, USA (Figures 9.1, 9.2). Mellichamp (2009) and McPherson and Schnell (2011) reviewed its classification, morphology, nomenclature, and biogeography. An anthocyanin-free form, *D. californica* f. *viridiflora* B. Rice, is known from only one site, in which a few plants grow with many of the typical form (McPherson and Schnell 2011). No other infraspecific taxa have been named in *Darlingtonia*.

9.2.2 *Heliamphora*

In *Heliamphora* (Figure 9.3) few synonyms or infraspecific taxa exist. Maguire (1978), Steyermark (1984), Berry et al. (2005), and McPherson et al. (2011) comprehensively reviewed its classification, identification, morphology, and biogeography. The most recent review lacks an identification key, making identifications difficult. Nine species have been described since Berry et al. (2005) published a key to the species.

Currently, *Heliamphora* comprises 23 species and one non-autonymic variety. All are endemic to the Guayana Highlands of southern Venezuela and small portions of adjacent Brazil and Guyana (Figure 9.2). Since 2000, 14 new taxa have been described and three of Maguire's (1978) varieties have been raised to species rank (McPherson et al. 2011). According to the current taxonomy, all *Heliamphora* taxa are narrow or extremely narrow endemics (McPherson et al. 2011). For example, *H. ceracea* and *H. uncinata* are known from only one or very few populations. Even the species with the largest geographic distributions—*H. exappendiculata*, *H. heterodoxa*, *H. minor*, *H. neblinae*, *H. pulchella*, and *H. tatei*—have small ranges. Current knowledge of taxonomic diversity in *Heliamphora* likely is incomplete, as extensive portions of the Guayana Highlands are botanically

Naczi, R. F. C., *Systematics and evolution of Sarraceniaceae*. In: *Carnivorous Plants: Physiology, ecology, and evolution*. Edited by Aaron M. Ellison and Lubomír Adamec: Oxford University Press (2018).
 DOI: 10.1093/oso/9780198779841.003.0009

Figure 9.1 (Plate 8 on page P7) *Darlingtonia* morphology, wild-grown plants. (**a**) *D. californica* pitcher and flower, California, USA. (**b**) *D. californica* population in seepage, Oregon, USA. (**c**) *D. californica* plants in fruit, California, USA. (**d**) *D. californica* pitchers in late-afternoon light, Oregon, USA. All photographs by R. Naczi.

unexplored. Furthermore, McPherson et al. (2011) illustrate three potentially new taxa that they did not name because of incomplete knowledge of their morphology.

Past taxonomic controversies in *Heliamphora* (Maguire 1978, Steyermark 1984) have been resolved (McPherson et al. 2011). Both Maguire and Steyermark reached different taxonomic conclusions because of the difficulty of delimiting morphologically plastic taxa, overlap among diagnostic traits, and relatively few specimens. Recent treatments based on more specimens and extensive field observations of living plants use additional characters, including the shape, relative size, and carriage of hoods ("nectar spoons"). By describing several new species as segregates of previously known species, the confusingly broad range of morphologies previously assigned to certain species, e.g., *H. heterodoxa* and *H. neblinae*, has been narrowed and clarified. Maguire (1978), Steyermark (1984), and McPherson et al. (2011) all use morphological characters to diagnose taxa. Thus, their use of the morphological species concept is apparent, albeit not explicitly stated in their treatments.

The taxonomy of McPherson et al. (2011) is as yet untested because of recent taxonomic innovation in *Heliamphora* and the substantial difficulties involved with conducting field work in the Guayana

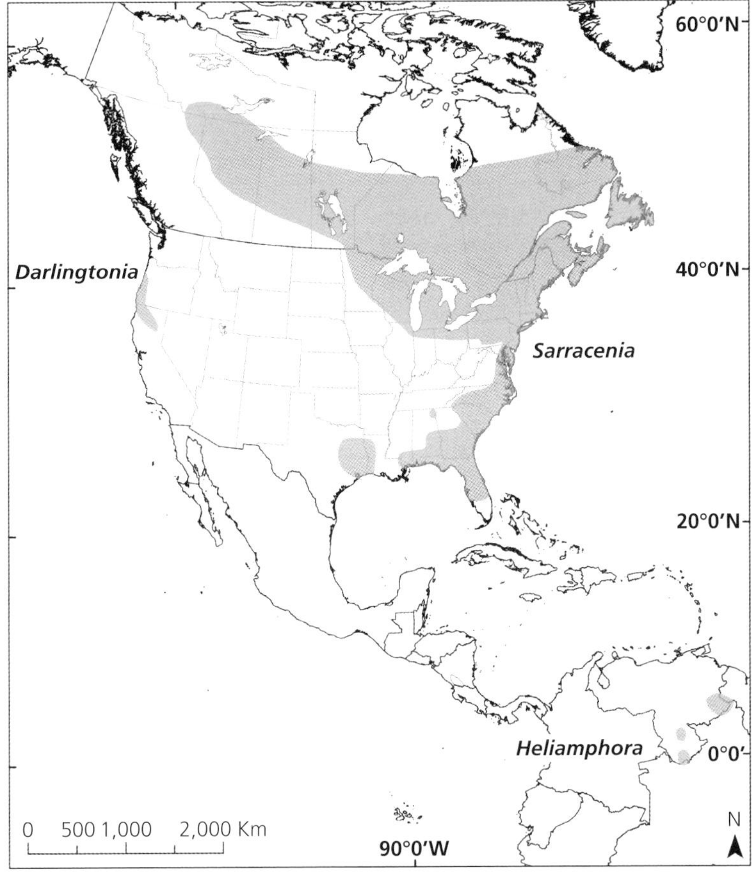

Figure 9.2 Geographic distribution of Sarraceniaceae, native populations only.

Highlands. Although it appears well supported by morphological data, the breadth of morphological variation within taxa, amount of morphological overlap among taxa, and influence of environmental factors on species limits remain incompletely documented and understood. Open questions include: are some species split too finely? Are some recently described species actually hybrids or environmentally induced variants of previously known species? Because independent tests of the current *Heliamphora* classification are lacking and more work on species delimitations is needed, taxonomy of this genus is likely to change.

9.2.3 *Sarracenia*

Sarracenia (Figure 9.4) includes 11 species native to portions of eastern and northern North America (Figure 9.2). Mellichamp and Case (2009) and McPherson and Schnell (2011) reviewed its classification, identification, morphology, nomenclature, and biogeography. Mellichamp and Case (2009) include an identification key.

Some *Sarracenia* species have been known for a long time; *S. purpurea* was known during pre-Linnaean times, and most of the species were described by the early 1900s. Exceptions are *S. alabamensis* and *S. rosea*, segregates of *S. rubra* and *S. purpurea*, respectively, both published within the past 20 years. Two subspecies were named in the 1970s: *S. rubra* ssp. *gulfensis* D.E. Schnell and *S. rubra* ssp. *wherryi* D.E. Schnell. Numerous varieties and forms have been described since the late 1990s (McPherson and Schnell 2011). Most recently, molecular analyses revealed two main, geographically correlated lineages within *Sarracenia alata*, which Carstens

Figure 9.3 (Plate 9 on page P8) *Heliamphora* taxonomic and morphologic diversity, wild-grown plants. (**a**) *H. heterodoxa*, pitchers and inflorescence, Gran Sabana, Venezuela. (**b**) *H. minor* plant, Auyán-tepui, Venezuela. (**c**) *H. minor*, upper portions of two inflorescences, Auyán-tepui, Venezuela. (**d**) *H. minor* population (inflorescences with pale flowers visible above surrounding foliage) in bog, Auyán-tepui, Venezuela. (**e**) *H. elongata*, population in bog, Ilú-tepui, Venezuela. (**f**). *H. elongata*, plant with flowers, Ilú-tepui, Venezuela. All photos by R. Naczi.

and Satler (2013) suggest might be distinct species. Morphological differences correlated with these two subclades are unknown, and no one has yet named a segregate species from *S. alata*. Whether the amount of divergence between lineages of *S. alata* is truly sufficient to designate separate species remains to be determined.

In nearly every species of *Sarracenia*, anthocyanin-free color forms have been discovered and named (McPherson and Schnell 2011). Other than their lack of red pigments, plants of the anthocyanin-free forms are identical to typical plants. In nature these forms are rare, though several are known from multiple localities. In addition, anthocyanin-free individuals co-occur with the much more numerous, anthocyanin-producing, typical forms of the same species. Using standard breeding experiments, Sheridan and Mills (1998) revealed inheritance of the anthocyanin-free expression in *Sarracenia* likely is controlled by a single locus with two alleles, the anthocyanin-free one being recessive.

Several taxa are treated differently by different authors; synonymy is extensive. One point of controversy concerns pitcher color morphs. McPherson and Schnell (2011) accept varieties in several *Sarracenia* species based solely on pitcher color patterns. The appropriate rank for these color morphs, if they should even be named, is form, not variety (Ellison

Figure 9.4 (Plate 10 on page P9) *Sarracenia* taxonomic and morphologic diversity, wild-grown plants. (**a**) *S. oreophila* pitchers and flowers, Alabama, USA. (**b**) *Sarracenia alabamensis* ssp. *wherryi* plants in fruit, Alabama, USA. (**c**) *S. purpurea* pitchers, anthocyanin-free form (left) and typical form (right), Michigan, USA. (**d**) *S. flava* population in wet pineland, Florida, USA. (**e**) *S. rosea* plant, Florida, USA. (**f**) *S. leucophylla* plants (left) syntopic with *S. flava* plants (right) in wet pineland, Florida, USA. All photos by R. Naczi.

et al. 2014; R.F.C. Naczi *unpublished data*). Justifications for the rank of form are that color morphs: differ from typical conspecific plants usually only in color patterns; usually occur in populations with typical color morphs; and usually occur sporadically in different parts of the geographic range of the species, mostly uncorrelated with geography. These are the same criteria for treating anthocyanin-free morphs as forms.

The naming of color morphs as varieties is part of a larger problem by McPherson and Schnell (2011). They use a subjective and intuitive species concept, rather than an objective, data-driven one. For example, McPherson et al. (2011: 27) suggest that "[t]he supposed distinguishing characteristics of *Sarracenia alabamensis*, *S. jonesii*, and *S. rosea* are outweighed by taxonomically more important characteristics that unify these plants to *S. rubra* and *S. purpurea* … " This statement ignores data and analyses demonstrating multiple morphological discontinuities used in concert with the morphological species concept to segregate each of these species from previously described ones (Case and Case 1976, Naczi et al. 1999). Future taxonomic delimitations should be based on formal analysis of all available data.

Remaining taxonomic controversy concerns the classification of various species complexes. Within the *S. purpurea* species complex, three points are controversial—recognition of subspecies within *S. purpurea*, recognition of *S. purpurea* var. *montana*, and recognition of *S. rosea* as a species. The latter two are disagreements about the merit and rank of taxa, but the first is more complex.

The vast geographic range of *S. purpurea*, spanning 30 degrees of latitude and 70 degrees of longitude, is the largest in the family (Naczi et al. 1999, latitude adjusted to include range of only *S. purpurea*, s.s.). Geographically correlated morphological variation would be expected within such a large range; does this variation merit taxonomic recognition? Two subspecies of *S. purpurea* are recognized by Mellichamp and Case (2009) and McPherson and Schnell (2011), but other authors do not accept the subspecies (Bell 1949, McDaniel 1971, Ellison et al. 2004). The hypothesis that *S. purpurea* is one species exhibiting a continuum of morphological variation that should not be divided into subspecies has not been tested comprehensively; morphometric and phylogeographic analyses are lacking. Morphological characters used by some authors to recognize subspecies in *S. purpurea* vary clinally and continuously across its geographic range (Naczi et al. 1999, Ellison et al. 2004). Molecular phylogenetic analyses have not settled the matter because of insufficient sampling (Ellison et al. 2012, Stephens et al. 2015b).

Another taxonomic controversy in the genus concerns *Sarracenia rubra* and its segregates. Mellichamp and Case (2009) and McPherson and Schnell (2011) agree on the number of taxa recognized but disagree on some of their ranks. Recent phylogenetic analyses using molecular data reveal the *S. rubra* complex to be polyphyletic (Ellison et al. 2012, Stephens et al. 2015b): *S. alabamensis* ssp. *alabamensis* is in a separate lineage from *S. rubra*, which supports the segregation of *S. alabamensis* from *S. rubra*.

The final taxonomic controversy concerns the status of *S. minor* var. *okefenokeensis* D.E. Schnell and *S. psittacina* var. *okefenokeensis* S. McPherson & D.E. Schnell. Bell (1949) and McDaniel (1971) concluded that these varieties represented only extremes in the continua of variation in their respective species. McPherson and Schnell (2011) did not address these hypotheses, and the taxonomic merit of these varieties remains uncertain.

9.3 Phylogenetic relationships

9.3.1 Fossils

The fossil record has not yet provided a reliable source of data for inferring relationships or establishing the age of Sarraceniaceae. The only fossil species reported to have affinities with Sarraceniaceae is *Archaeamphora longicervia* Hongqi Li from an early Cretaceous formation (≈125 million years ago) in Liaoning province, northeastern China (Li 2005). The fossils are suggestive of pitchers of *Sarracenia purpurea* or *Heliamphora* in lateral view, especially the ventricose portion connected by a column-like constriction to a hood-like distal appendage. However, the suggestion that these fossils "suggest a relationship to Sarraceniaceae" (Li 2005) is problematic because their age is much closer to the time at which the angiosperms diverged rather than the much later likely time of origin of Sarraceniaceae (Ellison et al. 2012). Based on similar but more recently discovered fossils, Wong et al. (2015) suggest that the structures interpreted by Li (2005) as pitchers more likely are arthropod-induced galls on the leaves of the extinct conifer *Liaoningocladus boii* G. Sun.

9.3.2 Morphological evidence for relationships of Sarraceniaceae

Family-level relationships. Ascertaining relationships of Sarraceniaceae from morphology has been difficult. The family possesses leaves that have been modified uniquely into distinctive tubular pitchers that function as carnivorous traps. Some authors (e.g., Cronquist 1981) suggested Sarraceniaceae are closely related to Nepenthaceae (Chapter 5) based on superficial similarity of their pitchers. However, pitchers comprise the whole leaf in Sarraceniaceae, but only its distal portion in Nepenthaceae. Sarraceniaceae pitchers also differ from those of Cephalotaceae (Chapter 10) in several ways, especially in the unridged lip and the adaxial wing that is distally single in the Sarraceniaceae.

Reproductive features are difficult to align, too, since little about Sarraceniaceae flowers or fruits are unusual, at least at the macromorphological scale. However, careful study of suites of floral features has led some botanists to suggest affinities of the

Sarraceniaceae with other flowering plants. Intuitive approaches based primarily on floral evidence led several botanists to suggest the Sarraceniaceae as the sole family within Sarraceniales (Thanikaimoni and Vasanthy 1972, Maguire 1978, Dahlgren 1980, Takhtajan 1980). Thanikaimoni and Vasanthy (1972) aligned Sarraceniaceae with the families Ranunculaceae and Papaveraceae. Maguire (1978) avoided placing Sarraceniales near any other flowering plants. Dahlgren (1980) placed the Sarraceniales together with the Cornales, Ericales, and two other orders in the superorder Corniflorae. Takhtajan (1980) placed Sarraceniales with Papaverales and Ranunculales in the subclass Ranunculidae. Other botanists suggested affinities of the Sarraceniaceae with Theales (DeBuhr 1975b, Thorne 1992).

Three studies of family-level relationships included explicit phylogenetic analyses of Sarraceniaceae based primarily on morphological evidence from floral features. Anderberg (1992) concluded Sarraceniaceae was sister to many families now placed in several orders, including Apiales, Asterales, Cornales, Ericales, and Gentianales. Hufford (1992) found Sarraceniaceae to be sister to Loasaceae, and Keller et al (1996) placed it in the Ericales. The overall consensus includes only two alternatives. One has Sarraceniaceae evolving relatively early in the angiosperm evolutionary tree (Thanikaimoni and Vasanthy 1972, Takhtajan 1980), near Ranunculales as presently circumscribed, which includes both Papaveraceae and Ranunculaceae (APG IV 2016). The other considers Sarraceniaceae to be evolutionarily more derived, in or near Ericales or Cornales (DeBuhr 1975b, Dahlgren 1980, Thorne 1992, Keller et al. 1996).

Generic relationships. No morphology-based phylogenetic analyses of infrafamilial relationships within Sarraceniaceae have been published. However, most who have considered phylogenetic relationships within the family have suggested that *Darlingtonia* and *Sarracenia* are more closely related to each other than either genus is to *Heliamphora* (e.g., Thanikaimoni and Vasanthy 1972, Maguire 1978, Kubitzki 2004; but see McPherson 2007 for the alternative hypothesis that *Sarracenia* is sister to *Darlingtonia* + *Heliamphora*). Chrtek et al. (1992) went further, splitting the Sarraceniaceae into Heliamphoraceae (*Heliamphora* only) and Sarraceniaceae, s.s.

Infrageneric relationships. Uphof (1936) divided *Sarracenia* into two groups, section *Sarracenia* (which he called section *Decumbentes*, but nomenclatural rules specify that this group must be called section *Sarracenia* because it includes *S. purpurea*, the type species of *Sarracenia*) for *S. psittacina* and *S. purpurea*, and section *Erectae* Uphof for the remaining species of the genus. Although he did not name groups within *Sarracenia*, Bell (1949) regarded *S. psittacina* and *S. purpurea* as comprising a group; and subdivided the species of Uphof's section *Erectae* into the "closely related" *S. alata* (as *S. sledgei*), *S. flava*, and *S. oreophila*, and a group with the remainder of the species. Chrtek et al. (1992) divided *Sarracenia* into the same groups as Uphof (1936), but called them subgenus *Sarracenia* (*S. psittacina* and *S. purpurea*) and subgenus *Sarraceniella* Chrtek, Slavíková & Studnička.

Hypotheses of infrageneric relationships in *Heliamphora* are recent and incomplete. Suggested relationships for some of the new species are based on morphological features shared with other species (e.g., Wistuba et al. 2002, 2005, Carow et al. 2005).

9.3.3 Molecular evidence for relationships of Sarraceniaceae

The field of molecular systematics has provided the breakthroughs necessary to infer relationships of Sarraceniaceae based on explicit data and analyses.

Family-level relationships. The first molecular phylogenetic analysis involving Sarraceniaceae, based on DNA sequences of the *rbcL* gene, found Sarraceniaceae to be monophyletic and to include all three genera traditionally placed in the family, *Darlingtonia*, *Heliamphora*, and *Sarracenia* (Albert et al. 1992). This was the first explicit demonstration of monophyly for the family; previously, the monophyly of Sarraceniaceae had been assumed or implied. This analysis also placed the Sarraceniaceae as sister to *Roridula* (Roridulaceae; Chapter 10) and in a larger lineage with Ericales (Albert et al. 1992). Though some earlier botanists had suggested affinities of the Sarraceniaceae with Ericales (§9.3.2), the sister-group relationship with the glandular-leaved Roridulaceae was a surprise.

The next molecular phylogenetic analysis included *rbcL* plus ITS sequences, and again found Sarraceniaceae monophyletic and sister to Roridulaceae (Bayer et al. 1996). Albach et al. (2001), based on phylogenetic analyses of four genes (18S rDNA, *atpB*, *ndhF*, and *rbcL*), found *Sarracenia* sister to a lineage containing Actinidiaceae and Roridulaceae. The sister-group relationship of Sarraceniaceae to Actinidiaceae + Roridulaceae also was supported by Anderberg et al. (2002) and Bremer et al. (2002), who used, respectively, five and six molecular regions, four of which did not overlap. Later analyses with increased molecular and taxonomic sampling found the same relationship with high levels of statistical support (Schönenberger et al. 2005, Löfstrand and Schönenberger 2015).

Generic relationships. Every study that has sampled all three genera of Sarraceniaceae (Albert et al. 1992, Bayer et al. 1996, Neyland and Merchant 2006, Ellison et al. 2012, Löfstrand and Schönenberger 2015) found *Darlingtonia* sister to *Heliamphora* + *Sarracenia*, usually with strong statistical support. These relationships are consistent with separate analyses of nuclear, plastid, and mitochondrial partitions of the genome (Ellison et al. 2012). This result is another unexpected outcome from molecular systematics since morphology-based hypotheses posited either that *Heliamphora* is sister to *Darlingtonia* + *Sarracenia* or that *Sarracenia* is sister to *Darlingtonia* + *Heliamphora* (§9.3.2).

Infrageneric relationships. Infrageneric relationships in *Sarracenia* remain incompletely known. In the two most recent, best-sampled phylogenetic analyses, low levels of sequence divergence stymied attempts to infer many relationships within the genus (Ellison et al. 2012, Stephens et al. 2015b). Stephens et al. (2015b) not only extensively sampled molecular regions but also used target-enrichment to enhance sampling of variable regions of DNA. Stephens et al (2015b) used a total of 128,110 base pairs of nuclear DNA, only 4% of which were variable and parsimony-informative in *Sarracenia*, and 42,031 base pairs of plastid DNA, only 0.5% of which were variable and parsimony-informative.

Despite low support and incomplete resolution of relationships among taxa, Ellison et al. (2012) and Stephens et al. (2015b) agree that the *S. rubra* complex is polyphyletic and that *S. flava* and *S. psittacina* are sisters. Neither analysis supports hypothesized sister-group relationship for *S. psittacina* and *S. purpurea* (Uphof 1936, Bell 1949, Chrtek et al. 1992). However, Ellison et al. (2012) and Stephens et al. (2015b) disagree about relationships among *S. alata*, *S. leucophylla*, *S. minor*, *S. oreophila*, and *S. rosea*.

Infrageneric relationships in *Heliamphora* remain largely unresolved. The chief problem has been the difficulty of sampling the species; the two studies incorporating the greatest number of *Heliamphora* species (Ellison et al. 2012, Löfstrand and Schönenberger 2015) sampled only six and seven species, respectively. Using nearly non-overlapping sets of molecular markers, both Ellison et al. (2012) and Löfstrand and Schönenberger (2015) identified two main clades: one containing *H. minor*, *H. neblinae*, and *H. pulchella*; the other containing *H. heterodoxa* and *H. nutans*. However, because so few individuals of so few species have been sampled, these infrageneric relationships should be regarded as provisional.

9.3.4 Molecular divergence time estimation

The first dated phylogeny for the Sarraceniaceae and molecular divergence time estimates indicate that the stem-group of Sarraceniaceae had originated by mid-Eocene, ≈49 million years ago (Mya), with a range of 44–53 Mya (Ellison et al. 2012). Subsequent analyses, with more robust sampling of extant and fossil taxa, estimate the origin of the stem to be older, ≈70 Mya (range 34–79 Mya; Tank et al. 2015) or 88 Mya (85–91 Mya; Magallón et al. 2015).

Though no Sarraceniaceae fossils are known (§9.3.1), a few fossils from within the crowns of the sister groups Actinidiaceae + Roridulaceae have been described. *Parasaurauia allonensis* J.A. Keller, Herend. & P.R. Crane is an extinct member of Actinidiaceae known from late Cretaceous flowers and fruits found in Georgia, USA (Keller et al. 1996). The geologic formation in which *P. allonensis* occurs currently is estimated to be of late Santonian age (Konopka et al. 1998), ≈84 Mya (Ogg and Ogg 2008). Roridulaceae are known from two fossil leaves from western Russia, estimated to be 35–47 Mya (Sadowski et al. 2015). Thus, based on the fossil record of the sister-group, the origin of stem

Sarraceniaceae appears to be older than 84 Mya (late Cretaceous), substantially older than estimated by Ellison et al. (2012) but in line with the 88 Mya estimated by Magallón et al. (2015). The calibrating fossils used by Magallón et al. (2015) did not include *Parasaurauia* and thus these fossils did not influence their analyses, and their estimate of 88 Mya for the origin of stem Sarraceniaceae seems reasonable.

The divergence time estimates of Ellison et al. (2012) are the only infrafamilial ones available for the Sarraceniaceae. According to these estimates, *Darlingtonia* diverged from *Heliamphora* + *Sarracenia* 25–44 Mya, probably in the late Eocene. *Heliamphora* and *Sarracenia* then diverged 14–32 Mya, probably in the late Oligocene. *Heliamphora* began to diversify 5–14 Mya, during the late Miocene. *Sarracenia* appears to have begun diversifying considerably more recently: 2–7 Mya, probably during the Pliocene.

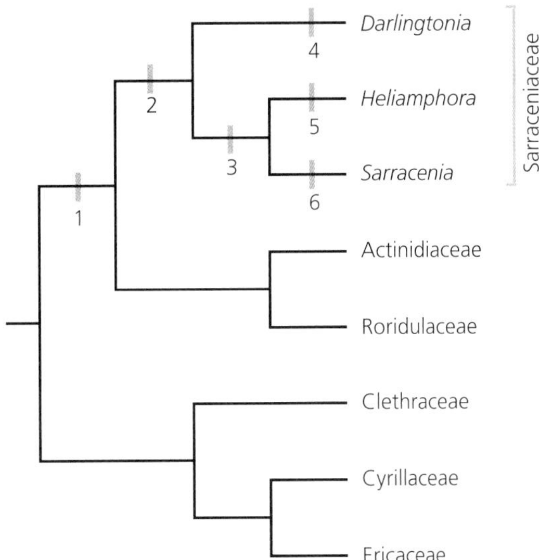

Figure 9.5 Consensus phylogenetic tree of Sarraceniaceae and their closest relatives (Schönenberger et al. 2005, Ellison et al. 2012, Löfstrand and Schönenberger 2015). Numbered bars identify key branches for use with Table 9.1.

9.3.5 Interpreting morphology in light of molecular phylogeny

The molecular phylogeny for Sarraceniaceae and close relatives provides the opportunity to understand morphology of this group better than had been possible in the past. Multiple studies sampling different gene regions provide a consensus phylogeny (Fig. 9.5) with a high degree of confidence (Schönenberger et al. 2005, Ellison et al. 2012, Löfstrand and Schönenberger 2015). Because the consensus is relatively recent, interpretations of morphology have begun only within the past few years. Identification of morphological synapomorphies is important because they diagnose clades independently from molecular data.

Robustly supported as the sister-group to Sarraceniaceae is the clade of Actinidiaceae + Roridulaceae. The ≈360 species of Actinidiaceae inhabiting temperate and tropical eastern Asia and the Neotropics are vines, trees, or shrubs with flat, usually ovate leaves (Dressler and Bayer 2004). The two species of Roridulaceae are endemic to South Africa. The flat, linear or narrowly lanceolate leaves of these carnivorous shrubs bear numerous long-stalked glands that produce viscous droplets (Conran 2004b, Sadowski et al. 2015; Chapter 10).

Morphologically, species of Actinidiaceae and Roridulaceae appear to have little in common with Sarraceniaceae. Yet, ancestral state reconstructions of floral features (Löfstrand and Schönenberger 2015) reveal three putative synapomorphies diagnosing a clade comprised of Sarraceniaceae, Actinidiaceae, and Roridulaceae (Table 9.1). These synapomorphies are: petals proximally thicker than the sepals (vs. the plesiomorphic petals thinner or equal to sepals in thickness); polystemony (vs. stamens less than or equal to the number of corolla lobes); and presence (vs. absence) of a nucellar hypostase, a dark area at the base of the embryo sac (Anderberg et al. 2002).

Ancestral state reconstructions identify at least one and possibly two synapomorphies for Sarraceniaceae. Anthers invert from introrse to extrorse at the beginning of anthesis, inversion "late type B" (vs. the plesiomorphic anthers not inverted; Löfstrand and Schönenberger 2015). Abaxial anther attachment (with a switch to proximal in *Darlingtonia*) may be synapomorphic for the Sarraceniaceae, although the state polarity of this character is uncertain (Löfstrand and Schönenberger 2015). R.F.C. Naczi (*unpublished data*) has identified six additional plausible synapomorphies: plants herbaceous (vs. the plesiomorphic woody); leaves are pitchers (vs.

Table 9.1 Morphological synapomorphies of the clade containing Sarraceniaceae, the Sarraceniaceae, and genera of Sarraceniaceae. Branches are numbered as in Figure 9.5. Synapomorphies identified through ancestral state reconstructions (Löfstrand and Schönenberger 2015) and phylogenetic analysis in preparation ("additional;" R.F.C. Naczi *unpublished data*).

Branch No.	Clade	Synapomorphies
1	Sarraceniaceae + (Actinidiaceae + Roridulaceae)	*from ancestral state reconstructions*: • petals proximally thick • polystemony • nucellar hypostase
2	Sarraceniaceae	*from ancestral state reconstructions*: • late anther inversion type B • possibly, anther attachment abaxial *additional*: • plants herbaceous • leaves ascidiate • flowers pendant • sepals glabrous • pollen grains each with 4 or more colpi • seeds transverse
3	*Heliamphora* + *Sarracenia*	*from ancestral state reconstructions*: • ovary summit not depressed *additional*: • sepals abaxially with sessile glands
4	*Darlingtonia*	*from ancestral state reconstructions*: • proximal anther attachment • late anther inversion type C *additional*: • pitchers longitudinally twisted • fishtail appendage • ovary distally widened • ovary truncate • seeds stipitate
5	*Heliamphora*	*from ancestral state reconstructions*: • styles united *additional*: • sepals 2 • petals persistent • ovary pubescent • seeds winged
6	*Sarracenia*	*additional*: • bracts immediately adjacent to sepals • sepals broadly ovate • pollen grains each with 6–10 colpi • styles forming umbrella-like expansion • ovary tuberculate

flat and nontubular); pendant flowers in which the peduncle or pedicel adjacent to the flower is curved ≈180 degrees at anthesis (vs. flowers erect or on only a slightly curved stalk); glabrous sepals (vs. pubescent); pollen grains each with 4 or more colpi (vs. with 3 colpi); and seeds transverse so that they are oriented with their long axes perpendicular to the long axis of the gynoecium in mature fruit (vs. seeds oriented with their long axes parallel to the long axis of the gynoecium).

Only two morphological apomorphies are known to support the sister-group relationship of *Heliamphora* and *Sarracenia*. These two are: ovary summit not depressed (vs. the plesiomorphic depressed; Löfstrand and Schönenberger 2015); and sepals abaxially with sessile glands (R.F.C. Naczi *unpublished data*).

Several morphological synapomorphies are apparent for each of the three genera. Ancestral-state reconstruction yielded two synapomorphies for *Darlingtonia*: proximal anther attachment (vs. the plesiomorphic adaxial attachment); and anthers inverting at the beginning of anthesis but in an uncertain direction (inversion "late type C" vs. the plesiomorphic anthers not inverted; Löfstrand and Schönenberger 2015). Additional synapomorphies of *Darlingtonia* appear to be pitchers longitudinally twisted (vs. untwisted); the distal portion of the pitcher bearing a deeply lobed fishtail appendage (vs. distal portion of pitcher lacking a fishtail appendage); ovary wider distally than proximally (vs. ovary wider proximally); ovary truncate at summit (vs. ovary tapering to summit); and seeds stipitate (vs. seeds lacking stipes; R.F.C. Naczi *unpublished data*).

For *Heliamphora*, united styles (vs. the plesiomorphic partially distinct styles) is a synapomorphy from ancestral state reconstructions (Löfstrand and Schönenberger 2015). Additional synapomorphies include two sepals (vs. five); petals persisting through fruiting (vs. deciduous); pubescent ovary (vs. glabrous); and winged seeds (vs. unwinged; R.F.C. Naczi *unpublished data*).

Sarracenia has five morphological synapomorphies (R.F.C. Naczi *unpublished data*): bracts immediately adjacent to the sepals (vs. the plesiomorphic bracts separated from the sepals); broadly ovate sepals (vs. ovate or lanceolate sepals), pollen grains each with 6–10 colpi (vs. three colpi); styles forming an umbrella-like expansion (vs. styles filiform); and tuberculate ovary (vs. smooth ovary).

Future work will undoubtedly lead to discovery of additional synapomorphies; vegetative characters have been little studied in this regard. In addition, polarity of certain character states is unclear, making some characters presently unavailable for discerning synapomorphies. Mode of anther dehiscence (poricidal in *Heliamphora*, but longitudinal in *Darlingtonia* and *Sarracenia*); carpel number (three in *Heliamphora* but five in *Darlingtonia* and *Sarracenia*); and number of flowers per plant (multiple in *Heliamphora* but single in *Darlingtonia* and *Sarracenia*) are all characters that differ among genera but are constant within them, and thus potentially valuable for phylogenetics.

9.4 Evolutionary patterns and processes

9.4.1. Patterns

Estimates of divergence times (Ellison et al. 2012), recent molecular phylogenetic analyses (Ellison et al. 2012, Löfstrand and Schönenberger 2015), and identification of putative morphological synapomorphies (Löfstrand and Schönenberger 2015; R.F.C. Naczi *unpublished data*; §9.3.5) present a well-supported pattern of three well-separated, quite divergent genera. However, the species of *Heliamphora* and *Sarracenia* are very closely related within each genus. In the phylogenetic trees, relatively large amounts of sequence divergence separate the genera, but only small amounts of sequence divergence are present among the species. Speciation within *Heliamphora* and *Sarracenia* appears to have been relatively recent and rapid (Ellison et al. 2012).

9.4.2 Chromosome number variation

Chromosome number is markedly different among genera but apparently invariant within *Heliamphora* and *Sarracenia*, similar to patterns revealed by molecular and morphological evidence. The chromosome number of *Darlingtonia* is $2n = 30$ (Bell 1949), whereas *Heliamphora* is $2n = 42$, reported from *H. nutans* (Kress 1970), *H. heterodoxa* (Kondo 1972b), and *H. minor* (Kondo 1973). Nearly all species of *Sarracenia* have chromosome counts, and are $2n = 26$ (Bell 1949, Hecht 1949). Chromosome counts of $2n = 26$ also exist for three *Sarracenia* hybrids: *S. leucophylla* × *S. purpurea*, *S. purpurea* × (*S. leucophylla* × *S. purpurea*), and *S. flava* × *S. leucophylla* (Tjio 1948, Hecht 1949).

9.4.3 Genetic diversity

Genetic diversity within *Sarracenia* has been characterized in a few illuminating cases, but similar studies of *Heliamphora* or *Darlingtonia* are lacking. Genetic diversity of *S. alabamensis*, *S. jonesii*, *S. leucophylla*, *S. oreophila*, and *S. rubra* ssp. *rubra* throughout their geographic ranges has been characterized using allozymes (Godt and Hamrick 1996, 1998, Wang et al. 2004). Diversity of *S. alabamensis*, *S. jonesii*, and *S. oreophila* also has been examined with DNA-based analyses (Furches et al. 2013a). Among these, *S. oreophila* and *S. jonesii* consistently show the lowest values of genetic diversity, whereas *S. leucophylla* has the highest; *S. alabamensis* also has high genetic diversity (Wang et al. 2004). Comparisons with a wide range of other, non-*Sarracenia*, narrowly distributed species indicate that levels of genetic diversity in *Sarracenia* are comparable to or greater than that of other narrowly distributed species (Wang et al. 2004). In fact, genetic diversity values of *S. alabamensis* and *S. leucophylla* far exceed values of the comparison group of narrowly distributed, noncarnivorous plants. It is unknown if observed levels of genetic diversity have limited *Sarracenia* evolution, or if these levels pose limits for future evolution. However, ongoing and future reductions in population sizes and numbers may pose serious threats to the genetic capacity and survivability of *Sarracenia* species by diminishing genetic variation within them (Godt and Hamrick 1998, Furches et al. 2013a).

Sarracenia species are self-compatible (Schnell 1976). The complex consequences of inbreeding on *Sarracenia* genetic structure, population size, and long-term survival have received little study. Sheridan and Karowe (2000) demonstrated that outcrossing is important in maintaining viable populations of *S. flava*, apparently through maintaining genetic variation. Self-pollinated plants produced significantly fewer seeds, seeds with depressed germination, and shorter pitchers than outcrossed plants. Offspring of crosses between plants from sites with small populations produced significantly taller pitchers than offspring of intra-site crosses (Sheridan and Karowe 2000).

Based on data from intentional introductions of *S. purpurea* into many different localities throughout the world where it is not native (e.g., Schwaegerle 1983, Walker 2014), inbreeding in this species appears to have consequences that are very different from those observed in *S. flava*. These introductions usually are of only very few plants, but most introduced populations expand dramatically through inbreeding and seed production, not vegetative reproduction (Schwaegerle 1983, Walker 2014). One well-documented example is an introduced population in Ohio, USA that was initiated in 1912 with a single individual and grew to *c.* 157,000 plants by 1978 (Schwaegerle 1983). Similar population expansions that have occurred in other sites have led to *S. purpurea* being regarded as an invasive plant (e.g., Walker 2014).

These repeated dramatic population expansions by *S. purpurea* defy expectations of inbreeding depression. The introduced Ohio population had low levels of genetic variability, yet experienced dramatic growth (Schwaegerle and Schaal 1979). Analyses of Swiss populations failed to detect inbreeding depression on number of seeds produced (Parisod et al. 2005). In fact, the authors found evidence for outbreeding depression, exhibited by a greater reduction in seed weight with increasing genetic distance of the parent plants (Parisod et al. 2005).

Apparently, *S. purpurea* can expand its populations successfully via inbreeding, and low genetic diversity appears not to limit population viability, at least within a short time period. Perhaps *S. purpurea* has undergone selection for inbreeding (Parisod et al. 2005). Recent analyses suggest that self-pollinating species like *Sarracenia* generally have larger ranges than obligately outcrossing ones (Grossenbacher et al. 2015), but apparently all *Sarracenia* species can self, and range sizes vary from very small (*S. jonesii*) to huge (*S. purpurea*). Documentation of genetic variability, its interrelationships with population growth, and its consequences for inbreeding or outbreeding depression require more study before we can understand the complex genetic bases of population dynamics in Sarraceniaceae.

9.4.4 Hybridization

Hybridization in *Sarracenia* is a popular topic; several reviews of the morphology, nomenclature, frequency, and geographic distribution of natural *Sarracenia* hybrids have been published (Bell 1949, 1952, Bell and Case 1956, McDaniel 1971, Mellichamp and Case 2009, McPherson and Schnell 2011, Neyland et al. 2015). Further evidence of the great interest in *Sarracenia* hybrids, fueled especially by horticultural popularity, is the fact that nearly every first generation (F_1) hybrid has been named with a binomial, as have several backcrosses (Bell 1952, Nelson 1986, Neyland et al. 2015).

Moore (1874) recorded the artificial hybridization of cultivated *S. flava* and *S. leucophylla*, and the first reports of natural hybrids in *Sarracenia* were by Manda (1892) and Harper (1903). The most recent reviews mention ≈20 natural F_1 hybrids (Mellichamp and Case 2009, McPherson and Schnell 2011, Ellison et al. 2014, Neyland et al. 2015). Hybrids of *Sarracenia* usually are morphologically intermediate between their parents, making them easy to recognize when the parents are distinctive, but more challenging to discern when similar. Hybrids are especially difficult to recognize in cases of backcrossing (Bell 1952).

Natural hybrids have been reported for every *Sarracenia* species except *S. oreophila* (Bell 1949, Ellison et al. 2014); every *Sarracenia* species that is known to co-occur with at least one congener is involved in hybridization. The greatest number of different hybrids occurs in the region of greatest sympatry among *Sarracenia* species—southern Alabama, northern Florida, and southern Georgia (Bell 1952)—where up to five species co-occur at the same locality (Folkerts 1982; R.F.C. Naczi *unpublished data*). Though most hybrids are rare, a few, such as *S. alata* × *S. leucophylla*, are common (Bell 1949, Bell and Case 1956, McDaniel 1971). *Sarracenia* species appear to be completely interfertile and hybrids are quite fertile, possibly fully fertile (McPherson and Schnell 2011). Backcrossing is reported by several authors, with hybrid swarms sometimes resulting (e.g., Bell 1949, 1952, McDaniel 1971). Introgression was suspected by some authors (Bell 1949, Bell and Case 1956, Folkerts 1977), a hypothesis later supported by a genetic analysis of plants at a site supporting syntopic populations of *S. alabamensis* ssp. *wherryi*, *S. alata*, *S. leucophylla*, and hybrids among them (Furches et al. 2013b). In addition, Ellison et al. (2012) detected a likely case of introgression involving *S. purpurea* var. *montana*. Some authors report that hybrids are more frequent in disturbed areas (McDaniel 1971, Folkerts 1977, 1982). In short, natural hybridization in *Sarracenia* is relatively frequent and sometimes even locally extensive.

Given the extensive syntopy and interfertility of *Sarracenia* species, hybrids are much less frequent in nature than expected. Chromosome number—constant in the genus—does not appear to preclude hybridization. Instead, some other isolating mechanism(s) evidently reduces the incidence of hybridization. Geography appears to be the strongest pre-zygotic isolating mechanism; allopatric populations do not interbreed (Bell 1952). Different blooming times among the species also present a pre-zygotic barrier to hybridization. Although some phenological overlap does occur among syntopic species (Bell 1952), they tend to differ in color, size, and height of flowers, differences that probably provide additional isolating mechanisms (McDaniel 1971, Folkerts 1982). Species also appear to exhibit microhabitat specialization, resulting in less intermingling than otherwise might occur (McDaniel 1971). Post-zygotically, some hybrids appear to have reduced fitness relative to their parents (Folkerts 1982).

Natural hybrids between *Heliamphora* species were discovered comparatively recently (Baumgartl 1993). In the most recent review of natural hybridization in *Heliamphora*, McPherson et al. (2011) report 11 suspected F_1 hybrids. Although unconfirmed as hybrids to date, their morphology usually is intermediate between, and they usually were found in close proximity to, the suggested parents. In some localities, hybrids are locally abundant, such as in the case of *H. purpurascens* × *H. sarracenioides*, and backcrosses are suspected, too (McPherson et al. 2011). Assuming hybridization occurs, reports of natural hybridization in *Heliamphora* are noteworthy especially because multiple species hybridize, and hybridization occurs at many sites in environments that appear to be undisturbed by humans.

Understanding of the role of natural hybridization in the evolution of Sarraceniaceae is in its infancy. Hybridization indisputably occurs and interspecific gene flow has been reported in a group of co-occurring *Sarracenia* (Furches et al. 2013b). However, it remains unknown whether any species of Sarraceniaceae are of hybrid origin. The possibility of *Heliamphora arenicola* being a hybrid is mentioned by its authors and supported by its morphology being intermediate between *H. elongata* and *H. nutans* (McPherson et al. 2011). Also unknown is the effect of hybridization on identity of species: does hybridization cause genetic erosion of existing species?

9.4.5 Heterochrony

Heterochrony is the evolutionary change in timing or rate of events in the development of an organism, relative to ancestral development (Alberch et al. 1979, Li and Johnston 2000). Through heterochrony, development of organs in a more recent species may be accelerated or delayed relative to more ancestral species. In this way, heterochrony can cause relatively abrupt morphological changes in closely related species. Though heterochrony has not been studied in Sarraceniaceae, morphological evidence indicates it may have occurred in *Sarracenia* and *Heliamphora* (R.F.C. Naczi *unpublished data*).

Though the flowers are very similar, pitchers of *S. psittacina* are remarkably divergent from those of all other *Sarracenia* species. The pitchers of *S. psittacina* have oblique or vertical orifices (rather than horizontal in the other species); the hoods nearly completely surround the orifices (vs. raised above the orifices or only partially surrounding them); the pitcher interior is lined for most of its length with long, rigid, intermeshing hairs (vs. the upper portion of the pitcher devoid of hairs); and the volume of the pitchers is quite small (vs. volume considerably greater). Within *Sarracenia*, every one of these characteristics is unique to *S. psittacina*, with one exception: every one of these seemingly anomalous traits occurs in juvenile pitchers of multiple, possibly all, species of *Sarracenia* (Lloyd 1942, R.F.C. Naczi *unpublished data*).

The expression of otherwise juvenile morphological traits in adults of *S. psittacina* raises the possibility that *S. psittacina* expresses these traits through heterochrony (R.F.C. Naczi *unpublished data*). The heterochronic process that produces morphologies in descendants that resemble juveniles of ancestors is called paedomorphosis (Alberch et al. 1979). Paedomorphosis is little known in plants, and the hypothesis of heterochrony resulting in a paedomorphic *S. psittacina* remains to be tested. Recent phylogenetic analyses, in which *S. psittacina* has a relatively derived position (Ellison et al. 2012, Stephens et al. 2015b), do provide some support for the paedomorphic *S. psittacina* hypothesis.

A parallel situation occurs in *Heliamphora*. The pitchers of *H. sarracenioides* resemble juvenile pitchers of other *Heliamphora* species (Carow et al. 2005, McPherson et al. 2011), although its flowers are similar. Thus, *H. sarracenioides* may have diverged morphologically because of paedomorphosis.

Alternatives to heterochrony could explain the divergent vegetative morphology of *S. psittacina* and *H. sarracenioides*. For example, each species may be highly divergent because of a large degree of anagenesis and accumulated autapomorphies in its evolution. The role of heterochrony, if any, in the evolution of Sarraceniaceae remains to be determined through future research.

9.4.6 Evolution of the Sarraceniaceae pitcher

One of the long-standing mysteries of Sarraceniaceae evolution has been the evolution of the pitcher. Previous authors speculated that pitchers evolved from planar, possibly peltate, leaves in which the margins inrolled and fused (Juniper et al. 1989; Chapter 18, §18.5.6). Discovery of the sister-group, Actinidiaceae + Roridulaceae, has not helped solve the mystery since no extant members of those families are known to have pitchers or pitcher-like leaves. Recent developmental studies of the *S. purpurea* pitcher reveal that differential cell division patterns in the leaf primordium produce the tubular leaf morphology of a pitcher (Fukushima et al. 2015).

9.4.7 Historical biogeography

Evolution of the tripartite disjunction of Sarraceniaceae genera (Fig. 9.2), and of species within *Heliamphora* and *Sarracenia*, have long interested biogeographers. Juniper et al. (1989), Bayer et al. (1996), McPherson et al. (2011), and Ellison et al. (2012) present hypotheses for evolution and biogeographic history of the three genera, and Stephens et al. (2015b) for species of *Sarracenia*. Ellison et al. (2012) hypothesize that stem-group Sarraceniaceae originated in South America, and ancestral species migrated to occupy much of North America. Then, an east–west disjunction occurred in the North American range, separating ancestral *Darlingtonia* from ancestral *Heliamphora* + *Sarracenia*, probably due to mid-continental aridification. Subsequently, a disjunction separated South American ancestors from those in North America, which diversified to become *Heliamphora* and *Sarracenia*, respectively.

One open question about the historical biogeography of the family is the role of long-distance dispersal (Juniper et al. 1989, Bayer et al. 1996). Obligately symbiotic arthropods of Sarraceniaceae offer a clue to answering this question. Species of the mite genus *Sarraceniopus* (Histiostomatidae) are restricted to pitchers of Sarraceniaceae (Fashing and OConnor 1984). *Sarraceniopus* species, most still undescribed, occur in all Sarraceniaceae taxa sampled to date: *Darlingtonia*, all species of *Sarracenia*, and four species of *Heliamphora* (R.F.C. Naczi *unpublished data*). The aspects of *Sarraceniopus* most relevant for Sarraceniaceae biogeography are the thin cuticles of these aquatic and semi-aquatic mites. Dispersing instars (deutonymphs) desiccate within hours of leaving pitchers if they do not reach the interior of another pitcher (R.F.C. Naczi *unpublished data*). The obligate co-association of *Sarraceniopus* with Sarraceniaceae and the rapid desiccation of mites upon exiting pitchers support the divergence of Sarraceniaceae through vicariance, rather than long-distance dispersal (R.F.C. Naczi *unpublished data*).

9.5 Future research

The most fundamental aspect of the Sarraceniaceae—its taxonomic diversity—is far from settled. Rigorous tests of existing taxonomic hypotheses, with explicit applications of species concepts that reflect the biology of these plants, clearly are needed. A major effort to stabilize taxonomy in the family should consist of additional phylogenetic analyses. Phylogenetic relationships need to be inferred with more taxonomic and molecular sampling, and increased statistical support, using molecular and morphological data.

Morphological evolution of Sarraceniaceae needs attention, too. Little is known about how morphological diversity has been achieved, especially vegetative diversity. Particularly pressing is the long-standing question of how the pitcher has evolved from ancestors lacking pitchers (see also Chapter 18, §18.5.6).

Historical biogeography and speciation in the family merit further work. The central question for historical biogeographers is how and when the current disjunctions among Sarraceniaceae genera occurred. Answering this question will be helped through integrating phylogenetics with paleofloristics, paleoclimatology, and geology (especially plate tectonics). The speciation process in Sarraceniaceae is poorly understood. The importance of such factors and processes as genetic diversity, inbreeding, hybridization, and heterochrony should be investigated more intensively, and should use both comparative and experimental methods.

CHAPTER 10

Systematics and evolution of small genera of carnivorous plants

Adam T. Cross, Maria Paniw, André Vito Scatigna, Nick Kalfas, Bruce Anderson, Thomas J. Givnish, and Andreas Fleischmann

10.1 Introduction

The Droseraceae, Lentibulariaceae, Nepenthaceae, and Sarraceniaceae are noteworthy for their diversity and long botanical history (Chapters 4–9), but carnivory also has evolved in six small, species-poor or even monogeneric plant families—Drosophyllaceae and Dioncophyllaceae (Caryophyllales); Cephalotaceae (Oxalidales); Roridulaceae (Ericales); and Byblidaceae (Lamiales)—and in at least one species each from genera in the Bromeliaceae, Eriocaulaceae (Poales), and Plantaginaceae (Lamiales) in which carnivory is not a synapomorphy (Figures 10.1, 10.2). These groups contain some of the most unique carnivorous plants: species reliant upon digestive mutualisms (*Roridula* spp.), genera in which carnivory is not a universal trait (*Brocchinia, Catopsis, Paepalanthus*), species with complex ecological requirements and biotic interactions (*Byblis* spp., *Cephalotus follicularis, Drosophyllum lusitanicum*), and species with unique ecologies and life history (*Philcoxia* spp., *Triphyophyllum peltatum*).

Most carnivorous species in these families appear to be paleoendemics: systematically isolated taxa with relatively ancient origins that are often ecological specialists (Stebbins and Major 1965). Almost all were likely to have once been widespread but now exhibit relictual ranges and narrow, fragmented distributions (Figure 10.3). Isolated and refugial distributions often are a characteristic of paleoendemics, and the reason such taxa are often considered to be "on the way to extinction" (Stebbins and Major 1965). *Brocchinia*, in contrast, appears to include neoendemic species (§10.3.2).

10.2 *Brocchinia*

10.2.1 Life history, morphology, and systematics

Brocchinia is a genus of ≈20 species of perennial rosette-forming herbs in the monogeneric subfamily Brocchinioideae of the Bromeliaceae (Givnish et al. 1997, 2014a). Species vary strikingly in vegetative stature (0.05–>3.0 m tall), growth form (impounding vs. non-impounding rosettes, impounding treelets, non-impounding shrubs, and one vine-like form), and mechanism of nutrient capture. Two species of *Brocchinia* are carnivorous: *B. reducta* and *B. hechtioides. Brocchinia reducta* is relatively short (≈30 cm tall), with steeply inclined, bright yellow leaves forming a cylindrical rosette, whereas *B. hechtioides* produces tall and slightly splayed rosettes (≤1 m; Figure 10.1a). The inner surfaces of the rosettes in both species are covered with a fine, waxy dust that attracts insect prey (mainly Hymenoptera) and precipitates them into the rain-filled tank (Givnish et al. 1984, 1997, Gaume et al. 2004). Flowers are minute and hermaphroditic with imbricate petals and vary in color from white or cream to orange (Figure 10.2a). Flowers appear to be insect-pollinated, although it has been suggested that bats visit the small, capitate inflorescences of the noncarnivorous *Brocchinia uaipanensis* (Varadarajan and Brown 1988).

Cross, A. T., Paniw, M., Scatigna, A. V., Kalfas, N., Anderson, B., Givnish, T. J., and Fleischmann, A., *Systematics and evolution of small genera of carnivorous plants*. In: *Carnivorous Plants: Physiology, ecology, and evolution*. Edited by Aaron M. Ellison and Lubomír Adamec: Oxford University Press (2018). © Oxford University Press. DOI: 10.1093/oso/9780198779841.003.0010

Figure 10.1 (Plate 11 on page P10) Carnivorous representatives of Bromeliaceae, Byblidaceae, Cephalotaceae, Dioncophyllaceae, Drosophyllaceae, Plantaginaceae, and Roridulaceae. (**a**) *Brocchinia hechtioides* (Bromeliaceae) on the summit of Auyán-tepui, Venezuela. (**b**) *Catopsis berteroniana* (Bromeliaceae) growing epiphytically in the Gran Sabana, Venezuela. (**c**) *Byblis lamellata* (Byblidaceae) growing among low shrubs in the Kwongan heathland north of Perth, Western Australia. (**d**) *Cephalotus follicularis* (Cephalotaceae) on a moss sward over exposed granite in southwest Western Australia. (**e**) *Triphyophyllum peltatum* in the rainforest understory of Sierra Leone. (**f**) *Drosophyllum lusitanicum* from Mediterranean heathland near Algeciras, Spain. (**g**) *Philcoxia rhizomatosa* growing on white sand in Brazil. (**h**) *Roridula dentata* from the shrubby vegetation of the fynbos, South Africa. Photographs **a–f** and **h** by Andreas Fleischmann; **g** by André Vito Scatigna.

Brocchinia was formerly placed in subfamily Pitcairnioideae, but recent studies indicate that Pitcairnioideae is paraphyletic and contains the embedded subfamilies Tillandsioideae and Bromelioideae (Givnish et al. 1997, 2011, 2014a). Brocchinoideae diverged from the common ancestor of all other bromeliads 19–23 million years ago (Mya) and extant lineages within *Brocchinia* began diverging from each other ≈13 Mya (Givnish et al. 2011, 2014a). Variation in chloroplast DNA restriction-sites identified four major lineages within *Brocchinia*: the Prismatica, Melanacra, Micrantha, and Reducta clades (Givnish et al. 1997). The first two clades are successively sister to other species of *Brocchinia*; Micrantha and Reducta are sister to one another. Almost all members of the latter two clades are tank rosettes in which the leaf bases overlap so tightly that they impound rainwater, which accumulates within the tank and generally keeps the leaf bases wet in the perhumid climate atop tepuis (Givnish et al. 1997). Except for *B. paniculata*, all members of the Prismatica and Melanacra clades lack tanks.

10.2.2 Carnivory

Brocchinia has evolved more different mechanisms of nutrient capture than any other genus of angiosperms, including carnivory, ant-fed myrmecophily, N-fixation, tank epiphytism (presumably involving the breakdown of fallen vegetable detritus), possible mutualisms with frogs, and "normal" nutrient uptake by roots in some terrestrial or saxicolous species (Givnish et al. 1997, Givnish 2017). These include all the nutrient-capture mechanisms seen across the Bromeliaceae except for the

Figure 10.2 (Plate 12 on page P11) The flowers of carnivorous representatives of Bromeliaceae, Byblidaceae, Cephalotaceae, Dioncophyllaceae, Drosophyllaceae, Plantaginaceae, and Roridulaceae. (**a**) *Brocchinia hechtioides* (Bromeliaceae). (**b**) The inflorescences of *Catopsis berteroniana* (Bromeliaceae). (**c**) *Byblis gigantea* (Byblidaceae). (**d**) *Cephalotus follicularis* (Cephalotaceae). (**e**) *Triphyophyllum peltatum*. (**f**) *Drosophyllum lusitanicum*. (**g**) *Philcoxia bahiensis*. (**h**) *Roridula dentata*. Photographs **a–d** and **h** by Andreas Fleischmann; **e** by Jan Schlauer; **f** by Maria Paniw; **g** by André Vito Scatigna.

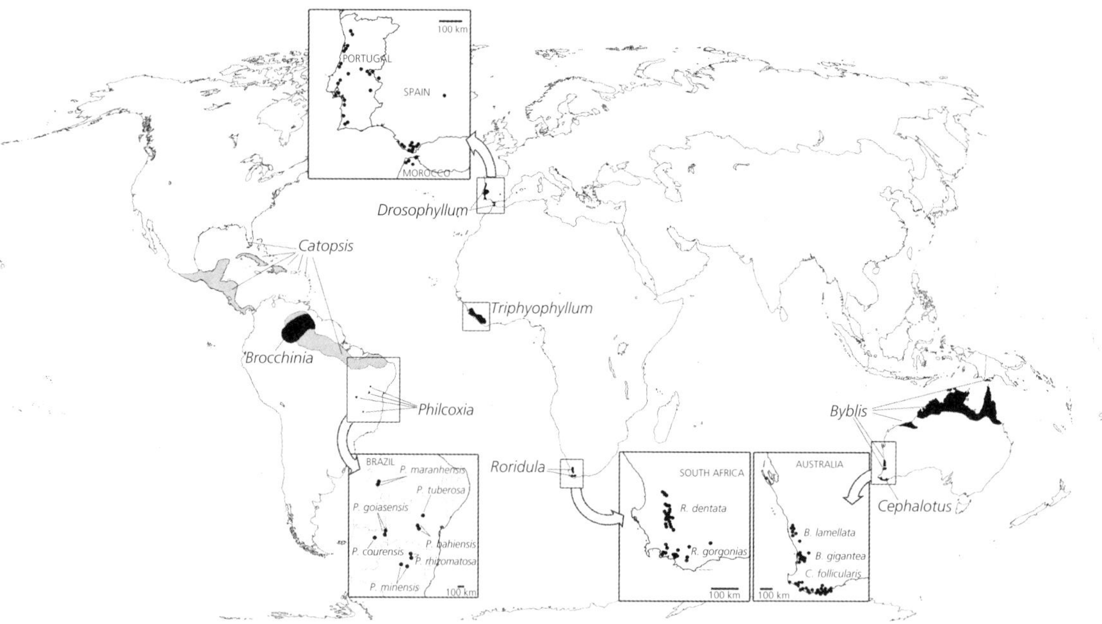

Figure 10.3 Ranges of *Brocchinia* (modified from Givnish 1989, Givnish et al. 2014a), *Catopsis* (following Givnish 1989, Givnish et al. 2014a), *Byblis* (Lowrie 2013 for Australia and based on herbarium records for New Guinea), *Cephalotus* (CHAH 2006, Lowrie 2013), *Drosophyllum* (Müller and Deil 2001), *Roridula* (following B. Anderson 2006), *Triphyophyllum* (modified from Airy Shaw 1951), and *Philcoxia* (Carvalho and Queiroz 2014, Scatigna et al. 2017). Map drawn by Andreas Fleischmann.

extreme atmospheric habit seen in some *Tillandsia* species (highly reduced growth form and dense coverage with highly absorptive leaf trichomes; e.g., Spanish moss, *T. usneoides*, and ball moss, *T. recurvata*).

All *Brocchinia* species with novel nutrient-capture mechanisms have impounding leaf bases, forming a tank from which the live trichomes on the leaf bases can absorb water and nutrients from prey, excreta, or fallen vegetable debris (Givnish et al. 1997). The trichomes of *B. reducta* are large (≈6100 μm²) and dense (≈200/mm²), and cover a larger area (≈12%) of the leaf base than other members of the genus (Givnish et al. 1984). The trichomes of *B. hechtioides* are smaller (≈2500 μm²), but still occupy a relatively large surface area of the leaf base (≈6%; Givnish et al. 1997). Additionally, *Brocchinia* is one of only three carnivorous lineages (the others being *Catopsis* and *Paepalanthus*; §10.3, §10.4) in which carnivory is present but not synapomorphic. This implies a recent evolution of the carnivorous habit in these three genera, in which character-state transitions may be inferred more easily than in taxonomically isolated, uniformly carnivorous genera or families (Givnish et al. 1984, 1997).

The tank fluid of the two carnivorous *Brocchinia* species is highly acidic (pH ≈4) and emits a sweet, nectar-like odor. The taller tanks of *B. hechtioides* appear to capture mainly bees and wasps, whereas ants are the predominant prey of *B. reducta* (Givnish et al. 1997). The greater dependence on flying insects by taller carnivorous plants, and on crawling insects by shorter plants, parallels a pattern observed for *Sarracenia* in the southeastern United States (Givnish 1989). Radioactive labeling studies conducted in the field show that the live trichomes on the leaf bases are capable of absorbing cations and amino acids (and thus organic N) at high rates (Givnish et al. 1984). Phosphatase activity has been noted on the surfaces next to trichomes on the leaf base, although the activity was much weaker than present in all other pitcher plants screened (Płachno et al. 2009a). The outer cell walls of the trichomes have an unusual labyrinthine structure with pores large enough to pass small protein particles (Owen et al. 1988, Owen and Thompson 1991). Unlike trichomes in Tillandsioideae species, those in the Brocchinioideae die after desiccation, which probably accounts for why *Brocchinia* has not evolved the atmospheric habit (Givnish 2016).

Larvae of inquiline flesh-flies (*Sarcophaga* spp.) and mosquitoes (*Wyeomyia* spp.) live in the tank fluid of both carnivorous *Brocchinia* species (Givnish et al. 1997), as does the recently described tree frog *Tepuihyla obscura*, at least on Chimanta-tepui (Kok et al. 2015), raising the possibility of additional nutrient inputs via frog excretions. The carnivorous bladderwort *Utricularia humboldtii* occasionally grows within the tanks of *B. hechtioides* (Givnish et al. 1997, A. Cross *unpublished data*).

Additional high-resolution phylogenetic studies are needed to elucidate the specific steps leading from the tank habit to carnivory, particularly as the tank epiphytic habit of *B. tatei* (which occasionally grows terrestrially, with heterocystous N-fixing cyanobacteria forming a plug) appears derived from a carnivorous ancestor (Givnish et al. 1997).

10.2.3 Distribution, habitat, and conservation

Brocchinia is endemic to the Guayana Shield (Figure 10.3) and the species occur mostly on the tepuis (sandstone table mountains) and nearby sand plains of southern Venezuela, with limited extension into adjacent parts of Colombia, Brazil, and Guyana (Smith and Downs 1974). A few taxa occur on granitic outcrops. *Brocchinia reducta* is restricted to the Altiplano of the Gran Sabana and the *tepuis* that emerge from it in southeastern Venezuela and nearby Guyana (Smith and Downs 1974), whereas *B. hechtioides* grows on most *tepuis* in southern Venezuela in the states of both Amazonas and Bolivar (Givnish et al. 1997). Both species grow in wet sandy savannas, on exposed sandstone, or in peaty meadows and bogs.

The presence of charcoal remains in many *tepui* bogs and *Stegolepis*-dominated meadows suggest that these burn with some regularity. Both *B. hechtioides* and *B. reducta* have been observed to survive ground fires, possibly because of their rainwater-filled tanks (Givnish et al. 1997). Fire may play an important role in promoting the carnivorous, ant-fed, and N-fixing species by volatilizing much of the N present in the ecosystem. Although both species occupy large ranges in relatively remote areas, their conservation status has not been assessed.

10.3 *Catopsis*

10.3.1 Morphology and systematics

Catopsis is a genus of 19 epiphytic species in the epiphytic bromeliad subfamily Tillandsioideae (Givnish et al. 2014a, Luther 2014). Among these, only *C. berteroniana* currently is considered to be carnivorous (Fish 1976, Frank and O'Meara 1984). *Catopsis berteroniana* is urn-shaped, with erect, bright-yellow leaves and a dense layer of highly reflective wax on the outer (abaxial) leaf surfaces facilitating prey attraction and capture (Gaume et al. 2004). It often occurs on the uppermost branches of shrubs and small trees (Figure 10.1b). Flowers of *Catopsis* are minute, insect-pollinated, hermaphroditic or dioecious, and are white to orange (Figure 10.2a). Molecular analyses place *C. berteroniana* as sister to *C. paniculata*, with both sister to the remainder of the genus (Gonsiska 2010).

10.3.2 Carnivory

Fish (1976) argued that *Catopsis berteroniana* was carnivorous because its dense waxy coat, highly reflective in the ultraviolet range, made it invisible against the sky. This argument was not compelling: the plant is highly visible in the remainder of the visual spectrum, and waxy dust on the outer leaf surfaces simply might reduce light interception and transpiration (Givnish et al. 1984). However, the large numbers of dead insects in the leaf axils of *C. berteroniana* (Fish 1976), the potential attractive value of the bright yellow and white leaves, the denial of insect foothold by its waxy covering (Gaume et al. 2004), and a $\delta^{15}N$ isotopic signature consistent with N input from animals (Gonsiska 2010) all suggest that this species is carnivorous. Carnivory in congeners with heavy waxy coats (e.g., *C. morreniana, C. subulata*) also should be investigated.

10.3.3 Distribution, habitat, and conservation

Catopsis berteroniana is distributed widely from the southern tip of Florida into the Caribbean, Central America, Colombia, Ecuador, Venezuela, and Brazil (Figure 10.3). The species is almost exclusively epiphytic and occurs in a wide range of wooded habitats. Although *C. berteroniana* occupies an extensive range, its conservation status has not been assessed.

10.4 *Paepalanthus*

The evolution of carnivory recently has been proposed for a third genus in the Poales: *Paepalanthus*, a genus of approximately 380 species in the Eriocaulaceae (a family of some 1200 species of predominantly Neotropical affinity; Stützel 1998). Only *P. bromelioides* (subg. *Platycaulon*) is thought to be carnivorous (Nishi et al. 2013). It occurs in *campos rupestres* on nutrient-poor soils over quartzite, and is endemic to the Serra do Cipó in the Serra do Espinhaço of Bahia state in southeastern Brazil (Figueira et al. 1994, Trovó et al. 2013). *Paepalanthus bromelioides* produces unusually large, bromeliad-like leaf rosettes with a central tank containing an acidic, mucilaginous liquid, and insects may be attracted to the rosettes by ultraviolet reflection by the leaves or in search of food or shelter (Nishi et al. 2013). The leaves are coated with a slippery wax and the leaf bases bear hydrophilous trichomes that appear to absorb nutrients (Jolivet and Vasconcellos-Neto 1993, Figueira et al. 1994, Nishi et al. 2013). Although 67% of nitrogen taken up by *P. bromelioides* is assimilated through its roots, ≈27% of total N inputs is derived from the excreta of associated invertebrate predators (especially spiders; Nishi et al. 2013). *Paepalanthus bromelioides* is not thought to secrete proteolytic enzymes (Figueira et al. 1994), and bacteria do not appear to contribute significantly to N cycling in the tank (Nishi et al. 2013), suggesting that the digestion of prey by invertebrate predators provides simple nitrogenous compounds that are absorbed by the leaf trichomes (Nishi et al. 2013). At least one of these predators, the web spider *Alpaida quadrilorata* (Araneidae), appears to be strictly associated with *P. bromelioides* where the plant occurs. On the basis of these observations, Nishi et al. (2013) proposed that *P. bromelioides* is carnivorous by means of digestive mutualism, as is *Roridula* (§10.8). Future studies should examine the absorptive capacity of the leaf trichomes in *P. bromelioides*, and explore the role of ultraviolet reflectivity in the leaves as a possible insect attractant.

10.5 *Drosophyllum*

10.5.1 Life history, morphology, and systematics

Drosophyllum lusitanicum is a short-lived, woody, perennial subshrub endemic to the Western Mediterranean (Garrido et al. 2003). Recruitment in natural habitats largely occurs from the soil seed bank following fire (Paniw et al. 2017b). It has a rosette growth form (Figure 10.1f), typically producing each season an additional secondary branching stem for each rosette from a woody rootstock (Correia and Freitas 2002, Paniw et al. 2017b). Rosettes comprise 10–20 erect, narrowly linear leaves with thread-like tips that are downwardly circular when in bud and can reach a length of 25 cm when fully developed (Correia and Freitas 2002). Leaves are densely covered by stalked sessile and pedunculate glands (Ortega-Olivencia et al. 1995), the former producing digestive enzymes and the latter secreting carbohydrate-rich mucilage (Lloyd 1942, Bertol et al. 2015a). As in the closely-related *Triphyophyllum* (§10.6), the glands of *Drosophyllum* are multicellular, multi-layered, and vascularized (Lloyd 1942).

Flowering typically takes two years to occur, with each rosette producing a single flowering scape of 4–7 flowers arranged in cymose, bracteate inflorescences with a pseudocorymbose appearance (Ortega-Olivencia et al. 1995, Salces-Castellanos et al. 2016). Flowers are hermaphroditic, actinomorphic, and pentameric (Figure 10.2f). They possess a dichlamydeous perianth consisting of five concrescent sepals at the base, and five large, free, sulfur-yellow petals. The stamens occur in two whorls of five and the gynoecia have five stigmas (Ortega-Olivencia et al. 1995). Individuals flower from March to July, with a peak in April or May (Ortega-Olivencia et al. 1995, Salces-Castellanos et al. 2016). *Drosophyllum lusitanicum* is highly autogamous (Ortega-Olivencia et al. 1995), although pollinators may play a significant role in population dynamics by assisting self-pollination and hence increasing seed production to replenish the persistent seed bank (Salces-Castellanos et al. 2016). Pollinator–prey conflict (Chapter 22, §22.2) appears not to exist (Bertol et al. 2015a).

The root system of *D. lusitanicum* is well developed and consists of a long, strongly branched, woody tap root with xeromorphic features (Carlquist and Wilson 1995, Adlassnig et al. 2005b). It may be 1-cm thick at its base, up to 40-cm long, and comprises ≈25% of the total dry biomass of the plant (Adamec 2009a, Paniw et al. 2017a). These tap roots initially were believed to be critical for water acquisition (Juniper et al. 1989), but recent studies indicate that root water supply is not sufficient to fulfil plant transpiration demands (Adlassnig et al. 2006, Adamec 2009a). Instead, plants maintain water balance with a thick leaf cuticle that reduces transpiration (Adamec 2009a) and by absorbing water from dew through a highly hygroscopic mucilage (Adlassnig et al. 2006).

Until the early 1990s, *Drosophyllum* was placed in Droseraceae as an outgroup to the *Drosera* clade (Albert et al. 1992) based on morphological similarities in its complex, multicellular flypaper traps (Juniper et al. 1989, Albert et al. 1992). Recent phylogenetic analyses using *rbcL*, sequences of the chloroplast *matK* gene, or a combination of several genetic markers, place *Drosophyllum* in the monogeneric Drosophyllaceae (Williams et al. 1994), sister to the Dioncophyllaceae + Ancistrocladaceae clade; all three families form a sister clade to the Nepenthaceae (Meimberg et al. 2000, Cameron et al. 2002, Cuénoud et al. 2002, Heubl et al. 2006, Renner and Specht 2011). Further strong support for the placement of *Drosophyllum* outside the Droseraceae was provided by extensive analyses of the chloroplast gene *rpl2*, which is interrupted by an intron in *Drosophyllum* but not in the Droseraceae (Meimberg et al. 2000).

10.5.2 Carnivory

Drosophyllum lusitanicum uses a passive flypaper trapping mechanism and captures insects using carbohydrate-rich adhesive mucilage produced by large, stalked glands (Darwin 1875, Adlassnig et al. 2010). Little is known about the chemical composition of the mucilage, but the sticky leaves are very efficient at attracting insect prey (Bertol et al. 2015a), which are digested by enzymes secreted by small, sessile glands (Adlassnig et al. 2006). The lack of lateral roots in adult plants indicates poor potential for soil nutrient uptake (Adamec 2009a) and recent investigations indicate that *D. lusitanicum* relies heavily on nutrients assimilated from prey for growth (Paniw et al. 2017a).

10.5.3 Distribution, habitat, and conservation

Drosophyllum lusitanicum is endemic to Mediterranean heathlands in the southwestern Iberian Peninsula and northern Morocco (Figure 10.3) and exhibits a restricted range and reduced genetic diversity likely resulting from range constriction following past climatic changes (Garrido et al. 2003, Heubl et al. 2006, Paniw et al. 2015). Preliminary population genetic analyses using microsatellite markers indicate that populations possess very limited intrapopulation diversity, but very high interpopulation variation (Paniw et al. 2014), probably exacerbated by the disjunction of populations through habitat fragmentation (Garrido et al. 2003) and poor seed dispersal (Ortega-Olivencia et al. 1995, 1998).

Contrary to the common association of carnivorous plants with wet habitats (Juniper et al. 1989; Chapters 2, 18), *D. lusitanicum* occurs in fire-prone heathlands on dry, nutrient-poor sandstone soils. The xeromorphic features of its roots are an adaptation to its habitat, and *D. lusitanicum* exhibits significant physiological adaptation to prolonged seasonal (summer) drought (Garrido et al. 2003, Adlassnig et al. 2006). It is a short-lived, fire-responsive species with distinct life-history adaptations to recurrent fire, and individuals are outcompeted by surrounding shrubs in as little as 3–4 years after fire (Paniw et al. 2017c). The formation of extensive soil seed banks facilitates population persistence; seedling emergence is stimulated by fire-derived cues such as heat and smoke in a similar fashion to southwest Australian *Byblis* (Correia and Freitas 2002, Cross et al. 2013, Paniw et al. 2015, 2017b, M. Paniw and A. Cross *unpublished data*), and re-sprouting shrubs in early post-fire habitats facilitate the successful establishment of *Drosophyllum* seedlings (Paniw et al. 2017b).

Most extant *D. lusitanicum* populations occur in secondary habitats where small-scale human disturbances such as mechanical uprooting or slashing to create fire breaks have replaced the role of fire in vegetation clearance (Garrido et al. 2003, Paniw et al. 2017c). Although fire-suppression policies and the large-scale degradation of natural habitats threaten many populations (Correia and Freitas 2002), small-scale human disturbances may play a critical role in sustaining several populations (Paniw et al. 2015). However, detailed demographic analyses comparing populations in fire-disturbed heathlands and human-disturbed secondary habitats have demonstrated that repeated disturbances such as vegetation slashing increases the mortality of mature plants and also may reduce reproductive success (Adlassnig et al. 2006, Paniw et al. 2017c).

Drosophyllum lusitanicum does not appear in the European Red List of Vascular Plants despite its phylogenetic and ecological uniqueness and declining populations (Bilz et al. 2011). Populations are largely protected in southern Spain, where the species is listed as vulnerable in the Andalusian Red List of Threatened Plants (BOJA 1994); it is considered rare in Morocco, where it is protected locally (Fennane and Ibn Tattou 1998); and it does not have formal protection in Portugal, where populations are declining rapidly (Correia and Freitas 2002).

10.6 *Triphyophyllum*

10.6.1 Life history, morphology, and systematics

Triphyophyllum is a monotypic genus comprising the single species *Triphyophyllum peltatum*, a tropical woody liana reaching 40–50 m in length in pristine rainforest, or growing as a scandent shrub in disturbed habitats (Airy Shaw 1951, Menninger 1965, Green et al. 1979). The stems are terete and glabrous, upright and self-supporting in juvenile plants and flexuous and scrambling in mature climbing vines. Old stems, covered by blackish-brown bark, can reach up to 10 cm in diameter at the base. In contrast to most other carnivorous plants, but similar to the related *Drosophyllum lusitanicum*, its root system is well-developed (Green et al. 1979).

Triphyophyllum is developmentally and seasonally heterophyllous, producing three morphologically and functionally distinct types of leaves depending on age and season: strap-shaped and spirally arranged juvenile leaves (20–35-cm long) with slightly undulate margins formed on juvenile plants 7–50-cm tall and on lateral short branches of the vine (Figure 10.1e); filiform, glandular carnivorous leaves, also produced mainly by young plants but occasionally occurring on the lateral spur shoots of mature lianas (Figure 10.1e); and the typical, hooked, mature oblong-elliptic leaves (6–17-cm long) with widened laminae and an apical pair of strong, backward pointing hooks that form only on climbing shoots with prolonged internodes (long shoots; Menninger 1965, Green et al. 1979). The juvenile leaves only photosynthesize, but the mature hooked leaves both photosynthesize and support the climbing vine, clinging or twisting around nearby twigs and branches. All three leaf types are caducous, with an articulating petiole base that remains attached to the stem (Airy Shaw 1951, Green et al. 1979).

The carnivorous leaves of *Triphyophyllum* resemble those of the related *Drosophyllum* (Drosophyllaceae), both in the unique reversed (outwardly-pointing) circinate vernation pattern and in the anatomy of the stalked and sessile glands (Lloyd 1942, Airy Shaw 1951, Green et al. 1979, Marburger 1979). Carnivorous leaves are held erect (1–6 can be present on a single shoot; Green et al. 1979), whereas the two types of noncarnivorous leaves are patent, spreading horizontally from the stem axis. The laminae of the carnivorous leaves is fully or partially reduced (the lowermost part of the latter "transitional leaves" is expanded blade-like, to the midrib; Green et al. 1979) to form a thread-like leaf 15–30-cm long that is densely covered by vividly red-colored, long-stalked, and sessile glands (Airy Shaw 1951, Menninger 1965, Green et al. 1979, Marburger 1979).

The stalked glands are multicellular, multilayered, and vascularized with both xylem and phloem (Metcalfe 1952, Green et al. 1979, Marburger 1979) and occur in two size classes. The larger ones may be up to 3-mm long, with a hemispherical gland head up to 1-mm diameter. Although these glands have been reported to be the largest in the plant kingdom (Green et al. 1979), glandular emergences with a gland head at least as large can be found on the resinous hypanthium of some species of the noncarnivorous *Barbacenia* (Velloziaceae, Pandanales; A. Fleischmann *unpublished data*). The sessile glands usually are grouped around the bases of the stalked glands (Green et al. 1979, Marburger 1979). The lifespan of the carnivorous foliage is comparatively short, with glandular leaves remaining active for only 3–6 weeks before withering and being shed (Green et al. 1979, Rembold et al. 2010b).

The inflorescences are glabrous, few-flowered cymes formed on axillary shoots on the long shoots of the climbing vine. Flowers are actinomorphic and (tetra- or) pentamerous (Figure 10.2e), with five persistent sepals and five petals that are free to the base, 10 subsessile stamens on very short filaments and a (4-)5-carpellate superior ovary topped by apically branched styles (Airy Shaw 1951, Schmid 1964). The white petals are caducous and last for a single day (Bringmann et al. 2002); the flowers are sweetly fragrant (Menninger 1965, Bringmann et al. 2002). The fruit is a three- to five-valvate dehiscent, leathery capsule. The capsule splits open before the seeds are fully ripe; the seeds mature and increase in size while attached to the open capsule valves by the thickening funicle. This form of seed-ripening is unique among the angiosperms and appears to be associated with the large seed shape of *Triphyophyllum*; the vividly red seeds are disk-like with a huge, circular, membranous wing 4.5–10 cm in diameter (Airy Shaw 1951, Menninger 1965, Green et al. 1979, Bringmann et al. 2002).

Triphyophyllum is placed in the small angiosperm family Dioncophyllaceae ("non-core Caryophyllales", *sensu* APG IV 2016). Dioncophyllaceae is sister to Ancistrocladaceae, a family consisting of the entirely noncarnivorous liana genus *Ancistrocladus*, and both are consecutive sister to Drosophyllaceae and Nepenthaceae (Heubl et al. 2006). Based on their strange leaf morphology, heterophylly, and fruit characters, Dioncophyllaceae were considered "one of the strangest groups of plants to be found in the vegetable kingdom" by the author of the family (Airy Shaw 1951: 327).

10.6.2 Carnivory

Triphyophyllum peltatum was first regarded as a carnivorous plant by Green et al. (1979) based on the morphology of the glandular leaves, field observations of numerous insects adhering to them, and the presence of hydrolytic enzymes in glandular secretions. Menninger (1965) had speculated about possible carnivory, but cited field notes of the botanist F. Hallé from Ivory Coast who claimed that the caught insects "never are digested like in the case of *Drosera*" (p. 31). Bringmann et al. (2001) provided experimental proof of carnivory by demonstrating the uptake of organic matter through the glandular leaves of the plant: after feeding isotopically labelled alanine to the trapping leaves, the isotope label was detected in apical shoot parts and noncarnivorous leaves within two days, which required initial digestion and uptake of the amino acid from the carnivorous glands.

Neither the stalked glands nor the leaves of *Triphyophyllum* are capable of any movement upon stimulation, nor do the sessile glands show any activity until stimulated by prey (Green et al. 1979). In these respects, they are identical to the carnivorous leaves and glands of the related *Drosophyllum*. The acidic mucilage of *Triphyophyllum* is not very sticky and of low viscosity, but stimulation by struggling prey results in copious mucilage secretion from the stalked and sessile glands, thus quickly coating and suffocating captured insects (Green et al. 1979). Observations in Sierra Leone suggest that prey consists mostly of crepuscular or nocturnal beetles, but other insects, including ants, small wasps, termites, mosquitoes, and grasshoppers often are trapped in large numbers (Green et al. 1979).

In its natural habitat, carnivorous leaves usually are produced just before the height of the wet season (likely synchronized with the maximum occurrence of prey; Green et al. 1979, Juniper et al. 1989). However, single carnivorous leaves occasionally are formed at other times throughout the year (Green et al. 1979, McPherson 2008), often after pruning back the lianas (Schmid 1964, Menninger 1965, Green et al. 1979, McPherson 2008). In cultivation, although carnivorous leaves generally are produced in a regular, cyclic manner (Rembold et al. 2010b), they also can be fully absent throughout the life cycle (especially in fertilized specimens; Bringmann et al. 2002, A. Fleischmann *unpublished data*). Carnivory therefore appears not to be essential to complete the full life cycle (Bringmann et al. 2002). As carnivorous leaves are only formed during certain stages of development, *Triphyophyllum* can be considered as a "part-time carnivore" (Juniper et al. 1989). Although this has been interpreted as a partial regaining of carnivory after its initial evolutionary loss in Dioncophyllaceae (Heubl et al. 2006), the suggestion of part-time carnivory in *Triphyophyllum* as a "phylogenetic snap-shot" on the way toward loss of carnivory seems both more reasonable and phylogenetically parsimonious (Fleischmann 2010, 2011a). This would imply a transitional stage from carnivorous ancestors such as the phylogenetically close Drosophyllaceae to noncarnivorous lineages such as the sister genera *Dioncophyllum* and *Habropetalum*, and sister family Ancistrocladaceae (Fleischmann 2010, 2011a; Chapter 3).

10.6.3 Distribution, habitat, and conservation

Triphyophyllum peltatum has a very narrow range in tropical West Africa (Figure 10.3), where it occurs in a small area in Sierra Leone, Liberia, and Ivory Coast, ranging from sea level up to ≈700 m altitude in the Loma Mountains of Sierra Leone (Airy Shaw 1951, Green et al. 1979). It grows in tall, dry-evergreen primary rainforest, but also in old secondary forest or disturbed habitats, usually in shaded conditions in the coppice of larger trees (Deighton in Airy Shaw 1951, Green et al. 1979, McPherson 2008). It prefers nutrient-poor, iron-rich, shallow, acidic soils consisting mainly of decomposed laterite pebbles and clay (usually overlying gneissic rock) that are waterlogged during the rainy season but well-drained during the dry season (Cooper and Record 1931, Green et al. 1979). Once widespread and common, the flexible woody stems of the liana were used to make ropes (Airy Shaw 1951). Although the species has not been assessed by the IUCN, *T. peltatum* faces a high risk of extinction in large parts of its range because of deforestation and anthropogenic habitat degradation. Not a single one of the sites in the Western Area Peninsula Forest Reserve near Freetown, Sierra Leone that were studied in 1972 by Green et al. (1979) could be relocated 34 years later, as most of the primary or secondary forest habitats had been deforested or turned into oil-palm plantations (Munro 2009, S. McPherson and A. Fleischmann *unpublished data*).

10.7 *Cephalotus*

10.7.1 Morphology and systematics

Cephalotus follicularis is the only species within the monogeneric Cephalotaceae. It is a small, herbaceous, rhizomatous perennial forb that forms low-lying rosettes of both lanceolate noncarnivorous leaves and complex prey-trapping pitchers (Figure 10.1d). Mature pitchers are 2–5-cm long and 1–2-cm wide and are green in shaded conditions and burgundy-red when exposed to full sunlight (Arber 1941, Juniper 1989). Pitchers possess two small outer keels and a large central double keel thought to provide structural stability, a toothed peristome with inward-facing hooks, and a large lid or operculum (Arber 1941). The lid of *C. follicularis* is unique among pitcher plants in being ventral, meaning that the high development of the ventral system of pitcher venation leads to its development, rather than the convergence of veins toward the midrib (Arber 1941, Juniper et al. 1989). The teeth along the peristome do not contain nectar glands and an additional three teeth set above the outer keels that are more developed are thought to add further structural stability.

Clusters of small, white, hermaphroditic flowers are produced at the end of a single long scape in early summer (Figure 10.2d), and are visited by small generalist insect pollinators (Conran 2004a). Each individual flower has six free sepals as lobes, no petals, and two whorls each of six stamens (Hamilton 1904, Conran 2004a). Anthers are distinct, introrse and cruciform, with a unique connective dorsal protrusion of large cells thought to act as "pseudonectaries" (Endress and Stumpf 1991). The gynoecium consists of six free carpels housed within superior ovaries that alternate with the inner stamens and sepals (Conran 2004a). Seeds have a small, linear embryo enclosed in a large endosperm, are contained within a hairy dispersal unit and are dispersed by wind or water upon dehiscence (Cronquist 1981). Seedling morphology is notable in that a non-vascularized extension of the hypocotyl grows back into the follicle, potentially representing a food reserve for the developing seedling (Conran and Denton 1996). *Cephalotus* possesses a chromosome number of $2n = 20$ (Keighery 1979).

Cephalotaceae were originally placed in the Saxifragales, Rosales, or their own order Cephalotales (Bessey 1915, Takhtajan 1997). Current molecular evidence places the family within the Oxalidales (APG 2016). Its nearest relatives are considered to be members of the Brunelliaceae and Elaeocarpaceae, but significant morphological differences with these families indicate an early divergence for *Cephalotus* (estimated at ~66 Mya; Heibl and Renner 2012). The geographic disjunction and phylogenetic isolation of *Cephalotus* from other genera producing pitcher traps indicates that carnivory evolved independently in Cephalotaceae.

10.7.2 Carnivory

The pitchers of *C. follicularis* open toward the apex of the leaf. Fine hairs covering the ridges and lid likely act as a pathway to draw prey toward the trap opening (Joel 1988). The lid is lifted, creating a small opening, and has multiple transparent patches thought to reflect light that attracts and disorients prey (Arber 1941, Parkes and Hallam 1984). Pitchers also possess glands on the outer and inner walls that secrete nectar, and the large inward-facing teeth around the peristome and an inner rim inside the pitcher mouth are obstacles preventing prey escaping from the small pool of fluid in the base of the pitcher (Parkes and Hallam 1984). Enzyme secretion and nutrient absorption occur from two sets of digestive glands in the pitcher wall. Prey consists primarily of the endemic ant *Iridomyrmex conifer*, but also includes small flies, wasps, arachnids, and molluscs. Pitcher size does not seem to affect prey capture, as even the smallest mature pitchers often contain prey. Although *C. follicularis* assimilates large amounts of its total potassium and iron intake from its prey, it is not reliant exclusively upon carnivory and can persist through periods of prey scarcity (Schulze et al. 1997, Pavlovič 2011).

10.7.3 Distribution, habitat, and conservation

The range of *C. follicularis* is a small area in the southwest corner of Western Australia centered on Albany (Figure 10.3), with isolated populations occurring from Nannup in the west to Manypeaks in the east. Although it is likely that the species once was distributed more or less continuously throughout a larger range, land clearing for agricultural expansion and climatic change have restricted the species to a handful of disjunct populations predominantly in *Homalospermum firmum/Callistemon glaucus* peat-swamp thickets. Only ≈2100 ha of this habitat remain in the greater Albany region, of which less than 500 ha lies within conservation reserves (Department of Parks and Wildlife *personal communication*). It appears that *C. follicularis* disperses poorly over long distances. Isolated populations have little genetic connectivity and habitat fragmentation has significantly reduced its genetic diversity (N. Kalfas *unpublished data*). Thus, it exhibits deep population structuring; distinct genetic groups have been identified along an east–west gradient, likely resulting from the geographical barrier of Mount Frankland and its associated ranges (N. Kalfas *unpublished data*). Recent surveys indicate that fewer than 25% of 114 historical locations recorded by the Australian Virtual Herbarium (AVH) remain (N. Kalfas *unpublished data*; CHAH 2006) and the decreasing rainfall predicted for the region by climate-change models may impact the long-term sustainability and persistence of the species (Pitman et al. 2004).

The pitchers of *C. follicularis* provide a dynamic habitat for the larvae of the wingless micropezid fly *Badisis ambulans*, which reside primarily under the rim of the pitchers and consume prey trapped in the pitcher fluid (McAlpine 1990, Adlassnig et al. 2011). *Badisis ambulans* is believed to have coevolved with *Cephalotus* and *Iridomyrmex conifer* (of which *B. ambulans* is a mimic; Yeates 1992) and the complex relationship between these three regionally endemic species parallels the complex biotic linkages in the southwest Western Australian floristic region.

10.8 *Roridula*

10.8.1 Morphology and systematics

Roridula is the only genus within the carnivorous plant family Roridulaceae, and consists of two South African endemic species, *R. dentata* and *R. gorgonias*. Both species are perennial and produce narrowly lanceolate leaves (40–90 mm) that are crowded at the branch tips (Figure 10.1h), although the leaves of *R. dentata* can be much smaller in winter (20–40 mm) when they catch fewer prey (B. Anderson *unpublished data*). The leaves are covered by prey-trapping glandular hairs, which produce extremely sticky, resinous glue. Glandular hairs are densest around the leaf margins, which are entire in *R. gorgonias* but toothed in *R. dentata*. *Roridula* has thin, woody stems, yet they are among the tallest free-standing carnivorous plants and can reach 2-m heights after a long fire-free period. Flowers are large and bright pink, with conspicuous yellow spheres at the base of each anther (Figure 10.2h). Pollen is released through a tiny apical pore in the anther of *R. dentata*, whereas the anthers of *R. gorgonias* dehisce laterally. *Roridula* anthers are unusual in that they swivel and change position when touched, but the function of both the yellow spheres and the swiveling action remains unknown. Both species have very poorly developed root systems, but they capture large amounts of prey.

Roridula has an unusually small genome of 0.76 pg (cf. 13.6 pg for other angiosperm taxa; Hanson et al. 2001, cf. Bennett et al. 1998). Reduced genome size is sometimes exhibited in species from phosphorus-impoverished habitats, as phosphorus is an essential component of repeated but dispensable DNA sequences (Hanson et al. 2001). Both *Roridula* species grow in very nutrient-poor soils.

Roridula frequently has been associated taxonomically with the similarly carnivorous but geographically distant *Byblis*. Both genera were grouped and placed in Droseraceae on the basis of their apparently similar traps (Bentham and Hooker 1865, Netolitsky 1926), in Byblidaceae (Domin 1922, Cronquist 1988), or in Roridulaceae (Marloth 1925, Kress 1970, Thorne 1983). Molecular analysis revealed that similarities between *Byblis* and *Roridula* resulted from convergence, and *Roridula* now is placed within the monogeneric Roridulaceae as a member of the Ericales (Chase et al. 1993). Roridulaceae is closely allied to the Actinidiaceae and the carnivorous Sarraceniaceae (Chase et al. 1993, Conran and Dowd 1993, Bayer et al. 1996; Chapter 9). These three families form a cluster that is basal to the Ericaceae.

The geographic isolation of *Roridula* from its closest relatives, none of which occur in South Africa, suggests that it is a paleoendemic genus (Linder et al. 1992). The recent discovery of fossilized Roridulaceae leaves from Eocene Baltic amber (dated to 35–47 Mya) suggests that *Roridula* once had a much broader distribution (Sadowski et al. 2015).

10.8.2 Carnivory

Roridula is very effective at trapping insect prey, but both species assimilate nitrogen predominantly through an obligately mutualistic relationship with *Pameridea* "bugs" (Hemiptera: Miridae; Ellis and Midgley 1996, Anderson and Midgley 2003, 2007, B. Anderson 2006; Chapter 26); each species of *Roridula* is tended exclusively by one of the two species comprising the genus *Pameridea* (Dolling and Palmer 1991). *Roridula gorgonias* is tended by *P. roridulae*, whereas *R. dentata* is tended by *P. marlothii*. *Roridula* relies upon *Pameridea* wastes for ≈70% of its total nitrogen uptake (Anderson and Midgley 2003) and *Pameridea* lays its eggs within the woody tissue of the plants and eats prey captured by the plants. High gene flow between populations within the same region indicates that *Pameridea* are capable of dispersing over short distances between *Roridula* populations (Anderson et al. 2004). However, the bugs cannot move between distant populations and it is likely that the geographic disjunction between, and subsequent divergence of, *Roridula* species resulted in co-speciation of *Pameridea* (Anderson et al. 2004).

10.8.3 Distribution and habitat

Roridulaceae is one of five families endemic to the Cape Floristic Region (CFR) of South Africa, a global biodiversity hotspot harboring over 9000 species in an area of only 78,555 km^2 (Goldblatt and Manning 2000). *Roridula* occurs on very nutrient-poor soils supporting fire-prone fynbos vegetation, but unlike the other families characteristic of fynbos vegetation (e.g., Proteaceae, Restionaceae, and Ericaceae), Roridulaceae consists of just a single genus and two species (Obermeyer 1970b). The two species of *Roridula* are geographically separated by ≈70 km (Obermeyer 1970b, Carlquist 1976): *R. gorgonias* grows in the southern mountains of the CFR, whereas *R. dentata* grows in the northwest (Figure 10.3). Another gap of ≈50 km separates the northern and southern populations of *R. dentata*.

Roridula gorgonias grows in peaty seeps and marshy areas with permanent surface water, whereas *R. dentata* typically grows on sandy *vlakte* (flats) that remain drier in summer even though underground water may be plentiful (Carlquist 1976). Populations of both species occur in relative isolation and studies indicate that little or no gene flow occurs even between geographically close populations (Anderson et al. 2004). It is therefore likely that the dispersal of both seed and pollen occurs over very small distances, resulting in population divergence through local adaptation and genetic drift. The CFR has undergone a significant drying period between the Oligocene and Quaternary periods (35–1.8 Mya) that may have played a significant role in population fragmentation, reproductive isolation, and ultimately speciation within the genus (Anderson et al. 2004, B. Anderson 2006).

10.9 *Byblis*

10.9.1 Life history, morphology, and systematics

Byblis is a predominantly Australian-endemic genus of either annual or short-lived perennial carnivorous fibrous-rooted herbs or herbaceous subshrubs in the monogeneric family Byblidaceae. Individuals either regenerate each year from a subterranean rhizome or recruit each season from the soil seed bank. Most annual species are small (5–45 cm, occasionally to 60 cm) with simple erect or scrambling stems, whereas short-lived perennial species can become large, almost bushy plants up to 60-cm tall with multiple branched stems developing as secondary growths from a woody rootstock (Figure 10.1c). All *Byblis* species possess alternate, subulate-linear leaves (2–25 cm long) that are covered in both stalked and sessile sticky glandular hairs.

Byblis produces large and showy solitary axillary hermaphroditic flowers ≤4-cm diameter (Figure 10.2c), with cerise to pale-lilac, rarely white, obovate petals, and abaxially twisted stamens held on short filaments. Anther filament length is considered a diagnostic feature among the annual species (Lowrie 2013). *Byblis* flowers are thought to be buzz-pollinated by native bees and hoverflies, with pollen released by the anthers when they are vibrated by its pollinator's wings (Lowrie 1998, Conran et al. 2002). The size, color, and morphology of *Byblis* flowers have led to the hypothesis that they mimic flowers of the sympatric *Thysanotus* (De Buhr 1975a). Many of the ≈50 species of *Thysanotus* (Asparagaceae) co-occur with *Byblis* in southwest and northern Australia and exhibit similar flowering phenology. *Thysanotus* produce large and showy lilac flowers with abaxially twisted stamens that also are buzz-pollinated, although it is unknown whether their pollinators are conspecific with those of *Byblis* (Conran et al. 2002).

Fire is considered likely to play a significant role in the seed biology and population maintenance of most *Byblis* species (Cross et al. 2013, A. Cross *unpublished data*). Following fire, *B. gigantea* survives by regenerating from its deep, woody, perennial rootstock (Baird 1984), whereas *B. lamellata* re-sprouts from rhizomes and above-ground stems (Conran et al. 2002, Lowrie 2013). Germination in all species is enhanced by exposure to smoke-derived compounds (Lowrie 1996, A. Cross *unpublished data*), and extensive natural post-fire recruitment has been observed in the populations of most annual species (Cross et al. 2013).

Byblidaceae are placed in the Lamiales and formerly were considered to be close to Lentibulariaceae on the basis of shared morphology and embryology (Albert et al. 1992, Conran 1996). However, recent phylogenetic studies indicate that a close relationship between these two families is unlikely (Savolainen et al. 2000, Schäferhoff et al. 2010). Carnivory has evolved independently in Byblidaceae, which has an early-branching position in Lamiales far from the more derived Lentibulariaceae (Schäferhoff et al. 2010). Infrageneric phylogenetic analyses using molecular phylogenetic data and morphological traits indicate that the closely related *B. gigantea* and *B. lamellata* are sister to all tropical species, and that the annual northern taxa are divided into two well-supported sister clades: one containing *B. aquatica* and *B. filifolia*, the other containing *B. guehoi* and *B. liniflora*. *Byblis rorida* is a sister lineage to the less well-supported taxa (Fukushima et al. 2011). *Byblis pilbarana* appears to be related most closely to *B. rorida* on the basis of ecology and morphology (Lowrie 2013), and these two species are likely to comprise a sister clade to *B. guehoi* and *B. liniflora*.

10.9.2 Carnivory

The small glandular hairs of *Byblis* leaves are effective at capturing prey (Lloyd 1942), and the numerous sessile glands show evidence of being both digestive and absorptive (Bruce 1905). All *Byblis* are considered to be carnivorous, and phosphatase is produced by glands in the leaves (Płachno et al. 2006). Ongoing studies examining the isotopic signature of numerous *Byblis* species indicate that plants assimilate significant amounts of N and P from captured prey (L. Skates, *unpublished data*). The mucilage produced by the glands on *Byblis* is relatively weak, but all above-ground parts of the plant except for the corolla are covered in glands that trap small- to medium-sized flying insects.

Byblis show a relationship with the predatory bugs of the genus *Setocoris* (Hemiptera: Miridae; Lowrie 1998, 2013, Conran et al. 2002), which live upon the plants and feed on captured prey (Hamilton 1903, Lowrie 1998). Although the relationship between *Byblis* and *Setocoris* remains unexplored, it is likely that these insects assist in nutrient turnover as has been demonstrated for *Pameridea* bugs on *Roridula*.

10.9.3 Distribution, habitat, and conservation

Eight species of *Byblis* are currently recognized (Lowrie and Conran 1998, 2007, Conran et al. 2002, Lowrie 2013), including six annual species from the semiarid to humid tropical region of northern Australia (*B. aquatica, B. filifolia, B. rorida, B. guehoi, B. liniflora*, and *B. pilbarana*) and two perennial species (*B. gigantea* and *B. lamellata*) from Mediterranean southwestern Australia (Figure 10.3). *Byblis liniflora* is the only taxon whose distribution extends outside of Australia, into southern New Guinea.

All northern Australian species grow in seasonally wet or damp habitats such as open savannas on alluvial sediments (*B. liniflora*), herbfields and seepage areas on sandstone and sandstone-derived soils (*B. filifolia, B. rorida, B. aquatica*), or in open sandy soil patches between low native tussock grasses (*B. guehoi, B. pilbarana*). Most of these species are widespread and not threatened: *B. aquatica* ranges from Darwin to Berry Springs in the Northern Territory; *B. filifolia* throughout northern Western Australia and western Northern Territory; *B. rorida* scattered throughout the northern Kimberley; *B. liniflora* patchily distributed throughout monsoonal northern Australia; and *B. pilbarana* in the northwest Pilbara region of Western Australia. In contrast, *B. guehoi* is known only from a small area on the Dampierland Peninsula, near Beagle Bay Mission, and is therefore a species of potential conservation significance. *Byblis lamellata* grows on deep sands in fire-prone dry heathlands in the Geraldton Sandplains, and although the species may be impacted by current and future mining activities, it currently is not threatened (Conran et al. 2002). *Byblis gigantea* is restricted to seasonally waterlogged *Leptospermum*/Restionaceae low scrub on the Swan Coastal Plain (Speck and Baird 1984, Cross et al. 2013) and undoubtedly was more widespread in the past. Land clearing and development have significantly reduced the range of the species to only a handful of disjunct populations (Conran et al. 2002, Cross et al. 2013).

10.10 *Philcoxia*

10.10.1 Morphology and systematics

Philcoxia (Plantaginaceae) is a small and unusual genus of annual or perennial carnivorous herbs (Figure 10.1g). Plants have an underground stem, sometimes much reduced, rhizomatous, or tuberous, holding laxly to densely arranged rosettes of tiny (1–7 mm in diameter) peltate leaves at or slightly below the soil surface (Taylor et al. 2000, Fritsch et al. 2007, Fleischmann 2012b, Pereira et al. 2012, Carvalho and Queiroz 2014, Scatigna et al. 2015). Leaves are usually covered by a thin layer of sand grains stuck on stalked, capitate glandular hairs, which secrete a sticky mucilage and trap and digest nematode prey (Pereira et al. 2012). Although it is difficult to distinguish underground stems from petioles in some species (Fritsch et al. 2007, Scatigna et al. 2015), the well-developed underground structures of *P. bahiensis*, *P. minensis*, *P. rhizomatosa*, and *P. tuberosa* suggest that these species are long-lived perennials, whereas *P. goiasensis* and *P. maranhensis* have delicate and reduced stems and are probably annuals (Taylor et al. 2000, Scatigna et al. 2016a, 2017). Stem development is highly variable in *P. courensis* and further anatomical studies on the genus are required. *Philcoxia* produces simple or branched racemose inflorescences up to 30-cm tall with a fractiflex axis, bearing numerous white to purple flowers (Scatigna et al. 2016a). Flowers are tubular and personate with a spurless lamialean corolla (Figure 10.2g), and have only two stamens with monothecous anthers (Taylor et al. 2000, Fritsch et al. 2007).

The genus comprised just three species when it was first described less than 20 years ago, but the number of species increased to seven in the last four years (Carvalho and Queiroz 2014, Scatigna et al. 2015, 2017). *Philcoxia* was first placed in Scrophulariaceae (*s. l.*) with affinity either with Scrophularieae (now Scrophulariaceae *s. str.*, following APG IV 2016) or Gratioleae (now a tribe of Plantaginaceae *sensu* APG III 2009). Fritsch et al. (2007) and Schäferhoff et al. (2010) provided phylogenetic evidence for the inclusion of the genus within Gratioleae, but left its specific placement in question. Scatigna et al. (*unpublished data*) propose that *Philcoxia* is sister to *Stemodia stellata*, a species endemic to the Brazilian quartzite outcrop, and also demonstrated interspecific relationships within the genus implicating geography as a probable driver of speciation.

10.10.2 Carnivory

The restriction of *Philcoxia* to nutrient-poor habitats, the morphology of the glandular-pubescent leaves, and the presence of dead soil nematodes (or their chitin cuticle leftovers) stuck to leaf surfaces led to speculation about carnivory in the genus (Fritsch et al. 2007). Although the carnivorous nature of *Philcoxia* was questioned by Taylor et al. (2000), and an initial test for carnivory done on greenhouse-grown plants of *P. minensis* was negative (Fritsch et al. 2007), recent detailed studies have provided more compelling evidence that the genus obtains nutrition directly from captured prey (Pereira et al. 2012). Phosphatase enzyme activity has been detected in the leaf glands of *P. minensis* and *P. rhizomatosa* (Pereira et al. 2012, Scatigna et al. 2015), and isotopic analysis indicates that nitrogen originating from nematode prey is assimilated into plant tissues as a direct result of enzymatic processes (Fleischmann 2012b, Pereira et al. 2012).

10.10.3 Distribution, habitat, and conservation

All seven currently described species of *Philcoxia* are endemic to sandy, comparatively dry and nutrient-poor open areas of *cerrado* (savanna), *campos rupestres* (savanna-like vegetation, associated with quartzite and ironstone outcrops and a nutrient-poor soil) or *caatinga* (Brazilian tropical dry forest) in the Brazilian states of Bahia, Goiás, Maranhão, and Minas Gerais (Figure 10.3). Although the conservation status of *Philcoxia* has not been assessed by the IUCN, the listing of all species as Critically Endangered has been proposed on the basis of isolation, fragmentation, and continuing decline of suitable habitat through inappropriate fire management, forestry, cattle grazing, and diamond mining activity (Guimarães et al. 2014, Scatigna et al. 2015, 2016a, 2016b, 2017). Only the distribution of *P. minensis* and *P. tuberosa* are completely encompassed by protected areas.

10.11 Future research

The small carnivorous genera covered are some of the most poorly studied carnivorous plant groups, so there are many, diverse opportunities for future research about them. A thorough examination of population genetics is needed for all nine genera, and especially for the increasingly at-risk species *C. follicularis, D. lusitanicum*, and *T. peltatum*. Such studies should be complemented with field research and modeling of their distribution, environmental niches, and spatial dynamics. Similarly, research into the conservation status and threats to all species is an urgent priority. Little to nothing is known about the reproductive ecology and recruitment dynamics of *Brocchinia, Catopsis, Paepalanthus, Philcoxia*, or *Triphyophyllum*, and further examination of mechanisms facilitating prey attraction, prey capture, and nutrient acquisition are needed for *Paepalanthus* and *Philcoxia*. The mechanisms by which prey digestion and nutrient uptake occur in poorly studied groups should be examined in detail (e.g., Poales). Foliar and root morphology, growth and development, genomics, and nutritional status should be compared between carnivorous and non-carnivorous species in families where carnivory is not ubiquitous. Studies using modern molecular techniques are needed to resolve taxonomy and develop well-resolved phylogenies for *Byblis* and *Philcoxia*. Finally, the ecological significance of the relationships noted between several of these species and associated invertebrates represent interesting areas deserving of future research: *Byblis* and *Setocoris* bugs, *Cephalotus* and *Iridomyrmex conifer* ants, and *Roridula* and the spiders *Peucetia nicolae* and *Synema marlothi* (Chapter 26).

CHAPTER 11

Carnivorous plant genomes

Tanya Renner, Tianying Lan, Kimberly M. Farr, Enrique Ibarra-Laclette, Luis Herrera-Estrella, Stephan C. Schuster, Mitsuyasu Hasebe, Kenji Fukushima, and Victor A. Albert

11.1 Introduction: flowering plant genomes with a twist

Carnivorous flowering plants are angiosperms that have specialized over evolutionary time to exploit metazoan prey for nutritive purposes (Chapters 1, 3). Although carnivorous plants have unique combinations of biochemistries, morphologies, and physiologies, we are only beginning to understand whether the underlying adaptations for these features also are unique at the genome level (the genome is the sum of all protein-coding and non-coding DNA in an organism). The recent complete sequencing of high-quality nuclear genomes for *Utricularia gibba* and *Cephalotus follicularis* is the first step toward examining the genomic basis of the evolution of botanical carnivory.

Like other angiosperms, the evolutionary landscape of carnivorous plant genomes has been influenced heavily by a combination of polyploidy events and upsurges in non-coding DNA. All living angiosperms descend at their base from one polyploidy event, and thereafter the most species-rich lineage (the eudicots, ≈75% of all angiosperms) underwent a complex triplication event prior to its evolutionary diversification (Soltis et al. 2009). Whole genome duplication (WGD) events copy every gene within a genome, and even though not all duplicated copies are retained evolutionarily following such events, many new or subdivided gene functions can develop each time a polyploidy event occurs (Van de Peer et al. 2009, Sankoff and Zheng 2012).

Genome size is superficially related to the number of such doublings that have occurred, but is more the result of proliferations of non-coding DNA, such as transposable elements (TEs), in various ways and in different lineages. The proportion of genomic space occupied by genes is not significantly greater for a hypothetical, pre-polyploid 25,000-gene genome than it is for a post-polyploid net-30,000-gene genome (following its diploidization), but the two can differ dramatically in overall size because of "blooms" of copy-paste TEs (Bennetzen and Wang 2014, Zhao et al. 2015). Such is the case for *U. gibba*; it contains essentially the same number of genes as *Vitis vinifera* (grape), *Solanum lycopersicum* (tomato), and even *C. follicularis*, despite having a total haploid nuclear genome size of 100 million bases (Mb), as opposed to 500 Mb in *V. vinifera*, 825 Mb in *S. lycopersicum*, and at least 1600 Mb in *C. follicularis* (Fukushima et al. 2017, Lan et al. 2017). Moreover, genome sequencing of these species has revealed that *V. vinifera* and *C. follicularis* genomes have not duplicated further since the ancient triplication event, but the tomato lineage has experienced a second triplication of its own, and in the case of *U. gibba*, at least two lineage-specific genome doublings occurred since sharing last common ancestry with *V. vinifera* and *C. follicularis* (Fig. 11.1). Present-day genome size and ancient genome duplication status are not fundamentally correlated.

Another process that occurs continuously in angiosperm genomes is local duplication of genes, often on a single gene-by-gene basis (Panchy et al. 2016). Such small-scale duplication processes

Renner, T., Lan, T., Farr, K. M., Ibarra-Laclette, E., Herrera-Estrella, L., Schuster, S. C., Hasebe, M., Fukushima, K., and Albert, V. A., *Carnivorous plant genomes*. In: *Carnivorous Plants: Physiology, ecology, and evolution*. Edited by Aaron M. Ellison and Lubomír Adamec: Oxford University Press (2018). © Oxford University Press. DOI: 10.1093/oso/9780198779841.003.0011

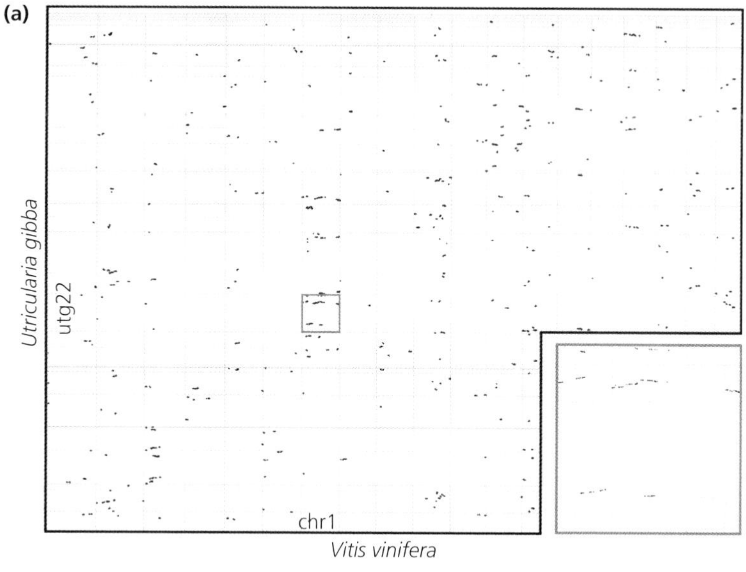

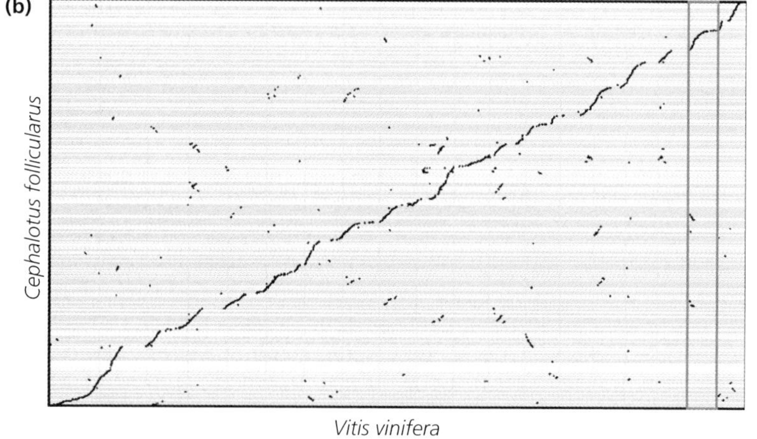

Figure 11.1 Syntenic path alignments (SPA) for (**a**) *Utricularia gibba* (*y*-axis) and *Vitis vinifera* (*x*-axis), and (**b**) *Cephalotus follicularis* (*y*-axis) and *V. vinifera* (*x*-axis). Inset in (**a**) expands a region of the SPA where three syntenic blocks within an entire chromosome (assembled as genome fragment utg22) match *V. vinifera* chromosome 1. In (**b**), a region of the SPA (surrounded by a gray rectangle) highlights an entire *V. vinifera* chromosome (*x*-axis) against multiple *C. follicularis* scaffolds (*y*-axis). Three internally syntenic blocks exist in both genomes due to the shared ancient gamma hexaploidy that occurred at the base of the core eudicots. In this comparison, the diagonal represents orthologous gene pairs between *C. follicularis* and *V. vinifera*, and the two smaller blocks of homologs below the diagonal, the left-hand side of the *V. vinifera* chromosome, represent paralogous gene pairs. SPAs constructed using CoGe, a comparative genomics platform (Lyons et al. 2008).

generate a different set of possible functional evolutionary possibilities for angiosperms. When we seek to understand differences between genomes of carnivorous plants and their noncarnivorous relatives, we must always keep in mind these two distinct processes: WGD and local gene duplication. In this chapter, we review salient research and provide new insights on nuclear genomes of carnivorous plants. We also review the status of our understanding of their organellar genomes; these are the far smaller (130 Kb–4 Mb) genomes that exist in the chloroplast and mitochondrial organelles of all plants. Insights into these genomes and the far more functionally diverse nuclear genomic components of plants can be drawn from rich comparative analyses afforded by considerable advances in DNA sequencing over the last decade and deep taxonomic sampling of angiosperm lineages.

11.1.1 Nuclear genome sequencing and assembly efforts for carnivorous plants

Next-generation sequencing has focused heavily on the Lentibulariaceae (Lamiales) and *Cephalotus* (Oxalidales), although the carnivorous habit has evolved independently at least three additional times within the Ericales (Sarraceniaceae), Caryophyllales (Dioncophyllaceae, Nepenthaceae,

Droseraceae), and Poales (Bromeliaceae; Chapter 3). Lentibulariaceae genomes are incredibly diverse in size, ranging from perhaps 60 to 1,500 Mb, so they have been a focus both of general efforts to understand genome evolution and of targeted research aimed at understanding genomic features specific to carnivorous plants (Greilhuber et al. 2006, Veleba et al. 2014). Genome sequencing does have technological limits. Until very recently, short-read shotgun sequencing approaches were standard. These projects, based on the sequencing of millions of random, short pieces of DNA (e.g., 100–500 bases at a time), have yielded helpful genome assemblies in terms of discovery of overall "gene space," but do not capture the full content of the highly repetitive, non-coding fractions of genomes. For example, the published genome of *U. gibba* based on a short-read assembly is approximately 82 Mb in length (Ibarra-Laclette et al. 2013), but Lan et al. (2017) discovered an additional 20 Mb of DNA using new, long-read sequencing technology. Long-read technologies (e.g., PacBio single-molecule sequencing) now regularly provide random DNA reads > 10,000 bases, which permits far easier deconvolution of repetitive regions of genomes such as TE families. Instead of such identical or near-identical repeats collapsing on one another and confounding the *in silico* assembly process, sequencing through them on single DNA molecules resolves them more simply. Thus, the first published genome of *U. gibba* (Ibarra-Laclette et al. 2013) was fragmented into small pieces: half of the assembly represented pieces > 100 Kb long and half of them less (a statistic called "N50"), but its newer PacBio genome has an N50 of genome fragments equal to 3.4 Mb (Lan et al. 2017). The reliability of our evolutionary analyses increases enormously with genomes like these that are much more contiguous.

Fragmented genomes also limit evolutionary research by introducing unknown biases in gene numbers and identities, coding sequence completeness, non-coding DNA content, and structural context of all genome features relative to each other. For example, the PacBio assembly of *U. gibba* gained some 2,000 genes that predominantly lay within the repeat-rich parts of the genome that previously had been unsequenced (Lan et al. 2017). Unfortunately for comparative studies within Lentibulariaceae, only genome fragment N50s < 100 Kb are currently available for *Genlisea* species (Leushkin et al. 2013, Vu et al. 2015). As such, low gene numbers reported for several *Genlisea* genomes that were even poorer in their contiguity than the first *U. gibba* genome assembly should be considered unreliable, and as unreliable as their entire assembly lengths, which were claimed to be even shorter than the first *U. gibba* short-read genome assembly (Leushkin et al. 2013, Vu et al. 2015).

Not all short-read genome assemblies suffer from poor fragment contiguity, but far greater effort and expense are required to get them up to par with long-read assemblies. Short-read assemblies also can be combined with long-read gap filling. On the other hand, current long-read sequencing methods have a considerably higher per-base error rate than short-read methods. By combining both technologies, users can reap the scaffolding benefits of long reads with the lower error rates (and higher depth of coverage) of short reads for ensuring confidence in base calls. This combined sequencing method was carried out for the *Cephalotus* genome (Fukushima et al. 2017). Although the *C. follicularis* genome comprises a massive 1.6 Gb, the fragment-size N50 for the its genome assembly was about three times longer than for the first *U. gibba* assembly, which was based on short reads alone (Ibarra-Laclette et al. 2013, Fukushima et al. 2017). Most other carnivorous plants are predicted to have large genome sizes, some (e.g., *Nepenthes* spp.) even larger than *C. follicularis*, so long-read single-molecule assemblies will be vital for reliable comparative studies (Greilhuber et al. 2006, Knápek 2012, Jensen et al. 2015).

11.2. Genome evolution

11.2.1 *Utricularia gibba* has a dynamic genome

The bladderworts, *Utricularia*, comprise one of two largest and most diverse carnivorous plant genera (Chapter 8). Through use of their bladder-like traps, prey are captured and digested via hydrolytic enzymes (Chapters 16, 19, 25). Using PacBio SMRT third-generation sequencing technology, a revised high-quality *U. gibba* nuclear genome was assembled and annotated (Lan et al. 2017). This additional sequencing permitted the assembly of several complete chromosomes, including one ≈8.5 Mb in length. Although the new genome sequence contained considerable repetitive DNA that had escaped the earlier ~82 Mb short-read assembly (Ibarra-Laclette et al. 2013), its genome size is still considerably reduced relative to its ancestors. Compared to other angiosperms, this shrinkage was not accompanied by an appreciable reduction in overall gene number. Rather, even the short-read *U. gibba* assembly showed that it had experienced a 1.5% net gain of conserved genes previously thought to be single-copy in *Arabidopsis thaliana* (thale cress), *Oryza sativa* (rice), *Populus trichocarpa* (poplar), and *V. vinifera*. Therefore, the principal cause of genome shrinkage in *U. gibba* was contraction of non-coding DNA.

Maximum likelihood estimates of gene-family dynamics showed that *U. gibba* has both higher gene gain and loss rates compared with *A. thaliana*, *S. lycopersicum*, *V. vinifera*, and *Mimulus guttatus* (yellow monkeyflower) genomes, all while having undergone genome size reduction. Although this pattern provides evidence for 'sloughing off' of many genes newly acquired through WGD or small-scale duplication, adaptive retentions of new gene duplicates also must have occurred. Evidence for the adaptive maintenance of duplicates can be found, for example, in the form of *U. gibba*-specific gene-family expansions related to processes that could be functionally associated with carnivory, such as nutrient acquisition and stress response (§11.4.1).

Separating the polyploid versus tandemly derived duplicate portions of the *U. gibba* genome is important to understanding the species' genome adaptive landscape. The new PacBio assembly has allowed for these different duplicate fractions to be conservatively distinguished (Lan et al. 2017). Fifty-four homeologous (i.e., homologous, yet polyploid) block pairs that descend from the most recent *U. gibba* WGD event were identified using syntenic analysis. Lan et al. (2017) then were able to reconstruct the nine-chromosome pre-polyploid ancestor of the modern *U. gibba* genome. Many large-scale inversion events could be inferred to have post-dated this WGD. An additional reconstruction was then performed on this ancestral genome to derive an earlier, six-chromosome ancestor that existed immediately prior to *U. gibba*'s second-most recent polyploidy event. Lan et al. (2017) could not resolve the third WGD event reported by Ibarra-Laclette et al. (2013), yet microsynteny analyses revealed many examples of eight (or more)-to-one syntenic block relationships with the *V. vinifera* genome. However, some of these relationships may date all the way to the gamma paleohexaploidy, which *Utricularia* shares with *V. vinifera*; in such cases, there may be 1–2 additional *V. vinifera* blocks that could map to the 8:1 or greater *Utricularia:V. vinifera* examples studied.

By closely examining the duplicate block pairs from the most recent WGD event, Lan et al. (2017) assessed the degree of gene loss experienced by each subgenome following polyploidization (i.e., its "fractionation"). Results of this analysis indicated a pattern of deletion bias characteristic of subgenome dominance inherited through a polyploidy event. Subgenome dominance refers to the condition where one genome contributing to a polyploidy event tends to retain more homeologous genes than the other, which fractionates less over time, likely because of gene-expression dominance (Sankoff and Zheng 2012). Gene-expression dominance in turn may be reflected by different TE distribution patterns in the new subgenomes, whereby greater densities of TEs and their epigenetic silencing near genes can influence gene expression negatively (Cheng et al. 2012, 2016).

In *U. gibba*, this bias in fractionation was matched by subgenome expression dominance, and on dominant blocks, fewer single nucleotide polymorphisms (SNPs) on average across all DNA sequences. This overall lesser DNA variation on dominant blocks likely reflects stronger purifying selection in and surrounding genes, as would be expected for homeologs taking on the greatest share of expression. Such subgenome dominance appears to derive most often through allopolyploidization events, wherein the genome of one parent species shows less epigenetic repression surrounding its gene space. Therefore, the most recent *U. gibba* WGD likely was an allopolyploidization event, including hybridization of two ancestral species, after which time a third species with novel phenotypic traits may have been generated instantly. Furthermore, evidence of heterogeneous patterns of heterozygosity in the *U. gibba* genome that do not match the boundaries of the observable polyploid blocks suggests broad but homoploid outcrossing events following the most recent WGD (Lan et al. 2017). As aquatic *Utricularia* can undergo vegetative propagation to produce genetically identical progeny, strongly heterozygous genotypes could represent states of particularly adaptive "frozen" heterozygosity, similar to those observed in populations of clonal *Populus* and unisexual hybrid vertebrates (Lampert and Schartl 2008, Macaya-Sanz et al. 2016).

11.2.2 Selection for genome size reduction in the Lentibulariaceae

Some investigators have argued that genome size reduction in Lentibulariaceae may have resulted from some adaptive factors favoring genome size decrease (Veleba et al. 2014). An argument based on the nutrient limitations imposed by carnivory does not hold in any obvious way, since many carnivorous plant genomes, including other Lentibulariaceae, are very large, and all arguably live under more-or-less similar phosphorus and nitrogen stress (Chapters 2, 17–19).

There are no well-studied cases in plants where selection for genome size reduction has been demonstrated. Initial work on the *Arabidopsis lyrata* genome and PCR-based indel comparisons of multiple *Arabidopsis* accessions suggested selection for deletions, but the results could not be repeated when massively parallel shotgun sequencing was used instead of DNA fragment amplification (Hu et al. 2011). Ibarra-Laclette et al. (2013) did not dismiss the idea of some selection pressure possibly favoring genome size reduction in *U. gibba*, but they showed that this process would not be required under random genetic drift-biased population genetic conditions (i.e., low effective population size). Random drift theoretically will enhance effects of any underlying molecular mechanistic biases in genomic change, and the most prevalent of these in plants appears to be unchecked retroelement copy-paste processes and non-adaptive genome size increases (Lynch 2007). If, however, *U. gibba* has particularly effective epigenetic silencing mechanisms to inhibit retrotransposon proliferation, but error-prone DNA repair capabilities that favor deletions of DNA over insertions (as is known in *Arabidopsis*), then random genetic drift causes genome size to decrease passively (Ibarra-Laclette et al., 2013). Regardless, essential gene space in the *U. gibba* genome will be preserved by purifying selection pressure, even if there is no positive selective pressure for DNA loss.

11.2.3 Adaptive evolution through gene duplication is largely limited to small-scale events in *Cephalotus follicularis*

The large *C. follicularis* genome (§11.1.1) reveals no evidence for further polyploidy events since it diverged from the last common ancestor it shares with *Utricularia*. When polyploidy events occur, all DNA in an ancestral nucleus is duplicated (or triplicated, etc.), and mechanisms of DNA deletion often incur great losses thereafter. In *C. follicularis* this deletion process may have been milder than in *U. gibba*, and although there are more predicted genes in the former species (36,503 versus 29,666), the principal contributor to the genome size of *C. follicularis* is massive proliferation of TEs; these essentially parasitic elements account for about 77% of the entire genome (Fukushima et al. 2017).

Although *C. follicularis* is the only member of Oxalidales with a sequenced genome, it can be reasonably assumed through its own duplication status that the basic polyploid state for the lineage is like all other rosids, which share the ancient gamma hexaploidy event with asterids. With the absence of further WGDs in *C. follicularis*, adaptive evolution through gene duplication largely has been limited to local, small-scale events. Still, the functional diversification afforded by these ongoing duplications has presented *C. follicularis* with considerable fodder for its evolution of the carnivorous habit and other lineage- or species-specific features.

11.3 Contribution of whole genome duplications to functional diversity

In both animals and plants, duplicates descended from polyploidy events tend to share transcription factor function. Polyploidy events generate entire genomic complements of precisely dosage-balanced duplicate genes that divergent selection pressures can act upon to generate phenotypic diversity. Research on model plant genomes supports the theory that modular, dosage-sensitive functions, including transcriptional regulation, are over-represented among duplicates that descend from polyploidy events, whereas duplicates deriving from small-scale, local duplication events are more likely to survive and adapt for dosage-responsive functions, for example in enhancing biosynthetic (e.g., secondary metabolic) pathways (Freeling 2009, Chae et al. 2014). Most duplicated genes, of both kinds, return to single copy after one redundant duplicate becomes pseudogenized.

To examine adaptive genetic features in *U. gibba* that descend from its polyploid duplicates, we examined enrichments of gene functional annotations among homeologs maintained in syntenic blocks that descend from *U. gibba*'s lineage-specific WGDs. Such duplicates mostly are enriched for transcriptional regulatory functions, and as expected based on earlier studies, Lan et al. (2017) found highly similar results for likewise analyzed *Arabidopsis thaliana* WGD duplicates. However, within general transcription regulatory function, no particular biological processes were themselves specifically enriched. For *Arabidopsis thaliana*, only response to jasmonic acid was over-represented among syntenic duplicates, suggesting that broad functional ranges of transcription functions are retained after WGDs.

We examined polyploid duplicates in the *C. follicularis* genome using the same methodology (*unpublished data*). We could verify that transcription factor function was similarly enriched among surviving syntenically related duplicates, but the polyploid event that gave rise to the homeologs was far more ancient, those genes having descended from the gamma paleohexaploidy event at the base of all core eudicots. The transcriptional enrichment phenomenon is neither age-restricted nor taxonomically biased; we obtained similar results with *V. vinifera*, which is 1:1 in ploidy level with *C. follicularis*.

11.4 The adaptive roles of small-scale gene duplication events

Unlike syntenic homeologs retained after WGD events, local, small-scale duplicates in the modern *U. gibba* and *C. follicularis* genomes (recovered as a byproduct of syntenic block discovery; Lan et al. 2017) are enriched for many secondary metabolic functions, including specific functions that could be anticipated for a carnivorous plant (Table 11.1). Tandemly duplicated *A. thaliana* and *V. vinifera* genes discovered in the same manner (Lan et al. 2017) were similarly enriched for secondary metabolic activities, as anticipated based on earlier results, but in many cases their activities were distinctively different than in the *U. gibba* or *C. follicularis* genomes.

11.4.1 *Utricularia gibba* small-scale gene duplication events

The *U. gibba* genome, while tiny, descends from an ancestral Lentibulariaceae genome that was considerably larger in size, up to 1.5 Gb. Gene duplicates that survived the considerable deletion pressure during genome evolution of *U. gibba* must have persisted under greater purifying selection pressure than in the larger genomes of most other angiosperms. Such a deletion-prone genome should be particularly revealing as to how duplicates generated by WGD vs. tandem events differentially survived deletion.

The *U. gibba* genome reveals large expansions of defense-related gene families, in addition to genes involved in nutrient acquisition and stress response (Lan et al. 2017). Among these are tandem arrays of cysteine protease genes with trap-specific expression that evolved within a protein family already known to be involved in carnivory in other species (Schulze et al. 2012, Libiaková et al. 2014, Buch et al. 2015; Figure 11.2). The cysteine protease gene family includes papains, which were characterized in papaya and thought to have evolved as part of protection from herbivorous insects (Konno et al. 2004). These enzymes also are prevalent endopeptidases in the digestive fluid of *Dionaea*, where they are regulated in part by jasmonates (Libiaková et al. 2014). Careful genome analysis showed that the *U. gibba* genes duplicated locally both prior to and following its most recent WGD event but, through considerable gene deletion, have become nearly totally restricted to one specific subfamily of cysteine proteases. These findings strongly suggest

Table 11.1 Gene ontology (GO) functional enrichment analysis of tandem duplicates in *Cephalotus follicularis* and *Utricularia gibba*, with the top 10 significantly enriched GO categories listed. *P* values resulting from Fisher Exact Tests after Bonferroni correction for multiple tests are shown (Lan et al. 2017). "Senescence-associated vacuole" under *U. gibba* identifies the tandemly duplicated cysteine proteases discussed in the text and shown in Figures 11.2 and 11.3. Some GO categories clearly overlap, such as oligopeptide transport, dipeptide transport, dipeptide transporter activity, and oligopeptide transporter activity.

	GO term	GO description	*P* values
Cephalotus follicularis	GO:0,016,042	Lipid catabolic process	3.84×10^{-7}
	GO:0,098,655	Cation transmembrane transport	6.81×10^{-7}
	GO:0,005,618	Cell wall	7.23×10^{-7}
	GO:0,046,527	Glucosyltransferase activity	5.21×10^{-7}
	GO:0,008,194	UDP-glycosyltransferase activity	5.48×10^{-7}
	GO:0,015,020	Glucuronosyltransferase activity	5.56×10^{-7}
	GO:0,004,497	Monooxygenase activity	7.2×10^{-7}
	GO:0,019,825	Oxygen binding	8.5×10^{-7}
	GO:0,006,952	Defense response	9.72×10^{-7}
	GO:0,016,757	Transferase activity, transferring glycosyl groups	1.08×10^{-6}
Utricularia gibba	GO:0,010,282	Senescence-associated vacuole	7.48×10^{-8}
	GO:0,015,238	Drug transmembrane transporter activity	1.20×10^{-7}
	GO:0,006,857	Oligopeptide transport	1.45×10^{-7}
	GO:0,042,938	Dipeptide transport	1.74×10^{-7}
	GO:0,042,936	Dipeptide transporter activity	1.74×10^{-7}
	GO:0,006,855	Drug transmembrane transport	2.77×10^{-7}
	GO:0,016,705	Oxidoreductase activity, acting on paired donors, with incorporation or reduction of molecular oxygen	3.25×10^{-7}
	GO:0,005,764	Lysosome	3.29×10^{-7}
	GO:0,019,825	Oxygen binding	3.89×10^{-7}
	GO:0,015,198	Oligopeptide transporter activity	4.26×10^{-7}

that diverse, related enzymes known from model plant species have become superfluous during *U. gibba* genome evolution. Furthermore, molecular evolutionary analyses showed that some of the amino acid divergence among *U. gibba* cysteine protease tandem duplicates was influenced by positive selection pressures and located near the functionally important substrate-binding cleft of these proteins (Figure 11.3; Lan et al. 2017).

Additional enriched functions among *U. gibba* tandem duplicates that have trap-enhanced expression include peptide transport, potentially involved in intercellular movement of broken-down prey proteins; ATPase activities for bladder-trap acidification and transmembrane nutrient transport; hydrolase and chitinase activities for breakdown of prey polysaccharides; and cell wall dynamic components that may be involved in active bladder movements.

11.4.2 Small-scale gene duplication events in *Cephalotus follicularis*

Gene families were determined *in silico* for the *C. follicularis* and *U. gibba* genomes, and for several noncarnivorous plants. When analyzing the entire duplicate gene space, maximum-likelihood gene gain and loss analysis detected lineage-specific expansion of almost 500 gene families in the *C. follicularis* genome. Among these, functional enrichment analysis highlighted two *C. follicularis*-specific families containing purple acid phosphatases (PAPs), which are well-known enzymes among carnivorous plants (Clancy and Coffey 1977, Sirová et al. 2003, Płachno et al. 2006; Chapter 16). Of interest were gene families shared between distantly related, independently evolved carnivorous plants. Functional enrichment analysis of gene families seen only in *C. follicularis* and *U. gibba* highlighted a *Cephalotus–Utricularia* pair of RNase T2 enzymes, which are, like the PAPs, well known protein components of trap secretions (Okabe et al. 2005b, Stephenson and Hogan 2006, Nishimura et al. 2013). Among gene families restricted to *C. follicularis* was one composed of 10 genes encoding dihydropyrimidinases, which are involved in the recycling of nitrogen from nucleic acids into general nitrogen metabolic pathways (Zrenner et al. 2009). Nitrogen is one of the

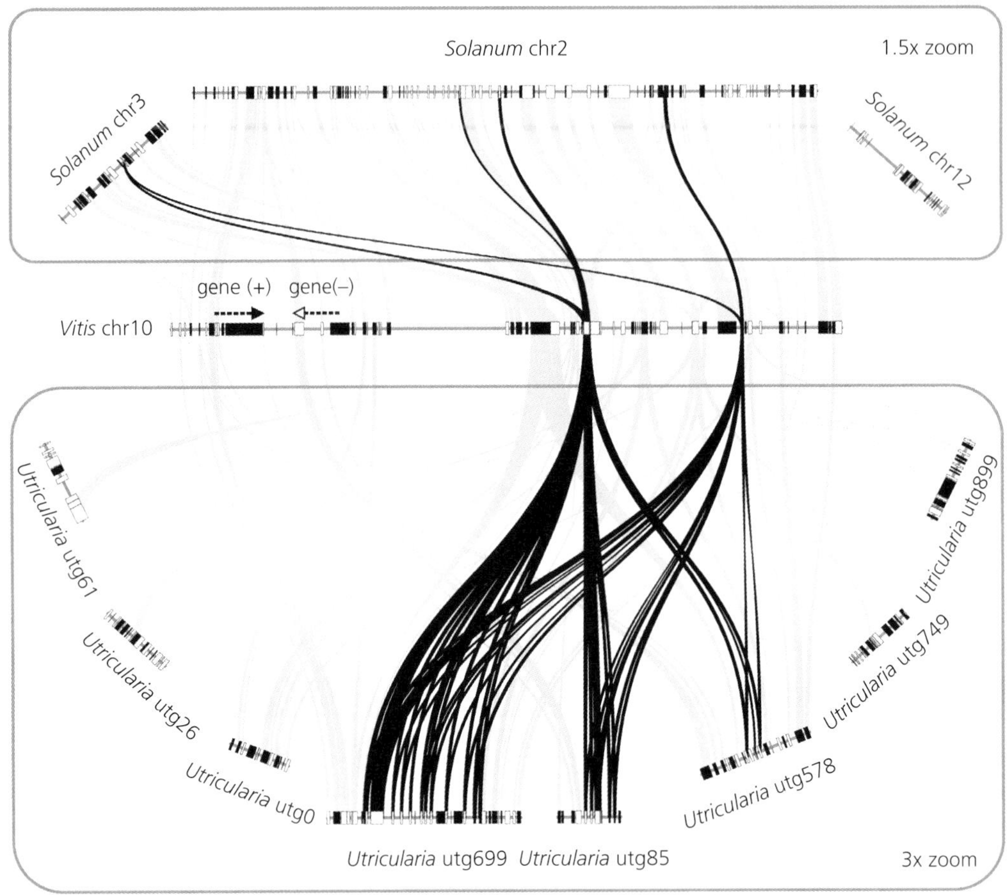

Figure 11.2 Syntenic relationships between *Vitis vinifera*, *Solanum lycopersicum*, and *Utricularia gibba* regions containing tandemly duplicated cysteine protease genes. Some parts of these tandem arrays clearly pre-existed in the pre-polyploid ancestral genomes of *U. gibba*, with further tandem duplications having occurred since these events, together increasing functional potential for its carnivory. A typical ancestral region in *V. vinifera* can be traced to up to three regions in *S. lycopersicum* (genome triplication) and up to eight regions in *U. gibba* (where up to three WGDs are possible). Black connecting lines highlight matching cysteine proteases in the selected regions; genes otherwise syntenic are shown in gray. Figure adapted from Lan et al. (2017).

primary limiting nutrients for carnivorous plants (Hanslin and Karlsson 1996, Wakefield et al. 2005, Farnsworth and Ellison 2008; Chapters 17–19).

The waxy epidermis of carnivorous pitfall traps promotes prey capture and prevents them from escaping (Juniper et al. 1989). Certain cytochrome P450 (CYP) genes belonging to a clade containing *Arabidopsis* genes involved in wax and cutin biosynthesis (CYP86 and CYP96A; Bak et al. 2011) showed expansion in membership in *C. follicularis* and pitcher-predominant expression. Like the *U. gibba* cysteine proteases, these genes appear to be duplicated tandemly in the *C. follicularis* genome. Tandem gene location and pitcher-preferential expression likewise were discovered for duplicates of wax ester synthase (WSD1, At5g37300; Li et al. 2008). Furthermore, three aspartic proteases were identified from the *C. follicularis* digestive fluid proteome, and the gene family encoding them also was found to have expanded in the gene gain and loss analysis. Indeed, three tandemly duplicated clusters of aspartic protease genes were discovered that contain both pitcher-preferential and constitutively expressed genes.

The duplicate gene space of *C. follicularis* is considerable, but it is limited mostly to duplicates

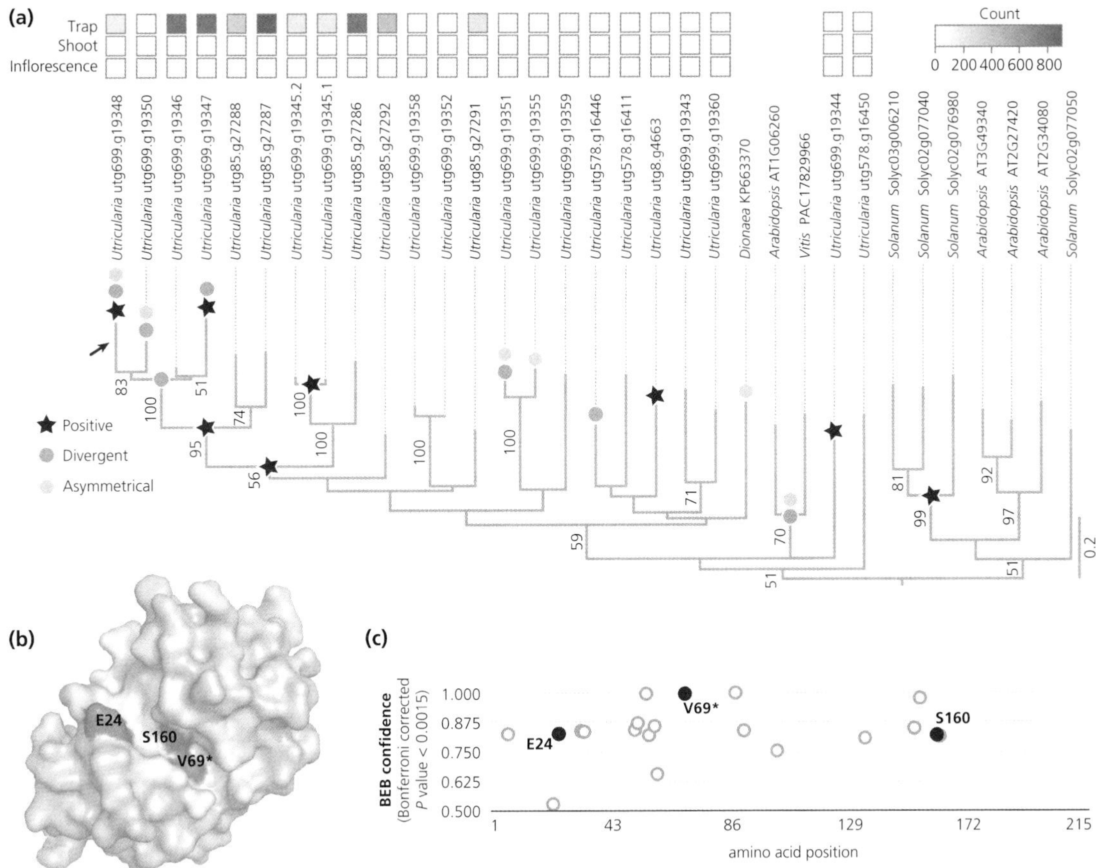

Figure 11.3 Molecular and structural evolutionary analysis of *Utricularia gibba* cysteine proteases suggests adaptive protein evolution accompanying WGD and tandem duplication events. (**a**) Best-scoring tree from maximum-likelihood based searches with bootstrap support (BS) values ≥50 indicated at branches. Symbols on branches indicate significant evidence for positive selection (black star), heterogeneous selective pressures (dark gray circle), or asymmetrical sequence evolution (light gray hexagon) as determined using PAML (Yang 2007). Heat-map above the phylogeny shows trap-dominant expression of *U. gibba* homologs, based on transcriptome data (Ibarra-Laclette et al. 2011). (**b**) Protein homology surface model for the catalytic domain of *U. gibba* utg699.g19348 based on the enzyme structure of *Dionaea muscipula* (Risør et al. 2016) shows that three of the amino-acid sites (E24, V69, S160) under positive selection (arrow in **a**) are within five amino acids of known *D. muscipula* functional residues, where they line the substrate-binding cleft. (**c**) Plot of *U. gibba* utg699.g19348 amino-acid sites under positive selection, with the three sites highlighted in (**b**) marked as black circles (BEB confidence >0.82, Bonferroni corrected *P* value < 0.0015). Figure adapted from Lan et al. (2017).

descending from small-scale, tandem events. We also examined *C. follicularis* tandem duplicates for functional enrichments in the same manner that Lan et al. (2017) examined them for *U. gibba* (Table 11.1). The most significantly over-represented functional category among these duplicates was lipid catabolism (Bonferroni corrected *P* value = 3.84×10^{-7}), which reflected many lipase genes, including those encoding GDSL lipases and patatin-like lipases. This function also was an expanded category among *U. gibba* tandem duplicates, where four genes showed greatly trap-enhanced RNA expression. Moreover, GDSL lipases were discovered in the trap fluid proteome of *Nepenthes* (Rottloff et al. 2016), and because of this and trap-specific expression of several GDSL lipases (Fukushima et al. 2017), we speculate that *C. follicularis* GDSL lipases also may have a carnivorous function.

The second-most significantly enriched functional category among small-scale duplicates was cation transmembrane transport (Bonferroni corrected *P* value = 6.81×10^{-7}), and genes falling under this annotation included oligopeptide and phosphate transporters. Oligopeptide transporters are known constituents of carnivorous physiology in other species (Schulze et al. 1999), and were over-represented among *U. gibba* tandemly duplicated genes. Phosphate transport is clearly an important function for carnivorous plants, given their P-limited environmental preferences (Chapter 2, 17, 19). Taken together, duplication and transcriptional specialization of tandemly duplicated genes in *C. follicularis* may have been of critical importance in the evolution of its carnivorous habit, just as it was for *U. gibba*.

11.5 Evolutionary rates and gene loss in *Utricularia gibba*

Genes in families with no predicted *U. gibba* member were collected and annotated for gene ontology (GO) and functional categories using the Kyoto Encyclopedia of Genes and Genomes (KEGG) to explore the functional consequences of genes that are putatively missing in the *U. gibba* genome yet present in other angiosperm genomes (i.e., *A. thaliana, M. guttatus, P. trichocarpa, Prunus persica, Theobroma cacao, V. vinifera*). We then compared the collections of GOs and KEGGs against those for the entire *A. thaliana* genome to obtain significant enrichments (*unpublished data*). The question asked here is essentially: what would an *Arabidopsis* plant be missing, and therefore be hindered by, if it lacked genes from gene families missing in *U. gibba*? This does not directly answer how the absence of certain functions in *U. gibba* has affected its evolution, but the exercise does suggest hypotheses. Genes putatively missing from the *U. gibba* genome may have been lost since it diverged from the last common ancestor shared with these other species because they are no longer needed, or such losses could directly reflect molecular or morpho-physiological phenotypes in *U. gibba*.

11.5.1 ROS scavenging and DNA repair

The method by which *Utricularia* traps its prey may have played a key role in the dynamics of its genome evolution. The suction traps of *Utricularia* build up a strong, negative internal pressure via active ion transport (Chapters 8, 14). In response to mechanostimulation, they relieve this pressure, and in turn, rapidly suck in prey. The energy demand for this active process likely is substantial, and supporting experiments *in vivo* have demonstrated that respiratory rates are greater in *Utricularia* traps than in vegetative tissues (Adamec 2006). However, increased rates of oxygen metabolism can have detrimental effects to plant cells by promoting the formation of reactive oxygen species (ROS). ROS include chemical species such as peroxides, superoxides, and singlet oxygen, which cause oxidative damage to biomolecules. In fact, ROS are involved in various mutational processes, such as base excision and DNA strand breaks (Bleuyard et al. 2006, Roldán-Arjona and Ariza 2009). Genes coding for ROS detoxification enzymes are expressed at significantly higher levels in *U. gibba* traps than in other tissues, suggesting a correlation between trap action, ROS production, and scavenging (Ibarra-Laclette et al. 2011). Thus, trap-induced ROS could be responsible for increased rates of nucleotide substitution in *U. gibba* and its extreme reduction in genome size.

The KEGG category base excision repair is enriched among annotated genes predicted to be absent in the *U. gibba* genome. Among such gene families, we found that several members of the peroxidase superfamily that are expressed in various tissues in *A. thaliana* are missing in *U. gibba* (KEGG Phenylpropanoid biosynthesis; AT3G50990, AT3G17070, AT2G18140, AT2G18150, AT5G19890, AT5G19880, AT1G49570, AT5G66390, AT1G44970, AT4G36430; *unpublished data*). One peroxide gene of interest is *PA2* (AT5G06720), which encodes a peroxidase with diverse roles in defense response and other oxidative stresses (Choi and Hwang 2012). Peroxidases are well known for their ability to catalyze the decomposition of hydrogen peroxide (a ROS highly damaging to DNA), and it is conceivable that gene family loss may negatively affect ROS detoxification in *U. gibba*.

Even more striking is the loss of genes involved in DNA repair. These include five out of the 15 DNA glycosylases present in the *A. thaliana* genome (InterPro ID: IPR011257; AT1G13635, AT5G44680, AT3G12710, AT3G47830, AT1G75090), four of which are further annotated to include methyladenine glycosylases (InterPro ID: IPR005019; AT1G13635, AT5G44680, AT3G12710, AT1G75090), for which *A. thaliana* has seven genes. Loss of poly (ADP-ribose) polymerase *PARP2* (AT4G02390) is of considerable interest due to its activity in response to DNA damage and involvement in repair of double strand breaks via non-homologous end-joining (Song et al. 2015). In addition to *PARP2*, two other such polymerases are present in the *A. thaliana* genome: *PARP1* (AT2G31320) directly interacts with *PARP2*, influencing its activity and abundance, whereas *PARP3* (AT5G22470) is important in seed viability and is likely involved in DNA repair during germination (Rissel et al. 2014, Song et al. 2015).

Additional DNA repair genes putatively lost from the *U. gibba* genome but not found within enriched GO or KEGG categories include two DNA ligases out of the eight present in the *A. thaliana* genome: DNA ligase 4 (AT5G57160), involved in telomere maintenance, and DNA ligase 6 (AT1G66730), involved in seed longevity and germination (Heacock et al. 2007, Waterworth et al. 2010). *U. gibba* is also putatively missing *UV REPAIR DEFECTIVE 3* (*UVR3*; AT3G15620), which is required for repair of UV-induced photolesions (6–4 photoproducts), which are highly mutagenic (Balajee et al. 1999, Li et al. 2010). Given the prevalence of ROS production in the bladders, it is unknown why gene family members involved in ROS detoxification and DNA repair have been lost. Previous transcriptomic and molecular evolutionary studies have suggested that ROS may increase nucleotide substitution rates (Ibarra-Laclette et al. 2011, Carretero-Paulet et al. 2015a), which could enhance gene turnover rates (Carretero-Paulet et al. 2015b).

11.5.2 Production of diploid gametes and the evolution of *Utricularia gibba* polyploidy

Loss of genes from meiotic errors that lead to diploid gametes may promote an environment in which genome duplication is more likely (Vanneste et al. 2014). *Utricularia gibba* is putatively missing (*unpublished data*) a member of the *JASON* gene family (AT1G06660), the product of which is associated with vesicles in an organelle band that maintains spindle position during chromosome division (Brownfield et al. 2015). *Arabidopsis thaliana* *JASON* mutants produce diploid male gametes, leading to triploid progeny that are generated by a defect in male meiosis II. Thus, it is conceivable that in losing *JASON*, *U. gibba* became primed for WGD. The entire *JASON* gene family has a close paralog that contains two *U. gibba* genes (unitig_8.g3738.t1 and unitig_22.g5291.t1) and one *A. thaliana* gene (*BARD1*; AT1G04030), but the latter has an altogether different function in root development. This ancient paralog may extend back to the gamma paleohexaploidy event or an ancient tandem duplication.

In keeping with a hypothesis that *U. gibba* may be prone to WGDs through loss of *JASON*, Ibarra-Laclette et al. (2013) hypothesized that the occurrence of WGDs may buffer against a state of "genomic collapse" incurred by runaway DNA deletion pressure. By the wholesale generation of duplicates genome-wide, scores of important functions could be buffered against loss-of-function genotypes, all at the same time, as opposed to here-and-there via small-scale duplication events.

11.5.3 Defense response

Putative losses of defense response genes are also apparent among enriched GO categories (*unpublished data*). These include *A. thaliana* resistance (R) genes *RPS5* (AT1G12220) and *RPM1* (AT3G07040), which confer resistance against plant pests such as the ubiquitous pathogen *Pseudomonas syringae* (Katagiri et al. 2002). Interestingly, a phytoalexin-deficient 4 (*PAD4*) gene (GO Triglyceride lipase activity; AT3G52430) known to be important for salicylic acid signaling and function in R gene-mediated and basal plant disease resistance also has been putatively lost (Cui et al. 2016). Finally, genes associated with disease susceptibility, such as *LOV1* (AT1G10920), are putatively missing (Lorang et al. 2012, Gilbert and Wolpert 2013). It is plausible that the microbial community associated with *Utricularia* may serve more in beneficial relationships than pathogenic ones, reducing the need for genes associated with defense (Srivastava et al. 2016; Chapter 25). Based on metagenomic analysis, *U. gibba* is known to harbor a diverse microbial assemblage associated with its traps, including a variety of *Pseudomonas* species that occur at high abundance (Alcaraz et al. 2016) and that catabolize numerous chemicals, including organophosphate compounds that could serve as an essential nutrient source (Horne et al. 2002).

11.5.4 Essential nutrient transport and enzyme activity

Losses of genes related to the uptake and transport of essential nutrients is puzzling, but many of these also are closely associated with root development and physiology; they may not be needed in the rootless *U. gibba*. *PHO2*, involved in phosphate starvation response (GO Phosphate ion transport; AT2G33770) also apparently is missing from the *U. gibba* genome, as is purple acid phosphatase 16 (*PAP16*; AT3G10150), which has serine/threonine phosphatase activity (Secco et al. 2012, Pant et al. 2015). At least two nitrate transporters (*NRT1:2*) (GO Metabolic processes, AT1G2704 and GO Integral component of membranes, AT5G62730) are absent, as are the nitrate efflux transporter *NAXT*, mainly expressed in the cortex of adult roots (GO Metabolic processes; AT3G45650), and a NIN-like protein 7 (*NLP7*), which modulates nitrate sensing and metabolism (GO Metabolic processes; AT4G24020) (Segonzac et al. 2007, Menz et al. 2016). Of these, *NAXT* was identified as missing by Ibarra-Laclette et al. (2013). An ammonium (NH_4^+) transporter, *DELTA TONOPLAST INTEGRAL PROTEIN* (*DELTA-TIP*; AT3G16240), is highly expressed in flowers, shoots, and stems, but is putatively absent in *U. gibba* (Kirscht et al. 2016). Transporters of nitrogenous bases and amino acids also are contracted; the purine transporter *PURINE PERMEASE 5* (*PUP5*; AT2G24220), and three amino acid permease transporters *AAP1*, *AAP6*, and *AAP8* (AT1G58360, AT5G49630, AT1G10010), also are putatively missing (Gillissen et al. 2000, Okumoto et al. 2002).

11.5.5 Auxin response

Auxin is a key growth regulator in plants that controls many aspects of development. Loss of genes in these pathways hypothetically could be associated with the unusual Bauplan of *U. gibba* (Chapter 8). The AUX/IAA gene family is large, with many members induced by auxin (Reed 2001). Ibarra-Laclette et al. (2013) found increased numbers of *U. gibba* AUX/IAA genes relative to *A. thaliana* and *S. lycopersicum* occurring in four lineages of AUX/IAA genes, but they also reported gene loss. Among the gene families apparently absent in *U. gibba* include the loss of some lineage I AUX/IAA genes, for which there is one *A. thaliana* homolog (*IAA29*; AT4G32280), and lineage III, which includes three *A. thaliana* homologs; in *U. gibba*, *IAA6/SHY1* (AT1G52830) and *IAA19/MSG2* (AT3G15540) have been lost. Of these, *IAA19* regulates hypocotyl growth responses and lateral root formation (Tatematsu et al. 2004), *IAA29* plays a role in hypocotyl elongation (Shimizu et al. 2016), and *IAA6* mutants have a dominant leaf curling phenotype and shortened hypocotyls (Krogan and Berleth 2007).

There also are genes related to auxin efflux and response that are putatively missing in *U. gibba* (*unpublished data*). These include *PIN6*, which is directly involved in catalyzing cellular auxin efflux and required for morphogenesis in the early lateral root primordium in *A. thaliana* (AT1G77110) (Simon et al. 2016), and *PIN8*, which encodes an auxin transporter with strong expression in male gametophytes (AT5G15100) (Ding et al. 2012). The missing gene *AIL5* (AT5G57390), which encodes a member of the AP2 family of transcriptional regulators, is required to maintain high levels of *PIN1* expression at the periphery of the meristem and modulate local auxin production in the central region of the shoot apical meristem (SAM). *YUC11* (AT1G21430) also is putatively missing in *U. gibba* and plays an important role in the *A. thaliana* auxin biosynthetic pathway (Cheng et al. 2007).

11.5.6 Root and shoot morphogenesis and the transition to the aquatic habit

MADS-box genes are well known for their roles in floral organ identity, regulation of flowering time, and other aspects of reproductive development. However, MADS-box genes also are expressed in vegetative tissues, including the root. In *A. thaliana*, at least 50 MADS-box genes are expressed in root tissues. Of these, four members of the *AGL17*-like MADS-box clade that are root-specific in their expression (*AGL16*, *AGL17*, *AGL21*, and *AGL44*) are missing in *U. gibba* (Ibarra-Laclette et al. 2013).

However, only one of these MADS-box genes has been functionally characterized (*AGL44*); it is a component of a signaling pathway regulating lateral root growth in response to changes in external NO_3^- supply. Other genes putatively lost but important for root meristem growth and maintenance include *RGF6* (AT4G16515), involved in root meristem growth (Matsuzaki et al. 2010), and one of the nine functionally redundant root meristem growth factor genes in *A. thaliana*. *RGF* genes are expressed mainly in the root stem-cell area and the innermost layer of central columella cells, and are required for the maintenance of the root stem-cell niche and transit amplifying cell proliferation. *FEZ* (AT1G26870) and *SOMBRERO* (AT1G79580), which are part of a regulatory feedback loop involved in the orientation of cell division planes in *A. thaliana* root stem-cells (Willemsen et al. 2008), also putatively are missing (*unpublished data*), along with *TORNADO 2* (*TRN2/TET1*; AT5G46700), one of the 17 *A. thaliana* transmembrane tetraspanin genes that is required for the maintenance of both the radial pattern of tissue differentiation in the root and for the subsequent circumferential pattern within the epidermis (Wang et al. 2015). *DEFECTIVELY ORGANIZED TRIBUTARIES 3* (*DOT3*; AT5G10250), also putatively lost in *U. gibba* (*unpublished data*), has roles in primary root and shoot growth, as well as leaf venation patterning (Petricka et al. 2008). The loss of *WUS*-like *WOX5* (AT3G11260) is perhaps the most detrimental for root production; in *A. thaliana*, its protein product controls the root stem-cell niche and *WOX5* single mutants have reduced rooting rate (Hu and Xu 2016). The loss of *WOX5* is supported by previous analyses of high gene family turnover rates in *U. gibba* (Carretero-Paulet et al. 2015b).

Genes responsible for root hair growth and initiation also are apparently missing in *U. gibba* (*unpublished data*). These genes include an armadillo-repeat containing kinesin-related protein (AT3G54870), which plays a role during transition to root-hair tip growth (Lockhart 2014), various root hair specific (*RHS*) genes (AT3G10710, AT4G02270, AT4G22080, AT5G22410), and a root hairless (*RHL1*) gene involved in root-hair initiation (AT1G48380) (Schneider et al. 1998). Last, *GLABRA 2* (*GL2*; AT1G79840), a homeodomain protein that affects epidermal cell identity in root hairs, is missing (Balcerowicz et al. 2015).

In addition to the loss of root morphogenesis genes, several genes necessary for SAM development, maintenance, and organ polarity/patterning in *A. thaliana* are putatively missing in *U. gibba* (*unpublished data*). For example, *SHOOT APICAL MERISTEM ARREST 1* (*SHA1*; AT5G63780), which encodes a putative E3 ligase required for post-embryonic SAM maintenance in *A. thaliana* (Sonoda et al. 2007), is not present within the *U. gibba* genome. Members of the *YABBY* gene family specify abaxial cell fate in vegetative tissue (*YAB5*; AT2G26580) and polarity of floral organs (*INO*; AT1G23420) (Bowman 2000, Eckardt 2010, Simon et al. 2012), and *GRAND CENTRAL* (*GCT*; AT1G55325) is expressed during embryogenesis and regulates developmental timing and radial pattern formation (Gillmor et al. 2010). Both gene family members are putatively missing in *U. gibba*. Taken together, loss of genes important for root and shoot morphogenesis may be correlated with the absence of an obvious root in *U. gibba* and could also help to define its unique Bauplan suited to an aquatic habit.

11.6 Genomic insights into leaf patterning in *Cephalotus follicularis*

As part of the *C. follicularis* genome project, evolutionary changes in leaf developmental regulators were characterized to better understand the developmental switch between flat noncarnivorous leaves and carnivorous pitchers. Among transcription factors acting in the regulation of adaxial-abaxial polarity, *ASYMMETRIC LEAVES 2*, *AUXIN RESPONSE FACTOR 4* (*ARF4*), *YABBY 2* and *5*, and *WUSCHEL-RELATED HOMEOBOX 1* (*WOX1*) orthologs had higher expression levels in *C. follicularis* shoot apices that produce pitcher leaves (Fukushima et al. 2017). Differential deployment of these abaxial-adaxial patterning genes matches previous suggestions regarding the evolution of diverse leaf types in angiosperms (Fukushima and Hasebe 2014).

Gene family size differs substantially relative to *U. gibba* for some of these transcription factors (*unpublished data*). Trap ontogenies are different in the two species and neither one is more or less complex than the other in terms of developmental regulatory needs. For example, there is only one *WUSCHEL* (*WUS*) homolog in *C. follicularis*, as in most other eudicot species (e.g., *A. thaliana, S. lycopersicum, M. guttatus, V. vinifera*), whereas there are three paralogs in *U. gibba*. *WUSCHEL* controls the stem-cell niche of shoot apical meristems. As hypothesized in previous work (Ibarra-Laclette et al. 2013), the *WUS*-like subfamily expansion could be correlated with the specialized vegetative ontogeny of *U. gibba*. Similarly, *ARF4* has undergone paralog expansion in *U. gibba* to three genes, whereas only a single copy exists in *A. thaliana, C. follicularis*, and *V. vinifera*.

11.7 Evolutionary convergence of digestive enzymes

Carnivorous plants secrete digestive enzymes to degrade trapped prey (Chapter 16). Previous studies on several digestive enzymes of *Nepenthes* and *Drosera* species, *Dionaea muscipula*, and *C. follicularis* indicated that pathogenesis-related (PR) proteins were co-opted for digestive function and for preventing microbial colonization of digestive fluid (Okabe et al. 2005b, Hatano and Hamada 2008, 2012, Rottloff et al. 2011, Schulze et al. 2012, Buch et al. 2013, 2014, Michalko et al. 2013, Nishimura et al. 2013; Chapter 16).

To further investigate the origin and evolution of digestive enzymes, Fukushima et al. (2017) applied protein and transcriptome sequencing to obtain coding sequences for digestive fluid proteins arising from three independent origins of carnivory: *C. follicularis; Drosera adelae* and *Nepenthes alata*; and *Sarracenia purpurea*. In total, 35 coding sequences were identified among the four species (Figure 11.4). Phylogenetic relationships were inferred among the digestive fluid proteins (Figure 11.5) using previously identified enzyme sequences and homologs from carnivorous and noncarnivorous plants. When considering only biochemically confirmed digestive fluid proteins, glycoside hydrolase family 19 (GH19) chitinase, β-1,3-glucanase, PR1-like and thaumatin-like protein (all PR proteins; Mithöfer 2011), PAP, and RNase T2 showed orthologous relationships among carnivorous plants. This result was unexpected, because the plants have evolved carnivory independently and so the relationships could have been paralogous. This result suggests that orthologous genes, not closely related paralogous gene lineages, repeatedly have been co-opted for digestive functions in independent carnivorous plant lineages. Aspartic proteases, GH18 chitinases, and class III peroxidases, however, were recruited from paralogous gene lineages in the four species.

To predict possible ancestral functions of the various digestive-fluid proteins, Fukushima et al. (2017) analyzed expression patterns of closely related genes in *A. thaliana*. In most cases, these *A. thaliana* genes were up-regulated upon various stresses (biotic or abiotic). This finding suggests a general evolutionary trend, in which genes involved in plant carnivory were repeatedly co-opted from stress-responsive genes. This idea is consistent with earlier hypotheses that digestive enzymes originated from pathogenesis-related (PR) proteins in carnivorous Caryophyllales species (Mithöfer 2011, Renner and Specht 2012; Chapter 16). It remains unclear whether digestive enzymes have dual functions (i.e., both stress-response and carnivory), but evidence from *C. follicularis* and *N. alata* imply spatial expression optimization since the carnivory-related genes are expressed primarily in traps, not noncarnivorous laminate leaves.

Repeated deployment of similar genes in several independent evolutionary originations of carnivory may have similarly involved carnivory-specific selective pressures at the within-protein, amino acid level. Fukushima et al. (2017) developed a tree reconciliation method to test for such molecular convergence and found that GH19 chitinases, PAPs, and RNase T2 enzymes significantly accumulated convergent amino acid changes. In each case, significant molecular convergence was detected between *C. follicularis* and other carnivorous species included in analyses. To study the potential functional significance of the amino acid residues identified as convergent, Fukushima et al. (2017) mapped these sites onto three-dimensional protein models. The convergent sites did not correspond with, nor were they located near, residues essential for catalytic activity; instead they mapped onto exposed areas of the proteins. Although amino acid

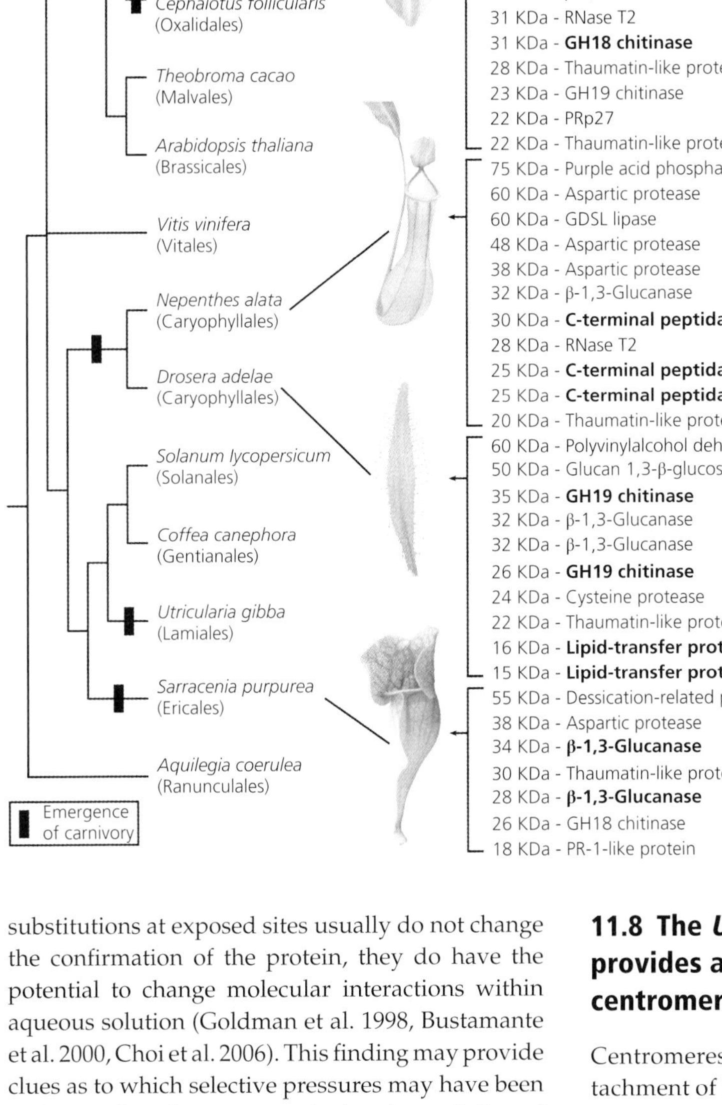

Figure 11.4 Consensus tree of plastid (Bell et al. 2010) and nuclear-based trees (Zeng et al. 2014, Wickett et al. 2014) showing phylogenetic relationships of plants that independently evolved the carnivorous habit. Digestive fluid proteins identified through proteomic analysis for each species tested are noted in brackets, with names in bold marking variants likely arising from the same gene. Figure adapted from Fukushima et al. (2017).

substitutions at exposed sites usually do not change the confirmation of the protein, they do have the potential to change molecular interactions within aqueous solution (Goldman et al. 1998, Bustamante et al. 2000, Choi et al. 2006). This finding may provide clues as to which selective pressures may have been acting on digestive enzymes during the evolution of plant carnivory. It is also important to note that the *C. follicularis* PAP and RNase T2 genes are located adjacent to one another (within a 40 kb interval) in the genome (Fukushima et al. 2017). This particular gene arrangement may have had an adaptive value, allowing for co-regulation such that gene expression is correlated for functions in plant carnivory.

11.8 The *Utricularia gibba* genome provides a look at complete plant centromeres

Centromeres are the chromosomal points of attachment of the kinetochores, which in turn recruit microtubules of the divisional spindle apparatus. Centromeres are usually highly repetitive regions in plant genomes, and such repeated sequence motifs make genome assembly difficult (§11.1.1). Plant centromeres often are composed of ≈100–725 bp tandem repeats, TEs, or a combination of the two. The PacBio genome of *U. gibba* was so highly contiguous that several complete chromosomes were

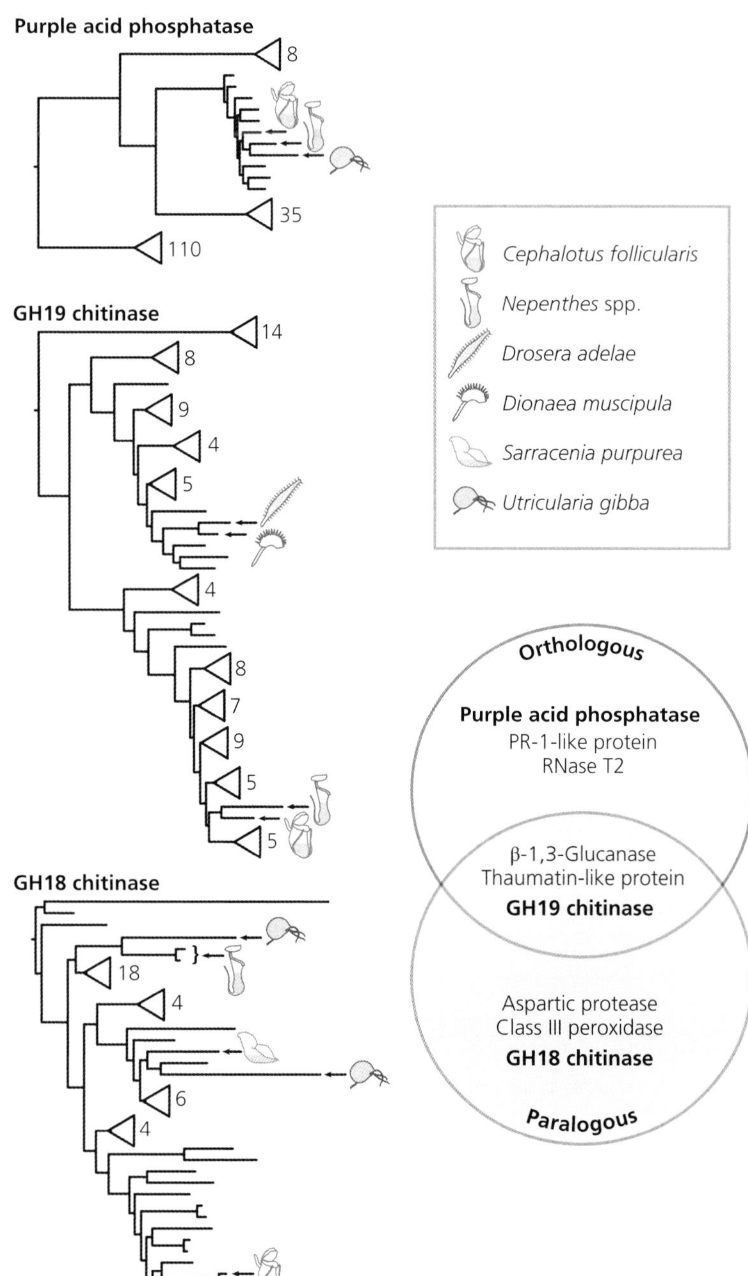

Figure 11.5 Phylogenies of carnivory-related genes. *Cephalotus follicularis*, *Nepenthes* spp., *Drosera adelae*, *Dionaea muscipula*, and *Sarracenia purpurea* cartoons denote sequences in the phylogenies with confirmed presence in digestive fluid. For these carnivorous plant sequences, ortholog–paralog relationships are indicated with a Venn diagram. Only gene tree relationships are shown for *Utricularia gibba* sequences, with presence in the proteome remaining to be tested. Gene numbers within collapsed clades are shown next to triangles. Figure adapted from Fukushima et al. (2017).

sequenced, from telomere to telomere, directly through their centromeres, permitting a detailed look at its centromeric structure (Lan et al. 2017). Since no high-copy tandem repeats were apparent in centromeric regions of chromosome-sized assembly fragments, a careful search was undertaken to discover putative centromeric TEs.

Although plant retrotransposon families in general are randomly dispersed, even in the densely gene-packed *U. gibba* genome, there are families such as the CRM centromeric chromoviruses that distinctly concentrate in centromeric regions. CRMs, a lineage of Ty3/gypsy retrotransposons, are known to be centromeric retrotransposons in the genomes of many species, including *Genlisea hispidula* and *G. subglabra* (Tran et al. 2015a). In a phylogenetic analysis, 55 sequences mined from the *U. gibba* genome grouped within the subgroup A CRMs (Neumann et al. 2011), which include many known centromere-specific CRMs. All but one of the *U. gibba* elements were resolved as a single, monophyletic CRM subfamily. To investigate the chromosomal localization of these *U. gibba* CRMs, Lan et al. (2017) plotted them on the chromosomes or partial chromosomal fragments of the genome assembly. Most *U. gibba* CRMs were located in the putative centromeric regions, but not all putative centromeres had CRM elements mapped to them, suggesting that other as-yet undetermined sequences help define *U. gibba* centromeres.

CRMs may play a functionally important role in stabilizing centromere structure (Topp et al. 2004, Slotkin and Martienssen 2007), although some investigators hypothesize that they are merely parasitic and tend to accumulate in recombination-poor (e.g., centromeric) regions where they can escape negative selection pressures (Gao et al. 2008). Since at least some putative centromeric regions in *U. gibba* lack CRMs or other high-copy centromeric tandem repeats, it seems that neither are universally critical for the maintenance of *U. gibba* centromere function. Further study of *U. gibba* centromere structure might use chromatin immunoprecipitation analysis of centromeric histone attachment points to locate those sequence motifs that are functional.

Centromere structure also will be very interesting to investigate in the Droseraceae once sufficient genomes are available to cover the diversity of that family. Most Droseraceae have holocentric chromosomes that lack primary constrictions on their arms; the kinetochores instead attaching along the entire axis (Junichi et al. 2011, Shirakawa et al. 2011, Veleba et al. 2017). *Dionaea, Drosophyllum*, and *Drosera regia* have chromosomes with clear primary constrictions, whereas *Aldrovanda* and the remaining *Drosera* species apparently do not (Shirakawa et al. 2011).

11.9 Additional nuclear genomes and transcriptomes of carnivorous plants

Genomic data have been published for a few other carnivorous plants. These include draft genomes for *Genlisea aurea, G. hispidula*, and *G. nigrocaulis*; *Drosera capensis*; and *Dionaea muscipula* (Leushkin et al. 2013, Vu et al. 2015, Butts et al. 2016a, 2016b, Hackl 2016). Although these genomes can allow for the identification and annotation of gene family members (e.g., aspartic and cysteine proteases), in most cases the assemblies do not capture highly repetitive regions and are too fragmented to address questions related to overall genome evolution.

Transcriptome data (also known as RNAseq) are available for carnivorous plants in varying quality; some of these data are considered "experimental" because various substrates were applied to induce expression of genes coding for digestive enzymes. For each study, differences in the presence of genes coding for digestive enzymes could be attributed to traps fed either prey or substrates prior to RNA extraction, the read depth of sequencing, or the type of sequencing platform.

Standardized methods for experimental setup, sample collection, and analysis of carnivorous plant transcriptomes also do not yet exist. For example, three studies that present *Dionaea muscipula* transcriptome data, some with focus on identifying digestive enzyme-coding genes, used very different approaches. One study used Roche 454 sequencing of RNA pooled from traps fed ants, a solution of coronatine (COR), and stimulated with filter paper saturated with urea, chitin, or water, and Illumina sequencing of RNA from traps stimulated with yellow meal-worm beetles (Schulze et al. 2012). A second used Illumina sequencing of RNA collected from freshly harvested flowers, petioles, and traps (Jensen et al. 2015). The third used Illumina sequencing of RNA collected from non-stimulated traps, petioles, roots, trap rims, and flowers, or those stimulated by live crickets or by spraying a 100 µM COR solution (Bemm et al. 2016). Various *Nepenthes* transcriptomes also are available, of which two are considered experimental and involve samples collected at various times after pitcher opening (Zakaria et al. 2015) or chitin treatment (Zakaria et al. 2016). Although it may be difficult to make comparisons between these various studies, much insight can be gained by analyzing which genes are differentially expressed within a given study in response to a given treatment (*e.g.*, transcripts coding for putative digestive enzymes; Chapter 16).

In addition to those sequenced for *U. gibba* and *C. follicularis*, transcriptomes currently are available for *Genlisea hispidula, G. nigrocaulis, Pinguicula vulgaris, Utricularia intermedia*, and *U. vulgaris* (Leushkin et al. 2013, Bárta et al. 2015, Vu et al. 2015). Results of comparisons between *U. vulgaris* and *U. gibba* revealed a number of similarities, including the presence or absence of root-associated genes (Bárta et al. 2015). *Sarracenia psittacina* and *S. purpurea* also have been sequenced, with the greatest amount of RNAseq data available for *S. purpurea* (Srivastava et al. 2011, Fukushima et al. 2017). Finally, transcriptomes for *Brocchinia reducta, C. follicularis, N. alata, Pinguicula agnata, P. caudata*, and *Roridula gorgonias* are available as part of the 1000 plants (1KP) initiative,[1] although information on tissue sampling is sometimes incomplete or RNA quantity or quality was compromised prior to library construction.

11.10. Organellar genomes

Complete chloroplast genomes from representatives of *Pinguicula, Utricularia*, and *Genlisea* have been sequenced to investigate plastome-wide changes in molecular evolution. Wicke et al. (2014) reported genome size reduction for *G. margaretae, P. ehlersiae*, and *U. macrorhiza* by up to 9%, mostly resulting from the independent loss of genes coding for plastid NAD(P)H dehydrogenase and alterations in the proportion of repeat DNA. The *ndh* complex is composed of eleven genes (*ndhA–ndhK*) encoded in the chloroplast and, together with four nuclear genes, they encode the thylakoid NAD(P)H dehydrogenase complex. Comparisons among aquatic and terrestrial forms of Lentibulariaceae suggest that aquatic species (e.g., *U. gibba* and *U. macrorhiza*) have retained functional copies of the eleven *ndh* genes, while terrestrial species (e.g., *G. margaretae, P. ehlersiae*, and *U. reniformis*), have either lost or truncated *ndhA–ndhK* (Silva et al. 2016a). These results suggest that *ndh* function may be dispensable in the terrestrial Lentibulariaceae. Some investigators have postulated that such gene loss might be a product of nutritional stress common to carnivorous plants or adaptations to a terrestrial environment (Silva et al. 2016a).

[1] <https://sites.google.com/a/ualberta.ca/onekp/>

Wicke et al. (2014) also demonstrated plastome-wide increases in substitution rates, particularly in *Utricularia* and *Genlisea*. Jobson and Albert (2002) first noticed such elevated substitution rates across all three genomes in the *Utricularia–Genlisea* lineage, focusing on only few loci. Additional early studies of other plastid-encoded genes demonstrated the same phenomenon (Müller et al. 2004), which was extended to large plastid, mitochondrial, and nuclear DNA sets by Ibarra-Laclette et al. (2013). Albert et al. (2010) argued that these increases in substitution rates could be related to ROS-related DNA damage. Although Ibarra-Laclette et al. (2013) did not report substantially elevated substitution rates for the *U. gibba* nuclear genome, preliminary analysis of the new PacBio assembly suggests a substitution rate greater than *V. vinifera* (Lan et al. 2017).

11.11 Future research

The bulk of genome research on carnivorous plants has focused on members of the Lamiales and Oxalidales, with the most complete genomes currently available for *U. gibba* and *C. follicularis*. Results from the minute *U. gibba* genome highlight important roles that tandemly duplicated genes generated from numerous small-scale events may play in the genomic architecture of a plant adapted for the carnivorous habit. While WGD events have received much attention as generators of diversity in other organisms, they may have played other adaptive roles (e.g., in ontogeny and morphology) during bladderwort evolution. Analysis of the *C. follicularis* genome detected presumed adaptive changes that may explain the evolution of carnivorous traits, including attraction, trapping, digestion, and nutrient absorption. Considering similar cases of convergent evolution shown for animal digestive enzymes (Stewart et al. 1987, Zhang 2006), major changes in nutritional strategy could impose a selective pressure strong enough to override evolutionary contingency in both plants and animals.

Functional consequences of genes that are putatively missing in the *U. gibba* genome, yet present in other angiosperm genomes, include gene families involved in ROS scavenging, DNA repair, defense, essential nutrient transport and enzyme activity, auxin response, root and shoot morphogenesis, and the production of diploid gametes, which may have influenced the evolution of *U. gibba* polyploidy. Losses of other genes could have influenced the unique physiology and rootless body plan of *Utricularia*.

Additional draft nuclear genomes and transcriptomes are available for carnivorous members of the Caryophyllales, Ericales, Lamiales, and Poales, but are limited in quantity and quality. Plastid genomes have been sequenced for each of the three Lentibulariaceae genera to investigate plastome-wide changes in molecular evolution. Additional genome sequencing is greatly warranted for the diverse array of carnivorous plant species, especially in combination with carefully planned transcriptome sequencing targeted at identifying genes expressed and associated with plant carnivory. Finally, a thorough assessment at the whole genomic level would greatly add to, and allow for, our understanding of convergent evolution not only among carnivorous plants, but among all plants.

PART III

Physiology, Form, and Function

CHAPTER 12

Attraction of prey

John D. Horner, Bartosz J. Płachno, Ulrike Bauer, and Bruno Di Giusto

12.1 Introduction

Because of the nutrient-poor soils they inhabit, carnivorous plants rely on a variety of animals, especially arthropod prey, to obtain limiting resources such as nitrogen (N), phosphorus (P), and potassium (K) (Juniper et al. 1989, Adamec 1997a, Ellison 2006; Chapters 3, 17–19). The plant traits associated with the acquisition of nutrients from animals constitute the carnivorous syndrome (Chapter 1). The ability to attract potential prey is one of these traits (Juniper et al. 1989, Adlassnig et al. 2010).

Juniper et al. (1989) recognized general mechanisms of prey attraction—visual, olfactory, and tactile cues, and rewards such as extrafloral nectar—but noted that "the efficiency of carnivory is apparently aided by special mechanisms of attraction. . . . Yet no behavioral study has examined this seemingly obvious fact" (Juniper et al. 1989: 74). Among the most significant advances during the last three decades in our understanding of prey attraction are behavioral studies that have examined attraction, detailed and sophisticated measurements of putative attractants, and experiments examining their roles in attracting prey. In this chapter, we review potential mechanisms by which carnivorous plants can attract prey, and summarize the evidence supporting these mechanisms. We ask whether attractants are universal among carnivorous plants; if multiple attractants are used; and if they vary over time (i.e., with trap age) and with environmental conditions. Anatomical details related to prey attraction are discussed in Chapter 13, whereas specialization by carnivorous plants for particular prey is discussed in Chapter 21.

12.2 Visual cues

12.2.1 Reflectance and absorption patterns

The visual attractants used by some carnivorous plants appear similar to those attracting pollinators to flowers. Joel et al. (1985) showed that *Sarracenia flava, Heliamphora nutans, Drosophyllum lusitanicum, Dionaea muscipula, Drosera capensis*, and *Pinguicula ionantha* exhibit distinctive patterns of ultraviolet (UV) reflectance and absorption. Some of these patterns derive from properties of the plant tissue, but others result from reflectance or absorption of nectar or digestive fluids.

Although UV patterns are lacking in some *Nepenthes* species (Joel et al. 1985), Moran (1996) found that peristomes of *N. rafflesiana* absorb UV light, leading to a strong contrast between the peristome and the rest of the pitcher, which reflects UV light. He went on to demonstrate experimentally that these patterns were important in prey attraction (Moran 1996). There is considerable interspecific variation in the ultraviolet and visible spectral reflectance characteristics, and the resulting contrast between peristome and pitcher body of *N. albomarginata, N. ampullaria, N. bicalcarata, N. gracilis, N. mirabilis* var. *echinostoma*, and *N. rafflesiana* (Moran et al. 1999). The differences in contrast among species were associated with the visual sensitivity maxima of insects and indicated species-specific efficacy of prey attraction and resource partitioning among *Nepenthes* species (Moran et al. 1999). Observed species-specific reflectance patterns also have been used as systematic characters. For example, Clarke et al. (2011) used differences in reflectance, among other characters, to distinguish *N. hemsleyana* (at the time named *N. baramensis*) from *N. rafflesiana*.

Horner, J. D., Płachno, B. J., Bauer, U., and Di Giusto, B., *Attraction of prey*. In: *Carnivorous Plants: Physiology, ecology, and evolution*. Edited by Aaron M. Ellison and Lubomír Adamec: Oxford University Press (2018).
 DOI: 10.1093/oso/9780198779841.003.0012

Sarracenia pitchers often possess or develop (usually) red veins that contrast with the most common base color (green) of the pitcher (Figure 12.1). In some cases the lines of contrasting color are associated with extrafloral nectar production, leading Joel et al. (1985) to refer to them as "nectariferous lines" that function similarly to nectar guides in flowers (Joel 1988). There is, however, no experimental evidence to date that these patterns serve to deceive pollinators and attract them as prey.

The pitchers of several *Nepenthes* and *Sarracenia* species possess fenestrations or areoles: translucent areas with relatively little pigmentation. Juniper et al. (1989) suggested that light passing through these fenestrations may confuse trapped insects, reducing the probability of their escape. In fact, the capture of *Drosophila* flies by *Nepenthes aristolochioides* was reduced substantially by masking the translucent areas in the dome of these plants (Moran et al. 2012). In contrast, masking the fenestrations of *S. minor* did not affect the number of flies or ants entering the pitchers, their retention time, or rate of capture (Schaefer and Ruxton 2014). However, more flies alighted on pitchers with unmasked fenestrations, suggesting that they act as long-range attractants. Neither Moran et al. (2012) nor Schaefer and Ruxton (2014) did their studies in the natural habitat of the plants. McGregor et al. (2016), working with natural populations, found that the effectiveness of the fenestrations of *S. minor* as a visual lure depends on the amount of incident sunlight reaching the plants. In a sunny site, masking a proportion of the areoles with indelible ink decreased total prey capture but had no effect on prey diversity. In contrast, altering the proportion of areoles on plants at a shady site did not affect total prey capture, but prey diversity was negatively correlated with the proportion of areoles masked.

Figure 12.1 Lines of contrasting color ("nectar guides" or "nectariferous lines") in a young pitcher of *Sarracenia alata*. These lines are lacking on newly opened pitchers but often become darker, more numerous, and more widespread on older pitchers. Photograph by Dr. Emma Hodcroft.

Color patterns also may be visible only under particular light conditions. Kurup et al. (2013) demonstrated that five species of *Sarracenia*, 14 species of *Nepenthes*, and *Dionaea muscipula*, but not *Drosera*, *Pinguicula*, or *Utricularia*, fluoresce when excited with 366 nm UV light, and suggested that this may act as a prey attractant. Masking the fluorescent regions of the pitchers reduced prey capture by *N. khasiana*; however, the same regions coincided with areas of nectar accumulation, and it is unknown whether the plant tissue or the secreted nectar was responsible for the fluorescence. It also remains unclear how the comparatively weak fluorescence might be perceived under natural daylight conditions when much stronger sunlight would mask it.

12.2.2 Red color as an attractant

Red color is particularly widespread among carnivorous plants and consequently has received much attention as a potential attractant. In addition to contrast, red coloration on traps may attract insects that can perceive red, or camouflage the traps from insects that lack red-light receptors (Briscoe and Chittka 2001). For example, *Nepenthes* pitchers painted red captured more Diptera than those painted green (Schaefer and Ruxton 2008). However, Bennett and Ellison (2009) pointed out that the study by Schaefer and Ruxton (2008) was not conducted in the plants' native habitat or with prey that would normally be encountered in the field. Schaefer and Ruxton (2008) also reported natural color variation within a population of *N. ventricosa*, but only measured the mean red and green coloration of pitchers, averaging out contrasts between pitcher and peristome that might be important for prey attraction. In an *in situ* study, Bennett and Ellison (2009) showed that the proportion of red coloration had no effect on the capture of flies or ants either in pitchers of *Sarracenia purpurea* or in painted artificial "pseudopitchers."

Several research groups working with different *Drosera* species found no evidence for the importance of color as a prey attractant in this genus. Using a combination of *D. rotundifolia* plants, artificial traps, and natural and artificial backgrounds, Foot et al. (2014) tested the effects of color on both attraction and camouflage (crypsis). Generally, red leaves captured more prey than green leaves, but in some cases red coloration deterred prey. Foot et al. (2014) concluded that red coloration may be confounded with other leaf traits and found no evidence that cryptic traps captured more prey. In a study on *D. brevifolia*, Potts and Krupa (2016) found that artificial sticky traps of different colors (red, blue, orange, and white) and leaves of *Drosera* caught similar numbers of insects per unit area per unit time. They suggested that, like spider webs, *Drosera* may capture sufficient prey without attractants. Jürgens et al. (2015) used model traps to investigate the effects of color, shape, position of leaves, and flower height on capture of prey and pollinators by *D. arcturi* and *D. spatulata*. They found that white artificial traps captured more prey than green ones, which in turn captured more than red. They suggested that, rather than attracting prey, red coloration may serve to exclude pollinating insects (which are often unable to perceive red) and thus protect them from being captured.

It is important to note that the colors of the artificial traps used in many of these studies poorly represent natural colors, and that the color perception of humans and insects differs significantly. The use of clear plastic laminates to protect paper or cardboard traps also might interfere with reflectance and absorption of both visual and ultraviolet wavelengths. Whereas some studies have verified the similarity in spectral reflectance (or absorption) of artificial and natural traps using spectrophotometers, others have not. Future studies should verify that reflectance and absorbance spectra are similar for artificial and natural traps.

Red coloration in leaves also might be entirely unrelated to carnivory. Red color is usually caused by anthocyanin, and the production of anthocyanins is influenced by light, ultraviolet radiation, and nutrient availability (Ichiishi et al. 1999, Close and Beadle 2003, Boldt et al. 2014). Therefore, red coloration predominantly may play a role in photoprotection or be an indicator of nutrient stress. Because carnivorous plants typically grow under high light conditions and in nutrient-poor soils, conclusions about the primary role of red coloration in these plants are difficult to draw. B. Molano-Flores and colleagues found that anthocyanin production by *Pinguicula planifolia* varied with water availability and vegetation height (i.e., the intensity of incident light), and that green leaves captured more prey than red leaves (B. Molano-Flores, *personal communication*).

12.3 Nectar rewards

All species of pitcher plants examined to date, including species of *Sarracenia* (Cresswell 1993, Cipollini et al. 1994, Green and Horner 2007), *Nepenthes* (Merbach et al. 2001, Bauer et al. 2008, 2009, Gaume et al. 2016), *Cephalotus* (Vogel 1998), and *Heliamphora* (Płachno et al. 2007b), possess extrafloral nectaries (EFNs). In *S. purpurea* and *S. alata*, nectar production is particularly high around the peristome (Cipollini et al. 1994, Green and Horner 2007). Nectar is also produced along nectariferous lines in *S. flava* (Joel et al. 1985) and *S. alata* (Green and Horner 2007). Measurable quantities of nectar have been detected on the top and bottom of the hood and elsewhere on the pitcher in *S. alata* (Green and Horner 2007). Nectar production in *S. purpurea* and perhaps other species varies throughout the day, and is depleted by insect foraging, making accurate measurements difficult (Deppe et al. 2000). Nectar production also varies with pitcher age, reaching a maximum approximately three weeks after *S. alata* pitchers open (Green and Horner 2007, Horner et al. 2012).

Among *Nepenthes* species, nectar production has been reported for *N. bicalcarata*, *N. gracilis*, *N. mirabilis* var. *echinostoma*, *N. rafflesiana*, *N. hemsleyana*, *N. lowii*, *N. macrophylla*, *N. rajah*, and *N. villosa* (Moran 1996, Merbach et al. 2001, Bauer et al. 2008, 2016, Chin et al. 2010, Moran and Clarke 2010). EFNs are scattered across the stems, laminae, tendrils, and pitchers. Higher densities of EFNs are found on the upper and lower surface of the pitcher lid and, in particular, lining the inner edge of the peristome. During pitcher development, the main location of nectar secretion is gradually shifted from the lamina, tendril, and outer surface of the pitcher in developing buds to the peristome and lower lid surface in open pitchers (Merbach et al. 2001). Peristome nectar production increases over the first two weeks after pitchers open in *N. rafflesiana* (Bauer et al. 2009). In this species, nectar not only serves as an attractant and reward, but also enhances the condensation of water from humid air on the peristome surface, thereby increasing the slipperiness of the trap and rate of prey capture (Bauer et al. 2008; Chapter 15). This function is a result of the hygroscopic properties of sugar. It is likely that the peristome nectar of other *Nepenthes* species, when produced in sufficient quantity and concentration (e.g., in *N. gracilis*), serves a similar function.

Generally, very little is known about the composition of the extrafloral nectar of carnivorous plants and how it varies with environmental factors or with the targeted prey. Few studies provide qualitative and quantitative analyses of the sugar or amino acid content which is crucial for attracting specific insects (Baker and Baker 1973, Heil 2011). For example, amino acid content often differs between floral and extrafloral nectars, and this allows for specific attraction of pollinators vs. defenders (Baker et al. 1978). In some cases, the nectar absorbs ultraviolet light, and Joel et al. (1985) suggested that the ultraviolet absorption of nectar pools in *S. flava* may indicate that the nectar contains amino acids. However, amino acids have so far only been reported in the nectar of *S. purpurea* (Dress et al. 1997).

Whether nitrogen limitation affects the amino acid content of extrafloral nectar in carnivorous plants is unknown. Sucrose:hexose ratios of nectar also can affect visitor attraction and discrimination between obligate and opportunistic defenders (Heil 2015). No study to date has shown how the nectar of insect-generalist species of carnivorous plants (e.g., *N. mirabilis* var. *echinostoma*) differs from the nectar of more specialized species. Specialists include *N. bicalcarata*, a myrmecophyte host for the ant *Camponotus schmitzi* that protects assimilatory organs, enhances prey capture, and aids in myrmecotrophy (Merbach et al. 1999, Bazile et al. 2012); and *N. lowii*, *N. macrophylla*, and *N. rajah*, each of which offers large quantities of extrafloral nectar to the mountain tree-shrew (*Tupaia montana*) in exchange for its nutrient-containing feces (Moran and Clarke 2010). Finally, differences in nectar composition and quantity have not been studied in different ontogenetic stages of pitchers, such as lower terrestrial pitchers of the rosette stage and upper aerial pitchers of the vine stage.

12.4 Olfactory cues

Scent has been recorded from pitcher traps (Juniper et al. 1989, Jürgens et al. 2009, Di Giusto et al. 2010, Wells et al. 2011), snap-traps (Jürgens et al. 2009, Kreuzwieser et al. 2014), and adhesive traps (Jürgens et al. 2009, Fleischmann 2016a). The attractive odor in some carnivorous plant traps is perceivable by humans, e.g., the honey-like odor of *Drosophyllum* (Lloyd 1942, Juniper et al. 1989, Bertol et al. 2015b), the flower-like odor of *S. flava* (Miles et al. 1975), and the sweet heavy fragrance of several *Nepenthes* species (*N. rafflesiana*, *N. undulatifolia*, *N. philippinensis*, *N. madagascariensis*) (Juniper et al. 1989, Moran 1996, Ratsirarson and Silander 1996, Di Giusto et al. 2008, Bauer et al. 2009, U. Bauer *unpublished data*). Others, including *N. albomarginata* that preys on blind termites (Merbach et al. 2002), may produce volatiles humans cannot perceive. However, detailed studies of the chemistry of volatiles have been performed for only a few species of carnivorous plants.

Among the most well-examined is *N. rafflesiana*, which, like other *Nepenthes* species, produces terrestrial pitchers as a juvenile and aerial pitchers when mature (Gaume and Di Giusto 2009). In *N. rafflesiana*, these two pitcher types differ not only in the quantity and diversity of captured prey (Moran 1996, Di Giusto et al. 2008) but also in the production of volatile compounds (Di Giusto et al. 2010). Upper pitchers emit a more complex odor with a larger spectrum of volatiles, resembling the bouquet of generalist flowers. Both lower and upper pitchers mainly capture ants, but upper pitchers catch a larger proportion of flying prey, including flower-visitors.

It is unknown whether *Nepenthes* traps have special osmophores that produce volatile compounds or whether the volatiles are secreted together with the nectar. Experiments with ants (*Oecophylla smaragdina*) and fruit flies (*Drosophila melanogaster*) have shown that the peristome of pitchers was the main site of both attraction and scent (Di Giusto et al. 2010). However, because the peristome has well-developed EFNs, it seems likely that they produce the volatile compounds; indeed, when isolated from the pitcher, the nectar itself is fragrant (U. Bauer *unpublished data*). Another source of volatile compounds may be the EFNs in the pitcher lid. The fluid in *Nepenthes* pitchers also has been reported to emit a sweet fragrance (Moran 1996, Moran et al. 1999), but the fluid alone is not very attractive to insect visitors (Di Giusto et al. 2008). It is unknown whether volatile compounds are produced by the digestive glands or whether the odor results from decaying pitcher contents or the fragrant nectar accumulating in the digestive fluid.

The production of volatile compounds in *Nepenthes* varies in at least two pitcher stages. Flower-like odors are produced by young, freshly opened pitchers that have captured few prey. Odor intensity and nectar production of *N. rafflesiana* increases simultaneously over the first week after pitcher opening (Bauer et al. 2009). Subsequently, putrefaction odors (volatile compounds produced by bacterial decomposition) are produced by older traps that have accumulated prey (Jürgens et al. 2009). Clarke (2001) noted that *Nepenthes* pitchers that are full of prey carcasses emit a special odor that attracts calliphorid and muscid flies. Because of this odor change, *Nepenthes* pitchers may attract different types of prey during different life stages: flower and fruit visitors at first and saprophagous insects later (Jürgens et al. 2009).

Nepenthes rajah "toilet" pitchers collect mammal feces (Chapter 26). They emit a sweet fruity odor that contains over 40 volatile compounds including hydrocarbons, alcohols, esters, ketones, and sulfur-containing compounds that attract mountain treeshrews and summit rats (*Rattus baluensis*) (Wells et al. 2011). These chemicals are common in fruit and flower odors, and their occurrence in pitchers may be an instance of convergent evolution of attractive traits.

Carnivorous plants with flypaper or sticky traps that, like *Nepenthes*, are members of the Caryophyllales (Chapter 3), also may use volatile compounds to attract prey. Few *Drosera* species have been reported to produce odors that are detectable by humans (Lowrie 2013, Fleischmann 2016a). Examples are the leaves of *Drosera slackii* (*D.* sect. *Drosera*) which emit a lemon-like scent; *D. fragrans* (*D.* sect. *Arachnopus* of the "*D. indica* complex") that produces a sweet honey-scent from outgrowths at the leaf bases that may function as osmophores (Lowrie 2013); and some specimens of *D. finlaysoniana* (also in *D.* sect. *Arachnopus*), which produce a strong, sweet, honey-like scent from their leaves but lack leaf-base outgrowths (Fleischmann 2016a). Because *D. finlaysoniana* attracts and catches a large number of Lepidoptera, it is probable that the fragrance of this species is fine-tuned to attract this group (Fleischmann 2016a). Interestingly, *Drosera* species that have scented leaves have scentless flowers, whereas species that do not have fragrant leaves have scented flowers (Fleischmann 2016a). These combinations may contribute to the avoidance of pollinator–prey conflicts (Chapter 22). Leaf-scent chemistry has been analyzed only for one species of *Drosera*, *D. binata*, whose leaves do not emit a human-perceptible scent (Jürgens et al. 2009). Only five volatile compounds were detected in this species, some of which also are characteristic of volatiles emitted from green leaves of typical noncarnivorous plants. It seems unlikely that these volatiles are involved in prey attraction (Jürgens et al. 2009).

Jürgens et al. (2009) and Kreuzwieser et al. (2014) independently analyzed the scent chemistry of *Dionaea muscipula* traps. Jürgens et al. (2009) found that *Dionaea* emits a detectable scent consisting mainly of benzyl alcohol and methyl salicylate, two compounds that commonly occur in the odors of flowers. In contrast, Kreuzwieser et al. (2014) found that *Dionaea* traps released over 60 volatile organic compounds, the majority of which had been identified previously in flowers or fruits. Kreuzwieser et al. (2014) also showed that starved fruit flies were strongly attracted to the volatile compounds released by *Dionaea*. They suggested that fruit flies looked for a food source in *Dionaea* traps because some of the volatile compounds that are released by the traps are common to ripe and rotten fruits.

The production of volatile compounds in pitcher traps of *Heliamphora*, *Darlingtonia*, and some *Sarracenia* species is associated with sweet-scented nectar (Juniper et al. 1989, Jaffe et al. 1995). The main sources of both nectar and scent in *Heliamphora* traps are nectar-spoons at the top of the pitcher (Płachno et al. 2007b, 2009b, Fleischmann 2016a). *Heliamphora tatei*, *H. neblinae*, and *H. chimantensis* produce a honey-like scent, whereas *H. sarracenioides* produces a notable chocolate-like scent (Fleischmann and McPherson 2010, Fleischmann 2016a). A sweet flower-like, honey-like, or fruit-like scent has been recorded from traps of *S. flava*, *S. alata*, *S. rubra*, *S. oreophila*, *S. leucophylla*, and *S. minor* (Miles et al. 1975, Slack 1979, Juniper et al. 1989, Jürgens et al. 2009, Fleischmann 2016a). According to Fleischmann (2016a), the production of scent in some of these species varies among populations and among individuals within a population. Further investigations are needed to determine whether these differences are genetically based, affected by season or variable weather conditions, linked to patterns of nectar secretion, or some combination of these.

Jürgens et al. (2009) analyzed the scent chemistry of *S. flava*, *S. leucophylla*, *S. minor*, and *S. purpurea* in detail using gas chromatography coupled with mass spectrometry (GC-MS). The first three species produced a high diversity of compounds, including some that are typical for flower or fruit scents, whereas the traps of *S. purpurea* produced weaker scents with fewer components that typically are emitted by green leaves. They also reported a difference between the odor composition of empty *S. purpurea* traps and traps that contained prey. Dimethyl disulfide (DMDS), which can attract saprophagous insects, was found only in traps that contained prey. Some of the volatiles emitted by prey-filled pitchers may not be produced by the plant itself but rather result from the decomposition of prey. Nevertheless, the plant may benefit from their production. For example, fluid from prey-filled pitchers of *Sarracenia alata* attracted more insects to artificial sticky traps than did deionized water (Bhattarai and Horner 2009). Jürgens et al. (2009) suggested that attractive syndromes in *Sarracenia* change over time in a similar fashion as in *Nepenthes*: young pitchers target flower-visiting insects but older pitchers attract carrion-feeders.

Ho and colleagues comprehensively analyzed the volatiles produced by *Heliamphora*, *Darlingtonia californica*, and 11 species of *Sarracenia* using GC-MS. They found some divergence of trap volatiles among species, but it was not as strong as that observed among the floral scents of the same species (W. Ho *personal communication*). They also found that traps of *S. flava* supplemented with prey significantly increase the emission of DMDS and several other compounds relative to plants without prey supplement. Electroantennograms were used to examine the response of calliphorid flies and bees to the compounds identified from flowers and traps. The results confirmed that saprophages and pollinators respond differently to floral and trap volatiles (W. Ho *personal communication*).

Among members of the Lentibulariaceae, scented traps have been recorded for *Pinguicula*, whose leaves emit a fungus-like odor (Lloyd 1942, Zamora 1995, Jürgens et al. 2009, Fleischmann 2016a). At least six species of this genus, including *P. primuliflora*, *P. lutea*, and *P. vallisneriifolia*, emit an odor that is detectable by humans (Fleischmann 2016a). The fungus-like odor is a strong attractant for fungus gnats (Diptera: Mycetophilidae and Sciaridae), which are the main prey of *Pinguicula* (Heslop-Harrison 2004, Fleischmann 2016a).

Finally, scented traps also are found among carnivorous Bromeliaceae. The leaves of *Brocchinia reducta* do not produce nectar, but emit a sweet nectar-like odor. *B. reducta* might mimic its nectar-producing neighbor, *Heliamphora heterodoxa* (Givnish et al. 1984, Juniper et al. 1989).

12.5 Acoustic attraction

At least one species of pitcher plant, *Nepenthes hemsleyana*, hosts roosting bats in its pitchers, and the plant receives nutrients from the bats in the form of feces (Chapter 26). M.G. Schöner et al. (2015a) demonstrated that the rear part of the inner pitcher wall in this species acts as an ultrasonic reflector, helping the bats to locate the pitchers in cluttered vegetation. The ultrasound reflection of *N. hemsleyana* pitchers was distinct from that of the closely related and sympatric *N. rafflesiana*, and the bats showed a clear preference for their host species. To our knowledge, this is the first report of an animal being acoustically attracted to a carnivorous plant.

12.6 Prey attraction in carnivorous plants with aquatic traps

Most studies on prey attraction have focused on terrestrial carnivorous plants, but a few have been performed on *Utricularia* and *Genlisea* spp. Although Luetzelburg (1910) suggested that carbohydrate-based mucilage produced by *Utricularia* traps may act as an attractant, Seine et al. (2002) concluded that attractants were used only by traps of terrestrial *Utricularia* species. Meyers and Strickler (1979) experimentally supported Darwin's (1875) hypothesis that the antennae and bristles on the trap of *U. vulgaris*, although not attractants *per se*, enhance capture success by guiding prey to the trap entrance. Sanabria-Aranda et al. (2006) and Manjarres-Hernandez et al. (2006) corroborated these results in other species of *Utricularia*. Sanabria-Aranda et al. (2006) also found that prey are lured to carbohydrates produced by *U. foliosa*. The quantity of carbohydrates on the traps was correlated to the abundance of periphyton (attached algae). They concluded that carbohydrate production was not part of a strategy to attract prey. Jobson and Morris (2001) showed that copepods were over-represented (and nematodes under-represented) in traps of *U. uliginosa* relative to their abundance in the surrounding soil, suggesting that the capture of prey by *U. uliginosa* is non-random. However, it is unclear whether the observed differences were due to differential attraction, capture success, or rate of decomposition, and Jobson and Morris (2001) did not identify any attractants.

It is still debated how *Genlisea* attracts prey. Barthlott et al. (1998) suggested that *Genlisea* traps may attract protozoa chemotactically. Studnička (2003a) hypothesized that *Genlisea* traps mimic soil interspaces that are partially filled with air and may attract oxygen-dependent microfauna. In support of this hypothesis, Adamec (2007b) measured radial oxygen loss and suggested that the oxygen released by traps may function as an attractant for prey. However, Płachno et al. (2008) suggested that prey move into *Genlisea* traps by accident, randomly wandering into small spaces filled with water. *Genlisea* may catch prey simply by providing many small openings that mimic the interspaces between soil particles, without the need for specific attractants.

12.7 Synergistic effects of multiple attractants

Most studies of prey attraction in carnivorous plants have examined only a single type of attractant. However, many species employ more than one means of attraction, which almost certainly act in synergy. Moran (1996) demonstrated an additive effect of high-contrast ultraviolet patterns and fragrance for prey attraction by *N. rafflesiana*. Pitchers with a combination of both traits captured more prey than traps with only one of these cues (Moran 1996). Therefore, the combination of attractive signals may be crucial for successful prey capture.

12.8 Temporal variation of attractive cues

Most studies of attractive cues have focused only on a particular period of trap development, for example, the first few weeks after trap opening. However, in many cases the quantity and quality of attractants change with trap age and prey accumulation. These changes may influence not only the quantity, but also the identity of attracted prey. Both Bauer et al. (2009) and Di Giusto et al. (2010) observed age-related changes in pitcher odor in

N. rafflesiana (§12.4). Nectar production, visual cues, and the rate of prey capture varied over the first six weeks after pitcher opening in *S. alata* (Green and Horner 2007, Horner et al. 2012). Although the rate of prey capture in *S. alata* is greatest approximately three weeks after pitchers open (Green and Horner 2007, Horner et al. 2012), prey capture continues at a lower rate over the life of the pitcher (J. Horner *unpublished data*). Prey capture in older pitchers also was observed for *S. purpurea* by Heard (1998), who found that pitchers continued to capture prey for up to two years.

The production of attractants also can change following prey capture. Changes in odors or volatiles emitted by pitchers containing prey have been observed for *Nepenthes* (Clarke 2001) and *S. purpurea* (Jürgens et al. 2009), and these may result in differences in the type of prey attracted over the lifespan of the trap.

Temporal variation of nectar secretion or composition also may have significant effects on the attraction of social insects, in particular ants recruiting *en masse* to food sources (Chapter 15). The nectar on the peristome of *N. rafflesiana* both attracts prey by providing a reward and facilitates prey capture by increasing surface wettability and aquaplaning of visiting insects (Bauer et al. 2008). Intermittent, wetness-based activation allows for the temporal separation of attraction of scout ants and trapping of recruited workers. It therefore may be a specific adaptation for the capture of ants that use mass recruitment (Bauer et al. 2015a; Chapters 13, 15).

12.9 Is production of attractants a crucial trait for carnivory?

Prey attraction has been termed a key trait of the carnivorous syndrome (Juniper et al. 1989 Adlassnig et al. 2010; Chapter 1), but it has been documented convincingly and consistently only for *Sarracenia* and *Nepenthes*. Putative attractants have been identified in a number of other species, but demonstration of their effectiveness is inconsistent. In some cases, the spectrum of captured prey hardly differs from the composition of arthropod species in the surrounding habitat. In others, there are substantial differences, suggesting that some carnivorous plants differentially attract (or capture) a subset of the available prey (Ellison and Gotelli 2009; Chapter 21).

Studies comparing natural and artificial traps in the field can help to shed light on the role of attractants. For example, leaves of *Drosophyllum lusitanicum* captured more insects (especially dipterans) than did artificial mimics (Bertol et al. 2015a). Similarly, the over-representation of certain taxa on plants compared to artificial traps may suggest the presence of attractants in *Pinguicula moranensis* (Alcalá and Dominguez 2003), *P. longifolia* (Antor and Garcia 1994), and *P. vulgaris* (Karlsson et al. 1987, 1994). However, these results should be interpreted cautiously because attractants have not yet been identified for these species. Furthermore, studies comparing natural traps and artificial mimics are problematic because artificial traps usually differ from natural ones in more than just the desired characteristics (§12.2.2).

Attractants have been confirmed for some carnivorous plants, but there are many carnivorous taxa for which no evidence of attractants has been found. Nonetheless, these plants also capture prey successfully. Because some plants capture insects and are capable of absorbing nutrients from the prey, but do not appear to produce attractants, specifying attractants as a defining trait of the carnivorous syndrome may need to be re-evaluated. Attractants simply may have escaped identification so far, but it is also possible that some species of carnivorous plants do not need attractants to capture prey. If prey are sufficiently abundant and regularly come into contact with traps by chance, attractants may be redundant. Moreover, as there usually is a cost associated with their production, there would be selective pressure against producing attractants.

12.10 Cost of attractants

The cost of attractants is important for at least four reasons. First, it has been hypothesized that carnivorous plants should be under strong selection to produce attractants so as to gain nutrients from captured insects. If carnivorous plants are under selection to produce attractants, then the cost of their production needs to be weighed against the benefits of prey capture. Attractants should be produced only when benefits exceed costs. Second, understanding the cost of attractants may help us understand the time-course and temporal variation of their production by those plants that produce them (§12.7). Third, understanding the costs of attractants may help us understand why some carnivorous plants do not appear to produce them. Finally, understanding the costs of attractants could contribute to our understanding of the evolution of botanical carnivory (Chapters 3, 18).

The production costs of attractants can be measured in several ways. One is the proximate metabolic cost of their construction and maintenance in terms of energy and nutrient requirements. Another valuable measure would be the ultimate cost of production: a reduction in growth and reproduction resulting from the reallocation of resources. Estimation of this ultimate cost should integrate not only factors such as the cost of enzymes used to produce attractants along with their concentrations and turnover rates but also the overlap between these resources and those limiting plant growth and reproduction (as Fagerstrom 1989 modeled for the cost of production of plant chemical defenses). For example, if growth and reproduction of a plant is limited by nitrogen or phosphorus (as in most carnivorous plants; Chapters 17–19), then the production of compounds containing only carbon, hydrogen, and oxygen (such as terpenoids, anthocyanins, and nectar that contains no amino acids) would have little negative impact on growth and reproduction. These compounds would have a low cost to the plant that could be offset easily by the benefits of enhanced prey capture.

12.11 Future research

Significant advances recently have been made in understanding how carnivorous plants attract prey or mutualists that provide essential nutrients to the plants. Nonetheless, our understanding of prey attraction in carnivorous plants is still relatively rudimentary.

The investigation of visual cues in most early studies, and even in some recent ones, has been primarily descriptive and based on human perception. Over the past 25 years, more research has used spectrophotometry to measure absorption or reflectance of both visible and ultraviolet light. Much more work is needed in this area. With modern qualitative and quantitative analytical methods and well-designed behavioral experiments, visual attractants may be discovered for additional carnivorous plant taxa. There are still wide gaps in our knowledge of the function and specificity of UV and visible guides for prey attraction. Behavioral assays with different insects will be essential to link descriptive studies of color spectra to attraction.

Similarly, many early descriptions of olfactory attractants were based on human perception of odors. Although there were a few early studies that used either GC alone or GC-MS, only recently has their use become routine in the analysis of scent. Their use is providing a much more accurate qualitative and quantitative understanding of volatiles emitted by carnivorous plants. However, more research is needed to elucidate causal relationships between volatile emissions and prey attraction. Electroantennograms are powerful tools with which to identify sensitivity of insects to individual volatile compounds or their mixtures. Behavioral tests (such as Y-tube tests) that present insects with a choice between a control and the compound(s) in question should be used to confirm attractive function.

To fully understand cost/benefit tradeoffs of nectar production, the composition of the nectar needs to be examined in detail. Both amino acid content and hexose:sucrose ratio are important factors determining attractive properties of nectar and its specificity of attraction, but neither have been studied in detail. Are the composition and patterns of extrafloral nectar secretion fine-tuned to the requirements of the prey of carnivorous plants? Moreover, general questions regarding the production of extrafloral nectar remain unanswered in most carnivorous plant species. What kinds of EFNs are involved (Chapter 13)? Do they reflect the large variety of EFNs observed in noncarnivorous plants? Temporal variation in the water content of nectar has been demonstrated to be important in prey attraction and capture in *N. rafflesiana*, and this aspect needs to be studied in a wider range of carnivorous plants, particularly the pitcher plants.

Combinations of cues (e.g., visual and olfactory) and rewards (e.g., nectar), and potential synergistic effects between them, have received far too little attention. For example, little is known about the interaction of UV patterns with other signals such as visible-light patterns, olfactory cues, and nectar rewards. In many cases, it is likely that combinations of cues are responsible for the attraction of prey. Weak or statistically insignificant results in prior research on attractants may have resulted from research focused on single cues, which may have little or no effect in isolation.

Temporal variation in cues is yet another area that requires further research. At this point, changes in attractants and in the type of prey attracted have been studied primarily over the first few weeks of trap activity in many species; very few studies have examined these aspects over the lifespan of the trap.

Finally, the adaptation of carnivorous plants to nutrient-poor soils likely influences the variety, composition, and quantities of attractants produced by these plants. Estimating the cost of production of attractants may provide insight into the evolution of attractants. Although the estimation of the cost of attractants will not be easy, it is a research area that promises to provide exciting new insights into the ecology and evolution of plant carnivory.

We emphasize that it is of the utmost importance that studies on prey attraction are conducted in the plants' native environment with the assemblage of natural prey. The production and emission of volatiles, nectar secretion, the synthesis of pigments, and acoustic background all vary with environmental factors. Attractants are likely to have been tuned through natural selection to the perceptions and preferences of prey, and laboratory experiments with different insects may identify spurious effects or miss real ones.

CHAPTER 13

Functional anatomy of carnivorous traps

Bartosz J. Płachno and Lyudmila E. Muravnik

13.1 Introduction

There is great variation in the anatomy of carnivorous traps that is related to their ability to attract, capture, kill, and digest prey, and to absorb the resulting mineralized nutrients. Many trap functions are carried out by glandular cells, organized as glandular hairs (trichomes), epidermis, or emergences (tentacles). Carnivorous plants have three main glandular systems: glands for attracting prey, which are primarily responsible for the production of nectar and olfactory attractants (Chapter 12); glands for trapping prey that produce mucilage, viscoelastic liquid, or resin; and digestive and absorptive glands. These systems may overlap as some glands have multiple roles. For example, the *Drosera* emergences trap, retain, and digest prey, whereas the digestive glands of *Nepenthes* produce not only enzymes but also antibacterial and anti-fungal substances (Hatano and Hamada 2008, Rottloff et al. 2009). In this chapter, we illustrate the diversity of trap design, methods of prey retention in different carnivorous plant lineages, and the evolution of digestive and absorptive systems, and discuss functional convergence in trap design across unrelated carnivorous plant taxa.

13.2 Nectar glands

Extrafloral nectaries (EFNs) have been described for all lineages of pitcher plants (Chapter 12).

13.2.1 Nectaries of the Sarraceniaceae

In *Sarracenia*, EFNs are located on, above, and below the peristome, along colored veins on external pitcher surfaces, on the hood, and on the upper (smooth) part of the pitcher (the "attractive zone" of Lloyd 1942). EFNs on the external surface and on the hood act as nectar guides leading to the peristome. The "*Sarracenia*-type" EFN consists of a barrel-shaped complex of 10–14 cells ≤ 60 μm in diameter that is sunken below the surface of the epidermis. The surface layer is made up of six cells: four peripheral and two central cells that taper toward the interior and have thick walls and dark contents. The subjacent layer consists of four cells, and the basal layer of two. Except for the basal cells, the whole gland body is isolated from the adjacent parenchyma by cutinized cell walls. Nectaries on the hood are slightly larger, comprising 2–4 surface layer cells and about 20 subjacent gland cells.

In *Darlingtonia*, nectar is secreted by EFNs on the fishtail appendage, peristome rim, and the pitcher's dome. EFNs on the peristome and interior surface of the dome are much larger than those on the outer surfaces. The large nectaries of *Darlingtonia* superficially resemble those of *Sarracenia*, but they are sunken below the surface of the epidermis and not visually conspicuous (Lloyd 1942).

Heliamphora has two types of EFNs: giant ones on the spoon and small ones on the outer surface of the pitcher, the wings, and the inner surface of the pitcher above the fluid level (Figures 13.1a, b). Both types are similar in structure to *Sarracenia*-type

Płachno, B. J., and Muravnik, L. E., *Functional anatomy of carnivorous traps*. In: *Carnivorous Plants: Physiology, ecology, and evolution*. Edited by Aaron M. Ellison and Lubomír Adamec: Oxford University Press (2018).
 DOI: 10.1093/oso/9780198779841.003.0013

EFNs (Lloyd 1942, Vogel 1998), but giant nectaries have an additional domed surface layer of ~50 apical cells. Both apical and central subjacent cells have fully cutinized radial walls, whereas peripheral cells have only partially cutinized walls. The cutin waterproofs the cell walls and prevents leaking between the nectary and adjacent parenchyma cells. Specialized cells with thickened flange-like walls control transport processes between the parenchyma and the nectary tissue.

The giant EFNs are the major source of nectar produced by *Heliamphora*. Some species produce so much nectar that it trickles from the spoon into the interior of the pitcher (McPherson 2007). *Heliamphora folliculata* has a unique nectar reservoir at the back of the spoon, where nectar accumulates before seeping out through a narrow channel at its front. The largest EFNs are found inside this nectar reservoir and on the surface of the nectar channel. The nectar cells of *H. folliculata* contain leucoplasts surrounded by endoplasmic reticulum, and vacuoles with large dark inclusions (Płachno et al. 2007b). The complex of endoplasmic reticulum and leucoplasts is characteristic of cells producing terpenes and phenols, indicating that the giant EFNs also produce volatile compounds (Chapter 12).

13.2.2 Nectaries of *Cephalotus*

Despite being unrelated to the Sarraceniaceae (Chapter 3), *Cephalotus* has anatomically similar, *Sarracenia*-type EFNs scattered across the outside of the pitcher, the inside of the lid, and the surfaces of the recurved teeth of the peristome. Glands of uncertain function are located on the smooth inner wall of the pitcher underneath the peristome. They are spherical, ≈50 µm in diameter, and comprise 25–30 cells arranged in several irregular layers. The largest glands of *Cephalotus* consist of ≤ 18 apical cells and 150–200 total cells. Twenty to 30 of these occur on each of a pair of prominent patches on the lower rear inner wall of the pitcher. These large glands are submersed and likely involved in digestion (Vogel 1998).

13.2.3 Nectaries of *Nepenthes*

EFNs of *Nepenthes* are found on the shoot axis, petiole, lamina, and tendril, on the outside of the pitcher, the pitcher lid, and along the inner margin of the peristome. Two main types of EFNs, disk-like and elongated ones, are produced by *Nepenthes* but differ in distribution, quantity, and size among the species (Merbach et al. 2001).

The ≈110-µm wide disk-shaped EFNs of the pitcher lid sit in pits and are structurally homologous to digestive glands (Parkes 1980, Vassilyev 2007). Each EFN contains a top layer of palisade secretory cells, two to three subjacent layers of secretory cells, an underlying "barrier layer" with cutinized anticlinal walls, and a single layer of basal cells. A vascular bundle containing one to three tracheal elements leads up to each nectary. In developing pitchers, before nectar begins to be secreted, the endomembrane system of the nectary's secretory cells is qualitatively and quantitatively very similar to that of digestive glands (Vassilyev and Muravnik 2007). Similarities include a well-developed rough endoplasmic reticulum, Golgi apparatus, and trans-Golgi reticulum producing coated vesicles. As the pitchers grow, these nectaries produce and discharge digestive fluid into the pitcher. Hydrolase is secreted, and a small quantity of polysaccharide mucilage, which large Golgi vesicles bring out from the cell, also appears in the periplasmic space. Once the pitcher is fully-formed and open and the EFNs are secreting nectar, synthesis and secretion of hydrolases cease but production and secretion of mucilage continues at a lower rate. Most *Nepenthes* species secrete nectar to attract prey, but EFNs are sparse or absent on the lid of *N. ampullaria*, a species that rarely captures prey but instead uses leaf litter as a source of nitrogen (Moran et al. 2003).

Morphology, anatomy, and ultrastructure of EFNs on the peristome (Figure 13.1c) differ from that of lid nectaries. Peristome nectaries are very large—in *N. khasiana*, for example, they are elongated, ≈320 µm high and ≈90 µm wide (Vassilyev 1977)—and sometimes have an extracellular cavity that serves as a nectar reservoir (Parkes 1980). A surface layer of palisade secretory cells covers several layers of parenchymal cells. The nectary is isolated from the peristome by a layer of barrier cells with cutinized anticlinal walls. A branch of a vascular bundle with one to three tracheal elements leads up to each nectary. In developing pitchers, the palisade and parenchyma cells are ultrastructurally nearly identical to typical mucilage-secreting

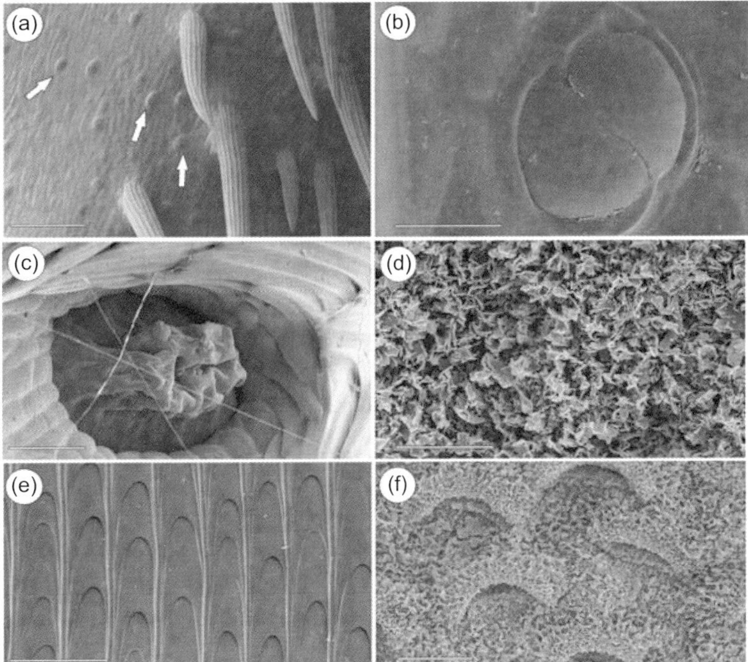

Figure 13.1 Micromorphology of several types of pitcher traps. **(a)** Small nectaries (arrows) and downward-pointing hairs on the inner wall of a *Heliamphora nutans* pitcher; bar = 200 µm. **(b)** Small nectary on the inner wall of an *H. nutans* pitcher; bar = 20 µm. **(c)** Nectary between the teeth of the inner edge of the peristome of *Nepenthes hurreliana*; bar = 50 µm. **(d)** Epicuticular wax crystal on the inner pitcher wall of *N. gracilis*; bar = 10 µm. **(e)** Microstructure of *N. tentaculata* peristome epidermal cells; bar = 50 µm. **(f)** Lunate cells covered by an epicuticular wax crystal in an *N. gracilis* pitcher; bar = 20 µm. Photographs by Ulrike Bauer.

glands, and include a slightly developed endoplasmic reticulum, a predominance of Golgi bodies that produce large secretory vesicles, and a periplasmic space filled with mucilage that may tightly seal the pitcher (Vassilyev 2007). In open pitchers, nectar is secreted, the number of Golgi bodies is reduced, and the mucilage in the periplasmic space is disappeared. In some species (e.g., *N. maxima*) nectaries in the peristome may form a ring of secretory tissue that is supplied with a ring of vascular bundles (Parkes 1980).

Secreted nectar completely moistens the peristome and forms a homogeneous thin film that makes the peristome extremely slippery for insects (Bauer et al. 2008). When the peristome is wet, this fluid film prevents the insects' tarsal adhesive pads from making close contact with the peristome surface, similar to how car tires aquaplane on a wet road. Freshly secreted peristome nectar of *N. rafflesiana* var. *typica* is hygroscopic and has a high sugar concentration (10–40%; W. Federle *unpublished data*). Spread on the completely wettable peristome surface, the nectar evaporates and crystallizes in the sun and wind but liquefies again at night as it absorbs water (Deppe et al. 2000). The role of nectar in prey capture by *Nepenthes* is remarkable because it involves a purely mechanical function of nectar (Chapter 15), but *Nepenthes* also may actively regulate the degree of wetting by adjusting the quantity of nectar secreted (Bauer et al. 2008).

In some *Nepenthes* species, nectar secreted by EFNs attracts ants that, in turn, protect the host against insect herbivores such as weevils that feed on pitchers (Merbach et al. 2007). The upper, uniquely shaped pitchers of *Nepenthes lowii* secrete a white sugar-rich substance at the tips of bristles under the lid. This substance, which is different from nectar, attracts birds (Clarke 1997).

13.3 Slippery surfaces of pitcher-plant traps and bromeliad tanks

Pitcher plants and carnivorous bromeliads have evolved sophisticated surface modifications to trap and retain prey. Adaptations for slipperiness can be found on virtually all interior surfaces of pitchers, including the lid and hood, peristome, and inner pitcher wall, and include epicuticular wax blooms, cuticular folds, and directional features formed by overlapping epidermal cells or inclined trichomes.

13.3.1 Epicuticular wax crystals

Wax blooms are widespread in *Nepenthes*, all genera of Sarraceniaceae (but not widespread in *Heliamphora*), and have been reported for *Brocchinia reducta* and *Catopsis berteroniana* (e.g., Givnish et al. 1984, Juniper et al. 1989, McPherson 2009). The epicuticular wax crystals in *Nepenthes, Brocchinia, Catopsis*, and *Darlingtonia* form a thick coating that is conspicuously white and lackluster (Juniper et al. 1989, Gaume et al. 2004), but those found in the remaining Sarraceniaceae form a less conspicuous "dusting" that covers the inner (usually hairy) pitcher surfaces (Jaffe et al. 1992, Poppinga et al. 2010). The wax crystal layer is reduced or absent in some *Nepenthes* species (Gaume and Di Giusto 2009, Bauer et al. 2012a).

The morphology of the wax crystals has not been described for all taxa and is best known for *Nepenthes.* In that genus, the epicuticular wax layer usually consists of a sponge-like matrix superimposed with a dense array of thin, upright-standing platelets (Figure 13.1d). Wax platelets are also found on the imbricate teeth of *S. leucophylla* pitchers (Poppinga et al. 2010). In contrast, the wax crystals on the inner surfaces of *Brocchinia* and *Catopsis* tanks are thread-like (Gaume et al. 2004). *Nepenthes gracilis* is unique among *Nepenthes* in having a wax-crystal layer on the underside of the pitcher lid that is involved in this species' unique trapping mechanism (Chapter 14). The micro-morphology of this layer is similarly unique, with the crystals forming distinct, pillar-shaped blocks arranged in irregular clusters on the lower lid surface (Bauer et al. 2012b; Chapter 15).

13.3.2 Teeth, folds, and ridges

Folds and ridges are found in all pitcher plants, but not in the carnivorous bromeliads. They often occur in conjunction with papillate cells and trichomes. In *Nepenthes* and *Cephalotus*, the peristome has a conspicuous, hierarchical ridge structure. In *Nepenthes*, macroscopic radial ridges are spaced 0.2–2-mm apart. These are superimposed with microscopic ridges formed by parallel rows of overlapping epidermal cells which are spaced 15–50-μm apart (Bauer 2010). In *Cephalotus*, there are peristome "teeth." Each tooth is covered in rows of tile-like, downward-oriented papillate cells that overlap like roof-tiles (Juniper et al. 1989). Finally, each papillate cell shows a delicate corrugation of fine cuticular folds along its length (Poppinga et al. 2010). A similar lengthwise striation is seen on the papillae and trichomes on the inner pitcher surfaces of *Sarracenia, Darlingtonia*, and *Heliamphora* (Juniper et al. 1989). In *Nepenthes* and some *Heliamphora* species, the ridged micropattern contributes to the high wettability of the trapping surfaces (Bohn and Federle 2004, Bauer et al. 2013; Chapter 15, §15.3.2).

13.3.3 Directional features

Like cuticular folds and ridges, directional features are found in pitcher plants but not carnivorous bromeliads. These features range from the macroscopic peristome teeth of *Nepenthes* and *Cephalotus* and the downward-pointing hairs of the Sarraceniaceae (Juniper et al. 1989) to the microscopic ridges and steps formed by individual epidermis cells. They provide a good foothold for insects in only one direction, and guide them toward the interior of the pitcher. The dense, downward-pointing hairs covering the inner walls of *Darlingtonia* and *Heliamphora* pitchers (Figure 13.1a) probably create a substantial barrier that prevent prey from escaping. Downward-pointing papillate cells with cuticular folds are found on the inner pitcher wall of *Sarracenia* pitchers and on the "funnel" structures inside *Cephalotus* pitchers (Juniper et al. 1989). The overlapping epidermal cells of the peristome of *Nepenthes* species form minute steps that descend toward the inner edge (Figure 13.1e; Bohn and Federle 2004) and lunate cells (Figure 13.1f) overhang the digestive glands on the inner pitcher wall (Gorb et al. 2004). Both peristome steps and lunate cells are oriented to create strongly directional surfaces that provide a foothold for insects moving into, but not out of, the pitcher.

13.4 Sticky glands of adhesive traps

Adhesive traps are known for *Pinguicula* (Lentibulariaceae), *Byblis* (Byblidaceae), and *Philcoxia* (Plantaginaceae) (all Lamiales); *Drosera, Drosophyllum, Triphyophyllum*, and a few species of *Nepenthes* (Caryophyllales); and *Roridula* (Ericales: Roridulaceae)

(Adlassnig et al. 2010, Bonhomme et al. 2011b, Pereira et al. 2012). Sticky trichomes that trap small Diptera also are found in the inflorescences of some *Utricularia* and *Genlisea* species, (Taylor 1989, Płachno et al. 2006, Adlassnig et al. 2010).

13.4.1 Mucilage glands of carnivorous Lamiales

In *Pinguicula* and *Byblis* (Lamiales) the mucilage glands are stalked hairs of epidermal origin. They have a similar architecture: each hair consists of a glandular head (glandular cells arranged in a circle); a neck cell with an endodermic cuticular incrustation of the lateral cell wall; one or two long stalk cells; and a basal cell (*Pinguicula*) or cells (*Byblis*). Mucilage is secreted at the tip of the hair and forms a mucilage drop (Lloyd 1942, Adlassnig et al. 2010). Mucilage is produced continuously in *Pinguicula* when the leaf is mature until senescence (Muravnik 1988). The mucilage in the cells of the glandular head of the trichomes of *Pinguicula* is formed in the Golgi apparatus and stored in vacuoles and between the plasma membrane and the cell wall (Muravnik 1988). Later, it is transported to the cell surface via the cell wall and cuticular discontinuities. Sessile hairs also produce gluey fluid in *Byblis*. The secretion of the stalked hairs is more viscous and probably stickier than the more liquid fluid of the sessile hairs that likely functions as a solvent for the digestive enzymes (Adlassnig et al. 2010).

Philcoxia has stalked capitate glandular trichomes on the upper leaf surfaces that produce sticky substances that tightly bind sand grains to the leaf surface and trap nematodes (Pereira et al. 2012; Chapter 10). These trichomes morphologically are similar to those of *Pinguicula* and *Byblis*, but they not only trap prey but also digest it. Although phosphatase activity has been recorded on trichome surfaces, it is unknown whether they also absorb nutrients from digesting prey or whether nutrients are absorbed through the leaf surface or via stomata.

13.4.2 Mucilage glands of adhesively trapping Caryophyllales

The glands of *Drosophyllum*, *Triphyophyllum*, and *Drosera* that produce adhesive substances are morphologically more complex but functionally similar to the mucilage glands of the carnivorous Lamiales (Green et al. 1979, Juniper et al. 1989, Adlassnig et al. 2010, Renner and Specht 2011, 2013). As in the mucilage glands of *Pinguicula*, mucilage of both *Drosophyllum* and *Drosera* is produced by a hypertrophic Golgi apparatus (Adlassnig et al. 2010).

The glands in the three genera of Caryophyllales with adhesive traps are large stalked emergences consisting of a multi-cellular stalk and a large glandular head. Although the mucilage glands of *Drosophyllum* and *Triphyophyllum* are structurally very similar, they appear to have evolved independently in the lineages leading to *Drosophyllum lusitanicum* and *Triphyophyllum* (Renner and Specht 2011). The common ancestor of Droseraceae, Drosophyllaceae, and Dioncophyllaceae probably had stalked glands with only xylem elements. The emergences of *Drosera* have a simple structure with only xylem elements in the stalk and few glandular cells in the head relative to *Drosophyllum* and *Triphyophyllum*, both of which have xylem and phloem elements in the stalk. The reduction or transformation of vascularized stalked emergences also appears to have occurred in a common ancestor of *Aldrovanda* and *Dionaea*.

All *Drosera* species form leaves with emergences. These are called “tentacles” because they move toward the prey and wrap over it similarly to the tentacles of a squid. The typical tentacle produces mucilage, but in some species (e.g., *D. rosulata*) the glandular tissue may be reduced (Poppinga et al. 2013a; Chapter 14). For many years, only one type of *Drosera* trap was considered, but we now know of a diversity of tentacle types in the genus, along with different types of traps and a range of morphological diversity that likely reflects adaptations for capturing different types of prey (Poppinga et al. 2012, 2013a, Hartmeyer et al. 2013). Some specialized tentacle types do not produce mucilage (snap-tentacles type T2 and type T3). The most interesting and unusual tentacles are of type T3, which are found only in *D. glanduligera*. These tentacles bend irreversibly within < 0.1 second and catapult walking prey into the center of the leaf (Hartmeyer et al. 2013, Poppinga et al. 2013a).

Most *Nepenthes* species rely on the slipperiness of their pitchers to capture prey (§13.2.3, 13.3.1), but the traps of a handful of species—*N. inermis*,

N. aristolochioides, *N. dubia*, *N. jamban*, *N. eymae*, *N. talangensis*, and *N. jacquelinae*—lack a waxy zone and instead contain a viscous and adhesive fluid specialized for capturing of flying insects (Gaume and Forterre 2007, Bonhomme et al. 2011b, Bauer et al. 2012a). The digestive glands of these species produce sticky compounds. A mixed trapping strategy is used by *N. rafflesiana* var. *typica*. Juvenile (lower) pitchers have a waxy zone, contain viscoelastic fluid, and trap primarily ants, Adult, upper pitchers also contain viscoelastic fluid that is deadly for flying insects but lack a waxy zone (Di Giusto et al. 2008, Gaume and Di Giusto, 2009). It is likely that adhesively trapping *Nepenthes* have diverged evolutionarily from other species in the genus, as the common ancestor of modern *Nepenthes* species is thought to have had relatively small pitchers with a fully developed wax crystal layer and narrow, symmetrical peristomes (Bauer et al. 2012a). However, it should be considered that some of the evolutionarily basal species (e.g., *N. pervillei*) have also very strongly viscoelastic fluid in the pitchers.

13.4.3 Resin emergences of carnivorous Ericales

Unlike carnivorous Lamiales and Caryophyllales, *Roridula* produces resin. The trap of *R. gorgonias* can be considered to have a hierarchical architecture consisting of the entire plant with a star-shaped leaf arrangement; individual glandular leaves; and three different types of sticky glands (long, medium, and short capitate glands; Voigt et al. 2009). All three of these glands (emergences) have the same morphology: a cellular stalk and a red glandular head that resembles the glands of *Drosera* (Adlassnig et al. 2010). But the different types of glands play different roles in capturing prey. The longest glands are the most flexible and least adhesive, and trap the prey. The shortest glands are less flexible and strongly adhesive and retain the prey.

13.4.4 Glands of other plants that entrap insects

Stylidium can trap small prey using mucilage produced from stalked hairs that are situated on the flower peduncles, pedicels, and sepals, and these glandular trichomes can be induced to digest protein (Darnowski et al. 2006). *Cleome droserifolia* (Cleomaceae), *Hyoscyamus desertorum* (Solanaceae), *Ibicella lutea*, and *Proboscidea parviflora* (Martyniaceae) all have glandular sticky leaves. However, these species do not appear to take up mineral nutrients from dead insects stuck to their leaves (Płachno et al. 2009a).

13.5 Suction traps and eel traps of the Lentibulariaceae

The common ancestor of *Genlisea* and *Utricularia* probably had delicate, bladder-like traps that were permanently opened and functioned in water or the water film on the soil surface. This basic trap type later evolved in two different directions leading to the closed suction trap of *Utricularia* that works on the basis of negative pressure, and the open eel-type trap of *Genlisea* that keeps prey moving in a single direction (Fleischmann 2012a; Chapters 14, 15).

13.5.1 The bladders of *Utricularia*

The *Utricularia* trap has long been considered to be one of the most fascinating and intricate structures in the plant kingdom (Lloyd 1942, Juniper et al. 1989). This suction trap is a small, hollow bladder that functions in water or within a wet substrate. Although all *Utricularia* species produce traps, trap count and size vary greatly among species. Generally, the traps are formed on different organs, including stolons, shoots, rhizoids, and phylloclades (leaf-like shoots) (Taylor 1989, Brugger and Rutishauser 1989, Rutishauser and Isler 2001, Reifenrath et al. 2006, Guisande et al. 2007). Taylor (1989) used the term "leaf" for practical reasons; however, we agree with Rutishauser (2016) that most *Utricularia* species do not produce true leaves. The location of a trap may be sectional or species-specific; for example, traps are formed only from the stem or from the peduncle base by some members of *U.* subg. *Polypompholyx* (Taylor 1989). Traps are produced soon after germination and in some species they are the second organs of a seedling (Płachno and Świątek 2010).

Utricularia traps range in size from 0.2 to as much as 12 mm (Lloyd 1942, Taylor 1989). The largest traps are in unrelated species: the amphibious *U. arnhemica* (*U.* sect. *Pleiochasia*) and the aquatic-epiphytic *U. humboldtii* (*U.* sect. *Iperua*) at 12 mm; and the aquatic *U. reflexa* (*U.* sect. *Utricularia*) at 8 mm (Taylor 1989, Adamec and Poppinga 2016). Generally, terrestrial species have smaller traps (typically 0.5–1.5 mm) relative to aquatic or amphibious ones (typically 1–6 mm long).

Some *Utricularia* species have dimorphic traps. The different-sized traps may allow the species to exploit prey of different size (Chapter 21). Trap-size dimorphism occurs in unrelated species, including *U. vulgaris*, which produces traps 1–5-mm long (Friday 1992) and *U. humboldtii*, which produces numerous small, terrestrial traps 1–1.5-mm long and few large aquatic traps 5–12 mm long (Taylor 1989). Polymorphism also exists in type of trap appendage or its entrance position (e.g., *U. volubilis, U. inaequalis, U. humboldtii*; Taylor 1989).

Most of our current knowledge of the structure and physiology of *Utricularia* traps is based on aquatic species of *U.* sect. *Utricularia* (Poppinga et al. 2016b; Chapter 19). A single trap is a hermetically closed, lens-shaped bladder with an entrance equipped with a trapdoor and a stalk that affixes the trap to the carrying organ (Figures 13.2a, b). There are three types of traps based on the position of the trap entrance and stalk: traps with a terminal entrance position, traps with a lateral entrance position, and traps with a basal entrance position (Taylor 1989; Poppinga et al. 2016b). When the trap is closed, the trapdoor rests on an enlarged part of the trap wall that is called the threshold, and *Utricularia* traps are divided into two major groups based on the posture of the trapdoor in relation to the threshold: species with a short tubular entrance (e.g., *U. vulgaris*) and species with a long tubular entrance (e.g., *U. bisquamata*).

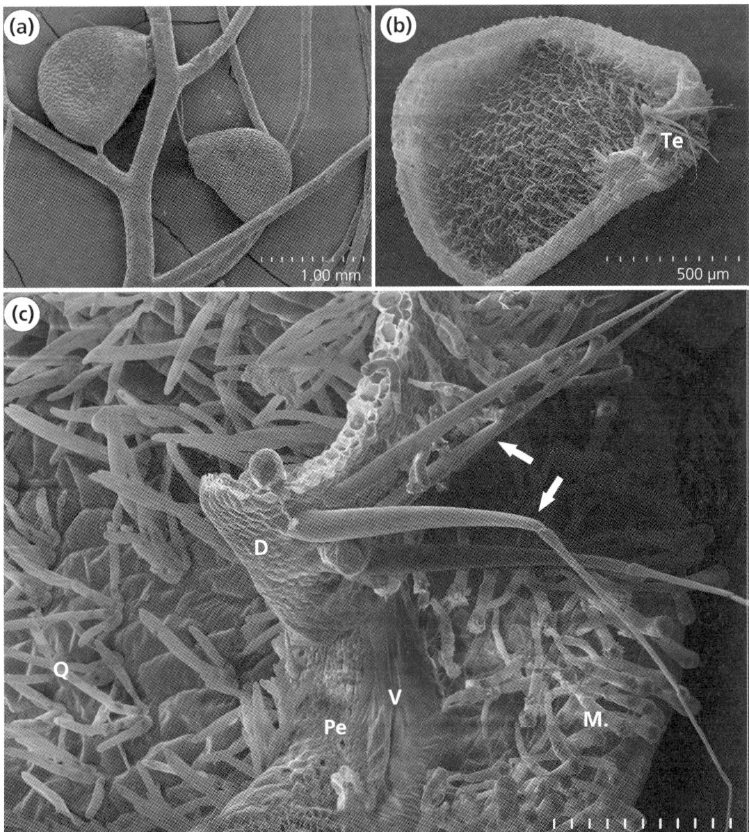

Figure 13.2 General bladder morphology of an aquatic *Utricularia aurea* from Cambodia ("*U. vulgaris* trap type"). **(a)** Scanning electron microscope image of whole traps; bar = 1 mm. **(b)** A longitudinal section of a *U. aurea* trap; trap entrance (Te); bar = 500 μm. **(c)** Higher magnification of the trap entrance; see four trigger hairs (arrows) on the trapdoor (D). On the threshold (T), there are glandular trichomes (pavement epithelium: Pe) that form the mucilage and velum (V) to seal the trap. On the external part of the entrance and the trapdoor, the mucilage trichomes occur (M), whereas quadrifids (Q) are on the internal trap surface; bar = 100 μm. Photographs by Bartosz J. Płachno.

The trap walls typically are flexible and consist of two to five cell layers depending on the species and the generic section (Lloyd 1942, Reifenrath et al. 2006, Płachno et al. 2015a), but traps of species in *U.* sect. *Pleiochasia* have comparatively stiff trap walls and a very stiff trapdoor (Płachno et al. 2015a).

A trapdoor typically consists of two layers of cells and can be compartmentalized into four different regions (hinge region, middle region, central hinge, and middle piece). The triggering mechanism is situated in the central hinge. In general, the traps are opened when a prey item touches the sensory bristles ("trigger" or "trip" hairs) that are situated on the trapdoor (in the *Utricularia vulgaris* trap type, see Figure 13.2c). There is wide structural diversity of the triggering mechanism of *Utricularia* traps. These include a group of non-glandular hairs (bristles) in aquatic species of *U.* sect. *Utricularia* (Figure 13.2c) and other sections, e.g., *U. humboldtii. Utricularia vulgaris* has four trigger hairs, but other aquatic species may have six. There are up to seven bristles in some species. The traps also fire spontaneously, without any mechanical stimulation, several times a day (Vincent et al. 2011a, Płachno et al. 2015a).

The triggering mechanism of aquatic *U. purpurea* and its relatives (*U.* sect. *Vesiculina*) differ from that of aquatic species in *U.* sect. *Utricularia* in being formed by a group of large glandular (mucilage-producing) hairs. Special mucilage hairs also may function as a trigger mechanism in *U. moniliformis* (*U.* subg. *Bivalvaria*) (Lloyd 1942). In contrast, most members of *U.* subg. *Polypompholyx* (except for *U. westonii*) lack triggering bristles (Taylor 1989). Sessile glands on the trapdoor may function as a trigger mechanism in species of *U.* sects. *Pleiochasia* and *Polypompholyx*, but data are scant. Regardless of trigger mechanism, there are some data that suggest that it may have only a small influence on the composition of captured prey and algae (Płachno et al. 2014b). Because algae and small prey are also trapped during spontaneous firing, this hypothesis needs to be tested on a larger sample of *Utricularia* species.

The traps of *U. multifida* and *U. tenella* (both *U.* sect. *Polypompholyx*) merit special attention because of their unique trap anatomy and morphology. Their trap walls are multilayered, the trapdoor is multi-cellular and thickened, and the trap appendages form two lateral tubular channels that are densely covered with hairs (Lloyd 1942). Reifenrath et al. (2006) suggested that these traps do not function with a suction mechanism, but Lloyd (1942) observed suction in *U. multifida* traps. In contrast to the robust morphology of *U. multifida* traps, traps of the species in *U.* sect. *Vesiculina* are delicate and thin, have a reduced threshold, and have special mucilage hairs as triggers (Lloyd 1942).

In most *Utricularia* species, there are two kinds of glandular hairs (glands) inside the traps: two-armed hairs on the threshold (bifids) and four-armed, "chromosome-shaped" hairs on the trap-wall surface (quadrifids) (Figure 13.3a). These glands have a similar architecture (Taylor 1989) and cell ultrastructure across the genus, but minor variations occur in the number of terminal cells. For example, species in *U.* sects. *Enskide, Benjaminia, Oligocista*, and *Stomoisia* have one-armed hairs on the threshold (with only one terminal cell) and bifid hairs on the trap-wall surface (Yang et al. 2009). The terminal cells in bifids and quadrifids are among the most specialized and complex plant cells (Fineran 1985). The internal glands have many functions, including enzyme production, nutrient absorption, and pumping water. Moreover, due to their high metabolism, they rapidly exhaust available oxygen in the trap (Adamec 2007b).

Utricularia traps not only trap and kill prey, but also host commensal bacteria, protozoa, and some species of algae (Chapters 19, 25). Limited data suggest that the physiology of traps changes as plants age and that carnivory becomes less important than nutrients derived from the food web of microbial commensals (Sirová et al. 2009; Płachno et al. 2012a).

13.5.2 The eel trap of *Genlisea*

The basic trap morphology is the same for all *Genlisea* species. This elongated trap (up to 18-cm long) is similar to the *Utricularia* trap in its early development, and consists of a stalk, a hollow bulb (digestive chamber) (Figure 13.3b), a tube (neck), and two helically twisted arms (Figure 13.3c) (Reut 1993, Płachno et al. 2007a, Fleischmann 2012a). The neck and arms are shaped like an inverted "Y." The trap mouth, which continues in both arms, is in the terminal part of the neck. This long and narrow mouth is divided into many small 400 × 180-µm openings (Barthlott et al. 1998) by giant hairs (called "prop-cells") (Figures 13.4a, b). Only small organisms can

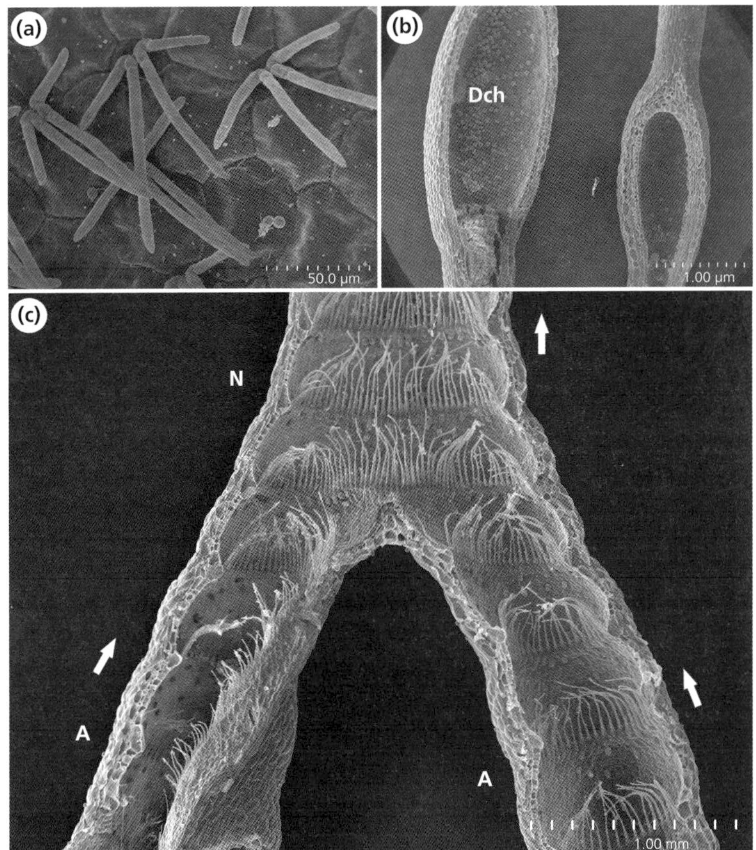

Figure 13.3 General trap morphology of *Utricularia* and *Genlisea*. **(a)** Quadrifids from a *Utricularia aurea* trap; bar = 50 µm. **(b)** Longitudinal section of the digestive chambers (Dch) of *Genlisea* spp. traps; bar = 1 mm. **(c)** Longitudinal section of the neck (N) and arms (A) of *Genlisea* traps, arrows show the direction of the movement of prey; bar = 1 mm. Photographs by Bartosz J. Płachno.

enter the trap, although sometimes larger prey such as long, thin annelids, also enter the traps (Płachno et al. 2005a).

Also in parallel with *Utricularia*, some *Genlisea* species have dimorphic traps in terms of size or position. For example, *G. nigrocaulis, G. aurea, G. margaretae*, and *G. oxycentron* produce filiform traps, which are long-stalked deep-soil traps that have a vertical position, and thick, short-stalked surface traps with a more or less horizontal position (Studnička 1996, Fleischmann 2012a; Chapter 7). The different traps probably help the plants capture different types of prey or are adaptations to different edaphic conditions. However, all species in *G.* subg. *Tayloria* along with *G. hispidula* and *G. subglabra* (both *G.* subg. *Genlisea*) produce monomorphic traps. Surface traps have a larger diameter but shorter arms with fewer windings than traps in deeper soil (Fleischmann 2012a).

The *Genlisea* trap also performs root-like functions such as water uptake, and glands in the trap may absorb water like root hairs. It remains an open question whether this process also can generate water flow in order to aspirate prey. Adamec (2003b) did not find active transport of water in *Genlisea* traps isolated from the plant. Moreover, the glandular trichomes ultrastructure in the arms and the distal part of the neck in *Genlisea* is different from that of barrier cells of *Utricularia* bifids (Płachno et al. 2007a, 2008). Together, these data suggest that *Genlisea* traps are passive and depend on detentive hairs (Figures 13.4a–c) that prevent prey from escaping by forcing them to move in only one direction: up the arms, through the neck, and finally into the digestive chamber, where it probably dies of suffocation (Adamec 2007b) or a high concentration of superoxide or other reactive oxygen species (Cao et al. 2015).

Digestive enzymes have been found both in the digestive chamber and in other trap parts

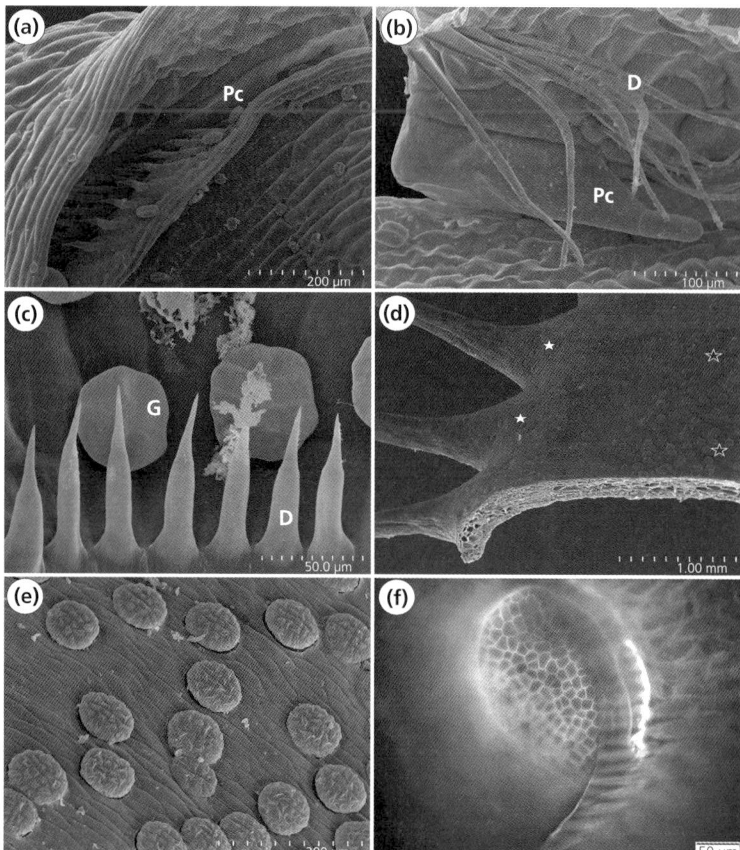

Figure 13.4 Trap micromorphology. (**a, b**). A trap opening in a *Genlisea* arm, including the giant non-glandular hair ("prop-cell:" Pc) that borders the opening and detentive hairs (D); bars = 200 μm and 100 μm, respectively. (**c**) Glandular trichomes (G) and detentive hairs (D) from the neck of a *Genlisea* trap; bar = 50 μm. (**d**) Parts of *Dionaea muscipula* blade, which are used to lure (white stars) and digest prey (black stars); bar = 1 mm. (**e**) Digestive glands of *D. muscipula* bar = 200 μm. (**f**) Digestive gland of *Nepenthes albomarginata* (autofluorescence observed in fluorescent microscopy); bar = 50 μm. Photographs by Bartosz J. Płachno.

(Heslop-Harrison 1975, Płachno et al. 2006). However, some of the digestive enzymes in *Genlisea* traps are produced by commensals (Płachno and Wołowski 2008, Cao et al. 2015). Caravieri et al. (2014) and Cao et al. (2015) found mostly anaerobic or facultative anaerobic bacteria in *Genlisea* traps. In *G. nigrocaulis* and *G. hispidula*, a diverse microbial community, including bacteria, protists of the SAR group (heterokont *Stramenopiles, Alveolata*, and *Rhizaria*), green algae, and microbial fungi live in traps (Cao et al. 2015). Thus, like *Utricularia, Genlisea* relies on a food web of trap commensals also in order to derive nutrients from its prey.

13.6 Fecal traps

The five main types of carnivorous traps—adhesive, pitfall, snap, eel, and bladders (Juniper et al. 1989, Król et al. 2012; Chapter 1)—are used to capture metazoan or protozoan prey. However, some carnivorous plants also "trap" plant detritus, including algae, pollen grains, and fungi hyphae (Moran et al. 2003, Peroutka et al. 2008). There are also two groups of carnivorous plants that are feces feeders. The first includes two species of *Roridula* that trap insects that are eaten later by symbiotic hemipterans (*Pameridea marlothii* on *R. dentata* and *P. roridulae* on *R. gorgonias*) (Ellis and Midgley 1996; Chapter 26); the host plant absorbs nitrogen from hemipteran feces. The tentacles of *Roridula* only produce resin to capture insects and they have no phosphatase activity (Płachno et al. 2006). Rather, nutrients are absorbed by the epidermal cells of the leaf blade through its cuticles, which have cuticular gaps (Anderson 2005, Płachno et al. 2009a).

The second coprophagous group of carnivorous plants includes three mountain giant *Nepenthes* species: *N. rajah, N. lowii*, and *N. macrophylla* (Clarke et al. 2009, Chin et al. 2010, Greenwood et al. 2011; Chapters 17, 26). Mammals visiting the large pitchers with their concave, reflexed lids consume nectar secreted on the lids and defecate into the pitcher. The enlargement of the pitcher orifice and lid angle matches the body size of tree shrews (Chin et al. 2010). The "toilet" pitchers of *N. lowii* are relatively ineffective arthropod traps (Chin et al. 2010) but *N. rajah* traps are more diversified. Its smaller pitchers catch arthropods, primarily ants, whereas its large pitchers are specialized for collecting feces. Another coprophagous species, *N. hemsleyana*, has a facultatively mutualistic interaction with bats (*Kerivoula hardwickii*), and has modified the morphology (narrow and cylindrical shape) and physiology (reduced liquid level) of the aerial pitcher to accommodate them (Grafe et al. 2011). Some modifications to the digestive gland physiology probably has occurred in coprophagous *Nepenthes* to absorb minerals from feces and urine. Thus, we propose a new type of carnivorous plant trap—"fecal traps," which are specialized in collecting mammal or bird feces as the main source of nutrients. This kind of trap has special characters in morphology and anatomy (e.g., a modification of the trap geometry; Chin et al. 2010) and also in the trap physiology both for attracting the animals that are the source of excrement and also to retain and use it.

13.7 Causes of prey death

Prey typically drown in pitcher fluid; suffocate in mucilage, viscid liquids, or resins; or are crushed and then suffocate in snap-traps. Prey die in pitcher fluid much more rapidly than they do in pure water (Adlassnig et al. 2011), and compounds such as naphthoquinones in the fluid of *Nepenthes* also may play a role in killing prey (Eilenberg et al. 2010, Raj et al. 2011, Buch et al. 2013). The acidity (pH) of carnivorous plant secretions also affects how prey die (Bazile et al. 2015). In addition to their role in the digestion of prey (Chia et al. 2004), free radicals may also accelerate prey death. Adamec (2007b) proposed that suffocation is the main cause of prey death in *Genlisea* and *Utricularia* traps.

Despite numerous observational studies, there have been only a few experiments aimed at understanding mechanisms of prey retention and death. Bazile et al. (2015) reported that pitcher fluid differed among *Nepenthes* species in both the capacity for insect retention and the time it took to kill them. In *N. hemsleyana* and *N. rafflesiana*, the time until death was affected by the age of the pitcher fluid. Death of prey was more rapid in older fluid, supporting the hypothesis that these plants produce additional compounds that help to kill prey (Chapter 15). In some carnivorous plants, prey is also killed by commensals, including mosquito larva in pitchers of some *Nepenthes* species and *S. purpurea*, *Camponotus schmitzi* ants in *N. bicalcarata*, and *Pameridea* hemipterans in *Roridula*.

13.8 Digestive and absorptive glands

Digestive and absorptive glands can be of epidermal origin, have a very simple architecture, and may consist of only a few (e.g., in *Genlisea*: Figure 13.4c; or in *Aldrovanda* and *Dionaea*: Figures 13.4d, e) or three cells (the internal trap glands of some *Utricularia* species), or be large structures with a complicated architecture (e.g., in *Nepenthes*: Figure 13.4f) that can also contain non-epidermal elements (e.g., in *Triphyophyllum, Drosophyllum, Drosera*) or epidermis (in *Sarracenia*). Regardless of their structure and origin, all digestive glands include: a terminal element (glandular cells that secrete digestive enzymes and absorb nutrients from dissolved prey bodies); a middle element (an endodermoid cell or cells that are heavily cutinized and have suberized lateral cell walls that are similar to a Casparian strip that creates a barrier for apoplastic transport); and a basal cell or cells (a reservoir element that connects the gland with other tissue cells; Juniper et al. 1989).

13.8.1 The terminal element and enzyme localization in digestive glands

The glandular cells at the periphery of the gland surface discharge hydrolytic enzymes and absorb nutrients; their cuticle needs to have discontinuities. The most universal cuticular discontinuities in these digestive glands are cuticular gaps—cutin-free wall regions. But cuticular pores occur in *Drosera* (Owen et al. 1988, Juniper et al. 1989, Owen and Thomson 1991, Muravnik 2005) and *Genlisea* (Płachno et al. 2005b). Such cuticular discontinuities also occur in the leaf epidermis in *Roridula* species, and are those places where the nutrients from excrement are absorbed (Anderson 2005). In some carnivorous plants, the cuticular discontinuities are formed only after a chemical stimulation of digestive glands (Joel et al. 1983). A common character of the glandular cells is a wall labyrinth: the occurrence of wall ingrowths bounded by plasmalemma.

The cell walls of the terminal cells can store digestive enzymes (Płachno et al. 2006), but large vacuoles of these cells are the primary location where these enzymes are stored (Król et al. 2012). These cells are rich in cytoplasm that has numerous organelles (especially many active mitochondria and extended endoplasmic reticulum), and thus are good models for cytological studies. Most researchers have concluded that digestive enzymes are produced by the endoplasmic reticulum; the proliferation of rough endoplasmic reticulum was observed in the glandular cells of *Drosera* after stimulation (Muravnik 2000, Adlassnig et al. 2005a, 2012).

In *Aldrovanda vesiculosa* and *Dionaea muscipula*, stimulation induced activation of the endomembrane system that was manifested by a complex of qualitative (Muravnik et al. 1995, Muravnik 2008) and quantitative (Muravnik 1996) characters. Aggregations of endoplasmic reticulum cisternae characteristic of unstimulated glands were dispersed after stimulation. Quantitative changes included growth of the volume densities of rough endoplasmic reticulum and Golgi apparatus.

It is probable that a granulocrine mode of enzyme secretion occurs in all carnivorous plants and that some evidence of holocrine secretion has been misinterpreted. The digestive enzymes are discharged via exocytosis, which involves membrane fusion of the coated vesicles of a trans-Golgi reticulum with lytic vacuoles or with the plasma membrane (Vassilyev and Muravnik 1988, Muravnik et al. 1995, Vassilyev 2005, Płachno et al. 2007a, Adlassnig et al. 2012, Król et al. 2012). Two types of digestive enzyme production can be identified: continuous acid hydrolase secretion in the immature and mature unstimulated digestive glands (e.g., *Drosophyllum*: Vassilyev 2005); and no acid hydrolase synthesis prior to stimulation (e.g., *Aldrovanda, Drosera*: Muravnik et al. 1995, Muravnik 1996, 2000).

Previous studies of the activity and distribution of enzymes in digestive glands were done either with destructive histochemical techniques or the substrate film method to localize proteases (Juniper et al. 1989). An enzyme-labeled fluorescence (ELF) assay was applied to detect phosphatase activity *in vivo* in glandular structures (Sirová et al. 2003). Płachno et al. (2006) used ELF to study traps of 46 species in 11 genera and proposed that phosphatases were the main digestive enzymes. Phosphatase activity was found in many cell compartments (cytoplasm, vacuoles, and cell walls) of all of species studied, with the exception of glands of *Roridula* and *Stylidium*, and an ELF assay confirmed that *Philcoxia* is a carnivorous plant (Pereira et al. 2012). Phosphatase activity also was detected in *Utricularia* traps growing in sterile culture *in vitro*. Commensal cyanobacteria and algae living in *Utricularia* traps also produced phosphatases (Płachno et al. 2006, Sirová et al. 2009). Finally, Adamec et al. (2011) used ELF to tag phosphatase and chitinase in the traps of several *Utricularia* species. Chitinases were rarely tagged on any inner glands but activity was observed on the surface of the antenna of six aquatic species.

13.8.2 Nutrient uptake and transport in the middle and basal elements

Owen et al. (1999) studied the possible pathways for nutrient transport in *Nepenthes alata* using lanthanum (for tracking apoplastic pathways) and 6(5) carboxyfluorescein (for tracking symplasmic pathways). The experiments with lanthanum revealed that the endodermoid cells of digestive glands block the apoplastic transport of water-soluble substances, and that transport into or out of the digestive gland must occur through the symplast (as the digestive gland is apoplastically isolated from other plant tissue). Experiments with lanthanum yielded similar results for *Utricularia* (Fineran and Gilbertson 1980) and *Brocchinia* (Owen et al. 1988), because all carnivorous plants have an endodermoid cell or cells in their digestive-absorption glands.

6(5)carboxyfluorescein was transported through glands from the pitcher lumen into the vascular endings immediately beneath the glands (Owen et al. 1999). Transport of 6(5)carboxyfluorescein from the petiolar vascular system to the glands and the pitcher lumen only occurred in immature pitchers. Owen et al. (1999) proposed that the functioning of the gland is developmentally regulated. Prior to reaching pitcher maturity, the primary function of the glands appears to be secretion, whereas after reaching pitcher maturity, the only function of the glands is nutrient absorption. These results add support to the hypothesis that old *Utricularia* traps (and glands) have a mainly absorption function (Sirová et al. 2009).

Juang et al. (2011) used food dyes and carboxyfluorescein diacetate to study nutrient transport in *Utricularia gibba*. They observed that symplasmic tracers were transported in the following order: through the quadrifids (terminal cells, stalks, pedestal cells, basal cells) and later through the surrounding epidermal cells, nearby phylloclades, and, last, through the stems. However, in aquatic *Utricularia* species traps, symplasmic transport also occurs in the opposite direction, from the phylloclades and stem to the trap wall and later to the quadrifids, because the plant provides organic C for the trap and trap commensals (Sirová et al. 2011).

Sakamoto et al. (2006) used fluorescent gelatine for an absorption study of *Aldrovanda vesiculosa* traps. After application of gelatine, they observed fluorescent structures with a granular appearance inside the glandular cells of the digestive glands. Some evidence of endocytosis was presented by Płachno et al. (2006), who detected phosphatase activity in the cytoplasm of *Utricularia* and *Drosophyllum* glands. Using fluorescent endocytosis markers, Adlassnig et al. (2012) demonstrated the occurrence of endocytosis in the glandular cells of *Nepenthes, Drosera, Dionaea, Aldrovanda, Drosophyllum*, and *Cephalotus*, but not in *Genlisea* or *Sarracenia*. Clathrin-dependent endocytosis most likely occurred in glands of both *Nepenthes* and *Drosera*. There are two types of glands at the bottom of the pitchers of *Cephalotus follicularis*: small glands (in each of the metamorphous stomata) and giant ones. But only the small glands are responsible for nutrient uptake; large glands secrete pitcher fluid or digestive enzymes. Both sessile and stalked emergences exhibit phosphatase activity and play a role in nutrient uptake in *Drosophyllum lusitanicum* (Płachno et al. 2006).

Adlassnig et al. (2012) concluded that nutrient uptake by carriers in some carnivorous plant species is supplemented by endocytosis (see also Schulze et al. 1999; Moran et al. 2010). This mechanism enables the absorption and intracellular digestion of whole proteins while at the same time saves energy because the plant does not need to secrete a large volume of digestive enzymes. They also proposed that Sarraceniaceae, which do not exhibit endocytosis, have evolved the capability of absorbing whole oligopeptides (Plummer and Kethley 1964), because the exocytosis of enzymes in their traps is dubious (exocytosis has to be compensated by endocytosis). Ammonium transporters are highly expressed in the glands of *Nepenthes* pitchers and NH_4^+ was the preferred nitrogen source for *Nepenthes*. Moreover, Moran et al. (2010) showed that lower rates of NH_4^+ uptake occurred in *N. ampullaria, N. bicalcarata*, which produce long-lived pitchers that capture few prey, relative to *N. rafflesiana*, which produces short-lived pitchers that have high capture rates.

13.9 Future research

Convergent evolution in carnivorous plants has occurred in trap types, and divergence has occurred due to prey specialization or to obtain new sources of nutrients such as feces or urine. Further insights into the anatomy, physiology, and evolution of traps and glands could be gained by determining: subcellular adaptations for carnivory and alternative nutrient pathways such as detritus or feces (e.g., enzyme secretion and nutrient absorption) in diverse carnivorous plant lineages; characteristics of nectar and olfactory substances in related and phylogenetically disparate carnivorous and noncarnivorous plants; comparative studies of aquatic, terrestrial, and epiphytic *Utricularia*; and experimental studies of the chemistry of compounds used to retain and kill prey. These studies could reveal further insights into the specific adaptations for trapping in different habitats and surrounding media. Generally, anatomical studies need to be linked more closely to physiological and ecological investigations of trapping. On the other hand, thorough comparative studies of the trap anatomy can help to shed further light on the evolution of plant carnivory. Updated comprehensive molecular phylogenies are urgently needed for this task.

CHAPTER 14

Motile traps

Simon Poppinga, Ulrike Bauer, Thomas Speck, and Alexander G. Volkov

14.1 Introduction

Motile traps are found in the genera *Aldrovanda* and *Dionaea* (snap-traps), *Drosera* and *Pinguicula* (adhesive traps with movable trap leaves and emergences), *Nepenthes* (passive-dynamic springboard traps), and *Utricularia* (suction traps). The traditional definition of moving traps as "active" and immobile traps as "passive" (Lloyd 1942) is somewhat blurred because the trap of *Nepenthes gracilis* moves, but this movement is externally driven and so it is still considered to be passive (Bauer et al. 2015b). Plants with active motile traps—snap-traps, motile adhesive traps including the catapulting flypaper traps of *D. glanduligera*, and suction traps—invest metabolic energy to produce movement, or for bringing the trap into a movable state (usually by mechanical pre-stressing of tissues). Plants with passive motile traps (springboard pitfalls) make no such investments, but instead exploit external energy sources. Note that the underlying "costs" of trap production and development (Chapter 18) are not considered here.

In our review of the physiology and biomechanics of carnivorous plant traps and their movements, we explain three general principles of movement actuation that are common to plants, carnivorous or not (reviewed by Forterre 2013, Poppinga et al. 2013b).

The first principle is that external mechanical forces (e.g., wind or rain) lead to elastic deformation of plant organs, causing passive motions. For example, plant stems bend in the wind and return to their original configuration when the wind stops.

The second principle is that water displacement between cells and tissues leads to differential expansion or contraction, thereby causing hydraulic movement. The movement velocity depends on the mechanical properties of the tissue, its permeability to water, and on the distance the water has to travel. Smaller structures principally move faster than larger ones when being actuated purely hydraulically. Two processes may be at work here. First, passive water transport, such as hygroscopic water uptake under wet environmental conditions or evaporative water loss under dry conditions, leads to swelling or shrinking of plant tissues. This does not require metabolic energy. Passive water transport typically actuates movements in dead plant structures such as seed capsules and cones, but is not known to play a role in carnivorous plant traps. Second, physiologically costly active water transport leads to cell turgor changes, causing either irreversible plastic cell deformation (growth) or reversible elastic cell deformation (e.g., in the guard cells of stomata).

The third principle is that the sudden release of elastic energy stored in cell walls and tissues as a result of (usually slow) passive or active hydraulic deformation processes can cause very fast motion.

14.2 Active motile traps

14.2.1 Snap-traps

Snap-traps are found in aquatic *Aldrovanda vesiculosa* and terrestrial *Dionaea muscipula*. Both species have mechanosensory structures for perceiving and processing stimuli, and motile zones that evoke the trapping motion after triggering (Sibaoka 1991). The snapping mechanics differ between these two species. The differences originally were hypothesized to reflect evolutionary adaptions to their different

Poppinga, S., Bauer, U., Speck, T., and Volkov, A. G., *Motile traps.* In: *Carnivorous Plants: Physiology, ecology, and evolution.* Edited by Aaron M. Ellison and Lubomír Adamec: Oxford University Press (2018). © Oxford University Press.
DOI: 10.1093/oso/9780198779841.003.0014

habitats and surrounding media (Poppinga and Joyeux 2011), but Poppinga et al. (2016a) showed that the kinematics and closure duration of the *Dionaea* trap are similar in air and water. Moreover, *Dionaea* plants can become seasonally inundated and the traps reportedly capture prey underwater (Bailey and McPherson 2012). Hence, the differences in snapping mechanics are unlikely to have evolved in response to the surrounding media, but instead may be adaptations to different prey. *Dionaea* typically catches large strong terrestrial arthropods one at a time (Gibson and Waller 2009), whereas *Aldrovanda* captures small to medium-sized prey such as zooplankton and mosquito larvae and also can capture several animals at once (Cross 2012a; Chapter 21).

The snap-trap of *Dionaea muscipula*. The *Dionaea* leaf consists of a basal green petiole and an apical, bilobed leaf blade forming a ≈2-cm-long snap-trap (Bailey and McPherson 2012). The bilaterally symmetric, doubly curved trapezoidal lobes are connected by a midrib. Each lobe bears 14–21 marginal teeth (Juniper et al. 1989). In the fully open state, there is a ≈80° angle between the lobes (Poppinga et al. 2016a). Prey are attracted to a band-like region with extrafloral nectaries on the inner trap surface, adjacent to the teeth (Chapter 12). The area between this region and the midrib contains numerous large digestive glands (Chapters 13, 16), and three or four mechanosensitive trigger hairs per lobe (Figure 14.1).

At room temperature (≈20 °C), touching a trigger hair at least twice within 20–30 seconds causes the trap to shut rapidly (Williams and Bennett 1982, Hodick and Sievers 1988). At higher temperatures (28–40 °C), a single touch entails closure. Each trigger hair has a circular constriction near its base that contains a single concentric row of mechanosensory cells connected to neighboring cells via multiple plasmodesmata. Their interior appears symmetrically polarized, with complexes of endoplasmic reticulum (ER) cisternae concentrated around vacuoles with polyphenolic contents at the basal and apical pole of each cell (Buchen et al. 1983, Juniper et al. 1989). Whereas the basal region of the hair consists of soft parenchymatous tissue, the distal part is reinforced by thick cell walls and very stiff (Buchen et al. 1983). When touched, the trigger hair bends mainly at the constriction, leading to deformation and stimulation of the sensory cells. As a result, mechanosensitive ion channels open, the cell membrane is depolarized, and a receptor potential is generated (Hodick and Sievers 1988, Sibaoka 1991, Król et al. 2006, Escalante-Pérez et al. 2011). During stimulation, the ER-vacuole complexes may act as pressure transducers with the phenolic contents supplying the ions necessary for electrical stimulus generation (Buchen et al. 1983, Juniper et al. 1989).

Voltage-gated ion channels then amplify the receptor potential and generate an action potential that is propagated through the trap tissue with constant amplitude, duration, and speed (Juniper et al. 1989, Volkov et al. 2008b, 2014). This process depends on cross-membrane fluxes of calcium, chloride, and potassium ions (Hodick and Sievers 1986, 1988, Sibaoka 1991, Volkov et al. 2008b, 2008d). Ion and water channel blockers (e.g., NaN_3) and uncouplers (e.g., 2,4-dinitrophenol [DNP]) can reduce or inhibit completely the response to physical and electrical stimuli (Hodick and Sievers 1989, Fagerberg and Allain 1991, Król et al. 2006, Volkov et al. 2008d) (Figure 14.1). Action potentials also are generated when the lobe epidermis or mesophyll cells are directly stimulated. The entire internal trap surface (but not the margins) is sensitive to touch, light, temperature, and electric or wounding stimuli (DiPalma et al. 1966, Hodick and Sievers 1988, Juniper et al. 1989, Trebacz and Sievers 1998).

Trap closure is all-or-nothing: there is no reaction to stimuli below a certain threshold and the speed of closing is independent of the strength (intensity) of stimuli above the threshold (Juniper et al. 1989, Volkov et al. 2007). At room temperature, a single stimulus is insufficient to achieve closure. The restoration of the resting potential after depolarization takes roughly 20–30 s. Because every stimulus causes a depolarization of equal magnitude, a second stimulus within this time frame has an additive effect. This "counting" or "memory" of mechanical perturbations by *Dionaea* may prevent "false" closing if leaves or soil particles land on the trap. The summation of receptor potentials has been demonstrated through repetitive application of small electrical stimuli (Volkov et al. 2007, 2008a, 2008b, 2008c, 2008d).

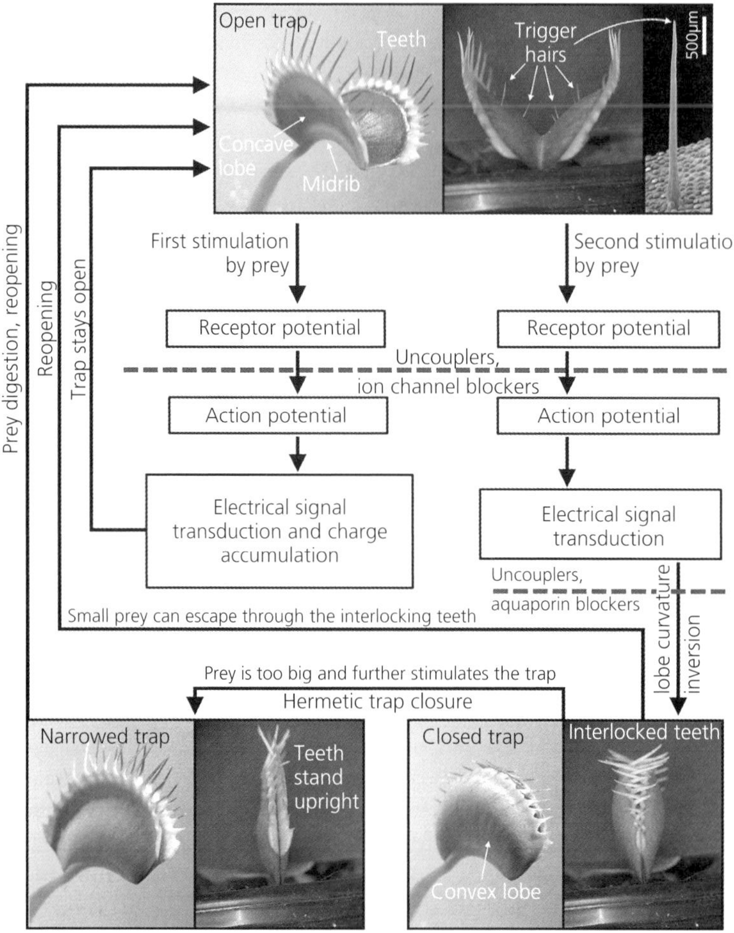

Figure 14.1 The snap-trap of *Dionaea muscipula*. Touching the trigger hairs of an open trap leads to the generation of receptor and action potentials. At least two consecutive mechanical stimuli are necessary for trap closure. Snapping comprises rapid curvature inversions of the two lobes. The electrophysiological processes of the sensory system and the physiological processes of trap lobe movement can be blocked biochemically. The closed trap is characterized by interlocking teeth. If small prey escape, the trap reopens. Prey that cannot escape further stimulate the trap, leading to hermetic closure with the teeth pointing more or less upwards. An enzymatic cocktail is released, prey are digested, and the trap re-opens once again. Sketch based on an original idea from Volkov et al. (2008b).

Drought stress can affect mechanoreception, signal transduction, and hence prey capture in *Dionaea* (Escalante-Pérez et al. 2011). Traps of water-stressed plants have lowered sensitivity and require three consecutive stimuli to close at room temperature. This makes sense because prey digestion (Chapter 16) is strongly water-dependent. Under dry conditions, the concentration of the phytohormone abscisic acid (ABA) is elevated in the traps. ABA is generally associated with stress responses in plants and plays an important role in water-balance by initiating the drought-induced closure of stomata (Roelfsema et al. 2004).

The closing of the *Dionaea* trap is very fast. It begins just 0.1–0.2 s after triggering (Volkov et al. 2008b, 2008d) and typically is completed within 0.1–0.5 s (Forterre et al. 2005, Volkov et al. 2007, 2008b, 2008d, Escalante-Pérez et al. 2014, Poppinga et al. 2016a). After triggering, the lobes first start to move slowly toward each other. It is not clear whether this initial movement is actuated purely hydraulically by water displacement processes between different layers within the lobes, or whether it is additionally supported by the relaxation of prestressed mesophyll tissue (Hodick and Sievers 1989, Fagerberg and Allain 1991, Colombani and Forterre 2011). Possible scenarios of hydraulic actuation include irreversible growth processes of cells of the outer lobe surfaces induced by acid-induced cell wall loosening (acid growth; Williams and Bennett 1982), or reversible deformation processes of cells in the different lobe layers due to the opening of water channels (aquaporins; Markin et al. 2008, Volkov et al. 2008b, 2011, Escalante-Pérez et al. 2014). The resulting deformation of the two lobes takes place mainly in directions perpendicular to the midrib, slowly driving the entire trap toward a closed state (Forterre et al. 2005). During closure, the amount of ATP in the lobes drops by ≈30% (Williams and Bennett 1982, Volkov et al. 2012) and their configuration as seen from the outside changes from concave to convex (Darwin 1875) (Figure 14.1).

Hydraulics alone cannot account for the rapid trap closure, and two hypotheses have been advanced to explain it. The first proposes that elastic curvature energy is stored in the lobes due to a difference in hydrostatic pressure between the hydraulic layers, and that its release drives the fast closing movement (Hodick and Sievers 1989, Markin et al. 2008). When water channels between the layers open, the pressure difference can equilibrate and the lobes rapidly relax to the closed configuration.

Alternatively, elastic instabilities may play a crucial role in trap closure. In a widely accepted model by Forterre et al. (2005), the concave lobe curvature acts as a geometrical constraint resisting the initial motion until the accumulated strain is released suddenly and the lobe rapidly flips into a convex state ("snap-buckling"). The amount of elastic energy released during the curvature inversion, and hence the trapping speed, increases with the dimensionless geometrical parameter $\alpha = \frac{L^4 K^2}{h^2}$, where L is the length of the lobe, K is its mean curvature in the open state, and h is its thickness. The parameter α equals the ratio between the energy barrier separating the two states of lobe curvature (the geometrical constraint) and the bending energy supplied by the change of the rest-state curvature (i.e., by the initial motion of the lobes), and depends only on trap morphology and geometry (Forterre et al. 2005, Forterre 2013). The model predicts that larger traps should snap shut faster than smaller ones, but this prediction was not confirmed experimentally (Volkov et al. 2008a, Poppinga et al. 2016a).

The *Dionaea* trap closes with considerable force. Volkov et al. (2013) measured a peak snapping force of 149 mN and a pressure between the narrow trap rims of 41 kPa. The force necessary for escaping from the closed trap reached 4 N, allowing little chance for prey to escape. If one lobe is cut off, the other lobe shows exaggerated bending toward the midrib because its movement is not blocked (Hodick and Sievers 1988, Fagerberg and Allain 1991). Differences in lobe morphology, stimulus processing, or movement actuation also may lead to asynchronous snapping in otherwise intact traps (Poppinga et al. 2016a). Seedling traps have smaller opening angles (≈50°) than adult plants and take much longer (up to several seconds) to close, probably because elastic instabilities play a minor, if any, role. As a result, seedlings may catch only relatively slow-moving prey (Poppinga et al. 2016a).

After snapping quickly, final closure occurs much more slowly because of internal dampening of the hydrated lobes (Forterre et al. 2005). The distance between the lobe edges immediately after snapping remains at 15–20% of the initial distance (Volkov et al. 2008b, 2008d). The teeth on the lobe margins interlock loosely, leaving numerous small gaps (Figure 14.1). This allows small animals to escape, ensuring that the plant does not initiate the metabolically costly digestion process for prey that yield little metabolic benefit. Only larger animals are retained and continue to stimulate the trigger hairs during their struggle to escape. This generates further action potentials (Lichtner and Williams 1977) and causes an increase in the cytosolic concentration of calcium (Escalante-Pérez et al. 2011). Together with the perception of N-containing substances (e.g., urea) from the prey, this initiates a further narrowing of the trap. Eventually the trap lobes flatten, causing the teeth to stand more-or-less upright on the lobe margins, and the trap closes hermetically (Figure 14.1). The lobes exert mechanical force up to 450 mN on the inside of the trap, corresponding to a maximum pressure of 9 kPa acting on the prey (Volkov et al. 2013). The formation of a sealed digestion chamber is then followed by the release of digestive enzymes (Escalante-Pérez et al. 2011; Chapter 16).

In most cases, the trap re-opens after prey is digested. A single trap can snap and re-open up to 12 times (Bailey and McPherson 2012). The re-opening process occurs over 1–2 days (Fagerberg and Howe 1996, Volkov et al. 2014, Poppinga et al. 2016a) and is controlled either by irreversible growth processes (Ashida 1934) or by hydrostatic pressure changes in the inner and outer tissue layers of the lobes (Markin et al. 2008). A noticeable bulge appears on each of the sealed lobes before the trap margins start to separate and the trap gradually re-opens (Fagerberg and Howe 1996). Reverse snap-buckling does not occur (Poppinga et al. 2016a).

The snap-trap of *Aldrovanda vesiculosa*. The aquatic *Aldrovanda vesiculosa* develops rootless floating shoots with whorls of six to nine leaves at each node. Each leaf consists of a flattened, air-filled petiolus that splits into three to eight bristles, and a leaf blade that is transformed into a 2.5–6-mm long, 4–7-mm wide snap-trap (Cross 2012a) (Figure 14.2). The two convex (as seen from the outside) trap lobes are attached to a midrib—an extension of the petiolus—and the whole trap is twisted approximately 90° counter-clockwise and tilted 30–40° backwards. As a result, one lobe is turned toward the bristles (the "bristle-side lobe"), whereas the other is turned away (the "free-side lobe") (Ashida 1934) (Figure 14.2a). The bristles are hypothesized either to serve as a mechanical barrier preventing objects other than prey from entering and triggering the trap, or to guide substrate-dwelling prey toward the trap (Cross 2012a). The twisted trap orientation toward the open water may help to maximize exposure to potential prey.

Each lobe is divided into a central and a peripheral region by a band-like enclosure boundary (Ashida 1934) (Figure 14.2a). The central, "three-layered" region consists of two epidermal cell layers on either side of a parenchymatous layer. Its interior surface has a multitude of trigger hairs and glands, which are especially densely arranged along the midrib. The peripheral, "one-layered" region is very thin, and consists only of two epidermal layers. Its inner surface is covered in four-armed (quadrifid) glands similar to those inside the *Utricularia* suction traps (Chapter 13). The lobe margin is bent inwards and densely lined with small, tooth-like projections.

Twenty to 40 trigger hairs are concentrated along the midrib and along the enclosure boundary of each trap (Lloyd 1942, Cross 2012a). The hairs are ≈1.5 mm long and 0.05 mm thick. A double articulation in their basal and median regions probably prevents fracture during trap closure (Darwin 1875). It is speculated that the hairs mimic filamentous algae and attract grazing prey (Cross 2012a). A single stimulation of a trigger hair is sufficient to effect trap closure (Lloyd 1942, Williams 1976). The characteristics of resting, receptor, and action potentials, and the patterns of signal transduction, all are similar to *Dionaea*. Upon triggering, Ca^{2+} ions are thought to enter the excitable cells of the three-layered region and cause a receptor potential (Juniper et al. 1989, Sibaoka 1991). Electrical and chemical stimuli also lead to trap closure (Ashida 1934). Trap sensitivity depends on trap age and water temperature, and spontaneous trap closure occurs when the water temperature is <10 °C or >40 °C (Czaja 1924).

Several consecutive phases of trap movement after triggering have been described (Ashida 1934, Lloyd 1942). Prey are captured during the first, rapid movement phase (trap shutting) (Figure 14.2). At 25 °C, traps can shut within 10 milliseconds, but

Figure 14.2 The snap-trap of *Aldrovanda vesiculosa*. Frontal views of a trap (**a**) and lateral views of the same trap (**b**) at the same points in time. The left images show the open trap before triggering; the middle images show the closed trap immediately after triggering with an electric charge; and the right images show the narrowed trap. Each lobe consists of a three-layered region and a one-layered region, which are fused by an enclosure boundary. The midrib connects the free-side lobe with the bristle-side lobe. During closure, the midrib notably bends, and its curvature is strongest when the trap is narrowed. At this stage, the free-side lobe has inverted its curvature, and the sickle-shaped cavity contains the caught prey.

this speed declines with decreasing temperatures. The inner epidermal cells of the three-layered region, which are located ≈0.15–0.25 mm from the midrib, constitute the "motor zone" responsible for the movement of the lobes. While the trap is open, the thick-walled motor cells are turgid and resist mechanical pressure exerted by the middle parenchymatous and outer epidermal layer. After triggering, turgor decreases (Ashida 1934, Sibaoka 1991) and the trap closes, probably aided by an acid growth response of cells on the outside of the lobe (Williams 1992). Both lobes move synchronously and smoothly without sudden accelerations or decelerations (Poppinga and Joyeux 2011). After closure, the small teeth on the trap margins interlock (Juniper et al. 1989).

It is still debated whether the shutting motion is supported by a release of stored elastic energy. The epidermis of the outer lobe is undulated when the trap is open but it is smooth in the closed state, indicating some degree of pre-stressing (Ashida 1934, Williams 1992). Moreover, the midrib is more or less straight when the trap is open and curved when the trap is shut (Figure 14.2b), indicating that it may be pre-stressed in the open state (Poppinga and Joyeux 2011). In contrast to *Dionaea*, the lobes of the *Aldrovanda* trap do not invert their curvatures during closure (Figure 14.2a), ruling out snap-buckling instability (Skotheim and Mahadevan 2005, Poppinga and Joyeux 2011). However, deformation of the motor cells alone cannot explain the high closure speed, which is thought to be achieved by mechanical coupling of the motor zone with the rest of the leaf, particularly the midrib. The trap geometry is probably adapted to amplify the bending of the midrib so that a small midrib movement results in a large displacement of the lobes (Poppinga and Joyeux 2011, Joyeux 2013).

The initial rapid trap closure leads to the formation of a cavity in which small prey can still move around freely, but continued stimulation of the trigger hairs initiates further narrowing (Ashida 1934; Figure 14.2). For approximately 30 minutes, both lobes continue moving toward each other until both enclosure boundaries almost touch. The free-side lobe then inverts its curvature in the one-layered region, forming a small, sickle-shaped cavity in the central trap zone that brings prey into contact with the digestive glands (Figure 14.2; Chapters 13, 16). The pressure continues to increase, and large animals may be crushed (Cross 2012a). The narrowing motion is caused by rising turgor resulting from water uptake, followed by elongation of the outer epidermal cells (Ashida 1934). The quadrifid glands also may pump water out of the trap (cf. *Utricularia*; Chapters 13, 16, 19), thereby sealing the trap tightly for digestion (Juniper et al. 1989). Without prey, closed traps stay in the narrowed phase for approximately 6–12 h until they reopen (Lloyd 1942).

Prey digestion lasts approximately 3–10 d (Czaja 1924, Cross 2012a; Chapter 16). Re-opening starts with curvature inversion of the free-side lobe before both lobes gradually open as the inner epidermal cells elongate (Ashida 1934). Opening is much slower than shutting or narrowing because of the mechanical resistance of the middle parenchymatous and outer epidermal cell layers, and because of the necessary water flow into the trap lumen during the inversion of the free-side lobe curvature. As with *Dionaea*, opening of the *Aldrovanda* trap involves irreversible trap growth.

14.2.2 Motile adhesive traps

Actively moving flypaper traps (Chapters 13, 15) are found in several species of *Drosera* and *Pinguicula*. Leaf laminae (*Pinguicula* and *Drosera*), emergences (*Drosera* tentacles), or both may move. Only a subset of *Pinguicula* and *Drosera* species uses movement for trapping, and movement patterns and speed vary greatly. Some *Drosera* species employ moving traps only during part of their ontogeny, which may be an adaptation to different prey targeted during different developmental stages (Poppinga et al. 2013a).

Motile *Drosera* traps. An insect struggling on a sticky *Drosera* leaf triggers movements of individual tentacles or the entire leaf. This not only hinders escape by bringing further glue-laden tentacles into contact with the prey (Chapters 13, 15), but also aids digestion by forming an "outer stomach" (Figure 14.3a; see Chapter 16). The animal eventually dies from suffocation due to clogging of its tracheae by mucilage (Juniper et al. 1989; Chapter 13). The speed of the tentacle movement depends on the species, the tentacle type, the intensity of stimulation, the plant's vigor, and on environmental conditions (Darwin 1875, Williams and Pickard 1972, Poppinga et al. 2013a). Bending can take as little as 75 ms in the catapulting tentacles of *D. glanduligera*, or up to several minutes as e.g., in *D. capensis* (Poppinga et al. 2012) (Figure 14.3a–d). Movements of the entire leaf tend to be even slower, typically lasting between several minutes and a few hours (Figure 14.3a).

Drosera tentacles are multicellular glandular emergences. They are typically several mm long, cylindrical in cross-section, and secrete sticky mucilage at the tip (Darwin 1875, Juniper et al. 1989, Seine and Barthlott 1993; Chapter 13). Bending begins a few seconds–minutes after stimulation and occurs in a distinct zone at the base of the tentacle. The site of mechanoreception is situated in stalk cells just below the constricted tentacle head, where the greatest deformation occurs during mechanical stimulation (Darwin 1875). The stalk cells are homologous to the receptor cells of *Dionaea* (Williams and Pickard 1972, Williams and Spanswick 1972, Williams 1976). Mechanical triggering

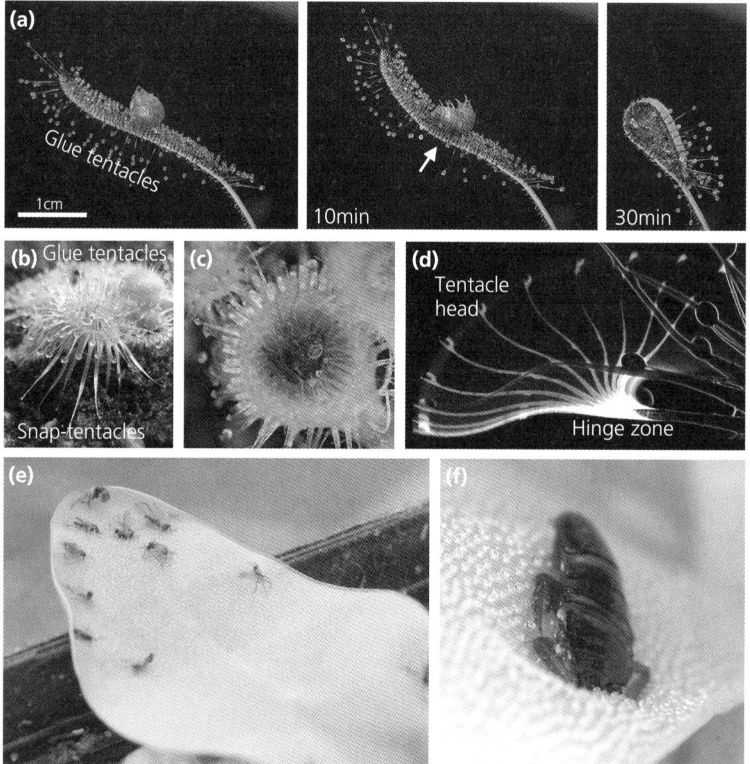

Figure 14.3 The motile traps of *Drosera* and *Pinguicula*. (**a**) A time-lapse recording of a manually fed *D. capensis* trap leaf. Ten minutes after prey is deposited on the leaf, the glue tentacles have bent upwards to hinder prey escape (arrow), and the leaf has begun to bend. Thirty minutes later, the leaf has completely wrapped around the animal, and digestion begins in the outer stomach. Afterwards, the leaf uncurls. (**b**) *D. glanduligera* possesses marginal, mucilage-free snap-tentacles in addition to the glue tentacles that are developed more centrally on the trap leaf. (**c**) Prey (here, a manually placed fruit fly) are catapulted onto the leaf of *D. glanduligera* and then slowly drawn within the concavity by sticky tentacles. (**d**) A high-speed recording of the motion of a *D. glanduligera* snap-tentacle (grayscale inverted for clarity). The period between the first two steps of the motion, which begins on the bottom left of the image, is 20 milliseconds, and the period between the other motion steps is 5 milliseconds. Bending takes place at the hinge zone, whereas the apical tentacle part with the head does not undergo bending deformation. (**e**) The leaf margin of *P. grandiflora* curls inwards as a response to the presence of stuck prey (see also Figure 6.3h). (**f**) A depression forms on the leaf lamina of *P. hirtiflora* at the place where the prey is stuck, leading to a sinking of the prey into a digestive bath. Figures (**b**), (**c**), and (**d**) are modified from Poppinga et al. (2012).

generates a graded receptor potential with amplitude approximately proportional to the intensity of the stimulus. Once a threshold value is reached, a series of short action potentials travels rapidly from the receptor site to the bending zone. They are transmitted along the excitable outermost two cell layers of the cylindrical tentacle stalk (Williams and Spanswick 1972). The tentacle moves, probably due to differential growth processes, as long as a threshold number of action potentials reach the bending zone (Williams 1976). The tentacles on the leaf margins are typically longer, more sensitive to mechanical stimulation, and show a stronger bending response than the central tentacles. Stimulation of marginal tentacles alone does not entail any other trap movements (Darwin 1875). Stimulation of central tentacles leads to membrane potential oscillations and jasmonate accumulation at the area of stimulation, entailing a delayed motion of (further) non-stimulated marginal tentacles (Nakamura et al. 2013, Krausko et al. 2017). Wounding of the trap leaf petiolus also leads to tentacle movement (Krausko et al. 2017).

Several tentacle types with distinct morphologies have been described (Seine and Barthlott 1993, Poppinga et al. 2013a). Especially noteworthy are the snap-tentacles found on the leaf margins of various species (Hartmeyer and Hartmeyer 2010). Snap-tentacles move much faster than normal glue tentacles. They are not cylindrical, but bilaterally symmetric, and have a raised gland that does not produce glue. Similar to glue tentacles, snap-tentacles are triggered by mechanical stimulation of the tentacle head, and move by deformation of a distinct bending zone. In *D. glanduligera*, movement occurs ≈400 ms after stimulation and lasts 75 ms. Prey are literally catapulted onto the sticky leaf surface, where mechanosensitive and motile glue tentacles are situated (Poppinga et al. 2012; Figures 14.3b, c). Theoretically, the rapid motion of snap-tentacles could be achieved by pure hydraulic actuation due to their small size. However, elastic instabilities caused by pre-stressing of tentacle tissue also may contribute. *Drosera glanduligera* snap-tentacles function only once, a phenomenon that could be explained by irreversible cell-buckling effects resulting from the high bending speed (Poppinga et al. 2012). In contrast, the snap-tentacles of some species with slower movements (several seconds–minutes) function repeatedly and reset within a day. Some small species of *Drosera* sect. *Bryastrum*, however, possess very fast snap-tentacles that bend within a fraction of a second but function repeatedly (Hartmeyer and Hartmeyer 2015).

Entire leaves move slowly when live prey or prey extracts are placed on the laminae (Darwin 1875, Nakamura et al. 2013; Figure 14.3a), often leading to the development of an "outer stomach" (Darwin 1875; Chapter 16). Leaf movements differ among species and can entail the formation of depressions and deflections of the leaf margins (e.g., *D. rotundifolia*; Darwin 1875), curving and folding of leaves (e.g., *D. capensis*; Bopp and Weber 1981, Bopp and Weiler 1985), or spiral curling (e.g., *D. regia*). The observed kinematics further depend on the location of the prey. For example, Bopp and Weber (1981) found that *D. capensis* leaves bend at the site of prey attachment, and the degree of bending was determined by the distance between prey and leaf tip. Leaf bending originally was thought to be based on auxin-mediated growth processes (Bopp and Weber 1981, Bopp and Weiler 1985, Juniper et al. 1989), but Nakamura et al. (2013) and Mithöfer et al. (2014) reported prey capture-induced accumulation of jasmonates in the leaf tissue and showed that these phytohormones alone can trigger leaf bending. Jasmonates are universally involved in asymmetric growth, herbivore defense, and wounding responses.

Motile *Pinguicula* traps. The slow leaf movements of *Pinguicula* spp. still are poorly understood. No scientific advances have been made on the physiology, functional morphology, or biomechanics of these traps since Juniper et al. (1989). Only some *Pinguicula* species (e.g., *P. vulgaris*) show movement responses within hours or days after prey capture (Darwin 1875), but others never move (e.g., *P. gypsicola*). The diet of some *Pinguicula* may consist mainly of pollen and other plant debris (Karlsson et al. 1994).

Pinguicula leaves are covered with sessile digestive glands and stalked, glue-producing trapping glands (Chapter 13). They trap mostly small dipterans, especially fungus gnats. In some species, mechanical stimulation in combination with nitrogenous matter leaching from prey induces an upward bending of the leaf margins (Figures 6.3h,

14.3e). When prey is placed near the center of the leaf, the leaf margins may curl toward the center. These movements are based on differential growth processes (Lloyd 1942) and may boost digestion by bringing additional digestive glands into contact with the prey (Juniper et al. 1989). This may be further aided by the formation of a depression of the leaf lamina around the prey (Figure 14.3f). The development of this cavity results from turgor loss of the basal cells of the glands and of neighboring epidermal cells (Heslop-Harrison and Knox 1971). As a consequence, prey is immersed in a digestive pool. Once digestion is completed, the leaf unfolds again, and the cavities disappear within approximately 24 hours (Darwin 1875, Lloyd 1942).

14.2.3 Suction traps

Suction traps are the fastest carnivorous plant traps and unique to *Utricularia* spp. (Figures 14.4, 14.5; Chapters 8, 13, 19). Prey animals are sucked into the trap ("bladder") within half a millisecond, which is ≈20 times faster than the *Aldrovanda* snap-trap, 150 times faster than the snap-tentacles of *Drosera glanduligera*, and >200 times faster than the *Dionaea* snap-trap. Most work so far has focused on the relatively large traps of aquatic species (*U.* sect. *Utricularia*, comprising ≈16% of the genus), as they are comparably easy to observe. How the traps of non-aquatic species—approximately 80 % of the genus—function remains largely unknown. Our description of functional morphology, physiology, and biomechanics of *Utricularia* suction traps therefore focuses on aquatic species (*U. vulgaris* trap type; Lloyd 1942).

The architecture of bladders of the *U. vulgaris* trap type is uniform, with trap diameters ranging from 0.5–6 mm. Some species (e.g., *U. vulgaris*) have bladders that are dimorphic in size and shape (Chapter 13). The bladders are hollow, water-filled, lentil-shaped vesicles with flexible lateral walls (Figure 14.4a, 14.4b, 19.4). The reinforced

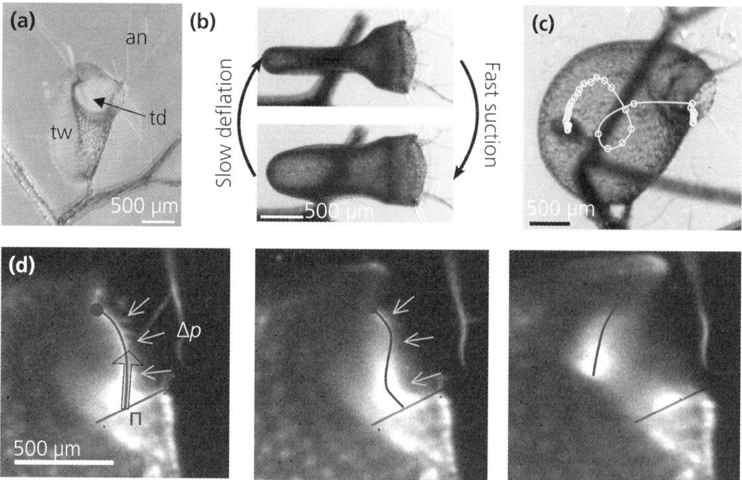

Figure 14.4 General morphology and kinematics of the *Utricularia vulgaris* type of suction trap. (**a**) The trap (bladder) is a hollow vesicle with flexible, lateral trap walls and an entrance region that is closed watertight by the trapdoor. Antennae are crucial for attracting prey and for leading them toward the trapdoor. (**b**) The bladderwort trap works in two phases. First, the trap slowly deflates as water is pumped out of the lumen, and elastic energy is stored in the deformed walls. Second, triggering entails trapdoor opening and relaxation of the trap walls, accompanied by a very fast suction of water and prey. (**c**) High-speed recording of the capture of a small crustacean. The path of the prey during suction is indicated, showing that the animal loops inside the trap. (**d**) Trapdoor kinematics, as visualized by high-speed cinematography under laser sheet fluorescence microscopy. The trapdoor possesses an outward curvature when the trap is ready to fire (left). The force exerted on the door due to the water pressure difference (Δp) between the trap interior and exterior is counterbalanced by friction forces (π) on the threshold. Also, the door resists inwards buckling due to its double curvature. Once the pressure difference is high enough (middle), small mechanical perturbations on the trigger hairs, which protrude from the trapdoor (see Figure 14.5), are sufficient to entail trapdoor buckling. The door is now unlocked and swings open rapidly as water flows in rapidly (right). A critical negative pressure inside the trap leads to spontaneous firing. Abbreviations: an = antennae, td = trapdoor, tw = trap wall. All images modified from Vincent et al. (2011a).

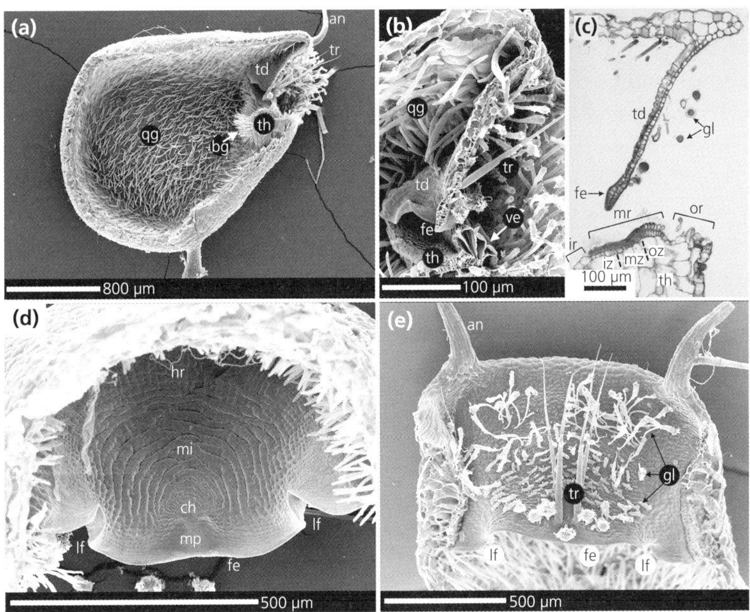

Figure 14.5 Functional morphology of the *Utricularia vulgaris* type of suction trap. (**a**) Scanning electron micrograph of a longitudinal section of an *U. vulgaris* trap. The quadrifid and bifid glands, and the threshold, trapdoor, and trigger hairs are visible. The door rests at an approximate 90° angle to the threshold, and the entrance forms a short tube. (**b**) Scanning electron micrograph of the trap entrance. Note the cut trapdoor with its free edge, the trigger hairs, and the velum on the threshold. (**c**) Light microscopic image of a 10-μm-thick longitudinal section of the trap entrance. The trapdoor consists of two cell layers. The three regions of the threshold, as well as the three zones of the pavement epithelium are visible. (**d**) The inner trapdoor surface is strikingly compartmentalized. The concentric constrictions are hypothesized to act as pre-folds for channeling the process of door buckling and unbuckling (see also Figure 14.4d). Two lateral folds on the free edge help in positioning the door on the pavement epithelium when it is closed, and, by unfolding, in the process of door opening. (**e**) The outer door surface is, in contrast, not compartmentalized. Note the multitude of glands and the long trigger hairs. Abbreviations: an = antennae; bg = bifid glands; ch = central hinge of the door; fe = free door edge; gl = glands of unknown function; hr = hinge region of the door; ir = inner region of the threshold; iz = inner zone of the pavement epithelium; lf = lateral door folds; mi = middle region of the door; mp = middle piece of the door; mr = middle region of the threshold (= pavement epithelium); mz = middle zone of the pavement epithelium; or = outer region of the threshold; oz = outer zone of the pavement epithelium; qg = quadrifid glands; td = trapdoor; th = threshold; tr = trigger hair(s); tw = trap wall; ve = velum. **(a)** modified from Poppinga et al. (2016b); **(d)** and **(e)** modified from Vincent et al. (2011a).

entrance forms a short tube and is kept watertight by a trapdoor (Lloyd 1942). The functional morphology of the trapdoor is highly sophisticated (Poppinga et al. 2016b). It consists of two cell layers (Figures 14.5b, 14.5c) and is only 20–40 μm thick. The cells of the outer layer are uniform. In contrast, the inner layer is compartmentalized into several distinct regions, including a central hinge and, immediately below, the so-called middle piece (Lloyd 1942, Vincent et al. 2011b; Figure 14.5d). Several trigger hairs protrude from the outer surface of the trapdoor in this central region (Figures 14.5b, 14.5e). Further antennae and bristles are located at the entrance region (Figures 14.4a, 14.4b, 14.5a, 14.5b). These appendages are crucial for leading prey toward the entrance (Chapter 12).

The hinge region connects the trapdoor to the arched upper end of the entrance. The lower trapdoor edge rests on the collar-like threshold, forming an angle of approximately 90° (Figures 14.5a–14.5c). The threshold can be subdivided into three regions. The outer region bears stalked glands of unknown function and the inner region extends into the trap lumen (Lloyd 1942, Poppinga et al. 2016b; Figure 14.5c). Sandwiched in between is the middle region which is the part where the free edge of the trapdoor rests. It is characterized by a pavement epithelium formed by short glandular cells, and is subdivided into an outer, middle, and inner zone. The cells of the outer zone are bloated like balloons and are characterized by exfoliated cuticles. They form the so-called velum (Figure 14.5b) which is, in combination with secreted mucilage, responsible for keeping the trap watertight. The outer and middle zones together form a large bump; and the adjacent inner zone forms a separate smaller bump. Both elevations run along the whole pavement epithelium, and the furrow between them holds the free edge of the trapdoor securely in place while the trap is closed. Lateral folds of the free door edge may help to position the door accurately in the furrow (Lloyd 1942, Vincent et al. 2011b; Figures 14.5b–14.5e), and probably add displacement space into the lumen when the trap fires.

Two types of glands are found inside the trap. Almost the entire inner surface is covered with four-armed glands (quadrifids) that probably secrete the digestive enzymes (Chapters 13, 16). Two-armed glands (bifids) near the trap entrance presumably pump water out of the trap continuously (Sasago and Sibaoka 1985a, 1985b, Juniper et al. 1989, Adamec 2011d). This assumption is based on the observation that water bubbles form around the entrance region when the trap is submersed in paraffin oil (Sasago and Sibaoka 1985a). The energy-demanding pumping process generates a negative hydrostatic pressure of approximately –12 to –16 kPa inside the trap (Sydenham and Findlay 1973, Sasago and Sibaoka 1985a, Singh et al. 2011). Once this pressure difference is reached, the outward flow may be equalized by inward flows due to trap wall permeability and trapdoor leakage (Adamec 2011d, Vincent et al. 2011a). Vincent et al. (2011b) estimated a Young's Modulus of 5–20 MPa for the trap body, similar to fully turgescent parenchymatous tissue. Trap deflation takes at least 15 minutes. During deflation, the lateral trap walls deform and curve inward (Figure 14.4b). This leads to storage of elastic energy as the trap becomes increasingly prestressed. Meanwhile, the trapdoor is held in place by the furrow on the threshold, and stabilized by its convex curvature (Figure 14.4d).

The ultrafast firing of the trap is triggered when prey (typically small crustaceans) touch one of the trigger hairs (Figures 14.4c). It is not yet fully understood whether the underlying process is purely mechanical, with the trigger hairs acting as levers, or whether electrical signaling (as in *Dionaea*, *Aldrovanda*, and *Drosera*) is involved. Masi et al. (2016) measured electrical responses in *Utricularia* trap cells but did not demonstrate signal transduction. Attempts to trigger *Utricularia* traps electrically have not been successful (e.g., Sydenham and Findlay 1973), and low temperature, ion channel blockers, and cytochrome oxidase inhibitors affect trap deflation, but not triggering (Adamec 2012a). These findings strongly support a purely mechanical trigger mechanism.

A mechanical buckling scenario sufficiently explains how the mechanical sensitivity of the trapdoor increases with decreasing internal pressure until very small perturbations (i.e., prey touching the trigger hairs) suffice to trigger door opening. As water is continuously pumped out of the trap, negative pressure builds up inside, putting the domed trapdoor under increasing tension. Eventually, the deflection of a single trigger hair on the surface is sufficient to induce buckling and a rapid inversion of the door curvature (Vincent et al. 2011b; Figure 14.4d). The deformation is initiated by bulging of the central hinge and middle piece and further amplified by numerous concentric constrictions on the inner door surface (Figure 14.5d) so that it progressively spreads across the rest of the door. The resulting inversion of the door curvature unlocks the free edge, and the door swings open within 0.5 ms (Poppinga et al. 2016b; Figure 14.4d). As the pressure equalizes, the trap walls relax and the bladder suddenly expands, sucking in water and prey with velocities of up to 4 m/s and accelerations of up to 2800 *g*. Phases of very high acceleration during onsets of suction are immediately followed by phases of similarly high deceleration (max.: –1900 *g*) inside the bladders, leading to immobilization of the prey which then dies (Poppinga *et al.* 2017). The rapid influx of water makes escape virtually impossible, and vigorous turbulence may cause prey to swirl inside the trap (Figure 14.4c). Further investigations are needed to determine whether these turbulences aid in prey retention.

In contrast to the *Dionaea* trap, both the initial inversion of the *Utricularia* trapdoor and the reversion to the convex configuration involve buckling. This allows the trapdoor to re-close extremely quickly, within approximately 2.5 ms. The hinge region of the trapdoor may promote unbuckling by acting as a spring (Lloyd 1942). The trapdoor remains insensitive to mechanical triggering until the negative pressure inside the trap has reached a threshold value. Immediately after suction, the lateral trap walls remain slightly concave, indicating that they still store some elastic energy and that the pressure difference between interior and exterior has not yet fully equilibrated (Lloyd 1942, Poppinga et al. 2016b). In contrast, piercing the trap (e.g., with a fine needle) leads to fully relaxed trap walls. This indicates that the trapdoor re-closes as soon as the resetting force exceeds suction force. This fast closing may prevent the escape of prey, and minimize the outflow of nutrient-rich water. Prey die due to anoxia (Adamec 2007b) and are enzymatically digested (Chapters 13, 16, 19). Remarkably, the trap can continue to capture prey while digesting.

Traps also may fire spontaneously when a critical negative pressure inside the trap is reached (Adamec 2011d, 2011f, Vincent et al. 2011b). The mechanical relaxation might help to avoid fatigue of the trap tissue (Poppinga et al. 2016b). Spontaneous firings can occur at regular intervals, randomly, or in bursts (Vincent et al. 2011a). The mode of spontaneous firing is characteristic for each individual trap and is species-independent. Small differences in door positioning on the pavement epithelium may contribute to this variation (Poppinga et al. 2016b). The critical pressure for spontaneous firings (–3.9 to –44.9 kPa) varies among species (Adamec and Poppinga 2016). Small algae, protozoa, bacteria, or detritus may be sucked into the bladder during spontaneous firing events and add to the nourishment of *Utricularia* (Koller-Peroutka et al. 2015).

The trap diversity among bladderwort species from various habitats is extraordinary. Many non-aquatic species have elongated tubular trap entrances with lower door-to-threshold angles of only ≈20–60°, and large amounts of mucilage are speculated to aid in door fastening (Lloyd 1942). The respective trapdoor movements also vary. For example, in some species (e.g., *U. uniflora*) no door curvature change takes place prior to opening, and the free door edge detaches from the threshold in a comparably slow manner (Westermeier et al. 2017). Trigger-hair morphology also is highly variable, and some species do not have trigger hairs at all. Westermeier et al. (2017) could not stimulate *U. multifida* traps and they speculated that these traps may function as lobster traps similar to those of *Genlisea* (Reifenrath et al. 2006; Chapter 15). Spontaneous firings also were not observed in *U. multifida*, further corroborating the assumption that it employs passive (i.e., non-motile) traps.

14.3 The passive motile trap of *Nepenthes gracilis*

Nepenthes pitchers are typically passive pitfalls that trap prey using slippery surfaces and retentive fluids (Chapters 12, 15). The roof-like pitcher lid is widely thought to prevent fluid dilution and prey loss during heavy rains. *N. gracilis* is the only species known to directly use the pitcher lid for capturing prey with an externally powered passive movement. The impact force of rain drops leads to a deformation of part of the pitcher, causing a rapid downward movement and subsequent oscillation of the pitcher lid which is crucial for prey capture (Figure 14.6). The passively induced lid movement depends on environmental physical factors (rain) and hence cannot be actively controlled by the plant.

The unique trapping mechanism of *N. gracilis* relies on a combination of surface and material adaptations. Similar to the inner pitcher wall of many *Nepenthes*, its lower lid is covered with epicuticular wax crystals (Chapters 12, 15); however, the crystal morphology is strikingly different (Figure 15.2g). In contrast to the delicate platelets found on the inner pitcher wall, the lid wax forms dense clusters of flat-topped pillars. These provide sufficient grip for insects to walk upside down under the lid and harvest nectar from highly prolific nectaries, but not enough to withstand severe perturbation. Coating the lower lid surface with an anti-slip polymer caused a significant reduction of captured prey in the field (Bauer et al. 2012b).

The mechanical properties of the *N. gracilis* lid similarly are well-adapted to its trapping function (Bauer et al. 2015b). The lid is very stiff and will break before it bends significantly. However, the rear of the pitcher just below the lid attachment is

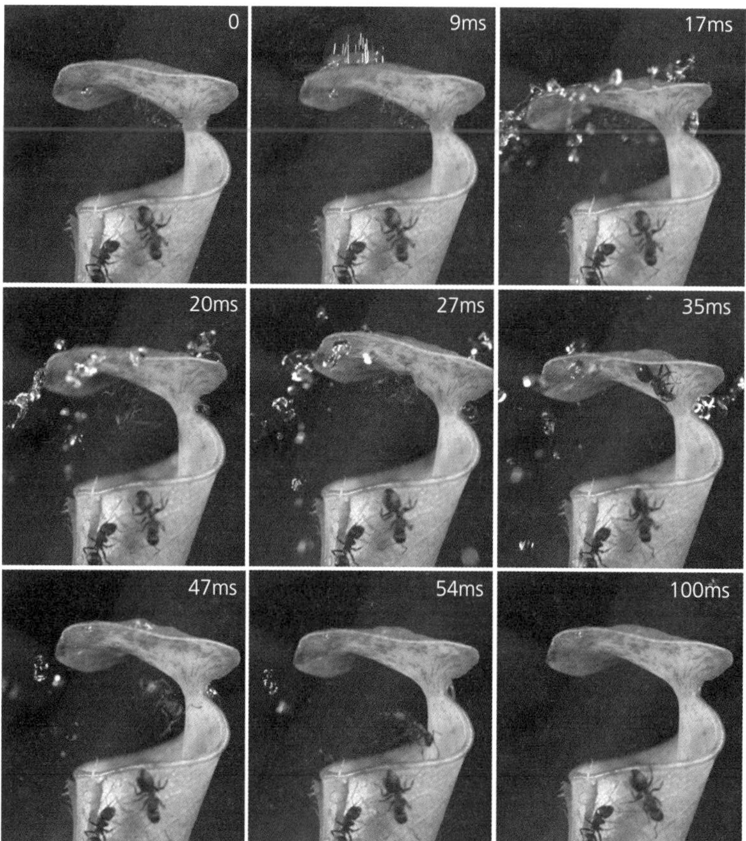

Figure 14.6 The passive motile trap of *Nepenthes gracilis* illustrated with single frames from a high-speed recording of a trapping sequence from a laboratory feeding experiment. In the first image, the ant that will later be trapped is situated at the underside of the pitcher lid and searches for nectar. Other ants can be seen on the outer pitcher surface. At $t = 9$ ms, the splash of the impacting water drop can be seen on the upper lid surface. Afterwards, the lid moves rapidly downward and subsequently oscillates. The ant loses its grip and falls down into the pitcher, which is filled with digestive and retentive fluid.

flexible. Upon impact of a rain drop, this region deforms and acts as a hinge about which the stiff lid pivots. During the initial downstroke, the lid tip can reach a top speed of 1.5 m s^{-1} and a maximum acceleration of approximately 30 *g*. Velocity and acceleration increase linearly from the base of the lid to its tip, confirming that the lid does not bend but instead functions as a torsional spring that responds to drop impacts with highly stereotyped damped oscillations (Bauer et al. 2015b). The combination of specialized material and surface properties provides *N. gracilis* with a highly effective additional trapping mechanism that may be partly responsible for the great ecological success of this widespread and abundant species. The importance of the lid for prey capture is further corroborated by the fact that *N. gracilis* allocates a much higher proportion of its attractive nectar to the lower lid surface than other investigated species (Bauer et al. 2012b).

14.4 Future research

Motile traps of carnivorous plants are ideal candidates for in-depth investigations of plant movements. Although their general kinematics and underlying physiological processes have been investigated in detail, the precise mechanisms are not yet fully resolved. Particularly on cellular and molecular level, many processes remain poorly understood (Forterre 2013). Open questions include: how do biological signals such as action potentials affect the post-stimulatory mechanical responses (e.g., turgor changes); how is active water transport maintained (e.g., out of *Utricularia* traps) and how do water channels work; which channels regulate ion transport across membranes, and how do the mechanical and electrical properties of cells and cellular constituents affect this? Commercially available loss-of-function mutants in *Dionaea* offer an exciting opportunity to study these mechanisms (Bailey and McPherson 2012, Poppinga et al. 2013b).

In addition to controlled laboratory experiments, it is important that trap movements and trapping mechanisms are investigated in the field. These can reveal if trap movements are fine-tuned to exploit prey behavior; how plants avoid being triggered by rain, wind, or debris; how trap sensitivity is modulated by environmental factors; and whether some animals are adapted to avoid traps or escape from them. The ecological implications of motile traps have barely been studied and are likely to provide exciting and surprising discoveries in the future.

Motile traps also can inspire technical devices such as structures that autonomously respond to the environment (Chapter 20). The interest in plant movement for the development of biomimetic applications is constantly increasing. For example, the trapping movements of carnivorous plants have become important sources of inspiration for "elastic architecture" such as biomimetic façade-shading systems (Schleicher et al. 2015). Furthermore, plants in general and carnivorous plants in particular provide ample inspiration for resilient hinge-free movements that function reliably over long time spans while requiring minimal maintenance.

CHAPTER 15

Non-motile traps

Ulrike Bauer, Reinhard Jetter, and Simon Poppinga

15.1 Introduction

Some carnivorous plants have trapping mechanisms that are not based on plant movement, but instead use stickiness, slipperiness, or direction-dependent mechanical obstructions to capture prey. Movement-independent trap types include adhesive "flypaper" traps, pitfall traps (but cf. Chapter 14, §14.3), and eel traps. Most non-motile traps employ complex mechanisms combining two or more movement-independent mechanisms (Rice 2007). For instance, the pitfall traps of *Nepenthes* often combine slippery surfaces with a sticky, viscoelastic trap fluid, and the flypaper traps of *Drosera* and *Pinguicula* regularly combine sticky mucilage and glues (Chapter 13) with prey-triggered leaf and tentacle movement (Chapter 14).

In general, successful trapping depends on a combination of prey attraction, capture, and retention. Juniper et al. (1989) distinguished between initial capture and prey retention as separate tasks, but also noted that one trap component can serve both purposes. For example, both the sticky mucilage of *Drosera* and *Drosophyllum* and the waxy inner surface of many *Nepenthes* pitchers serve to catch prey and to retain it. The peristome of *Nepenthes* pitchers attracts insects by means of nectar secretion and contrasting UV reflectance patterns (Chapter 12) while also playing a key role for trapping (Chapter 13).

The simplest adaptations for prey capture affect trap orientation, shape, and geometry. Traps are often oriented to maximize their exposure to potential prey (Juniper et al. 1989). Examples include the broadened, flat leaves of many *Drosera* and *Pinguicula* species, the outward-oriented pitchers on *Cephalotus* rosettes, and the pitchers of *Darlingtonia* that are basally twisted so that each grows in a slightly different direction relative to others on the same rosette. Pitchers of *Sarracenia* and *Nepenthes* that specialize in capturing flying prey tend to be taller and more slender than those targeting crawling prey (Juniper et al. 1989). Moran (1996) found that the slender, funnel-shaped upper pitchers of *N. rafflesiana* capture more flying prey than its short, ovoid ground pitchers, but he also noted differences in how each pitcher type attracted prey. Furthermore, narrow tubular pitchers directly aid in the capture of flying prey. As prey try to fly out of the tubular section of the pitcher, they create downward air currents that hinder their escape. The magnitude of this effect depends on the shape and diameter of the pitcher tube, and on the location of the insect within it (Iosilevskii and Joel 2013).

In their review of trapping mechanisms, Juniper et al. (1989) emphasized micromorphology of pitcher and eel traps, and glue biochemistry of flypaper traps. Actual trapping mechanisms, if even discussed, largely were inferred from morphological data. The only systematic experimental study of trapping mechanisms done before 1989 was the series of simple but ingenious experiments by Knoll (1914) on wax crystal-covered surfaces of *Nepenthes* and of noncarnivorous *Iris germanica*. In contrast, a large number of detailed investigations into the trapping mechanisms of pitcher and flypaper traps have been done in the last twenty years. This increased interest in the function of non-motile traps has both helped to deepen our understanding of previously known trapping mechanisms and led to the discovery of

Bauer, U., Jetter, R., and Poppinga, S., *Non-motile traps*. In: *Carnivorous Plants: Physiology, ecology, and evolution*. Edited by Aaron M. Ellison and Lubomír Adamec: Oxford University Press (2018). © Oxford University Press.
DOI: 10.1093/oso/9780198779841.003.0015

entirely new ones (Bohn and Federle 2004, Bauer et al. 2012b, 2013). As a result, pitcher plants in particular have gained the attention of physicists and engineers, inspiring the development of novel biomimetic anti-adhesive and self-cleaning surfaces (Koch and Barthlott 2009, Wong et al. 2011; Chapter 20). In this chapter, we review how carnivorous plants use glues, slippery surfaces, and directional (anisotropic) features to trap and retain prey. Mechanisms of attraction are reviewed in Chapter 12; anatomy and morphology of traps are reviewed in Chapter 13.

15.2 Sticky traps and trap glues

Trap glues often are multifunctional and may contain digestive enzymes (Chapters 13, 16) or act as attractants (Chapter 12). In typical flypaper traps, prey (predominantly arthropods) are lured to the aerial trap leaves where mucilage-producing glands are located (Darwin 1875, Lloyd 1942, Juniper et al. 1989; Chapter 13); *Philcoxia* uniquely possesses below-ground trap leaves that capture nematodes (Pereira et al. 2012; Figure 15.1a). The mucilage glands typically are raised on stalks—the

Figure 15.1 Sticky traps and trap glues. (**a**) Scanning electron micrograph showing details of the subterranean *Philcoxia minensis* trap leaf. Numerous stalked glands are visible on the surface. The image also shows several nematodes (the main prey of this species) and a sand grain; bar = 200 μm (Modified from Pereira et al. 2012 and reproduced with permission from *PNAS*). (**b**) *Drosera rotundifolia* trap leaves with prey. (**c**) *D. capensis* tentacle with mucilage drop; bar = 200 μm. (**d**) *Drosera* mucilage is characterized by high extensional viscoelasticity and forms long threads when touched and pulled away from the tentacle; bar = 500 μm. (**e**) *Triphyophyllum peltatum* trap leaf with glue-laden tentacles and a captured blowfly. (Image from Rembold et al. 2010b and reproduced with permission from the *Carnivorous Plant Newsletter*). (**f**) A *Roridula gorgonias* plant has captured multiple flies on its extremely sticky leaves. Three sizes of glandular trichomes occur along the leaf margins. (**g**) A young *Drosophyllum lusitanicum* leaf, showing outward circinate vernation, mushroom-shaped stalked glands, and trap mucilage. (Photograph by Anja and Holger Hennern, and used with permission). (**h**) Leaves of *D. lusitanicum* covered in small flying prey. (Photograph by Anja and Holger Hennern, and used with permission). (**i–k**) A fly on the surface of *Nepenthes rafflesiana* pitcher fluid wraps itself in long filaments of the highly viscoelastic liquid; bar = 3 mm. (Images from Gaume and Forterre 2007).

"tentacles" of *Byblis*, *Drosera* (Figure 15.1b–d), *Roridula* (Figure 15.1f), and *Triphyophyllum* (Figure 15.1e) (Chapter 13, §13.4). Prey get stuck to the glue, struggle, and try to escape, thereby contacting even more glue, and eventually suffocate as their tracheae get plugged. However, some animals can move freely and live as commensals on adhesive traps (Chapters 1, 13, 23, 26).

The glues of *Drosera* and *Drosophyllum* (and probably those of less well-studied *Byblis*, *Nepenthes*, *Pinguicula*, *Philcoxia*, and *Triphyophyllum*) are sugar-based (Vintéjoux and Shoar-Ghafari 2000, Adlassnig et al. 2010). In all but very arid conditions (Volkova and Shipunov 2009), the *Drosera* glue drop is typically tens of μm in diameter (Figure 15.1c), corresponding to a nL fluid volume (Erni et al. 2011). The glue itself is a homogeneous, aqueous solution (pH 5) of an acidic polysaccharide composed of a D-glucurono-D-mannan backbone with alternating monosaccharide side groups (Rost and Schauer 1977, Gowda et al. 1982, 1983). It constitutes a natural hydrogel, consisting of a flexible, fibrous polysaccharide nano-network that allows for large stretching deformation (Huang et al. 2015) and is characterized by high viscosity, capillary thinning, and extensional viscoelasticity (Erni et al. 2011). These rheological properties are susceptible to pH changes and extreme temperatures. Freezing, thawing, and raising the temperature to 80 °C all lead to an irreversible loss of the glue's viscoelasticity (Rost and Schauer 1977). The fresh mucilage can be drawn out into long slender threads (Figure 15.1d). Elastic forces need to be overcome to stretch the long-chain polymers, conferring high resistance to extensional flows. The movement of struggling prey is too fast for these elastic forces to relax, and the glue behaves like an elastic band.

The glue of *Drosophyllum lusitanicum* is similarly composed of the monomers arabinose, galactose, xylose, rhamnose, glucuronic acid, and ascorbic acid. However, it is more acidic (pH 2.5–3) than the *Drosera* glue (Juniper et al. 1989, Adlassnig et al. 2010), and emits a noticeable honey scent (Chapter 12). In further contrast to *Drosera*, the mushroom-shaped glands of *Drosophyllum* carry glue drops even when the relative humidity falls below 40% (Adlassnig et al. 2006, Figure 15.1g). The glue is highly hygroscopic (Darwin 1875), probably helping this xeromorphic plant to harvest water from fog (Adamec 2009a). Unlike the glue of *Drosera*, that of *Drosophyllum* cannot be drawn out into threads and is easily detached from the glands. Larger and heavier prey "slide" down the trap leaf, thereby coming into contact with even more glue, whereas small prey get firmly stuck to the trap (Rice 2007; Figure 15.1h).

The glandular tips of *Roridula* tentacles secrete a water-insoluble, viscoelastic resin (Bruce 1907; Chapter 13; Figure 15.1f) that consists of lipophilic aliphatic esters and carboxylic acids (Simoneit et al. 2008, Voigt and Gorb 2008, Frenzke et al. 2016). Its high desiccation resistance may be an adaptation to periodically dry environments (Voigt and Gorb 2010a). The resin also stays fully functional under wet conditions, even when submersed for 24 h (Voigt et al. 2015). It is extremely sticky, so that plants are often covered in trapped insects of considerable size and mass (Voigt and Gorb 2008). In addition to the adhesive properties of the glue, the hierarchical architecture of the tentacles plays a crucial role in prey capture (Voigt et al. 2009), with tentacles of three different lengths acting sequentially. First, long, slender, and very flexible tentacles make comparably weak adhesive contact with the prey. The still mobile animal, in its struggle to escape, touches more of the long tentacles, until it also comes into contact with a second type of tentacles. These are thicker, shorter, and four times stiffer than the long ones, and produce an adhesive that is 50% stickier. Finally, the animal sticks to the short and thick tentacles that are almost 50 times stiffer than the longest ones and produce the stickiest glue. The animal exhausts itself in its attempts to escape and eventually dies.

Roridula does not produce its own digestive enzymes (Marloth 1925), but relies on a mutualistic relationship with mirid bugs for prey digestion (Chapter 26). The bugs feed on the trapped insects and then defecate on the leaves. The mineral nutrients contained in the feces are absorbed through the thin, porous cuticle of the leaves (Ellis and Midgley 1996). A combination of locomotive adaptations and a thick coating of anti-adhesive epicuticular grease enables the bugs to walk on the trapping leaves without getting stuck (Voigt and Gorb 2008, 2010b).

Large amounts of water-insoluble mucilage also are secreted on the outer surfaces of leaves, inflorescences, and traps of some species of *Genlisea* and *Utricularia*. The biological function of these mucilages has not yet been investigated in detail (Fleischmann 2012a, Poppinga et al. 2016b).

Finally, the viscoelastic fluid in the pitchers of some *Nepenthes* (Nepenthaceae) species must also be considered a trap glue, despite its predominantly retentive function. Initial trapping is largely based on anti-adhesive surfaces (§15.3). Prey slip and fall into a fluid pool with both retentive and digestive properties. Gaume and Forterre (2007) showed that the viscoelastic relaxation time of the trap fluid of *N. rafflesiana* exceeds the typical timescale of the leg movements of struggling prey. As a consequence, prey become entangled in multiple sticky fluid threads, and eventually drown and die (Figure 15.1i–k). Gaume and Forterre (2007) found no effect of surface tension for prey retention in *N. rafflesiana*, but Armitage (2016a) showed in laboratory experiments that pitcher-dwelling bacteria reduce the surface tension of *Darlingtonia californica* pitcher fluid, leading to improved prey retention. Bazile et al. (2015) investigated the effects of fluid viscoelasticity and pH on prey capture and retention in four *Nepenthes* species. Fluid pH affected mainly survival times, with more acidic fluid killing prey more quickly. Retention efficiency increased exponentially with viscoelasticity, higher levels of which were necessary for effective retention of ants relative to flies.

15.3 Anti-adhesive surfaces

The pitfall traps of *Brocchinia, Catopsis, Cephalotus, Darlingtonia, Heliamphora, Nepenthes*, and *Sarracenia* all rely predominantly on slippery surfaces for prey capture and retention, often aided by directional surface features (Chapter 13, §13.3). Whether a surface is more effective for initial capture or prey retention depends on two factors: its position within the trap and the presence or absence of attractive features, particularly nectar (Chapter 12). Generally, surfaces at or close to the trap opening tend to play a more important role in initial trapping, whereas features deep inside the trap are more heavily involved in retention. However, attractive components such as nectaries often extend deep into the pitcher interior, e.g. in *Heliamphora* (Juniper et al. 1989, Bauer et al. 2013; Chapter 12), and insects such as mosquitoes are attracted to the pitcher fluid for breeding (Beaver 1983) and may get trapped in viscoelastic *Nepenthes* fluid during oviposition.

The anti-adhesive properties of pitfall trap surfaces are based on four general principles. First, microscopic roughness greatly reduces the available contact area for adhesive pads on the tarsi of insects (Scholz et al. 2010). The same effect prevents adhesive tape from sticking to fine-grained sandpaper. At the same time, the tarsal claws cannot interlock with the plant surface if its surface roughness is sufficiently fine-scale. This principle has been experimentally validated for *Nepenthes* and is likely to also play a crucial role on the trapping surfaces of *Cephalotus, Darlingtonia*, and *Sarracenia*. Second, easily detachable wax crystals (Chapter 13, §13.3.1) may contaminate the adhesive pads of insects walking on the surface, making the insects more likely to fall (Federle et al. 1997, Markstädter et al. 2000). This has been reported for *Brocchinia, Catopsis*, and *Nepenthes* (Gaume et al. 2004), and is likely to work in a similar way in other species with trap surfaces covered in epicuticular wax crystals. Third, directional features such as the downward-pointing hairs or papillae of the Sarraceniaceae and *Cephalotus*, and the overlapping peristome cells and downward-oriented lunate cells of *Nepenthes*, allow claws to interlock on the way into the pitcher, but not in the opposite direction (Gaume et al. 2002, Gorb et al. 2004, Bauer and Federle 2009, Bauer et al. 2013; Chapter 13, §13.3.3). And fourth, highly wettable (superhydrophilic) surfaces lead to the formation of stable water films under humid or rainy conditions (Bohn and Federle 2004, Bauer and Federle 2009, Bauer et al. 2013; Chapter 13, §13.2.3). Experimental studies on the detailed trapping mechanisms to date have focused almost exclusively on the genus *Nepenthes*.

15.3.1 Wax blooms

Epicuticular wax crystal coatings fulfill multiple ecological functions in all major plant groups (Barthlott 1989) and are often extremely slippery for insects (Jeffree 1986). They also occur in all genera, but not all species, of carnivorous bromeliads and pitcher plants, excepting *Cephalotus*. Wax crystal layers usually are located on interior surfaces of the trap above the digestive fluid. Juniper et al. (1989) described a thick coating of filamentous wax crystals in *Brocchinia reducta*, noting that insects not only slip on these surfaces, but also get entangled in the fine waxy threads. Gaume et al. (2004) and Poppinga et al. (2010) showed a similar wax crystal structure on the adaxial leaf surface of *Catopsis berteroniana* (Figure 15.2a). Platelet-shaped wax crystals are found in *Sarracenia* (Figure 15.2b), *Darlingtonia*, and *Nepenthes* (Figure 15.2c, 15.2d), but they are produced by *Darlingtonia* only when it grows under high-light conditions (Juniper et al. 1989, Poppinga et al. 2010). Wax blooms also are apparent in some *Heliamphora* species (notably *H. ceracea* and *H. macdonaldae*;

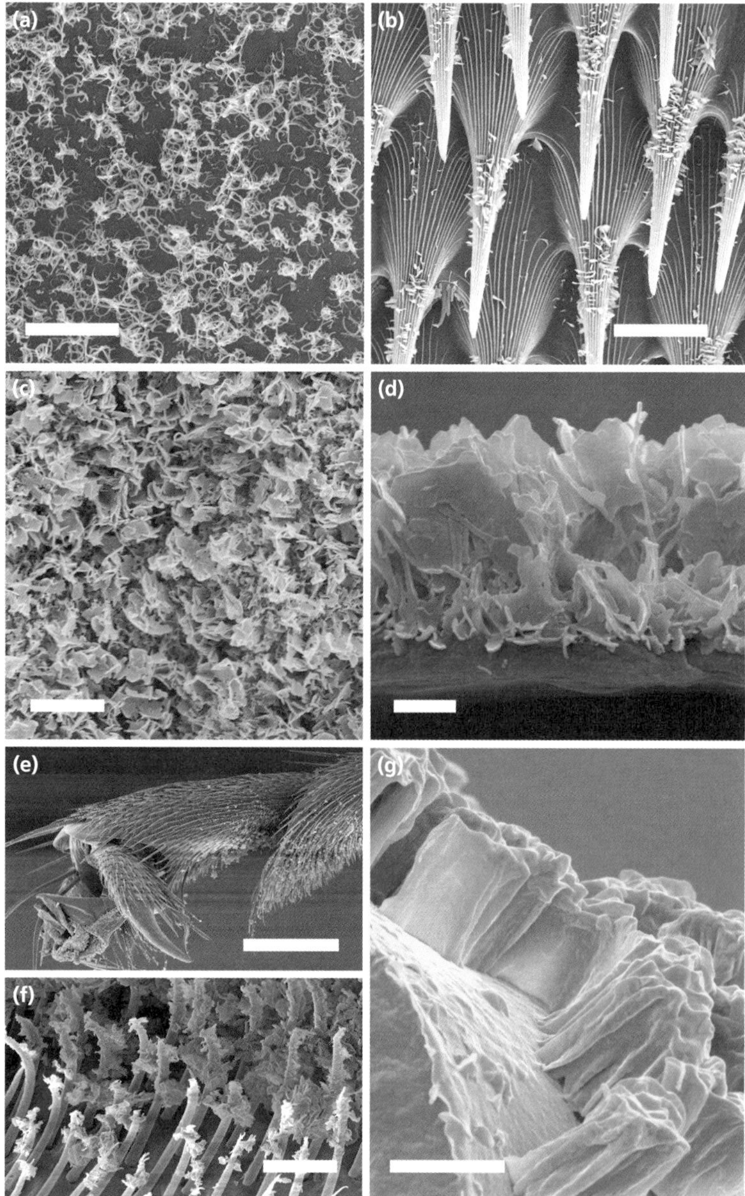

Figure 15.2 Wax blooms and their effect on insect locomotion. (**a**) Thread-like wax crystals on the inner surface of a *Catopsis berteroniana* leaf; bar = 20 μm (modified from Poppinga et al. 2010 and reproduced with permission from CSIRO Publishing). (**b**) Wax scales and cuticular folds (see also Figure 15.3a) on the imbricate cells of a *Sarracenia leucophylla* pitcher; bar = 20 μm (modified from Poppinga et al. 2010 and reproduced with permission from CSIRO Publishing). (**c–d**) Upright-standing wax platelets on the inner pitcher wall of *Nepenthes gracilis* in top (**c**) and side (**d**) views; bars = 5 μm (**c**) and 1 μm (**d**). (**e–f**) Severe wax contamination on adhesive pads and pretarsal hairs of a weaver ant after trying to climb the waxy inner wall of a *Nepenthes* pitcher; bars = 100 μm (**e**) and 10 μm (**f**); (Micrographs by Walter Federle, and used with permission). (**g**) Pillar-like wax crystals on the lower pitcher lid surface of *Nepenthes gracilis*; bar = 2 μm.

McPherson et al. 2011), but the crystalline ultrastructure has not yet been examined in this genus.

The majority of *Nepenthes* species employ wax crystals during at least some stages of their ontogeny, and a recent comparative analysis suggested that the presence of epicuticular wax crystals on the entire pitcher wall from fluid to rim is the ancestral character state within the genus (Bauer et al. 2012a). Ontogenetically early ground pitchers often show wax crystals throughout the pitcher wall, whereas they are reduced or absent in upper pitchers of the same species (Gaume and Di Giusto 2009). The ultrastructure of inner-wall surface waxes in *Nepenthes* is very uniform across all species examined to date, further supporting a common evolutionary origin.

Generally, the epicuticular wax layer consists of a sponge-like lower matrix with a superimposed dense array of very thin, upright-standing platelets with variable orientation (Figure 15.2d). Juniper et al. (1989) proposed that the wax platelets would be oriented downward, more or less parallel to the inner pitcher wall, but both SEM and AFM images consistently show platelets protruding perpendicularly from the underlying matrix (Riedel et al. 2003, Gaume et al. 2004, Scholz et al. 2010). The platelets are connected to the matrix via thin stalks that break easily. Detached platelets contaminate adhesive pads of insect tarsi in experimental trials (Juniper and Burras 1962, Gorb et al. 2005; Figure 15.2e, 15.2f). However, the extent to which wax crystals may break under natural trapping conditions remains controversial. Scholz et al. (2010) observed that the wax platelets of *N. alata* pitchers were remarkably stable, and did not detect significant wax contamination on the tarsal pads of ants or stick insects after they had slipped off the surface naturally. These authors also showed experimentally that a reduction of the available contact area because of microscopic roughness, at a scale similar to the *Nepenthes* wax, is by itself sufficient to impede insect adhesion.

Nepenthes gracilis is the only species known to produce morphologically distinct wax crystals on the surfaces of two different pitcher parts (Bauer et al. 2012b). The inner wall is coated with wax platelets similar to those of other *Nepenthes* species, whereas the underside of the pitcher lid carries clusters of coarser, pillar-shaped wax blocks (Figure 15.2g). This surface is far less slippery than the inner wall, and insects can walk upside down under the lid and harvest the generously secreted nectar. However, when the lid is perturbed, for example by the impact of a rain drop, the insect loses its footing and falls into the pitcher (Bauer et al. 2015b; Chapter 14, §14.3). When the wax crystal layer is removed experimentally, the *N. gracilis* lid loses its trapping function (Bauer et al. 2012b).

The ultrastructure of epicuticular wax crystals is largely determined by their chemical composition, although environmental influences during crystal formation may affect their structure to a lesser extent (Koch and Ensikat 2008). Riedel et al. (2003) found that the epicuticular wax crystals on the inner wall of *N. alata* consist of a mixture of aliphatic compounds: very long-chain (C_{22+}) aldehydes account for approximately 60% of the mixture; triacontanal (43%) is the most abundant constituent. The remaining 40% of the wax mixture include alcohols, fatty acids, esters, and a small amount of alkanes. The slippery epicuticular crystals are formed by accumulation of particularly high amounts of triacontanal (together with other wax aldehydes) near the pitcher wall surface, likely leading to spontaneous crystal formation. In contrast, the underlying intracuticular wax contains higher proportions of alcohols (31%), esters, and pentacyclic triterpenoids, but only 28% aldehydes and 9% triacontanal.

The chemical composition of epicuticular waxes on inner pitcher walls of *N. albomarginata, N. khasiana*, and three further hybrids is nearly identical to that of *N. alata* (Riedel et al. 2007), suggesting very similar mechanisms of crystal formation and hence wax biosynthesis in most, if not all, *Nepenthes* species that rely on slippery waxes to capture prey. Aldehyde-based crystals of similar morphology also have been described in noncarnivorous plants, including rice (Haas et al. 2001), suggesting convergent evolution of slippery surfaces in very distantly related plant lineages.

15.3.2 Cuticular folds

Microscopic surface roughness can be achieved not only by epicuticular wax blooms, but also by fine-scale folds of the cuticle itself. Such cuticular folds, usually arranged parallel to each other, are found ubiquitously across all genera of pitcher plants except the carnivorous bromeliads (Chapter 13, §13.3.2). The scale, density, and location of folds on the pitcher vary greatly between genera. In *Sarracenia, Darlingtonia*, and *Cephalotus*, strikingly similar downward-pointing imbricate cells line the upper sections of the pitchers. Each of these shark tooth-shaped cells is delicately striated by fine, ridge-like cuticular folds (Figure 15.3a). Their function has not yet been studied experimentally in pitcher plants, but cuticular folds hamper attachment of beetles in several noncarnivorous plants (Prüm et al. 2012). We hypothesize that these folds reduce the available contact area for tarsal adhesive pads of insects in a manner similar to the wax platelets in *Nepenthes* pitchers.

Broader, more widely spaced cuticular folds also are found on the surface of downward-pointing trichomes in *Heliamphora* (Figure 15.3b) and *Darlingtonia*. For *Heliamphora*, it has been proposed that these ridges help to increase the wettability of the hairs (Bauer et al. 2013). Even larger-scale cuticular ridges mark the centerlines of the elongated epidermal cells on the *Nepenthes* peristome (Figure 15.3d), creating a microscopic radial pattern of parallel ridges and grooves that enhances the wettability of the surface (§15.3.4).

15.3.3 Directional (anisotropic) surfaces

All pitcher plants other than carnivorous bromeliads have various types of directional surface structures. Frequently, different anisotropic features are found in different parts of the trap, or even in combination on the same tissue. All have in common that they provide a secure grip for an insect's tarsal claws in only one direction. Examples of such directional structures include the tooth-like imbricate cells of *Sarracenia* (Figure 15.3a), *Darlingtonia*, and *Cephalotus*, and the trichomes of *Heliamphora* (Figure 15.3b, 15.3c, 15.3e) and *Darlingtonia*. Similar downward-pointing trichomes are also found on the inside of the hood of many *Sarracenia* pitchers (Juniper et al. 1989). In *H. nutans*, the trichomes densely cover the inner pitcher wall, from its top

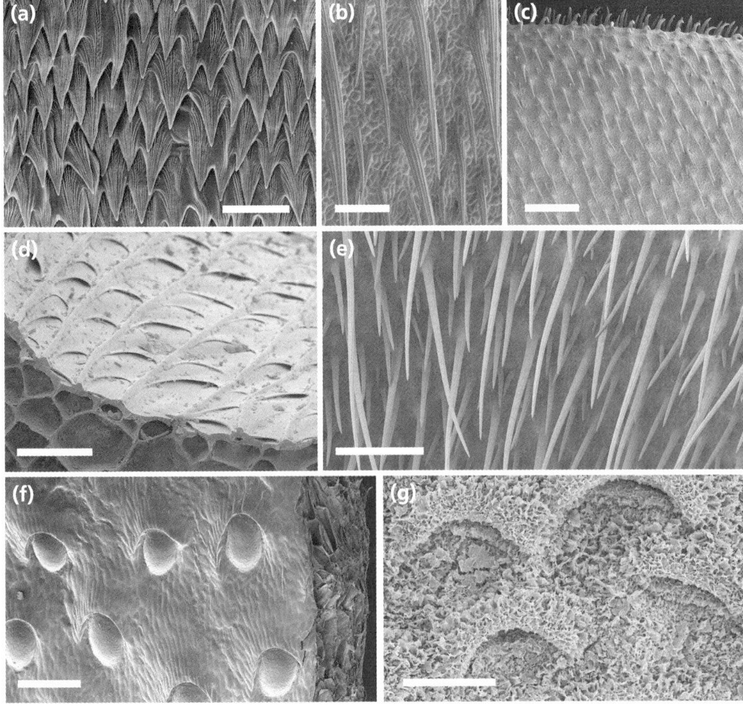

Figure 15.3 Cuticular folds and anisotropic surfaces. (**a**) Imbricate cells on the inner pitcher surface of a *Sarracenia* hybrid are finely striated by cuticular folds (see also Figure 15.2b); bar = 50 μm. (**b**) Cuticular folds on the downward-pointing trichomes of *Heliamphora nutans*; bar = 200 μm. (**c**) The upper part of the *H. nutans* inner pitcher wall is densely covered in short, downward-pointing, highly wettable hairs; bar = 500 μm. (**d**) Cuticular folds forming the small-scale ridges on the *Nepenthes* peristome. This image also shows the tile-like steps formed by overlapping epidermal cells. Both steps and ridges are crucial for the high wettability of the surface; bar = 50 μm. (**e**) Long trichomes in the lower region of the hairy zone of a *H. nutans* pitcher; bar = 500 μm. (**f**) Hooded digestive glands on the inner wall of a *Nepenthes* pitcher, providing a foothold for claws only in downward direction; bar = 200 μm. (**g**) Lunate cells in the waxy zone of a *N. gracilis* pitcher; bar = 20 μm.

down to just above the maximum fluid level that is determined by a small drainage hole in the pitcher wall. The trichome length gradually increases from ≈0.2 mm at the top (Figure 15.3c) to >1.5 mm at the bottom of the hairy section (Figure 15.3e). At the same time, the incline of the wall steepens from top to bottom, approaching vertical in the lower half of the hairy section. Bauer et al. (2013) found that ants could walk relatively safely on the short trichomes of the upper region, but the surface became increasingly treacherous as they moved deeper into the pitcher. Ants invariably fell once they reached the lower half of the hairy section. Experimental studies on the anti-adhesive function of the trichomes of other species of Sarraceniaceae as yet are lacking.

Downward-pointing trichomes with slightly different morphology also are found in the lower part of *Sarracenia* and *Heliamphora* pitchers, where they are permanently submerged in the digestive liquid. The exact function of these hairs is unknown, but they probably play a role in prey retention, either by directly obstructing attempts to climb the pitcher wall or by encouraging disoriented prey to move deeper into the pitcher. The latter effect can commonly be observed in *Nepenthes* pitchers, where the hooded glands of the digestive zone provide unidirectional footholds for claws, causing struggling insects to drag themselves deeper into the pitcher (U. Bauer *unpublished data*).

Nepenthes pitchers not only possess hooded glands (Figure 15.3f)—usually located below the fluid surface but in some species extending all the way up to the peristome—but also characteristic lunate cells within the upper zone of the inner wall where the slippery epicuticular wax crystals are found (Figure 15.3g). The function of these cells is not currently known, but they do not appear to be secretory. The anisotropic structure of the lunate cells allows insect tarsal claws to grip only in a downward direction (Gaume et al. 2002, 2004, Gorb et al. 2004). The unidirectional orientation of the lunate cells may be an adaption for prey retention, but it is unlikely to contribute much to initial prey capture. Knoll (1914) proposed that the projections of the lunate cells may have a jolting effect on insects, causing them to fall away from the wall instead of sliding down the surface. This could make prey more likely to fall backwards onto the fluid surface, thereby improving the retention efficiency of the trap. Systematic experimental studies are needed to confirm this hypothesis.

The secretions of the glandular zone also may aid in retention of at least some prey taxa. Gaume et al. (2002) found that the fresh glandular surface of *N. alata* impeded the locomotion of ants. The surface appeared to be sticky, making it difficult for ants to detach their feet while prompting extensive cleaning behavior. Gorb et al. (2004) measured significantly lower friction forces for *Pyrrhocoris apterus* bugs on fresh relative to dried glandular surfaces of *N.* ×*ventrata*, but found no measurable effect for *Calliphora vicina* blowflies. The glandular surface also has been described as readily wettable by the same authors. More research is needed to investigate whether aquaplaning effects (§15.3.4; Chapter 13, §13.2.3) also play a role on the glandular surface.

15.3.4 Wettable (superhydrophilic) surfaces

The trapping function of fully lubricated, superhydrophilic surfaces is a relatively recent discovery (Bohn and Federle 2004). It is astonishingly effective and has been found in all *Nepenthes* species examined to date and in one species of *Heliamphora* (Bauer et al. 2013). When the *Nepenthes* peristome is dry, it is completely safe for insects to walk on (Lloyd 1942, Gaume et al. 2002), but it becomes extremely slippery as soon as it gets wet. The extraordinary wettability of the peristome (Figure 15.4a) probably results from a combination of a hydrophilic surface composition and an enhancing effect of the complex hierarchical surface topography. Microscopic roughness amplifies the chemical properties of a surface; a familiar example of this effect is the water-repellent, self-cleaning lotus leaf whose properties are based on surface hydrophobicity combined with a hierarchical micropattern (Barthlott and Neinhuis 1997). The opposite effect occurs on the *Nepenthes* peristome: an applied water droplet spreads instantly and forms a thin stable film on the surface. This water film prevents the tarsal adhesive pads of insects from making contact with the underlying surface and causes them to slip, leading Bohn and Federle (2004) to term the

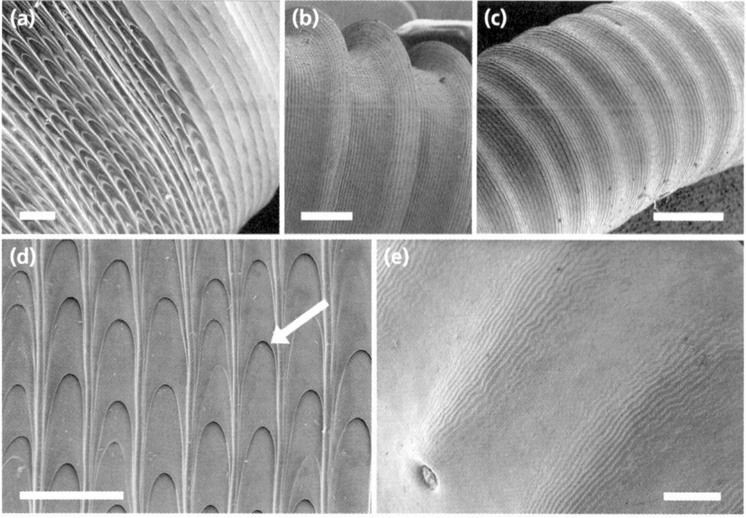

Figure 15.4 Microtopography of the *Nepenthes* peristome. Arrow points toward the interior of the pitcher. (**a**) Environmental scanning electron micrograph of a *N. alata* peristome. The two macroscopic grooves on the far right are covered by a continuous water film; bar = 100 µm. (**b–c**) Typical hierarchical ridge structure. The lower pitcher of *N. lowii* (**b**) shows more acute macroscopic ridges and deeper macroscopic grooves than the *N. muluensis* peristome (**c**); bars = 500 µm. (**d**) Detail of the microscopic ridges with steps (arrow) in between. The steps promote directional water transport from the inside to the outside of the peristome; bar = 50 µm. (**e**) Detail of the peristome of an *N. lowii* upper pitcher, showing only faint remnants of the typical ridge structure. The upper pitchers of this species are not effective insect traps and collect mammalian feces instead; bar = 500 µm.

phenomenon "insect aquaplaning" when they first discovered it in *N. bicalcarata*.

The microstructure of the *Nepenthes* peristome, although to some degree variable between species, is universally characterized by a radial ridge pattern. Most species show hierarchical ridges on two distinct length scales (Figure 15.4b, 15.4c). The larger, macroscopic ridges are visible with the naked eye. Water spreads along these ridges, filling grooves between them before spreading to neighboring grooves. This confines water spreading and may result in partially slippery peristomes when only a few individual drops are applied. Further, it limits the exposed surface area of a given volume of water on the peristome, reducing evaporation rates and prolonging periods of slipperiness. The grooves vary in depth among species and can be shallow and indistinct (e.g., *N. muluensis*; Figure 15.4c) or semicircular in cross-section and delineated by acute ridges (e.g., lower pitchers of *N. lowii*; Figure 15.4b). The effect of different groove cross-sections on wettability and water-spreading speed has not yet been studied, but it is widely accepted that the grooves aid water spreading via capillary effects (Bauer and Federle 2009).

Superimposed on this macroscopic ridge structure is a pattern of microscopic ridges and grooves. The latter are shallow in cross-section and are formed by parallel rows of elongated epidermal cells. The cells overlap like tiles on a roof, creating a series of sloped steps toward the interior of the trap (Figure 15.4d). Similar to lunate cells, these steps render the peristome surface strongly anisotropic, providing good footholds for tarsal claws only when an insect is traveling into the pitcher. Using a series of experimental manipulations of ant feet and subsequent friction force measurements, Bohn and Federle (2004) showed that the directional topography is effective against the claws, whereas the lubricating water layer incapacitates the adhesive pads of the insects. In addition, the steps promote directional wetting, aiding water to spread outward and upward, against gravity, on the peristome. The tile-like arrangement of cells within each groove creates a microcavity at the top end of each cell, overhung by a sharp edge formed by the bottom end of the next cell. In top view, this edge runs roughly semicircularly across the groove before tapering downwards along the adjacent ridges. Water gets pinned at sharp edges, preventing it from flowing to the other side, an effect similar to that seen when overfilling a glass. The same effect hampers downward spreading of water on the peristome. On the way up, however, the water gradually rises to and above the level of the next step while filling the groove and cavity below, allowing it to overflow easily into the next compartment (Chen et al. 2016). In the natural setting of a more or less vertically oriented peristome, this effect is likely to be counterbalanced by gravity and capillary forces because the grooves narrow toward the inner peristome edge. The combination of all three effects probably ensures that the peristome is evenly wetted, independent of where water is deposited.

The microscopic groove pattern appears to be highly conserved across *Nepenthes* species, except for *N. lowii*, for which only the lower pitchers possess a peristome with this hierarchical ridge pattern. The upper pitchers have a much reduced peristome with a largely smooth surface and only faint remnants of a microscopic ridge structure locally above each nectar gland (Figure 15.4e). This reduced peristome is neither wettable nor slippery for insects, and upper pitchers of *N. lowii* have been shown to be specialized for collecting feces of tree shrews (*Tupaia montana*) instead (Clarke et al. 2009; Chapter 13, §13.6; Chapter 21).

The wetness-induced slipperiness of the *Nepenthes* peristome is paralleled in pitchers of the South American *Heliamphora nutans*. Bauer et al. (2013) showed that its hairy inner pitcher wall also is highly wettable, and water spreads easily on the surface. In contrast to *Nepenthes*, the water appears to spread uniformly in all directions. Wetting markedly increases the trapping efficiency of *H. nutans*, particularly in the upper section of the pitcher that provides a secure foothold when dry, but becomes extremely slippery when wet. Superhydrophilicity is an unusual trait for densely hairy surfaces which are often strongly water-repellent. The parallel cuticular folds on the trichomes (Figure 15.3b) of *H. nutans* are similar in dimension to the microgrooves on the *Nepenthes* peristome and are thought to be crucial for its wetting properties.

15.4 Mechanical obstructions

Beside microscopic surface features (§15.3.3), larger-scale mechanical barriers for prey capture and retention also are common in pitfall or eel traps. Examples include the overhanging funnel region of *Cephalotus* (Figure 15.5a), the peristome of *Nepenthes* (Figure 15.5b), and the long trichomes inside *S. psittacina* pitchers (Figure 15.5c). The overhanging geometry and downward-pointing teeth of the *Nepenthes* peristome may aid in the retention of captured prey (e.g., Lloyd 1942, Juniper et al. 1989), but at least ants can cross this barrier freely in both directions if the peristome surface is dry (U. Bauer *unpublished data*).

Long inward-pointing trichomes, commonly present in the retention zones of *Cephalotus, Darlingtonia, Heliamphora* (Figure 15.5d), and *Sarracenia* pitchers, also contribute to prey retention. The pitchers of *Sarracenia psittacina* grow horizontally, and their interior surface is densely covered with such trichomes (Figure 15.5c). These traps have been described as lobster or eel traps (Lloyd 1942, Cheek 1988). Prey can enter the trap by pushing through the hairs, but are unable to reverse direction against their incline. Eel traps are also found in *Genlisea* species. Their subterranean leaves are achlorophyllous and, like roots, provide both anchorage and nutrient absorption from the soil (Adamec 2008c). They also function as traps for attracting, capturing, and digesting prey (Darwin 1875). The general architecture of the trap is homogeneous among the species: an inverted Y-shape, with a terminal digestive vesicle, and a tubular neck and helical arms equipped with rows or funnel-like rings of stiff, unidirectional bristles (Fleischmann 2012a; Chapter 13, §13.5.2). Prey enter the trap through openings at the branching region and along both helical arms. Like the *S. psittacina* hairs, the bristles inside the trap only allow prey to move further inwards, toward the digestive vesicle.

15.5 Ecological implications of wetness-activated trapping mechanisms

The discovery of wetness- or rain-activated trapping mechanisms in *Nepenthes* and *Heliamphora* fundamentally changed our understanding of how these traps work and interact with their

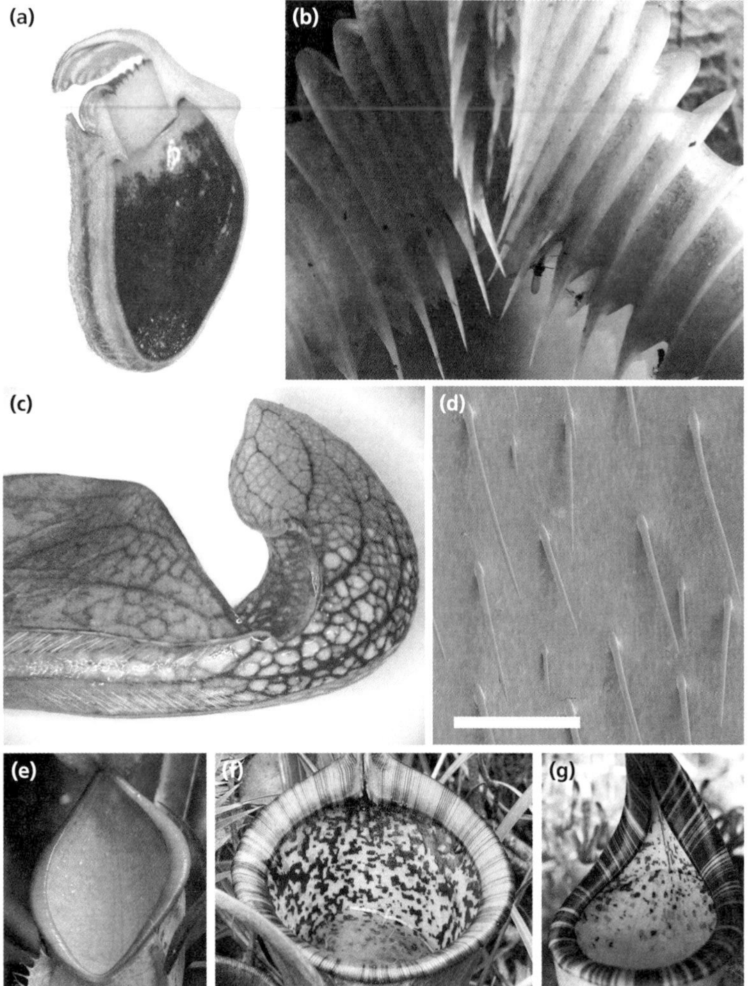

Figure 15.5 Diversity of physical barriers in pitcher traps. (**a**) Cross-section of a *Cephalotus follicularis* pitcher showing the peristome, funnel, and main pitcher body with accumulated prey. (**b**) The overhanging architecture and downward-pointing teeth of many *Nepenthes* peristomes are thought to present an escape barrier for captured prey, as well as preventing theft. (**c**) Eel trap of *Sarracenia psittacina*, with long, inward-pointing hairs inside the narrow tubular pitcher. (**d**) Downward-pointing trichomes in the permanently submerged retentive zone of a *Heliamphora nutans* pitcher; bar = 500 μm. (**e–g**) Variations in the extent of the waxy zone in three species of *Nepenthes*. (**e**) *N. tentaculata* with a waxy layer covering the entire inner pitcher wall above the fluid. (**f**) *N. attenboroughii* free of wax crystals. (**g**) Upper pitcher of *N. fusca* showing a largely wax-free inner wall with a small remnant wax crystal patch in the upper rear section. In lower pitchers of the same species, the entire inner wall above the fluid is covered in wax crystals.

environment. Previously, non-motile traps were believed to be static not only with regards to the absence of movement (but see Chapter 14, §14.3), but also in effectiveness over time. Only long-term changes of capture rates with trap age had been reported. Fish and Hall (1978) observed peak capture rates in *Sarracenia purpurea* pitchers approximately two weeks after pitcher opening. After that, prey capture gradually decreased over the remaining lifespan of the pitcher, but it was unclear whether these changes were caused by changes in attractiveness, trapping efficiency, or both. Bauer et al. (2009) also observed age-dependent changes in prey capture in *Nepenthes rafflesiana*. Prey-capture rates during the first week after pitcher opening coincided not only with an increase in nectar production (attractiveness), but also with increased trapping efficiency due to changes in peristome wettability and resulting slipperiness.

In addition to these gradual changes, however, the wetness-activated slipperiness of both the *Nepenthes* peristome and the *Heliamphora* inner pitcher wall also leads to dramatic short-term variability in trapping efficiency. This has been studied for *N. rafflesiana*, but similar effects likely occur in other, mainly peristome-trapping, *Nepenthes* species, and in other pitcher plants with wettable slippery surfaces that occupy habitats with pronounced short-term variability in ambient humidity.

Nepenthes rafflesiana in Northern Borneo typically inhabits open shrublands and forest edges at low elevations. These habitats experience large diurnal variations of temperature and humidity, and pitcher peristomes generally are dry during the hotter hours of the day. Traps become wet when it rains or when rising humidity leads to condensation on the peristome. Nectar secreted onto the peristome facilitates condensation because its high sugar content renders it strongly hygroscopic (Bauer et al. 2008; Chapter 12, §12.3). During dry periods, the peristome is safe for insects to visit, and provides a rich source of nectar (Bauer et al. 2008). This enables ants, the predominant prey of *N. rafflesiana* (Moran 1996, Adam 1997), to establish foraging trails to the pitchers. Many ant species show a pronounced division of labor with dedicated scouts constantly searching for new food sources. The temporary ineffectiveness of the *N. rafflesiana* peristome ensures that scouts visiting during dry periods can safely sample nectar from the peristome, and return to recruit nest-mates to the trap. When the conditions turn wet later, the established foraging trail ensures that more ants fall prey to the now slippery pitcher (Bauer et al. 2015a; Chapter 12, §12.7).

A similar pattern is expected for the rain-driven lid trapping mechanism of *N. gracilis* (Chapter 14, §14.3) because the prey of *N. gracilis* is as ant-dominated as that of *N. rafflesiana*. In addition, *Nepenthes gracilis* also may benefit from the habit of small and delicate insects to seek shelter from heavy rain on the underside of leaves. A generally poor trap performance (Tan 1997), or the formation of localized slippery patches on partially wetted peristomes (§15.3.4) also may promote scout survival and ant recruitment.

Only the genus *Nepenthes* is known to have considerable interspecific variation in functional trap features. Wax crystals may cover the entire inner pitcher wall above the digestive fluid (Figure 15.5e), the wax layer may be confined to the uppermost section of the pitcher wall (Figure 15.5g), or it may be absent altogether (Figure 15.5f). Peristome size and geometry, and fluid viscoelasticity also are variable. Differences occur not only between species, but often also between upper and lower pitchers of the same species (Gaume and Di Giusto 2009, Bonhomme et al. 2011b, Bauer et al. 2012a, Benz et al. 2012). Moreover, Bauer et al. (2012a) and Bonhomme et al. (2011b) both noted the existence of distinct trapping syndromes with mutually exclusive combinations of morphological or physiological adaptations. For example, enlarged peristomes and viscoelastic fluids are typically found in the absence of wax crystals. Moran et al. (2013) found strong correlations between the trapping syndromes of 94 *Nepenthes* species and climate variables linked to their geographical distribution. Peristome- and viscoelasticity-dependent trapping strategies were confined to perhumid regions, whereas wax crystals dominated in drier and more seasonal areas.

15.6 Future research

Much progress has been made in unraveling trapping mechanisms of non-motile traps in the past decades, but large knowledge gaps still remain. Most research to date on adhesive, flypaper traps has focused on the physiology of mucilage production and the chemical composition of the mucilage. Comparatively few studies have examined the detailed biomechanics of prey capture, or the physical properties of the trap glues. The development of new portable devices for measuring fluid viscoelasticity in the field (Collett et al. 2015) should provide further insights into the rheology of trapping secretions, and force measurements and high-speed video recordings could be used to quantify effects of viscoelastic secretions on insect locomotion and prey retention.

Prey capture mechanisms by plants with eel traps remain speculative. For example, it is still unclear whether *Genlisea* traps are completely passive or whether the plants generate physiologically costly water currents to actively suck up prey (as does *Utricularia*; Chapter 14, §14.2.3). Fleischmann (2012a) comprehensively summarized arguments for both scenarios, visualized trap water currents with ink, and highlighted immobile trap contents such as soil debris and non-ciliate algae. Future experiments, including careful observations of water displacement inside the traps and the targeted inhibition of active water transport processes, are necessary to determine how *Genlisea* traps really work.

A great deal of knowledge has accrued over the past two decades about pitcher-plant traps, but studies have focused almost exclusively on *Nepenthes* (but see Bauer et al. 2013 for comparable work on *Heliamphora nutans*). The hairy surfaces found in pitchers of *Heliamphora*, *Sarracenia*, and *Darlingtonia* are morphologically similar, but circumstantial observations suggest that not all of them are equally wettable. Systematic investigations of surface microstructures and wettability in Sarraceniaceae are needed to fully understand possibly divergent functions. Conversely, striking morphological similarities of *Cephalotus* pitchers, when compared with both *Nepenthes* and *Sarracenia,* strongly suggest evolutionary convergence of trapping mechanisms across all three families of pitcher plants. However, the trapping mechanism of *Cephalotus* pitchers has not yet been studied in any detail, so evolutionary convergence remains but an intriguing hypothesis.

We now have some understanding of various trapping mechanisms, but very little is known about either the evolution or the ontogeny of the underlying structures. The taxonomic and morphological diversity of *Nepenthes* provides an ideal opportunity for comparative morphological and phylogenetic studies, and the few studies of the evolution of trapping mechanisms have focused on this genus. Bonhomme et al. (2011b) investigated the occurrence of wax layers and viscoelastic trap fluids in 23 species and related these traits to the altitudinal distribution of the species. They also performed retention experiments with insects, but unfortunately did not analyze their results in a phylogenetic context. Further studies are needed to establish whether differences in viscoelasticity are partially responsible for variations of prey spectra between ground and aerial pitchers, or between sympatric *Nepenthes* species. The chemical composition underlying the viscoelastic properties of the *Nepenthes* pitcher fluid also remains to be investigated.

Benz et al. (2012) and Bauer et al. (2012a) independently did comparative phylogenetic analyses of *Nepenthes* pitcher traits, including peristome size and shape, and presence and extent of the wax crystal layer on the inner wall. The latter study comprised a total of 60 species and presented strong evidence for the divergent evolution of distinct trapping strategies within the genus. Ancestral state reconstruction suggested that the common ancestor of modern *Nepenthes* species had pitchers with narrow peristomes that were symmetrical in cross-section, and a full-length wax crystal layer. Further comparative studies of the trapping mechanisms and prey spectra of diverse *Nepenthes* species, and also across the Sarraceniaceae, are needed to refine these initial conclusions. Updated comprehensive molecular phylogenies are needed to place the results in a phylogenetic context.

Hardly anything is known about the ontogeny of trapping surfaces. The few ontogenetic studies on *Cephalotus* and *Nepenthes* focused on macroscopic pitcher development (Arber 1941, Froebe and Baur 1988, Owen and Lennon 1999) or physiological and trapping-related traits of already opened pitchers (Bauer et al. 2009). More studies are needed to unravel the development of the complex directional surface structures such as the peristome micropattern inside developing pitcher buds, and the genetic underpinnings controlling these developmental processes. Our understanding of the development of the epicuticular wax crystal layers is similarly limited. The mechanisms of wax biosynthesis in epidermal cells are reasonably well understood for the genetic model plant species *Arabidopsis thaliana* (Samuels et al. 2008), but the function and regulation of the biochemical machinery that produces the specific wax compounds involved in trapping mechanisms of carnivorous plants remain unknown.

The study of diverse interactions between carnivorous plants and their prey organisms, and with the environment, is fascinating and has prompted a number of astonishing discoveries over the past two decades. In recent years, engineers have increasingly taken interest in anti-adhesive plant surfaces as an inspiration for the development of self-cleaning, self-repairing, or non-sticky surfaces and paints (e.g., Wong et al. 2011, Barthlott et al. 2017; Chapter 20). Future advances in the study of surface development and genetic determination of surface patterns could open up novel approaches to pesticide-free, mechanical crop protection, and the viscoelastic trap glues of Droseraceae, Lentibulariaceae, *Triphyophyllum*, and *Roridula* may inspire the development of environmentally friendly, sugar- or resin-based elastic adhesives. The increased scientific interest in plant biomechanics and the continuing discovery of new trapping mechanisms and new species of carnivorous plants likely will provide further exciting opportunities for cross-disciplinary research and tantalizing practical applications.

CHAPTER 16

Biochemistry of prey digestion and nutrient absorption

Ildikó Matušíková, Andrej Pavlovič, and Tanya Renner

16.1 Introduction

The carcasses of captured prey have to be decomposed before carnivorous plants can take up and use the essential nutrients bound within them. Previous studies on the biochemistry of prey digestion were restricted to analyses of typical lytic enzymes (Juniper et al. 1989). In the last 20 years, the endogenous production of many digestive enzymes has been demonstrated, corresponding genes have been isolated, and products analyzed (Adlassnig et al. 2011). More comprehensive views of endocrine biology combine biochemistry with ultrastructural studies (e.g., Bemm et al. 2016). Most recently, whole genome analyses (Chapter 11), proteomics, and transcriptomics have broadened our knowledge of carnivorous plant biochemistry still further.

The evolution of botanical carnivory reflects metabolic tradeoffs (Chapter 18), and exploitation of adaptable plant morphology and anatomy (Chapters 13, 15) under particular ecophysiological conditions (Renner and Specht 2013, Pavlovič and Saganová 2015). In this chapter, we review general mechanisms of prey digestion that appear to be conserved across genera, and features specific to individual genera of carnivorous lineages within the Caryophyllales (*Drosera, Dionaea, Nepenthes*) that may contribute to the success and efficiency of carnivory. Finally, we summarize available evidence addressing the hypothesis that the proteins involved in prey decomposition evolved from proteins related to defense against herbivores and pathogens.

16.2 Composition of the digestive fluid

Specialized multicellular secretion glands produce fluids that smother, kill, or digest prey, and provide a solution for assimilation of released nutrients. Saccharides in, and viscoelasticity of, secretions also retain prey captured by some pitcher plants and all carnivorous plants with adhesive traps (Gaume and Forterre 2007; Chapters 13–15). The anion and element composition of the fluid is poor in nutrients and consists mainly of ions K^+, Na^+, Ca^{2+}, and Mg^{2+} (e.g., Buch et al. 2013 for *Nepenthes*), but the properties and the protein content differ among genera. After prey capture and the release of nitrogenous compounds, the fluid begins to acidify (Gallie and Chang 1997, An et al. 2001). Early findings identified HCl as the source of acidification (Juniper et al. 1989), the secretion of which is regulated by a membrane proton H^+-ATPase in *Nepenthes* (An et al. 2001).

Soon after prey capture, digestive glands secrete different naphthoquinone derivatives, such as plumbagin or droserone. These metabolites come into contact with electron-transferring flavin enzymes including NAD(P)H-dehydrogenases or oxidases (diaphorases) on the cell membrane of the prey, and generate reactive oxygen species that precondition the carcass for proteolytic degradation (Galek et al. 1990, Chia et al. 2004, Eilenberg et al. 2010). Peroxidases also are likely to be involved; nine different enzymes were detected in digestive fluid of *Dionaea* (Schulze et al. 2012), and others in pitchers of five *Nepenthes* species (Lee et al. 2016, Rottloff et al. 2016). To avoid detrimental

Matušíková, I., Pavlovič, A., and Renner, T., *Biochemistry of prey digestion and nutrient absorption*. In: *Carnivorous Plants: Physiology, ecology, and evolution*. Edited by Aaron M. Ellison and Lubomír Adamec: Oxford University Press (2018).
 DOI: 10.1093/oso/9780198779841.003.0016

effects of oxidation, traps actively produce oxidative scavengers and inhibitors of programmed cell death (Bemm et al. 2016).

The major mass of the prey body is decomposed in the traps by a cocktail of hydrolytic enzymes (Table 16.1), which either are stored in sub-cellular compartments and released following stimulation by prey or are synthesized *de novo* when needed. Despite numerous cytochemical studies (reviewed in Juniper et al. 1989), these enzymes were considered to be of microbial origin. Indeed, both microbe-derived enzymes and autolytic ones from the prey itself play important roles in recycling their nutrients, especially for *Byblis, Brocchinia, Catopsis, Sarracenia, Darlingtonia, Heliamphora*, and *Utricularia*, all of which produce few if any digestive enzymes themselves (Adlassnig et al. 2011). More recently, plant-governed hydrolytic decay was demonstrated using studies on sterile carnivorous plants. Enzymes either are already present in prey-free traps (e.g., *Nepenthes, Utricularia*) or are secreted in response to stimuli from captured prey (e.g., *Dionaea*). Ultrastructure studies revealed that the endocrine system of the *Dionaea* trap consists of three layers of cells with tripartite functional morphology; the outer two layers produce and secrete hydrolases (Bemm et al. 2016).

Proteins of digestive fluid have been studied using high-throughput techniques applied to *Dionaea* (Schulze et al. 2012, Bemm et al. 2016) and several species of *Nepenthes* (Hatano and Hamada 2012, Rottlof et al. 2016). The proteins described from these species to date vary with the chemical nature of prey (Matušíková et al. 2005, Pavlovič et al. 2014, Lee et al. 2016) and trap ontogeny (Biteau et al. 2013). Proteases have been studied most extensively, but chitinases that partly destroy the exoskeleton of insects, phosphatases, and nucleases also have been described (§16.2.2–16.2.5). Recent molecular tools have enabled the identification of proteins that are less abundant in digestive fluid, but their role in prey digestion remains hypothetical or unknown. For example, Rottloff et al. (2016) described two basic lipid transfer proteins (NmLTP1 and NmLTP2) in *Nepenthes* sterile secretion fluid, whereas Bemm et al. (2016) identified a lipid transfer protein in *Dionaea* traps that may increase permeability of prey membranes or assist lipases in decomposing lipoproteins and wax, thereby increasing availability of the inner parts of prey and facilitating uptake of metabolites. Thaumatin-like proteins with similar hypothesized functions have also been described (Rottloff et al. 2016, Fukushima et al. 2017, Krausko et al. 2017).

16.2.1 Proteases

Hydrolysis of proteins, both inside (endopeptidase) and at the ends (exochitinase) of peptide chains, has been described repeatedly within traps (Juniper et al. 1989). As for other digestive enzymes, the origin of these proteases—endogenously by the plant itself or exogenously by bacteria or other organisms living within the traps (Chapters 23–26)—has been controversial (Juniper et al. 1989). Some of the proteolytic activity in digestive fluid has been assigned to bacteria, but molecular evidence now supports strongly the involvement of plant-derived proteases in digestion.

Athauda et al. (2004) were the first to purify, characterize, and sequence two aspartic proteases (nepenthesin I and II) from the pitcher fluid of *N. distillatoria*. As with previous studies done on crude pitcher fluid or partially purified enzymes, they found a pH optimum of 2–3, temperature optimum at 50–60 °C and complete inhibition by pepstatin, indicating that the enzymes belong to the typical pepstatin-sensitive aspartic proteases. Nepenthesin I has six potential N-glycosylation sites and the presence of carbohydrate may make nepenthesin I more stable than nepenthesin II; indeed, temperature stability of non-glycosylated recombinant nepenthesin I is lower than of the natural enzyme (Kadek et al. 2014b). The complete amino acid sequences of nepenthesin I and II were deduced by cloning and nucleotide sequencing of cDNA clones obtained from *N. gracilis* (Athauda et al. 2004). Nepenthesins contain a signal sequence for secretion and are expressed as zymogens capable of autoactivation in an acidic environment by cleavage of the pro-sequence. Nepenthesin I and II lack the plant-specific insertion typical of plant vacuolar aspartic proteases, but have an approximately 22-residue insertion, which contains four cysteine residues and was named the nepenthesin-type AP specific insertion (NAP). Recently, Lee et al. (2016) identified three new nepenthesins (III–V) sharing these characteristics.

Table 16.1 Enzymes of plant origin detected in the digestive fluid of carnivorous plants (—indicates unstudied for that genus; if a genus is not mentioned then no data currently are available). Results of histochemical and transcriptomic analyses were not included, as further evidence is required to determine relevance to digestive processes.

Genus	Nuclease	Protease	Phosphatase	Chitinase	Other carbohydrate-hydrolyzing enzymes	Other enzyme types	References*
Caryophyllales							
Dionaea	S1/P1 Nuclease 1,2 RNase T2 (RNS1) DNase	Cysteine protease C1A (SAG12) Aspartic protease 1,2 Serine carboxypeptidase S10, 49 (SCPL49)	Acid phosphatase PAP20, PAP27	Chitinase ATEP3 Chitinase VF1	β-1,3-Glucanase (BGL2) Thioglucosidase Thaumatin-like protein	Peptide-N4-asparagin amidase A Osmotin-like protein Pathogenesis-related protein Lipid transfer protein Peroxidase LysM-containing protein	1–7
Drosera	DNase RNase	Cysteine protease Aspartic protease	Acid phosphatase	Chitinase class I	β-1,3-glucanase α-Galactosidase Thaumatin-like protein	Esterase Lipid transfer protein 1,2	8–15
Drosophyllum	—	—	Phosphatase	—	—	Peroxidase	10
Nepenthes	RNase Endonuclease ENDO2	Aspartic protease Serine carboxypeptidase III Serine carboxypeptidase S10 Prolyl endoprotease Serine carboxypeptidase-like 204,751	Protein phosphatase C2 (NmPP1) Serine/threonine phosphatase (NmPP1)	Chitinase class I, III, IV	α-Glucosidase β-Glucosidase 7-like protein β-1,3-Glucanase 1,2 α-Galactosidase 1,2 β-Galactosidase 1–3 β-Xylosidase Thaumatin-like protein	Peroxidase 27 Peroxidase N Lipid transfer protein 1, 2 Cationic peroxidase 1 Esterase/lipase PR-1 protein	16–20
Ericales							
Sarracenia	RNase DNase	Aspartic protease	Neutral phosphatase Acid phosphatase	Chitinase GH18	β-1,3-Glucanase Thaumatin-like protein		15, 21
Lamiales							
Utricularia	—	Aminopeptidases	Phosphatase	β–Hexosaminidase (chitinase)	α-Galactosidase β-Galactosidase	—	22
Oxalidales							
Cephalotus	RNase T2	Aspartic protease	Purple acid phosphatase	Chitinase GH18,19	β-1,3-Glucanase Thaumatin-like protein	Class III peroxidase	2, 7, 15

*References. 1. Schulze et al. (2012); 2. Takahashi et al. (2009); 3. Takahashi et al. (2011); 4. Paszota et al. (2014); 5. Libiaková et al. (2014); 6. Bemm et al. (2016); 7. Nishimura et al. (2013); 8. Okabe et al. (1997); 9. Okabe et al. (2005b); 10. Juniper et al. (1989); 11. Matušíková et al. (2005); 12. Takahashi et al. (2012); 13. Michalko et al. (2013); 14. Krausko et al. (2017); 15. Fukushima et al. (2017); 16. Hatano and Hamada (2008); 17. Rottloff et al. (2011); 18. Hatano and Hamada (2012); 19. Lee et al. (2016); 20. Rottloff et al. (2016), 21. Gallie and Chang (1997); 22. Sirová et al. (2003).

The amino acid sequence identity of nepenthesins with ordinary aspartic proteases is remarkably low (≈20%). Phylogenetic analyses have revealed that nepenthesins form a novel subfamily of aspartic proteases characterized by high cysteine residue content and the NAP insertion (Athauda et al. 2004). Based on inhibitory studies, Takahashi et al. (2009) described aspartic protease in the digestive fluid of two carnivorous Caryophyllales (*Dionaea, Drosera capensis*), and of *Cephalotus follicularis*, that inhibits proteolytic activity of oxidized insulin B chain at low concentrations of pepstatin. Following earlier nomenclature, these were described as dionaeasin, droserasin, and cephalotusin, respectively (Takahashi et al. 2009, 2012). Aspartic proteases in *Cephalotus* recently were confirmed by comparative proteomics (Fukushima et al. 2017). Schulze et al. (2012) obtained complete amino acid sequences of two dionaeasins (dionaeasin-1 and dionaeasin-2) from *Dionaea*. Both dionaeasins contain an active site residue, signal sequence, propeptide, and NAP specific insertion similar to nepenthesin (Schulze et al. 2012). Based on partial amino acid sequence and homology searches, droserasin was identified in the digestive fluid of *D. capensis* (Krausko et al. 2017), and several additional aspartic proteases were annotated in its genome (Butts et al. 2016a). Aspartic protease also was recently identified in *Sarracenia purpurea* (Fukushima et al. 2017).

In contrast, the major endopeptidase in the digestive fluid of *Dionaea* differs from nepenthesin. Robins and Juniper (1980) suggested it was a papain-like enzyme; over 30 years later, Takahashi et al. (2011) obtained its partial amino acid sequence, which is homologous with cysteine peptidases in the papain family of peptidases (hence its name, "dionain"). Cysteine peptidases share a common catalytic mechanism involving a nucleophilic cysteine thiol in a catalytic triad or dyad. The activity of dionain is inhibited by the cysteine peptidase E-64 inhibitor and optimal activity is achieved at pH 5.4–6.0 (Risør et al. 2016). Schulze et al. (2012) obtained full amino acid sequences of four dionains (named dionain1–4). The alignment with papain from papaya illustrates that its reactive site residues and cysteine residues that form disulfide bonds in papain are conserved dionains. A conserved motif for the pH-dependent autoactivation of cysteine proteases is present in dionain, suggesting that autoactivation is a part of the dionain activation mechanism at low pH (≈3.6; Risør et al. 2016). The active site of the zymogen is sterically blocked by a reverse-oriented substrate-like propeptide and its catalytic activity is gained upon acidification by the loosening of pro-domain contacts. The catalytic activity is directed toward proteins (e.g., myosins) that have a high content of basic polar amino acids (lysine and arginine with high nitrogen content), facilitating uptake of essential nitrogen compounds (Risør et al. 2016). Crystal structures of nepenthesin-1 and dionain-1 have been resolved to 2.8 Å and 1.5Å, respectively (Fejfarová et al. 2016, Lee et al. 2016, Risør et al. 2016).

Cysteine proteases often act cooperatively with aspartic proteases. Based on inhibitory studies, Takahashi et al. (2012) predicted the occurrence of cysteine protease in *D. indica*. Krausko et al. (2017) identified partial amino acid sequences for at least three cysteine proteases in *D. capensis* and Fukushima et al. (2017) did likewise for *D. adelae*, indicating that *Drosera* uses both aspartic and cysteine peptidases in digestion. A survey of the *D. capensis* genome expanded the list of enzymes to seven aspartic proteases and four cysteine proteases (Butts et al. 2016a, 2016b). Stephenson and Hogan (2006) cloned and characterized a cysteine proteinase from a *N. ventricosa* pitcher. Although they found inhibition of proteolytic activity by the cysteine proteinase inhibitor E-64 *in vitro*, cysteine proteinase has never been identified in *Nepenthes* digestive fluid (Hatano and Hamada 2008, 2012, Lee et al. 2016, Rottloff et al. 2016). Lee et al. (2016) identified but did not fully characterize a prolyl-endoprotease (neprosin) in *Nepenthes* digestive fluid.

Exopeptidases catalyze the cleavage of terminal or penultimate peptide bonds, releasing a single amino acid or dipeptide from the peptide chain. Depending on whether the amino acid is released from the amino or the carboxy terminal, exopeptidases are classified further as aminopeptidases or carboxypeptidases, respectively; both types were detected in *Nepenthes* and *Dionaea* (Robins and Juniper 1980). Carboxypeptidases usually are classified into one of several families, depending on whether they contain metal, serine, or cysteine residue in the active site. Serine carboxypeptidase (SCP) is relatively abundant in *Dionaea* digestive fluid, and sequence analysis revealed that it belongs to the S10 family of serine proteases, which are active only at low pH (Schulze et al. 2012). Four putative serine carboxypeptidases (NmSCP3, NmSCP20, NmSCP47, and NmSCP51) were identified in digestive fluid of *Nepenthes,* all belonging to the S10 family of serine proteases. The molecular masses of NmSCP3, NmSCP20, and NmSCP47 range from 52.4 to 55.3 kDa, and the proteins themselves contain several glycosylation sites (Rottloff et al. 2016). The catalytic activity of SCPs is induced through a charge-relay system involving a catalytic triad composed of an aspartic acid residue linked to a histidine and a serine.

Several proteolytic activities have been detected but molecular analysis has confirmed the presence of only three main endopeptidases in digestive fluid: aspartic peptidase, cysteine peptidase, and prolyl peptidase; and one exopeptidase, serine carboxypeptidase. The proteases can act cooperatively in strongly acidic (aspartic proteases) and less acidic (cysteine proteases) conditions. As the pH of the digestive fluid changes over time, different enzymes might have different optimal activities.

16.2.2 Phosphatases

Phosphatases (phosphomonoesterases) are common plant enzymes with low substrate specificity that catalyze the hydrolysis of phosphate esters. They are produced in significant quantities to mobilize phosphate from prey carcasses (Juniper et al. 1989). Płachno et al. (2006) used fluorescence labeling to study phosphatase activity in digestive glands of 47 carnivorous plant species, mostly Lentibulariaceae. Phosphatase production in all types of glands all over the traps appeared as a common feature within *Genlisea* and aquatic *Utricularia*. Phosphatases are the most abundant, constitutively present enzyme in *Utricularia* traps (Sirová et al. 2003, Płachno et al. 2006) and are involved in prey hydrolysis, ion transport, and pumping water out of the traps (Płachno et al. 2006).

Phosphatase activity also is present in different types of glandular and epidermal cells of *Aldrovanda, Cephalotus, Dionaea, Drosophyllum, Drosera,* and *Nepenthes* (Płachno et al. 2006, 2009a, Pavlovič et al. 2014). Phosphatase production also has been found in hydathodes and in secretory cells within the inflorescence of *Pinguicula*, suggesting the presence of an additional carnivorous organ in some species (Płachno et al. 2006).

Phosphatases contribute to digestion of nematodes by *Philcoxia minensis* (Pereira et al. 2012), and high phosphatase activity occurs in the glands of *Byblis liniflora* and *Roridula gorgonias* (Płachno et al. 2006). The invariably high phosphatase activity found in traps in various carnivorous plant taxa is consistent with high efficiency of P uptake estimated from model prey (Adamec 2002, Płachno et al. 2009a, Pavlovič et al. 2014). Although there is some phosphatase activity in the leaves of *Ibicella lutea, Proboscidea parviflora,* and *Cleome droserifolia*, the released nutrients are not absorbed by these apparently noncarnivorous plants (Płachno et al. 2009a).

Molecular studies of phosphatases are rare. The acidic, possibly metal-coordinated serine/threonine phosphatase NmPP1 of *N. mirabilis* and the nucleotide pyrophosphatase/phosphodiesterase NmNPP1 of *N. sanguinea* are the first putative phosphatases identified for any *Nepenthes* species (Rottloff et al. 2016). Acid phosphatases also were detected for *Dionaea* using transcriptome annotation (Schulze et al. 2012), and through a combination of profile-based single peptide annotation and high-throughput screening proteomics (Bemm et al. 2016). Fukushima et al. (2017) found phosphatases in digestive fluid of *Cephalotus* and *N. alata* using comparative genomic and proteomic analyses.

16.2.3 Chitinases

Chitinases break down the chitin-rich exoskeleton of arthropod prey, releasing bound nitrogen and allowing other digestive enzymes to penetrate and degrade internal tissues. Chitinase gene families are organized into five classes based on sequence similarities and underlying homology. Classes I, II, and IV (family GH19) share a homologous catalytic domain, whereas classes III and V (family GH18) are more similar to fungal and bacterial chitinases than to plant chitinases and have additional lysozyme activity (Grover 2012). Class I chitinases are divided further into two subclasses: subclass Ia with a carboxyl terminal extension (CTE) that targets the chitinase to the plant vacuole; and subclass Ib without a CTE that renders the chitinase extracellular (Neuhaus et al. 1991). Early studies showed that the concentrated trap fluid from *Nepenthes* and *D. peltata* digests colloidal chitin, but the plant origin of the acting enzymes was doubted (Juniper et al. 1989). Similarly, chitinase activity of field-grown *D. whittakeri* and *D. binata* was attributed to microorganisms or insects. Involvement of plant chitinases in prey digestion first was demonstrated for *D. rotundifolia* grown *in vitro* (Matušíková et al. 2005) and subsequently in the sterile liquid of closed *N. khasiana* pitchers (Eilenberg et al. 2006). Since the 1980s, ten chitinases have been identified and characterized to some extent in digestive fluid of *Nepenthes* spp. and *Dionaea*. Stimulation by prey induces glandular secretion of VF chitinase-I of *Dionaea*. This chitinase is one of the most abundant proteins in its digestive fluid during the first three days of prey digestion (Paszota et al. 2014, Bemm et al. 2016). Functionality of this enzyme in the proteolytically aggressive environment is facilitated by high proline content in the exposed loop regions and many glycosylation sites. Structurally, VF chitinase-I belongs to the class I family of chitinases that plays a role in defense against pathogens in noncarnivorous plants. It probably is assisted by two other chitinases that are less abundant and might have different but complementary chitinolytic functions (Schulze et al. 2012).

Multiple but structurally different chitinases also act in *Nepenthes* (Eilenberg et al. 2006, Lee et al. 2016). Eilenberg et al. (2006) isolated four class I chitinases (Nkchit1b-1, Nkchit1b-2, Nkchit2b-1 and Nkchit2b-2) from *Nepenthes* that differed in their gene exon-intron structure. The two Nkchit1b genes share exon preservation and a proline-rich region with eight putative glycosylation sites. They are active only in the secretory regions of the pitcher after chitin induction. In contrast, the vacuolar Nkchit2b, similar to the three different chitinases in vegetative tissues of *Drosera* (Libantová et al. 2009), is expressed constitutively with a possible role in protection against pathogens or in morpho-physiological processes. Class III (Hatano and Hamada 2008, 2012, Rottloff et al. 2011, 2016, Ishisaki et al. 2012a, Lee et al. 2016) and class IV chitinases from *Nepenthes* (Hatano and Hamada 2008, Ishisaki et al. 2012b, Lee et al. 2016, Rottloff et al. 2016) also are well characterized. The structural features of chitinase genes are tied closely to particular functions and have served as evidence for the evolution of the carnivorous syndrome (§16.4; Chapter 1).

16.2.4 Nucleases

Ribonucleases (RNases) and deoxyribonucleases (DNases) release nitrogen from nucleic acids in prey. Ribonucleases first were detected and enzymatically characterized in the field from open pitchers of *N. gracilis*, *N. ampullaria*, and *N. rafflesiana*, and later from the transverse walls of the glands of *N. khasiana* and *Pinguicula* (Juniper et al. 1989). All RNases characterized to date belong to the so-called S-like ribonucleases that are induced by other plants in response to phosphate starvation, wounding, or senescence. However, they are hypothesized also to have evolved as a constitutive expression in carnivorous traps as an adaptation to nutrient-poor habitats (Okabe et al. 2005b). The sticky digestive fluids of *Cephalotus*, *Dionaea*, and *Drosera adelae* contain an abundant amount of the S-like RNases CF-1, DM-1 (Nishimura et al. 2013), and DA-1 (Okabe et al. 2005a, Nishimura et al. 2013). In the secretome of *Dionaea*, additional nucleases were induced by insect stimulation (Bemm et al. 2016).

Recombinant protein analyses were made for DM-1, CF-1, *Aldrovanda* AV-1, *N. bicalcarata* NB-1, and *S. leucophylla* SL-1 (Nishimura et al. 2014), and for native DA-1 (Okabe et al. 2005a). These enzymes are acidophilic, with a pH optimum of 3.5–4.5. They also are insensitive to EDTA chelator and share a similar optimum temperature, although NB-1 also tolerates high temperatures. In *Aldrovanda* and *S. leucophylla*, the nuclease enzyme is probably not secreted into the trap (Adlassnig et al. 2011, Nishimura et al. 2014). The structures of S-like RNases isolated from carnivorous plants have conserved amino acid residues at positions 85 (Asn), 111 (Ser), 115 (Ser), and 135 (Glu) (Nishimura et al. 2014). Kinetic assays of the recombinant proteins suggest that these residues function in maintaining the histidine residues at amino acid positions 66 and 124 that are used in RNA cleavage and maintenance of the conserved active site of the enzyme in the most functional conformation (Nishimura et al. 2014). The native DA-1 digests poly(A), poly(I), poly(U), but not poly(C) (Okabe et al. 2005b). Protein structure, optimum pH values, and the kinetic curves of these enzymes all appear to be more closely tied to functional similarities of insect trapping and digestion than to deep homology. Sophisticated mechanisms regulate activity of these RNases (§16.3).

Fewer data exist for DNases. High enzyme activity in *Dionaea* was detected after three days of secretion, with maximum activity at pH 4.5–6.0 (Juniper et al. 1989). Another endonuclease was identified and partially purified by column chromatography from *D. adelae*. This enzyme cuts double-stranded DNA at pH 3.5–5.0 and requires divalent cations for maximum activity (Okabe et al. 1997).

16.2.5 Carbohydrate-digesting enzymes

Additional nutrient sources may include glucan-rich airborne organic particles from pollen grains, fungal spores, seeds, or plant detritus. Direct uptake of prey-derived organic carbon is of crucial importance for aquatic carnivores when CO_2 and light are limiting (Adamec 1997b). A β-1,3-glucanase in tentacle glue of *D. rotundifolia* degrades laminarin, a β-(1–3)-glucan polymer with β-(1–6)-branches, and the products are taken up by the trap tissue (Michalko et al. 2013). β-1,3-glucanases also were identified in *Cephalotus, N. alata*, and *S. purpurea* (Fukushima et al. 2017). Other putative glucanases (NkGluc1 and NaGluc2) have been detected from different species of *Nepenthes* (Hatano and Hamada 2008, 2012, Rottloff et al. 2016) and *Dionaea* (Schulze et al. 2012, Bemm et al. 2016). The observed glucan-hydrolase activity in secretions also might result from action of other enzymes like the thaumatin-like protein (TLP), a sweet-tasting protein originally regarded as a salt-induced component in osmotically stressed plant cells. TLPs have been identified in digestive fluid of *N. mirabilis, N. alata, N. gracilis, N. sanguinea* (Hatano and Hamada, 2008, Rottloff et al. 2011, Fukushima et al. 2017), *Dionaea* (Schulze et al. 2012), *Cephalotus, Sarracenia* (Fukushima et al. 2017), *D. capensis*, and *D. adelae* (Fukushima et al. 2017, Krausko et al. 2017).

Hatano and Hamada (2012) assembled molecular evidence from *Nepenthes* digestive fluids for the presence of other carbohydrate-hydrolyzing enzymes, including α- and β-glucosidases (NmAGluc1, NmBGluc7), α- and β-galactosidases (NmAG1-2, NmBG1-3), and β-D-xylosidase. The role of these enzymes in traps is unclear. In addition to facilitating digestion and absorption of nutrients, they might adjust the saccharide composition and thereby the sweetness, viscoelasticity, or retention capacity of the digestive glue (Gaume and Forterre 2007). The involvement in decomposition or assimilation of sugars, however, rarely has been discussed except for a few works on *Nepenthes* and *Aldrovanda* (reviewed by Juniper et al. 1989), *Utricularia* (Sirová et al. 2003), and *D. capensis* (Adlassnig et al. 2012).

16.3 Regulation of enzyme release and activity in traps

Lytic enzymes give rise to a mixture of degradation products. Enzyme release is tightly coupled with processes of nutrient absorption, and transporters contributing to these processes have recently been isolated and characterized for *Dionaea* and *Nepenthes* (Chapter 17). Because terrestrial carnivorous plants grow in nutrient-poor soil and their photosynthesis is not very efficient, the production of enzymes likely represents a significant "cost" (Pavlovič and Saganová 2015). Regulation of digestive enzyme activity may help to reduce these costs, and both electrical and hormonal networks are involved in regulating their production. A huge remodeling of the "transportome" in prey-stimulated glands of *Dionaea* involves mostly proteins on the plasma membrane that transport metallic cations, nitrogen, sulfur, phosphate, and different cations, amino acids, peptides, and nucleotides (Bemm et al. 2016). Many of these transporters also are expressed in roots.

16.3.1 Enzyme induction

In the traps of *Dionaea* and *Aldrovanda*, the digestive process is fully inducible. Digestive fluid is not produced before the plant receives a stimulus indicating that prey has been captured. When such a stimulus is perceived, a temporary digestive cavity is formed (Juniper et al. 1989). *Dionaea* uses electrical signals not only for trap movement but also for induction of enzyme secretion. Mechanical stimulation of trigger hairs on *Dionaea* generates action potentials (APs; Chapter 14). After rapid closure secures the insect prey, the struggling of the entrapped prey in the closed trap results in a generation of further APs, which cease when the prey stops moving. This results in further closure and tightening of the trap into a tightly appressed ("narrowed") state and secretion of digestive fluid (Libiaková et al. 2014). Gland cells that receive a series of at least three APs increase their cytoplasmic calcium level (Escalante-Pérez et al. 2011). Escalante-Pérez et al. (2011) also hypothesized that a jasmonate phytohormone *cis*-12-oxophytodienoic acid (OPDA) is responsible for secretion of digestive fluid, and Böhm et al. (2016) showed that a jasmonic acid (JA) signaling pathway is activated after the second AP and mechanical stimulation-induced accumulation of JA and OPDA in the trap (Escalante-Pérez et al. 2011, Libiaková et al. 2014). After > 5 APs, this JA signaling pathway activates transcription of genes responsible for prey digestion (dionain, serine carboxypeptidase, and chitinase) and nutrient absorption (sodium channel); expression is proportional to the number of mechanical stimuli and APs (Böhm et al. 2016).

All of these observations are consistent with immunodetection of dionain in digestive fluid after mechanostimulation of trigger hairs, indicating that regulation occurs at the level of transcription (Libiaková et al. 2014). But the fact that the trap remains closed for several days with dead prey, which does not elicitate any APs, indicates that other stimuli from prey also must be sensed (Darwin 1875). Chemical stimuli from digested prey can induce accumulation of JA, its isoleucine conjugate (JA-Ile), and OPDA, which are correlated with abundance of dionain in digestive fluid (Libiaková et al. 2014). A possible link between chemical stimuli and high jasmonate concentration is depolarization of the membrane potential resulting from ion uptake in digestive glands (Scherzer et al. 2015). Indeed, Böhm et al. (2016) reported a correlation between membrane depolarization and the relative expression level of sodium channels by *Dionaea*.

Nishimura et al. (2013) documented inducible expression of S-like ribonucleases (DM-1) in response to mechanical and protein stimulation and they suggested the presence of a specific factor—likely a jasmonate—that triggers expression. Expression of VF chitinase–I markedly increases in response to chitin. In noncarnivorous plants, chitin sensing often involves receptor-like kinases such as LysM, and a receptor-like kinase is up-regulated by *Dionaea* in response to insect stimulation (Bemm et al. 2016). Thus, *Dionaea* has at least three sensing mechanisms that help to optimize enzyme production: mechanostimulation; stimulation by inorganic ions; and chitin stimulation (Figure 16.1).

16.3.2 Combinations of constitutive and inducible production of enzymes

In contrast to *Dionaea*, pitcher plants maintain permanent levels of digestive fluid and sticky-trapped plants produce mucilage secretion without prey stimuli. However, insect stimuli induce additional secretion of pitcher fluid (Juniper et al. 1989) and mucilage secretions of *Drosera capensis* produce few enzymes in its resting state. By delivering a mechanical stimulus to the marginal tentacle, a series of APs are triggered and are propagated to its base (Krausko et al. 2017). Within the first two hours, low levels of jasmonates accumulate and trigger the secretion of phosphatases and proteases (Nakamura et al. 2013, Mithöfer et al. 2014, Krausko et al. 2017). The response is much stronger when chemical signals (e.g., ammonium) from prey are detected (Pavlovič et al. 2014, Krausko et al. 2017).

Chitinase activity markedly increases in *D. rotundifolia* leaf exudates upon application of chitin, but not proteins (Matušíková et al. 2005). In contrast to *Dionaea*, however, the S-like RNase of *D. adelae* (DA-1) appears to be constitutively expressed, but only in glandular tentacles. An unmethylated promoter in the tentacles is responsible for this tissue specificity (Okabe et al. 2005b, Nishimura et al. 2013). Nishimura et al. (2013) also documented constitutive expression of the S-like RNase (CF-1) gene in *Cephalotus*. Its expression is restricted to digestive glands and is triggered by an unknown transcription factor not present in normal photosynthesizing leaves. Regulatory mechanisms of DA-1, CF-1, DM-1 differ from one another and are uncorrelated with species-level phylogenetic relationships.

Nepenthes also maintains a permanent level of digestive fluid and closed, developing pitchers have enzymes present in digestive fluid (Eilenberg

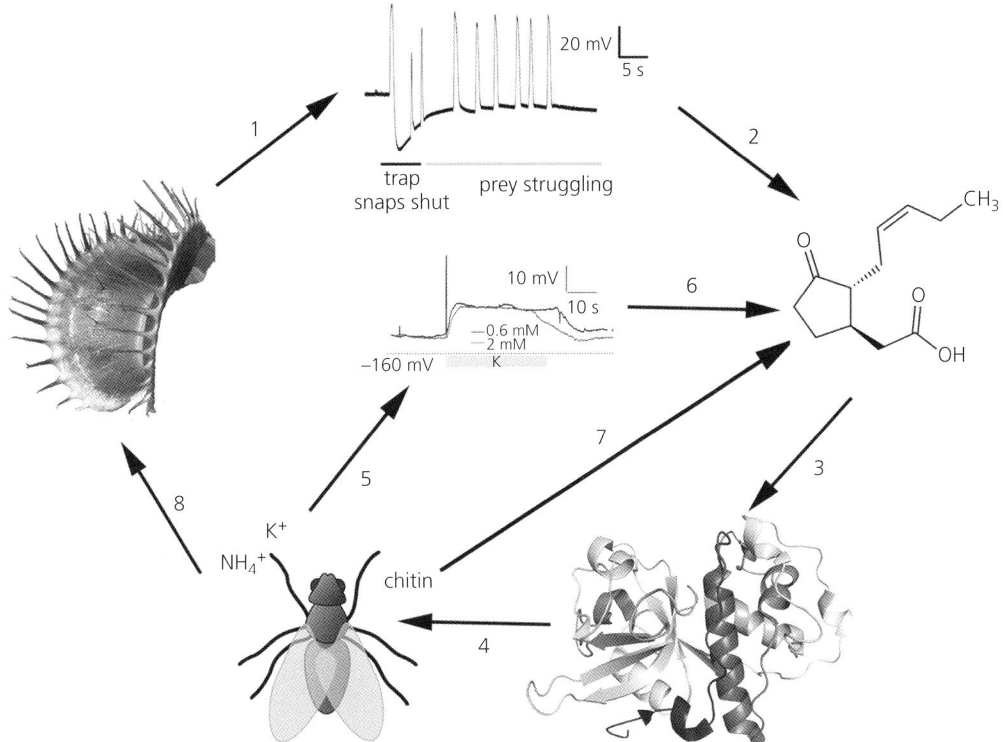

Figure 16.1 Regulation of enzyme secretion in *Dionaea*. Movement of prey stimulates the trigger hairs protruding from the upper leaf epidermis. Mechanosensitive ion channels are activated and generate an action potential. Two action potentials activate the trap to shut in a fraction of second. Struggling of the entrapped prey generates further action potentials (1), which activate the jasmonic acid signaling pathway (2) and expression of digestive enzymes (3) in a dose-dependent manner (4). Ions released from the prey depolarize the membrane potential in digestive glands (5) and may induce even higher and sustained accumulation of jasmonates (6) and enzymes (3) for completion of digestion (4). Chitin released from prey also activates accumulation of jasmonates (7) and hydrolytic enzymes (3). Nutrients gained from prey increase plant fitness (8). Figure based on results of Libiaková et al. (2014), Scherzer et al. (2015), Risør et al. (2016), Böhm et al. (2016), and Bemm et al. (2016).

et al. 2006). Production of enzymes in *Nepenthes* likely is ontogenetically regulated and their production is enhanced after pitcher opening even if prey is not captured (Biteau et al. 2013). This latter observation suggests that *Nepenthes* are adapted to slow, continuous prey capture and some of the enzymes are constitutively expressed.

Eilenberg et al. (2006) found that the chitinase gene Nkchit2b is constitutively expressed in *N. khasiana* and encodes for a vacuolar enzyme. However, this species also produces digestive enzymes when it receives stimuli from prey. Chitin can induce transcription of the class I chitinase gene Nkchit1b in *N. khasiana* (Eilenberg et al. 2006), and the expression of class III acid endochitinase is induced when *Nepenthes* captures prey (Rottloff et al. 2011, Yilamujiang et al. 2016). *Nepenthes* plants probably release selective chitinases depending not only on the structure of available substrate, but also on their ability to act when the pH of the pitcher fluid changes (Ishisaki et al. 2012b). The transcription and secretion of nepenthesin in digestive fluid also increases in response to prey capture and the role of jasmonates in an induction of proteolytic activity was also confirmed in *Nepenthes* (Yilamujiang et al. 2016). The RNase NvRN1 from *N. ventricosa* also was expressed in pitcher tissue and, for unknown reasons, in flowers (Stephenson and Hogan 2006).

Expression of proteases, RNases, nucleases, and phosphatase also is induced in the fluid of *S. purpurea* by the addition of reduced nitrogen, nucleic acids, or proteins, suggesting that hydrolase expression may be induced upon perception of the appropriate chemical signal. Fukushima et al. (2017) provide some molecular evidence supporting the existence of plant-derived enzymes in *Sarracenia*. In summary, enzymes are produced both constitutively and inductively in traps of carnivorous plants, expression is regulated developmentally, and it is tissue specific (Gallie and Chang 1997, Rottloff et al. 2011, Biteau et al. 2013, Nishimura et al. 2013). Enzyme production in aquatic *Utricularia* appears only to be constitutive, but this needs further investigation (Sirová et al. 2003).

16.3.3 Enzyme activity

Whether an enzyme is always present or induced, its maximum activity is dependent on pH. After prey capture, the pH of *N. alata* pitcher fluid declines within a day (An et al. 2001). This is probably triggered by prey-derived ammonium that induces the H^+-ATPase membrane proton pump Na-PHA1 in the secretory glands (An et al. 2001). For example, ammonium concentration in digestive fluid increases when released from insect hemolymph and by the action of glutamine deaminase (An et al. 2001, Scherzer et al. 2013). Acidification also may be induced by the activity of an ammonium transporter, as was hypothesized by Scherzer et al. (2013) for *Dionaea*. Homologous transporters have been immunolocalized in digestive glands of *Nepenthes* (Schulze et al. 1999).

Acidification may induce a rapid autoactivation mechanism of nepenthesin and other aspartic and cysteine proteases with an autocatalytic removal of propeptide, converting the inactive form of the enzymes into their active form within one hour (Kadek et al. 2014b). It is unknown whether autoactivation occurs in the trap tissue or the digestive fluid. Either way, once the enzyme is in its active form, pH affects enzyme activity through the state of ionization of acidic or basic amino acids. If the state of ionization of amino acids in a protein is altered, then the ionic bonds that help to determine the three-dimensional shape of the protein can be altered. If commensal organisms aid in digestion (Chapters 23–26), *Nepenthes* maintain a higher pH of the digestive fluid (e.g., *N. bicalcarata*; Moran et al. 2010).

16.4 Evolution of digestive enzymes and their regulatory mechanisms

The digestive enzymes in traps generally favor acidic conditions, have a broad range of substrate specificity, and are both rendered stable and protected from activity of proteases by glycosylation (Buch et al. 2014, Rottloff et al. 2016). Many chitinases, glucanases, thaumatin-like proteins, RNases, and PR-1 proteins (Buch et al. 2014) are strongly structurally and functionally homologous to a group of common plant defense enzymes involved in pathogenesis.

Co-option is an evolutionary mechanism that facilitates the emergence of new functions. It is especially common in plants, for which many examples exist, including the co-option of polarity gene networks for the evolution of reproductive organs and transcription factors in the regulation of floral symmetry (Rosin and Kramer 2009, de Almeida et al. 2014). Increasing evidence suggests that defense mechanisms used against herbivores and pathogens have been co-opted to function in the digestion of prey.

Plants use an array of defensive chemicals and proteins to protect themselves against microbial pathogens, herbivorous insects, and other environmental stressors. These include pathogenesis-related (PR) proteins that are induced in response to microbe-associated molecular patterns or insect herbivory, and those that are constitutively expressed in tissues vulnerable to attack (Savatin et al. 2014). In addition to serving protective roles, PR proteins appear to be important players in prey digestion for carnivorous plants (Renner and Specht 2013). Within the past decade, various PR proteins have been identified in *Nepenthes* pitcher-fluid proteomes and *Dionaea* transcriptomes, and recovered from the genomes of other carnivorous Caryophyllales (Hatano and Hamada 2008, 2012, Renner and Specht 2012, Schulze et al. 2012, Böhm et al. 2016, Butts et al. 2016a). Chitinases, glucanases, and proteases have been studied in the greatest detail within an evolutionary framework and investigated to determine the extent co-option has played in their functional diversification.

Besides protein similarity, the signaling pathway leading to carnivory also was probably co-opted from plant-defense signaling pathways. In plant-defense mechanisms, changes in membrane potentials or modulation of ion fluxes at the plasma membrane level are among the earliest cellular responses to biotic and abiotic stresses, and later elements of the signal transduction pathway are represented by a network of phytohormones, mainly jasmonates (Maffei et al. 2007, Wasternack and Hause 2013). The link between changes in membrane potential (action and variation potentials) and jasmonate accumulation is an increased cytosolic Ca^{2+} concentration, which is sensed by Ca^{2+}-sensor proteins such as calmodulin, calmodulin-like, and calcineurin B-like proteins, or Ca^{2+}-dependent protein kinases (Fisahn et al. 2004, Maffei et al. 2007). JA-Ile is the only jasmonate for which the molecular basis of its gene-regulatory activity has been elucidated. After JA-Ile binds to the coronatine-insensitive 1 receptor (COI1), it mediates the ubiquitin-dependent degradation of jasmonate ZIM domain (JAZ) repressors resulting in the activation of JA-dependent gene expression encoding proteins involved in plant defense (Wasternack and Hause 2013). JA-Ile readily accumulates in *Dionaea, Drosera*, and *Nepenthes* in response to mechanical and chemical stimuli (Nakamura et al. 2013, Libiaková et al. 2014, Mithöfer et al. 2014, Yilamujiang et al. 2016, Krausko et al. 2017) and JAZ and COI1 proteins have been identified in *Dionaea* (Böhm et al. 2016). Carnivorous plants probably employ the same signaling pathway in prey digestion. After rapid changes of electrical potentials, cytosolic Ca^{2+} is increased, and the jasmonates in prey-stimulated traps accumulate and induce expression of carnivory-related genes (Escalante-Pérez et al. 2011, Nakamura et al. 2013, Pavlovič and Saganová 2015). The similarity of these signaling pathways is further supported by the observation that wounding and insect oral secretions can induce enzyme secretion and tentacle bending in *D. capensis* (Mithöfer et al. 2014, Krausko et al. 2017). Another plant hormone involved in plant resistance to biotrophic pathogens is salicylic acid, which, however, does not induce expression of carnivory-related genes (Matušíková et al. 2005, Krausko et al. 2017).

16.4.1 Subfunctionalization of class I chitinases for defense and digestion

Plant chitinases (PR protein families 3, 4, 8, and 11) play an important role in defense against fungal pathogens and herbivory by insects. Their activity consists of hydrolyzing β-1,4-glycosidic linkages between *N*-acetylglucosamine oligomers in chitin, a major constituent of the fungal cell wall and arthropod exoskeleton. Demonstrated to inhibit fungal growth *in vitro* and enhance resistance to fungal and insect attack, it is not surprising that chitinases have been discovered in all plant genomes sequenced to date, including the moss *Physcomitrella patens*.

Compelling evidence suggests that class I chitinases involved in plant defense have been co-opted to function in carnivorous Caryophyllales. Molecular evolutionary studies of this clade have revealed that class I chitinase genes primarily used for pathogenesis response by ancestral noncarnivorous plants duplicated and diverged during the evolution of carnivory. Functional divergence of class I chitinases is supported by the separation of subclasses Ia and Ib chitinases into distinct phylogenetic clades (defense and digestion) (Renner and Specht 2012; Figure 16.2a). Through subsequent co-option, class I chitinases could have been subfunctionalized for pathogenic response (subclass Ia) and carnivory (subclass Ib).

Additionally, there is evidence that selection pressures acting on subclass Ib chitinases have shifted as pathogenesis responses enabled a role specific to carnivory, affecting structure and function (Renner and Specht 2012). In modeling *N. khasiana* subclass Ia and Ib chitinases, an amino acid substitution was identified at a positively selected site in subclass Ia (Phe276; Figure 16.2b). Because this particular site is located within the substrate-binding cleft, changes at the amino acid level could greatly affect substrate binding, activity, and functionality. This observation is supported further by site-directed mutagenesis studies in *Arabidopsis* and *Zea* at a site positionally homologous to Phe276, in which chitinase activity was modified when phenylalanine was substituted for an essential tyrosine important for chitin degradation (Verburg et al. 1993). These results also support the hypothesis that differential selection is driving the process of subfunctionalization in class I chitinases.

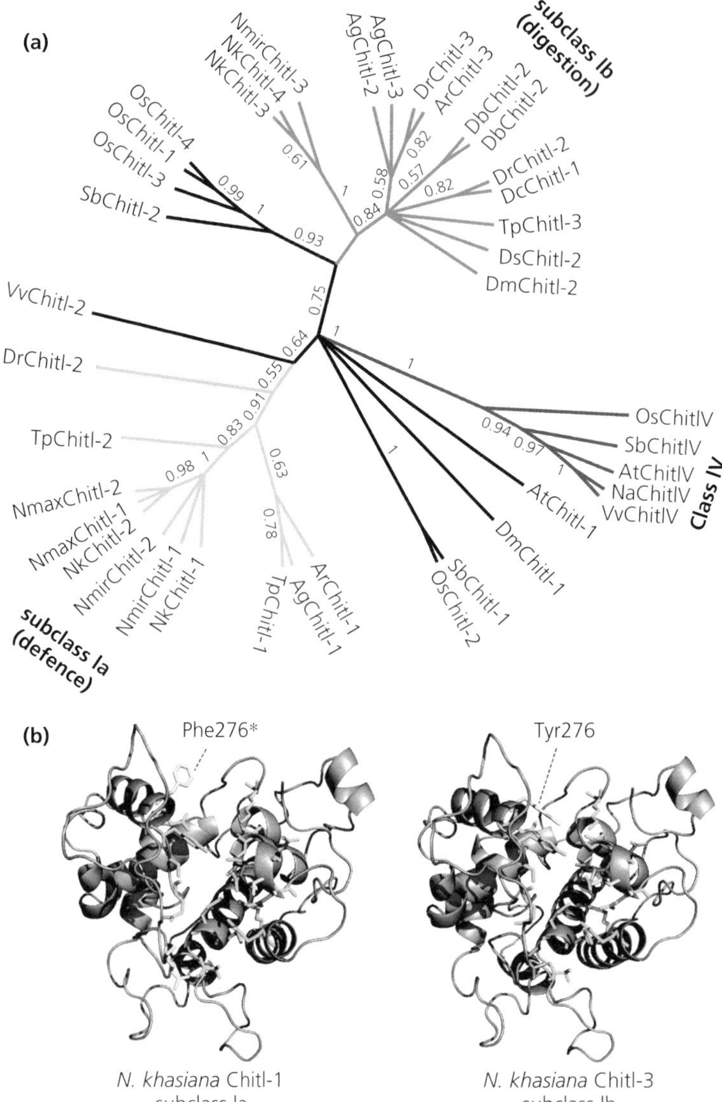

Figure 16.2 Evolution of chitinases with roles in plant carnivory. (**a**) The 50% majority-rule phylogenetic tree inferred from Bayesian analyses of angiosperm chitinases, with posterior probabilities indicated at nodes. Two distinct clades comprised of carnivorous Caryophyllales chitinases are evident in the phylogeny: (1) subclass Ia, used in defense and (2) subclass Ib, used in digestion. (**b**) Three-dimensional models of *Nepenthes khasiana* subclass Ia (NkChitI-1) and Ib (NkChitI-3) chitinases in association with NAG and water. In subclass Ia, residue Phe276 (asterisk) disrupts interaction with the substrate and is a site under positive selection. Phylogeny and models based on previous analyses of class I chitinases (Renner and Specht 2012).

16.4.2 Evolution and expression of class III chitinases

Phylogenetic studies of class III chitinases have been limited to *Nepenthes* (Rottloff et al. 2011). Similar to class I, class III chitinases are induced by abiotic stress, up-regulated in response to fungal pathogens, and may be developmentally regulated (Grover 2012). In *Nepenthes*, class III chitinase is present within the digestive fluid of both closed and open pitchers, yet the enzyme is also up-regulated in response to prey in both the glands and their surrounding tissues (Rottloff et al. 2011). These results suggest that *Nepenthes* class III chitinases have a broader expression pattern and an evolutionary history different from the class I chitinases. Rottloff et al.'s (2011) results suggest that class III is represented by a single gene in *Nepenthes* that codes for an enzyme with roles in both defense and carnivory, as opposed to subfunctionalization of chitinase activity.

Protein-specific divergence events are not evident in the phylogenetic reconstruction, which includes class III sequences from eight species of *Nepenthes*. Instead, the phylogenetic relationships are fairly congruent with a species tree reconstructed from a single chloroplast molecular marker (Meimberg et al. 2001). Gene duplication does not seem to be the major mechanism by which *Nepenthes* class III chitinases have evolved to function in carnivory.

The presence of multiple chitinase classes associated with digestive fluids of trap tissues may indicate synergistic roles in prey digestion that could be influenced by differential patterning in temporal and spatial expression. For example, class I, III, and IV chitinases have been discovered in the digestive fluid or identified in the traps of *Nepenthes*, *Drosera*, and *Dionaea* under varying conditions (§16.2.3). Differences in the relative expression of chitinase genes can be influenced strongly by multiple factors, including the presence or absence of prey or prey-like substrates, the method of stimulation, timing of the experiment, and types of tissues collected.

16.4.3 Evolution and expression of class V *β*-1,3-glucanases

The β-1,3-glucanases form a large and diverse gene family in plants, represented by five classes of glucanase. Expression and phylogenetic analyses suggest that ancestral functions of the gene family likely were involved in cell division and cell-wall remodeling (Doxey et al. 2007). Because of alterations in gene structure and expression, some β-1,3-glucanases are extracellular and function in pathogen response.

Glucanases also have been suggested to play a role in decomposition of prey. A β-1,3-glucanase gene (*DrGln1*) was isolated from *D. rotundifolia* by a genome walking approach, and phylogenetic reconstructions provided strong support for evolutionary relatedness to the class V β-1,3-glucanases (Michalko et al. 2017). Further, *DrGln1* forms a strongly supported clade with some homologs derived from carnivorous Caryophyllales that represent sequences either expressed in response to prey (Schulze et al. 2012) or active within the digestive fluid of *Nepenthes* pitchers (Hatano and Hamada 2012). At least for *Dionaea* and *Nepenthes*, the class V β-1,3-glucanases could represent a unique example of non-defense related genes that have been co-opted for the carnivorous syndrome. *DrGln1* is expressed in vegetative tissues but not in flowers and digestive glands, and encodes for a functional enzyme when expressed in transgenic tobacco. Detailed analyses of the supposed promoter both *in silico* and in transgenic tobacco suggest that this glucanase plays a role in development. Specific spatiotemporal activity was observed during seed germination. Later during growth, the *DrGln1* promoter was active in marginal and sub-marginal areas of apical true leaf meristems of young tobacco plants. The results suggest the isolated glucanase gene is regulated endogenously, possibly by auxin.

16.4.4 Evolution and specificity of proteases

Functional and evolutionary studies for genes encoding other digestive enzymes are incomplete, but with increasing amounts of sequence data this is expected to change. Sequence comparison, molecular modeling, and network analyses already have revealed potential functional divergences among carnivorous plant proteases (Butts et al. 2016b). This work has led to the discovery of variations in their stability, pH optima, substrate specificity, and cleavage patterns (Athauda et al. 2004, Butts et al. 2016a). Despite the universal distribution of proteases with various functions, those used in digestion have adapted extracellular activities. Among noncarnivorous plants, cysteine proteases have been shown to be under diversifying selection, coevolving with pathogens as a means of plant defense. These proteases may have been co-opted for roles in botanical carnivory, such as the cysteine proteases active in digestive fluid. Functional differences among these proteases are likely adaptive, and allow for a broad range of prey substrates to be digested.

16.5 Future research

Although available data on the chemistry of digestive secretomes are biased toward *Dionaea* and *Nepenthes*, they nevertheless provide a general picture on the mode of prey decay. However, these data are still fragmented. Available proteomic tools will have to be altered to deal with unique features of digestive proteins, such as low frequency of Lys/Arg residues (Lee et al. 2016). Although strong convergent evolution accounts for similar components of digestive enzymes among different lineages of carnivorous plants (Fukushima et al. 2017), research on other carnivorous genera should reveal the nature of secretome diversity and discovering novel digestive enzymes is likely. The unique properties of nepenthesin unexpectedly have proven to be useful for hydrogen/deuterium exchange mass spectrometry used more widely for studies of protein dynamics, structure, and interactions (Rey et al. 2013, Kadek et al. 2014a). The general and peculiar features of plant carnivory across different taxa will require not only studying the linkages between processes leading from prey perception to digestion and absorption of nutrients, but also analysis of the interplay of enzymatic and non-enzymatic fluid components during digestion. The role of electrical and jasmonate signaling, and possible hormonal cross-talk deserve special attention. These goals will require application of modern analytical techniques to analyses of the complex molecules that comprise digestive fluids.

High-throughput genome sequencing coupled with experimental transcriptomics should pinpoint genes induced in response to prey and help to elucidate processes by which gene families involved in botanical carnivory have evolved. Functional evolution among carnivorous plants of defense proteins such as chitinases has already been supported with experimental and phylogenetic data. Research on the evolution of digestive enzymes should be expanded and take advantage of the rapidly increasing amounts of molecular data (Chapter 11).

The microbiology of the trap fluids is another topic that deserves further attention. Discovering the roles of microorganisms in digestion and nutrition, in addition to the ways they interact with the plant, will require information on the biodiversity, abundance, and metabolic rates of microbial flora within traps (Chapters 23–26). The role of prey autolysis in digestion likely has been underestimated.

Finally, tests for a cost/benefit model of the evolution of prey digestion mechanisms should also consider expected changes of the climate and address how outcomes of this model are likely to change in response to new extremes of temperature, water shortage, or biotic stress (Chapter 28).

CHAPTER 17

Mineral nutrition of terrestrial carnivorous plants

Lubomír Adamec and Andrej Pavlovič

17.1 Introduction

The term "mineral nutrition of plants" includes processes of mineral nutrient uptake by plants from the ambient medium, nutrient translocation within the plant, incorporation of mineral nutrients to plant metabolism and physiological functions, and mineral nutrient reutilization (recycling) from senescent organs to young ones. Knowledge of mineral nutrition of terrestrial carnivorous plants is based on studies of only ≈70 species (i.e., <10% of all carnivorous plants), published in ≈100 papers that have appeared since the 1950s.

The topic of mineral nutrition was covered only fragmentarily by Juniper et al. (1989). Although there were fewer than two dozen publications on nutrient dynamics prior to 1988, mineral nutrition long has been thought to be the key determining process leading to the evolution of botanical carnivory (Givnish et al. 1984, Juniper et al. 1989, Adamec 1997a, 2011c, Ellison 2006). Since 1990, there has been a rapid rise in papers and reviews on this topic. Adamec (1997a) comprehensively reviewed the literature published mainly in the decade after Juniper et al. (1989), whereas Ellison (2006) focused more narrowly on reviewing effects of added prey or mineral nutrients on growth and tissue nutrient contents with an eye toward identifying nutrient limitations. More recently, Adamec (2011c) reviewed another decade of research. He identified the primary physiological effect of foliar capture of prey to be the stimulation of nutrient uptake by roots.

In this chapter, we review the current understanding of (eco)physiological processes associated with mineral nutrition of terrestrial carnivorous plants, with additional attention to papers having been published after 1989 and to integrative studies of *Nepenthes*. From an evolutionary perspective, the costs and benefits of carnivory are likely to be similar in terrestrial and aquatic species (Ellison and Adamec 2011; Chapter 18), but aquatic plants, with their different ecophysiological traits are discussed in Chapter 19. In highlighting differences between carnivorous and noncarnivorous plants, we illustrate how ecophysiological studies of the former can stimulate research on the latter.

17.2 Ecophysiological traits in stressful habitats

The majority of terrestrial carnivorous plants grow in wet or waterlogged peat or sandy soils where they encounter persistently unfavorable conditions. These soils are mostly acidic (pH 3–6; Juniper et al. 1989, Adamec 1997a) but some are neutral or slightly basic (pH 6.9–8.3 for *Pinguicula* spp. [Adamec 2015b]; 6.1–7.7 for "ultramafic" *Nepenthes* spp. [van der Ent et al. 2015]). The substrates usually contain a high proportion of slowly decomposing organic matter or clay particles and are hypoxic or anoxic. In wet soils, decomposition of organic matter may lead to an accumulation of toxins (H_2S or S^{2-}) and to a low redox potential. Under these conditions, iron and manganese may solubilize and become toxic to plant roots,

Adamec, L. and Pavlovič, A., *Mineral nutrition of terrestrial carnivorous plants*. In: *Carnivorous Plants: Physiology, ecology, and evolution*. Edited by Aaron M. Ellison and Lubomír Adamec: Oxford University Press (2018).
 DOI: 10.1093/oso/9780198779841.003.0017

whereas other microelements (e.g., Mo) may become unavailable (Crawford 1989). However, some *Nepenthes* and *Drosera* species grow in ultramafic soils that have extremely elevated levels of heavy metals (Co, Cr, Zn, Ni, Fe, Mn; van der Ent et al. 2015). Substrates that support carnivorous plants may contain high total levels of macronutrients, but there is a 100–1000-fold difference between available and total macronutrient levels in most bog and fen soils. The available content of N, P, K, Ca, and Mg in bogs and fens are normally in the range 1–30 mg kg^{-1} dry weight (DW; Adamec 1997a) and carnivory is one mechanism by which plants overcome this very low macronutrient availability (Juniper et al. 1989).

The normal uptake of water and nutrients by roots is impeded by the low nutrient availability in wet, anoxic soils. Carnivorous plants generally are "stress-strategists" (*sensu* Grime 1979) that grow slowly: the median relative growth rate (RGR) of *Drosera, Dionaea, Sarracenia, Genlisea*, and *Utricularia* is only 0.020–0.025 g g^{-1} d^{-1} (range 0.003–0.15 g g^{-1} d^{-1}; Adamec 2011c, Ellison and Adamec 2011, Kruse et al. 2014). Carnivorous plants do not require a high supply rate of mineral nutrients from soils to support low growth rates and they also reutilize them efficiently from senescent organs (Dixon et al. 1980, Adamec 1997a, 2002, Butler and Ellison 2007).

Carnivorous plants characteristically have weakly developed root systems (Juniper et al. 1989, Adamec 1997a, 2005a); root:total biomass ratio ranges from 3.4% to 23% (Adamec 1997a, 2002). Short, weakly branched roots usually tolerate anoxia and related stressors in wet soils (Adamec 2005a) and are able to regenerate easily (Adamec 1997a). Although roots usually are less important than foliar traps for nutrient acquisition, the roots are still quite active. Their aerobic respiration rate and water exudation rate per unit biomass are comparable to or greater than those reported for roots of noncarnivorous plants (Adamec 2005a). Roots of carnivorous plants rely on an aeration diffusive mechanism supported by exodermal diffusive barriers (Adamec 2005a, 2011c).

17.3 Nutrient content and stoichiometry

Generally, the medians and quartiles of foliar N and K content in terrestrial carnivorous plants are ≈2× lower than those for noncarnivorous plants, whereas the values for P are more-or-less equivalent (Table 17.1). Foliar N, P, and K content of terrestrial carnivorous plants are most similar to values found in evergreen trees and shrubs (Ellison 2006, Ellison and Adamec 2011). Terrestrial carnivorous plants have considerably lower foliar/shoot macronutrient content than aquatic ones; for N, P, and Ca, this could be partly due to inclusion of prey in traps of aquatic species (Adamec 1997a).

Tissue nutrient content also depends markedly on the age (or position) of the organ or organs that are harvested for nutrient analysis (Adamec 1997a, 2002), which rarely are described in publications. Thus, tissue nutrient content alone is an unreliable measure of nutrient uptake; total nutrient accumulation and plant growth rate should be estimated simultaneously (Adamec 1997a, 2002, Farnsworth and Ellison 2008). The very low foliar N and K contents in terrestrial carnivorous plants (Table 17.1) are associated with very low growth and net photosynthetic rates (Ellison 2006, Ellison and Adamec 2011; Chapter 18), suggesting that the "critical" contents of foliar N and P limiting their growth could be ≈2× lower than for noncarnivorous plants (2.0% N, 0.10% P; Ellison 2006, Ellison and Adamec 2011). We note that the values tend to be lower for *Sarracenia* and *Nepenthes* (Ellison 2006, Ellison and Adamec 2011) than for the Droseraceae (Adamec 1997a, 2002, Pavlovič et al. 2007, 2009, Farnsworth and Ellison 2008).

Table 17.1 Tissue nutrient content (median % dry weight; lower and upper quartiles in parentheses) of nitrogen (N), phosphorus (P), and potassium (K) in leaves or shoots of all plants of various ecological groups, terrestrial carnivorous plants, and aquatic carnivorous plants. Data from Ellison (2006) and Ellison and Adamec (2011).

Plant category	N	P	K
All plants (n = 397–1973)	1.81 (1.22–2.41)	0.10 (0.04–0.17)	1.86 (1.16–3.03)
Terrestrial carnivorous (n = 51–94)	0.88 (0.70–1.21)	0.11 (0.07–0.18)	0.91 (0.57–1.21)
Aquatic carnivorous (n = 43–54)	2.40 (1.90–2.81)	0.22 (0.12–0.92)	1.55 (1.20–1.92)

Nutrient stoichiometry also can provide information on which nutrients limit growth; available data suggest co-limitation of carnivorous plant growth by N + P or N + P + K (Ellison 2006). These data suggest at least two selective advantages for capturing animal prey (Adamec 2011c). First, animals are a relatively rich source of some mineral macronutrients (e.g., N, 9.9–12.1; P, 0.6–1.47; K, 0.15–3.18; Ca, 0.1–4.4; Mg, 0.094% DW of insect and other arthropod prey; Adamec 2011c). The tissue N and P contents of prey carcasses are ≈5–10× higher than those in shoots or traps (Table 17.1), whereas the K content is comparable. Although uptake of N and P from animal prey likely was a key evolutionary innovation, much of the nutrients in prey tissue is unavailable (Dixon et al. 1980, Hanslin and Karlsson 1996, Adamec 2002, Płachno et al. 2009a, Pavlovič et al., 2014; Chapter 16). Second, most captured prey species are abundant but relatively small, and they are frequent visitors to plants as pollinators or herbivores (the "predictable prey-capture hypothesis"; Müller et al. 2004). Prey-capture effectiveness varies >10-fold among individual plants (Thum 1989a, 1989b, Karlsson et al. 1994).

17.4 Mineral nutrient economy

Foliar nutrient uptake from prey and root nutrient uptake from the soil; mineral nutrient reutilization from senescing shoots (nutrient resorption efficiency); and stimulation of root nutrient uptake by foliar nutrient uptake together make up the mineral nutrient economy. Efficiency of mineral nutrient uptake from prey by traps has been well studied, but virtually no data have been published on the nutrient uptake affinity and capacity of carnivorous plant roots. We hypothesize that root uptake affinity in nutrient-poor habitats is relatively high, but uptake capacity of these slow-growing plants is very low. The consequence of these traits indicates that roots take up mineral nutrients even from very barren soils but the low uptake rate supports only slow growth. Finally, we note that the ecological importance of carnivory in the field depends primarily on what proportion of mineral nutrients (as seasonal nutrient gain or consumption) are acquired directly from prey for seasonal growth (Adamec 1997a, 2011c, Ellison and Gotelli 2001).

17.4.1 Mineral nutrient uptake from prey

N, P, K, S, Na, Ca, Mg, Fe, Mn, Zn, and Co have been found to be taken up by traps from prey or mineral nutrient solutions (Adamec 1997a, Adlassnig et al. 2009), but the ecophysiological importance of absorption of a given macro- or micronutrient from prey depends on its uptake efficiency. To date, nutrient uptake efficiency from insect prey has been estimated only in several studies; the greatest attention has been focused on N. Greenhouse-grown *Drosera erythrorhiza* fed fruit flies (*Drosophila*) took up 76% of the total fly N (Dixon et al. 1980); much of the residual N in the flies was locked up in insect chitin. N uptake of chitin by *D. capensis* was ≈4–8 times less than N uptake of bovine serum albumin (Pavlovič et al. 2016). In greenhouse-grown *D. capillaris* and *D. capensis*, the percent uptake efficiency of N, P, K, and Mg from fruit flies and mosquitoes was relatively efficient (43–62% N, 61–97% P, 60–96% K, 57–92% Mg), whereas uptake of Ca was not and depended greatly on insect tissue Ca content (Adamec 2002, Pavlovič et al., 2014).

Similar values (56–65% N, 59–67% P) were reported for greenhouse-grown *D. closterostigma* fed fruit flies (Karlsson and Pate 1992). Płachno et al. (2009a) reported similar values of uptake efficiency from fruit flies for greenhouse-grown *Drosophyllum lusitanicum* and *Roridula gorgonias* (33–47% N, 62–75% P, 44–86% K, 33–39% Mg), but uptake efficiency was zero for four "proto"-carnivorous plants (*Proboscidea parviflora, Ibicella lutea, Cleome droserifolia, Hyoscyamus desertorum*) with hairy glandular leaves. N uptake efficiency of 39–51% from fruit flies was estimated for three subarctic European *Pinguicula* species and *D. rotundifolia* in a greenhouse, but was only 29–41% in field-grown plants (Hanslin and Karlsson 1996). In three greenhouse-grown *Sarracenia* species, the mean ^{15}N uptake efficiency from fruit flies was 46%, but was 66% from $^{15}NH_4NO_3$ supplied directly to the pitchers (Butler et al. 2008). However, only 9.5% of absorbed prey-^{15}N and 6.6% of absorbed NH_4NO_3-^{15}N were translocated to roots and rhizomes in *S. purpurea*, suggesting that prey-N might be absorbed in organic form (Karagatzides et al. 2009) and translocated more to below-ground organs in preference to mineral forms. Underground leaves (sticky traps) of *Philcoxia minensis*

(Plantaginaceae) absorbed 15% of total ^{15}N from labeled nematodes as prey in a greenhouse during 48 hours (Pereira et al. 2012).

In conclusion, the uptake efficiency of P, K, and Mg from prey can be much greater than that of N but field-based values can be significantly lower than those obtained in greenhouse studies. Significant uptake of mineral nutrients from model prey is a more reliable test of suspected plant carnivory than enzyme secretion alone (Płachno et al. 2009a; Chapters 12, 16).

17.4.2 Mechanism of nutrient uptake from prey

Mechanisms of mineral and organic nutrient uptake from prey include both membrane carriers and endocytosis. Adlassnig et al. (2012) used fluorescent markers to demonstrate endocytosis in traps of *Aldrovanda, Drosera, Dionaea, Drosophyllum, Nepenthes,* and *Cephalotus*, but not in *Sarracenia* or *Genlisea*. They proposed that nutrient uptake by carriers was supplemented by endocytosis, which enabled absorption and intracellular digestion of proteins and other nutrients. Moreover, traps of many species efficiently take up soluble organic substances (mainly amino acids) from prey or solutions applied (Juniper et al. 1989, Adamec 1997a, Karagatzides et al. 2009) but the ecological importance of this carbon uptake is unclear. In field-grown *S. purpurea*, simultaneous addition of NH_4NO_3 and amino acids to the pitchers decreased amino acid uptake threefold (Karagatzides et al. 2009). Regardless of the transport mechanisms, a good deal of N can thus be readily absorbed from prey as amino acids but this strongly interferes with the uptake of mineral N forms.

When *Dionaea* traps were fed milled insects or pretreated with coronatine (a substitute for insect stimulation), Scherzer et al. (2015) found a marked, high-affinity for K^+ uptake by the traps resulting in the decrease in $[K^+]$ in the trap suspension from 30 mM to 10 μM after only a few days. A half-saturation $[K^+]$ of 60 μM was calculated for the K^+ uptake from K^+-selective microelectrode measurements. Unstimulated traps, however, did not exhibit any high-affinity K^+ uptake. Furthermore, digestive glands of stimulated *Dionaea* traps expressed and up-regulated the potassium channel ortholog DmKT1 and the high-affinity H^+-driven K^+ transporter DmHAK5. The same calcium sensor kinase (CIPK23) was found to regulate both transporting proteins. Thus, stimulated *Dionaea* traps use K^+ uptake from prey both as a high-affinity, pH-dependent K^+/H^+ symporter and a low-affinity, high capacity K^+ transporter so that the resultant K^+ uptake efficiency from prey can be extremely high (see also Adamec 2002 for *Drosera* spp.).

17.4.3 Mineral nutrient reutilization

Reutilization (i.e., recycling, nutrient resorption efficiency) of mineral nutrients is very efficient, ranging from 33–99% for N, 51–98% for P, and 41–99% for K (Adamec 1997a, 2002; Table 17.2), minimizing losses of these nutrients from senescing organs. Mean reutilization efficiencies of N (70–75%) and P (75–80%) in leaves or shoots of carnivorous plants are 20–25 percentage points greater than in co-occurring noncarnivorous plant (Aerts et al. 1999). In contrast, highly variable Mg reutilization and even negative Ca reutilization have been reported for both types of plants.

17.4.4 Leaf–root nutrient interaction

A marked stimulation of absorption of nutrients by roots as a consequence of foliar uptake of nutrients from prey is an ecophysiological peculiarity of mineral nutrition in *Drosera* and *Pinguicula* (Hanslin and Karlsson 1996, Adamec 1997a, 2002). Plants fed either insects or mineral nutrient solutions in various growth experiments grew rapidly and accumulated much more mineral nutrients in biomass (≈1.6–27× more for N, P, K, Ca, and Mg relative to unfed control plants) than they could have acquired from prey alone. In three field-grown *Pinguicula* species, the stimulatory effect on roots depended directly on the amount of prey fed to the plant (Hanslin and Karlsson 1996). The extent of this stimulation is several times greater for K, Ca, and Mg than for N and P under natural conditions. However, the mechanism of the stimulation of root uptake by prey capture remains unknown. For example, PO_4 taken up by leaves from a phosphate solution stimulates root nutrient uptake (Karlsson and Carlsson 1984), but the contribution of other nutrients (especially N) is unexplored.

Table 17.2 Mineral nutrient reutilization efficiency (% of initial content;—no data) by terrestrial carnivorous plants.

Species	Organ	N	P	K	Ca	Mg	Reference*
Drosera capensis[†]	Leaf	82	92	79	−88	−23	1
D. peltata[†]	Leaf	78	86	41	−187	−45	1
D. scorpioides[†]	Leaf	70	87	98	−32	33	1
D. rotundifolia	Shoot	81	—	—	—	—	4
D. erythrorhiza	Leaf	79	88	56	25	63	2
D. erythrorhiza	Tuber	95	98	99	−56	83	2
D. erythrorhiza	Leaf	94	—	—	—	—	3
D. erythrorhiza	Stem	99	—	—	—	—	3
Dionaea muscipula[†]	Leaf	80	51	99	−119	−75	1
Three *Pinguicula* spp.	Shoot	58–97	—	—	—	—	4
Three reproductive *Pinguicula* spp.	Leaf	33–70	—	—	—	—	6
Three non-reproductive *Pinguicula* spp.	Leaf	38–74	—	—	—	—	6
Sarracenia purpurea	Leaf	56	75	89	—	—	5
S. alata	Leaf	42	—	—	—	—	7

*References: 1. Adamec (2002); 2. Pate and Dixon (1978); 3. Dixon et al. (1980); 4. Hanslin and Karlsson (1996); 5. Small (1972); 6. Eckstein and Karlsson (2001); 7. Horner and Schatz (2016).
[†]Values include correction for biomass decrease in senescent organs.

In three *Drosera* species in a greenhouse experiment, slightly greater root lengths could explain only ≈17% of the uptake stimulation; the higher theoretical uptake rate of roots per unit root biomass accounted only for 15–30%, but the greater root biomass explained 70–85% of the effect (Adamec 2002). The total root biomass of the fed plants was markedly greater than that of the unfed controls, but the proportion of root biomass to the total biomass of fed plants decreased slightly. However, root aerobic respiration rate was unchanged and the stimulatory effect on the roots was uncorrelated either with root or shoot mineral nutrient contents. Because a smaller proportion of K, Ca, and Mg are taken up from prey, stimulation of root uptake of these cations from soil should be very important for nutrient budgets of carnivorous plants in the field.

17.4.5 Seasonal nutrient gain

The ecological importance of carnivory depends mainly on what proportion of mineral nutrients (as seasonal nutrient gain) are acquired directly from prey and used for seasonal growth (Adamec 1997a, 2011c, Ellison and Gotelli 2001). Based on measured rates of seasonal prey capture and an average 76% efficiency of nutrient uptake from prey (Dixon et al. 1980), the calculated values of this proportion vary intraspecifically, interspecifically, and seasonally (Table 17.3). Moreover, real nutrient uptake efficiencies may differ considerably among plant species and the prey they capture (Adamec 1997a, 2002), and the overall average efficiency, leading to overestimation of seasonal N gain but an underestimation of P and K gains.

Seven to 100% of the seasonal N and P gain, a much smaller proportion of K (1–16%), and likely even less Ca and Mg is obtained from prey. The seasonal N gain in non-flowering individuals of three *Pinguicula* species was ≈2× higher than in flowering ones, and the values of the most successful individuals were two to four times higher than the mean (Karlsson et al. 1994). The 100% seasonal N and P gain from prey in *Sarracenia purpurea* (Table 17.3; Chapin and Pastor 1995) reflects the low mineral N uptake efficiency of *S. purpurea* roots (Butler and Ellison 2007). Prankevicius and Cameron (1991) estimated a surprisingly high bacterial dinitrogen

Table 17.3 Mean or range (%) of seasonal mineral nutrient gain coming from carnivory in the field.

Species	N	P	K	Reference*
Pinguicula vulgaris	26–40	36	7–16	1, 2, 3
P. alpina	8–14	12–19	1.3–1.9	1, 2, 3
P. villosa	7–15	6–10	3–12	1, 2, 3
P. vallisneriifolia	44	44	5.5	4
Drosera rotundifolia	63	95	1.1	5
D. intermedia	92	100	1.6	5
D. erythrorhiza	11–17	—	—	6
D. erythrorhiza	100	100	2–3	7
Sarracenia purpurea	100	100	—	8

*References: 1. Karlsson (1988); 2. Karlsson et al. (1994); 3. Hanslin and Karlsson (1996); 4. Zamora et al. (1997); 5. Thum (1988); 6. Dixon et al. (1980); 7. Watson et al. (1982); 8. Chapin and Pastor (1995).

fixation rate in *S. purpurea* pitchers that could account for 3.8× the total seasonal plant N budget. Although the importance of this N_2 fixation in the pitcher cavity for direct plant N uptake cannot be assessed, the fixed N may be released from senescent leaves (Butler and Ellison 2007).

Seasonal N gain from prey has been estimated for many species using natural ^{15}N distribution in potential prey and plant shoots. No seasonal N gain was found for seven Australian rosette *Drosera* species, but N gain was 12–32% in the rosette-forming *D. erythrorhiza*; 37–57% in erect low species (*D. huegellii, D. menziesii, D. stolonifera*); 49–65% in erect high species (*D. gigantea, D. heterophylla, D. marchantii*); 35–87% in vine species (*D. gigantea, D. pallida, D. subhirtella*); 29–64% in *D. rotundifolia* and *D. intermedia*; 21% in semi-terrestrial *Polypompholyx multifida*; 26–47% in *Cephalotus follicularis*; 76% in *Darlingtonia californica*; 54–68% in *Nepenthes mirabilis, N. albomarginata*, and *N. rafflesiana*; 46–75% in *Dionaea muscipula*; and 40–70% in *Roridula dentata* and *R. gorgonias* via their digestive mutualism with hemipterans (E.D. Schulze et al. 1991, W. Schulze et al. 1997, 2001, Moran et al. 2001, Ellison and Gotelli 2001, Anderson and Midgley 2003, Millett et al. 2003; Chapters 12, 14). Prey-derived N was preferentially allocated to flowers (63% gain) and leaves (51%), but much less to roots (29%), of *D. rotundifolia* (Millett et al. 2003). Among different populations, N gain was correlated significantly with shoot N content. Prey capture thus supports flowering, seed set, and production of larger trapping area of leaves (as positive feedback). However, in three *Pinguicula* species growing in the field, the mean direct N uptake from extra added prey was only 39% of the total increased N; the remaining 61% was taken up indirectly from the soil when root uptake was stimulated by prey capture (Hanslin and Karlsson 1996; §17.4.4).

17.5 Growth effects

Because greenhouse experiments simplify field conditions (e.g., lack of competition, reduced mortality, predictable watering), results from them generally reflect only potential physiological abilities and growth consequences of uptake of mineral nutrients by leaves from prey or by roots from soil. All terrestrial species studied to date can grow satisfactorily *in vitro* in natural peaty soils without either supplemental prey or soil fertilization. However, foliar application of a mineral nutrient solution (Karlsson and Carlsson 1984, Adamec 2002) increased growth of *Drosera* as much as did prey additions, lending support to the hypothesis that N and P absorbed from prey, but not organic substances, were crucial for growth (Juniper et al. 1989, Adamec 1997a, 2011c).

Depending on experimental conditions and species, supplemental prey or soil fertilization increased the growth rate and mineral nutrient uptake two- to five-fold (reviewed by Adamec 1997a, Ellison 2006). However, species-specific differences suggest that there are three main ecophysiological groups (Adamec 1997a). Growth rate of "nutrient-requiring" species (most *Drosera* spp., most *Pinguicula* spp., all *Sarracenia* spp.) depends on both soil and leaf nutrient supply; their root nutrient uptake rate also may be stimulated by foliar uptake of nutrients. In contrast, "root–leaf nutrient competitors" (some Australian *Drosera* species from drier areas, *Pinguicula villosa*, and *Dionaea muscipula*) depend on both nutrients accumulated from captured prey and from root uptake, but their growth enhancement is usually lower than among the nutrient-requiring species because of competition between root and foliar nutrient uptake. Last, "nutrient-modest" species (some Australian pygmy *Drosera* and *P. vallisneriifolia*) have roots with a very low nutrient uptake capacity and rely almost entirely on prey capture for their nutrient budget.

Field studies have revealed the ecological importance of carnivory for growth and development in different microhabitats with varying environmental conditions and suites of other interacting species. Relevant interactions include not only competition, mortality, robbing of prey by opportunistic predators (kleptoparasites, kleptobionts), but also the washing out of either entire prey carcasses or dissolved nutrients by heavy rains (Adamec 1997a). Growth in the field with or without supplemental prey was comparable in parallel greenhouse experiments (Adamec 1997a, Ellison 2006). Moreover, prey addition experiments done in the field have shown repeatedly that much more prey than would normally be captured can be used for growth

(Thum 1988, Chapin and Pastor 1995, Hanslin and Karlsson 1996, Zamora et al. 1997, Farnsworth and Ellison 2008). Field studies also have shown that growth increase of particular individuals always depends on the amount of captured prey; prey capture of even closely spaced individuals of the same species could vary by more than an order of magnitude (Thum 1989a, 1989b, Karlsson et al. 1994). Variability in prey capture among individuals likely results in local differences in individual plant size (Thum 1988, Schulze and Schulze 1990, Zamora et al. 1997).

In their classic cost/benefit model, Givnish et al. (1984) predicted that plant carnivory would be beneficial only in nutrient-poor soils. To test this model, Ellison (2006) pooled data from 29 studies that tested the influence of prey and soil fertilization on growth in either greenhouse or field conditions. His meta-analysis found a significantly positive effect of prey capture on growth ($P = 0.02$) but no effect of soil fertilization ($P = 0.15$) or soil nutrient × prey interaction ($P = 0.81$). Thus, the effect of mineral enrichment of natural peaty soils may not always lead to growth increase (cf. Svensson 1995), and an efficient use of prey is not limited to nutrient-poor soils. It is not possible to generalize from these experiments because the level of mineral enrichment varied greatly among the studies and also could have been excessive (Adamec 1997a). Field experiments also revealed that prey capture was much more important for seedlings and young plants than for adult ones. Generally, prey capture by juveniles is limited, but in successful individuals it supports much faster growth and attainment of maturity, prolific flowering, and seed set (Thum 1988, Zamora et al. 1997, Thorén and Karlsson 1998, Hatcher and Hart 2014). In a positive feedback loop, the successful, larger individuals are better competitors (Wilson 1985) and capture larger prey. There are no indications that capture of prey in adult plants is allocated selectively to reproduction, but it does accelerate the time by which an individual plant reaches the minimum size necessary for flowering.

An alternative way to quantify growth benefits of carnivory expresses increase in plant growth as a function either of nutrients absorbed from captured prey or an application of mineral solution onto the traps in greenhouse or field growth experiments. Using published data (Karlsson and Carlsson 1984, Thum 1988, 1989b, Chapin and Pastor 1995, Adamec 2002) the biomass increase (treatments minus controls) for 13 *Drosera* species, *P. vulgaris*, and *S. purpurea* per amount of N absorbed from prey (in mg mg^{-1}) ranged from 21 to 5020; for P, 206–13,000, and for K, 170–10,400. Based on usual values of shoot nutrient content these species (Table 17.1), the values far exceeding 70–100 for N, 800–1200 for P, and 70–120 for K indicate stimulation of root nutrient uptake by prey capture (§17.4.4).

17.6 Effects of mineral nutrition on expression of carnivorous traits

Expression of carnivorous traits in *Drosera rotundifolia, Dionaea muscipula*, and *Sarracenia purpurea* depends at least in part on the availability of mineral nutrients (mainly N) in the soil or from atmospheric deposition. Leaves of *D. rotundifolia* grown outdoors reduced their "stickiness" (as a measure of mucilage production by tentacles; Chapter 12) by 45% after their peaty soil was fertilized (Thorén et al. 2003). This species also showed increased long-term mortality in a *Sphagnum*-dominated raised bog following application of 2 or 4 g N (as NH_4NO_3) m^{-2} yr^{-1} (Redbo-Torstensson 1994). On 16 ombrotrophic bogs across Europe with background N deposition rates of 0.1–2.3 g N m^{-2} yr^{-1}, Millett et al. (2015) found that the seasonal N gain by *D. rotundifolia* shoots from prey decreased linearly from 54% to 20% with increasing N deposition while the shoot N content and total plant N amount doubled. Increasing soil N availability increased the proportion of N absorbed by roots and shifted the limiting nutrient for growth from N to P.

Feeding milled insects to greenhouse-grown *Dionaea muscipula* significantly decreased the trap:petiole biomass ratio in young leaves of autumnal plants with large traps, indicating a feedback regulation of trap construction by prey-derived nutrients (Kruse et al. 2014). Similarly, *D. muscipula* fed milled insects accumulated 113% more total N than unfed controls after five weeks, indicating a 48% N uptake efficiency from prey, but increased the root:shoot ratio by 51% (Gao et al. 2015). Supplemental feeding also decreased the trap:petiole ratio in young leaves by 2.75×, reflecting the doubled N content of fed plants relative to unfed controls. Large amounts of NH_4^+ and glutamine were absorbed by *Dionaea* roots during a one-day application of 2-mM NH_4NO_3 + 1-mM glutamine solution to the peaty soil (an increase of 22–29% relative to the controls). Unfed fertilized plants absorbed 55% of their total increased N amount as NH_4^+, 40% of N as glutamine, but only 5% as NO_3^-, illustrating a surprisingly high uptake capacity of these roots and a marked preference for NH_4^+ (Adamec 1997a). These results are in good agreement with the known expression of the NH_4^+ transporter DmAMT1 in *Dionaea*; constitutive DmAMT1 expression was 2.5× higher in roots than in traps. However, Scherzer et al. (2013) found that NH_4^+ uptake by *Dionaea* traps was inducible and stimulated during prey digestion. One-day pretreatment of roots of *Dionaea* plants with 0.1- or 0.5-mM NH_4Cl significantly reduced the uptake affinity of root cells whereas their uptake capacity was unchanged (Gao et al. 2015).

Sarracenia purpurea relies on a food web of organisms inhabiting its water-filled pitchers to break down the prey and a microbiome to mineralize it (Chapter 24). Most *Sarracenia* species produce wide pitchers, albeit with narrow keels (wings), for prey capture and nutrient uptake as well as leaves with wide keels but greatly reduced tubes (phyllodia) that have higher photosynthetic rates (Ellison and Gotelli 2002, Farnsworth and Ellison 2008). In a poor fen, Ellison and Gotelli (2002) fed open pitchers of *S. purpurea* with solutions of NH_4Cl, NaH_2PO_4, or both at two concentration levels for three seasons and measured the morphology of newly produced leaves. The production of pitchers or phyllodia depended on N, but not P, addition to pitchers. These authors also found a significant linear association between the relative keel size of plants growing in 26 bogs or fens throughout New England (USA) whose N deposition rates ranged from 0.85 to 2.7 g N m^{-2} yr^{-1}. *Sarracenia purpurea* is thus very sensitive to external addition of mineral N from deposition either directly to its pitchers or to the adjacent peat and, in line with the cost/benefit model (Chapter 18), produces phyllodia in response to increasing N addition. The relative abundance of phyllodia can be used for monitoring the quantity of N deposition (Ellison and Gotelli 2002). In a transplant experiment between an acidic bog and a neutral fen differing in their N soil status, Bott et al. (2008) confirmed that the pitcher–phyllode plasticity depended on soil chemistry and N shoot content. However, pitcher N and P contents of field-grown *S. purpurea* plants doubled when fed supplemental houseflies but pitcher morphology was unchanged (Wakefield et al. 2005). It appears that growth of *S. purpurea* populations in bogs is N limited and that direct uptake of mineral, but not organic, N either from roots or pitchers regulates expression of carnivorous traits of developing leaves.

17.7 Mineral nutrition of *Nepenthes*

Nepenthes pitcher plants have become a suitable object for mineral nutrition and cost/benefit studies because their leaves are clearly differentiated into a photosynthetic lamina and a pitcher trap. By comparing tissue nutrient content in eight lowland *Nepenthes* species with those of co-occurring noncarnivorous plants, Osunkoya et al. (2007) concluded that *Nepenthes* pitcher plants were more N- than P-limited. Overall, the mean assimilation leaf N content of 0.81% in lowland heath ("kerangas") and peat-swamp inhabiting *Nepenthes* species studied was significantly lower than the 1.16% in the leaves of co-occurring noncarnivorous plants of any growth form (Osunkoya et al. 2004). However, the mean leaf contents of P (0.19%) and K (1.08%) in *Nepenthes* were similar to co-occurring noncarnivorous plants (0.12%, 1.05%, respectively). P and K contents of the *Nepenthes* leaves and pitchers also were similar to other terrestrial carnivorous plants (Adamec 1997a, Ellison 2006). *Nepenthes* pitchers had significantly lower N, lower or equal P, and higher K contents than their corresponding photosynthetic laminae (Osunkoya et al. 2007, Pavlovič et al. 2007, 2009, Karagatzides and Ellison 2009, Pavlovič and Saganová 2015).

We hypothesize that the lower N content of *Nepenthes* pitchers is a consequence of pitcher specialization for trapping and digestion rather than photosynthetic carbon assimilation. The most abundant protein in green plant cells is the carboxylation enzyme Rubisco and 20–30% of the total leaf N is incorporated into it (Evans 1989, Feller et al. 2008). Indeed, the Rubisco content and corresponding photosynthetic capacity (A_{max}) in *Nepenthes* pitchers are the lowest among all pitcher-plant genera (Pavlovič et al. 2007, 2009, 2011b, Karagatzides and Ellison 2009, He and Zain 2012, Pavlovič and Saganová 2015). The higher K content is hypothesized to be associated with the greater osmoregulatory activities linked with transfer of ions and nutrients in pitcher walls relative to the leaf lamina (Osunkoya et al. 2007). However, Pavlovič and Saganová (2015) found elevated K content in *Nepenthes* pitchers mainly in the peristome and the lid, failing to support Osunkoya et al.'s (2007) hypothesis.

Van der Ent et al. (2015) investigated nutrient content of the obligate ultramafic highland species of *Nepenthes* in Borneo's Kinabalu National Park. There, the ultramafic soils have high levels of Mg, Ca, Co, Cr, and Fe. This was partially reflected in the foliar chemistry of *N. rajah*, *N. villosa*, *N. edwardsiana*, and *N. burbidgeae*, which had higher foliar Mg, Ca, and Na relative to highland *N. lowii* and *N. macrophylla* growing on non-ultramafic soils. However, contents of other trace elements (Co, Cr, Fe, Mn, Ni) that are high in ultramafic soils were low in *Nepenthes* foliage. The low contents of Al, Ca, Co, Mn, and Ni in foliage of *Nepenthes* relative to co-occurring noncarnivorous plants growing on ultramafic soils provides additional evidence for active exclusion of metals by *Nepenthes*. Finally, the foliar N contents were similar for *Nepenthes* and co-occurring species, whether or not they were growing on ultramafic soils (0.86–1.0%).

Similarities in mineral nutrition between lowland and highland *Nepenthes* outweigh their differences (Osunkoya et al. 2007, van der Ent et al. 2015). Both groups had higher or comparable foliar P and K contents relative to co-occurring noncarnivorous plants, but lower or comparable foliar N contents. *Nepenthes* is strongly N-limited where prey abundance and capture is high; P and K are absorbed more readily from prey than N (Pavlovič et al. 2014). Indeed when *Nepenthes* was prey-deprived for 18 weeks, there was a stronger negative effect on P than N contents (Moran and Moran 1998).

Nonetheless, *Nepenthes* plants depend strongly on prey for N; insect prey contributes 54–68% of total foliar N in *Nepenthes*. Young *Nepenthes* leaves with closed pitchers are a major sink for N, which they obtain mainly from insects captured by older pitchers (Schulze et al. 1997, Moran et al. 2001). Schulze et al. (1999) used *in situ* hybridization for localization of NH_4^+ transporter expression (NaAMT1) in head cells of the digestive glands of *N. alata*. This transporter is homologous to that found in *Dionaea* traps (Scherzer et al. 2013; Chapter 16). Moran et al. (2010) investigated NH_4^+ uptake in three *Nepenthes* species using vibrating ion-selective microelectrodes and found that the nutrient sequestration strategy was mirrored by pitcher physiology. Species producing long-lived pitchers with low prey-capture rates (*N. ampullaria*, *N. bicalcarata*) had lower rates of NH_4^+ uptake than *N. rafflesiana*, a species producing short-lived pitchers with high capture rates. As expected, the highest NH_4^+ uptake rate was detected in the digestive zone of the pitchers.

Nepenthes lowii, *N. rajah*, and *N. macrophylla* have evolved specialized nutrient acquisition strategies: mutualistic associations with small mammals that defecate into the pitchers (Clarke et al. 2009, Chin et al. 2010, Grafe et al. 2011; Chapters 12, 15), which obtain substantial amount of N and other nutrients from the feces (Chapter 26). Other species (*N. bicalcarata*) cooperate with ants in prey capture (Bazile et al. 2012) or collect leaf litter (detritivory; *N. ampullaria*) into their pitchers (Moran et al. 2003, Pavlovič et al. 2011b; Chapter 15). Foliar N content increased in both young and mature leaves of *N. talangensis* as a result of N uptake from prey (Pavlovič et al. 2009). However, in other studies, N contents were observed to increase either slightly or not at all, while total plant N increased due to increased biomass (foliar N content × DW: Moran and Moran 1998, Pavlovič et al. 2011b). It seems that if the amount of captured prey is low, *Nepenthes* preferentially invests N from captured prey into production of new foliage instead of increasing foliar N (Moran and Moran 1998). However, if prey capture exceeds a threshold for relatively slow growth, foliar N content and total plant biomass increase together. Plant photosynthesis is then increased either per unit leaf mass or total leaf mass (Pavlovič et al. 2009). Soil fertilization with mineral solution can increase leaf foliar N content and photosynthetic rate, indicating that roots of *Nepenthes* have the potential for rapid nitrogen uptake; however, fertilized plants stop producing pitchers (Pavlovič et al. 2010b). As for other pitcher-plant genera, the possible stimulation of root mineral nutrient uptake by foliar nutrient uptake in *Nepenthes* has not been investigated. In conclusion, some ecological nutritional strategies in *Nepenthes* indicate a partial retreat from carnivory but carnivory in all species is retained, though its efficiency might be reduced.

17.8 Nutritional cost/benefit relationships of carnivory

In their microhabitats, terrestrial carnivorous plants grow together with many noncarnivorous ones and both groups experience similar ecological conditions (Chapter 2). Carnivory is thus only one of many possible adaptations to wet, nutrient-poor soils. Givnish et al. (1984; Chapter 18) predicted that carnivory would evolve among plants growing in sunny, moist, and nutrient-poor habitats when the marginal photosynthetic benefits of carnivory exceeded its marginal costs. However, the emphasis on photosynthesis and C-acquisition as the primary benefit of carnivory likely has been overestimated. Terrestrial carnivorous plants as stress-tolerating species have low A_{max} and corresponding RGRs, and their capacity to increase their growth rates appears to be limited (§17.2., Adamec 2002, 2011c, Farnsworth and Ellison 2008). However, the physiological cause of the low photosynthetic rates remains unknown. Because photosynthesis commonly increases in well-fed relative to starved plants (Pavlovič and Saganová 2015) and foliar N and P contents are low, the slow growth rates of carnivorous plants are limited at least partially by a shortage of mineral nutrients. We therefore suggest that the primary physiological benefit of carnivory is acquisition of N and P needed for increase of photosynthesis and essential growth processes in shoot apices and young foliar tissues, as well as for stimulation of nutrient uptake by roots (Adamec 2011c).

The ecological cost/benefit relationships of carnivory could be expressed better in terms of marginal gain of limiting mineral nutrients coming from carnivory or the efficiency of mineral nutrient investment in traps (Adamec 2011c). The basic mineral nutrient cost of carnivory is the investment of some amount of mineral nutrients in trap production. A portion of this amount (M_1) is lost from senescent biomass; the rest (M_2) is reutilized. Traps gain M_3 mineral nutrients from prey over their lifespan. The M_3:M_1 ratio characterizes the direct nutritional benefit (and efficiency) of carnivory for each nutrient. Moreover, due to the stimulation of root nutrient uptake, an additional proportion of mineral nutrients (M_4) is taken up by roots from the soil; this represents the indirect ecophysiological benefit of carnivory. The M_4:M_1 ratio specifies the indirect nutritional benefit of carnivory, the sum $M_3 + M_4$ the total ecophysiological benefit of carnivory in terms of nutrient gain, and $(M_3 + M_4)/M_1$ is the total nutritional benefit of carnivory. When this latter value for a given nutrient >1, carnivory will be ecologically beneficial (Adamec 2011c).

To what extent does this model reflect reality? For *Drosera* and *Pinguicula* species, the total nutritional benefit of carnivory at high prey capture rates should be within 2–10 for N, P, and K, ≈1 for Mg, but only ≈0.5 for Ca; the direct values for all nutrients (or total ones at low prey capture) should be much lower (<1). All values decline with increasing soil nutrient content, illustrating that direct mineral nutrient benefits from carnivory are usually several times lower than indirect benefits due to stimulation of nutrient uptake by roots. To be nutritionally beneficial, carnivorous plants simultaneously need to capture prey efficiently, maximize nutrient uptake from prey, and minimize nutrient losses in senescing traps.

This "nutritional" model of carnivory does not deny the outcomes of the "photosynthetic" model (Givnish et al. 1984; Chapter 18) as it is still unclear which is more important: root nutrient uptake stimulation as an indirect ecophysiological benefit of carnivory or photosynthetic benefit? However, the nutritional model (Adamec 2011c) to date has received much less attention, probably due to common evidence for increase of photosynthetic rates as a function of prey capture.

17.9 Future research

Experimental prey additions have revealed that prey-derived N is allocated mainly to growth of new (young) leaves. Both positive and negative interactions occur between mineral nutrient uptake by roots and traps; the signs of these interactions are species-specific. Uptake of organic substances from prey by traps could regulate important processes of mineral nutrition in whole plants although the direct ecological significance of this uptake appears to be marginal. Overall, carnivory in itself, although it creates only a part of purposeful adaptations of carnivorous plants, is almost indispensable for naturally growing ones.

Future research on the ecophysiology of growth and nutrition of terrestrial carnivorous plants could profitably address a number of key questions. First, given the phylogenetic diversity of carnivorous plants and the structural diversity of their traps and other traits, is it possible to define botanical carnivory on a common basis of mineral nutrient uptake from animal prey? Second, with respect to growth effects of carnivory, what are the primary and secondary physiological effects of growth enhancement by carnivory? In particular, is the positive growth effect caused by stimulation of cell divisions in shoot apices of juvenile tissues? What is the role of tissue nutrient content?

Along these same lines, the mechanism by which foliar uptake of nutrients stimulates root uptake of nutrients remains unknown. Is it mediated by increased allocation of photosynthates or mineral nutrients from leaves or traps to roots? Does it represent primary or secondary effects of utilization of prey? In a related vein, what is the significance of direct uptake of organic nutrients (e.g., amino acids) by carnivorous plants? And finally, what is the affinity and capacity of root uptake by carnivorous plants relative to noncarnivorous ones for a range of different mineral nutrients?

CHAPTER 18

Why are plants carnivorous? Cost/benefit analysis, whole-plant growth, and the context-specific advantages of botanical carnivory

Thomas J. Givnish, K. William Sparks, Steven J. Hunter, and Andrej Pavlovič

18.1 Introduction

Most plant species are consumed, in whole or part, by animals acting as herbivores, pollinators, or seed dispersers, but carnivorous plants have turned the ecological tables and consume animals as prey. Carnivorous plants thus interact in unique ways with animals that serve as competitors for prey, digestive symbionts, food guards, butlers, kleptoparasites, or even prey mutualists and sources of nutrients via excreta (Chapters 21–26 and below). More importantly, carnivorous plants—by absorbing mineral nutrients from animals via costly traps that attract, capture, and/or digest prey—have gained the ability to live and compete successfully in nutrient-poor environments, but at the expense of reduced competitive ability elsewhere (Givnish et al. 1984, Ellison and Gotelli 2002, Ellison and Adamec 2011, Pavlovič and Saganová 2015; Chapter 2).

Darwin (1875) devoted only a few lines to the ecological value of botanical carnivory, stating that "[t]he absorption of animal matter from captured insects explains how *Drosera* can flourish in extremely poor peaty soils . . . considering the nature of the soil where it grows, the supply of nitrogen would be extremely limited, or quite deficient, unless the plant had the power of obtaining this important element from captured insects" (pp. 14–15). Darwin (1875) focuses almost entirely on the structure and workings of the traps in different groups, with little said about the functional consequences of carnivory. The beneficial impacts of prey capture on growth and seed production by *Drosera* were first documented three years later by his son, Francis Darwin (1878).

Givnish et al. (1984) addressed the ecological value of carnivory in detail for the first time with a cost/benefit model for the evolution of carnivory in plants. This model explained why carnivorous plants are common in habitats that are not only nutrient-poor but also sunny and moist, and why plants might adjust their allocation to carnivory in different circumstances. This model has stimulated a large amount of research over the past 32 years, and provided a conceptual framework for many studies of ecophysiology, ecological distribution, resource allocation, and evolution of carnivorous plants (Givnish 1989, Adamec 1997a, Ellison and Gotelli 2002, Ellison 2006, Anderson and Midgley 2007, Gibson and Waller 2009, Ellison and Adamec 2011, Clarke and Moran 2016; Chapters 2, 3, 10, 17, 19).

Here we re-assess this model, discuss how its predictions can be extended in a number of important ways, and evaluate its assumptions and predictions in terms of what is now known about photosynthesis, respiration, relative growth rate (RGR), and resource allocation in carnivorous versus noncarnivorous plants, nutrient limitation and stoichiometry, and adaptation to different kinds of prey.

Givnish, T. J., Sparks, K. W., Hunter, S. J., and Pavlovič, A., *Why are plants carnivorous? Cost/benefit analysis, whole-plant growth, and the context-specific advantages of botanical carnivory*. In: *Carnivorous Plants: Physiology, ecology, and evolution*. Edited by Aaron M. Ellison and Lubomír Adamec: Oxford University Press (2018). © Oxford University Press.
DOI: 10.1093/oso/9780198779841.003.0018

18.2 The cost/benefit model for the evolution of plant carnivory

Traps in carnivorous plants provide an alternative source for the nutrients absorbed by roots in other plants, but they involve additional energetic costs, including production of attractive nectars or aromas (Chapter 12), secretion of digestive enzymes (Chapters 13, 16), absorption of nutrients (Chapter 16), trap activation or resetting (Chapter 14), and reduced carbon gain associated with production of photosynthetically inefficient leaves (Givnish et al. 1984, Givnish 1989, Adamec 2006, 2010a, 2010c, Hájek and Adamec 2010, Ellison and Adamec 2011, Pavlovič and Saganová 2015). Botanical carnivory—the attraction, capture, or digestion of prey, and subsequent uptake of nutrients from dead prey resulting in increased plant growth or reproduction (Chapter 1)—should be favored whenever the energetic benefits of a small investment in carnivory exceed the costs (Figure 18.1).

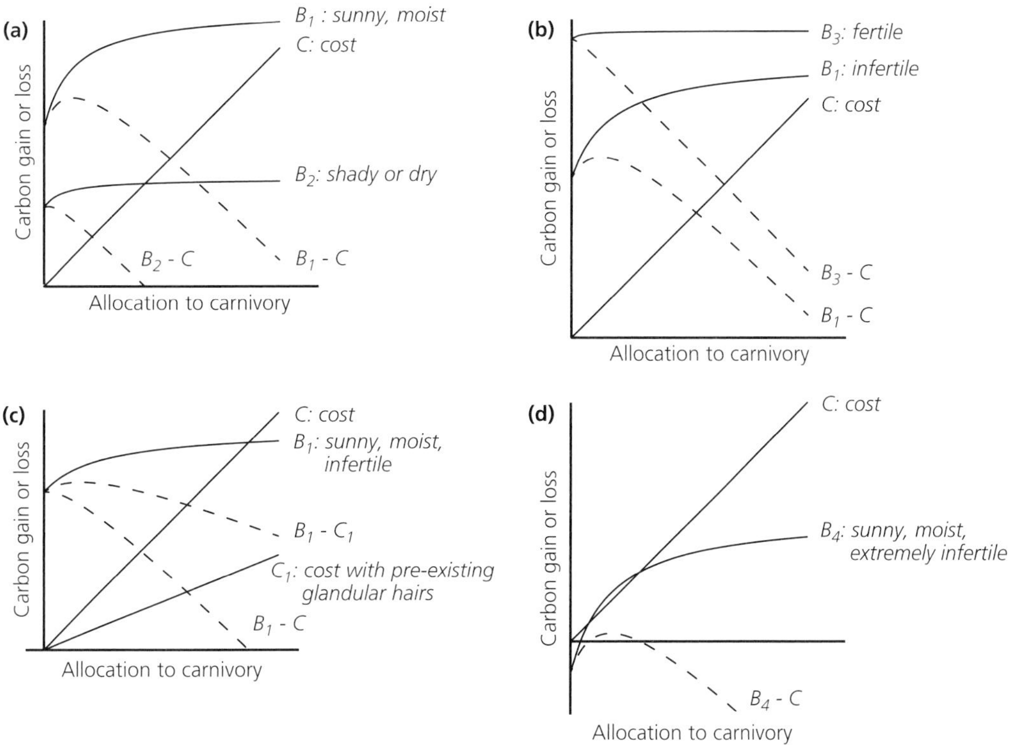

Figure 18.1 Cost/benefit model for the evolution of carnivory in plants, plotting photosynthetic benefits and costs against different levels of investment in carnivorous adaptations under different conditions. (**a**) On nutrient-poor sites, the rate at which photosynthesis (*B*) increases with investment in carnivory (as a result of greater nutrient supply) should be higher, and show less tendency to plateau, in well-lit and moist microsites (B_1) than where light or water more strongly limit carbon uptake (B_2). Dashed lines show net difference between photosynthetic benefit and cost (*C*) of obtaining nutrients through carnivory. Carnivory should evolve whenever the benefit of a small investment in carnivory exceeds its own cost; that is, when $dB/dx > dC/dx$, or when the net profit curve $B–C$ slopes upward near $C = 0$. (**b**) On sunny, moist, nutrient-rich sites, photosynthetic rates (B_3) should be elevated relative to otherwise similar but nutrient-poor sites where carnivory can make little or no increment to photosynthesis ($dB/dx \approx 0$), so carnivory is unlikely to be favored. (**c**) In lineages in which glandular hairs are already present and co-used in carnivorous mutants, the cost curve (C_1) should be reduced, favoring the evolution of carnivory. (**d**) On bare rock or extremely sterile soil, noncarnivorous ancestors should show negative carbon balance; even under well-lit and -watered conditions, nutrients gained from a small investment in carnivory are unlikely to result in positive carbon balance, even if a greater, optimal investment would. Thus, adaptation to extremely infertile sites via carnivory is unlikely to arise directly, but instead in steps, with small initial investments favored on sunny, moist, somewhat infertile sites, leading to plants being able to survive on, and later become adapted to, extremely infertile sites.

18.2.1 The benefits of carnivory

The cost/benefit model assumes that advantages in energy capture translate into advantages in intra- and interspecific competition that lead to gains in fitness. It also assumes that any initial trap structure (or subsequent refinement thereof) is a simple phenotypic step from ancestral forms (Chapter 3); no matter what the energetic or fitness advantage of carnivory, it is extremely unlikely that a single mutation or set of mutations encoding a complex trapping mechanism could arise all at once. Finally, it also assumes that investment in carnivory is scalable once traps have evolved; this is straightforward, given the modular construction of plants and resulting ability to vary trap number, size, and nectar and digestive enzyme secretion rates once traps have evolved. Given these assumptions, selection within species or competition among species should favor carnivory whenever the initial marginal benefit obtained by carnivory (measured as increased rates of photosynthesis per unit leaf mass or whole-plant growth) exceeds the marginal cost (measured in the same units of carbon) of constructing traps (Givnish et al. 1984, Givnish 1989; Figure 18.1).

Based on first principles, carnivory might provide four kinds of energetic benefits that would enhance whole-plant carbon gain: (1) an increased rate of photosynthesis per unit leaf mass; (2) an increased rate of conversion of photosynthate to new leaf tissue; (3) reduction in allocation of photosynthate to unproductive roots as nutrient absorption by trap leaves increases; or (4) partial replacement of autotrophy with heterotrophy (Givnish et al. 1984, Givnish 1989). Uptake of N, P, or other nutrients via carnivory (Chapters 17, 19) could result in increased concentrations of Rubisco or other photosynthetic proteins in new leaf tissue (Pavlovič et al. 2016), increased stores of nutrients to produce such proteins and convert stocks of carbohydrates into new leaves, or decreased demand for nutrients supplied by roots.

Increased photosynthesis. Many studies show that feeding enhances the growth of carnivorous plants, especially under nutrient-poor conditions (Chapters 17, 19), but few have tested the proposed elevation of leaf nutrient levels and photosynthesis by feeding (Hypothesis 1). The strongest support for Hypothesis 1 is provided by Farnsworth and Ellison (2008) for ten *Sarracenia* species and by Pavlovič et al. (2014) for *Drosera capensis*. Gao et al. (2015) supported Hypothesis 2 by showing that fed *Dionaea* produce larger petioles and smaller traps. Comparative data indicate that carnivores often have substantially smaller root systems than non-carnivores in the same habitat (Darwin 1875, Brewer 2003, Brewer et al. 2011)—3.4 to 23% of total biomass for the few species studied (Adamec 1997a)—providing circumstantial support for Hypothesis 3. However, some experiments have shown increased root growth and nutrient uptake following prey capture (Hanslin and Karlsson 1996, Adamec 1997a, 2002, Lenihan and Schulz 2014, Gao et al. 2015).

Heterotrophy or autotrophy? There is little support for Hypothesis 4. Terrestrial plants appear to obtain mostly nutrients, not carbon, from carnivory (e.g., Chandler and Anderson 1976, Adamec 1997a), although some carbon uptake does occur, perhaps via absorption of amino acids (Dixon et al. 1980, Rischer et al. 2002). Fasbender et al. (2017) showed that *Dionaea* uses prey-derived amino acid carbon to fuel respiration. However, the situation may be different for aquatic *Utricularia*, which can grow in complete darkness if supplied with a carbohydrate-rich medium (Harder 1970). Terrestrial species of *Utricularia* allocate < 9% of their biomass to leaves, calling into question whether they are fully autotrophic (Porembski et al. 2006). Michalko et al. (2013) showed that *Drosera rotundifolia* secretes a β-1, 3-glucanase that can cleave plant glucans and lead to absorption of simple sugars. The magnitude of carbon uptake via this pathway, however, has yet to be quantified and may be quite low, because insects do not contain glucans. Glucanases in carnivorous plants have traditionally been viewed as defenses against microbial pathogens (Juniper et al. 1989, Hatano and Hamada 2008, Schulze et al. 2012).

18.2.2 Benefits vary with environmental conditions

Let us consider plants of a given size—with an initial fixed biomass in leaves and roots—and ask how allocation of energy to carnivory would affect their net growth and, thus, competitive ability and fitness

(Givnish et al. 1984). As the amount of energy x invested in carnivory (e.g., traps, digestive enzymes) per gram of leaf mass rises (the cost curve $C(x)$ in Figure 18.1), prey capture, nutrient absorption, and effective rate of photosynthesis per unit leaf mass (g C g^{-1} leaf s^{-1}) should also increase, rising linearly initially and then plateauing with increasing investments in carnivory (the benefits curve $B(x)$ in Figure 18.1). The initial increase could result from (a) an increase in the absolute rate of photosynthesis; (b) an increase in the rate of conversion of carbon skeletons to new leaf tissue because of the higher availability of nutrients provided by carnivory; or (c) a decline in the fractional allocation of energy to unproductive but nutrient-absorbing roots (Givnish et al. 1984, Givnish 1989). In retrospect, we think that (c) is unlikely to alter the curve $B(x)$, and instead should discount the net cost of carnivory, decreasing the slope of the cost curve $C(x)$.

Photosynthetic enhancement. The amount by which photosynthesis (or conversion rate of photosynthate to leaf tissue) can be enhanced by increased nutrient input depends on ecological conditions. The effective rate of photosynthesis is unlikely to rise unless nutrients are in short supply and limit photosynthesis or the conversion of photosynthate into new leaf tissue, so the greatest benefit is expected on nutrient-poor sites. The usual increase in the growth of carnivorous plants when supplied with prey on nutrient-poor substrates disappears if nutrient availability is increased by fertilizing the substrate itself (Ellison 2006). Growth is not photosynthesis, but these results are consistent with (and analogous to) how soil fertilization leads to enhanced rates of photosynthesis and whole-plant rates of growth in noncarnivorous plants (e.g., Reich et al. 2003, Ellison 2006, Drenovsky et al. 2012). The well-documented rise following supplemental feeding in the concentrations of N and P in leaf tissue of terrestrial carnivorous plants (Chandler and Anderson 1976, Christiansen 1976, Karlsson and Pate 1992, Chapin and Pastor 1995, Wakefield et al. 2005, Ellison 2006, Farnsworth and Ellison 2008, Pavlovič et al. 2014) implies an increased rate of photosynthesis, based on the rates of carboxylation, electron transport, and net photosynthesis increasing with leaf N and P concentration across plants worldwide (Walker et al. 2014).

If factors such as light or moisture are in limited supply, they should limit photosynthesis and the extent to which additional nutrients provided by carnivory can elevate carbon gain (Givnish et al. 1984). On economic grounds, we expect lower levels of moisture or light to reduce optimal stomatal conductance and, thus, maximum photosynthetic rates (A_{max}) at a given level of leaf N concentration and mesophyll photosynthetic capacity (Givnish and Vermeij 1976, Cowan and Farquhar 1977, Wong et al. 1979), flattening the A_{max} vs. [N] response. Lower moisture supply should favor thicker leaves with more mass per unit area (Givnish 1979), resulting in more internal self-shading and longer average distances for CO_2 to diffuse from the stomata, also flattening the A_{max} vs. N response. As expected, photosynthesis per unit leaf mass at a given leaf N concentration is lower in drier habitats around the world, and increases less rapidly with leaf N concentration (Wright et al. 2005).

Plateauing of benefits. As the amount of energy x devoted to carnivory rises, the benefit curve $B(x)$ should plateau, perhaps because of saturation in prey capture efficiency (e.g., a Type-II functional response), but certainly as factors other than nutrients limit photosynthesis or the conversion of photosynthate into new leaf tissue (Givnish 1989). For example, $B(x)$ would level off if increased nutrient supply increased the rate of conversion of photosynthate into new leaf tissue at a constant nutrient concentration but not the nutrient concentration per unit leaf mass (Givnish 1989). As nutrient uptake increases with investment in carnivory, the rate at which new leaves can be produced should depend less on limiting nutrients (e.g., N, P) and more on the availability of carbon skeletons. The latter should depend on the availability of light and water, so the conversion rates should rise most quickly and plateau more slowly in well-lit, moist, nutrient-poor areas (Givnish et al. 1984).

The difference between the benefit and cost curves in Figure 18.1 can be used to predict whether carnivory should be favored in a given environment, and if so, what the optimal level of investment in carnivory would be. Carnivory should evolve if benefits rise faster than costs at low levels of investment in carnivory: if $dB/dx > dC/dx$ or, equivalently, if the initial marginal benefit of carnivory exceeds its marginal cost; or the initial slope of the benefit curve is steeper than that of the cost curve. The optimal level of investment in carnivory should occur where the difference between the benefit and cost curves is maximized: $dB/dx = dC/dx$.

18.3 Predictions of the cost/benefit model

Thirteen specific predictions emerge from the cost/benefit model.

18.3.1 Carnivory is most likely to evolve and be favored ecologically in habitats that are sunny, moist, and nutrient poor

Sunny, moist, and nutrient-poor conditions are most likely to increase the initial steepness of the benefit curve; carnivory should evolve when nutrients alone limit photosynthesis, either directly or via effects on other processes (e.g., cell division) that limit photosynthesis (Figure 18.1a). This prediction does not depend on potential benefits of carnivory associated with decreased root function in anoxic soils (§18.3.10). Fertile substrates should elevate substantially the benefit curve but largely eliminate the marginal benefits of investment in carnivory, flattening $B(x)$ and working against the evolution and competitive ability of carnivorous plants (Figure 18.1b).

Many habitats are poor in nutrients. Wet soils *per se* are likely to be deficient in nitrate due to anoxia—O_2 diffuses 10,000 times more slowly in water than in air—reducing nitrogen fixation by free-living or symbiotic prokaryotes at pO_2 < 40–50 kPa (Serraj and Sinclair 1996), while denitrifiers remain metabolically highly active and release of nitrogen from dead plant tissue is greatly slowed (Vitousek and Howarth 1991). Ombrotrophic bogs are precipitation-fed and thus deficient in bedrock-derived P; consequently, they are also deficient in N because of the high energy requirements and P demands of N-fixers (Vitousek and Howarth 1991), and to the slow decomposition of dead plant remains. Ancient, highly leached soils and uplifted marine sands are likely to be deficient in both P and N (Chadwick et al. 1999), accounting for highly infertile sites in the boreal zone, the Atlantic and Gulf Coastal Plains of the USA, *tepuis* and adjacent sand plains of the Guayana Shield, and ancient sandy soils of South Africa and southwestern Australia, all hot-spots of carnivorous plant diversity (Givnish 1989). Frequent burning can impoverish soil fertility further by volatilizing leaf N and leading to partial losses of other nutrients from ash via leaching or run-off (Givnish et al. 1984). Paradoxically, calcareous spring-heads and highly calcareous soils may be P-deficient but extremely rich in cations because of P complexing with Ca, especially as a result of the degassing of CO_2 and co-precipitation of P with $CaCO_3$ as insoluble calcite around spring-heads (Boyer and Wheeler 1989).

18.3.2 Epiphytism works against carnivory and favors myrmecotrophy

Perches on tree boles and branches are nutrient-poor and often sunny, but also are episodically dry and thus unlikely to support carnivorous plants (Givnish et al. 1984, Benzing 1990). Sunny perches are more likely to support ant-fed plants than carnivores because of reduced rates of evaporation in the internal chambers of ant-fed myrmecophytes versus the active secretion of nectar or digestive fluids in carnivores (Thompson 1981, Givnish et al. 1984). For bromeliads, Benzing (1990) argued that lower costs would permit ant-fed plants to tolerate more shade, and that lax, nearly horizontal-leaved taxa (e.g., *Nidularium*) would be adapted best to the low light and high humidity of forest understories, and obtain added nutrients from fallen leaf debris.

18.3.3 Optimal investment in carnivory in terrestrial plants should increase toward the sunniest, moistest, most nutrient-poor sites

Sunny, moist, and nutrient-poor conditions should increase the initial steepness of the benefit curve and shift the optimum level of investment in carnivory higher (Figure 18.1a). This prediction can help account for both spatial patterns and seasonal variation in the expression of carnivorous traits, including clines in pitcher-plant form and seasonal heterophylly—the production of noncarnivorous leaves at particular times of year.

18.3.4 Optimal trap mechanism and form should depend on tradeoffs associated with environmental conditions, prey type, and trap type

Included here are variations in the form of sticky traps in *Drosera* and degree of leaf curling in *Pinguicula*, the advantage of snap-traps versus sticky traps in *Dionaea*, and the tradeoffs between wide-mouthed pitchers with wettable peristomes versus narrow-mouthed pitchers with waxy scales and viscoelastic fluids in *Nepenthes* from rainier versus drier areas. Also pertinent are the conditions that favor carnivory versus other strategies (e.g., N fixation, myrmecotrophy, coprophagy, and detritus capture) that can provide alternative sources of limiting resources (Givnish et al. 1984, Givnish 1989, Benzing 1990, Bonhomme et al. 2011a, Bazile et al. 2012, C.R. Schöner et al. 2015).

18.3.5 Carnivorous plants should have low photosynthetic rates and RGRs

The advantage of carnivory lies not in absolutely high photosynthetic rates, but rather in an increase in benefits minus costs per unit leaf mass. The poor, wet soils likely to favor carnivory should, by themselves, lead to low photosynthetic rates and RGR. Unless a given investment in carnivory were to yield a substantially greater return of nutrients than the same energy invested in roots, carnivory is unlikely to reverse this situation. If there were such a large return/cost advantage of carnivory versus roots, most plants would have already become carnivorous.

18.3.6 Rainy, humid conditions or wet soils favor carnivores by lowering the costs of glandular secretion or permitting passive accumulation of rainwater

Pitcher plants and functionally similar bromeliads (*Brocchinia hechtioides, B. reducta; Catopsis berteroniana*) typically are found in areas of heavy rainfall and where the ratio of precipitation to evapotranspiration is high (Givnish 1989). The benefit of heavy rainfall and high humidity is seen most easily for species that impound rainwater (e.g., *Heliamphora, Sarracenia purpurea, Brocchinia, Catopsis, Paepalanthus*), but such conditions also should reduce the amount of fluid that others (e.g., *Cephalotus, Darlingtonia, Nepenthes*) must secrete. Potential origins of carnivorous pitchers as hydathodes (§18.5.6) also would be facilitated by wet soils. Root pressure should be investigated as a mechanism for maintaining abundant glandular secretions. A dry atmosphere or soil would greatly increase the costs of maintaining glandular secretions in plants with sticky traps. There is a tension between increased supplies of soil moisture and decreased rates of evaporation on the one hand, and increased rates at which glandular secretions are washed away on the other, which might work against plants with sticky traps in areas of heavy rainfall (Givnish 1989).

18.3.7 Possession of defensive glandular hairs should facilitate the evolution of carnivory

In essence, the pre-existence and co-use of glandular hairs would reduce the cost curve and favor carnivory for smaller increments to photosynthesis and the benefits curve (Figure 18.1c). This would favor one of Darwin's proposed pathways to carnivory—and one that was almost surely the basis of the evolution of carnivory in five of ten carnivorous clades (Chapter 3)—but do so on economic grounds, adding to Darwin's (1875) functional analogies.

18.3.8 Fire over infertile substrates favors carnivory

Fire increases light availability, and often soil moisture, because of reductions in above-ground biomass and leaf area. Fire also volatilizes N and reduces the availability of this element in the long term (Christiansen 1976). These shifts all tend to elevate and steepen the benefits curve (Figure 18.1a) and thus favor carnivory. Brewer et al. (2011) argue that carnivores may never have an edge over noncarnivores, and that fire favors them simply by opening competition-free space (Chapter 2). But this ignores the advantage in local competition (or at least physiological tolerance) that carnivorous plants have on such cleared sites relative to noncarnivores, and the edge in regional competition obtained by reproducing on such sites. In essence, Brewer hypothesized that carnivorous plants can have an advantage only where other plants cannot grow, and that there is no set of conditions, slightly more fertile, slightly less open, where they continue to have a growth advantage. This seems unlikely, but no experiment has yet tested this proposition.

18.3.9 The ability of carnivorous plants to grow on bare rock or sterile sands must have evolved in stepwise fashion

A progenitor with no alternative source of nutrients would have negative carbon balance (respiration, no photosynthesis) at zero investment in carnivory, so any initial small investment in carnivory still would leave plants with negative growth (Figure 18.1d). However, if a plant evolved carnivory on a slightly more fertile substrate, and then increased its investment to the optimal level there (Figure 18.1a), it might then be able to survive on substrates with zero substrate fertility.

18.3.10 Anoxic or toxic soils should favor carnivory on open, moist sites

Anoxic or toxic soils render root function costly; shifting part or all nutrient uptake to trap leaves above-ground would effectively lower the cost curve via reduced allocation to roots, and thereby favor the evolution of carnivory and higher optimal levels of allocation to carnivory. However, this hypothesis ignores the possibly complex interplay of allocation to leaves, roots, and traps and their impact on photosynthesis per unit leaf mass and whole-plant growth. To assess these tradeoffs, we present a simple, two-dimensional model to explore the impact of allocation to roots versus traps, which predicts increased allocated to traps and decreased allocation to roots on sites with unfavorable root environments (Box 18.1).

Box 18.1

We ignore tissue respiration and mortality, assume that all tissues have the same construction cost, and further assume that the fractional allocations f_L, f_R, and f_C to leaf tissue, root tissue, and carnivory are constant and sum to 1. Let the water supply available per unit leaf mass $W = f_R \times \alpha_R / f_L$, where α_R measures the rate of water uptake per unit root mass, a function of root physiology and available water per soil volume. Let the supply of the most limiting nutrient $N = (f_R \times \beta_R + f_C \times \beta_C)/f_L$, where β_R measures the rate of nutrient uptake per unit root mass and β_C measures the rate of nutrient uptake per unit investment in carnivory. The β values reflect the physiology of roots and traps and the availabilities of the most limiting nutrient in soil versus air. Under these conditions, the parameter measuring the exponential rate of growth in whole-plant mass is $f_L \times A(W, N)$, where A_{mass} is photosynthetic rate per unit leaf mass, modeled as a compound Michaelis-Menten process, $A_{mass} = A_{max} \times W/(W + k_W) \times N/(N + k_N)$. The k values measure the extent to which additional water and nutrients can enhance photosynthesis; the smaller k is, the more rapidly A_{mass} rises with the supply of each resource and less of each resource is required to saturate photosynthesis.

The optimal allocation strategy (f_R, f_C) will always involve intermediate allocations to roots and leaves ($0 < f_R, f_L <$ 1), given that only roots can provide water (this may not be valid for pitfall plants like *Brocchinia reducta*) and that zero growth would result if no energy or all energy were allocated to leaves. Low soil fertility, anoxia, or soil toxins that reduce root efficiency would reduce β_R enough so that it should favor nutrient capture by above-ground trap leaves and decreased allocation to roots.

This analysis is a "toy" model, in that the parameters have not been adjusted to reflect the behavior of real plants. Nevertheless, it can be used to show that carnivorous plants are most likely to evolve when nutrient availability to roots is low and that to traps is high, and when photosynthesis is responsive to nutrient increments and little root allocation is needed to saturate water supply. As expected, optimal allocation to roots (f_R) increases as α_R declines, while optimal allocation to carnivory (f_C) increases as k_N increases and β_C declines; at the transition to carnivory, increases in β_C result in a large increase in allocation to carnivory and large decrease in allocation to roots (Figure 18.2).

Brewer et al. (2011) argued against carnivory evolving on wet soils through a mechanism that involved increasing photosynthetic rates because carnivorous plants have comparatively low photosynthetic rates (Ellison 2006); this misinterprets the cost/benefit model, which focuses on marginal, not absolute gains. Brewer et al. (2011) also suggested that carnivorous plants might be absent from dry sites not because such sites limit the benefits of carnivory, but because the shallow root systems of carnivores would exclude them from such sites. This claim involves a conundrum: do carnivorous plants have shallow root systems because they often grow in wetlands with anoxic

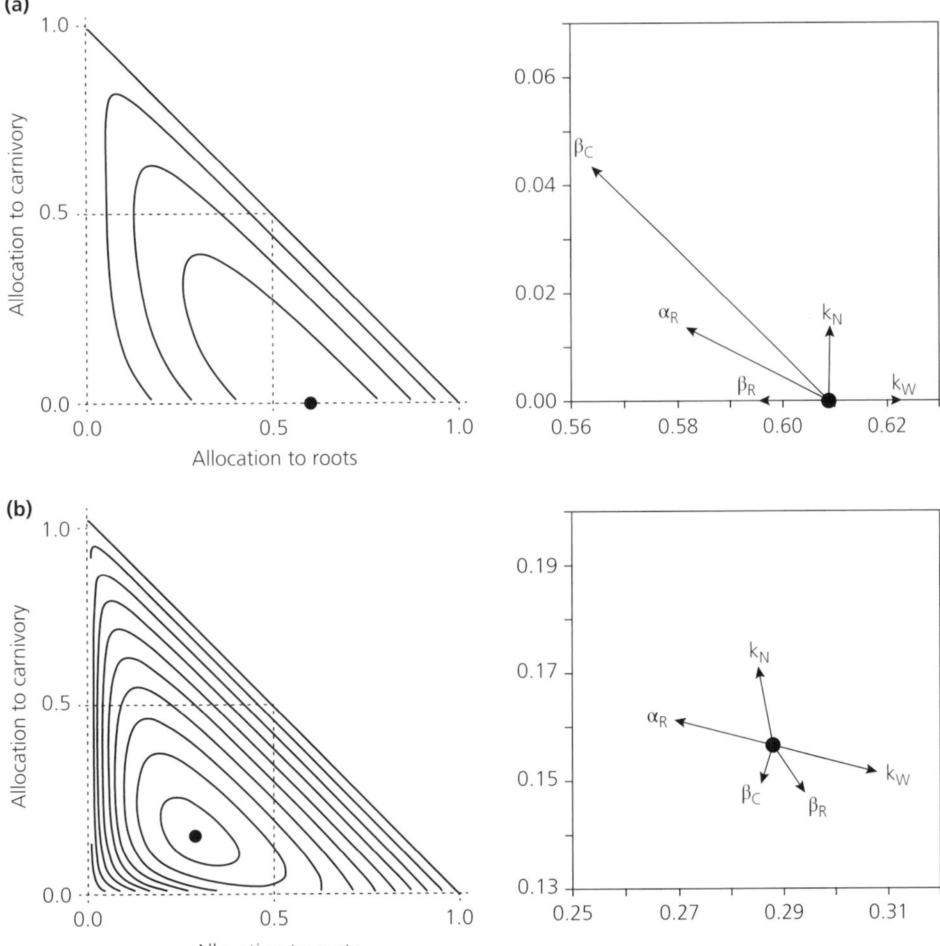

Figure 18.2 Two-dimensional cost/benefit model, plotting contours of the exponential coefficient of whole-plant growth rate against allocation to carnivory (f_C) and roots (f_R). Shown are contours of relative growth rate for (**a**) an edge optimum ($f_C = 0$) and (**b, c**) two internal optima (f_C, $f_R > 0$) (see Box 18.1). The dot in each left panel indicates the precise position of the optimum. Arrows in the enlargements (right) indicate shifts in each optimum resulting from a 20% increase in the value of the indicated parameters. For internal optima, increases in the effective delivery of the limiting nutrient per unit investment in carnivory (β_C)—or decreases in the same per unit investment in roots (β_R)—favor increased allocation to carnivory and decreased allocation to roots. A similar pattern may apply to some edge optima if the parameters favor a shift to carnivory. Greater water returns per unit root investment (α_R) favor increased root investment. Increased values of k_W decrease the value of additional small increments of water supply and result in decreased root allocation and increased allocation to carnivory; the opposite results from increased values of k_N, which decrease the value of additional small increments of nutrient supply. Parameter sets (α_R, β_C, β_R, k_N, k_W) are (1.5, 3, 1.51, 1, 1.05) for the edge maximum, and (2.87, 3, 0.42, 0.26, 0.49) and (1.29, 0.92, 0.08, 0.57, 0.16) for the internal maxima in (**b**) and (**c**).

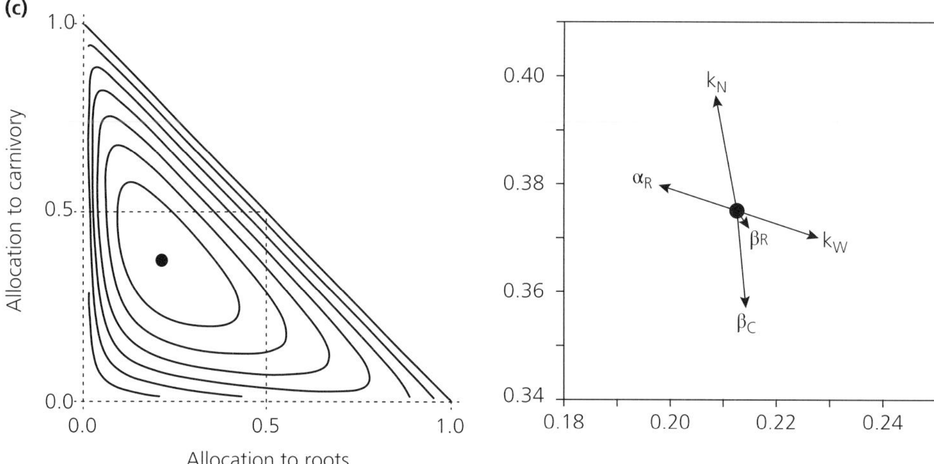

Figure 18.2 (*Continued*)

soils, or do their shallow root systems restrict them to wetlands?

Given that several carnivorous plants (including at least *Drosophyllum, Byblis lamellata, Nepenthes pervillei*, and pygmy sundews) have extensive root systems to exploit rock fissures and deep, well-drained sands (Juniper et al. 1989, Conran et al. 2002, Adlassnig et al. 2005b), it appears likely that the shallow root systems of other carnivores were shaped by anoxic soils, rather than being an inevitable consequence of the carnivorous habit. Brewer et al. (2011) also argued that in wetlands, carnivory simply may be an adaptive alternative to the production of deep roots with extensive aerenchyma, and found that all carnivores in a Mississippi wetland lacked aerenchyma. However, this view ignores the negative impact of sodden soils on N availability (§18.3.1). Further, the absence of aerenchyma is simply not characteristic of all carnivores: *Darlingtonia californica* and many species of *Drosera, Pinguicula*, and *Sarracenia* have gas-filled intercellular spaces in their root cortex (Adlassnig et al. 2005b).

18.3.11 Growth co-limitation by multiple nutrients may favor the paradoxical increase in root investment seen in carnivorous plants that have recently captured prey

For a single nutrient, the two-dimensional model outlined in Box 18.1 would favor decreased root allocation whenever increases in prey density or capture efficiency elevate β_C. However, if two or more nutrients co-limit plant growth, and prey capture elevates the level of one of them, it could favor increased root growth and nutrient capture. This would be even more likely if one or more of the co-limiting nutrients were retrieved more cheaply from the soil than from prey ($\beta_C > \beta_R$). But increased root growth and metabolism also would likely lead to increased uptake of the nutrient(s) retrieved more efficiently from prey, perhaps explaining the paradoxical finding in some feeding experiments that carnivores absorb more of a limiting nutrient than are contained in the prey captured (Hanslin and Karlsson 1996, Adamec 1997a, 2002).

Without nutrient complementarity, the two-dimensional model would argue against any simple positive feedback involving prey capture leading to increased root allocation (e.g., Ellison and Gotelli 2001): if prey capture in a particular site is a better source of a single limiting nutrient ($\beta_C > \beta_R$), then prey capture should mainly stimulate more trap production. Adamec (1997a, 2002) argues that animal prey are such rich sources of N and P, and such poor sources of K and Mg, that prey capture might favor increased root growth to absorb K and Mg (Chapter 17). This intriguing idea could be tested by modeling the uptake and benefits of several different nutrients while implicitly assuming that all nutrients mentioned essentially co-limit growth (cf. Tilman 1982).

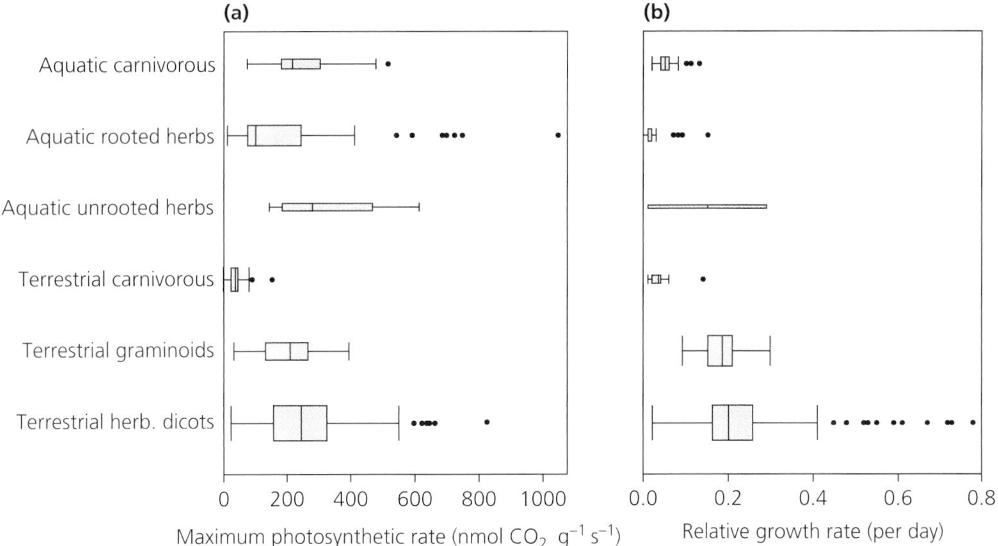

Figure 18.3 (**a**) Maximum photosynthetic rates per unit leaf mass (nmol CO_2 g^{-1} s^{-1}) for species in six growth-form categories of herbaceous plants. Boxes indicate median rates (center horizontal line), upper and lower quartiles (limits of grey boxes), upper and lower deciles (horizontal lines delimiting ends of vertical lines), and individual observations beyond the latter range. Box width is proportional to sample size, from $n =$ 8 for unrooted, submersed noncarnivores to $n = 141$ for terrestrial, herbaceous, noncarnivorous dicots. Note low rates for terrestrial carnivores vs. terrestrial noncarnivores; aquatic carnivores have rates comparable to those for aquatic noncarnivores. (**b**) Relative growth rates (RGR, g g^{-1} day^{-1}) for the same growth forms. Boxes as above, with sample sizes ranging from $n = 2$ for unrooted, submersed noncarnivores to $n = 208$ for terrestrial, herbaceous, noncarnivorous dicots. Note the low values of RGR for terrestrial carnivores vs. noncarnivores. Growing conditions were not controlled across either dataset. Redrawn from Ellison and Adamec (2011).

18.3.12 Paradoxically, in aquatic carnivorous *Utricularia*, harder, more fertile waters should favor greater investment in traps

Harder waters contain more cations, and often more P and N, but they also contain more CO_2 when in equilibrium with the atmosphere, and a greater pool of bicarbonate ions from which additional CO_2 can emerge as some is absorbed by photosynthesis (Hanson et al. 2006). Lakes also are often supersaturated in CO_2, presumably because of decomposing organic matter and dissolved inorganic carbon (DIC) delivered by springs (Hanson et al. 2006, Adamec 2008d, 2012b). Because of this large variation in CO_2 availability across lakes, and the much slower diffusibility of CO_2 in water than in air, photosynthesis by submersed aquatic plants can be strongly limited by CO_2 availability. Although there are more mineral nutrients (cations) in ponds with harder waters, and sometimes more PO_4-P and NO_3-N (e.g., Adamec 2008d), we suspect that the dominant effect setting optimal allocation to costly traps is CO_2 availability. The greater $[CO_2]$ is, the greater should be the benefit in realized photosynthesis for a given increment to leaf nutrients via prey capture, favoring greater investment in traps (Adamec 2015a; Chapter 19).

18.3.13 Soil anoxia or extreme infertility militate against tall, woody plants and may restrict carnivory to short, mostly herbaceous plants

Woody plants, because of their secondary thickening, lack aerenchyma linking roots to leaves to air, and thus typically are excluded from sodden, nutrient-poor soils. Extreme soil infertility favors carnivory but not tall, woody plants (Givnish 2003, Givnish et al. 2014b). Together these considerations help explain why carnivory is restricted mostly to short, herbaceous plants.

18.4 Assumptions of the cost/benefit model

The modified cost/benefit model (Figures 18.1, 18.2) has five important assumptions: (1) the costs of carnivory, including the low photosynthetic capacity of trap leaves, are substantial relative to photosynthetic inputs; (2) as allocation to traps increases or prey density rises, prey capture should also increase; (3) benefits of carnivory include an increased rate of photosynthesis per unit leaf mass, an increased rate of conversion of photosynthate to new leaf tissue, or a reduction in photosynthate allocation to unproductive roots; (4) the marginal net benefits of carnivory should peak and then decline with increasing investments in carnivory as factors other than nutrients limit photosynthesis; and (5) prey capture should result in carnivorous plants having a higher growth rate than noncarnivores in the same microsites.

18.4.1 Costs of carnivory

The costs of carnivory can be substantial. *Drosera* from southwestern Australia allocate 3–6% of total photosynthesis to mucilage production alone (Pate 1986). *Darlingtonia californica, Sarracenia purpurea*, and two species of *Drosera* have photosynthetically inefficient trap leaves with lower photosynthetic rates per unit leaf mass at a given% leaf N content and specific leaf mass (SLM, g m^{-2}), and a lower N content at a given SLM, than across noncarnivorous plants worldwide (Ellison and Farnsworth 2005). Ellison (2006) documented significantly lower rates of photosynthesis per unit leaf mass and whole-plant growth in carnivorous plants than in any of the other growth forms tabulated from a global database (Figure 18.3).

Although these data suggest either an opportunity cost (Givnish et al. 1984) or little increase in photosynthesis associated with carnivory, such interpretations neglect the fact that few of the noncarnivorous species included in Figure 18.3 occur in the same extremely unproductive habitats as the carnivorous ones. To identify differences between species that result from differences in traits rather than environments, it is necessary to control for ecological distribution. Pavlovič et al. (2007, 2009) and Pavlovič and Saganová (2015) avoided this problem by studying variation between lamina (leaf) and trap (pitcher) tissues in three species of *Nepenthes*, each grown under identical greenhouse conditions; their findings clearly identify an opportunity cost associated with trap production. Leaves outperformed pitchers in photosynthesis per unit area for most levels of light availability and internal CO_2 concentration (c_i), apparent photochemical quantum yield (ϕ_{PSII}), stomatal density and conductance, tissue N and P concentration, and the concentrations of chlorophyll and carotenoids (Figure 18.4). Leaf tissue also had higher concentrations of Rubisco, consistent with the higher initial slopes of the A–c_i curve.

18.4.2 Allocation to carnivorous structures

The structural and opportunity costs of carnivory in trap construction are slightly smaller than estimates based solely on biomass allocation. A survey of 23 terrestrial carnivorous species showed that traps have a significantly lower construction cost per unit mass (CC_{mass} = 1.29 ± 0.20 g glucose/g dry mass) than leaves or laminae (1.41 ± 0.14 g glucose/g DM) of the same species under the same conditions; the latter costs were similar to those of leaves of 267 noncarnivorous species across a wide variety of habitats (Karagatzides and Ellison 2009). Slightly lower CC_{mass} and much lower photosynthetic rates result in a long payback time for traps (Karagatzides and Ellison 2009).

Aquatic species of *Utricularia* allocate up to 61% of their vegetative mass to traps, which have dark respiration rates R_d (mmol kg^{-1} dry mass h^{-1}) two to three times higher than leaves, and maximum photosynthetic rates A_{max} 7–10 times lower (Adamec 2006, 2008d). As a result, traps have high construction costs per plant, and high maintenance costs and low photosynthetic capacities per unit of bladder mass: R_d/A_{max} is 0.50–1.40 in traps but only 0.036–0.082 in leaves. High rates of trap respiration in *Utricularia* probably are related to the costs of pumping water from the trap (Adamec 2006, 2011f) and may underlie the evolution of a highly unusual,

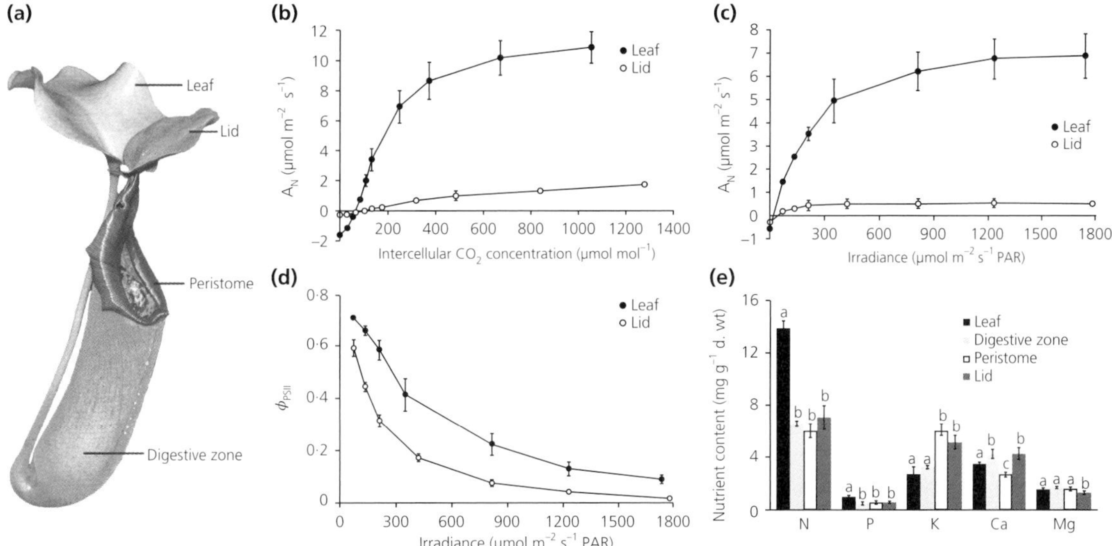

Figure 18.4 Data bearing on the cost/benefit model for evolution of carnivory in *Nepenthes*, based on comparisons between the pitcher (the lid being a flat part of the pitcher) and the leaf. (**a**) Leaf and pitcher of *Nepenthes truncata*. (**b**) A vs. c_i photosynthetic response curves. (**c**) Light response curves. (**d**) Light responses of effective photochemical quantum yield of photosystem II (ϕ_{PSII}). (**e**) Elemental composition of the leaf vs. different portions of the pitcher. Data shown are means ± SE ($n = 5$); different letters indicate significant differences among tissue types ($P < 0.05$; one-way ANOVA). Redrawn from Pavlovič and Saganová (2015).

energy-efficient mutation in the sequence of cytochrome *c* oxidase in *Utricularia* (Jobson et al. 2004).

Other continuing costs of carnivory, including nectar, mucilage, and enzyme secretion (Chapter 12); digestion and nutrient resorption (Chapters 3, 16); elemental allocation (Chapters 17, 19); electrical and jasmonate signaling in *Dionaea* and *Drosera* (Chapters 14–16); and resetting of aquatic trap leaves in *Utricularia* (Chapter 14) may be substantial, but many of these remain unquantified (Pavlovič and Saganová 2015). The peristome of *Sarracenia purpurea* secretes ≈70 μg cm^{-2} h^{-1} of carbohydrate as nectar (Deppe et al. 2000). Pitcher plants (*Cephalotus*, *Nepenthes*, Sarraceniaceae) secrete abundant nectar to attract prey, often ants (Givnish et al. 1984, Givnish 1989); observations on *Sarracenia alata* (Horner et al. 2012) and experiments on *S. purpurea* (Bennett and Ellison 2009) indicate that nectar secretion is critical in determining the rate of prey attraction and capture.

Respiration rates of traps often are very high during periods of high metabolic activity, including water pumping from *Utricularia*, rapid trap closure and prey retention in *Dionaea* (Pavlovič et al. 2010a, 2011a), and rapid tentacle movement in *Drosera* (Adamec 2010a). Costs of pumping water from recently triggered *Utricularia* bladders may be increased further by spontaneous triggering of those bladders in the absence of animal prey, which can occur ≈15–40 times during a typical trap's three-week lifetime (Adamec 2011f; Chapter 14). In two *Utricularia* species, 20–25% of newly fixed carbon is secreted into the trap fluid, including simple sugars (which may help feed symbiotic microbiota; Chapter 25), lactic acid, and phosphatases (Sirová et al. 2010, 2011). In aquatic *Aldrovanda* and *Utricularia*, the mineral costs of carnivory exceed 50% of total plant K and P, which may be related to the high energetic and elemental costs of pumping water from the traps that lead to very high concentrations of K in bladders (3.7–8.7% dry mass) and needs for ATP (Adamec 2010c).

The costs to *Nepenthes* of producing waxes (0.02–0.61 μg cm^{-2}) composed of aldehydes and viscoelastic fluids composed of long-chain polysaccharides are not yet known (Riedel et al. 2007, Bonhomme et al. 2011b). During the triggering and closure of

Dionaea traps and the generation and propagation of action potentials, leaf respiration spikes to at least ten times the background rate, while non-photochemical quenching rises, and apparent quantum yield (ϕ_{PSII}) and photochemical quenching fall, each in successive waves (Pavlovič et al. 2010a, 2011a, Pavlovič and Saganová 2015). During trap closure, 29% of cellular ATP is lost (Jaffe 1973). During the digestive phase, respiration rates of *Dionaea* traps more than double, while net photosynthetic rates in bright light fall by roughly 20% (Pavlovič and Saganová 2015).

Given the high costs of prey digestion, water pumping, and electrical signaling, Pavlovič and Saganová (2015) proposed that the inducible rather than constitutive nature of these processes is key to the evolution of rapidly closing traps and plant-mediated digestion. Jasmonates, well-known plant defense hormones, are involved in this inducibility and their accumulation can also affect photosynthetic reactions (Krausko et al. 2017). In noncarnivorous plants, jasmonates act as signals to redirect the gene expression and biosynthetic capacity from photosynthesis and growth to defense (Chapter 16), a significant allocation cost for plants that may be offset by the fitness benefit of not incurring these costs when defense is not needed. In carnivorous plants, jasmonates induce production of digestive enzymes, which are pathogenesis-related proteins (Buch et al. 2015, Bemm et al. 2016), and their accumulation also can affect photosynthetic reactions in *D. capensis* (Krausko et al. 2017). Given the long time between meals for snap-traps of *Dionaea* (≈23 days; Gibson and Waller 2009), the savings from inducible carnivory might be large.

18.4.3 Prey capture increases with allocation to carnivory

Increased mucilage production, droplet size, and gland density increase the apparent rate of prey capture in *Pinguicula vallisneriifolia* (Zamora 1995). Prey capture in *Sarracenia alata* increases with trap size (Green and Horner 2007, Bhattarai and Horner 2009) and nectar secretion rate (Horner et al. 2012). Larger traps of *Dionaea* capture larger prey at roughly the same rate as smaller traps capture smaller prey, so larger traps capture far more biomass per unit time (Gibson and Waller 2009). *Dionaea* releases a cocktail of more than 60 volatile organic compounds (VOCs), including terpenes, benzenoids, and aliphatics, that increase attraction of *Drosophila* under lab conditions (Kreuzwieser et al. 2014). Field studies in Borneo have shown a similar value of VOCs in attracting flies and ants to *Nepenthes rafflesiana* pitchers (Di Giusto et al. 2010). Absence of benzenoids in pitchers near ground level led to the attraction mainly of ants.

Comparisons among *Nepenthes* species highlight a complex series of tradeoffs among energy allocation to a large peristome, slippery waxes, and viscoelastic trap fluids (Chapters 12, 15). In species without viscoelastic fluids, prey capture increases with allocation to trap waxes, whereas species with viscoelastic fluids capture more prey than species dependent on slippery waxes alone; investments in these two mechanisms are negatively correlated (Bonhomme et al. 2011b).

Small investments in carnivory can yield intermittent but predictable bonanzas. The peristome of several *Nepenthes* is extremely slippery when wet by rain, fog, or nectar, but not when dry (Bauer et al. 2008; Chapters 12, 15). The inefficiency of traps during dry periods allows scout ants to recruit large numbers of workers to extrafloral nectaries (EFNs); large batches of workers then can be captured under wet conditions (Bauer et al. 2015a). Continuous experimental wetting of the peristome increases the number of non-recruiting prey but decreases that of ants, with trapping shifting from batches to individuals. A wettable peristome thus appears to be an adaptation for capturing ants or other social insects (Bauer et al. 2015a; Chapter 15). Waxy zones below the peristome are effective under both wet and dry conditions; the latter are likely in the seasonal lowlands. Capture rates of ants by *Nepenthes* increase with extrafloral nectar and slippery wax walls (Gaume et al. 2016), whereas termite capture increases with the presence of a rim of edible trichomes (Merbach et al. 2002) and symbiotic association with ants. Capture of flying insects increases with pitcher aperture and presence of alluring odors.

The unique status of *Nepenthes bicalcarata* as both a carnivorous plant and a myrmecophyte long has been puzzling (Beccari 1904, Givnish 1989; Chapters 23, 26). It invests in domiciles in the form of swollen tendrils for its ant partner, *Camponotus schmitzi*, and its EFNs prove the ants with food. *Camponotus schmitzi* can run safely over the slippery trapping surfaces, and dives into the pitchers to retrieve prey (Clarke and Kitching 1995, Merbach et al. 2007). The benefits of this kleptoparasitism are unclear, but the plant gains two advantages from investing in ants. First, ants attack and prey on inquiline dipteran larvae that feed on prey captured by the pitcher, preventing their escape from traps and loss of prey-derived nutrients (Bonhomme et al. 2011a, Scharmann et al. 2013). Second, the ants clean the peristome surface and maintain its ability when wet to "aquaplane" prey into the pitcher fluid (Thornham et al. 2012; Chapters 12, 15); the presence of *C. schmitzi* increases prey capture by 45%. Fungal hyphae contaminate the peristomes of ant-free older pitchers; experimental contamination of clean peristomes with starch also greatly reduced capture efficiency. Peristomes of ant-colonized pitchers were cleaned and returned to high capture efficiency in about one week. *Nepenthes bicalcarata* has unusually long-lived pitchers, which may have driven its unusual investment in what we might term ant butlers (Thornham et al. 2012).

18.4.4 Benefits of carnivory

Experimenting with *Drosera rotundifolia*, Francis Darwin (1878) provided the first demonstration that prey capture increased the growth and reproduction of carnivorous plants. Pavlovič and Saganová (2015) recently showed that feeding increased the photosynthetic rates of carnivorous plants in 16 of 19 species spread across *Aldrovanda*, *Dionaea*, *Drosera*, *Nepenthes*, and *Sarracenia*. In most terrestrial carnivorous plant species, increases in photosynthetic rates are positively correlated with increases in leaf N or P concentration and whole-plant growth (Ellison 2006, Pavlovič and Saganová 2015; Chapter 17). Comparable findings for aquatic carnivores are rare and inconsistent (Adamec 2000, 2008d, Adamec et al. 2010c; Chapter 19). Ellison and Adamec (2011) argued that this difference might simply reflect the greater methodological challenges of working with *Utricularia*'s tiny bladders.

Early studies that failed to show an impact of prey capture on photosynthetic rate (e.g., Méndez and Karlsson 1999, Wakefield et al. 2005) measured gas exchange on fully expanded leaves that had been fed for a short period, in which up-regulation of photosynthesis might not have occurred for several reasons. In contrast, measurements of photosynthesis on newly produced leaves (e.g., Farnsworth and Ellison 2008) generally find up-regulation following feeding. This result is consistent with increased allocation of nutrients in newly formed leaves in many carnivorous genera (Schulze et al. 1997, Butler and Ellison 2007, Kruse et al. 2014).

Increased flowering and seed production as a consequence of feeding has been reported in several terrestrial species (e.g., Darwin 1878, Karlsson and Pate 1992, Thorén and Karlsson 1998, Pavlovič et al. 2009; Chapter 17). The question remains whether these phenomena are caused by increased photosynthesis and plant growth, or direct limitation of reproduction by certain critical nutrients, not carbon (Givnish et al. 1984). Increased growth after feeding also may reduce the time required to achieve minimum flowering size (Pavlovič et al. 2009).

18.4.5 Plateauing benefits of carnivory

A meta-analysis of 29 studies demonstrated a significant positive effect of feeding on growth ($P < 0.02$), but no significant effect of nutrient additions ($P = 0.15$) or nutrient × prey interaction ($P = 0.81$), showing that additional nutrients contributed by carnivory are less valuable when plants are growing on more nutrient-rich substrates (Ellison 2006). Direct evidence of a plateauing of the photosynthetic benefits of carnivory, however, is sparse. Increased feeding of ten *Sarracenia* species in a greenhouse led to a rapid rise to a maximum A_{mass} ($r^2 = 0.51$, $P < 0.02$) and fluorescence ratio F_v/F_m ($r^2 = 0.74$, $P < 0.0015$) (Farnsworth and Ellison 2008; Figure 18.5). Additional studies should be conducted to test whether this effect applies to a wide range of carnivores.

18.4.6 Growth advantage of carnivorous plants

Experiments supporting a growth advantage of carnivorous versus noncarnivorous plants have not been done, perhaps because they would require a unique combination of physiological and ecological approaches under field conditions for both types of plants. The studies that come closest to supporting this assumption are those that show a significant

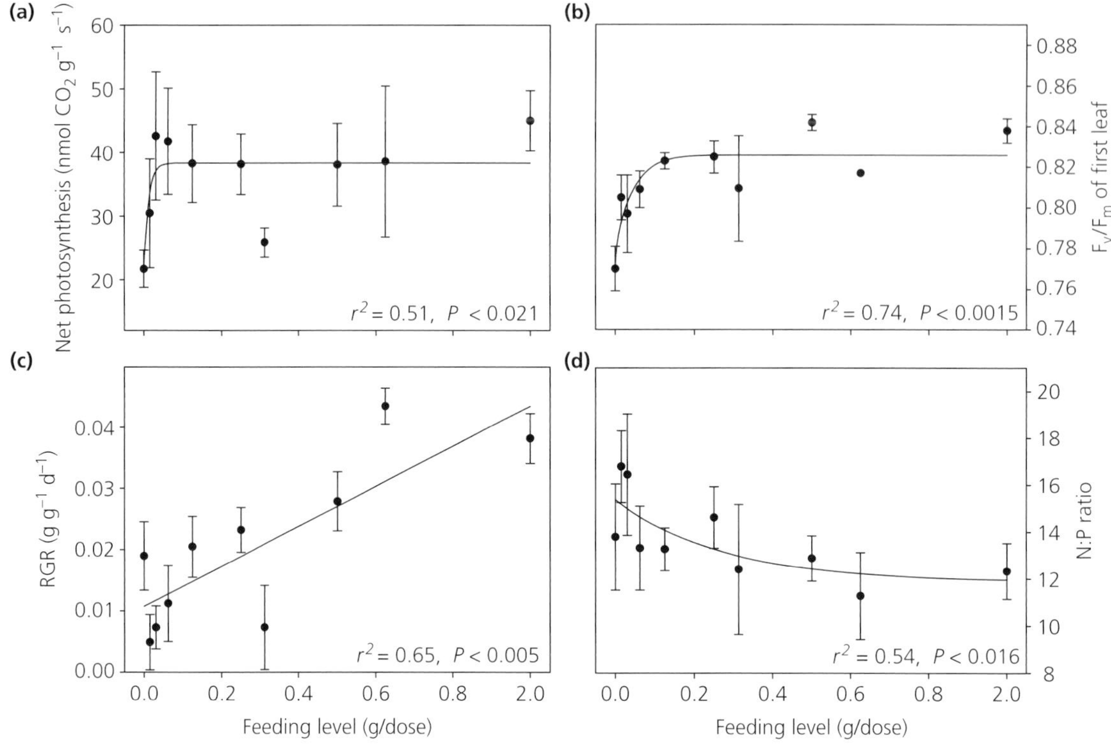

Figure 18.5 Responses of (**a**) net photosynthesis per unit leaf mass, (**b**) fluorescence ratio F_v/F_m of the first leaf, (**c**) RGR (g g^{-1} day^{-1}), and (**d**) N:P ratio to experimental feeding in ten *Sarracenia* species (n = 12/species). Points are means ± SE for all plants at each feeding level and all variables showed significant responses across species to feeding ($P < 0.05$, ANOVA). Curves for (**a**) and (**b**) are non-linear regressions fit to a three-parameter exponential rise to a maximum [$y = y_0 + a(1-e^{-bx})$]; (**c**) is a linear regression; and (**d**) is a non-linear fit to a three-parameter exponential decay ($y = y_0 + ae^{-bx}$). Redrawn from Farnsworth and Ellison (2008).

advantage of fed carnivores over unfed carnivores in photosynthesis or growth under the same lab or field conditions (§18.4.4, 18.4.5). For example, van der Ent et al. (2015) found that five montane *Nepenthes* species had similar leaf N but higher leaf P contents relative to 34 co-occurring noncarnivorous plants. This study needs to be followed up with investigations of gas exchange, whole-plant growth, and competitive interactions.

18.5 Tests of predictions of the cost/benefit model

18.5.1 Botanical carnivory is most likely in nutrient-poor, sunny, and moist habitats

The great majority of carnivorous plants in fact grow in sites that are nutrient-poor, sunny, and moist, at least during the growing season (Heslop-Harrison 1978, Thompson 1981, Lüttge 1983, Givnish et al. 1984, Juniper et al. 1989, Midgley and Stock 1998, McPherson 2010, Pereira et al. 2012, Nishi et al. 2013). Such habitats include bogs and wet tundra, open sites on moist to wet sands and other highly leached soils, fire-swept substrates, bare rock, and oligotrophic ponds and streams.

There are few exceptions to this rule. *Drosophyllum lusitanicum* actively grows on arid sites during the dry Mediterranean summer, but has a relatively extensive, deep root system that may tap groundwater (Adamec 1997a, 2009a, Adlassnig et al. 2005b) and hygroscopic glandular secretions that may allow substantial water uptake from fog (Adamec 2009a). A few shade-loving *Drosera* (*D. adelae, D. prolifera, D. schizandra*) inhabit the understories of Queensland rain forests, but they are only weakly carnivorous, possessing few glandular tentacles per leaf or slow-moving to stationary tentacles (Givnish et al. 1984, McPherson 2008). Other *Drosera* (e.g., *D. erythrorhiza, D. falconeri*) occur on calcareous sands

(Adlassnig et al. 2005b). *Nepenthes* vines mostly inhabit forest openings on nutrient-poor soils, but several grow on base-rich calcareous or serpentine soils; few species are epiphytic or grow under closed canopies (Givnish et al. 1984, Clarke and Moran 2016). *Darlingtonia* also is endemic to serpentine soils (Ellison and Farnsworth 2005). Most *Utricularia* are aquatic or grow on moist open ground in highly oligotrophic sites (Adamec 1997a), but a few species grow in hard, cation-rich waters (e.g., the Florida Everglades, McCormick et al. 2011) or tolerate shade (Adamec 2008a); 12 of ≈240 species are epiphytic in cloud forests (Fleischmann 2015a; Chapter 8).

Substantial numbers of species (>30) of *Pinguicula* occur on calcareous substrates, especially in México but also at mid- to high latitudes in Europe (Basso 2009). Karlsson and Carlsson (1984) found that growth of *Pinguicula vulgaris* on a calcareous mire substrate was more limited by P. This result is consistent with our expectations: calcareous soils may be base-rich but P available to plants may be in short supply (§18.3.1). Several *Pinguicula* species in México and Europe occupy partly to densely shaded microsites, with the distribution of some apparently representing an adaptive compromise between sunny but dry and insect-poor microsites and shady but moist and insect-rich microsites (Zamora 1995, Zamora et al. 1998, Alcalá and Domínguez 2003, 2005).

Seasonal growth or expression of carnivory explains other apparent exceptions. Several tuberous *Drosera* occupy extremely nutrient-poor but semiarid upland sites in southwestern Australia, but are active mostly during the moist winter and spring (Erickson 1978) and thus are not an exception to predictions. Similarly, butterworts in México stop producing sticky mucilage during the dry season, when few insects are present (Alcalá and Domínguez 2005). The single species of *Triphyophyllum* undergoes its carnivorous phase in the understory of seasonally waterlogged forests on shallow lateritic soils in West Africa, with carnivorous individuals seen both under canopy openings and dense shade (Green et al. 1979, A. Fleischmann *personal communication*).

Based on tissue stoichiometry, most terrestrial carnivorous plants appear to be limited by N, P, or both N and P (Figure 18.6). In contrast, aquatic *Aldrovanda* and *Utricularia* may be limited by K, which may reflect their heavy investment of K in below-water traps and water pumping (Adamec 2010c), lack of K recycling from old tissues, and low [K] in some waters (Ellison and Adamec 2011; Chapter 19). Two points should be made regarding these inferences. First, the N:P:K stoichiometry of leaf tissue shown is downstream of elemental inputs via carnivory. Data for a limited number of terrestrial species indicate that nutrient inputs from prey enhance both foliar N and P contents by small to quite large amounts (Ellison 2006), but the effect appears reversed in aquatic *Aldrovanda* and several *Utricularia* species (Adamec 1997a, 2000, 2008d). Second, N:P:K stoichiometry only indicates relative amounts of growth limitation by these elements (Olde Ventertink et al. 2003), not the total amount by which plant growth could be increased by *ad libitum* nutrient additions. This total limitation is likely to be quite large for carnivorous plants, but it has never been measured.

18.5.2 Carnivorous epiphytes should be rare but myrmecophytic epiphytes should be more common

Tree branches and boles are nutrient-poor, but often are shady and moist or sunny and dry. Thus the cost/benefit model would predict that epiphytes rarely should be carnivorous.

Of the nearly 800 species of carnivorous plants, only 17 (2%) are epiphytic, including four *Nepenthes* (Chapter 5), twelve *Utricularia* (all in *U.* sect. *Orchidoides*; Chapter 8), and one *Catopsis* (Chapter 10). This compares with an estimated 9% of epiphytic species among all the nearly 300,000 species of angiosperms (Zotz 2013). If carnivory were randomly distributed among angiosperms, we would expect there to be ≈70 carnivorous epiphytes, four times more than are known.

True epiphytism among carnivorous plants may be even rarer than these numbers suggest, as many carnivorous epiphytes live in wet microsites that differ little, functionally, from those typically occupied by aquatic bladderworts. Five of the 12 epiphytic *Utricularia* species often grow as emergent aquatic plants in open microsites, frequently

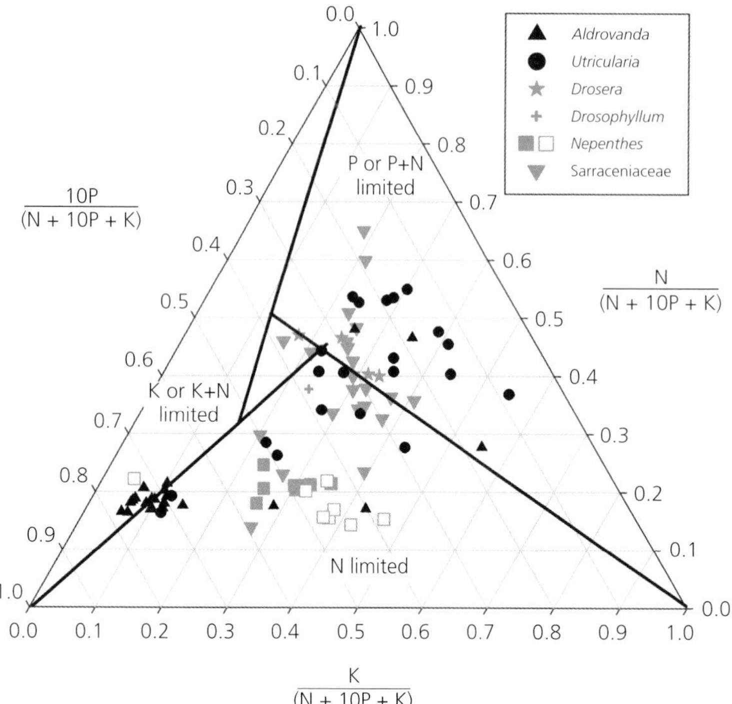

Figure 18.6 Tissue stoichiometry for aquatic submersed carnivorous plants (black) and terrestrial carnivorous plants (grey). Solid symbols indicate entire plants or traps; open symbols indicate leaves (laminae) measured separately on *Nepenthes*. Sarraceniaceae here includes *Darlingtonia* and *Sarracenia*. Dark diagonals separate regions of N, P or P + N, and K or K + N limitation following Olde Venterink et al. (2003). Redrawn from Ellison and Adamec (2011).

living in the tanks of sunlit bromeliads, including the carnivorous *Brocchinia hechtioides* and *B. reducta*. A single truly epiphytic species (*U. quelchii*) studied by Porembski et al. (2006) allocated only 15% of its biomass to leaves, just above the average of 8.1 ± 1.5% (SD) in six small terrestrial *Utricularia* species and well below the average of 59.1 ± 1.0% in three aquatic species. Even more remarkably, the leaf:trap biomass ratio for epiphytic *U. quelchii* was 0.50, much smaller than the ratios of 3.9 ± 1.3 for aquatic species and 32.4 ± 17.6 for terrestrial species. This raises the question of whether *U. quelchii* and the few other epiphytic species growing on densely shaded, mossy boles and branches can obtain carbon autotrophically via photosynthesis or heterotrophically via carnivory. *Utricularia* can grow in the dark when provided with sucrose (Harder 1970), and the tiny leaves of many annual terrestrial species suggests they might be heterotrophic. These terrestrials produce so few traps, however, that heterotrophy seems unlikely (Porembski et al. 2006). Such is not the case for epiphytic species like *U. quelchii*, which should be investigated as possible heterotrophs or mixotrophs.

As predicted by cost/benefit analysis, myrmecotrophy is far more common in epiphytes than carnivory. Almost all the ≈200 ant-fed plants originally tallied by Thompson (1981) are epiphytes, mostly occupying sunny or partly shaded perches; many more species (e.g., in *Tillandsia*) would be added to this list if it were recompiled today. At least 93 more species—all epiphytes—receive nutrients from ant gardens growing around their roots, as well as possibly other services, including seed dispersal and protection against herbivores (Orivel and Leroy 2011).

18.5.3 Investment in carnivory by terrestrial plants should increase toward the sunniest, moistest, most nutrient-poor sites

One confirmation of this prediction is the cline in the relative size of the photosynthetic keel versus the carnivorous tube of the northern pitcher plant (*Sarracenia purpurea*) across a nitrogen deposition gradient in New England (Ellison and Gotelli 2002; Figure 18.7a). Across 26 bogs, carnivorous pitchers were smaller while photosynthetically more efficient keels were larger where the concentration of ammonium in the pore water (and, presumably, in precipitation) was greater. The two exceptions to this rule apparently involved cases where tissue N:P stoichiometry suggested that P rather than N was limiting (Ellison and Gotelli 2002). Within a single bog remote from coastal industrial areas, plots experimentally sprayed with different concentrations of NH_4NO_3 developed pitchers that exhibited a similar pattern to that seen across the N deposition gradient (Figure 18.7b). Pitchers in another bog that were fed N and P solutions exhibited a regular shift toward relatively large keels and small tubes with increasing N (Figure 18.7c). As expected, pitchers with larger keels had higher photosynthetic rates. Similarly, feeding *Sarracenia* (Weiss 1980) and fertilizing *Nepenthes* (Pavlovič et al. 2010b) often suppresses trap formation and facilitates phyllode production. Variation in relative tube and keel size also occurs in response to natural variation in macronutrient availability in different wetland types (Bott et al. 2008). Feeding *Dionaea* decreases the trap:petiole ratio and red trap pigmentation (Gao et al. 2015).

Among the three *Drosera* species native to the New Jersey Pine Barrens, leaves of *D. rotundifolia* growing on partly shaded hummocks are horizontal and broad, with a relatively small trap surface. In contrast, leaves of *D. filiformis*, which grows on wet sand and *Sphagnum*, are vertical and thread-like with a relatively large trap surface. *Drosera intermedia* is intermediate in leaf shape, relative trap volume, and likelihood of shading (Givnish 1989). In *Drosera rotundifolia* grown in outdoor plots, Thorén et al. (2003) documented significant declines in leaf stickiness—reflecting the amount of mucus secreted by the tentacles and the polysaccharide concentration of the secretions—with experimental shading and additions of nutrients, with nutrients having a larger effect. These patterns are consistent with the cost/benefit model for carnivory, although Thorén et al. (2003) interpreted them in light of the carbon/nutrient-balance theory of Bryant et al. (1983). However, the quantitative pattern of secretion of the anti-microbial agent 7-methyljuglone contradicted the carbon/nutrient-balance theory.

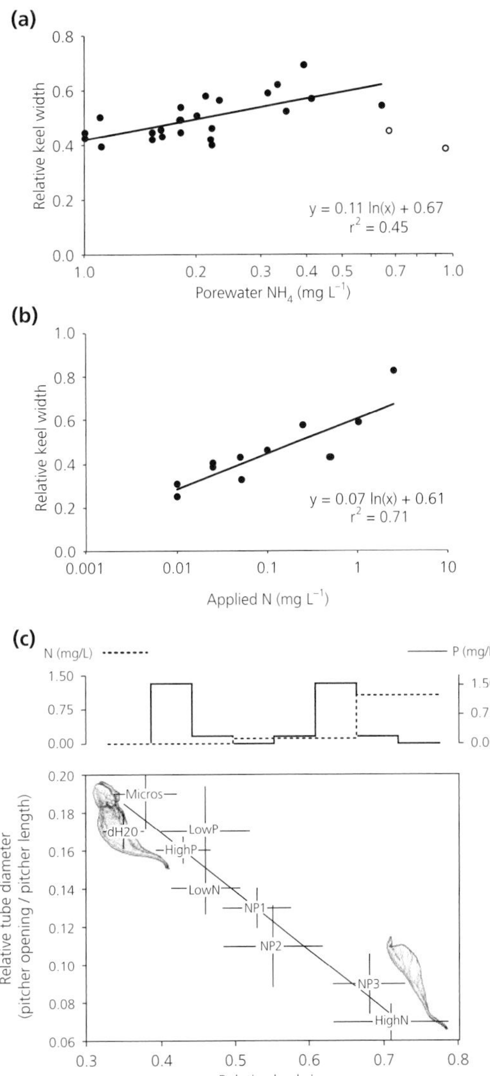

Figure 18.7 (**a**) Allocation to photosynthesis vs. carnivory measured by relative keel width [keel width/(keel width + tube width)] in *Sarracenia purpurea* growing in 26 bogs along an anthropogenic nitrogen deposition gradient (measured as pore-water NH_4) in Massachusetts and Vermont, USA. RKW (relative keel width) = 0.11 × ln [NH_4] + 0.67 ($r^2 = 0.45$, $P < 0.001$), excluding two Vermont outliers that appear to be P-limited (hollow symbols). (**b**) Relative keel width as a function of the concentration of N experimentally sprayed on the substrate at Molly Bog, Vermont, USA: RKW = 0.07 × ln [N] + 0.61 ($r^2 = 0.71$, $P < 0.001$). (**c**) Relative keel width and tube diameter as a function of the concentrations of N and P (upper panel) regularly added to the tank fluid of 90 plants at Hawley Bog, Massachusetts, USA. Insets show form of leaves at either end of the nutritional gradient. Redrawn from data provided by Ellison and Gotelli (2002).

Mucilage secretion in *Pinguicula vallisneriifolia* increased from deeply shaded to sunny microsites in Spain; leaves varied from curled and more secretory in sun to nearly flat and dry in shade (Zamora et al. 1998). During the Mediterranean summer, sunny sites offer maximum amounts of light, while shaded sites offer maximum amount of moisture and prey, shaping allocation to carnivory and resulting in highest reproduction in intermediate sites. Similarly, peak fitness of *P. moranensis* growing in México occurred in the middle of a light–water–prey gradient, where selection for, and investment in, capture glands was greatest in N-poor environments (soil and prey supplies). Selection for, and investment in, digestive glands was greatest in N-rich environments (Alcalá and Domínguez 2005).

Many carnivorous plants do not produce traps, or produce leaves with a lower allocation to carnivory, during unfavorable seasons when factors other than nutrients may limit growth. Such patterns of seasonal heterophylly also appear to support the predictions of the cost/benefit model (Givnish et al. 1984, Givnish 1989). *Sarracenia flava*, *S. oreophila*, and *S. leucophylla* develop trapless phyllodes during late summer droughts (Christiansen 1976, Weiss 1980). *Cephalotus* develops phyllodes during winter; *Dionaea* produces leaves with broader photosynthetic petioles and smaller traps in winter as well (Slack 1979). Many Mexican *Pinguicula* lose their carnivorous leaves and sprout succulent leaves, or die back to resting stages to survive the winter drought (Alcalá and Domínguez 2005). *Nepenthes* may not develop pitchers in excessively dry or shady sites (Slack 1979). The seasonal flush of glandular leaves in juvenile *Triphyophyllum* occurs at the beginning of the rainy season, as the soil becomes waterlogged and insect abundance may peak, and before the heavy rains likely to wash away glandular secretions (Green et al. 1979).

18.5.4 Form and function of traps depends on tradeoffs associated with environmental conditions and prey type

Studies on *Nepenthes* over the past decade provide some of the most compelling examples illustrating this prediction. Most *Nepenthes* occur in forests edges and openings over highly infertile soils and prey on arthropods, mainly ants (Juniper et al. 1989, Clarke 1997, Clarke and Moran 2016), using one or more of three trapping mechanisms, involving epicuticular waxes, a wettable peristome, and viscoelastic pitcher fluid (Moran et al. 2013; Chapters 12, 15, §18.4.3).

Bonhomme et al. (2011b) proposed that the nature of insect prey helped determine which of these mechanisms was most effective, and showed that waxier pitchers were more effective at capturing ants, that viscoelastic fluids were more effective against flies than ants, and that flies were more prevalent than ants in cloud forests at higher elevations. Moran et al. (2013) argued instead that climate shaped pitcher form and trapping mechanism, showing that perhumid cloud forests were strongly associated with large peristomes (with a large limb inside the pitcher), waxless pitchers, and viscoelastic fluid, whereas more seasonal lowlands were associated with small peristomes, waxy pitchers, and non-viscoelastic fluid. Gaume et al. (2016) proposed that both climate and insect prey shape pitchers: waxy traps and cylindrical pitchers with narrow apertures are associated with seasonal lowland areas and mainly capture ants and termites, whereas funnel-shaped traps with wide apertures, viscoelastic fluids, and aromatic traps dominate perhumid cloud forests and mainly capture flying prey. Wettable peristomes also can be highly effective in the seasonal lowlands, based on mass recruitment to nectaries during dry periods, when the pitchers are "safe," followed by mass captures under wet conditions (Bauer et al. 2012a).

Three *Nepenthes* species with giant pitchers at high elevations—*N. lowii*, *N. macrophylla*, *N. rajah*—have become partly to largely dependent on the feces of tree shrews and rats for N capture (Clarke et al. 2009, Chin et al. 2010, Greenwood et al. 2011; Chapters 15, 26). All three have large, concave lids that secrete abundant nectar and are held at nearly right angles to the pitcher; the distance from the lid glands to the front of the pitcher orifice precisely matches the length of the head plus body of the tree shrew (*Tupaia montana*) to ensure capture of feces when it visits the pitcher to feed from the lid. Feces account for 57–100% of foliar N in *N. rajah*, but all three species continue to capture arthropod prey presumably because of its superiority as a source of N (9.8% of insect mass versus 4.9% for feces).

Nepenthes ampullaria, with its unusually short, wide-aperture pitchers with highly reflexed lids at ground level, is adapted to detritivory and obtains >35% of leaf N from fallen leaves that collect in its pitchers, despite N content of those leaves being only 1.2% (Moran et al. 2003, Pavlovič et al. 2011b). It is one of three species with highly specialized nutrient-capture strategies adapted to densely shaded conditions under closed-canopy heath forests and peat swamps. The others are *N. bicalcarata*, with ant butlers that keep its long-lived pitchers functional (§18.4.3), and *N. hemsleyana*, which serves as a bat roost and obtains substantial amounts of its foliar N in that way (Grafe et al. 2011, C.R. Schöner et al. 2015; Chapter 15). Presumably, these three species can tolerate densely shaded conditions because they have substantially or greatly reduced secretion of nectar and tank fluid (especially in *N. hemsleyana* and *N. ampullaria*; C.R. Schöner et al. 2015), because the costs of bat–plant mutualism are low for each partner (C.R. Schöner et al. 2015), or because ant guarding and long-lived pitchers in *N. bicalcarata* increase the efficiency of those pitchers and reduce their cost of replacement (Thornham et al. 2012).

Darwin (1875) suggested that the rapidly closing snap-traps of *Dionaea* and *Aldrovanda* facilitate escape of small prey through the spaces between the marginal teeth. This feature, together with digestion in an enclosed chamber, may have arisen to capture larger, and consequently more rewarding, prey than could be immobilized by the sticky traps ancestral to Droseraceae. Gibson and Waller (2009) supported this hypothesis using a simple economic model and data on the distribution of body masses of insect prey and escape from both snap traps and sticky traps as a function of body size. Larger plants with larger traps should have a very large advantage in growth, making early prey capture especially important. Prey capture enhances long-term photosynthesis in *Dionaea* (Kruse et al. 2014), but in the short term leads to a spike in respiration and reduced photosynthesis associated with trap closure, prey retention, and digestion (Pavlovič et al. 2010a, 2011a), leading Pavlovič and Saganová (2015) to argue that the inducibility of costly enzyme secretion in plants with active traps may be crucial to their evolution (§18.4.2).

Finally, the return on carnivory in *Roridula dentata* is strongly context-dependent, and varies non-linearly with the density of hemipterans (*Pameridea*) that cruise its leaves, eating trapped insects and defecating on leaves. At very low and very high hemipteran density, growth of *Roridula* is negative, reflecting little or no N input due to the absence of bugs or their abandonment of host plants after consuming all prey (Anderson and Midgley 2007). Variation in the density of *Pameridea* predators or competitors may thus determine the value of carnivory in *Roridula*.

Other forms of trophic interactions also may facilitate the origin or maintenance of carnivory. Joel (1988) and Givnish (1989) independently proposed that pitcher plants in the genera *Cephalotus*, *Nepenthes*, and *Sarracenia*—all of which prey primarily on ants—may actually be mutualists with their prey. Plants in nutrient-poor habitats are long on carbohydrates but short on N, P, and other mineral nutrients, whereas ants in the same environment may be relatively long on nutrients, because of their predation on insects, but short on carbon. Given the large amount of nectar provided by ant-specialist pitcher plants, and the few individuals from colonies that are lost to pitchers, it might pay ant colonies to sacrifice the occasional sister into the well for all the sugar gained, just as it might pay plants to provide ants with large amounts of sugar if they obtain enough nutrients from the occasional ant prey to elevate photosynthesis by a greater amount. Moon et al. (2010) demonstrated experimentally that *Sarracenia minor* earns a net benefit from its interactions with ants; the question remains whether a net benefit also accrues to the ants. If these ant specialists are mutualists with their prey, there would be a positive feedback between the potential benefits from carnivory and increasing prey density—an outcome not originally envisioned by either Joel (1988) or Givnish (1989).

18.5.5 Carnivorous plants should have low photosynthetic rates and RGR

Meta-analyses support these predictions (Ellison 2006, Ellison and Adamec 2011). Both terrestrial and aquatic carnivorous plants have lower rates of photosynthesis and whole-plant growth than most other types of plants, but aquatic carnivorous plants have rates of both that are similar to or exceed those of aquatic noncarnivorous plants (Adamec 2013; Figure 18.3). Initially, these comparisons led Ellison (2006) to conclude they were evidence against the cost/benefit model. However, they are not; the model only predicts that carnivory will evolve if it produces a net increase in growth, not a high absolute growth rate. The key question, therefore, is how the photosynthetic rates and RGRs of fed and unfed carnivorous plants would differ from one another and from noncarnivorous species growing in the same nutrient-poor habitat. None of the comparisons presented by Ellison (2006) or Ellison and Adamec (2011) are controlled in this way, and no such experiments yet have been done.

18.5.6 Rainy, humid conditions or wet soils favor carnivorous plants by lowering the costs of glandular secretion or allowing passive accumulation of rainwater

The occurrence of the great majority of carnivorous plants under rainy and humid conditions or on wet soils, and the growth of most pitcher plants in areas of high rainfall and low evaporation rate (Givnish 1989, Moran et al. 2013) are consistent with this prediction. No studies using modern hydraulic approaches have examined how secretory glands and traps are connected to a carnivorous plant's hydraulic system, and whether any carnivores exhibit root pressure, although Adamec (2005a) did demonstrate water exudation by roots in 12 carnivorous plant species.

The hydathode hypothesis. Root pressure and water exudation might have played a key role in the evolutionary origin of pitchers in *Nepenthes* and possibly other groups. Biologists have long recognized that different clades of carnivorous plants have evolved either from ancestors with glandular hairs (Darwin 1875) or from ancestors with peltate leaves that capture and hold rainwater in areas with heavy precipitation (Baillon 1870, Franck 1976; Chapter 3). Available data suggest that five of the ten carnivorous plant clades (carnivorous Caryophyllales, *Roridula* in Ericales, and Byblidaceae, Lentibulariaceae, and *Philcoxia* in indumentum-rich Lamiales; Chapter 3) evolved from ancestors with glandular hairs. Two others—*Cephalotus* and Sarraceniaceae—are natural candidates for the peltate-leaf pathway, but none of their relatives have peltate leaves, and Fukushima et al. (2015) provide detailed genetic and developmental evidence that the *Sarracenia* trap leaf was not derived from a peltate form and instead reflects shifts in the orientation of the planes of cell division. We hypothesize that pitchers in these two groups, and in *Nepenthes*, are derived from terminal hydathodes in ancestral taxa. Hydathodes can allow plants to refill embolized xylem vessels under root pressure without flooding the leaf mesophyll and thereby adversely affecting photosynthesis (Feild et al. 2005). Often, hydathodes lie at the end of secondary or tertiary veins; they are inherently tubular in nature, providing a key step in the evolution of pitchers, although in almost all species the hydathodal tubes themselves are quite small and flattened.

In *Nepenthes*, the secondary veins converge toward the leaf apex and terminal tendril; a young pitcher begins as a slightly concave depression in the tendril tip (Owen and Lennon 1999). Similarly, in Sarraceniaceae and *Cephalotus*, young pitchers begin as tiny tubular excrescences of the leaf tip. One can imagine an evolutionary pathway in which a terminal hydathode or gland, fed by converging secondary veins, attracts minute prey that drown and are then digested by bacteria, leading to nutrient uptake and a fitness advantage, favoring the evolution of a larger and more complex trap derived from the hydathode. This proposal is close to Hooker's (1859) appendicular theory. Seedling leaves of *Nepenthes* have a depression at the end of the tendril terminating the leaf, and the adult leaves and stem are beset with hydathodes (Cheek and Jebb 2001). In both *Nepenthes* and *Cephalotus*, the tank fluid is supplied by hydathodes inside the trap (Adlassnig et al. 2011). The hydathode hypothesis seems plausible, and a first test of it would be to determine if *Cephalotus* and *Nepenthes* exhibit root pressure, and whether guttation occurs in the terminal tendril of the seedling leaves of *Nepenthes*.

18.5.7 Possession of defensive glandular hairs facilitates the evolution of carnivory

Five carnivorous clades are derived from ancestors with glandular hairs (§18.5.6). An additional two to

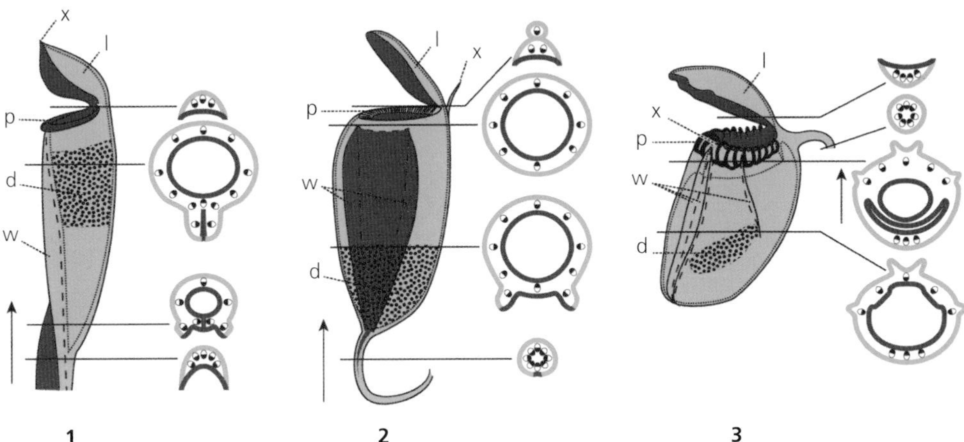

Plate 1. (Figure 3.2 on page 25). Convergent evolution of foliar pitchers in unrelated taxa. External pitcher appearance, surfaces, and schematic cross sections of **1**. *Sarracenia*; **2**. *Nepenthes*; and **3**. *Cephalotus*. All have pitcher leaves of epiascidiate ontogeny but of fundamentally different morphology and anatomy. Arrows: position and growth direction of shoot axis, axis in cross sections located below; *d*: digestive (glandular) zone inside pitcher; *w*: wing (ala) of pitcher outer surface; *p*: peristome; *l*: lid; *x*: true leaf apex; green: abaxial surface; red: adaxial surface; partially black-and-white filled ellipses: main vascular bundles in cross section, black: xylem, white: phloem. Pitchers not shown in correct size relations. Pitfall traps of the carnivorous monocots (Chapter 10) are made up of the entire rosette, while in the taxa illustrated here, each pitcher is derived from a single leaf. Illustration by Jan Schlauer.

Plate 2. (Figure 3.3 on page 25). Convergent and homologous evolution of adhesive traps in carnivorous plants. The peculiar outwardly circinate vernation of (**a**) *Drosophyllum* also is observed in the related *Triphyophyllum* (**b**), both in the emerging carnivorous (top) and noncarnivorous leaves (bottom; both Caryophyllales). (**c**) This is paralleled in the unrelated *Byblis* (illustrated by *B. aquatica*; Lamiales). *Drosera* (Droseraceae, Caryophyllales; illustrated by (**d**) *Drosera tracyi* and (**e**) *D. capensis*) and (**f**) *Pinguicula* (early-branching Lentibulariaceae, illustrated by *Pinguicula heterophylla*), have similar active flypaper traps, but with inward circination. In both, the leaves of several species are motile upon stimulation by prey. Photographs by Andreas Fleischmann.

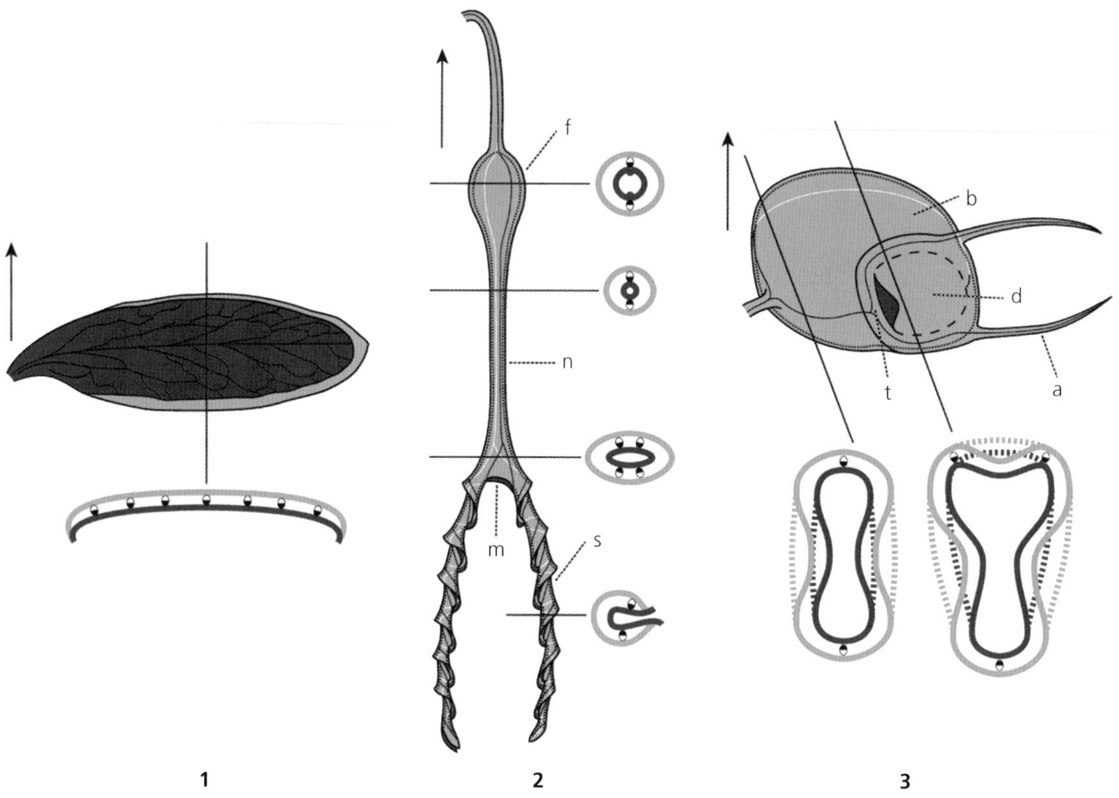

Plate 3. (Figure 3.7 on page 38). Homology of tissues among traps in Lentibulariaceae. External appearance, surfaces, and schematic cross sections of **1**. *Pinguicula*; **2**. *Genlisea*; and **3**. *Utricularia* (before suction and after door opening, dashed lines in cross sections indicate shape after suction). Arrows: position and growth direction of shoot axis, axis in cross sections located below; *f*: flask; *n*: neck; *m*: mouth; *s*: spiral arm; *b*: bladder; *t*: threshold; *d*: door; *a*: antenna; green: abaxial surface; red: adaxial surface; partially black-and-white filled ellipses: vascular bundles in cross section, black: xylem, white: phloem. Traps not shown with true size relations. Illustration by Jan Schlauer.

Plate 4. (Figure 3.8 on page 40). Loss of carnivory in carnivorous Caryophyllales (**a–e**) and Lamiales (**f**). (**a**) Normally developed *Drosera caduca* (Droseraceae) leaves consist of an enlarged petiole, the lamina greatly reduced to often a single, apical tentacle, or fully absent. This *Drosera* produces carnivorous foliage only in juvenile plants and after dormancy, but for the largest part of its life, it is a noncarnivorous sundew. (**b**) An almost entirely eglandular, noncarnivorous, naturally occurring mutant of *Drosera erythrorhiza*. (**c**) The predominant habit of the part-time carnivorous liana *Triphyophyllum peltatum* is noncarnivorous (shoot with the double-hooked climbing leaves shown). (**d**) The post-carnivorous Ancistrocladaceae (illustrated by *Ancistrocladus abbreviatus* from Sierra Leone); the inset shows the typical Mettenian glands that link it to Nepenthaceae. (**e**) *Nepenthes lowii*, a coprophagous rather than carnivorous pitcher plant (Chapter 26). (**f**) Shoots of the aquatic rheophyte *Utricularia neottioides* from Brazil usually lack traps almost entirely. Photograph (**b**) by Kingsley Dixon, all others by Andreas Fleischmann.

Plate 5. (Figure 4.5 on page 54). Morphological diversity of *Drosera*. (**a**) The annual therophyte *D. sessilifolia* (*D.* sect. *Thelocalyx*). (**b–d**) Pygmy sundews of *D.* sect. *Bryastrum*: (**b**) in most species, the flower rivals the vegetative part in size, illustrated here by *D. barbigera*; (**c**) dormant stipule bud, illustrated here by *D. androsacea*; (**d**) vegetative propagation by gemmae, illustrated here by *D. pedicellaris*. (**e**) *Drosera ordensis* of the woolly sundews, *D.* sect. *Lasiocephala*. (**f**) The dichotomously branched lamina of *D. binata* (*D.* sect. *Phycopsis*). (**g**) Filiform leaves of *D. spiralis* (*D.* sect. *Brasiliae*). (**h, i**) A remarkably similar geophyte habit evolved in parallel in some Australian tuberous, illustrated here by (**h**) *D. zigzagia* (*D.* sect. *Ergaleium*), and members of the South African *D.* sect. *Ptycnostigma*, illustrated here by (**i**) *D. cistiflora*. Photographs (**a–e, g, i**) by Andreas Fleischmann, (**f**) by Robert Gibson, and (**h**) by Laura Skates.

Plate 6. (Figure 6.1 on page 71). Morphological diversity and growth types of *Pinguicula*. A lithophytic habit with strap-shaped leaves has evolved in parallel in all three subgenera: (**a**) *P. megaspilaea* (*P.* subg. *Isoloba*); (**b**) *P. mundi* (*P.* subg. *Pinguicula*); (**c**) *P. calderoniae* (*P.* subg. *Temnoceras*). Those three species flower from the carnivorous rosettes. In contrast, (**d**) flowering from the noncarnivorous winter rosettes in *P. rotundiflora* (frequent in Mexican members of *P.* subg. *Temnoceras*); (**e**) sprouting hibernaculum of *P. grandiflora* (*P.* subg. *Pinguicula*). (**f**) *P. alpina* with hibernaculum formed in the center of a rosette of carnivorous leaves. (**g**) The (facultative) therophyte *P. lusitanica* has small leaves with strongly involute margins. (**h**) In most species, such as this *P. jarmilae*, the leaf margins are motile, enrolling over caught prey. (**i**) Species with thread-like leaves and revolute margins, such as this *P. heterophylla*, cannot move. Photographs (**b, e**) by Aymeric Roccia, (**c**) by Fernando Rivadavia, and remaining photos by Andreas Fleischmann.

Plate 7. (Figure 7.1 on page 82). Vegetative and generative morphology of *Genlisea*. (**a, d**) *Genlisea flexuosa* (*G.* subg. *Tayloria*), with monomorphic traps. (**b, e**) *G. margaretae* (*G.* sect. *Recurvatae*) with dimorphic traps (note the short, thick surface traps and the longer, filiform deep-soil traps). (**c, f**) *G. repens* (*G.* sect. *Genlisea*), the sole species with a prostrate stoloniferous habit. (**g**) The unique multiple-circumscissile capsule dehiscence pattern of *G.* subg. *Genlisea*, illustrated here by *G. hispidula*. Photographs by Andreas Fleischmann.

Plate 8. (Figure 9.1 on page 106). *Darlingtonia* morphology, wild-grown plants. (**a**) *D. californica* pitcher and flower, California, USA. (**b**) *D. californica* population in seepage, Oregon, USA. (**c**) *D. californica* plants in fruit, California, USA. (**d**) *D. californica* pitchers in late-afternoon light, Oregon, USA. All photographs by R. Naczi.

Plate 9. (Figure 9.3 on page 108). *Heliamphora* taxonomic and morphologic diversity, wild-grown plants. (**a**) *H. heterodoxa*, pitchers and inflorescence, Gran Sabana, Venezuela. (**b**) *H. minor* plant, Auyán-tepui, Venezuela. (**c**) *H. minor*, upper portions of two inflorescences, Auyán-tepui, Venezuela. (**d**) *H. minor* population (inflorescences with pale flowers visible above surrounding foliage) in bog, Auyán-tepui, Venezuela. (**e**) *H. elongata*, population in bog, Ilú-tepui, Venezuela. (**f**). *H. elongata*, plant with flowers, Ilú-tepui, Venezuela. All photos by R. Naczi.

Plate 10. (Figure 9.4 on page 109). *Sarracenia* taxonomic and morphologic diversity, wild-grown plants. (**a**) *S. oreophila* pitchers and flowers, Alabama, USA. (**b**) *Sarracenia alabamensis* ssp. *wherryi* plants in fruit, Alabama, USA. (**c**) *S. purpurea* pitchers, anthocyanin-free form (left) and typical form (right), Michigan, USA. (**d**) *S. flava* population in wet pineland, Florida, USA. (**e**) *S. rosea* plant, Florida, USA. (**f**) *S. leucophylla* plants (left) syntopic with *S. flava* plants (right) in wet pineland, Florida, USA. All photos by R. Naczi.

Plate 11. (Figure 10.1 on page 121). Carnivorous representatives of Bromeliaceae, Byblidaceae, Cephalotaceae, Dioncophyllaceae, Drosophyllaceae, Plantaginaceae, and Roridulaceae. (**a**) *Brocchinia hechtioides* (Bromeliaceae) on the summit of Auyán-tepui, Venezuela. (**b**) *Catopsis berteroniana* (Bromeliaceae) growing epiphytically in the Gran Sabana, Venezuela. (**c**) *Byblis lamellata* (Byblidaceae) growing among low shrubs in the Kwongan heathland north of Perth, Western Australia. (**d**) *Cephalotus follicularis* (Cephalotaceae) on a moss sward over exposed granite in southwest Western Australia. (**e**) *Triphyophyllum peltatum* in the rainforest understory of Sierra Leone. (**f**) *Drosophyllum lusitanicum* from Mediterranean heathland near Algeciras, Spain. (**g**) *Philcoxia rhizomatosa* growing on white sand in Brazil. (**h**) *Roridula dentata* from the shrubby vegetation of the fynbos, South Africa. Photographs **a–f** and **h** by Andreas Fleischmann; **g** by André Vito Scatigna.

Plate 12. (Figure 10.2 on page 122). The flowers of carnivorous representatives of Bromeliaceae, Byblidaceae, Cephalotaceae, Dioncophyllaceae, Drosophyllaceae, Plantaginaceae, and Roridulaceae. (**a**) *Brocchinia hechtioides* (Bromeliaceae). (**b**) The inflorescences of *Catopsis berteroniana* (Bromeliaceae). (**c**) *Byblis gigantea* (Byblidaceae). (**d**) *Cephalotus follicularis* (Cephalotaceae). (**e**) *Triphyophyllum peltatum*. (**f**) *Drosophyllum lusitanicum*. (**g**) *Philcoxia bahiensis*. (**h**) *Roridula dentata*. Photographs **a–d** and **h** by Andreas Fleischmann; **e** by Jan Schlauer; **f** by Maria Paniw; **g** by André Vito Scatigna.

Plate 13. (Figure 19.1 on page 257). Monomorphic shoots of *Aldrovanda vesiculosa* bearing regular whorls of leaves with snapping traps, each 4–5 mm long.

Plate 14. (Figure 22.2 on page 296). Pollinators and methods for investigating the importance of insects for the reproductive success in carnivorous plants. (**a**) *Utricularia multifida* with halictid bee, near Perth, Western Australia (photograph by T. Carow); (**b**) *Drosera arcturi* with *Melangyna novaezelandiae* (Syrphidae) probing pollen, Mt Cook, New Zealand (photograph by A. Jürgens); (**c**) *Drosera liniflora* and the hopliinid beetle *Anisonyx ursus*, Baineskloof, South Africa (photograph by A. Fleischmann); (**d**) *Heliamphora sarracenioides* with the carpenter bee *Xylocopa* sp. buzzing pollen from the poricidal anthers, Ptari-tepui, Venezuela (photograph by A. Fleischmann); (**e**) Pair of scanning electron micrographs of *D. anglica* tetrads on *Toxomerus marginatus* (Syrphidae) (photographs by G. L. Murza); (**f**) Pollinator exclusion experiment testing autonomous selfing in *D. arcturi* (photograph by A. Jürgens).

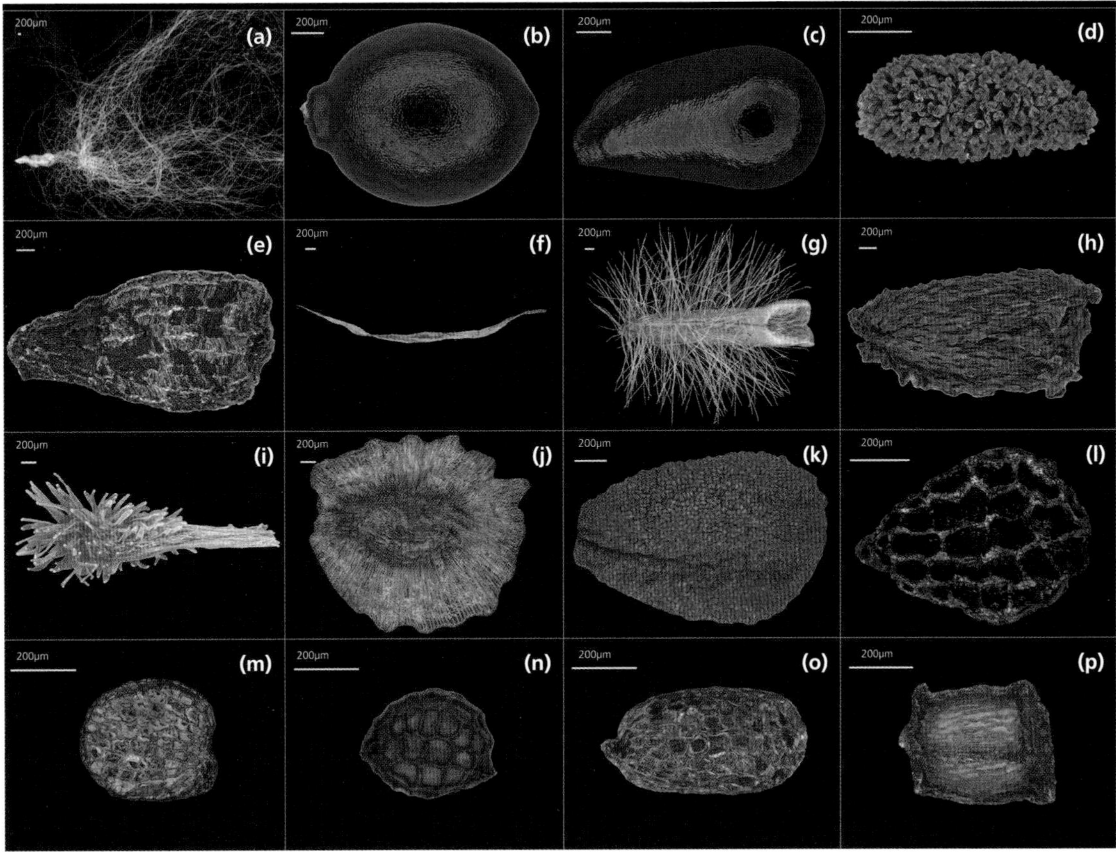

Plate 15. (Figure 22.3 on page 306). Comparative seed morphology of representative carnivorous plants in the Poales (**a**), Caryophyllales (**b–f**), Oxidales (**g**), Ericales (**h–k**), and Lamiales (**l–p**). (**a**) *Catopsis berteroniana*; (**b**) *Aldrovanda vesiculosa*; (**c**) *Dionaea muscipula*; (**d**) *Drosera intermedia*; (**e**) *Drosophyllum lusitanicum*; (**f**) *Nepenthes mirabilis*; (**g**) *Cephalotus follicularis* (fruit [indehiscent follicle, which acts as a diaspore] shown); (**h**) *Roridula gorgonias*; (**i**) *Darlingtonia californica*; (**j**) *Heliamphora heterodoxa*; (**k**) *Sarracenia rubra*; (**l**) *Byblis filifolia*; (**m**) *Philcoxia minensis*; (**n**) *Genlisea hispidula*; (**o**) *Pinguicula vulgaris*; (**p**) *Utricularia vulgaris*. All bars = 200 μm. All images by D.R. Symons.

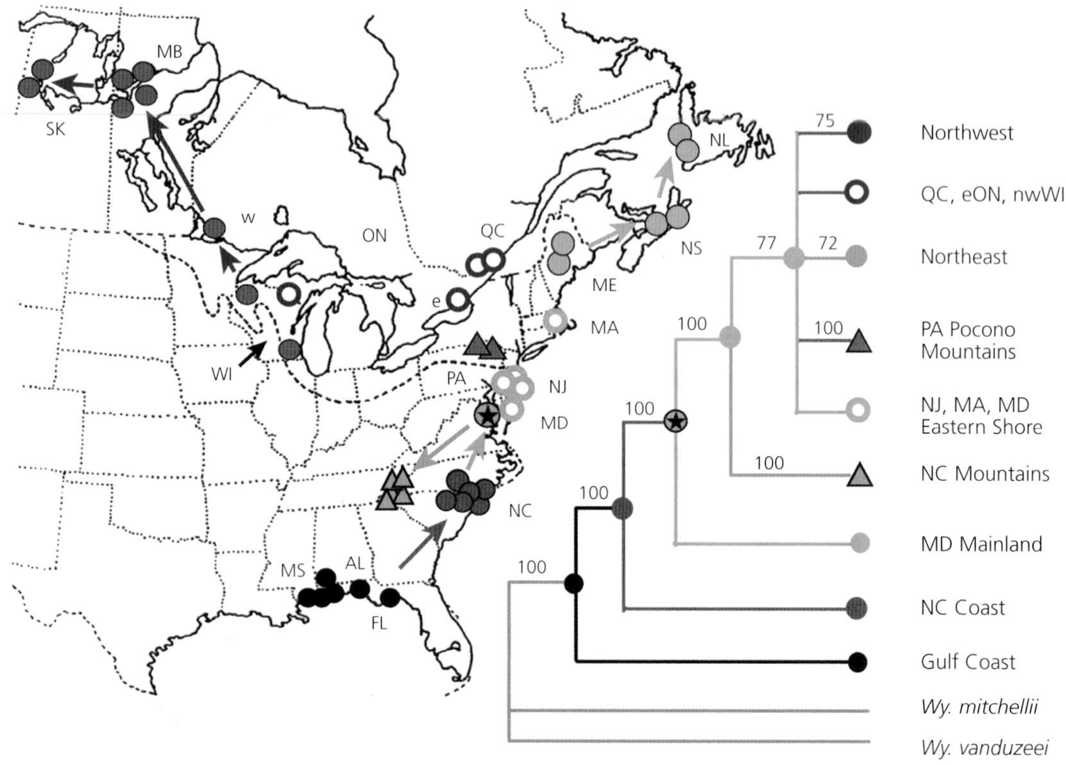

Plate 16. (Figure 24.1 on page 335). Phylogeography of *Wyeomyia smithii* based on the combined 46-population tree. Arrows indicate directions of range expansion. Maximum extent of the Laurentide Ice Sheet at the last glacial maximum is plotted as a dotted line. Two-letter abbreviations identify each state or province. USA states: AL—Alabama; FL—Florida; MA—Massachusetts; ME—Maine; MD—Maryland; MS—Mississippi; NC—North Carolina; NJ—New Jersey; PA—Pennsylvania; WI—Wisconsin. Canadian provinces: MB—Manitoba; NL—Newfoundland and Labrador; NS—Nova Scotia; ON—Ontario; QC—Quebec; SK—Saskatchewan. Triangles represent mountain populations in the southern Appalachians of North Carolina (green) or in the Pocono Mountains of Pennsylvania (brown). The filled orange symbol with the star indicates the basal mainland Maryland population (MD2). Numbers associated with each node or branch tip represent maximum parsimony bootstrap support for that clade. There are no support numbers for the QC, eastern ON (eON) and northwest WI (nwWI), or the NJ, MA, and Eastern Shore MD populations because they do not constitute a monophyletic grouping. Adapted from Merz et al. (2013).

Plate 17. (Figure 26.5 on page 369). *Pameridea* bugs attacking a fly trapped on the sticky surface of a *Roridula* leaf. The bug in the center of the picture is using its piercing mouthparts to suck fluids from the body of the fly.

Plate 18. (Figure 26.6 on page 371). Unidentified orb-web spider (Araneidae) attacking a *Pameridea* bug on a leaf of *Roridula dentata*.

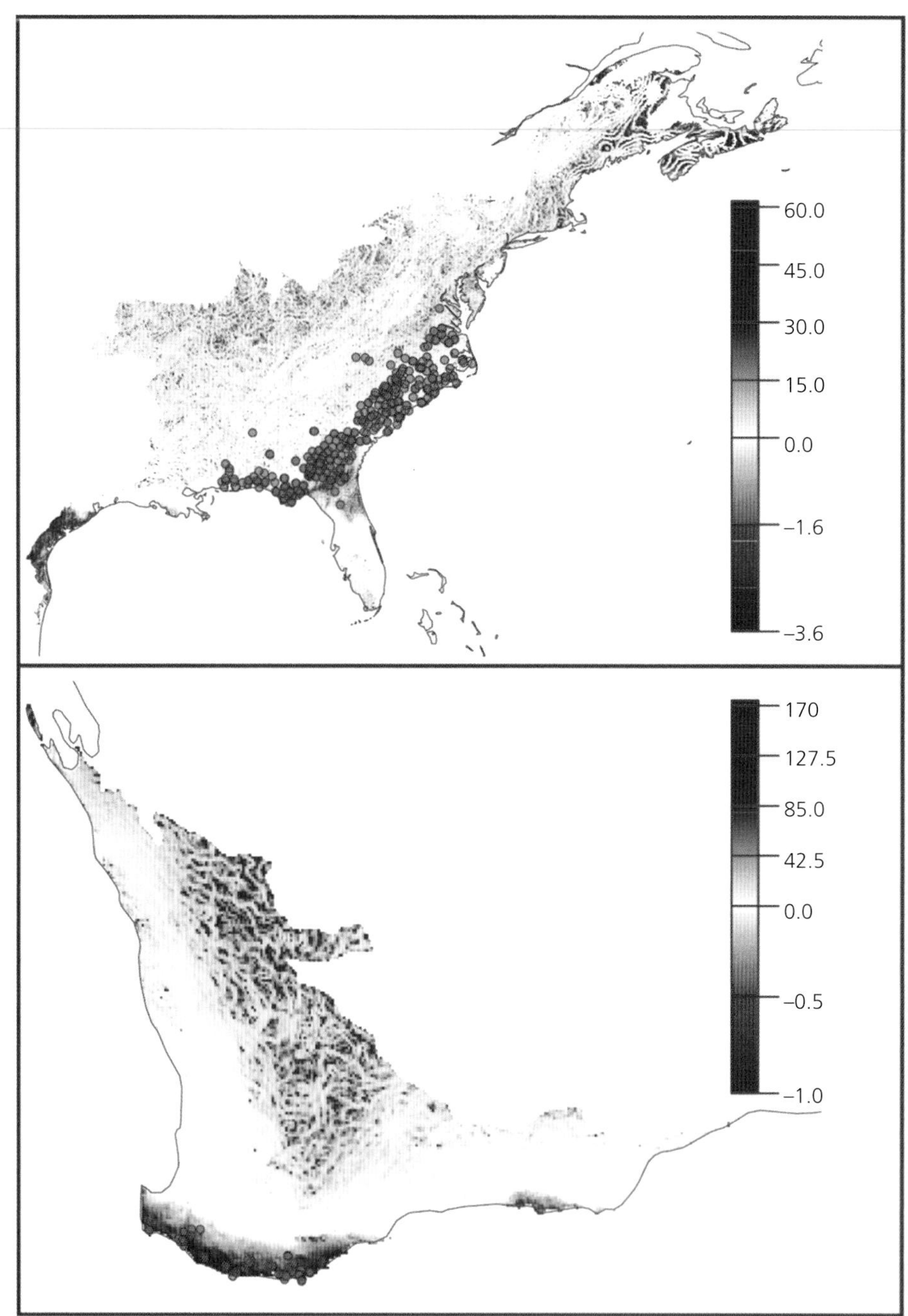

Plate 19. (Figure 28.4 on page 401). Maps of bioclimatic velocity (km yr^{-1}) between current climate and year 2050 for (top) *Sarracenia flava* and (bottom) *Cephalotus follicularis* averaged across 32 future climate scenarios. Black circles represent occurrence locations used in model fitting. For *S. flava*, most locations where populations have been observed are projected to experience increases in habitat suitability and low velocity declines only at southernmost locations. In contrast, *C. follicularis* is projected to experience relatively low velocity decreases in suitability across all locations where populations have been observed.

three origins (in Poales) involve ancestors with eglandular but absorptive leaf hairs. A pre-existing investment in hairs would have reduced the energetic threshold for carnivory. In addition, to the extent that the evolution of sticky hairs also may protect carnivores from herbivores, it would increase selection for their evolution. A field experiment involving the removal of secretory glands from the leaves of *Pinguicula moranensis* showed that glandless plants sustained 18 times more damage than control plants (Alcalá et al. 2010). Even if some of the observed effect is due to trauma increasing the attractiveness of glandless plants, this study suggests that much of the benefit this carnivore obtains is through a reduction in herbivory. A similar effect is seen in the ant-specialist *Sarracenia minor*, in which increased ant visitation independent of their capture as prey reduced herbivory and pitcher mortality, and increased the number of pitchers per plant (Moon et al. 2010).

18.5.8 Fire over infertile soils favors carnivorous plants

Many carnivorous plants are associated strongly with nutrient-poor, moist habitats that burn frequently (Givnish 1989). Fire should favor carnivores on such sites by volatilizing N from burnt tissue, and increasing light levels and soil moisture by removing competing plants and foliage (Givnish 1989). Other forms of disturbance over sterile substrates, including logging, lake drainage, or fractures or landslides that expose bare rock, also can favor carnivores without volatilizing N. However, fire also can increase levels of many cations and P, raising the question of whether the greater supply of such nutrients would select against botanical carnivory, or instead enhance its benefits through nutrient complementarity (Givnish 1989). The possibility of the latter is reinforced by the degree to which fire enhances the abundance of N-fixing plants in prairies and pinelands (Towne and Knapp 1996), although wet anoxic soils where carnivorous plants grow are unfavorable for N-fixing legumes.

18.5.9 Gradual evolution of carnivory is essential in extreme habitats

Small allocations to carnivory probably would not provide enough nutrients to maintain plant life when there is essentially no P or N in the substrate. Presumably, the origin of carnivores like *Brocchinia reducta* or *Dionaea muscipula* that can grow on substrates with essentially no nutrients occurred through their initial colonization of nutrient-poor substrates, subsequent optimization (and thus, increase) in allocation to carnivory, and eventual colonization of more-or-less nutrient-free substrates. Testing this hypothesis will require simulation studies with properly parameterized versions of our new model for the evolution of carnivory (Box 18.1) or genetic manipulations that reduce allocations to carnivory.

18.5.10 Anoxic or toxic soils should favor carnivory on open, moist sites

This hypothesis is supported by the widespread occurrence on saturated or toxic soils of carnivorous plants with small root systems. At least fifteen species of *Nepenthes* are restricted to serpentine outcrops in northern Borneo, Palawan, and parts of Wallacea (Clarke and Moran 2016), and *N. vieillardii* is restricted to serpentine outcrops in New Caledonia (Kurata et al. 2008). Van der Ent et al. (2015) found that the serpentine *Nepenthes* from Mt. Kinabalu and Mt. Tambuyukon were heavy-metal excluders; their restriction to serpentine may reflect lowered costs of heavy-metal exclusion because of their carnivory, limited root systems, and the open nature of the vegetation over serpentine. In northern California and southern Oregon, *Darlingtonia californica* is endemic to serpentine (Ellison and Farnsworth 2005), where it co-occurs with another serpentine endemic, *Pinguicula macroceras* ssp. *nortensis*, and non-endemic *Drosera*. In western Newfoundland, *Sarracenia purpurea* grows luxuriantly on open serpentine gravel with water not far below the surface (T. J. Givnish *unpublished data*).

18.5.11 Co-limitation of growth by multiple nutrients may favor the paradoxical increase in root investment by carnivorous plants that recently have captured prey

This hypothesis provides the only plausible explanation for this puzzling phenomenon. An approach to testing this idea would be to feed plants on a series of defined media that are lacking one or more nutrients, and determine whether additional root growth occurs only in those circumstances in which the elements available from prey and soil are complementary.

18.5.12 Harder, more fertile waters should favor greater investment in traps by *Utricularia*

Knight and Frost (1991) found that the number of traps in *Utricularia macrorhiza* increased with water hardness across nine lakes in northern Wisconsin. They quantified water hardness as specific conductance, a measure that increases with cation concentration in the water column and, most likely, bicarbonate concentration, [CO_2], total dissolved inorganic carbon (DIC), and possibly [P] weathered from the substrate (Hanson et al. 2006). Their experiments did show that water chemistry, not prey supply, was responsible for this pattern but direct measurements of DIC and [CO_2] were not provided. In two of the lakes, variation in bladder numbers per leaf reflected the proportional allocation of biomass (Knight and Frost 1991).

Adamec (2008a) found that the percentage of biomass allocation to traps (T) in *U. australis* in 29 Czech fishponds and bog or fen pools was correlated significantly with only one external factor: dissolved CO_2: T = 14.33 [CO_2] + 32.21 ($r^2 = 0.33$, $P < 0.001$; Figure 18.8). In moving from oligotrophic to eutrophic waters in this landscape, dissolved [CO_2] increased two-fold (from 0.26 ± 0.20 mM to 0.54 ± 0.56 mM), while NO_3-N also increased two-fold (from 3.6 ± 4.1 to 8.0 ± 7.1 µg/L), NH_4-N increased four-fold (from 27 ± 10 to 102 ± 135 µg/L), and PO_4-P increased more than six-fold (from 11.7 ± 5.2 to 71.3 ± 86.1 µg/L). However, no measure of available N or P showed a significant correlation with trap allocation under two-tailed *t*-tests ($r = -0.09$, –0.19, and –0.3 for NO_3-N, NH_4-N, and PO_4-P, respectively; all $P > 0.05$). As expected, shading or feeding *Utricularia vulgaris* stopped trap formation (Englund and Harms 2003), whereas trap allocation in *U. foliosa* increased as water [NO_3] declined along an Amazonian creek (Guisande et al. 2004). Adamec (2008a, 2015a) argued that proportional biomass allocation to traps is set by the balance of a positive effect of water [CO_2] and a negative effect of tissue N (or P) concentration, which he viewed as a homeostatic mechanism without reference to optimal energy capture. Alternatively, Adamec's data may be an exemplar of optimal allocation to carnivory in an aquatic carnivore and a mechanism for generating and maintaining a favorable pattern of allocation.

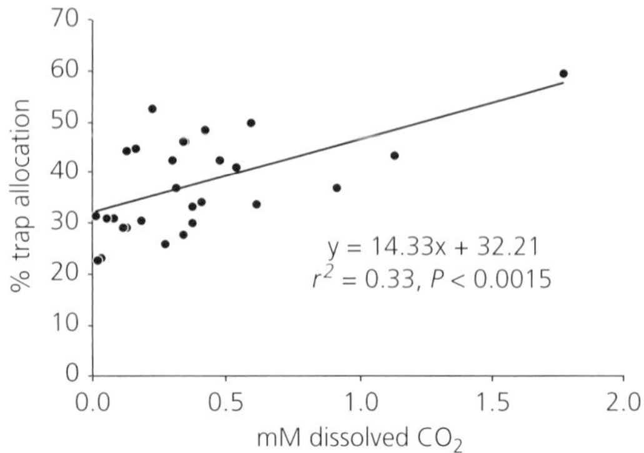

Figure 18.8 Percent allocation of biomass to traps as a function of dissolved [CO_2] across 29 Czech sites of *Utricularia australis* studied by Adamec (2008a). Line is least-mean-squares regression,% trap allocation = 14.33 × mM dissolved CO_2 + 32.21. Original data kindly made available by Lubomír Adamec.

18.5.13 Soil anoxia or extreme infertility makes tall, woody carnivores impossible

Almost all carnivorous plants are relatively short herbs. Perhaps the simplest explanation for this prediction is that the wet soils that favor carnivory (§18.5.1, §18.5.6, §18.5.10) often disfavor woody plants. However, the decrease in leaf-specific hydraulic conductance with height in woody plants (Mencuccini 2003) also might be important. That decrease would militate against gigantic carnivores by decreasing water potential and selecting for small leaves with low photosynthetic rates (Koch et al. 2004, Ishii 2011) that are unlikely to be increased by nutrient inputs via carnivory. Thus, the restriction of the carnivorous phase to relatively short juveniles (≈1 m tall) in *Triphyophyllum* should be re-examined and the role of decreasing water potential in limiting the photosynthetic rate with increasing height of adult vines studied.

18.6 Future research

The cost/benefit model for the evolution of carnivorous plants provides a qualitative explanation for patterns in their distribution, allocation to traps, variation in trap mechanism, association with growth form, low rates of carbon uptake and whole-plant growth, and ecological characteristics relative to plants with other "non-standard" mechanisms of nutrient capture, including myrmecotrophy and nitrogen fixation.

In some ways, it is remarkable that Darwin himself did not advance something analogous to the cost/benefit model. On July 21, 1875, Alfred Russel Wallace—co-creator of the theory of natural selection—wrote to Darwin:

> *Many thanks for your kindness in sending me a copy of your new book. . . . The account of* Utricularia *is most marvellous, and quite new to me. I'm rather surprised that you do not make any remarks on the origin of these extraordinary contrivances for capturing insects. . . . I daresay there is no difficulty, but I feel sure they will be seized on as inexplicable by Natural Selection, and your silence on the point will be held to show that you consider them so!. . . Here are plants which lose their roots and leaves to acquire the same results by infinitely complex modes! What a wonderful and long-continued series of variations must have led up to the perfect "trap" in* Utricularia, *while at any stage of the process the same end might have been gained by a little more development of roots and leaves, as in 9,999 plants out of 10,000!*

A day later, Darwin responded to Wallace, pointing out that he had in fact traced an evolutionary pathway to carnivory in Droseraceae, a proposal that continues to inspire research (e.g., Gibson and Waller 2009). But Darwin did not respond to the thrust of Wallace's suggestion—that plants might have responded to shortages of soil nutrients by increased allocation to roots or decreased allocation to leaves, rather than evolve carnivory—a suggestion which, if considered carefully, would almost surely have led Darwin to a consideration of the relative costs and benefits of carnivory.

Perhaps the one direction in which our own responses to this central question might be improved over the coming years is to produce a fully quantitative model for the evolution of carnivory, and compare quantitative predictions of optimal plant form and allocation with reality. Studies by Zamora et al. (1998), Ellison and Gotelli (2002), Alcalá and Domínguez (2005), Adamec (2008a), Gibson and Waller (2009), Bonhomme et al. (2011b), and Gaume et al. (2016) are all steps in this direction. Developing quantitative optimality models, especially for *Utricularia* or *Nepenthes* species that have a clear distinction between photosynthetic and carnivorous organs, and comparing them with alternative approaches (e.g., Adamec 2015a; Chapter 19) would enable the most powerful tests of cost/benefit analysis (Mäkela et al. 2002). Experiments comparing photosynthesis and whole-plant growth by non-carnivorous plants with those of carnivorous plants in the same field locations and with and without simultaneous access to prey are a very high priority. Such experiments will provide the critical data needed to test many of the predictions of the cost/benefit model.

CHAPTER 19

Ecophysiology of aquatic carnivorous plants

Lubomír Adamec

19.1 Introduction

The ≈60 submersed aquatic or amphibious species of *Aldrovanda* and *Utricularia* (*U.* sects. *Pleiochasia, Avesicaria, Avesicarioides, Lecticula, Utricularia,* and *Vesiculina*) are strictly rootless and take up mineral nutrients from the ambient water (and/or sediment) and captured prey only via their trap-bearing shoots. Their traps are among the fastest moving organs found within the plant kingdom (Juniper et al. 1989, Vincent et al. 2011b; Chapter 14). Most ecophysiological studies on aquatic carnivorous plants postdate Juniper et al. (1989). They have morphological and physiological features that are very different from their terrestrial counterparts (e.g., Colmer and Pedersen 2008, Ellison and Adamec 2011; Chapter 17), and historically have been comparatively under-studied (Juniper et al. 1989, Adamec 1997a, Ellison and Adamec 2011).

Adamec (1997a, 2011c) comprehensively reviewed the ecophysiology of most aquatic carnivorous plants, and Guisande et al. (2007) reviewed ecophysiological traits and cost/benefit relationships of *Utricularia*. Adamec (2011a) reviewed the ecophysiology of aquatic *Utricularia* traps, emphasizing the role of microbial trap commensals, whereas Poppinga et al. (2016b) focused on morphological and functional aspects of *Utricularia*. Ellison and Adamec (2011) reviewed functional differences between terrestrial and aquatic carnivorous plants with an emphasis on growth rate, photosynthesis, shoot mineral content, and cost/benefit relationships. This chapter focuses on results published since 2011, and includes not only peculiarities of the habitats of aquatic carnivorous plants, but also growth, photosynthesis, mineral nutrition, and regulation of investment in carnivory, peculiarities of *Utricularia* traps, and turion ecophysiology. Four strictly rheophytic and virtually unstudied species (Taylor 1989)—*U. neottioides, U. oliveriana, U. rigida,* and *U. tetraloba*—are excluded from this review.

19.2 Habitat characteristics

Aquatic carnivorous plants usually grow together with vascular noncarnivorous plants in shallow standing or slowly running humic (dystrophic) waters. In temperate regions, a partly decomposed, nutrient-poor litter of reeds and sedges or *Sphagnum*-based peat usually accumulates in these waters. This slowly decomposing litter gradually releases mineral nutrients, humic acids, tannins, and CO_2, and the waters often are rich in free [CO_2] (0.1–1 mM; Adamec 1997a, 1997b, 1999, 2007a, 2008a, 2009b, 2010b, 2011c, 2011g; global median 0.30 mM [interquartile range 0.14–0.92 mM] in Adamec 2012b). Typical humic waters are usually poor in minerals such as N (NH_4+, NO_3^-), P (the concentration of both is commonly 5–25 μg L^{-1}), and sometimes K (<0.5 mg L^{-1}). In waters not impacted by human activity, concentrations may even be several times lower (Adamec 1997a, 2008a, 2009b, Guisande et al. 2007). Concentrations of SO_4^{2-}, Ca, Mg, and Fe are usually >1 mg L^{-1} and do not limit plant growth. In bogs, fens, swamps, backwaters, lakes, and fishponds, where most aquatic carnivorous plants grow in mildly or medium humic, oligo-mesotrophic waters, they can tolerate much

Adamec, L., *Ecophysiology of aquatic carnivorous plants.* In: *Carnivorous Plants: Physiology, ecology, and evolution.*
Edited by Aaron M. Ellison and Lubomír Adamec: Oxford University Press (2018). © Oxford University Press.
DOI: 10.1093/oso/9780198779841.003.0019

higher total concentrations of humic acids and tannins (very dark waters; worldwide median 11 mg L^{-1}, usually 5–20 but even >60 mg L^{-1}; Adamec 2012b) than noncarnivorous ones. Other potentially unfavorable factors of dark humic waters are shading, cooling, harmful concentrations of humic acids and tannins, and low pH.

A [CO_2] > 0.15 mM is considered the primary factor supporting vigorous growth and propagation of stenotopic *A. vesiculosa* and many *Utricularia* species (Adamec 1999, 2011c, 2015a). Low concentrations of dissolved oxygen (within 0–12 mg L^{-1} but commonly 2–6 mg L^{-1}) occur at many sites, sometimes with a marked daily oscillation (Adamec 1997b, 1999, 2007a, 2010b, Guisande et al. 2000, 2004, Adamec and Kovářová 2006). Most species grow in soft to moderately hard (total alkalinity 0.2–2 meq L^{-1}), acid or neutral waters (global median pH = 6.3, interquartile range 5.7–7.0; Adamec 2012b), but some temperate-zone species also may occur in hard, alkaline waters (pH 8–9.3; Adamec 1997a, 2009b). The widespread *U. australis* and *U. minor* can grow across a range of five pH units (Adamec 2011c). Unlike most terrestrial carnivorous plants, many aquatic species also grow well under very shaded conditions (<5% of total irradiance; Adamec 2008a).

19.3 Morphology

Rootless aquatic carnivorous plants float freely below the water surface or are weakly attached to loose sediments as submerged or partly amphibious types; firmly attached species (*U. resupinata*) are exceptional (Taylor 1989, Guisande et al. 2007). Most species have a linear, modular shoot structure consisting of nodes with filamentous leaves and tubular, fragile internodes. Exceptions include *Aldrovanda*, whose leaves are arranged in true whorls, and the rosette-shaped *Utricularia volubilis*. The majority of linear-shoot species have homogeneous (monomorphic), non-differentiated green shoots with traps (Figures 19.1, 19.2). Several species (e.g., *Utricularia intermedia, U. floridana*) have dimorphic shoots differentiated into green photosynthetic ones (usually bearing only a few traps) and pale carnivorous (trapping) ones with many traps (Figure 19.3). These species are intermediate in body plan between the aquatic *Utricularia* species with monomorphic shoots and terrestrial species (e.g., *U. uliginosa*; Taylor 1989). However, morphologically and developmentally, the distinction between stems, branches, air shoots, and leaves is not clear; all these organs might be considered homologous and they exhibit a great morphological and functional plasticity (Sattler and Rutishauser 1990).

Figure 19.1 (Plate 13 on page P11) Monomorphic shoots of *Aldrovanda vesiculosa* bearing regular whorls of leaves with snapping traps, each 4–5 mm long.

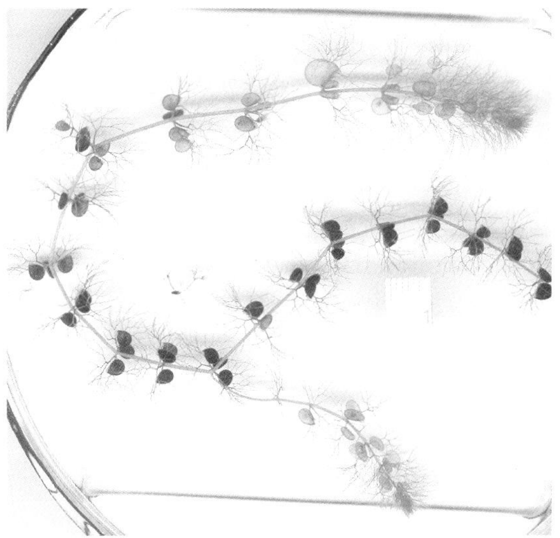

Figure 19.2 Monomorphic shoots of *Utricularia reflexa* from Zambia photosynthesize but also bear dozens of suction traps. Bars (ticks) = 1 mm.

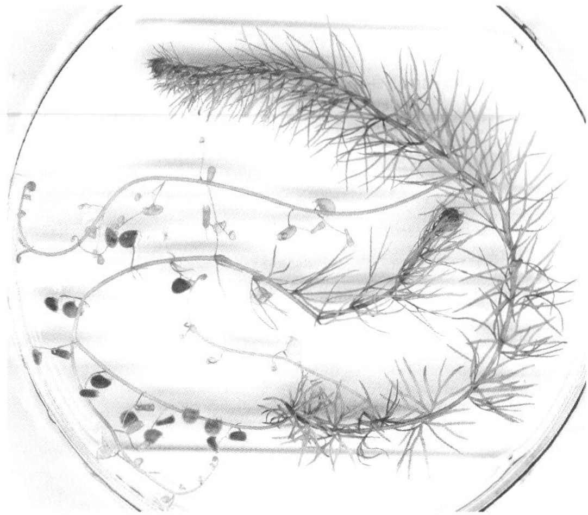

Figure 19.3 Dimorphic shoots of *Utricularia intermedia* are differentiated into green photosynthetic shoots without traps and pale carnivorous shoots bearing dozens of traps, each up to 5 mm long.

19.4 Growth, mineral nutrition, photosynthesis, and respiration

Rootless aquatic carnivorous plants grow very rapidly, exhibiting rapid apical shoot growth and frequent shoot branching in nutrient-poor but CO_2-rich habitats. Besides carnivory, these species efficiently take up nutrients from water and reutilize nutrients from senescing shoots, enabling them to access scarce supplies of mineral nutrients and photosynthesize rapidly (Kamiński 1987a, Kosiba 1992, Friday and Quarmby 1994, Adamec 2000, 2008a, 2008d, 2014, Englund and Harms 2003).

19.4.1 Growth

Aquatic carnivorous plants have growth characteristics that differ substantially from terrestrial ones (Adamec 2011c, Ellison and Adamec, 2011). Adult aquatic carnivorous plants maintain approximately constant length of main shoots throughout the season: they show very rapid apical shoot growth but their basal shoot segments age and die at about the same rate (a "conveyer-belt" system of shoot growth). New biomass is allocated only into branching or flowering. Under favorable conditions, apical shoot growth rate of *Aldrovanda* is 1.0–1.2 whorls per day; 1.2–4.2 nodes per day in field- or 0.9–4.4 nodes per day in greenhouse-grown *Utricularia* species (Friday 1989, Adamec 2000, 2008b, 2009b, 2010b, 2015a, Adamec and Kovářová 2006, Adamec et al. 2010), but only 0.25 nodes per day in *U. purpurea* (Richards 2001). Despite its very high apical growth rate, the relative growth rate (RGR) of *U. australis* approaches zero simultaneously (Adamec 2009b), and its conveyer-belt growth may give it competitive release from epiphytic algae (Friday 1989). Zeatin cytokinins predominate apically in shoots of *Aldrovanda* and *U. australis* and decrease basipetally along the shoots, but auxin concentrations are more homogeneous throughout (Šimura et al. 2016). These results suggest that rapid polar growth and distinct physiological polarity is correlated with a polar gradient of cytokinins.

Whereas most terrestrial carnivorous plants grow slowly and tolerate environmental stress (*sensu* Grime 1979; Chapter 2), the high RGRs reported for several aquatic species under favorable conditions suggest they are more like ruderals (*sensu* Grime 1979; Chapter 2). The mean doubling time of biomass (i.e., ln 2/RGR) of *Aldrovanda* is only 8.5–28.7 days in field or culture, and 4.9–40.5 days for six *Utricularia* species (Table 19.1). Pagano and Titus (2007) reported a two- to threefold RGR increase in three *Utricularia* species with increasing [CO_2]. Comparable or higher doubling times were recorded by Nielsen and Sand-Jensen (1991) and Pagano and Titus (2007) for biomass in dozens of rooted submersed, noncarnivorous species (medians 9.2, 18.7; range 6.4–53.3 days, respectively). In contrast, corresponding values for terrestrial carnivorous plants are much greater (median 28–35, range 21–104 days; Ellison and Adamec 2011; Chapter 17).

A further important growth characteristic is the production of shoot branches that later develop into new individuals (Adamec 1999, 2011e). In *Aldrovanda*, the branches always develop into new plants, but in some aquatic *Utricularia* species, initiated branches (mainly axillary buds) may not develop into mature plants (Adamec 1999, 2009b, 2011e). The number of branches per shoot or the total branch:total plant biomass ratio generally is considered the principal growth parameter and a criterion for plant vigor and propagation rate; they also reflect the suitability of a habitat for plant growth (Kamiński 1987a, Adamec 1999, 2000, 2009b,

Table 19.1 Mean or median doubling time (range) of biomass or total shoot apices of aquatic carnivorous plants. Comparative data on 12 rooted and submerged noncarnivorous species are given in each of the last two rows.

Species	Growth conditions	Doubling time (days)	Reference*
A. vesiculosa	outdoors	28.7 (12.8–44.5)	1
A. vesiculosa	greenhouse	8.5 (6.4–10.6)	2
A. vesiculosa	field	24.0 (12.9–34.9)	3
A. vesiculosa	field	14.4 (8.4–21.5)	4
U. australis	greenhouse	4.9 (4.7–5.2)	2
U. australis	field	19.1 (9.1–33.2)	4
U. stygia	field	9.2	5
U. intermedia	field	6.6	5
U. macrorhiza	greenhouse	18.4 (12.4–29.5)	6
U. macrorhiza	greenhouse	24.8 (13.1–36.5)	7
U. geminiscapa	greenhouse	28.6 (22.4–34.7)	7
U. purpurea	greenhouse	40.5 (23.1–57.8)	7
noncarnivorous species (rooted)	greenhouse	18.7 (9.9–53.3)	7
noncarnivorous species (submerged)	greenhouse	9.2 (6.4–34.7)	8

*References: 1. Adamec (2000); 2. Adamec et al. (2010); 3. Adamec (1999); 4. Adamec and Kovářová (2006); 5. Adamec (2010b); 6. Pagano and Titus (2004); 7. Pagano and Titus (2007); 8. Nielsen and Sand-Jensen (1991).

2010b, 2011c, 2011e, 2015a, Adamec and Kovářová 2006, Adamec et al. 2010).

In some species, branching rate (number of internodes between two branches) is regular under some conditions (Table 19.2), suggesting both species specificity and ecological regulation (in *U. australis*). If *U. australis* grows under optimal conditions, regular branching by apically initiated branches occurs with frequent axillary budding along main shoots (Adamec 2011e) and the apparent branching rate is commonly 1–3 internodes per branch. Branching frequency (branching rate/apical shoot growth rate) specifies the time needed for initiating successive branches on the shoot and is a good measure of RGR; field-grown *Aldrovanda* and *U. australis* have similar branching frequencies (4.7–5.5 days per branch; Adamec and Kovářová 2006). In these species, however, competition occurs between growth of the main shoot and branches: the apical growth rate of branches (nodes per day) is ≈70–83% of that

Table 19.2 Mean (standard error of the mean) branching rates (number of internodes between two branches).

Species	Shoot type	Branching rate	Reference*
A. vesiculosa	Main shoot	6.2 (range 3–11)	1
A. vesiculosa	Main shoot	5.3±0.2	2
U. stygia	Photosynthetic	12.2±0.4	3
U. stygia	Carnivorous	6.7±0.2	3
U. stygia	Photosynthetic	12.2±0.2	4
U. intermedia	Photosynthetic	16.8±0.4	3
U. intermedia	Carnivorous	5.9±0.1	3
U. intermedia	Photosynthetic	17.7±0.4	4
U. australis	Main shoot	22.1±1.2; 10.5±0.4	5
U. australis	Main shoot	14.5±1.2; 15.1±1.7	2
U. australis	Main shoot	5.9±0.3; 22.6±1.1	6
U. australis	Branches	6.6±0.1	6

*References: 1. Adamec (1999); 2. Adamec and Kovářová (2006); 3. Adamec (2007a); 4. Adamec (2010b); 5. Adamec (2009b); 6. Adamec (2011e).

of the main shoot (Adamec 1999, 2011e). Similarly, formation of flowers in *Aldrovanda* competes with branching (Cross et al. 2016).

19.4.2 Mineral nutrition

Although rootless aquatic carnivorous plants grow in mineral-poor habitats, their macroelement composition is similar to rooted aquatic noncarnivorous plants (Adamec 1997a, Ellison and Adamec 2011). Their tissue nutrient content (% DW) in young shoots is usually in the ranges 1.0–4.0 for N; 0.12–0.50 P; 1.5–5.0 K; 0.15–3.0 Ca; 0.2–0.7 for Mg, even at low prey availability. Regardless of a marked polarity of tissue N, P, and Ca along shoots and large differences between leaves and traps (Adamec 2000, 2008a, 2008b, 2010b, 2014, 2016), mean shoot content of N, P, K, Ca, and Mg in aquatic carnivorous plants is ≈1.5–3-fold greater than that in terrestrial carnivorous plants (Ellison and Adamec, 2011) and likely reflects much faster growth and the absence of mechanical tissues in aquatic species. The marked polarity of tissue N, P, Ca (N and P increase apically; Ca increases basally) suggests very efficient N and P reutilization from senescing shoot segments (Friday and Quarmby 1994, Adamec 2000, 2008a, 2014, 2016). *Aldrovanda* reutilizes 60–90% N and 61–72% P from senescing shoots, and >90% for both elements in autumnal *Aldrovanda* shoots forming turions (Adamec 2000). *Utricularia* species reutilize 27–57% N and 44–77% P (Table 19.3); the reutilization efficiency differs between shoots and traps of *U. reflexa*. However, very low or even negative K reutilization in senescent shoots by aquatic carnivorous plants contrasts with the 41–99% reutilization of this element by terrestrial carnivorous plants (Chapter 17, §17.4.3), including five terrestrial *Utricularia* species (Adamec 2014). Zero K reutilization occurs even when *Aldrovanda* and *U. australis* are grown without prey at very low [K^+] (<4 μM; Adamec 2016).

In sum, aquatic carnivorous plants lose only a relatively small amount of the N and P in their senescent shoots, but almost all the K, Ca, and Mg. However, their N and P reutilization efficiency is still 10–30 percentage points lower than that found for terrestrial carnivorous plants (Chapter 17, §17.4.3). For unknown reasons, no reutilization of K appears to be common for all submersed aquatic plants (Adamec 2014). Considerable differences in tissue nutrient content occur also between shoots and traps in aquatic carnivorous plants. In both *Aldrovanda* and *Utricularia*, N, Ca, and Mg usually are greater in photosynthetic shoots than in the traps, whereas the opposite applies for P and K suggesting a considerable "mineral" cost of carnivory (Adamec 2008a, 2010c, 2014).

Table 19.3 Mineral nutrient reutilization efficiency (% of the initial content). All values assume a 17.3% decrease in biomass of senescent organs (Adamec 2016). Negative values indicate higher nutrient content in senescent organs after correction. Means or ranges of values from different experiments are shown;—no data.

Species	Growth conditions	N	P	K	Reference*
Aldrovanda	outdoors	90	72	11	1
Aldrovanda	greenhouse	60	61–62	5–33	2
U. australis	greenhouse	28–37	50–60	−8.5–2	2
U. australis	Field	57	77	8	3
U. vulgaris	Field	48	—	—	4
U. purpurea	Field	42	54	12	5
U. reflexa—shoot	Greenhouse	27	44	−25	6
U. reflexa—traps	Greenhouse	6	61	−63	6

*References: 1. Adamec (2000); 2. Adamec (2016); 3. Adamec (2008a); 4. Friday and Quarmby (1994); 5. Moeller (1980); 6. Adamec (2014).

Aquatic carnivorous plants grow in oligo-mesotrophic waters where NH_4^+ usually predominates over NO_3^-. Like most aquatic plants growing in dystrophic waters, *Aldrovanda* and some aquatic *Utricularia* species preferentially take up NH_4^+ from diluted NH_4NO_3 solutions (Adamec 1997a, 2000, 2016, Fertig 2001). NH_4^+ uptake by *Aldrovanda* apical shoot segments was the same as that by basal parts, but PO_4 uptake by apical segments was double that of basal parts. At ecologically relevant concentrations, NH_4^+ uptake on a molar basis by *Aldrovanda* shoots exceeded that of PO_4 13–40-fold and *U. australis* 11–19-fold (Adamec 2000, 2016). Prey capture significantly decreased NH_4^+ uptake but increased PO_4 uptake by *Aldrovanda* shoots, whereas there was no comparable effect of prey capture for *U. australis* (Adamec 2016). K^+ uptake from 83 μM K^+ by apical shoot segments (or photosynthetic shoots in *U. stygia*) in *Aldrovanda* and four aquatic *Utricularia* species (grown without NH_4^+) in light was 2.5–3.5-fold higher than that by basal shoot segments (or carnivorous shoots in *U. stygia*); K^+ uptake by whole shoots was the average of the two (Adamec 2016). That *Aldrovanda* takes up K^+ only by the basal segments (Adamec 2000) or that K^+ uptake may be significantly negative (Adamec 2016) suggests that uptake of NH_4^+ interferes with uptake of K^+ (Szczerba et al. 2008). The interference is strongest in *Aldrovanda*.

Numerous data show marked effects of prey on growth, both in culture and in the field (Adamec 2011c). Overall, feeding leads to longer shoots, greater biomass, faster apical shoot growth, greater RGR, and increased branching, and prey capture usually has a greater effect on *Aldrovanda* than on aquatic *Utricularia* species. However, it is unclear how the growth effects of carnivory are induced, because tissue N and P contents in apical or young shoot segments in prey-fed plants were lower than in unfed plants (Adamec 2000, 2008d, 2011a). The hypothesis that N and P absorbed from prey preferentially supports essential growth processes associated with cell division in shoot apices (Adamec 2008d) was supported only in *Aldrovanda*, but not in *U. australis* and *U. bremii* (Adamec 2011a).

Aquatic carnivorous plants may gain a substantial amount of mineral nutrients from prey, but only Friday and Quarmby (1994) have quantified N uptake efficiency from prey by any of these species. In *U. vulgaris* fed on mosquito larvae, they estimated N uptake efficiency to be at least 83% of total prey N, and ≈52% total plant N was obtained from the prey. P was also taken up rapidly from the prey. N uptake efficiency from prey in *Utricularia* traps may be even higher than that in terrestrial species (Chapter 17, §17.4.3).

The proportion of seasonal (daily) N and P gain obtained from prey is an important ecological parameter associated with mineral nutrition. The proportion of seasonal N gain from carnivory in *U. macrorhiza* is ≈75% (Knight 1988). However, in robust *U. foliosa* growing at a barren site with extremely low prey availability in Florida, the mean proportion was only 0.9% N and 3.5% P (Bern 1997). Koller-Peroutka et al. (2015) used ^{15}N to estimate a mean seasonal N gain from animal prey of 68–100% for *U. vulgaris, U. minor*, and *U. australis* in Austria. As only 4–10% traps usually captured animal prey and the traps mainly captured pollen grains and unicellular algae, these values may overestimate the contribution of prey to the plants' mineral nutrition. The capture of pollen and algae was correlated significantly with plant DW and length, but pollen capture was not correlated with ^{15}N in shoots.

Adamec (2011c) modeled the proportion of daily N and P gain that could be obtained from a daily capture of one small *Cyclops* (DW 25 μg) by *Aldrovanda* and *U. australis*. In rapidly growing plants (plant biomass doubles in 15 days), *Cyclops* makes up 15% of the daily N and 4% of the daily P gain by the smaller *Aldrovanda*, but only 0.62% N and 0.56% P gain by the larger *U. australis*. Under unfavorable conditions, when plants are not growing but maintaining constant biomass *Cyclops* can account for all the daily N and 16% P gain for *Aldrovanda* and 1.8% N and 2.6% P gain for *U. australis*. In a greenhouse experiment, *Aldrovanda* obtained 73% N, 49% P, and 100% K and Mg of its total nutrient content from ostracods relative to unfed controls (Adamec et al. 2010). In contrast, *U. australis* obtained all its N, but no P, K, or Mg from diaptomid prey. As in terrestrial species, the ecological importance of N and P uptake from prey depends primarily on the quantity of captured prey. Capture of prey in aquatic species is thus one of the most important determinants of their rapid growth and is especially important for their successful propagation.

19.4.3 Photosynthesis and respiration

Maximum net photosynthetic rate (A_{max}) per unit dry (DW) or fresh weight (FW) in aquatic carnivorous species with linear shoots (eight species, 40–160 mmol O_2 kg^{-1} FW h^{-1}) is comparable to or even higher than the highest values found in submersed noncarnivorous species (30–110 mmol kg^{-1} FW h^{-1}; Adamec 1997b, 2006, 2011b, 2011c, 2013, Ellison and Adamec 2011). Such high values have been found even in old leaves of *U. australis* and *U. vulgaris* without functional traps (Adamec 2013). Chlorophyll-based A_{max} of *U. australis, U. vulgaris*, and *U. purpurea*, which all have linear shoots, is significantly greater than that of three rhizomatous/rosette species (*U. dichotoma, U. resupinata, U. volubilis*), but only marginally higher when calculated based on FW or DW. Very high A_{max} appears to be a prerequisite for the rapid growth observed in *Utricularia*, as the rapid, permanent decay of senescent shoot segments causes a great loss of structural and non-structural carbohydrates (Adamec 2000), and concomitantly high maintenance costs of traps (§19.7.).

In amphibious species with dimorphic shoots, A_{max} of emergent photosynthetic shoots increases considerably relative to submerged ones (Colmer and Pedersen 2008). Traps of aquatic *Utricularia* species are physiologically very active organs and have high dark respiration rates (RD). Simultaneously, they incur high photosynthetic (metabolic, energetic) costs because of relatively lower A_{max}. In seven aquatic *Utricularia* species, RD of traps (5.1–8.6 mmol O_2 kg^{-1} FW h^{-1}) was 1.7–3.0 times higher than that in leaves, whereas A_{max} in photosynthetic leaves exceeded that in the traps (5.2–14.7 mmol kg^{-1} h^{-1}) 7–10-fold (Adamec 2006, 2014). Such high RD:A_{max} ratios (50–140%) for *Utricularia* traps illustrates their high maintenance and photosynthetic costs: in *U. stygia* and *U. intermedia* with dimorphic shoots, the trap RD could account for 34–44% of the total plant respiration, and 63% in

U. australis with monomorphic shoots (Adamec 2006, 2007a, 2011c). However, in *U. macrorhiza*, mean trap RD was only about 10% higher than leaves, but trap A_{max} in lake water was 41–67% of leaves (Knight 1992). For *Aldrovanda* traps, it reached 67% (Adamec 1997b).

Aquatic carnivorous plants usually grow in waters with high $[CO_2]$ >0.1 mM. All species tested so far use only CO_2 for photosynthesis (Adamec 1997a, 1997b, 2011c, Adamec and Kovářová 2006, Pagano and Titus 2007); the apparent slight use of HCO_3^- by *U. australis* induced by growing at pH ≈9.2 is ecologically unimportant (Adamec 2009b). In several species growing in the field or culture, CO_2 compensation points range from 1.5–13.2 μM (Adamec 1997a, 1997b, 2009b, 2011c, Adamec and Kovářová 2006, Pagano and Titus 2007); similar to the 1.5–10 μM reported for noncarnivorous species (Maberly and Spence 1983). CO_2 compensation points in 17 culture-grown species or accessions of *Utricularia* and *Aldrovanda* (mean 5.3, range 1.9–13.6 μM) are similar to those found in these species growing *in vitro* (mean 5.2, range 2.5–8.8 μM; Adamec and Pásek 2009). The positive relationship between CO_2 compensation point and $[CO_2]$ in culture water suggests flexible ecological regulation of CO_2 affinity. However, in *U. australis* growing at 17 sites of different trophic levels in the Czech Republic, CO_2 compensation points ranged from 0.7–6.1 μM (mean 2.6 μM) but is correlated significantly neither with water chemistry factors nor capture of prey (Adamec 2009b).

Both *Aldrovanda* and *U. australis* fed zooplankton grew outdoors significantly faster than unfed ones (Adamec 2008a). Feeding increased A_{max} by 59% in *Aldrovanda* but decreased it by 25% in *U. australis*. CO_2 affinity was unchanged following feeding in *Aldrovanda* but decreased from 5.2 to 9.2 μM in *U. australis*, while RD values were unaffected by feeding in both species. Thus, the hypothesis of stimulation of photosynthesis by prey capture (Givnish et al. 1984) is supported for *Aldrovanda* but not supported for *U. australis*. Carnivory could partly compensate for photosynthetic CO_2 uptake, but the uptake of organic carbon from prey has never been quantified, although it occurs (Adamec 1997a; §16.2.5). Nevertheless, organic carbon uptake from prey may be ecologically important when $[CO_2]$ is limiting. Field-grown *Aldrovanda* also grew at pH>9.0 when catching numerous prey (Adamec 1999), and greenhouse-grown *U. vulgaris* fed on prey grew better and branched more only at pH values of 7.6–9.1 (Kosiba 1992).

19.5 Trap ecophysiology of aquatic *Utricularia*

Utricularia suction traps are hermetically closed bladders functioning on the basis of negative pressure (Fig. 19.4; Sydenham and Findlay 1973, Juniper et al. 1989, Guisande et al. 2007, Vincent et al. 2011b, Singh et al. 2011; Chapter 14, §14.2.3). Unlike the traps of other species, solutes and suspended particles aspirated from the ambient water are retained in the lumen until the senescent trap disintegrates. Four types of glands (hairs) occur inside or outside the traps: numerous and large internal quadrifid and bifid glands are crucial for trap physiology (Juniper et al. 1989, Guisande et al. 2007; Chapter 13, §13.5.1).

19.5.1 Water flow

Almost all of the biophysical knowledge of *Utricularia* trap functioning derives from four classic

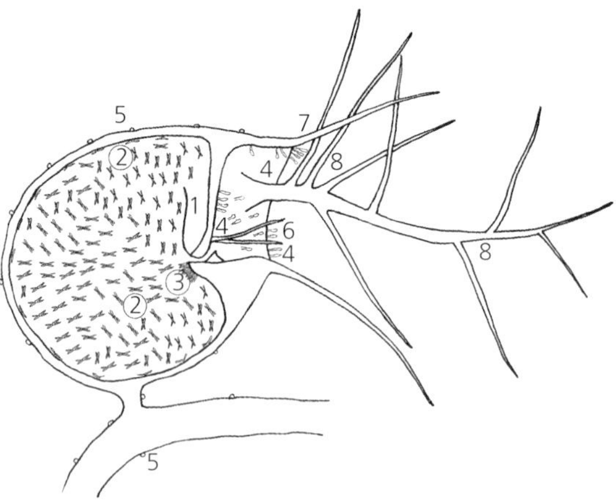

Figure 19.4 Schematic longitudinal section through a trap of *Utricularia* with glands and other structures (modified by J. Vrba after Juniper et al. 1989). 1: trapdoor; 2: quadrifid glands; 3: bifid glands; 4: stalked mucilage glands; 5: spherical sessile glands; 6: trigger hairs; 7: rostrum; 8: antennae.

studies (Sydenham and Findlay 1973, Sasago and Sibaoka 1985a, Adamec 2011c, 2011g). A negative pressure of about −16 kPa (−0.16 bar) relative to the ambient water is maintained inside a fully set trap. When a prey animal touches sensory hairs situated on the trapdoor, it opens, the small animal is aspirated into the trap, and the door shuts again. This process of firing is completed within 10–15 ms (Sydenham and Findlay 1973). Immediately after firing, the negative pressure inside the trap is zero, but pressure is restored by the gradual removal of ≈40% of the water from the fired trap until the original concave shape is reached. This first phase of trap resetting lasts ≈25–30 min, by which time the trap is ready to fire again, but full resetting lasts much longer.

Pumping water out of the traps is an active process associated with ATP consumption. Aerobic respiration inhibitors added to the trap fluid block the water-pumping and trap-narrowing processes. Bifid glands attached close to the trapdoor take part in water pumping so that water is exuded from the pavement epithelium close to the door. Electrophysiological measurements have suggested that Cl^- ions are taken up actively from the trap fluid by the bifid glands and, due to their movement, osmotically drag water molecules (Sasago and Sibaoka 1985a, Adamec 2011c). Monovalent cations (Na^+, K^+) accompany the Cl^- fluxes, while divalent cations (Ca^{2+}, Mg^{2+}) inhibit them. The final part of the water pathway is still unclear, but water is probably expelled from the cells of the pavement epithelium through a very leaky plasmalemma by turgor pressure (for solutes <600 Daltons; Adamec 2011c). Traps also can pump out water when immersed in liquid paraffin oil or in moist air (Sasago and Sibaoka 1985a, Adamec 2012a).

High-speed cinematography revealed that the basis of the reversible trapdoor opening and closing is a rapid curvature inversion called "buckling and unbuckling" (Singh et al. 2011, Vincent et al. 2011a, 2011b; Chapter 14, §14.2.3). Because of the negative pressure inside the trap, the curved trapdoor is metastable and any above-threshold mechanical stimulation (mediated by prey through the trigger hairs) triggers the curvature inversion (buckling) leading to trap opening and water inflow. When the negative pressure is gone, the trapdoor spontaneously returns to its initial curvature, closing the door again. The negative pressure inside the traps is thus an essential component part of the opening and closing mechanism. The complete process of trap firing lasts only 2.5–5 ms in several species.

New stimuli for studying *Utricularia* trap ecophysiology have arisen from the discovery of spontaneous firing in *Utricularia* traps by high-speed cinematography of intact shoots (Vincent et al. 2011a, 2011b) and linear-position sensor measurements of excised traps (Adamec 2011d, 2011f). Spontaneous firings occurred 0.3–2.4 times during the 24-hour resetting period and the mean time between two spontaneous firings varied from five to 16 hours. Spontaneous firings quantitatively resembled mechanically stimulated ones (trap thickness changes as firing and resetting rate: Adamec 2011f). In two *Utricularia* species, Vincent et al. (2011a) distinguished temporal patterns of spontaneous trap firings as metronomic, random, and bursting. Spontaneous firings may function as a "safety valve," protecting the integrity of the trapdoor or walls. After firing, traps commonly pump water out for at least 5–10 hours until a steady-state is reached (Adamec 2011f), and recent data suggest that traps always are pumping water out (Adamec 2011d, Vincent et al. 2011a). Thus, water either recirculates constantly through some leaks under the trapdoor or through the walls, or the mechanism of water pumping becomes thermodynamically inefficient at high negative pressure.

Adamec and Poppinga (2016) gradually applied negative pressure inside cut off aquatic *Utricularia* traps to determine the critical negative pressure (CNP) at which the traps (located in air) fire and aspirate an air bubble (as in spontaneous firings). In 15 aquatic *Utricularia* species representing four generic sections, mean CNP values ranged from −7 to −35 kPa. The average in all 20 species or variants tested was −20 ± 2 kPa. In 13 species or variants in section *Utricularia*, the CNP was −17 ± 2 kPa, significantly higher than that of two species (*U. dichotoma*, *U. volubilis*) in sect. *Pleiochasia* (−34 ± 1 kPa). The high intraspecific variability in CNP could represent the three types of spontaneous firings (Vincent et al. 2011a).

Adamec (2011d) compared taxonomic aspects and basic functional characteristics of trap firing

(firing and resetting rate as "trap efficiency") of 13 aquatic *Utricularia* species. These species differed significantly in their firing (3.7–4.2×) and resetting (10–24×) rates per unit trap thickness or length and on an absolute scale. Overall, traps of very common species (*U. australis, U. stellaris, U. inflata*) showed the greatest firing and resetting rates. The magnitude of firing and resetting per unit trap thickness or length was correlated negatively with both trap thickness and length. Smaller and narrower traps are thus more effective at trap firing and resetting than larger traps. Firing or resetting characteristics were the same in unfed and prey-fed traps of *U. reflexa* for both stimulated and spontaneous firings.

The occurrence of a distinct lag-period after trap firing in two accessions of aquatic *U. dichotoma* suggests that water-pumping regulation varies among generic sections (Płachno et al. 2015a). Traps of three accessions of two species of *U.* sect. *Pleiochasia* possess 2–4 cell layers in their trap walls (Płachno et al. 2015a), can be stiffer than those in other sections, and their trap efficiency characteristics differed significantly from those in 12 species of sect. *Utricularia*. The magnitude of firing per unit trap length of species in *U.* sect. *Pleiochasia* averaged 1.7–2× less and their resetting rate per unit trap length 3–6× less than species in sect. *Utricularia* (Adamec 2011d, Płachno et al. 2015a). However, no correlation between the number of cell layers in trap walls and the trap characteristics in *U.* sect. *Pleiochasia* was found.

Last, biophysical diversification of *Utricularia* traps into three types and different trap sizes could contribute to a diversification of prey capture in environments with various prey availability. The ratio between the stiffness of the lateral trap walls and that of the trapdoor is a primary determinant of trap efficiency (Vincent et al. 2011b) and could vary phylogenetically and for different prey. New biophysical findings indirectly support the physical (not electrophysiological) concept of *Utricularia* trap triggering (Adamec 2012a).

19.5.2 Prey digestion

The pH of trap fluid in four aquatic *Utricularia* species is independent of prey digestion and equals 5.0 ± 0.1 (Sirová et al. 2003). Although several types of hydrolytic enzymes have been described from *Utricularia* traps using biochemical and cytochemical methods (protease, esterase, acid phosphatase; Juniper et al. 1989; Chapter 16, §16.3), only the activity of phosphatases at pH 4.7 was determined to be biologically significant by an *in situ* analysis of prey-free trap fluid in aquatic *Utricularia* (Sirová et al. 2003) and trap age was the key factor in the pattern of phosphatase production. Trap activities of α- and β-glucosidases, β-hexosaminidases, and aminopeptidases at pH 4.7 usually were lower by one or two orders of magnitude but were higher at the same pH in the culture water. Thus, a greater part of the enzyme activity, excepting phosphatases, enters traps from ambient water after they fire.

More generally, enzyme activity is independent of prey capture (digestion) and is not inducible by prey or loading of N and P salts into the traps (Sirová et al. 2003, Adamec 2011g). Due to a very efficient total N uptake from prey in *U. vulgaris* traps (Friday and Quarmby 1994), proteinaceous N as the main N source from prey also must be digested and absorbed effectively. The absence of aminopeptidases (proteases) in traps could be compensated for by the autolysis of dead prey tissues, yet the discrepancy between the almost complete lack of protease activity in trap fluid and the presence of large secretory vesicles (Golgi apparatus) rich in proteases in quadrifid glands (Vintéjoux and Shoar-Ghafari 2005) still persists. As proteases commonly act on a very diverse set of substrates, it is possible that the commonly used microfluorimetric method identifies only a small number of the proteases that actually are present in digestive fluid.

A similar discrepancy exists between the invariably high phosphatase activity found in trap fluid and a very low activity of enzyme-labeled fluorescence (ELF) of phosphatase usually detected on the surface of quadrifid glands (Sirová et al. 2003, Adamec 2011g). Some of this discrepancy might be explained by methodical limitations of ELF. Furthermore, it is not clear what proportion of enzyme activity in the filtered trap fluid is produced by trap glands alone and the various trap commensals. Nevertheless, the consistently high trap-fluid phosphatase activity found in all *Utricularia* species studied to date implies that P uptake from prey or detritus might be more important than that of N for these plants.

19.5.3 The role of trap commensals

Commensal microorganisms (mainly bacteria, algae [*Euglena*], ciliates, rotifers) occur and propagate in the traps of all aquatic *Utricularia* species studied to date (Richards 2001, Peroutka et al. 2008; Chapter 25). However, their role in trap functioning and possible benefit for the plants are still rather unclear. Moreover, perfect *Utricularia* traps usually capture relatively few animal prey in typical barren, dystrophic waters (usually only 5–50% during their lifespan) even though a high abundance of commensals invariably occur in all prey-free traps (Friday 1989, Richards 2001, Adamec 2008a, 2009b, Peroutka et al. 2008, Sirová et al. 2009, Koller-Peroutka et al. 2015). Some of these commensals could, to varying extents, act as digestive mutualists, breaking down prey with their own enzymes (Richards 2001, Płachno et al. 2006, Sirová et al. 2009). For aquatic *Utricularia* species in barren waters with low trapping efficiency (e.g., *U. purpurea*), commensal communities in traps could be more beneficial than the trapping of prey alone (Richards 2001).

In prey-free traps, which have aspirated detritus or phytoplankton from the ambient water during spontaneous firings, a miniature microbial food web may develop (Sirová et al. 2009). Its main components are bacteria, Dinophyta, ciliates, and rotifers. Similar food webs occur in *Sarracenia* and *Nepenthes* pitchers (Chapters 23, 24). In filtered fluids from prey-free traps of two field-grown *Utricularia* species, Sirová et al. (2009) detected high concentrations of soluble organic carbon (60–310 mg L^{-1}), organic N (7–25 mg L^{-1}), and soluble P (0.2–0.6 mg L^{-1}). However, the total content of C, N, and P in the trap fluid, including mainly the particulate form (commensal organisms and detritus), was several times greater (in mg L^{-1}): C, 632–1570; N, 21–81; P, 0.9–4.2. The contents usually increased with trap age and were correlated with commensal biomass. On the basis of phospholipid fatty acid analysis of the trap commensal biomass, a complex microbial food web was revealed and bacteria formed >75% of the viable microbial biomass. Very rapid turnover of organic substances also occurs in *U. reflexa* traps (Šimek et al. 2017).

In a two-day ^{13}C-labeling experiment using *U. australis* and *U. vulgaris*, a large proportion of newly fixed CO_2 was allocated from shoot bases to shoot apices and mature shoot segments (Sirová et al. 2010). Total C allocation in plant tissues rapidly decreased with increasing age of the shoot segments, but the ratio of C exuded into the trap fluid to that in plant tissues increased markedly with age: twice as much newly fixed C was allocated to trap fluid relative to plant tissue in the oldest analyzed segments. Overall, 20–25% of newly fixed C was allocated into the trap fluid. Sirová et al. (2011) further showed that these organic exudates fuel respiration of the heterotrophic microbial commensals inside prey-free traps. Up to 30% of total dissolved organic C in trap fluid is easily metabolized compounds (mainly glucose, fructose, sucrose, and lactate, at 2–56 mg L^{-1} each) and the proportion of these compounds and their microbial use decreases with increasing mineral nutrient supply (N, P) and trap age. The total concentration of 46 analyzed organic compounds in trap fluid was 37–260 mg L^{-1}. Concentrations in the range 9–78 mg L^{-1} were found in three other *Utricularia* species but shaded plants exhibited lower values (Borovec et al. 2012).

Most of the C in the commensal organisms in the trap fluid is provided by the plant (Sirová et al. 2009, 2010, 2011) and this extensive C supply appears to be an important additional maintenance (energetic) cost of traps. Because the A_{max} of aquatic *Utricularia* is very high (Adamec 2006, 2013; §19.4.3), the plants can afford such "gardening" of commensals. The concentration of organic compounds in the trap fluid also depends on species, photosynthetic conditions, and water chemistry. High concentrations of organic acids and amino acids in the fluid could lead to the observed pH values ≈5 (Sirová et al. 2003), indicating a high buffering capacity. The plants also could gain growth-limiting N and P from phytoplankton, pollen, and detritus decomposed in prey-free traps "in trade" for exuded organic compounds (Richards 2001, Koller-Peroutka et al. 2015). Aquatic *Utricularia* species growing in oligotrophic habitats with low prey availability might better be considered to be bacterivorous or detritivorous rather than carnivorous.

N_2 fixation by cyanobacteria also has been shown to occur on the outer trap surface of *U. inflexa* (Wagner and Mshigeni 1986) and it also could occur inside the traps and provide the traps with N. Sirová et al. (2014) estimated $^{15}N_2$ fixation rate in shoots and traps of *Aldrovanda* and aquatic *Utricularia* and suggested that N-fixation made only a small contribution to seasonal plant N gain. Higher activity in older traps and shoots implied a key role of the periphyton in N-fixation. The reason for limited N_2 fixation could have been the high concentration of NH_4^+ -N (2.0–4.3 mg L^{-1}) found in the trap fluid of *Utricularia*.

Recent data on trap operation specifies in part how much growth-limiting N and P enters the traps from the ambient medium and becomes a substrate for the microbial commensals. Spontaneous trap firings and constant water recirculation (§19.5.1) could lead to a substantial N and P gain for the traps, especially for plants in barren waters that otherwise trap few prey. Adamec (2011d, 2011g) constructed a simple nutrient budget model, in which a theoretical accumulation of N and P from the ambient natural water (Adamec 2008a) was compared with the total contents of N and P estimated for prey-free trap fluid in two *Utricularia* species (Sirová et al. 2009). Modeled total N and P accumulated in prey-free traps very slowly (Adamec 2011d, 2011g). For a 10-day old, prey-free trap, it would take 40–70 more days to accumulate the measured N or 15–23 days for P if there were only spontaneous firing and constant water recirculation mechanisms operating. However, the modeled rate of N and P accumulation could have been underestimated as the aspirated ambient water on the trap surface also contained phytophilous zooplankton (unlike the "bulk" water). Adamec (2011d, 2011g) also calculated that the total N and P amount found in the trap fluid represents only ≈3.5% of total plant N and ≈1.2% of total P.

Several conclusions can be drawn from this model (Adamec 2011d, 2011g). N and P inputs from ambient water are low, they cannot account for the total N and P amounts in the prey-free trap fluid, and no net N or P uptake occurs from trap fluid. Therefore, the traps exude N and P to enhance the microbial community and inoculation of young traps by microorganisms could stimulate the traps to exude N, P, and organic C. Prey-free *Utricularia* traps derive no clear nutritional benefit from the trap commensal community and the trap microorganisms behave more like parasites than commensals. Any nutritional benefit for the plant occurs only when it traps animal prey. The model is supported by the finding of high NH_4^+ concentration in the trap fluid (demonstrating very low uptake affinity) of two *Utricularia* species (Sirová et al. 2014), which strongly contrasts with the ≈1000× higher uptake affinity of *Utricularia* shoots growing in very barren waters: <0.4 μM for NH_4^+ and <0.1 μM for phosphate (e.g., Adamec 2009b).

19.5.4 Oxygen regime and trap respiration

An absence of O_2 is consistently measured in the fluid of excised and intact prey-free traps of *Utricularia* bathed in an oxygenated medium, regardless of trap age and irradiance (Adamec 2007b). Permanent anoxia inside the traps is interrupted by trap firing (prey capture or spontaneous firing), but only for short time periods; aerobic respiration of the inner glands and trap walls is so high that any available O_2 is exhausted within 10–40 minutes after trap firing. The traps can partly restore the negative pressure within 30 minutes, which requires high amounts of energy derived from aerobic respiration and is inhibited by respiration inhibitors (Sydenham and Findlay 1973, Sasago and Sibaoka 1985a, Adamec 2012a).

It is unclear how the traps or glands provide sufficient ATP energy for their demanding functions under anoxia, although a mitochondrial mutation of cytochrome *c* oxidase found in *U. gibba* should provide greater energetic power for the traps (Laakkonen et al. 2006). These authors hypothesized that mitochondrial proton pumping was decoupled from electron transfer and provided ATP energy during the short aerobic period after trap firing. Such decoupling would allow the traps to increase power output during periods of need, but resulted in a 20% decrease in total energy efficiency of the respiratory chain. Moreover, it seems improbable that the traps provide most of their ATP energy need from anaerobic fermentation. Evidently, the inner trap glands possess an extremely high O_2 affinity (<0.5–1 μM; cf. 0.3–1.1 μM reported for leaves of terrestrial plants by Laisk et al. 2007) and use the permanent O_2 influx from fine intercellular spaces in the trap walls connected with the shoot. The above-mentioned cytochrome *c* oxidase mutation also can explain such a high O_2 affinity. Finally, transcriptomic global gene expression analysis in *U. gibba* confirms that traps significantly over-express genes involved in respiration (Ibarra-Laclette et al. 2011). In sum, it appears that the extremely low $[O_2]$ in the trap fluid is a result of a functional compromise: it must be very low (<15–30 μM) to reliably kill the captured prey but higher than the threshold (0.4 μM) for effective aerobic respiration. Anoxia causes captured prey to die of suffocation in traps, while all trap commensals are adapted to facultative anoxia (Adamec 2007b).

19.6 Regulation of investment in carnivory

As an investment in carnivory, traps in aquatic *Utricularia* species incur costs in structure (organic biomass), photosynthesis (decrease of A_{max}), energy (consumption for trap operation, metabolism, and maintenance), and minerals (total amount of minerals in the traps or that lost in senescent traps; Ellison and Adamec 2011). In aquatic *Utricularia* species, the proportion of traps to the total plant biomass is usually 10–65%, but this structural cost is regulated flexibly by the plants to minimize all costs according to habitat factors, especially water chemistry, prey capture, and irradiance (Adamec 2011c, 2011g, 2015a). Although this regulation may be species-specific, increased mineral nutrient availability (water chemistry or prey capture) usually leads to decreased investment in carnivory in terms of trap number per leaf, mean trap size (weight), or proportion of trap biomass. Similarly, the regulation of investment in carnivory by *Aldrovanda* is at best very weak (Adamec et al. 2010).

Much attention has been focused on the regulatory role of shoot nutrient content in *Utricularia* species. The number of traps per leaf in *U. foliosa* is inversely proportional to shoot N and P content (Bern 1997). Out of all nutrient factors measured for field-grown *U. australis*, only N content in apical and young adult shoot segments was significantly, and negatively, correlated with trap proportion (Adamec 2008a). This result was consistent with the suggestion of Guisande et al. (2004) that ambient N sources are the key factor regulating trap proportion and supports Adamec's (2008a, 2011c) "nutrient hypothesis" that all external factors that decrease tissue N content in young shoots (poor prey capture, low $[NH_4^+]$, high $[CO_2]$, etc.) increase trap production in young shoots and *vice versa*. This negative feedback also contributes to stabilizing the tissue mineral contents of other nutrients. However, in *U. vulgaris*, Kibriya and Jones (2007) found a central regulatory role for P, whereas both N and P played the regulating role in *U. foliosa* (Bern 1997).

As prey capture and photosynthesis support plant growth, growth rate itself also is a component of this endogenous regulatory system. Other indirect data indicate that changes of trap proportion in aquatic *Utricularia* are regulated photosynthetically (Bern 1997, Englund and Harms 2003, Adamec 2008a; Chapter 18): at low A_{max} (low irradiance or $[CO_2]$), trap proportion is low or even zero, indicating that photosynthetic regulation outweighs nutrient regulation. Supporting this hypothesis, three aquatic *Utricularia* species grown at high $[CO_2]$ increased trap proportion 3–82-fold, whereas there was much less effect of prey removal (1–4.7-fold increase in trap proportion; Adamec 2015a). The trap proportion in all species was negatively correlated with shoot N and P contents but positively correlated with mean trap DW. Trap numbers per leaf node DW (without trap DW) between the variants varied only 1.2–3.1×, whereas mean trap DW varied 2.7–249×.

More generally, under favorable CO_2 and light conditions, prey capture is associated more with high trap proportions in aquatic *Utricularia* (positive feedback) than with apical shoot growth, although the trap proportion apparently does not depend on very low shoot N or P content (Adamec 2015a). At intermediate $[CO_2]$, shoot N and P contents are very variable and regulate trap proportion by negative feedback (nutrient regulation). Under poor photosynthetic conditions, apical shoot growth is supported preferentially and trap proportion is blocked by a shortage of photosynthates (photosynthetic regulation) and probably also by the very high shoot N and P content. All factors (including growth rate) that change shoot mineral content can regulate trap proportion, but photosynthesis outweighs nutrient regulation. Mean trap size (DW) is regulated much more than trap number per leaf but trap regulation is considerably species-specific.

19.7 Turions

Turions are dormant vegetative winter buds formed by perennial aquatic plants as a response to unfavorable ecological conditions including decreasing temperature, day-length, or nutrient availability (Bartley and Spence 1987). *Aldrovanda* and ten of 14 temperate *Utricularia* species form turions. Some (sub)tropical or temperate *Utricularia* species and *Aldrovanda* populations produce non-dormant, quiescent shoot apices ("winter apices").

Turions of aquatic carnivorous plants are tough and sturdy organs 1–25 mm in diameter, and are modified from shoot apices by extreme condensation of short, modified trap-free leaves at the end of the growing season (Figure 19.5). They are partly frost-resistant and protect fragile plant shoots from freezing. Turions are hardened by weak frosts and their frost hardiness is based on the shift from frost avoidance in non-hardened turions to frost tolerance. For example, dormant, non-hardened autumnal turions of *Aldrovanda* and seven *Utricularia* spp. exhibit extracellular freezing from −7.0 to −10.2 °C (leading to turion death), whereas outdoor-hardened turions of five species freeze at −2.8 to −3.3 °C and almost fully survive freezing (Adamec and Kučerová 2013). Because of their markedly increased frost tolerance, they can survive even −10 °C.

Figure 19.5 Ripe turions of *Utricularia vulgaris* ranging from 10 mm to 25 mm long.

Two dormancy states and their hormonal patterns were described in detail for turions of Canadian *U. macrorhiza* using bioassays (Winston and Gorham 1979a, 1979b): innate dormancy at the end of summer and imposed dormancy at the end of autumn when the turions can germinate and sprout under sufficient temperature and light conditions. These two dormancy states also occur in *Aldrovanda* turions (Adamec 2003a).

Turions of all aquatic carnivorous plants usually overwinter and break their innate dormancy in darkness or deep shade, under hypoxia or anoxia, while lightly covered by organic sediments. Two distinct ecophysiological strategies of their autumnal sinking and spring floating (rising) may be distinguished (Adamec 2003a, 2008b). All turions ripen at the water surface, overwinter at the bottom, and usually germinate and sprout at the water surface in warmer water. Only *Aldrovanda* turions have developed an active mechanism of sinking and rising, which is presumably caused by variable gas volume in the gas spaces of turion leaves; for rising, the evolved gas could come from respiration or fermentation (Adamec 2003a, 2008b). *Utricularia* turions are less dense than water and are dragged to the bottom by their decaying mother shoots. By early spring, the turions have separated from the shoots and rise to the water surface. A variable turion fraction in all species can overwinter on wet substrate or frozen in ice at the surface (Adamec 1999).

Turions are storage organs. In autumn, they accumulate starch, free sugars, reserve proteins, and lipids (Winston and Gorham 1979b, Adamec 2000, 2003a, Płachno et al. 2014a). Turions also represent storage organs for mineral nutrients (N, P) although storing nutrients presumably is less important than storing carbohydrates (Adamec 2010b, 2010d). The RD of turions of five aquatic carnivorous plants was 30–75% lower per unit FW and 80–85% lower per unit DW than adult shoots (leaves) of the same species at a standard temperature (Adamec 2008b, 2011d). In contrast, RD in non-dormant winter apices was comparable with that of their growing shoots (Adamec 2008b). True turion dormancy thus is associated strictly with low RD values. A great proportion of cyanide-resistant respiration (50–90%) was found in dormant turions. A_{max} of turions of *Aldrovanda* and four *Utricularia* spp. during imposed dormancy was negative or only slightly positive under standard conditions. However, both old and newly formed shoot segments in sprouting turions in all species exhibited as high an A_{max} as measured for adult plants (Adamec 2011b). This very high A_{max} combined with the storage functions of turions for N, P, S, and Mg contributes to rapid growth of sprouting turions and rapid production of standing biomass at the beginning of the growing season. In sum, ecophysiological traits of turions of aquatic carnivorous plants are the same as those of turions of aquatic noncarnivorous plants.

19.8 Future research

Traits and growth strategies of *Aldrovanda* and aquatic *Utricularia* are ecophysiologically quite dissimilar from those of terrestrial carnivorous plants. The principal growth traits in rootless aquatic carnivorous plants with linear shoots are associated with very steep physiological polarity along the shoots and require a combination of several unique ecophysiological processes. Carnivory markedly enhances growth and, for naturally growing species, is indispensable, but methodological difficulties associated with the aquatic medium have made it difficult to study the efficiency of mineral and organic nutrient uptake from prey in aquatic carnivorous plants.

Recent biophysical data on *Utricularia* traps and on interactions between traps and the microorganisms within them have forced a re-evaluation of classic views of both trap function and ecological characteristics of trap microorganisms: digestive mutualism in traps with prey but parasitism in prey-free traps. Despite the high costs of traps, all aquatic *Utricularia* species produce them. The benefit derived from the few traps that capture prey exceeds the costs of producing many traps that never catch prey. Finally, other than *Aldrovanda vesiculosa* and *Utricularia* species, there are no other rootless, submersed vascular plant species that grow in dystrophic, barren waters.

Further insights into ecophysiology of aquatic carnivorous plants will result from addressing a number of key areas (Ellison and Adamec 2011). First, which physiological processes are most important for growth enhancement, and are positive growth effects caused by stimulation of cell divisions in shoot apices? Defining the role of tissue N and P content in observed growth effects would be particularly useful. Second, what is the uptake affinity of shoots for mineral nutrients from water? Is shoot nutrient uptake from the ambient water stimulated by prey capture, and what is the efficiency of mineral and organic nutrient uptake from prey? Third, unlike terrestrial carnivorous plants but similar to all submersed aquatic plants, aquatic carnivorous plants do not reutilize K^+ in senescent shoots. What are the K^+ uptake characteristics (localization of uptake, affinity, uptake rates) of shoots of aquatic carnivorous plants, and are they similar to those in aquatic noncarnivorous plants?

Finally, further investigations of the commensal community will yield many new insights. Open areas of research include nutrient dynamics and interactions within traps with and without prey; regulation by the plant and its commensals in trap exudation of C, N, and P into the trap fluid; and the importance of phytoplankton, bacteria, and detritus as potential nutrient sources (N, P, K, Mg) for *Utricularia* in barren waters.

CHAPTER 20

Biotechnology with carnivorous plants

Laurent Legendre and Douglas W. Darnowski

20.1 Introduction

Biotechnology is defined as "the application of science and technology to living organisms, as well as parts, products and models thereof, to alter living or non-living materials for the production of knowledge, goods and services" (OECD 2016). Until recently, the only biotechnological processes linked to carnivorous plants involved the extraction of pharmaceutical substances or milk-clotting agents from total biomass or from the leaf phylloplane. The former still finds industrial applications and is covered in an abundant literature on *Drosera* (Banasiuk et al. 2012, Egan and van der Kooy 2013) and *Dionaea* (Gaascht et al. 2013). In this review, we focus on the many new products that are being developed in various industrial sectors thanks to the current understanding of the physiology of carnivorous structures. The biotechnology and phytopharmaceutical industries require plant biomass in quantities that cannot be supported by wild collection, and we review current methods for *in vitro* culture and genetic transformation of many carnivorous plant species.

20.2 Activity and production of pharmaceutical substances

20.2.1 Droseraceae and Nepenthaceae

Perhaps the most widely known pharmaceutical use of a carnivorous plant is "Drosera herba," a tincture of *Drosera rotundifolia* used to treat whooping-cough and other respiratory illnesses throughout Europe. A wide range of *Drosera* and *Nepenthes* species also are part of the traditional and commercial pharmacopeia of India, Malaysia, China, and Tibet (Table 20.1; Banasiuk et al. 2012). Ethnobotanical claims concerning anti-spasmolytic, anti-inflammatory, anti-oxidant, and anti-bacterial activities have been supported by modern cell and biochemical reaction-based experimental systems and have been extended to anti-cancer, anti-fungal, and crop-protection properties (Table 20.1).

Mystical power has been attributed to some carnivorous plants (Juniper et al. 1989), but bioassay-guided fractionation of plant extracts coupled to HPLC-DAD/MS/NMR-based analyses have revealed that pharmaceutical properties of *Drosera* and *Nepenthes* are derived from defensive secondary metabolites that are not required for fundamental developmental processes or carnivory and can, for the most part, be found in noncarnivorous species. Phytochemicals of *Drosera* and *Nepenthes* likely are homologous, and are shared with the other carnivorous Caryophyllales: *Dionaea, Aldrovanda, Drosophyllum,* and *Triphyophyllum*. These phytochemicals fall into three main classes: naphthoquinones, flavonoids, and phenolic acid derivatives (Figure 20.1a–c; Juniper et al. 1989, Banasiuk et al. 2012, Egan and van der Kooy 2013, Gaascht et al. 2013).

Naphthoquinones occurrence and activity. Since their discovery in *Drosera* over 50 years ago (Zenk et al. 1969), naphthoquinones (NQs) have attracted a lot of attention. They constitute a large class of natural substances that occur in fungi, lichens, algae, plants, and arthropods (Babula et al. 2009). All share a bicyclic naphthalene skeleton with C_1–C_2 or C_1–C_4 substitutions and are colored (yellow, orange,

Legendre, L., and Darnowski, D. W., *Biotechnology with carnivorous plants*. In: *Carnivorous Plants: Physiology, ecology, and evolution*. Edited by Aaron M. Ellison and Lubomír Adamec: Oxford University Press (2018).
 DOI: 10.1093/oso/9780198779841.003.0020

Table 20.1 Bioactivity of carnivorous plant extracts extracted with various methods (**bold** is most active; ° is least active) that have been supported by cellular or biochemical tests.

Species	Extraction	Bioactivity	References*
Dionaea muscipula	Chloroform	Anti-bacterial against the crop pathogen *Pectobacterium atrosepticum*; active metabolite: plumbagin.	1
	Methanol, chloroform	Anti-bacterial against four human pathogenic bacteria; most active secondary metabolite: quercetin.	2
	Organic synthesis	Anti-feedant on *Spodoptera litura*; hypothesized active metabolite: plumbagin.	3
Drosera aliciae	**Chloroform**, methanol	Inhibits RNA and protein synthesis of five human pathogenic bacteria.	2
	Methanol, chloroform,	Anti-oxidant against DPPH, FRAP.	2
	Methanol	Purified ramentaceone toxic to four tumor cell lines.	4
D. binata	Chloroform	Anti-bacterial against burn-wound *Streptococcus aureus* in planktonic and biofilm forms.	5
D. burmannii	Golden ash (*swarnabhasma*)	Used as a cure for memory loss, defective eyesight, infertility, overall body weakness, early aging, bronchial asthma, rheumatoid arthritis, diabetes mellitus, nervous disorder.	6, 7
D. capensis	**Chloroform**, methanol	Anti-bacterial against four human pathogenic bacteria.	2
D. indica	Macerated leaves	Cure for corns.	6, 7
	Golden ash (*swarnabhasma*)	Used as a cure for memory loss, defective eyesight, infertility, overall body weakness, early aging, bronchial asthma, rheumatoid arthritis, diabetes mellitus, nervous disorder.	6, 7
	Ethanol, water	Anti-cancer agent that reduces tumor weight, packed cell volume and viable cell count of lymphoma ascites-induced tumor in mice.	8
	Ethanol, water	Anti-cancer agent for DAL lymphoma, EAC carcinoma cell lines.	8
	Ethanol, water	Anti-oxidant agent for hydroxy radical, DPPH, superoxide scavenging, ABTS, chelating ability on Fe^{2+}, NO inhibition.	8
D. intermedia	**n-hexane**, Methanol, water°	Anti-microbial against seven bacteria and eight yeasts.	9
	n-hexane, methanol, water°	Anti-fungal against four food spoilage yeasts and five filamentous fungi producing mycotoxins; active metabolite: plumbagin.	10
	Methanol, n-hexane, water°	Anti-oxidant against TEAC, ORAC, Folin-Ciocalteu.	9
D. madagascariensis	Ethanol	Anti-inflammatory that inhibits human neutrophil elastase enzyme activity; active metabolites: quercetin, hyperoside, isoquercitin.	11
	Ethanol, water	Anti-inflammatory based on HET-CAM assay; most active metabolite: ellagic acid.	12
	Ethanol	Anti-spasmodic: inhibits cholinergic M3 and histamine M1 receptors in guinea pig ileum but not effective at contractile prostanoid receptors of guinea pig trachea; active metabolites: quercetin, hyperoside, isoquercitin.	11
	Ethanol-water mixture	Anti-spasmodic: inhibits muscarinic M3 receptor in guinea pig ileum; hypothesized metabolite: quercetin.	13
D. peltata	Golden ash (*swarnabhasma*)	Used as a cure for memory loss, defective eyesight, infertility, overall body weakness, early aging, bronchial asthma, rheumatoid arthritis, diabetes mellitus, nervous disorder.	6, 7
	Macerated leaves	Used as a cure for irregular menses, scrofula, old injuries, age-related disorders.	14
	Macerated leaves	Used to dispel gas, eliminate dampness, promote blood circulation, ease inflammations, cure bacterial infections.	15

(*continued*)

Table 20.1 (*Continued*)

Species	Extraction	Bioactivity	References*
	Chloroform, petroleum ether, diethylether, ethyl acetate, ethanol, water°	Anti-bacterial: agar gel diffusion assay against dental caries and periodontitis bacteria; active metabolite: plumbagin.	16
	Petroleum ether, chloroform, ethanol, n-butanol, water°	Anti-fungal against five plant pathogenic fungi; active metabolite: plumbagin.	17
	Ethanol-water mixture	Anti-spasmodic: inhibits muscarinic M3 receptor in guinea pig ileum; hypothesized metabolite: quercetin.	13
D. ramentacea	Ethanol	Anti-bacterial against *Bordetella pertussis*.	18 in 19
	Ethanol	Anti-spasmodic: inhibits bronchospasms induced in guinea pig.	20 in 19
	Ethanol	Anti-spasmodic in rabbit and guinea pig intestine.	18 and 21 in 19
D. rotundifolia	Tincture ("Drosera herba")	Used to treat whooping-cough, chronic bronchitis, asthma, phthisis.	19
	Ethanol, water	Anti-inflammatory: inhibits human neutrophil elastase; hypothesized metabolite: quercetin.	22
	Ethanol, water	Anti-inflammatory based on HET-CAM assay; most active metabolite: ellagic acid.	12
	80% ethanol	Anti-inflammatory: inhibits T-cell membrane-induced inflammatory gene expression in HMC-1 human mast cells.	23
	80% ethanol	Anti-bacterial based on agar disk diffusion assay with gram+ and gram− bacteria.	24
	Ethanol-water mixture	Anti-spasmodic: inhibits muscarinic M3 receptor in guinea pig ileum; hypothesized metabolite: quercetin.	13
	Ethanol, water	Anti-spasmodic: inhibits muscarinic M3 receptor of guinea pig ileum.	22
	Ethanol	Animal semen protectant: bovine spermatozoa mobility, viability, and superoxide scavenging activity during *in vitro* cell cultures.	25
D. spathulata	80% ethanol	Anti-inflammatory (inhibition of T-cell membrane-induced inflammatory gene expression in HMC-1 human mast cells).	23
D. tokaiensis	80% ethanol	Anti-inflammatory: inhibits T-cell membrane-induced inflammatory gene expression in HMC-1 human mast cells.	23
D. spp. in "Europe"	Tincture	Used as an aphrodisiac.	19
	Freshly squeezed leaf juice	Used to treat warts and corns.	19
	Ethanol	Anti-spasmodic: inhibits bronchospasms induced in guinea pig.	26 in 19
Nepenthes gracilis	**Hexane**, chloroform, ethyl acetate, ethanol, methanol, water°	Anti-fungal against six human pathogenic fungi; toxic to LLC-MK2 in rhesus monkey kidney epithelial cells; active metabolite: plumbagin.	27
N. khasiana	Pitcher fluid	Droserone and 5-O-methyldroserone toxic to *Candida albicans* and *Aspergillus* sp.	28
N. mirabilis	Methanol	Anti-bacterial: weakly active against *Staphylococcus aureus*.	29
	Methanol	Anti-oxidant: scavenges peroxyl radical; nine phenolic substances carry the activity once purified.	30
	Methanol	Anti-osteoporosis: suppresses tartrate-resistant acid phosphatase activity in NF-κB ligand-induced osteoclastic RAW 264.7 macrophage cells; active metabolites: nepenthosides A and B, and nine phenolic substances.	30
N. thorelii	Organic synthesis	Plumbagin, droserone, and 2-methylnaphthazarine present in roots toxic to malaria *Plasmodium*.	31

(*continued*)

Table 20.1 *(Continued)*

Species	Extraction	Bioactivity	References*
N. ventricosa; N × *maxima*	Hexane	Anti-fungal against eight plant pathogens; active metabolite: plumbagin.	32, 33
N. sp. from Malaysia	Powdered root	Used to treat stomach-aches and dysentery.	19
	Stem infusion	Used to treat fever.	19
Pinguicula spp. from Europe	Fresh leaf juice	Used to treat cattle wounds.	19
	Hog-lard	Used to treat for human wounds.	19
	White wine	Used to treat edema.	19
	Syrup	Used as a purgative and diuretic.	19
Sarracenia purpurea	80% ethanol, hot water	Increases glucose uptake by muscles and decreases glucose release from the liver; protects against diabetic neuropathy and reduces symptoms of Type-II diabetes; leaves more active than roots.	34—37
Sarracenia spp.		Used to treat smallpox, complaints of the liver, kidney, uterus, and stomach; laxative and diuretic.	19

*References. 1. Szpitter et al. (2014); 2. Krolicka et al. (2008); 3. Tokunaga et al. (2004); 4. Kawiak et al. (2011); 5. Krychowiak et al. (2014); 6. Ravikumar et al. (2000); 7. Reddy et al. (2001); 8. Asirvatham and Christina (2013); 9. Grevenstuk et al. (2009); 10. Grevenstuk et al. (2012a); 11. Melzig et al. (2001); 12. Paper et al. (2005); 13. Kolodziej et al. (2002); 14. Tibetan People's Publishing House (1971); 15. Chinese Herbal Medicine Compilation Group (1975); 16. Didry et al. (1998); 17. Tian et al. (2014); 18. Bezanger-Beauquesne (1954); 19. Juniper et al. (1989); 20. Ramanamajary and Botteau (1968); 21. Paris and Delaveau (1959); 22. Krenn et al. (2004); 23. Fukushima et al. (2009); 24. Kačániová et al. (2014); 25. Tvrda (2015); 26. Paris and Quevauvillier (1947); 27. Gwee et al. (2014); 28. Eilenberg et al. (2010); 29. Wiart et al. (2004); 30. Thanh et al. (2015); 31. Likhitwitayawuid et al. (1998); 32. Shin et al. (2007a); 33. Shin et al. (2007b); 34. Cieniak et al. (2015); 35. Harris et al. (2012); 36. Leduc et al. (2006); 37. Spoor et al. (2006).

or brown). They are very potent and have activities ranging from providing vitamin K to anti-microbial and anti-cancer toxins (Babula et al. 2009). All carnivorous Caryophyllales produce 1,4-NQs (Figure 20.1a; Zenk et al. 1969), whereas noncarnivorous Caryophyllales accumulate the red pigment betalain (Brockington et al. 2011). In the unrelated Bignoniaceae, Juglandaceae, Ebenaceae, and Plumbaginaceae, 1,4-NQs play defensive and allelopathic roles.

1,4-NQs undergo redox cycling after one- or two-electron bio-reduction in target cells (Babula et al. 2009), whereupon they produce reactive oxygen species, induce DNA cleavage, alkylate proteins, and inhibit topoisomerase II. They also disrupt microtubule networks by directly binding to tubulin. These activities are toxic and may be lethal. However, by directly inhibiting Nuclear Factor (NF)-κB signaling in many animal cell lines, they prevent the destructive action of carcinogens, inflammatory stimuli, and tumor necrosis factor (TNF)-α and can be used as anti-inflammatory, cancer preventive, or cancer curative agents (Gaascht et al. 2013).

Plumbagin is the primary NQ of most Droseraceae and Nepentheceae (Figure 20.1a). Its content is ≈2–3% (maximum of 5%) of the dry matter of *Dionaea* (Tokunaga et al. 2004), *Aldrovanda* (Adamec et al. 2006), and most *Drosera* species (Egan and van der Kooy 2013). Only *D. capensis* and *D. aliciae* have inverse content ratios of plumbagin and its topoisomer ramentaceone, and a few Australian *Drosera* sections accumulate only small amounts of both substances (Egan and van der Kooy 2013, Zenk et al. 1969). A dual plumbagin and ramentaceone chemotype is rare, but has been observed in *D. madagascariensis* and in some natural hybrids of parents with opposite chemotypes (Schlauer and Fleischmann 2016).

Plumbagin is also present in many species of Plumbaginaceae and Ebenaceae with medicinal value where it has attracted attention because of its potent anti-oxidant, anti-inflammatory, anti-cancer, anti-bacterial, anti-malarial, and anti-fungal

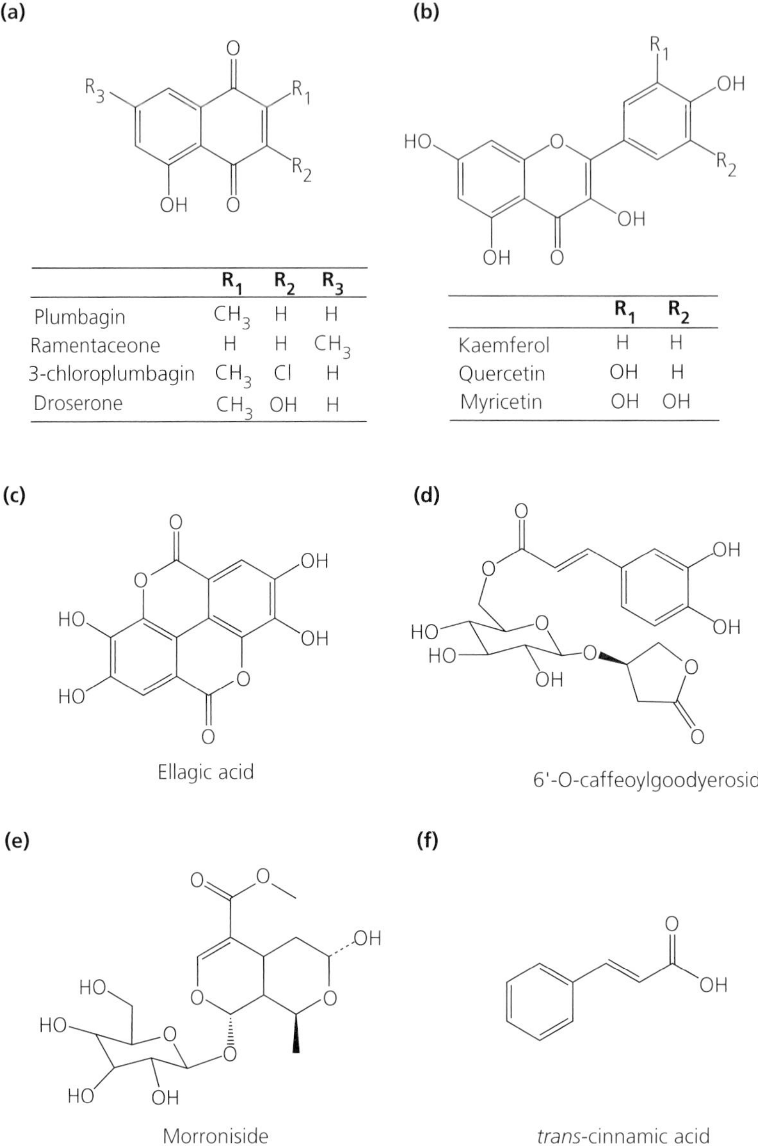

	R_1	R_2	R_3
Plumbagin	CH_3	H	H
Ramentaceone	H	H	CH_3
3-chloroplumbagin	CH_3	Cl	H
Droserone	CH_3	OH	H

	R_1	R_2
Kaemferol	H	H
Quercetin	OH	H
Myricetin	OH	OH

Figure 20.1 Examples of key phytopharmaceutical substances for each chemical class of secondary metabolite. (**a**) naphthoquinones, (**b**) flavonoids, (**c**) gallic acid derivatives, (**d**) goodyeroside derivative, (**e**) iridoid, (**f**) phenylpropanoid.

properties (Padhye et al. 2012, Sumsakul et al. 2014). Likewise, synthetic plumbagin (and other major *Drosera* and *Nepenthes* NQs) or purified plumbagin have biological activities similar to crude plant extracts, and bioassay-guided fractionation experiments have in some cases suggested that plumbagin was primarily responsible for the biological activity of the crude extracts (Table 20.1).

Minor NQs (e.g., droserone) consist of equal parts of glycosylated and unglycosylated NQ monomers, although a few dimers have been reported (Egan and van der Kooy 2013, Gaascht et al. 2013). NQs specific to *Drosera* or *Dionaea* are poorly studied, as are species-specific NQs in *Nepenthes* (Likhitwitayawuid et al. 1998, Aung et al. 2002, Rischer et al. 2002, Eilenberg et al. 2010, Thanh et al. 2015).

Industrial production of naphthoquinone. Plumbagin content in *D. capensis* and *D. natalensis* is half that of *Plumbago auriculata* (Crouch et al. 1990), and the slower-growing *Drosera* species appear to be less economically viable sources for plumbagin or ramentaceone. However, attempts have been made to increase NQ content in harvested material. One such attempt involved harvesting selected plant parts rather than whole plants, but NQs were found to accumulate throughout the plant body with only slightly higher contents measured in younger or reproductive organs of *Aldrovanda* (Adamec et al. 2006), *Drosera* (Egan and van der Kooy 2012), and *Nepenthes* (Babula et al. 2009), in agreement with the hypothesized defensive role of these substances. Despite large interspecific variations in NQ content, none of the many Droseraceae species analysed greatly surpassed the traditionally extracted species (Egan and van der Kooy 2013; Table 20.1). Intraspecific differences have received less attention; minor, albeit significant variability in NQ content occurs among three colored forms of *D. capensis* (Egan and van der Kooy 2012) and 13 populations of *D. rotundifolia* from northern Finland (Kämäräinen et al. 2003) for which interannual variability in NQ content exceeded interpopulation variability. *In vitro*-grown material also accumulated less NQ than either greenhouse-grown or wild harvested material (Egan and van der Kooy 2013). Overall, environmental factors appear to be more important determinants of NQ content than cultivar selection.

As with most other plant defense elements, NQ synthesis also can be elicited by substances derived from plant–pathogen interactions. For example, lysates of *Agrobacterium rhizogenes* are a powerful elicitor of NQ accumulation in *Dionaea* and *D. capensis* (2.6-fold for plumbagin and 1.9-fold for ramentaceone respectively) in four weeks (Krolicka et al. 2008). Yeast extract (0.5 mg L^{-1}) best elicited plumbagin accumulation in six days, in a search for elicitors for *D. burmannii*, and the nitrogenous sugar polymer chitosan also elicited a response (Putalun et al. 2010). Jasmonic acid (JA; Chapter 16) and salicylic acid (SA) also elicited responses in *Dionaea, D. burmannii*, and *D. capensis* (Krolicka et al. 2008, Ziaratnia et al. 2009, Putalun et al. 2010). NQ accumulation after JA and SA application on excised surfaces of shoot tips reaches its highest levels 48 hours after elicitation (Ziaratnia et al. 2009).

Chitin—the acylated form of chitosan—is a major surface polymer of fungi and arthropods and is a powerful elicitor of plumbagin accumulation in *Drosophyllum* cell-suspension cultures (Nahálka et al. 1998) and in *N. khasiana* pitcher fluid (Eilenberg et al. 2010). The elicitation turns the *Nepenthes* pitcher fluid red and the plumbagin is quantifiable easily with low-sensitivity detection techniques including NMR (Eilenberg et al. 2010). Unlike other elicitors, small chitin oligomers (degree of polymerization <7) elicit plumbagin accumulation without triggering undesirable defense responses such as hypersensitive cell death (Nahálka et al. 1998). Prey degradation frees eliciting chitin oligomers from their exoskeleton (Chapter 16), thereby amplifying plumbagin elicitation. Indeed, ramentaceone accumulation rises by 80% in *D. capensis* leaves upon prey capture (Egan and van der Kooy 2012). Enhanced production of NQs by prey-derived elicitors may be facilitated further by prey-derived nutrients; the feeding of L-alanine (a product of the proteolytic degradation of prey proteins by plant trap proteases) to *N. insignis* traps leads to the direct incorporation of 2-C units into plumbagin (Rischer et al. 2002).

Drosera capensis roots were genetically transformed with *Agrobacterium rhizogenes* in an attempt to boost NQ production, (Krolicka et al. 2010). Although transformation efficiency was only 10%, transformed roots ("teratomas") grew three times faster and produced 60% more ramentaceone. Combined with JA elicitation, ramentaceone production was enhanced seven-fold (Krolicka et al. 2010). Despite these successes, it is unknown if NQ production by carnivorous plants is economically profitable.

Biosynthesis of 1,4-NQ occurs via two separate pathways in plants. [^{14}C]-CO_2-labeling experiments revealed that plumbagin is produced via the acetate–polymalonate (polyketide) pathway in 16 species of Droseraceae and *Drosophyllum* (Durand and Zenk 1974). These results were confirmed by [^{13}C2]-acetate-labeling studies of *Triphyophyllum* (Bringmann et al. 2000) and by feeding [^{13}C2]- or [^{13}C3]-L-alanine to *in vitro*-grown *N. insignis* pitchers (Rischer et al. 2002). However, the gene products responsible for NQ biosynthesis in Droseraceae or Nepentheceae are neither known nor characterized.

Every step of the extraction process of accumulated NQs has been optimized (partially reviewed in Banasiuk et al. 2012). The highest extraction yields of plumbagin were obtained with ultrasound extraction of fresh *Drosophyllum* with methanol (Grevenstuk et al. 2008). Nevertheless, n-hexane extraction was preferred because it yielded purer extracts with a still high recovery rate (2.42 mg per g of fresh material). Purification of plumbagin from such extract was later optimized with Solid Phase Extraction (SPE) (Grevenstuk et al. 2012b), yielding 2.74 mg g^{-1} of 99% pure plumbagin from micropropagated *D. intermedia*—an 86.3% recovery rate from a scalable process. Though NQ glycosides are usually minor, their artificial de-glycosylation is feasible to enhance free NQ yield, as demonstrated using *D. spatulata* (Budzianowski et al. 1995).

Flavonoids and phenolic acid derivatives. There are several therapeutic issues attendant to replacing traditional *Drosera* extracts with purified NQs. First, NQs are not the only bioactive substances of Droseraceae and Nepenthaceae. Traditional extracts contain only small quantities of NQs because they are weakly soluble in water, are poorly extracted by traditional protocols, and have low bioavailability in whole animals. However, they contain a rich blend of flavonoids and phenolic acid derivatives (Figure 20.1b, c), and mostly glycosylated flavonols and gallic acid derivatives that include ellagic acid (Zehl et al. 2011, Egan and van der Kooy 2013, Braunberger et al. 2015) that represent 5.9% and 4.8% of *D. rotundifolia* and *D. anglica* dry matter, respectively (Zehl et al. 2011). A few studies of phenolic compounds also found members of these chemical classes in *N. mirabilis* and *N. gracilis* (Aung et al. 2002, Thanh et al. 2015). Flavonoids and gallic acid derivatives possess anti-cancer, anti-microbial, and anti-oxidant properties (Gaascht et al. 2013) and, in a few cases, have proven to be more active than NQs extracted from *Drosera* or *Nepenthes* (Table 20.1).

The therapeutic value of members of Droseraceae and Nepenthaceae may stem from the synergistic action of multiple, weakly active, components that include flavonoids, phenolic acid derivatives, and NQs (Eilenberg et al. 2010, Krolicka et al. 2008). Synergies have been underestimated in the literature because most of the phytomedical analyses are based on bioassay-guided fractionation strategies or the evaluation of single purified substances, and the use by pharmacopeia of NQ content to validate the quality of commercial extracts likely underestimates their effects (Egan and van der Kooy 2013). The general rarity of carnivorous plants in the field (Chapter 27) forces industrial firms to collect non-traditional species or extract elicited material. Either way, extracts are biased toward modified NQs over flavonoid/phenolic acid derivatives ratios (Krolicka et al. 2008, Kováčik et al. 2012) that may have higher toxicity and lower therapeutic value.

20.2.2 Sarraceniaceae

Bioassay-guided fractionation of an 80% ethanol *S. purpurea* leaf extract for enhancement of glucose uptake by C2C12 mouse muscle cells led to the identification of three kaempferol and quercetin glycoside derivatives (flavonoids—Figure 20.1b) (Asim et al. 2012). The same extract yielded caffeoylated and uncaffeoylated forms of goodyeroside (Figure 20.1d) that repressed glucose-6-phosphatase expression in H4IIE rat liver cells, a major control element of glucose release by these cells. Both of these activities lower blood glucose levels. In a separate analysis, bioassay-guided fractionation of a similar extract for inhibition of glucotoxicity in PC12 cells attributed an additional neuroprotective activity to quercetin-3-O-galactoside in case of low or high glucose toxicity and further revealed that the iridoid morroniside (Figure 20.1e) that was found to be inactive in the previous tests has similar activity (Harris et al. 2012). Taken together, these results support the traditional use of *S. purpurea* leaves by the healers of the Cree of Eeyou Instchee first nation in northern Quebec, Canada, as a cure for Type II diabetes (Table 20.1).

20.2.3 Lentibulariaceae

In line with their limited pharmaceutical usage (Table 20.1), the few phytochemical analyses of *Pinguicula* have revealed only minor quantities of *trans*-cinnamic acid (Figure 20.1f) and traces of some flavonoids and carotenoids (Juniper et al. 1989, Legendre 2000). The leaves of *Pinguicula* were, however, widely used in various parts of Great Britain and France, and are still currently used by the Laps in northernmost Europe to curdle, or thicken, milk (Legendre 2000). Milk curdling by *Pinguicula* leaves probably involves prey-digesting enzymes and mucilage-associated bacteria (Legendre 2000). Though most traps secrete prey-digesting proteases (Chapter 16) that potentially can mimic the rennet chymosin traditionally used to clot milk, only *Pinguicula* has such a recorded use. This may result from a more specific action of *Pinguicula* proteases on casein-κ to remove the casein macropeptide, the cause of the destabilization of the colloidal milk casein assemblages (micelles) that leads to their aggregation. *Pinguicula* leaves sequester their proteases intracellularly until their exocytosis is triggered by prey in a "one-off" secretory mechanism (Legendre 2000). As a result, a short contact with milk will extract only surface bacteria and mucilage whereas a longer contact would be necessary to induce exocytosis (i.e., extraction) of proteases.

20.3 Mass propagation

20.3.1 *In vitro* culture

Tissue culture largely has been worked out for *Drosera* and *Dionaea* (Banasiuk et al. 2012) to mass produce large quantities of a few species for, respectively, the phytopharmaceutical and horticultural industries. The capacity of tissue culture to multiply endangered species has been a driver to successfully develop *in vitro* micropropagation protocols for *Aldrovanda* (Adamec and Kondo 2002), *Drosophyllum* (Gonçalves and Romano 2005), *Cephalotus* (Tuleja et al. 2014), *Nepenthes* (Devi et al. 2013), *Pinguicula* (Saetiew et al. 2011, Grevenstuk and Romano 2012, Legendre 2012), *Utricularia* (Idei and Kondo 1998, Adamec and Pásek 2009), *Sarracenia* (Northcutt et al. 2012), *Darlingtonia* (Kim et al. 2006), *Heliamphora* (Kim et al. 2006), and *Genlisea* (L. Legendre *unpublished data*).

Starting material. Introduction of material *in vitro* has proved to be most successful from seeds because they withstand sterilization processes more easily and rarely contain endophytic microbial contaminants. Nevertheless, most studies start *in vitro* cultures with excised leaves, flower stalks, and, more rarely, shoot tips. Calcium hypochlorite (calcium bleach) as a sterilization agent is less damaging than sodium bleach because its salts are less prone to enter cells. Their action is enhanced by adding wetting agents (non-ionic detergents such as Tween-20), although the presence of detergent accelerates precipitation of calcium bleach, requiring longer washes. After introduction, *Dionaea* explants require activated charcoal (Teng 1999) in the supporting gel to absorb growth-inhibitory phenolic substances they liberate. Such addition is not necessary in subsequent multiplication steps.

Media. Comparison of the performance of different *Drosera* species on different basal mineral mixes revealed that a good compromise for growth performances and shoot multiplication rates was obtained on Murashige and Skoog (MS)-based medium. Satisfactory results also were obtained with Vacin and Went, Driver and Kunyuki Walnut, Fast, Lindemann, Reinert and Mohr, and Gamborg's B5 mineral media (Banasiuk et al. 2012). MS-based media are used for all other carnivorous plants, except *Aldrovanda* and *Utricularia*, for which Gamborg's B5 medium is used (Idei and Kondo 1998, Adamec and Kondo 2002). Numerous studies assessed the effect of diluting medium salts on the proliferative rate of various species and concluded that 50 to 25% MS formulations worked best (Banasiuk et al. 2012, Grevenstuk and Romano 2012, Northcutt et al. 2012, Devi et al. 2013, Tuleja et al. 2014). Growth is slowed and plants are stunted on undiluted media whereas plants may turn into crystal-type structures when dilution is too large. However, if moderately diluting macroelements prevents mineral toxicity, diluting complete mineral solutions is not advisable because it concomitantly dilutes microelements, vitamins, and iron. Keeping these later elements at their original concentration and diluting macroelements only improved the growth of *Cephalotus* (Tuleja et al. 2014) and *Pinguicula* (Legendre 2012). Modifying the iron supplement of MS medium boosted *Pinguicula* growth even further (Legendre 2012) but none of these modifications were tested for other species. pH titration with KOH led to better growth than titration with NaOH (Legendre 2012).

The improved performance of carnivorous plants on diluted published mineral solutions is in line with their adaptation to mineral-poor habitats, though direct causation is unlikely. Tissue culture media developed for carnivorous plants also are optimal for many small noncarnivorous and non-crop species. Further, even when diluted by a factor of four,

MS-basal salt solutions still contain macroelements (N, P, K) and Ca/Mg that are several orders of magnitude more concentrated than those found in the field. In part, this is because minerals added to a tissue-culture flask must be sufficient to sustain the enclosed plants until their next transplanting. A balanced ratio of ammonium and nitrate is also recommended *in vitro* because the preferred consumption of the former one by plants will lead to a pH decrease while the subsequent consumption of the latter one leads to a pH increase. The presence of both forms results in a "V-shaped" pH curve that eliminates drift toward dangerous values during a complete growth cycle.

In nature, biotic and abiotic conditions buffer these changes. The peat, clay, and sand in which carnivorous plants grow in the field retain minerals differently from agar gels. Regardless of their purity or grade, agar gels never are totally devoid of minerals. All of these factors are accounted for in the design of mineral solutions used for *in vitro* culture. Agar also will melt if pH is adjusted below 4.5 to mimic field conditions. Solid support elements are not needed for the freely floating *Aldrovanda* and aquatic *Utricularia* that grow particularly well *in vitro*. The absence of competition from neighboring plants and microorganisms may let these plants benefit from improved nutrition without having to make use of their unique adaptations for fast growth in dystrophic waters.

Aquatic carnivorous plants are strict CO_2 users for photosynthesis (Chapter 19), and thus good carbon availability is of foremost importance to sustain their fast growth (Adamec and Pásek 2009). Flasks are not airtight to sustain plant respiration but air flow is hindered to prevent flask contamination unless vent plugs—unknown in culture of carnivorous plants—are installed on the caps. Plant respiration and photosynthesis gradually increase the CO_2 to O_2 ratio in the flask head space because photosynthesis is slowed by low light and respiration is increased by the high sucrose content in tissue culture solutions. Though CO_2 contents in the flask head space have never been measured, the generally low pH values of 3.2–4.0 measured in used *in vitro* media of 12 *Aldrovanda* strains and *Utricularia* species suggest an intense H^+ efflux (Adamec and Pásek 2009). Despite different CO_2, sucrose, and light availabilities *in vitro* (and the absence of prey digestion to add to C acquisition; Chapter 19), plants of both genera exhibited the same CO_2 affinity (as estimated by measuring CO_2 compensation points) as these species grown outdoors or in aquaria (Adamec and Pásek 2009). However, this parameter varies among species grown *in vitro*.

Hormones and plant multiplication. When seeds are introduced *in vitro*, they germinate and grow into whole plants that can be subcultured. However, when shoot tips are introduced, they only grow into aerial plant parts unless roots are initiated ("organogenesis") to complete the plant structure. In the case of leaf fragments, embryos need to be generated ("embryogenesis") after dedifferentiation of clumps of cells that grow into a dedifferentiated and disorganized structure ("callus") if novel aerial plant parts are to develop. Both processes are hormonally regulated and can occur with hormones internal to the explant or can be encouraged by the external addition of natural or synthetic hormones. With the exception of *Nepenthes* (Devi et al. 2013), the addition of hormones is not necessary to induce embryogenesis and organogenesis in studied carnivorous plants, and both processes occur without prior callus formation (Bobák et al. 1995).

Leaf cuttings of *Dionaea, Drosera*, and *Pinguicula* naturally form new embryos and shoots. All rooted carnivorous plants also naturally produce adventitious roots on isolated shoot stems and will do so spontaneously *in vitro* on hormone-free media. Mature plants of most species will spontaneously surround themselves with daughter plants within a month or two via meristem division or the development of lateral buds. Multiplication rates can be quite impressive for most *Drosera* and *Pinguicula* species on hormone-free media, although even higher aerial shoot production rates are obtained with dilute additions of cytokinins including 6-benzylaminopurine (BA), kinetin, and zeatin. Faster rooting occurs following additions of auxins including 1-naphthaleneacetic acid (NAA) and indole-3-butyric acid (IBA) (Saetiew et al. 2011, Devi et al. 2013). Care should be taken when using cytokinins because high doses of BA and kinetin suppress shoot multiplication by, and may even become toxic to, *D. intermedia* (Rejthar et al. 2014) and

D. indica (Jayaram and Prasad 2007). In *D. burmannii*, they induce the formation of highly proliferative fasciated shoots (Liao and Ji 2014). In the case of the rootless, terrestrial, *U. praelonga*, differentiation rates of stems and leaves were affected by the addition of the hormone BAP (Idei and Kondo 1998).

Abiotic conditions and mineral nutrition. Increased shoot production by *Drosera* can occur in liquid medium (Kawiak et al. 2003) and temporary immersion has a greater impact than permanent immersion (Kopp et al. 2006). Etiolation and proper orientation of *Dionaea* flower stalk explants increased embryogenesis rates (Teng 1999). Trap differentiation can be induced preferentially in *U. praelonga* by low levels of KNO_3 (Ichiishi et al. 1999). Lowering the ratio of nitrate relative to ammonium retards growth and induces *Dionaea* and *D. spatulata* to accumulate red pigments (Ichiishi et al. 1999). Increased sucrose levels have similar effects on (Ichiishi et al. 1999), but do not affect shoot initiation of, *D. indica* (Jayaram and Prasad 2007) and *D. intermedia* (Rejthar et al. 2014).

To further accelerate biomass production, facilitate gene transfer, or study fundamental processes of plant cell biology, cultures of callus/cell suspension and protoplast have, respectively, been developed for *Drosophyllum* (Nahálka et al. 1998) and *Nepenthes* (Sweat and Bodri 2014). Four-hour incubation at 25 °C, 40 rpm in the presence of 0.5 M sorbitol, 5% cellulase, 0.5% macerozyme, and 0.3% pectolyase generated the highest *Nepenthes* protoplast yields with a 62.1% viability. These then formed cell walls but their division stopped and efforts to regenerate plants from individual *Nepenthes* protoplasts so far have failed (Sweat and Bodri 2014).

Growth. Terrestrial carnivorous plants grown *in vitro* are stunted and resemble small versions of outdoor-grown plants. Aquatic species tend to produce more numerous but smaller traps than they do in the field (Adamec and Pásek 2009), and their stems will fork spontaneously. Except for some large species, most can reach flowering stage *in vitro*. Stoloniferous *Pinguicula*, such as *P. gigantea* and *P. antarctica*, produce abundant stolons, and tuberous *Drosera* produce tubers that can be transplanted *in vitro* and *ex vitro*. Temperate *Pinguicula* that have a resting stage in nature will go dormant spontaneously if left non-transplanted long enough, and mother buds will become surrounded by daughter ones as they are in the field. They resume growth upon transfer to a new medium (Legendre 2012), although this may take a few weeks.

Genetic fidelity. Acclimation *ex vitro* of material cultured *in vitro* generates fertile plants for all species. Nevertheless, the high multiplication rate of *in vitro* cultures increases the probability of generating somaclonal (non-meiotic) mutations, and this probability is increased further by synthetic hormones and during embryogenesis. Genetic fidelity (stability) has been confirmed with RAPD for *D. aliciae* (Kawiak et al. 2011) and *Nepenthes* (Devi et al. 2013), but further investigation of chromosome counts and heterochromatin distribution patterns revealed increases in cytological variations in second and third generations of micropropagated *Nepenthes* that otherwise were morphologically similar (Devi et al. 2015). Morphologically distinct somatic variants of *P. ramosa* occasionally occur over several rounds of tissue culture (Legendre 2012).

20.3.2 Hydroponics

Hydroponics is another artificial method of accelerating plant growth and multiplication. It does not require complete asepsis and is less prone to somaclonal mutation. Using standard Knop mineral solution, high multiplication rates of several *Pinguicula* species were obtained with a tidal table apparatus and glass wool as plant support (Guitton et al. 2012). If tested and applied more widely, hydroponics could become a useful compromise between slow, standard cultivation and the more rapid, resource-intensive use of *in vitro* culture, meeting the needs of the horticultural and pharmaceutical industries.

20.4 Industrial products inspired by botanical carnivory

20.4.1 Production tools for recombinant proteins

The secretion of digestive enzymes by trap glands represents an attractive machinery that could be used to produce recombinant polypeptides of therapeutic interest, including insulin, interleukins, or growth factors. Currently, these substances are produced by microorganisms that, unlike plants or animals, are incapable of post-translationally modifying them to make them fully functional or protected from proteolysis during therapeutic use. Compared to animal-based heterologous expression systems, plants usually can be produced in large quantities at low cost. However, purification of recombinant polypeptides from plant material is challenging; it is hoped to be simpler from carnivorous plant secretions. Comparative pros and cons of such technological processes have been reviewed using *Sarracenia purpurea* as a model plant (Rosa et al. 2009) and have been patented (Biteau et al. 2012). Unfortunately, genetic transformation of carnivorous plants remains a bottleneck that limits such technological endeavors.

Expression of the luciferase gene has been observed after transformation of tissue-cultured *D. rotundifolia* leaves by *A. tumefaciens* strain C58C1 (Hirsikorpi et al. 2002). Success was low (17%) and selection of the transformed plants resulted from the co-transformation of the kanamycin-selection gene neomycin phosphotransferase. Transformation of roots with *A. rhizogenes* is feasible (Krolicka et al. 2010) but cannot be used for whole plants because it yields albino plants (Blehová et al. 2015). The *A. tumefaciens* protocol was improved to make it more routine and reliable (Blehová et al. 2015). Glutathione was added to the medium, plants were protected from light to reduce necrosis, and the use of an *Agrobacterium* strain that tolerates toxic NQs secreted by *Drosera* species is recommended. Co-transformation of the fluorescent protein GFP also was advised to assist the selection of transformants and weed out sports and chimeras.

20.4.2 Biomimetic materials

The fast, nastic movement of *Dionaea* (Chapter 14) is an endless source of inspiration for solving engineering problems in medicine, aeronautics, automotive technology, and architecture. Upon mechanical stimulation of trigger hairs, the change in hydraulic pressure-driven lamina cell volume induces leaf folding and subsequent capture of prey. To reduce the risk of prey escape, movement speed is not constant. The leaf snaps on the prey with an accelerating movement driven by buckling. Leaf lobes can harbor one of two stable shapes, whereas intermediate configurations are energetically unfavorable. The favoring of one stable configuration over the other depends on the respective pressures of the two surfaces of each trap lobe. This complex movement has been modeled mathematically and copied by man-made structures to make them change shape via accelerated snap-buckling upon sensing external stimuli (Guo et al. 2015).

As a direct demonstration that *Dionaea* trap movement can be mimicked by artificial structures, a *Dionaea* trap robot was recreated by assembling two sheets of ionic polymer-metal composites (Guo et al. 2015). Application of an electrical current to the conductive metal component forced the ionic polymer to change shape and the metal to snap-buckle. Such technology could be applied to cars, planes, or rockets that need pilot-controlled shape changes. But pilot and electricity-independent applications also could allow objects to change shape upon external stimuli perception without any need for batteries to be recharged regularly. For example, a thin polydimethylsiloxane (PDMS) layer was covalently layered on a thicker PDMS sheet punctured by a regular array of holes so that the thinner layer folded into domes above the holes. Upon application of biaxial compression loads, these domes buckled from concave to convex via a snap-through transition. Load thresholds and amplitudes of movements could be finely tuned by changing lens sizes and spacing. However, sensing does not have to be mechanical. Hydrogels that change volume when subject to changes in temperature, chemical, or light can be embedded in buckling structures either as circulating fluids or mixed with braided fibers with tunable stiffness and force generation capacity.

Research into the chemistry of these hydrogels is in its infancy and there is continuing interest in expanding the array of elements they can sense and to which they can respond. One development associates a photobase that acidifies upon light perception with a pH-sensitive gel that swells when pH drops. If such a hydrogel is introduced into a T-shaped, twisted, two-rod structure, light perception by the hydrogel forces the twisted stem to bend in the direction of the light. When coupled to photovoltaic plates, this mechanism helps them track and optimize their positioning toward a moving sun. Alternatively, they can be used as sun-activated, sun-protecting devices. If the photobase sensor is tuned to specific wavelengths, it will respond only to specific light stimuli.

The buckling structure that hosts hydrogels also is of interest. A variety of devices that snap-buckle within 12 ms to jump with strength like human muscle have been designed. Some are simple bistable helical ribbons that switch from one helical shape to the other upon hydrogel stimulation. Others have more complex shapes based on traditional origami designs (Li and Wang 2015) that only require hydrogel-hosting origami cells to be able to change volume. Stimulation-induced gel swelling forces the structure to change shape either from one three-dimensional configuration to another (e.g., between helical sheets of reverse chirality) or from a flat two-dimensional sheet to a three-dimensional structure. If multiple origami structures are integrated, additional stability forms are possible based on the network of interactions allowed among the various pressurized cells.

Dionaea trap movement is not the sole source of inspiration for technological products. Sundew mucilage has demonstrated its potential for tissue engineering (Zhang et al. 2010). Atomic force microscopy (AFM) has revealed that sundew mucilage sugar polymers are organized into nanofibers and nanostructures. Though highly elastic in their liquid form, they become stiff when dried to yield a surface on which PC12 neuron-like cells could adhere and grow.

The slippery zone of *Nepenthes* pitchers has inspired designers of ever-clean materials (Wang and Zhou 2016). The surface of this zone is made of downward-oriented lunate cells that are covered by a layer of epicuticular wax platelets (Chapter 13, §13.3.3; Chapter 15, §15.3). Hydrophobicity and capacity for self-cleaning from dust or other particles brought by wind are the two physical properties of this zone that ensure repeated capture of prey over the lifespan of the trap. Measurements of water-drop contact angle (free surface energy) and examination of three-dimensional microstructures of the slippery zones of three *Nepenthes* species suggested that their somewhat different hydrophobicities stemmed only from size differences of the lunate cells and wax platelets. These properties suggested a rationale for the development of biomimetic materials with hydrophobic and self-cleaning properties (Wang and Zhou 2016).

The *Nepenthes* pitcher also has been used for the green chemical synthesis of gold nanoparticles (GNPs). Among heavy metals, gold is notable for its distinctive and tunable surface plasmon resonance, and GNPs have found many applications in biomedical sciences including drug delivery, tissue/tumor imaging, photothermal applications, and identification of pathogens in clinical specimens. However, the synthesis of high quality GNPs from gold chloride is difficult. Uniformity of shape and size is of foremost importance for their physical behavior and toxic contaminants can preclude clinical applications. An aqueous extract from *N. khasiana* was used to form colloidal GNPs of improved size and shape uniformity, a capacity that was, at least in part, attributed to the anti-oxidant properties of the extract (Dhamecha et al. 2016). Compared to current industrial processes, the synthesis was inexpensive, non-toxic to the environment, and did not carry-over toxic substances. *Nepenthes*-formed GNPs also proved to be stable and compatible with blood components during a mouse fibroblast cell line (L929) viability assay.

20.5 Future research

Despite the long established use of *Drosera, Nepenthes*, and *Sarracenia* in herbal medicines around the globe, the active ingredients and mechanism of action of these traditional preparations are far from being understood. Only activities linked to single substances are known and synergistic effects among elements remain obscure and prevent the establishment of meaningful pharmaceutical standards. Understanding underlying mechanisms of synergies among metabolites also could lead to the development of innovative medicinal treatments. For example, NQs from carnivorous plants have been hypothesized to inhibit multidrug-resistance transporters whose function is to clear cell interiors of xenobiotic toxic substances (Gaascht et al. 2013). If this hypothesis is supported, these NQs may be able to enhance the activity of known therapeutic agents, allowing anti-cancer drugs to be used at lower doses with fewer side effects. Alternatively, non-toxic silver nanoparticles (AgNP) substantially potentiate the bactericidal activity of 3-chloroplumbagin and *D. binata* chloroform extracts to generate a healing cocktail devoid of toxicity against human keratinocytes (Krychowiak et al. 2014). Further combinations await exploration.

One goal of tissue culture is to generate large quantities of endangered plants for ornamental horticulture, but most research has focused on non-threatened plants that are easy to grow in greenhouses or outdoors. Legendre (2012) found that a rare, difficult-to-grow *Pinguicula* performed very poorly *in vitro*. Because many carnivorous plant species are threatened or endangered, research is desperately needed to improve nutrient media and develop protocols to grow them *in vitro* with their naturally associated microbiome. Establishing healthy breeding lines of carnivorous plants for the horticultural trade via the introduction of healthy explants *in vitro* (e.g., excised growing meristems or seeds of unopened seedpods) is a common deliverable of plant *in vitro* culture that also has not yet been tackled.

The current scant understanding of the unique physiological mechanisms of botanical carnivory already has led to innovation in a vast array of commercial fields. Further understanding of these processes should lead to still more innovation. A major drawback comes from the absence of efficient genetic transformation tools to conduct functional genomic studies (Chapter 11). To remove this constraint, several additional avenues to the traditional *Agrobacterium*-based leaf or root transformation protocols could be explored, such as the use of gene gun technology or the *Agrobacterium*-based floral dip-method.

PART IV

Ecology

CHAPTER 21

Prey selection and specialization by carnivorous plants

Douglas Darnowski, Ulrike Bauer, Marcos Méndez, John Horner, and Bartosz J. Płachno

21.1 Introduction

Carnivorous plants grow in a wide range of habitats, from wind-whipped mountain tops in the tropics to lowland bogs near the Arctic (Chapter 2). The wide range of habitats has an equally large range of potential prey, especially insects. Although Juniper et al. (1989: 74) asserted that "[n]ot one carnivorous species has been shown to be specific in choosing its prey," more recent evidence suggests otherwise, at least for some species. The primary evidence for selective predation would be a difference between the relative abundance of prey captured in traps and its relative abundance in the surrounding environment (Harms and Johansson 2000). Prey selectivity could have evolved to minimize competition for limited resources within populations and among sympatric species (Thum 1986, 1989a, Gibson 1991a, González et al. 1991, Ellison and Gotelli 2009) or between carnivorous plants and other predators, such as spiders (Jennings et al. 2010, Crowley et al. 2013). The latter can lead to conflicting goals in resource allocation. For example, production of glandular hairs by *Drosera* when prey are depleted by spiders is costly, but *Drosera* nonetheless should increase its investment in structures for attracting prey when competing for it (Crowley et al. 2013).

Prey composition also differs between terrestrial and aquatic genera, likely because of differences in trap morphology and mechanisms of attraction (Ellison and Gotelli 2009; Chapters 12–15). Different kinds of traps (e.g., motile sticky traps and snap-traps [Chapter 14]; non-motile pitfall traps [Chapter 15]) differ in their abilities to retain prey of contrasting size (Gibson 1991b). Differences in trap size and shape set limits to the size of prey that can be captured by *Sarracenia* or *Nepenthes* (Gibson 1991b, Chin et al. 2010, Gaume et al. 2016). These differences could lead to apparent prey selection as larger traps could catch either large and small prey or large prey alone, whereas small traps would be likely to capture successfully only smaller prey. Finally, differences in the chemical composition of mucilages among families (Chapter 12, 16) also could lead to differences in the size spectrum of prey (Adlassnig et al. 2010). In this chapter, we review evidence for prey selectivity that could arise from these and other differences in morphology, attractiveness, and rewards, and niche partitioning among co-occurring species (Table 21.1).

21.2 Prey selection by carnivorous plants with motile traps

21.2.1 *Aldrovanda*

The monospecific, aquatic *Aldrovanda* uses a snap-trap (Chapter 14) to capture prey ranging from very small crustaceans to occasional small tadpoles. The leaves that are modified into traps range in size from only a few mm when the plant grows in poor conditions to 10–15 mm when growing in ideal conditions. Non-living debris also could be

Darnowski, D., Bauer, U., Méndez, M., Horner, J. D., and Płachno, B. J., *Prey selection and specialization by carnivorous plants*. In: *Carnivorous Plants: Physiology, ecology, and evolution*. Edited by Aaron M. Ellison and Lubomír Adamec: Oxford University Press (2018). © Oxford University Press. DOI: 10.1093/oso/9780198779841.003.0021

Table 21.1. Evidence of prey selectivity for different genera of terrestrial carnivorous plants. Plain text indicates that trait has been identified as present, but no proof exists of its participation in prey selectivity; **bold** indicates positive proof of selectivity; underlined text indicates negative proof of selectivity; (parenthetical numbers) refer to references (at foot of table). Comparable data are unavailable for *Byblis* or *Roridula*.

Genus	Morphological traits	Visual signals	Odor	Rewards	Experimental tests
Brocchinia			Yes (27)		
Catopsis		UV (27)			
Cephalotus		UV (7)		Nectar (28)	
Darlingtonia	Fishtail appendage (1)	UV (6)			
Dionaea		UV (7); Fluorescence (9)	Yes (2); **Yes** (22)		No (23)
Drosera	**Architecture** (10)	UV (6, 7); No fluorescence (9); Red color (4)	Few (8); **Yes** (3)		No (14, 15, 19)
Drosophyllum		UV (7)	Honey-like (2, 7)		**Yes** (2)
Heliamphora		UV (7)	Yes (29)	Nectar (30)	
Nepenthes	**Translucent "windows"** (24); trap shape and size (25)	UV (6); **UV** (18); Fluorescence (9)	**Yes** (18)	**Nectar** (18); Edible trichomes (25)	
Pinguicula		UV (6, 7); No fluorescence (9)			No (11, 12, 13, 16); **Yes** (16, 17)
Sarracenia	**Orifice size** (5)	UV (7); Fluorescence (9); Fenestrations (21)	Yes (8)	**Nectar** (20)	**Yes** (26)

References: 1. Armitage (2016b); 2. Bertol et al. (2015b); 3. El-Sayed et al. (2016); 4. Foot et al. (2014); 5. Gibson (1991a); 6. Gloßner (1992); 7. Joel et al. (1985); 8. Jürgens et al. (2009); 9. Kurup et al. (2013); 10. Verbeek and Boasson (1993); 11. Antor and García (1994); 12. Zamora (1990b); 13. Zamora (1995); 14. Watson et al. (1982); 15. Jennings et al. (2010); 16. Karlsson et al. (1987); 17. Alcalá and Domínguez (2003); 18. Di Giusto et al. (2010) partial demonstration lacking olfactometer use; 19. Potts and Krupa (2016); 20. Bennett and Ellison (2009); 21. Schaefer and Ruxton (2014); 22. Kreuzwieser et al. (2014); 23. Hutchens and Luken (2009); 24. Moran et al. (2012); 25. Gaume et al. (2016); 26. Folkerts (1992); 27. Benzing (2000); 28. Joel (1988); 29. Jaffe et al. (1995); 30. Jaffe et al. (1992).

trapped by *Aldrovanda*, although the bristles along the edge of the leaves could minimize its collection. Only scant data are available for prey selectivity by *Aldrovanda*, perhaps because it is rare and difficult to cultivate.

Darwin (1875) listed only crustaceans and a few other arthropods as potential prey based on their occurrence in dried specimens of *Aldrovanda* from Australia and India. Adamec (2000) found that about 20–33% of all trapping leaves contained some prey during a feeding study of plants grown outdoors in culture and that prey trapping had significantly enhanced the growth of the plants. A range of small aquatic animals rest frequently on cultivated *Aldrovanda*, which preponderantly capture small snails and amphipods when offered a mixture of potential invertebrate prey (Chadwick and Darnowski 2002). Prey capture is related to which organisms rested on the plant and physical proximity likely affected the types of organisms that are caught.

21.2.2 *Dionaea*

The scientific name of *Dionaea muscipula* suggests it feeds on mice, whereas its common name implies it prefers flies, but neither is true. More often, its nearly or totally prostrate snap-traps ensnare ants foraging on the sandy soils where it grows, although Hutchens and Luken (2009) suggested that it was not selective in capturing prey. Kreuzweiser et al. (2014) proposed that *Dionaea* uses olfactory clues to attract potential prey—small *Drosophila* in their experiments—which could explain observations of *Dionaea* non-selectively trapping a mixture of local herbivores and frugivores.

21.2.3 *Utricularia*

Among its many species, *Utricularia* trap sizes range from 1–12 mm (Taylor 1989; Chapters 13, 14). The majority of data on prey capture come from only a few aquatic species in the derived subgenus

Utricularia, including *U. purpurea* (*U.* sect. *Vesiculina*) and several members of *U.* sect. *Utricularia* (Płachno et al. 2014b). A few recent studies address prey composition in other subgenera (Jobson et al. 2000, Jobson and Morris 2001, Seine et al. 2002, Płachno et al. 2014b).

Organisms and other objects recorded from traps of aquatic *Utricularia* (*U.* sect. *Utricularia*) include: bacteria (including cyanobacteria); protozoa; rotifers; cladocerans; copepods; annelids; insects; tardigrades; coelenterates; algae (mostly diatoms, green algae, euglenoids, desmids, and dinophytes); microfungi; pollen grains; fragments of fungi and plants; and debris (Mette et al. 2000, Gordon and Pacheco 2007, Peroutka et al. 2008, Sirová et al. 2009, Alkhalaf et al. 2011, Płachno et al. 2012b, Koller-Peroutka et al. 2015). The abundance and frequency of organisms inside the traps varies with trap age (Friday 1989, Sirová et al. 2009, Płachno et al. 2012b) and seasonal changes in its water reservoir (Guiral and Rougier 2007). For example, Friday (1989) showed experimentally the mean number of prey caught by traps of *Utricularia vulgaris* was largest in the six-day-old traps and declined sharply between the sixth and eighth days. Different prey taxa are digested at different rates, resulting in concentrations of remains of some species in traps and their consequent over-representation in datasets (e.g., crustaceans; Jobson and Morris 2001, Guiral and Rougier 2007). Larger prey also may be overestimated because they are easier to observe (Guiral and Rougier 2007). And not all of the organisms inside *Utricularia* traps are prey; some are commensals or kleptoparasites (Chapter 25).

Several authors (Mette et al. 2000, Gordon and Pacheco 2007, Peroutka et al. 2008, Sirová et al. 2009, Alkhalaf et al. 2011) have shown that aquatic *Utricularia* (*U.* sect. *Utricularia*) do not capture prey selectively but rather capture prey in proportion to the local abundance of phytophilous organisms. This result was supported further by the observations that *Utricularia* of the same species may prey upon different organisms in different habitats and that traps can fire spontaneously without any mechanical stimulation (Adamec 2011f, Vincent et al. 2011a; Chapters 13, 14).

In contrast, Harms (1999) showed that aquatic *Utricularia* in natural conditions captures prey non-randomly. Selective prey choice was observed in laboratory experiments with *Utricularia vulgaris* (Harms and Johansson 2000). In a choice experiment using a cladoceran and a copepod, more copepods were trapped than cladocerans. Prey behavior and interactions between prey also can affect prey selection. For example, Mette et al. (2000) found that *Utricularia* do not trap prey swimming in open water but instead predominantly capture those organisms that stay close to plants as phytophilous zooplankton.

Despite a different trapdoor mechanism (Chapters 13, 14), the prey of affixed aquatic *U. volubilis* (*U.* sect. *Pleiochasia*) traps is similar to that of *U.* sect. *Utricularia*. Algae are the most abundant group of all organisms in traps of *U. volubilis*, although only some of the algal species in the trap interior were killed (Płachno et al. 2014b). Selective or not, aquatic *Utricularia* most frequently captures crustaceans (Harms 1999, Mette et al. 2000, Guiral and Rougier 2007). In *U. inflata* and *U. gibba* traps, the density of cladocerans was higher than that in the water (Gordon and Pacheco 2007), but this result could reflect the accumulation of durable crustacean corpses.

Among terrestrial species, Seine et al. (2002) suggested that *Utricularia calycifida, U. livida*, and *U. sandersonii* specialized on protozoa, but they did not determine whether metazoans also could be attracted and trapped (cf. Jobson et al. 2000, Jobson and Morris 2001 for *U. uliginosa*). Darwin (1875) found metazoan prey in traps of terrestrial *Utricularia*. Jobson and Morris (2001) analyzed time series of organisms in the traps of *Utricularia uliginosa* and the surrounding soil and also did experiments on prey behavior and attraction. Their results illustrate that this species captures prey non-randomly. Nematodes are strongly under-represented in the traps relative to their abundance in the soil, whereas the copepod *Elaphoidella* are over-represented inside the traps.

The affixed aquatic *U. volubilis* (*U.* sect. *Pleiochasia*) and aquatic *Utricularia* in *U.* sect. *Utricularia* trapped similar animals, although *U. volubilis* has a different trapdoor mechanism from species in *U.* sect. *Utricularia*. Algae represented the dominant, most abundant group of all observed organisms in traps of *U. volubilis*. Some algal species remain alive within the trap interior whereas others are killed (Płachno et al. 2014b).

In conclusion, bladder age and size, prey availability, prey behavior, and size all determine prey selection at least in the more derived aquatic *Utricularia* species (Friday 1989, Harms and Johansson 2000, Guiral and Rougier 2007). Some aquatic *Utricularia* species (e.g., *Utricularia vulgaris*) have dimorphic traps that range from ≈1–5-mm long. Larger central traps attached near the foliar mid-rib have a wider aperture and a more capacious lumen than the more peripheral smaller traps (Friday 1991). It appears that different-sized traps may exploit different size ranges of prey. Variation in trap organization, size distributions, and the extent of epiphytic colonization of the traps all may affect the prey spectrum (Guiral and Rougier 2007).

21.2.4 *Drosera*

Following the distinction made between motile and non-motile traps (Chapters 14, 15), we include *Drosera* in this section, although motility varies among the nearly 250 species of *Drosera*. Many of the theoretical and empirical questions pertaining to prey selectivity could be addressed experimentally using *Drosera* as a model system. Many species occur sympatrically, and the sticky traps of *Drosera* are leaves covered by glandular hairs with some movement capacity (Chapters 13, 14). Competition for prey exists both within (Thum 1989a, Gibson 1991a) and between (Thum 1986, 1989a) *Drosera* species. Although prey composition has been described for only a handful of species, substantial variation has been reported (Figure 21.1). Although part of this variation appears to result from small-scale differences in habitat, there is growing evidence for direct selectivity of prey based on trap shape, size, and orientation (Bertol et al. 2015b).

Differentiation among *Drosera* species in growth form, leaf shape, glandular hairs (Chapters 12–15), or chemical composition of the trapping mucilage (Chapter 16) could indicate a separation of trophic niches, particularly if supported by empirical differences in prey composition. Upright sundew species significantly differ in their prey composition

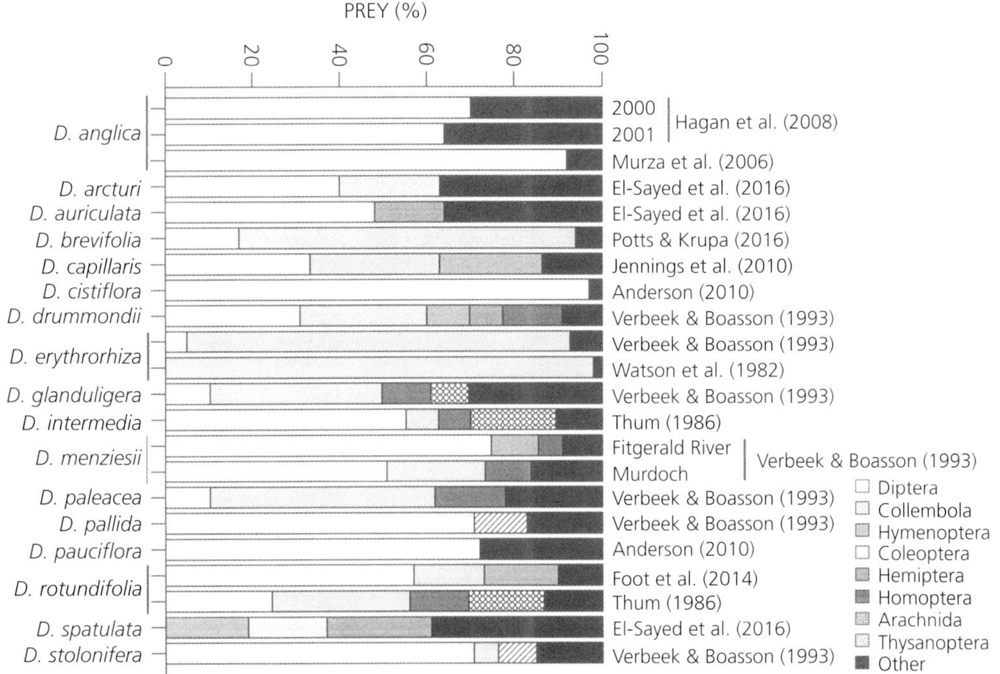

Figure 21.1. Diversity of prey captured by different *Drosera* species. Only orders or classes with >5% representation are separated; the remainder are pooled within "Other."

compared to more prostrate species, with the former capturing more aerial arthropods and the latter more terrestrial arthropods (Verbeek and Boasson 1993). Interspecific differences in leaf shape among sympatric *Drosera* species (Thum 1986, Hoyo and Tsuyuzaki 2013) or in the chemical composition of the trapping mucilage between *D. binata* and *D. capensis* (Adlassnig et al. 2011) have been reported, but none of them has been studied in relation to differential prey capture.

Similarly, interspecific differences in glandular hairs may be related to prey capture, but this hypothesis requires further investigation. Although all glandular hairs of carnivorous plants have similar structures (Adlassning et al. 2011; Chapter 13), 14 kinds of glandular hairs have been described in leaves and sepals of *Drosera* (Länger et al. 1995). The majority of *Drosera* species each have only one type of glandular hair, but some species have two to five types (Länger et al. 1995). At least four kinds of glandular hairs as emergences (tentacles; Chapter 13) have been described for *Drosera* (Poppinga et al. 2013a) that differ in their movement properties, presence of trapping mucilage, and arrangement in the leaves (Poppinga et al. 2013b).

Drosera may lure prey using visual (e.g., ultraviolet reflectance; Chapter 12) or olfactory cues, including plumbagin in *Drosera* (Egan and van der Kooy 2012, El-Sayed et al. 2016; Chapter 16). Volatiles are emitted by *D. binata*, but they are unlikely to play any role in prey attraction due to their low diversity, amount, and similarity to volatiles emitted by green leaves (Jürgens et al. 2009).

Indirect evidence based on observational data and inferences based on plant morphology suggest that *Drosera* species trap different kinds of prey and can attract a subset of the available prey (Table 21.1). Prey selectivity also can be demonstrated by comparing the identity of prey trapped by living plants with prey accumulated on artificial sticky traps with similar shapes and orientations placed nearby (Ellison and Gotelli 2009). Such tests suggest lack of prey selection by *D. erythrorhiza* (Watson et al. 1982), *D. capillaris* (Jennings et al. 2010), and *D. brevifolia* (Potts and Krupa 2016). There are two caveats associated with these types of experiments. First, negative results do not show lack of specialization, only a similar trapping with respect to a passive trap of similar size, shape, and orientation (Ellison and Gotelli 2009). Second, even if differences are observed, they do not reveal mechanisms by which prey are selectively trapped.

21.3 Prey selection by carnivorous plants with non-motile traps

21.3.1 *Genlisea*

Prey selectivity by *Genlisea* could be demonstrated either by comparison of the relative abundance of organisms in traps and the surrounding environment, or using laboratory experiments in which different types of organisms are offered to the plant. Artificial traps (models) also can be used to demonstrate attractiveness of plants to prey. Analysis of trap structure (e.g., trap morphology and size, size of trap openings versus body size of prey; Chapter 13) may be helpful in understanding prey selection. As always, it is important to distinguish between prey and commensals living inside the traps. For example, live nematodes (Studnička 1996, 2003b), algae (Płachno and Wołowski 2008), and microbes (Caravieri et al. 2014, Cao et al. 2015) have been reported from *Genlisea* traps.

Barthlott et al. (1998) showed that *G. aurea*, *G. violacea*, and *G. margaretae* can attract and trap the ciliate *Blepharisma americana* and absorb ^{35}S from digested prey. They also observed that protozoa are trapped in large amounts by *G. stapfii* in the field. Barthlott et al. (1998) suggested that *Genlisea* should be regarded as a protist specialist, but they did not determine whether metazoans also could be attracted and trapped. They also did not report detailed data of prey abundance inside the traps and surrounding soil.

Płachno et al. (2005a) showed experimentally that *Genlisea* attracted and trapped both annelids (*Chaetogaster* sp.) and protozoa (*Colpidium colpoda*, *Pseudomicrothorax dubius*), and they recorded nematodes, rotifers, annelids, tardigrades, crustaceans, and mites in the traps. Płachno et al. (2005a) concluded that prey trapped by *Genlisea* depended on the species of available organisms, but that the size of both the trap openings and the prey can affect whether or not prey are captured. Darnowski and Fritz (2010) showed that *G. filiformis* trapped relatively few prey of any type, but the prey encountered most frequently were ciliates and the copepods *Cyclops*. Darnowski and Fritz (2010) hypothesized that *Genlisea* preferentially preys on small crustaceans. Fleischmann (2012a) reported that the prey of *Genlisea africana* collected from Zambia consisted of copepods.

Demonstrations of selective predation also requires the analysis of available prey in natural conditions. Fleischmann (2012a) summarized all published data about organisms inside *Genlisea* traps, but most lack detailed information about available prey, are poorly analyzed, and do not account for differential rates of decomposition (e.g., protozoa, annelids versus arthropods with chitinous exoskeletons or delicate euglenophytes versus desmids and diatoms). The few reliable data suggest that *Genlisea* species are not specialized to use any particular prey but that the kind of prey in traps reflects the size of the traps and the potential prey that is available in the surrounding environment in which the plants are growing (Płachno et al. 2005a, Fleischmann 2012a). The role of trap dimorphism (Studnička 1996, Fleischmann 2012a; Chapters 13, 14) in prey capture by *Genlisea* remains an open question.

21.3.2 *Philcoxia*

Philcoxia was discovered in Brazil some decades ago but demonstrated to be carnivorous only recently (Pereira et al. 2012). It grows with its leaves below the surface of bright white silica sand outbreaks. The habitat of *Philcoxia* is flooded during the wet season when the plant is actively growing, during which time the leaf surfaces are covered with sticky trichomes. Pereira et al. (2012) showed that the plants exclusively trap nematodes.

21.3.3 *Drosophyllum*

This monotypic genus is endemic to the southwest Iberian Peninsula, where Bertol et al. (2015b) demonstrated experimentally using real plants and artificial traps that it preys selectively on small Diptera. *Drosophyllum* has two potential mechanisms to lure prey: ultraviolet reflection (Joel et al. 1985), and a sticky secretion that smells like honey (Bertol et al. 2015b).

21.3.4 *Pinguicula*

Pinguicula traps are compact rosettes of relatively small, glandular, often prostrate leaves. Some *Pinguicula* species may use ultraviolet absorption and reflectance patterns, variations in color, of scent to attract prey (Chapter 12). Karlsson et al. (1987) used two kinds of artificial traps to examine prey availability and selectivity by *P. vulgaris*, *P. villosa*, and *P. alpina* growing in different microhabitats. One trap mimicked the size, shape, and color of *Pinguicula* rosettes, the other was a transparent 2×2-cm piece of plastic. Both were coated with an odorless adhesive. More dipterans (suborder Nematocera) were captured by leaves of *P. vulgaris* than by artificial traps, indicating some attraction and selectivity. In contrast, prey capture by *P. alpina* and *P. villosa* was similar to capture by artificial traps (Karlsson et al. 1987).

In another study Karlsson et al. (1994) found that *P. vulgaris* again predominantly captured nematoceran flies, *P. villosa* captured predominantly Collembola, and the prey of *P. alpina* was more equitably split between Collembola and Nematocera. Because the three species occurred in different microhabitats and there were no estimates of prey availability within these microhabitats, it is unknown whether these differences reflect differences in prey availability or differences in attraction and selectivity.

Antor and García (1994) examined prey capture in the elongate hanging traps of *P. longifolia*. To assess prey availability, a series of artificial traps (7.8-cm diameter and coated with adhesive) were placed both within and outside the population. *Pinguicula longifolia* captured proportionately more Nematocera than were captured by sticky traps. Alcalá and Domínguez (2003) examined prey capture in *P. moranensis*. Prey availability was assessed using 8×10-cm cardboard traps coated with Tanglefoot™ glue, and plants captured proportionately more dipterans than did sticky traps. Last, Zamora (1995) used artificial traps made from wooden sticks that mimicked distal leaf size, shape, and aspect of *P. vallisneriifolia*. Again, leaves caught more nematoceran flies and aphids than the artificial traps. In contrast, the taxonomic diversity of prey capture by *P. nevadense* was similar to that accumulated on artificial traps (Zamora 1990b). In all studies of prey capture by *Pinguicula* in which prey size was examined, a difference was observed in the size of prey retained (most <2.0 mm, all species with a limit in prey size of 5 mm) (Gibson 1991b, Zamora 1995).

21.3.5 *Nepenthes*

More studies in the last 25 years on prey selectivity have been done on *Nepenthes* than on any other carnivorous plant. Researchers have demonstrated targeted attraction and trapping of particular prey taxa, and prey partitioning between ontogenetic stages of the same species as well. Other species have highly sophisticated adaptations for obtaining other nutrient sources, including mammal feces and leaf litter (Chapters 5, 15, 17, 26).

Field surveys have revealed that most *Nepenthes* species capture a broad spectrum of arthropods; ants commonly predominate (Erber 1979, Kato et al. 1993, Adam 1997, Rembold et al. 2010a). Moran et al. (2001) used stable isotope analysis to show that 68% of foliar nitrogen in *N. rafflesiana* was derived from *Crematogaster* ants. Kato et al. (1993) investigated ten Sumatran species and provided an early indication of consistent differences between prey spectra. Adam (1997) sampled prey from 18 Bornean *Nepenthes* species and found not only a striking diversity of prey, but also clear differences in prey assemblages between species growing in different habitats and at different altitudes. However, because he used one-off spot sampling and only looked at ordinal differences (except for ants; family Formicidae), he probably underestimated the amount of small and soft-bodied invertebrates in their diets.

Both Kato (1993) and Adam (1997) reported termites (Isoptera) as the dominant prey component in pitchers of *N. albomarginata*; Moran et al. (2001) subsequently showed that this species specifically targets termites as prey. This termite specialist was the first case of prey selectivity to be shown in *Nepenthes*, and remains one of the most convincing cases to date. Using stable isotopes, Moran et al. (2001) demonstrated that over 50% of foliar nitrogen in the studied plants was derived from termites. Merbach et al. (2002) showed that the conspicuous ring of white trichomes that distinguishes this species from all other *Nepenthes* is highly attractive to lichen-feeding termites in the sub-family Nasutitermitinae. Merbach et al. (2002) found pitchers in two distinct categories: empty ones bearing an immaculate white ring and pitchers with a grazed-down, brown ring that were full of termites, often quite literally full up to the rim.

Moran (1996) systematically studied attraction and prey capture in *N. rafflesiana* in Brunei and found that upper pitchers caught significantly higher numbers of flying prey than lower pitchers. This difference still persisted when upper pitchers were positioned at ground level. Upper and lower pitchers showed similar patterns of contrasting UV absorption and reflectance but Moran noticed differences in fragrance between the pitcher types, concluding that the fragrance serves as specific attractant for flying prey. Adam (1997) corroborated the higher proportion of flying prey in upper compared to lower pitchers. Di Giusto et al. (2010) showed this fragrance of upper pitchers resembles the odor bouquet of flowers, attracting generalist pollinators, and that the mere scent was enough to attract various insects. This specialized attraction (here more an extension of the prey spectrum) is associated with a modification of trapping mechanisms such as the loss of the inner waxy surface (Gaume and Di Giusto 2009) and an increase in the viscoelasticity of the digestive liquid, which has been shown to be particularly effective in retaining flying prey (Gaume and Forterre 2007; Chapter 15). *Nepenthes inermis* also appears to specialize on flying prey, and its pitchers also contain an exceptionally sticky, viscoelastic fluid (Salmon 1993).

Another strong indicator of prey specialization is found in *N. aristolochioides*. Like the unrelated North American *Darlingtonia californica, Sarracenia minor*, and *S. psittacina*, the pitchers of this species have a dome-shaped upper section with multiple translucent fenestrations, or "windows" (Chapter 12, §12.2.2). Small Diptera are a major component of the prey of *N. aristolochioides* in the wild. Moran et al. (2012) reported a three-fold reduction of captured *Drosophila melanogaster* flies when they covered the windows with a red filter. The similar pitcher morphology of *N. klossii* suggests that it might employ a similar prey capture strategy.

The predominant trapping mechanism of *Nepenthes* pitchers, insect aquaplaning on the wet peristome (Chapter 13, §13.2.3; Chapter 15, §15.3.4), might in itself constitute an adaptation for capturing ants. Bauer et al. (2015a) showed that the sudden shift between slippery (wet) and safe (dry) conditions in the natural environment increases ant recruitment and ultimately prey numbers in *N. rafflesiana* pitchers. The copious amounts of nectar secreted by many *Nepenthes* pitchers also might have evolved to attract ants, but no study yet has investigated the importance of nectar quantity or composition for prey selectivity in any *Nepenthes* species. Bennett and Ellison (2009) showed that nectar is the single most important attractant for ant visitors of *Sarracenia purpurea*.

The genus *Nepenthes* is unique in that it includes several species that have evolved away from carnivory and instead use their traps to exploit alternative nutrient sources. The first reported case was that of *N. ampullaria* which was shown to derive more than 35% of its foliar nitrogen from leaf litter (Moran et al. 2003). *N. ampullaria* is one of very few *Nepenthes* species that regularly grow under a closed forest canopy. Its pitchers form dense carpets on the forest floor. The pitcher lid is bent backwards so that the pitchers are open to the elements and readily accumulate leaf litter and other detritus falling from the forest canopy.

Clarke et al. (2009) and Chin et al. (2010) demonstrated that *N. lowii* and *N. macrophylla* attract mountain tree shrews (*Tupaia montana*) and capture their feces in wide, funnel-shaped upper pitchers (Chapter 13, §13.6). Stable isotopes revealed that mature *N. lowii* plants derive between 57 and 100% of their nitrogen from shrew feces. *Nepenthes rajah* was also shown to attract tree shrews and nocturnal rats (*Rattus baluensis*) and capture their feces (Greenwood et al. 2011, Wells et al. 2011). Pitcher size and geometry likely are the key adaptation for fecal capture in these species; Chin et al. (2010) showed that shrew body size exerts strong evolutionary pressure on minimum pitcher size, and only pitchers with an orifice matching or larger than the shrews' snout-vent length successfully capture feces. Fine-tuning of volatiles for attraction also has been reported (Wells et al. 2011), but tree shrews are also attracted to the nectar of generalist *N. gracilis* (Bauer et al. 2016).

Finally, the pitchers of the bizarre *N. hemsleyana* effectively reflect echolocation calls of tiny *Kerivoula hardwickii* bats that regularly are found roosting inside pitchers (Schöner et al. 2015). Additional adaptations to host the bats include an extraordinarily low fluid level and a specific pitcher shape with a distinctive "hip" that prevents the bats from slipping down into the digestive liquid (Lim et al. 2014). Grafe et al. (2011) demonstrated that *N. hemsleyana* gain approximately 34% of their foliar nitrogen from bat excrement.

21.3.6 *Sarracenia*

The pitcher-shaped leaves of *Sarracenia* form a rosette and arise from a short stem or rhizome. The leaves of different species vary from almost prostrate to completely erect, and from less than 10 cm to more than 70 cm in length. Most species have a hood that extends beyond the top of the pitcher, although in most species the hood does not occlude the pitcher opening. The inner portion of the pitcher can be subdivided into several zones, including a waxy zone near the top and a zone with downward-pointing hairs toward the bottom (Chapter 13, §13.3). All species of *Sarracenia* that have been examined produce nectar and a number of volatile compounds, and many exhibit variations in color including nectar guides or nectariferous lines and patterns of ultraviolet absorbance/reflectance (Chapter 12, §12.2.1). Some species also possess fenestrations or windows (Chapter 12, §12.2.2).

Only one study has simultaneously examined both prey availability and prey capture by *Sarracenia* (Folkerts 1992). She found that pitchers of *S. flava* captured fewer prey taxa than artificial traps. When several *Sarracenia* species occurred sympatrically, they consistently differed in pitcher size, prey numbers, and prey diversity. She concluded that differences in pitcher size are an adaptation for prey partitioning between sympatric species.

Data from several other studies have suggested that certain species specialize on particular groups of prey, but definitive tests are lacking. There are distinct differences between the prey composition of *S. minor* and *S. minor* var. *okefenokeensis* that occur within a few km of each other (Stephens et al. 2015a). Approximately 95% of the prey captured by *S. minor* consisted of ants in both spring and fall but prey captured by *S. minor* var. *okefenokeensis* consists primarily of beetles in the spring and a combination of beetles, flies, and ants in the fall. Since the two varieties do not co-occur in the same site, there could also be differences in prey availability.

The majority of prey items captured by *Sarracenia alata* in west-central Louisiana (USA) are ants (Bhattarai and Horner 2009, Horner et al. 2012), but the proportion of ants captured varied by site (Green and Horner 2007) and pitcher size. Pitcher size was confounded with season, however, so it is unclear whether the differences in prey capture were due to differences in pitcher size or differences in prey availability at different times of year. Similarly, Gibson (1991b) demonstrated that the maximum size of the prey captured was correlated with the maximum orifice size of ten species of *Sarracenia*.

21.3.7 *Brocchinia, Catopsis, Cephalotus,* and *Heliamphora*

The remaining genera of pitcher plants have various attractants (Table 21.1; Chapter 12), including odors (Jaffe et al. 1995, Benzing 2000), ultraviolet reflectance patterns (Joel et al. 1985, Benzing 2000), and nectar (Joel 1988, Jaffe et al. 1992), but their roles in prey attraction or selection are unknown. Experimental tests using artificial traps would provide useful data to understand prey capture and selectivity by these pitcher plants.

21.4 Future research

Even though prey capture is one of the most captivating aspects of carnivorous plant biology, careful studies of prey selectivity are uncommon. Future research in this area should focus on under-studied genera (e.g., *Aldrovanda, Genlisea*) and use artificial traps, censuses of the surrounding habitats, and consistent metrics (Darnowski 2016) to test for prey selectivity. The effect of sympatry—interspecific competition among co-occurring species—on prey selection and selectivity is of particular interest. Relatedly, it is of interest whether allocation to traps varies with prey availability as predicted by cost/benefit analysis (Chapter 18). Last, do carnivorous genera that no longer depend solely on carnivory for obtaining nutrients show any corresponding loss of prey specificity?

Future studies also need to address the best methods for assessing prey availability. Studies have used sticky traps, pitfall traps, and sweep netting. Each of these captures different subsets of insects so each gives a different view of prey availability (Norment 1987, Schoenly et al. 2007). Researchers also need to be mindful of plant-produced attractants and man-made chemicals that differ between living carnivorous plants and artificial traps. If a species produces attractants, then samples of available insects from co-located non-biological traps without attractants may still oversample prey attracted to the plants. For example, Gibson (1983) found that there was a positive relationship between insects captured on sticky traps and density of *S. flava* pitchers. This result suggested that *S. flava* produces attractants that also lured insects to the sticky traps. Artificial sticky traps rarely are analyzed to ensure that no odor (either attractant or repellent) is produced. If adhesives produce volatiles that either attract or repel insects (or particular groups of insects), this needs to be accounted for in estimates of prey availability. Finally, future studies also need to account for differential rates of digestion of prey. Not all prey are digested at equal rates, and some (e.g., wasps with large chitinous exoskeletons) leave much more obvious remains than soft-bodied prey such as slugs.

CHAPTER 22

Reproductive biology and pollinator-prey conflicts

Adam T. Cross, Arthur R. Davis, Andreas Fleischmann, John D. Horner, Andreas Jürgens, David J. Merritt, Gillian L. Murza, and Shane R. Turner

22.1 Introduction

Approximately 88% of flowering plants are pollinated by animals (Ollerton et al. 2011). Most rely on flower-visiting insects for pollination, and entomophily appears to be ubiquitous among carnivorous plants (Juniper et al. 1989). Pollination services are not free; plants invest substantial energy into conspicuous floral traits such as nectar, color, and scent to recruit, manipulate, or reward visiting animals to ensure reproductive success (successful pollination and subsequent seed production).

Reproductive success of carnivorous plants could be compromised if pollinators are captured, giving rise to the "prey/pollination paradox" (Juniper et al. 1989: 273) or "pollinator–prey conflict" (Zamora 1999) (henceforth, "PPC"). However studies to date have found little evidence for PPC (Murza et al. 2006, Anderson 2010, Meindl and Mesler 2011, A. Fleischmann *unpublished data*). Several authors have assumed that mechanisms have evolved to reduce or overcome PPC (e.g., Jürgens et al. 2015; El-Sayed et al. 2016), and features such as trap color, long trap-to-flower distances, and temporal separation of traps and flowers have been interpreted as adaptations to reduce the risk of pollinators landing on and being captured by trap leaves (e.g., Givnish 1989, Juniper et al. 1989, Anderson and Midgley 2001, Schaefer and Ruxton 2008, Bennett and Ellison 2009, Sciligo 2009, Horner 2014, El-Sayed et al. 2016).

Once pollination and fertilization have occurred, plants must produce seeds, which in turn must disperse into new locations and be able to survive until conditions are suitable for germination. Most carnivorous plants produce numerous small, lightweight (< 5 mg) seeds that germinate and establish in open or disturbed oligotrophic habitats (Chapter 2). Seeds can vary immensely in size and shape, and anatomical features of the seed coat (testa) significantly affect how they disperse and germinate (Baskin and Baskin 2014). Although vegetative propagation might play an important role in structuring populations of some carnivorous plants (Heslop-Harrison 1962, Chen et al. 1997), long-term persistence and recruitment into new habitats largely depend on seeds. Like seeds of ≈70% of all angiosperms, those of carnivorous plants exhibit some form of dormancy: an adaptive trait that prevents or inhibits germination until environmental conditions are suitable (Vleeshouwers et al. 1995). Dormancy is determined by physiological (PD), morphological (MD), or both physiological and morphological (MPD) properties of the developing embryo, the surrounding tissues, or their interactions (Baskin and Baskin 2014).

In this chapter we provide an overview of the reproductive biology of carnivorous plants, with a specific focus on breeding systems and pollinator specialization; the implications and methods by which species overcome potential PPC; and their seed dormancy and germination biology. These life-history traits are central aspects of conservation and restoration programs (Chapter 27) that depend on the genetic integrity of wild populations and the long-term storage of viable germplasm such as seeds.

Cross, A. T., Davis, A. R., Fleischmann, A., Horner, J. D., Jürgens, A., Merritt, D. J., Murza, G. L., and Turner, S. R., *Reproductive biology and pollinator–prey conflicts*. In: *Carnivorous Plants: Physiology, ecology, and evolution*. Edited by Aaron M. Ellison and Lubomír Adamec: Oxford University Press (2018). © Oxford University Press. DOI: 10.1093/oso/9780198779841.003.0022

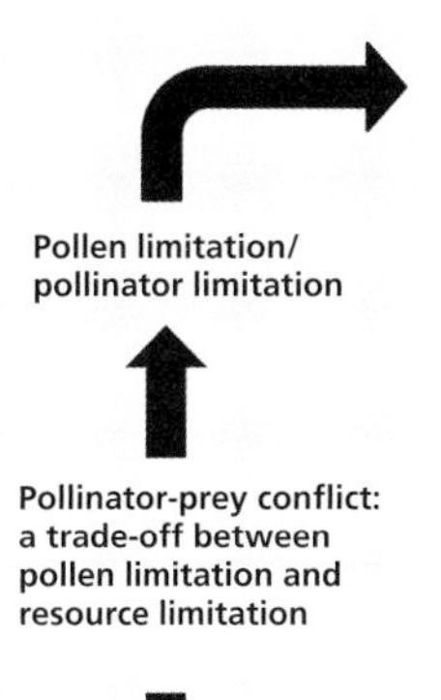

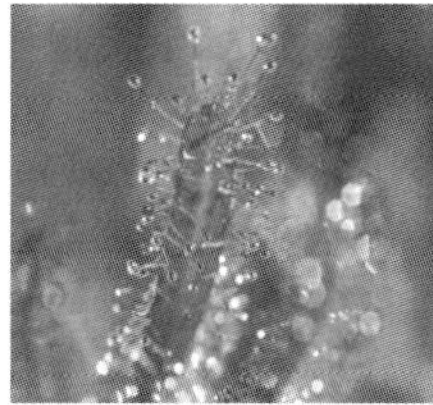

Figure 22.1 Consequences of the pollinator–prey conflict in carnivorous plants, which is a tradeoff between pollen limitation and resource limitation, are illustrated by *Drosera arcturi*, pollinated by *Melangyna novaezelandiae* on Mt Cook, New Zealand. (Photographs by A. Jürgens).

22.2 Pollinator–prey conflict

The existence of PPC and possible mechanisms to reduce it have been debated for several decades (Juniper et al. 1989, Jürgens et al. 2012). Several trait-based mechanisms have been proposed to ameliorate PPC (Figure 22.1), but few studies have investigated whether pollinators are ever trapped as prey or vice versa (Moran 1996, Zamora 1999, Murza et al. 2006, Anderson 2010, Fleischmann 2012a, 2016b, A. Fleischmann *unpublished data*). A more general problem is determining whether observed traits evolved to reduce PPC (adaptation), or whether their current function evolved in response to different selection pressures (exaptation).

22.2.1 Autogamy

Dependence on pollinators is a precondition for PPC (Jürgens et al. 2012), and one way to avoid PPC is to eliminate the need for pollinators through self-pollination (autogamy) and subsequent self-fertilization. Autogamy also may be an adaptation to reduce the cost of pollinator advertisement and associated rewards such as nectar (e.g. Hobbhahn et al. 2006). Jürgens et al. (2012) speculated that breeding system and independence from pollinators could affect the evolution of trapping efficiency; pollinator-independent plants not only could reduce floral rewards such as nectar but also could trap potential pollinators without reducing their fitness.

Pollen production often is reduced in plants that self-pollinate because of the higher efficiency of pollen transfer to stigmas, resulting in low pollen-to-ovule ratios (Cruden 1977).

Approximately 100 species of Lentibulariaceae, Droseraceae, and Drosophyllaceae are facultatively autogamous (Cross 2012a, Fleischmann 2012a, *unpublished data*, Salces-Castellano et al. 2016): *Aldrovanda vesiculosa*; nine annual species of *Pinguicula*; 20–40 species of *Utricularia* (predominantly annuals and some aquatics with cleistogamous flowers); seven annual species of *Genlisea*; ≈50 species of *Drosera*; and *Drosophyllum lusitanicum* (Olivencia et al. 1995). However, autogamy appears to be less related to PPC than to annual life-histories (some *Drosera*, *Byblis*, and aquatic *Utricularia*) and dynamic, highly seasonal habitats, where it ensures successful seed set in years with low pollinator activity or briefly favorable conditions for establishment (Fleischmann 2012a, 2016b). For example, the common occurrence of autogamy in aquatic *Utricularia* species and *Aldrovanda* mirrors a general trend found in submerged aquatics rather than being related to plant carnivory (Barrett 2015).

Autogamy also may cause inbreeding depression resulting in reduced mean population fitness (Sheridan and Karowe 2000, Sciligo 2009), and most angiosperms have evolved features to reduce or completely prevent self-pollination or self-fertilization. Similarly, most carnivorous plants have highly effective mechanisms to prevent selfing and favor outcrossing. These include the complex floral design of *Sarracenia* and *Darlingtonia* (Meindl and Mesler 2011); the bilabiate stigma of Lentibulariaceae covering the anthers (Fleischmann 2012a, 2016b); dichogamy in *Heliamphora* and certain *Drosera* species, and dioecy in *Nepenthes* (Renner 1989, Chen et al. 1997, A. Fleischmann *unpublished data*).

22.2.2 Specialization on pollinators and prey

Jürgens et al. (2012) emphasized that studies of PPC are best understood with respect to the importance of recorded flower visitors as pollinators (Figure 22.2). Despite its importance for evaluating PPC, the basic pollination biology of carnivorous plants is largely unexplored, as it is difficult to do detailed field studies in their often inaccessible habitats. Further, taxonomic resolution of pollinators and prey rarely is reported below the family level. For example, *Dolichopus* sp. (Diptera) visit flowers of *Drosera anglica*, but seven other Dolichopodidae spp. were trapped as prey (Murza et al. 2006). Similar taxonomic disparities are *in litt.* between trapped and flower-visiting Muscidae and Tachinidae, although members of these two families have been regarded solely as flower visitors in more recent studies of PPC in *Drosera*. In contrast, *Usia* sp. (Bombyliidae) visited the flowers of *Drosophyllum*, whereas *Bombylella atra* was recorded only as prey (Bertol et al. 2015b).

Trapping ineffective pollinators (e.g., infrequent or slow foragers that gather pollen but transfer little) or pollen feeders may provide nutritional benefits and reduce pollen discounting (Zamora 1999). For example, trapping of pollen feeders by *Drosera tracyi* is beneficial relative to the capture of seven large-bodied bee species that actually pollinate the flowers (Wilson 1995). Similarly, Murza et al. (2006) point out that the consequences of consuming potential pollinators of the autogamous *D. anglica* (Murza and Davis 2005) are small relative to the pollination deficit created by trapping thrips and beetles in shaded habitats by *Pinguicula vallisneriifolia* (Zamora 1999). Finally, a novel

Figure 22.2 (Plate 14 on page P12) Pollinators and methods for investigating the importance of insects for the reproductive success in carnivorous plants. (**a**) *Utricularia multifida* with halictid bee, near Perth, Western Australia (photograph by T. Carow); (**b**) *Drosera arcturi* with *Melangyna novaezelandiae* (Syrphidae) probing pollen, Mt Cook, New Zealand (photograph by A. Jürgens); (**c**) *Drosera liniflora* and the hopliinid beetle *Anisonyx ursus*, Baineskloof, South Africa (photograph by A. Fleischmann); (**d**) *Heliamphora sarracenioides* with the carpenter bee *Xylocopa* sp. buzzing pollen from the poricidal anthers, Ptari-tepui, Venezuela (photograph by A. Fleischmann); (**e**) Pair of scanning electron micrographs of *D. anglica* tetrads on *Toxomerus marginatus* (Syrphidae) (photographs by G. L. Murza); (**f**) Pollinator exclusion experiment testing autonomous selfing in *D. arcturi* (photograph by A. Jürgens).

mechanism for interspecific competition between carnivorous plants is suggested by the observation that a bombylid fly species that pollinates *Drosera trinervia* is a common prey of the sympatric *D. cistiflora* (A. Fleischmann *unpublished data*).

The spectrum of captured prey can be very broad, but some species use a combination of color, scent, and morphology to attract and capture particular prey (Chapters 12, 21). PPC is avoided in some species that produce different signals to guide prey and pollinators to leaves and flowers, respectively. For example, flowers and traps of *Drosera auriculata* are produced adjacent to one another, but specific volatiles (2-phenylethanol, 2′-aminoacetophenone, benzyl alcohol, benzaldehyde) distinguish flowers from leaves and application of the floral chemical blend alone to clear sticky discs attracted more flower-visiting types of insects (Syrphidae, Tachinidae, Muscidae) than recorded naturally on flowers (El-Sayed et al. 2016). The addition of trap-leaf odors reduced insect attraction, and typical prey (especially Nematocera and Hemiptera) were attracted to discs baited with both floral and trap-leaf odors more than to discs baited with trap scent alone. Discrimination between scents of flowers and trap leaves by guilds of potential-pollinating and prey-type insects reduces PPC (El-Sayed et al. 2016).

In contrast, two sympatric alpine species, *D. arcturi* and *D. spatulata*, lack fragrant volatiles to distinguish their floral and trapping organs (El-Sayed et al. 2016), and their generalist pollinators are frequently trapped by the leaves (Sciligo 2009, Jürgens et al. 2015). Similarly, trap leaves of *D. binata* are not scented and lack volatiles distinctive from green leaves of many non-*Drosera* species (Jürgens et al. 2009, Fleischmann 2016b). In general, scented sticky traps are infrequent among *Drosera* (Fleischmann 2016b), perhaps because they are "cost-prohibitive" (El-Sayed et al. 2016). Even in scented *Drosera* species (similar to the scent of *Sarracenia* traps), fragrance may be detectable only by a small percentage of potential pollinators or prey (Fleischmann 2016a).

22.2.3 Carnivorous traps that mimic flowers

Trap architecture, and the visual and olfactory features of the traps of some species, suggest they mimic flowers to attract flower-visiting insects as prey. The scents emitted from *Nepenthes rafflesiana* pitchers are effective cues for prey attraction, and upper pitchers produce olfactory cues that biochemically mimic flowers (Di Giusto et al. 2010). The trap-leaves of *Drosera finlaysoniana* from northern Australia are richly honey-scented, and the prey of a single plant comprised 35 adult Lepidoptera from five different species as well as many hundreds of Thysanoptera, Diptera, and other small insects (Fleischmann 2016a). Future studies should investigate flowering and pollination biology further in *Drosera* with scented leaves (e.g., *D. finlaysoniana* and *D. slackii*), and to analytically compare the scents of vegetative parts and flowers, particularly as flowers appear to lack obvious fragrance in species with scented leaves (Fleischmann 2016a).

22.2.4 Spatial separation of flowers and traps

The spatial separation of reproductive and trapping organs has been postulated to be a morphological solution to PPC. Flowers are spatially separated from trapping organs in many carnivorous genera (Juniper et al. 1989; Jürgens et al. 2012, 2015). The spatial separation of flowers and traps may reduce the risk that a pollinator approaching an inflorescence inadvertently falls into or lands on a trap. However, results of several studies suggest that long flowering stalks instead may have evolved to attract more flower visitors by increasing the visual floral display height rather than to reduce PPC (Anderson and Midgley 2001, Anderson 2010, Jürgens et al. 2015). This may explain why low-growing species occurring in very dense vegetation possess some of the longest inflorescences. Two extreme examples are *Cephalotus follicularis* (flowers held up to 60 cm above traps) and *Drosera hamiltonii* (flowers held up to 50 cm above traps). However, other species with long flowering stalks include terrestrial *Utricularia* and *Genlisea*, and aquatic *Utricularia* and *Aldrovanda* whose underground or submerged traps pose no risk to aerial pollinators (Anderson and Midgley 2001, Cross 2012a, Fleischmann 2012a). On the other hand, approximately one-third of all *Drosera* species—in a group often used as a standard example for spatial separation of traps and flowers—have flowers that are not widely separated from the traps (e.g., almost all tuberous *Drosera*; El-Sayed et al. 2016, A. Fleischmann *unpublished data*), and some species have almost sessile flowers displayed among the trapping leaves (Fleischmann et al. 2007, Rivadavia et al. 2009). Spatial separation of flowers and traps also may be achieved by positioning traps close to the ground, which not only may reduce the risk of PPC but also could lead to specialization on non-volant prey.

22.2.5 Temporal separation of flowering and trapping

PPC also may be avoided by the temporal separation of flowering and trapping. This might be achieved either by differences in seasonal phenology of flowering and trap activity or by operating traps and opening flowers during different times of the day. Species with seasonal dormancy, including *Sarracenia*, many heterophyllous Central American *Pinguicula*, and some tuberous *Drosera* have seasonally alternating display of traps and flowers (Rice 2011a, Lowrie 2013). These taxa usually flower immediately after resuming growth in winter or spring and prior to the emergence of new trapping leaves. Trap functionality in several *Nepenthes* and *Heliamphora* species seems to be associated with periods of high air humidity (Bohn and Federle 2004). These periods are probably times when winged pollinator activity is at its lowest. Nonetheless, pollinator and prey spectra rarely overlap in these species, most of which prey predominantly on ants. Differences in diel activity of flowers and traps are unknown. Those few *Drosera* with nocturnally open flowers have traps that capture nocturnal insects (Fleischmann 2016a), and nocturnal flowering appears to be associated with pollinator specialization among sympatric plant species (A. Fleischmann *unpublished data*).

22.3 Pollinator–prey conflict as a function of trap type

22.3.1 Sticky traps

Drosera. Approximately 50 species of *Drosera* are facultatively autogamous. Facultative autogamy generally occurs at the end of anthesis, or rarely in closed flower buds under unfavorable conditions, for a handful of predominately Northern Hemisphere species (A. Fleischmann *unpublished data*). Pollen-ovule ratios for *Drosera* are relatively low and within the ranges reported for cleistogamous or autogamous species (Takahashi 1988, Wilson 1995, Murza and Davis 2003). Only a single monophyletic lineage of *Drosera*, the Australian *D.* subgenus *Drosera* (Chapter 4) has obligatory xenogamous members with pollen self-incompatibility (Chen et al. 1997, A. Fleischmann *unpublished data*). However, there are multiple evolutionary reversals from self-incompatibility to self-compatibility in other lineages (A. Fleischmann *unpublished data*). Some members of the South African *D.* sect. *Ptycnostigma* (e.g., *D. cistiflora*) and *D. regia* use dichogamy (protandry) to overcome selfing. However, these species usually remain capable of facultative selfing at the end of anthesis (A. Fleischmann *unpublished data*).

Flowers of the majority of *Drosera* species lack odors perceptible to humans (Murza and Davis 2003, Gibson 2013, Fleischmann 2016a; Chapter 12, §12.4). However, fragrance is emitted by the flowers of some Australian *Drosera* subg. *Ergaleium* and five white-flowered members of *D.* sect. *Drosera* in South America (Fleischmann et al. 2007, Rivadavia et al. 2009, Gibson 2013, A. Fleischmann *unpublished data*). The flower fragrance of *Drosera* was thought to be produced by the petals (Gibson 2013), but it is now known that pollen is the osmophore (A. Fleischmann *unpublished data*). Insect visitors to *Drosera* flowers include bees (especially Halictidae and *Ceratina*), flies (especially Syrphidae), beetles (Scarabidae, Meloidae, Chrysomelidae), among others (Wilson 1995, Murza and Davis 2005, Murza et al. 2006, Anderson 2010, A. Fleischmann *unpublished data*). Although Lowrie (2001) reported that birds feed on *Drosera* nectar, this seems unlikely. *Drosera* flowers are nectarless (e.g., Murza and Davis 2003) and it is more likely that the birds are picking insects off the flowers (A. Fleischmann *unpublished data*).

Detailed investigations of pollinator–prey overlap in *Drosera* have to date involved six species, mostly generalist-pollinated alpine/boreal species with non-scented white flowers. Prey recorded from *D. anglica* in western Canada included over 109 arthropod species in 94 genera from 11 orders, but predominantly comprised the small, weak-flying adults of Ceratopogonidae and Chironomidae flies (Murza et al. 2006, Hagan et al. 2008). Only 1.7% of prey individuals captured during the flowering period belonged to any of the six Brachycera families recorded as floral visitors and none included particular species recorded as flower visitors (Murza 2002, Murza et al. 2006). The only overlap recorded was a single immature *Thrips* species. Separation of prey also has been reported for *D. arcturi*, *D. auriculata*, and *D. spatulata* from New Zealand (El-Sayed et al. 2016). Trapped insects included Diptera (40%) and Coleoptera (23%) for *D. arcturi*; Diptera (48%, of which 61% belong to 3 families of Nematocera) and Hemiptera (16%) for *D. auriculata*; and Hemiptera (24%), Hymenoptera (19%), and Coleoptera (18%) for *D. spatulata*. Families of Brachycera (mostly Syrphidae, Tachinidae, and Muscidae) comprised the majority of flower visitors, and represented only 4.3, 3.3, and 3.8% of prey trapped by these three species, respectively. Finally, Anderson (2010) reported Diptera constituted 72% and 97% of all prey captured by *D. pauciflora* and *D. cistiflora*, respectively, whereas Coleoptera (Scarabidae: Hopliinae, 64–66%; Chrysomelidae, 21–26%) were the predominant floral visitors. Pollinator-prey overlap was absent in *D. cistiflora* and only one chrysomelid beetle was trapped by *D. pauciflora*.

Notwithstanding the poor taxonomic resolution, flower visitors (potential pollinators) and prey of *Drosera* clearly are different. The lack of overlap between pollinators and prey implies little or no PPC, despite the absence of any seasonal separation between *Drosera* flowering and trapping (El-Sayed et al. 2016). Prey are readily captured by *Drosera* prior to flowering (Murza et al. 2006) or during flowering (Potts and Krupa 2016). Most *Drosera* populations have short, predictable daily flowering periods (Wilson 1995, Murza and Davis 2005), which may help attract dependable flower visitors and hence reduce their likelihood of capture.

Pinguicula. All *Pinguicula* have nectariferous spurred flowers, and some species have pollen-imitation and nectar guide marks (Fleischmann 2016b). Several small annual *Pinguicula* species are autogamous (Fleischmann 2016b), but cleistogamy, vivipary, or apomixis were not reported for the temperate *P. alpina*, *P. grandiflora*, *P. lusitanica*, or *P. vulgaris* (Heslop-Harrison 2004). All four of these species are self-compatible but vary in the degree of self-pollination: no self-pollination was observed in *P. grandiflora*; delayed self-pollination occurred in *P. alpina* and *P. vulgaris*; and self-pollination was prevalent in *P. lusitanica* (Heslop-Harrison 2004). Zamora (1999) notes that *P. vallisneriifolia* from Spain is pollen-limited and self-compatible, but spatial separation of male and female gametophytes (herkogamy) prevents spontaneous self-pollination. This species also displays both spatial and temporal overlap in flowering and trapping.

The main floral visitors to *P. vallisneriifolia* were bee flies (*Bombylius* spp.), bees (*Andrena* spp., *Anthophora* spp., *Bombus terrestris*, *Halictus* spp., *Lasioglossum* spp., *Osmia cornuta*), thrips, hawk moths, butterflies, calliphorid flies, and small beetles (Zamora 1999). The composition of floral visitors was influenced by microclimate, and controlled pollination experiments suggest that seed production in natural populations appeared to be limited by the availability of insect pollinators (Zamora 1999). All flower visitors were adult thrips, which also represented 77% of the prey of plants growing in unshaded areas (Murza et al. 2006). However, staphylinid beetles were the main floral visitors and prey of plants growing in the shade, although their capture was slower due to reduced retention capacity of the leaf mucilage (Zamora 1999). Thus, PPC appears to be significant for *P. vallisneriifolia*, but in a somewhat habitat-dependent manner (Zamora 1999).

Drosophyllum lusitanicum. This species has homogamous flowers (Olivencia et al. 1995); pollen germination and stigma receptivity occur before the flowers open. Bagging experiments confirmed a high capacity for spontaneous autonomy resulting in high fruit and seed set. During a study of potential PPC of *Drosophyllum lusitanicum* in Spain, both small (<5 mm) and large Diptera were reported as prey (Bertol et al. 2015b). One-third of prey was trapped nocturnally when leaves are not clearly visible, providing evidence that *Drosophyllum* also attracts prey with its honey-scented leaves (Chapter 12, §12.4). In contrast, predominant flower visitors were bees (*Panurgus* spp.: Andrenidae) and beetles (*Enicopus* spp.; *Homaloplia ruricola*), whereas flies (Syrphidae) were rare (Olivencia et al. 1995, Bertol et al. 2015b).

Other genera. *Byblis* spp. have poricidal anthers and are buzz-pollinated by native bees and hoverflies (Conran and Carolin 2004), but predominantly trap small Diptera (L. Skates, *unpublished data*). The flowers of *Triphyophyllum peltatum* are sweetly fragrant and produced on long shoots of the climbing vine that are spatially and temporally separated from the short-lived carnivorous leaves (Menninger 1965, Bringmann et al. 2002). *Roridula* are very effective at trapping even large Hymenoptera, and PPC may therefore have selected toward obligate autogamy; self-pollination of *Roridula* flowers is mediated by mutualist *Pameridea* bugs (Anderson et al. 2003).

Why is PPC rare among species with sticky traps? Several hypotheses have been proposed to explain the low PPC among species with sticky traps, even though many lack temporal or spatial separation between flowering and trapping. First, flower visitors may have sufficient strength to escape the sticky mucilage of traps, and they may avoid traps thereafter (*Drosera*: Anderson and Midgley 2001, El-Sayed et al. 2016; *Pinguicula*: Zamora 1999, *Drosophyllum*: Bertol et al. 2015b). However, when individuals of hoverflies known to pollinate *D. arcturi*, *D. auriculata*, or *D. spatulata* were placed individually on trapping leaves, up to 80–93% were successfully trapped (El-Sayed et al. 2016). Because very few of these hoverflies naturally are trapped as prey (van Achterberg 1973, Thum 1986, Murza et al. 2006, Bertol et al. 2015b, Fleischmann et al. 2016), they may be averse to landing on leaves and never actually contact the mucilage (El-Sayed et al. 2016).

Second, the position of erect leaves around a *Drosera* inflorescence may indirectly protect pollinators by obstructing flower visits (Jürgens et al. 2015). However, the greatest spatial separation between flowers and leaf-traps in *Drosera* occurs in non-climbing species with basal rosettes, and few tall or upright species possess lengthy leafless scapes (Anderson and Midgley 2001). Rather than long peduncles being an evolutionary response to protect pollinators from being captured, long scapes probably increase the general attraction of flower visitors (Anderson 2010, Jürgens et al. 2015). Indeed, Lowrie (2013: 492) notes that the long inflorescence of *D. hamiltonii* "often has to push its way up through a tangle of rather dense vegetation well before it can display its mature flowers in the open."

Last, difference in both color and scent between trapping leaves and flowers may reduce PPC. Whereas the red coloration of many *Drosera* leaves appears unimportant in the attraction of insect prey (Foot et al. 2014, El-Sayed et al. 2016, Potts and Krupa 2016), it may provide an important signal to insect pollinators and reduce the incidence of their capture (Jürgens et al. 2015). White discs mimicking the flowers of *D. arcturi* and *D. spatulata* were visited frequently by regular floral visitors of the two species (Jürgens et al. 2015), but were visited most often when placed in close proximity (5–10 cm) to red discs resembling leaves (El-Sayed et al. 2016). Additionally, pollinators visited pink discs with the floral scent of *D. auriculata* more frequently than green discs with trap scent, suggesting that visual plus olfactory cues aid in the discrimination between flowers and traps by some flower visitors (El-Sayed et al. 2016).

22.3.2 Pitfall traps

Available evidence suggests that like the species with sticky traps, those with pitfall traps also have little PPC.

Nepenthes. The dioecious flowers of *Nepenthes* are small, inconspicuous, and usually spatially separated from pitchers. The entomophilous greenish or claret-colored flowers secrete nectar and disperse pollen in tetrads (Kaul 1982, Adam 1998). The inflorescences of most species produce a fetid scent, especially from male individuals (A. Fleischmann, *unpublished data*), which may be involved in insect attraction (Juniper et al. 1989).

There appears to be little pollinator–prey overlap in *Nepenthes*. The flowers of *N. gracilis* in Sumatra are visited by pyralid moths at night and by calliphorid flies in the evening, whereas prey are mostly ants (Kato 1993). A similar pollination system was observed for *N. vieillardii* (Kato and Kawakita 2004). Diptera are the main flower visitors of *N. kinabaluensis*, *N. rajah*, and *N. villosa*, and Hymenoptera and Diptera visit flowers of *N. curtisii* and *N. reinwardtiana* in Borneo (Adam 1998). Diptera (Calliphoridae, Muscidae, and Syrphidae) are the main pollinators of *N. macfarlanei* in Malaysia (Chua 2000).

Sarracenia. Hymenoptera (especially bees and bumblebees) appear to be the main pollinators of all *Sarracenia* species. Pollinators visiting *Sarracenia* flowers must brush past one of the five stigmas, which are near the tips of the upside down umbrella-shaped stylar disc, to get into the chambered part of the flower where the nectar and stamens are located (Ne'eman et al. 2006). Although *Sarracenia* species are self-compatible, they rarely self-pollinate and are not apomictic (Thomas and Cameron 1986). Pollination in *S. purpurea* is primarily carried out by *Bombus affinis, B. bimaculatus, B. vagans*, the bee *Augochlorella aurata*, and the fly *Fletcherimyia fletcheri* (Thomas and Cameron 1986, Ne'eman et al. 2006), whereas prey are principally ants and flies but not bees (Cresswell 1991). *Bombus* spp. also seem to be the primary pollinators of *S. alata* (Horner 2014), and when pollinators were excluded from the flowers no ovules matured (J. Horner *unpublished data*). However, no *Bombus* species has ever been observed among the thousands of insects recorded as prey for *S. alata* (Green and Horner 2007, Bhattarai and Horner 2009, Horner et al. 2012). Sheridan and Karowe (2000) found that self-pollination had a significant negative effect on offspring quality and quantity in *S. flava*.

Pollinator–prey overlap therefore appears to be minimal or absent in *Sarracenia*, and PPC may be avoided by the significant temporal separation of flowering and trap production. Flowers matured and pollination was almost complete before pitchers matured and began capturing prey in two bogs studied in different years (Horner 2014). Empirical studies of the development and activity of flowers and pitchers have been published only for *S. alata* (Horner 2014). However, synchrony in flowering and trap activity is commonly observed in greenhouses and has also been documented for natural populations (J. Horner *unpublished data*). As this could result in PPC, it has been suggested that different volatiles produced by the flowers and traps of *Sarracenia* may assist pollinators in distinguishing between the two organs (Ho et al. 2016). Volatile emissions differed significantly between the flowers and traps of *Heliamphora, Darlingtonia californica*, and 11 species of *Sarracenia*, although some overlap was detected in the compounds emitted (Ho et al. 2016). Pollinators exhibited different responses to floral and trap volatiles compared with prey insects (W. Ho *personal communication*), suggesting that volatiles may be used as cues by pollinators and prey and contribute to the avoidance of PPC under circumstances in which both active flowers and traps occur. Although it is possible that these two mechanisms evolved due to selection to avoid PPC, it is equally likely that they represent mechanisms to maximize the effectiveness of pollination or prey capture.

Heliamphora. All *Heliamphora* species are dichogamous and protogynous: the stigma ripens before the anthers mature. The majority of species have nectarless flowers (Renner 1989), but two species have nuptial nectaries on their petaloid tepals (Fleischmann et al. 2009). *Heliamphora* have poricidal anthers and are buzz-pollinated by large bees (especially *Xylocopa* spp.) and bumblebees (Renner 1989; Fleischmann and McPherson 2010), but ants comprise the majority of captured prey (McPherson 2008). The only case of floral visitors frequently caught as prey by *Heliamphora* are the introduced honey bees (*Apis mellifera*). They were observed collecting pollen leftovers from the ripe anthers after previous visits by buzz-pollinating bees, and also were trapped frequently in the pitchers (Fleischmann and McPherson 2010).

Darlingtonia. Bees are thought to be the primary pollinators of the large, fragrant flowers of *Darlingtonia californica* (Meindl 2009, Meindl and Mesler 2011). The pollination biology of *D. californica* remains somewhat unresolved, and unlikely pollinators such as spiders have been proposed (Juniper et al. 1989, Nyoka and Ferguson 1999, Rice 2011a). Meindl and Mesler (2011) recorded the generalist solitary bee *Andrena nigrihirta* as the main floral visitor responsible for cross-pollination at two study sites in California. However, breeding experiments showed that *D. californica* can also be facultatively autogamous, potentially facilitated by spiders and bees (Meindl and Mesler 2011).

22.3.3 The suction traps of *Utricularia*

The vivid flowers and nectar-producing spur of *Utricularia* (Lentibulariaceae) are traits generally considered adaptations to insect pollination. More than 50 species of bees, butterflies, moths, and flies visited the small but colorful flowers of the mass-flowering, terrestrial *U. albocaerulea,*

U. purpurascens, and *U. reticulata* (Hobbhahn et al. 2006). Although low pollen-ovule ratios in these species imply autogamy and *U. purpurascens* and *U. reticulata* are highly self-compatible, insect vectors are required for pollen transfer because of the spatial arrangement of the reproductive organs (Hobbhahn et al. 2006). However, some species of *Utricularia* appear to be obligately autogamous (Khosla et al. 1998) and several aquatic species have reduced pollen fertility and mainly propagate vegetatively (Casper and Manitz 1975). *Utricularia australis*, for example, an amphiploid species of hybrid origin, is male-sterile and populations propagate only vegetatively (Casper and Manitz 1975, Taylor 1989, Araki and Kadono 2003). Isoenzyme analyses suggest clonal dominance and low seed recruitment in *U. australis* (Araki and Kadono 2003). The submerged or subterranean traps of *Utricularia* are so spatially separated from the aerial inflorescences that PPC is irrelevant in this genus (§22.2.4).

22.3.4 Snap-traps

Little information is available on the pollination biology of *Dionaea muscipula*, although Juniper et al. (1989) indicate that the species may be capable of self-pollination. The species produces small amounts of nuptial nectar (A.R. Davis *unpublished data*) and may rely upon insects for pollen transfer, and extrafloral nectaries also occur in the traps. The aquatic *Aldrovanda vesiculosa* appears to be obligately autogamous, with self-pollination occurring in a proportion of flowers (40–60%) that possess curved styles that contact the stamens upon flower closure (Cross 2012a). The small, white flowers of *A. vesiculosa* are unscented and open at the water surface only for very short periods (2–3 hours; Cross et al. 2016), and no pollinator has yet been observed for the species (Cross 2012a).

22.3.5 Eel traps

Bees are among the main pollinators of *Genlisea* spp., although the pollination biology of very few *Genlisea* species has so far been studied *in situ* (Fleischmann 2012a, Aranguren Díaz 2016). The role of bees as pollinators fits with floral features such as zygomorphic masked flowers or snap dragon-type blossoms, and nectar secretion inside a spur of the corolla. As with all Lentibulariaceae, *Genlisea* are herkogamous, although no self-incompatibility mechanism in the genus is apparent as experimentally self-pollinated flowers produce fertile seeds (Fleischmann 2012a). Autonomous self-pollination has also been observed in some small-flowered annual species of *Genlisea* (Fleischmann 2012a). As with *Utricularia*, spatial separation of traps and flowers eliminates PPC.

22.4 Seed morphology, germination biology, and seed dormancy

Once flowers have been pollinated and fertilization has been successful, seeds are produced. With the exception of the large and unique seeds of *Triphyophyllum peltatum*, the majority of carnivorous genera produce seeds that are small to minute (<5.0 mg per seed—Table 22.1), and in almost all cases these are smaller than the median seed size of other members of their respective orders (Moles et al. 2005). Increasing seed size appears to be linked to larger growth form (Moles et al. 2005), and average seed size and mass in carnivorous genera broadly increases along a gradient of growth form from the diminutive *Genlisea, Philcoxia, Drosera*, and *Utricularia* to larger species from genera such as *Drosophyllum, Nepenthes, Roridula*, and *Sarracenia* (Table 22.1).

Available data suggests that physiological dormancy (PD) is likely in all carnivorous plant genera that produce seeds with fully developed embryos, that is, all genera except *Paepalanthus, Roridula, Darlingtonia, Heliamphora, Sarracenia, Genlisea*, and *Utricularia* (Table 22.1). Seeds of some species in genera with PD germinate readily when sown (e.g., some *Drosera, Dionaea*), suggesting no dormancy is present. However, there is a possibility that these seeds may instead have non-deep PD that is yet to be detected. Since non-deep PD can be alleviated during storage, germination tests need to be conducted on freshly matured seeds to ensure that the dormancy state is assessed correctly at the time of dehiscence (Baskin and Baskin 2014). Of the genera with underdeveloped embryos, seeds of *Heliamphora* germinate readily several weeks after dehiscence and thus appear to have morphological dormancy (MD), whereas seeds of *Darlingtonia californica* and *Sarracenia* do not germinate unless stratified and therefore have morphophysiological dormancy (MPD) (Table 22.2). Further work is required to determine the dormancy type present in the seeds of *Genlisea, Paepalanthus, Roridula*, and whether *Utricularia* have MD or MPD.

PD and MPD are commonly alleviated by either cold (0–10 °C) or warm (20–30 °C) moist stratification for several weeks, dry after-ripening (warm [*c.* 20–35 °C] dry storage of seeds) for durations ranging from weeks to several years, dry after-ripening with intermittent pulses of wetting and re-drying, or short durations (minutes to hours) of

Table 22.1 Seed characteristics of carnivorous plant species with regard to: seed mass (mg), seed length and width (mm), embryo type, embryo development, and dispersal mechanisms (A = anemochory; B = barochory; H = hydrochory; Z = zoochory). ^—seed mass available for only a single representative species.

Order/Family/ Genus	Mass per seed (mg)	Seed length (mm)	Seed width (mm)	Embryo type	Embryo development	Dispersal mechanism	References*
Poales							
Bromeliaceae							
Brocchinia	?	3.0–4.0	0.5–1.0	Broad or linear	Fully developed	A, H	4, 9
Catopsis	0.924	4.0–6.0	0.6–1.0	Broad or linear	Fully developed	A, H	9, 15, 26
Eriocaulaceae							
Paepalanthus	?	?0.5	?0.5	?Basal	?Underdeveloped	A, H	11
Aldrovanda	0.82	1.2–1.6	0.8–1.0	Rudimentary	Fully developed	H, Z	19, 21
Droseraceae							
Drosera	0.006–0.247	0.1–3.0	0.1–2.0	Dwarf	Fully developed	B, A, H	3, 20, 22, 26, 27
Dionaea	0.351	1.5–2.5	1.0–2.0	Rudimentary	Fully developed	B	1, 20, 26
Caryophyllales							
Drosophyllaceae							
Drosophyllum	3.161–4.400	2.0–2.5	1.0–2.0	Linear	Fully developed	B	6, 24, 25
Nepenthaceae							
Nepenthes	0.074–0.259	3.0–25.0	0.1–1.0	Linear	Fully developed	A, B	2, 20, 26, 27
Dioncophyllaceae							
Triphyophyllum	450	50–120	50–120	Spatulate	Fully developed	A	5, 7
Oxidales							
Cephalotaceae							
Cephalotus	0.094	4.0–6.0	3.0–5.0	Linear	Fully developed	A, H	20, 26
Roridulaceae							
Roridula	1.136^	2.0–5.0	1.0–2.0	Linear	?Underdeveloped	B	12, 20, 26
Ericales							
Sarraceniaceae							
Darlingtonia	0.243	2.0–3.5	1.0–1.5	Linear	Underdeveloped	B, ?Z	20, 23, 26
Heliamphora	0.315^	2.0–4.0	2.0–3.0	Linear	Underdeveloped	A, H	17, 20, 26
Sarracenia	0.428–1.880	1.2–2.7	0.8–1.8	Linear	Underdeveloped	B	10, 20, 26, 27

(*Continued*)

Table 22.1 (*Continued*)

Order/Family/ Genus	Mass per seed (mg)	Seed length (mm)	Seed width (mm)	Embryo type	Embryo development	Dispersal mechanism	References*
Byblidaceae							
Byblis	0.054–0.311	0.3–1.1	0.3–0.5	Linear	Fully developed	B	13, 22
Lamiales							
Plantaginaceae							
Philcoxia	0.010^	0.3–0.5	0.3–0.5	?	?	A, H	8, 16, 26
Lentibulariaceae							
Genlisea	0.025^	0.1–0.6	0.1–0.4	Dwarf	Underdeveloped	B, A, H	18, 26
Pinguicula	0.018–0.047	0.4–1.0	0.1–0.6	Linear	Fully developed	B, A, H	3, 14, 26, 27
Utricularia	0.003–0.767	0.1–2.5	0.1–2.5	Dwarf	Underdeveloped	B, H, A, ?Z	3, 20, 26, 27

*References: 1. Smith (1931); 2. Corner (1976); 3. Dwyer (1983); 4. Varadarajan and Gilmartin (1988); 5. Takhtajan (1992); 6. Takhtajan (1996); 7. Bringmann et al. (1998); 8. Taylor et al. (2000); 9. Benzing (2000); 10. Ellison (2001); 11. Scatena and Bouman (2001); 12. Conran (2004b); 13. Conran and Carolin (2004); 14. Degtjareva et al. (2004); 15. Palací et al. (2004); 16. Fritsch et al. (2007); 17. Fleischmann et al. (2009); 18. Fleischmann (2012a); 19. Cross (2012a); 20. Baskin and Baskin (2014); 21. Cross et al. (2016); 22. Cross et al. (2013); A.T. Cross (*unpublished data*); 23. A.T. Cross (*unpublished data*); 24. Salces-Castellano et al. (2016); 25. M. Paniw (*unpublished data*); 26. S.R. Turner (*unpublished data*); 27. Royal Botanic Garden Kew (2017).

Table 22.2 Seed characteristics of carnivorous plants with regard to: dormancy class (ND = non-dormant; PD = physiological dormancy; MD = morphological dormancy; MPD = morphophysiological dormancy), light requirements for germination, dormancy alleviation treatment (WS = warm stratification; CS = cold stratification; AR = after-ripening; TP = thermic pulsing; SC = scarification), germination stimulants (E = ethylene; K = karrikinolide; S = smoke; N = nitrate), and longevity in seed banks.

Order/Family/Genus	Dormancy class	Light requirement	Dormancy alleviation treatment	Germination stimulant	Longevity in seed bank	References*
Poales						
Bromeliaceae						
Brocchinia	PD?	Yes	?	?	?Months	14
Catopsis	PD?	Yes	?	?	?Months	10, 14
Eriocaulaceae						
Paepalanthus	MD/MPD?	?	?	?	Months/?Years	14
Aldrovanda	PD	Yes	CS	E	Months/?Years	12, 15
Droseraceae						
Drosera	ND/PD	Yes	WS, CS, AR	S	Years/Decades	14, 18
Dionaea	PD	Yes	–	–	Months/?Years	9, 11
Caryophyllales						
Drosophyllaceae						
Drosophyllum	PD	No	TP, AR, SC	S	Years/Decades	20, 18
Nepenthaceae						
Nepenthes	ND/PD	Yes	AR, CS	?	Weeks; Months	3, 14
Dioncophyllaceae						
Triphyophyllum	PD	No	–	–	Weeks/Months	2, 6, 14
Oxidales						

(*Continued*)

Table 22.2 (*Continued*)

Order/Family/Genus	Dormancy class	Light requirement	Dormancy alleviation treatment	Germination stimulant	Longevity in seed bank	References*
Cephalotaceae						
Cephalotus	PD	No	–	S	Weeks/Months	14, 19
Roridulaceae						
Roridula	MD/MPD?	No	SC	S	?	14
Ericales						
Sarraceniaceae						
Darlingtonia	MD	Yes	–	–	?	14
Heliamphora	MD	Yes	–	–	?	14
Sarracenia	MPD	No	CS	S	Months/?Years	8, 14
Byblidaceae						
Byblis	PD	No	WS, AR	K, S, E	Years/Decades	16
Lamiales						
Plantaginaceae						
Philcoxia	?	?	?	S?	?Years	
Lentibulariaceae						
Genlisea	MD/MPD	Yes	AR	–	Months/?Years	14, 13
Pinguicula	PD	No	CS	–	Months	4, 5, 14
Utricularia	MD/MPD	Yes	CS, ?AR	E, N	Weeks/Years	7, 14, 1, 17

*References: 1. Kondo (1971); 2. Green et al. (1979); 3. Corker (1986); 4. Maas (1989); 5. Studnička (1993); 6. Bringmann et al. (1998); 7. Brewer (1999a); 8. Ellison (2001); 9. Schnell (2002); 10. Winkler et al. (2005); 11. Bailey (2008); 12. Cross (2012a); 13. Fleishmann (2012a); 14. Baskin and Baskin (2014); 15. Cross et al. (2016); 16. A.T. Cross (*unpublished data*); 17. A.T. Cross (*unpublished data*); 18. A.T. Cross (*unpublished data*); 19. D.J. Merritt (*unpublished data*); 20. M. Paniw (*unpublished data*).

exposure to higher temperatures between 60–120 °C (Bewley et al. 2013, Baskin and Baskin 2014). Many cool temperate species require a period of cold stratification for dormancy alleviation (Table 22.2), including all *Sarracenia* spp. and some species of *Drosera, Pinguicula,* and *Utricularia* (Baskin and Baskin 2014). Warm stratification alleviates dormancy in a number of species from Mediterranean climates (Table 22.2), including species of *Drosera* and *Byblis* (Cross et al. 2013; A.T. Cross *unpublished data*). After-ripening has been shown to alleviate dormancy in seeds of some species from subtropical regions with long, hot dry seasons (Table 22.2), such as *Byblis* (A.T. Cross *unpublished data*) and *Nepenthes mirabilis* (Corker 1986). Two techniques that can alleviate PD/MPD—cycles of wetting and drying and exposure of seeds to high temperatures for short periods—have not yet been tested widely on carnivorous plant seeds, although high temperatures readily promote dormancy loss in *Drosophyllum lusitanicum* seeds (A.T. Cross *unpublished data*).

Once dormancy has been alleviated, seeds remain quiescent until exposed to environmental conditions suitable for germination, which may include a requirement for an exogenous chemical stimulant such as smoke, nitrates, or ethylene (Baskin and Baskin 2014). Smoke products including aerosol smoke, smoke water, and karrikinolide (KAR_1) act as germination stimulants for many carnivorous species from fire-prone habitats (Table 22.2), particularly those of *Drosera* spp. and *Byblis* spp. (Cross et al. 2013, A.T. Cross *unpublished data*), and perhaps *Drosophyllum, Cephalotus,* and *Roridula* spp. Ethylene gas and nitrate (NO_3^-) are common germination stimulants for species with PD from freshwater habitats (Baskin and Baskin 2014, Cross et al. 2014, 2015), and a germination response to ethylene has been recorded for tropical *Byblis* (Cross et al.

unpublished data), *Aldrovanda* (Cross et al. 2016), and *Utricularia vulgaris* (A.T. Cross *unpublished data*). Nitrate appears to be a germination stimulant at least for some *Utricularia* spp. (Kondo 1971).

Many shorter-lived, fire-responsive species such as *Drosophyllum* and *Byblis gigantea* (Cross et al. 2013, Salces-Castellano et al. 2016, A.T. Cross *unpublished data*), and annual taxa from seasonally dry regions in genera such as *Byblis, Drosera, Philcoxia*, and *Utricularia* form a persistent soil seed bank (Table 22.2). It is uncertain to what extent species from relatively stable mesic habitats, including *Cephalotus, Darlingtonia, Dionaea*, and some *Sarracenia, Pinguicula*, and *Drosera* species form persistent soil seed banks. Species from tropical and humid environments rarely form persistent soil seed banks (Baskin and Baskin 2014), and it appears that the representatives of carnivorous genera in these regions usually produce seeds that are quite short-lived (*Brocchinia, Catopsis, Genlisea, Heliamphora, Nepenthes, Triphyophyllum, Utricularia*).

22.4.1 Bromeliaceae

Brocchinia spp. produce small, caudate, endospermous seeds (Table 22.1), with a homogeneous surface cellular pattern and dissimilar apical and basal ends (Varadarajan and Gilmartin 1988). Their broad or linear embryo occupies roughly one-quarter to one-third of the seed length (Benzing 2000) and is likely to be fully developed at maturity (Baskin and Baskin 2014). The seeds of *B. reducta* are somewhat flattened apically, and longitudinally divided into 4–8 finger-shaped basal lobes (Varadarajan and Gilmartin 1988). The seeds of *Brocchinia* appear suited to anemochory (wind dispersal) or hydrochory (water dispersal) on the basis of seed morphology (Varadarajan and Gilmartin 1988, Benzing 2000, Palací et al. 2004). No data are available on seed germination biology or seed longevity in the soil seed bank for *Brocchinia*, but given its tropical distribution, the seeds probably are short-lived.

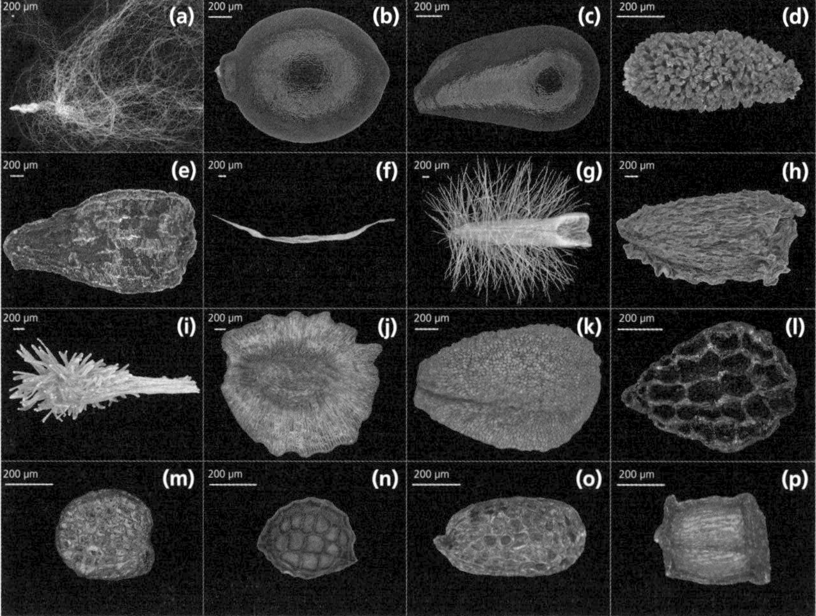

Figure 22.3 (Plate 15 on page P13) Comparative seed morphology of representative carnivorous plants in the Poales (**a**), Caryophyllales (**b–f**), Oxidales (**g**), Ericales (**h–k**), and Lamiales (**l–p**). (**a**) *Catopsis berteroniana*; (**b**) *Aldrovanda vesiculosa*; (**c**) *Dionaea muscipula*; (**d**) *Drosera intermedia*; (**e**) *Drosophyllum lusitanicum*; (**f**) *Nepenthes mirabilis*; (**g**) *Cephalotus follicularis* (fruit [indehiscent follicle, which acts as a diaspore] shown); (**h**) *Roridula gorgonias*; (**i**) *Darlingtonia californica*; (**j**) *Heliamphora heterodoxa*; (**k**) *Sarracenia rubra*; (**l**) *Byblis filifolia*; (**m**) *Philcoxia minensis*; (**n**) *Genlisea hispidula*; (**o**) *Pinguicula vulgaris*; (**p**) *Utricularia vulgaris*. All bars = 200 µm. All images by D.R. Symons.

Catopsis seeds are subglobose and endospermous (Fig. 22.3a), possibly with a fully developed broad or linear embryo (Table 22.1), and possess a large plumose appendage comprised of elongated cells (hairs with hooked ends) that extend from the seed apex (Benzing 2000, Palací et al. 2004). Seeds of epiphytic *Catopsis* appear well suited to anemochory, and the plumose appendages appear to be adapted specifically to disperse into aerial substrates (Benzing 2000). No data are available on seed germination biology or seed longevity in *C. berteroniana*. However, seeds of the epiphytic *C. sessilifolia* from humid montane forest in México have PD and appear to be desiccation sensitive (Winkler et al. 2005), and thus *C. berteroniana* seeds probably also have PD and are likely to be short-lived.

22.4.2 Eriocaulaceae

No data are available for the seeds or seed biology of *Paepalanthus bromelioides* (Table 22.1). Scatena and Bouman (2001) describe undifferentiated embryos in the seeds of members of *P.* sect. *Actinocephalus*, and thus *P. bromelioides* seeds may have MD or MPD. Other studied species of *Paepalanthus* produce very small seeds (<0.5 × 0.5 mm), but seed longevity in the soil seed bank remains completely unknown (Tarabini et al. 2001).

22.4.3 Droseraceae

Aldrovanda vesiculosa produces small, ovoid, endospermous seeds (Fig. 22.3b), with a small basal fully developed embryo (Table 22.1). Seeds possess a highly complex honeycomb-like exotesta hypothesized to provide lightweight structural support and assist in gas exchange (Cross 2012a, Cross et al. 2016), and are anemochorous or possibly zoochorous (animal dispersed) via water birds (Cross 2012a). Once released from the fruit, seeds sink to the sediment (Cross 2012a), and studies indicate that at least some seeds remain viable in the sediment seed bank for up to a year or more (Cross et al. 2016, L. Adamec *unpublished data*). Seeds of *A. vesiculosa* have PD, with dormancy alleviated by cold stratification followed by incubation at 25 °C; ethylene gas also acts as a germination stimulant under some conditions (Cross et al. 2016). Ethylene is produced by microbes in the sediment, usually in response to wetting and drying events (Cross et al. 2013), and germination of *A. vesiculosa* seeds probably occurs in spring or summer as the water temperature rises and light becomes more intense following winter stratification (Cross 2012a). The species most frequently grows in the shallow margins of wetlands where water levels are likely to fluctuate seasonally (Cross 2012a), and significant recruitment from the seed bank following the re-flooding of sediments that had been dry for weeks to months has been observed both in *ex situ* cultivated populations (A.T. Cross *unpublished data*), and at a natural site in Hungary (L. Adamec *unpublished data*).

Dionaea muscipula seeds are ovate to pyriform (Fig. 22.3c), with a small, fully developed embryo surrounded by copious endosperm and a smooth testa (Smith 1931, Dwyer 1983, S.R. Turner *unpublished data*). The seeds of *Dionaea* are dispersed by barochory (Table 22.1), and germinate within several weeks of dehiscence (Smith 1931, Schnell 2002, Bailey 2008). *Dionaea* seeds appear to have PD and are conditionally dormant at release. They appear to require light to germinate, and fail to germinate if covered with soil (Schnell 2002). The longevity of *Dionaea* seeds in the soil seed bank remains unknown. Pietropaulo and Pietropaulo (1986) found that seeds remained viable for less than 12 months when stored at low temperatures, and Roberts and Oosting (1958) reported germination slowly reduced to zero over 100 days of storage at room temperature. In contrast, Schnell (2002) reported that seeds stored in a sealed jar in a refrigerator remained viable for at least five years.

Seed morphology and biology of the diversity of *Drosera* species varies markedly. The seeds of *Drosera* can be fusiform, ovoid, linear, obovate, filiform, ellipsoid, subglobose, nail-like, or oblong, and the seed coat may be reticulate, papillose, pitted, or smooth (Dwyer 1983, Lowrie 2013, A.T. Cross *unpublished data*). Seed size and mass are highly variable (Table 22.1). Seeds possess a small but fully developed embryo surrounded by copious endosperm (Baskin and Baskin 2014, Cross et al. *unpublished data*). Seeds of many species have

PD, although those of a number of species appear to be ND or to have non-deep PD (Baskin and Baskin 2014, A.T. Cross *unpublished data*). Dormancy is alleviated in species from temperate regions by cold stratification (Grime et al. 1981, Burdic 1985, Crowder et al. 1990, Baskin et al. 2001), and in many species in *D.* sects. *Erythrorhiza* and *Stolonifera* from the Mediterranean southwest of Western Australia, by warm stratification (A.T. Cross *unpublished data*). After-ripening appears to alleviate dormancy in seeds of at least some species in *D.* sects. *Lasiocephala, Arachnopus,* and *Thelocalyx* from tropical northern Australia (A.T. Cross *unpublished data*). However, species from these groups appear to have deeply dormant seeds that can be difficult to germinate under *ex situ* conditions (Jayaram and Prasad 2006, A.T. Cross *unpublished data*). Most species in *D.* sect. *Ergaleium* appear to be ND, but some populations produce dormant seeds and are likely to have non-deep PD (A.T. Cross *unpublished data*).

Most *Drosera* seeds germinate readily once dormancy has been alleviated, but KAR_1 stimulates germination in at least some species from fire-prone habitats in southwestern Australia (A.T. Cross *unpublished data*). Further studies of seed dormancy and the germination biology of *Drosera* from sections such as *D.* sects. *Bryastrum* and *Drosera* are required. Seeds of most *Drosera* species are dispersed by barochory (e.g., the large unadorned seeds in *D.* sects. *Arachnopus, Arcturia, Erythrorhiza,* and *Stolonifera*). However, on the basis of size and morphology, the seeds of species in *D.* sect. *Drosera* such as *D. capensis* and the filiform seeds of species in *D.* sect. *Ergaleium* may be anemochorous (Fleischmann et al. 2007; Lowrie 2013). Several species from the Guayana Highlands of Venezuela appear to be hydrochorous, including the ombrochorous *D. felix, D. kaieteurensis,* and *D. solaris* (Fleischmann et al. 2007) and the hydrochorous *D. amazonica* from the northern Amazon basin (Rivadavia et al. 2009). Little information is available on seed longevity in *Drosera* from regions outside of Australia. However, most Australian species appear to form a persistent seed bank (Lowrie 2013, A.T. Cross *unpublished data*).

22.4.4 Drosophyllaceae

The seeds of *Drosophyllum lusitanicum* are pyriform and endospermous (Fig. 22.3e), with a small fully developed capitate embryo and a thick testa (Kubitzki 2003a, A.T. Cross *unpublished data*). Seeds are barochorous and disperse only short distances (Salces-Castellano et al. 2016). The species is a short-lived perennial (1–4 years; M. Paniw *unpublished data*) and has a persistent and extensive soil seed bank in its natural heathland habitats (Correia and Freitas 2002, Paniw et al. 2015, Salces-Castellano et al. 2016). Seeds of *Drosophyllum* have PD (Table 22.2), and dormancy is alleviated by exposure of seeds to high temperatures or after-ripening (A.T. Cross *unpublished data*). Scarification results in a high germination percentage (Gonçalves and Romano 2009, A.T. Cross *unpublished data*), suggesting that the embryo has a low growth potential (Baskin and Baskin 2014). Germination appears to be stimulated by smoke water and KAR_1, supporting observations of widespread recruitment following wildfire in natural populations (M. Paniw *unpublished data*, A.T. Cross *unpublished data*). Ziemer (2012) reports that *Drosophyllum* seeds retained viability and germinated after 22 years of storage in a refrigerator, and seeds may persist for several decades in the soil seed bank between disturbance events.

22.4.5 Nepenthaceae

Although seed size varies markedly within *Nepenthes* (Table 22.1), most species produce apically and basally elongated slender and filiform (truncate or ovoid) seeds (Fig. 22.3f), with a reduced testa (Corner 1976, Kubitzki 2003b). Seeds are endospermous (except at least *N. lowii*; Kaul 1982), with a minute but fully developed linear embryo (Corner 1976, Kubitzki 2003b). The seed morphology of all species, except the Seychelles endemic *N. pervillei* (in which the apical and basal extensions are almost completely reduced; Dwyer 1983), clearly implies anemochory (Ridley 1905, McPherson 2009). Few studies assessing the germination biology in *Nepenthes* have been undertaken, but it appears that seeds have PD or are occasionally ND and require light for germination (Green 1967, Ah-Lan and Prakash 1973, Wee 1978, Corker 1986). After-ripening alleviates dormancy in seeds of *N. mirabilis* (Corker 1986), and seeds of *N. gracilis* and *N. rafflesiana* appear to be ND or have weak PD (Green 1967, Ah-Lan and Prakash 1973, Wee 1978). The seeds of *N. gracilis* appear to be recalcitrant (desiccation sensitive *sensu* Farnsworth 2000), and it is likely that many *Nepenthes* from the humid tropics produce recalcitrant or very short-lived seeds.

22.4.6 Dioncophyllaceae

The seeds of *Triphyophyllum peltatum* are disciform and extremely large (Table 22.1), are borne on long, thickened funiculi, and ripen externally from the fruit (Bringmann et al. 1998). The embryo is fully developed (Baskin and Baskin 2014), large (≈10 mm in diameter), discoid obconicular or spatulate, and surrounded by copious endosperm (Takhtajan 1992, Bringmann et al. 1998). Most of the seed's surface area comprises a large papery wing, and seeds are dispersed considerable distances anemochorously (Bringmann et al. 1998). Light is not required for germination, which occurs several weeks to months after sowing (Bringmann et al. 1998). The seeds of *T. peltatum* therefore probably have PD, with germination delayed by low embryo growth potential (*sensu* Baskin and Baskin 2014). Seeds are released during the wet season in May and germinate at the peak of summer rains in July (Green et al. 1979). Since other species of tropical lianas from similar humid evergreen forest habitat produce recalcitrant seeds (Baskin and Baskin 2014), the seeds of *T. peltatum* are unlikely to be long-lived in the soil seed bank.

22.4.7 Cephalotaceae

Cephalotus follicularis produces hairy, indehiscent follicles (Fig. 22.3g), each containing a single, small, ovoid seed (Conran 2004a). The embryo is small and linear but fully developed (Table 22.1), and is surrounded by copious endosperm (Conran 2004a, D.J. Merritt *unpublished data*). Seeds are particularly lightweight relative to the average seed mass for Oxidales (Moles et al. 2005), and are likely to be anemochorous or hydrochorous with dispersal being facilitated by hairs. Seeds of *Cephalotus* appear to have PD and do not require light to germinate, but the dormancy-breaking requirements are not fully understood (D.J. Merritt *unpublished data*). Seeds have been observed to germinate readily on water agar without requiring cold stratification and germination appears to be stimulated by KAR_1, but the depth of seed dormancy appears to vary markedly between years (D.J. Merritt *unpublished data*). Seeds appear to be desiccation sensitive (D.J. Merritt *unpublished data*), and *Cephalotus* is unlikely to form a persistent soil seed bank.

22.4.8 Roridulaceae

The seeds of *Roridula* are small (Table 22.1), ellipsoid to pyriform (Fig. 22.3h), and have a thick and warty or smoothly reticulate testa (Conran 2004b). The linear embryo occupies ≈60% of the seed length (Diels 1928), is surrounded by a copious endosperm (Conran 2004b), and is probably underdeveloped (Baskin and Baskin 2014). Seeds are dispersed barochorously, but at dehiscence they possess a mucilaginous coating that may assist in zoochorous dispersal. Little empirical evidence is available on the seed dormancy and germination biology of either *Roridula* species, but both have a persistent soil seed bank and significant recruitment events occur only following fires (B. Anderson *unpublished data*). Since anecdotal evidence suggests that seeds of both species germinate several weeks after nicking of the seed coat (D'Amato 2013) and germination of *R. dentata* seeds appears to be cued by smokey water (A.T. Cross *unpublished data*), *Roridula* may have MPD with dormancy alleviated by after-ripening or exposure of seeds to high temperatures and germination stimulated by smoke-derived compounds.

22.4.9 Sarraceniaceae

Darlingtonia californica produces small, elliptic seeds with an elongated and truncated basal end and numerous short apical projections (Fig. 22.3i). Seeds have an underdeveloped linear embryo (Baskin and Baskin 2014). Some authors suggest a period of cold stratification prior to sowing seeds in cultivation (McPherson and Schnell 2011, D'Amato 2013), but seeds do not require stratification and germinate readily at cooler temperatures after several weeks (Schnell 2002). Seeds probably therefore have MD (Table 22.2). As *Darlingtonia* grows predominantly in seeps and streams in running water (Schnell 2002), the apical projections of the testa may facilitate hydrochory. However, it also has been postulated that these projections assist in zoochory (specifically by bears), and that long-distance zoochorous dispersal may explain the fragmented distribution of the species (Collingsworth 2015).

The seeds of *Heliamphora* are small and ovoid to obovoid (Fig. 22.3j), with an exotestal seed coat and often a large membranous wing (Fleischmann et al. 2009). They have a minute, underdeveloped, linear embryo surrounded by copious endosperm (Fleischmann et al. 2009, Baskin and Baskin 2014), but lack dormancy at maturity and germinate within several weeks of dehiscence (Pietropaulo and Pietropaulo 1986, D'Amato 2013). *Heliamphora* therefore appears to have MD (Table 22.2). Seeds most likely are dispersed anemochorously or hydrochorously, but little information is available on seed dispersal or seed longevity in *Heliamphora*.

Sarracenia produce numerous small, obovoid, endospermous seeds (Fig. 22.3k) that have an underdeveloped, linear embryo that fills one-third to half of seed volume (Ellison 2001, Baskin and Baskin 2014). Seeds of all species are dormant at maturity and have MPD (Table 22.1). Seeds require a period of cold stratification to alleviate dormancy, although the period of required stratification decreases markedly along a latitudinal gradient from up to six weeks in northern areas of the United States to only two weeks in the southeast (Ellison 2001). However, cold stratification alone resulted in the germination of only 40–80% of viable seeds tested in all species (Ellison 2001), suggesting that additional environmental cues are involved in regulating seed germination. Fire plays a role in the population dynamics of many *Sarracenia* (Weiss 1980, Barker and Williamson 1988, Brewer 1999c), and exposure of seeds to high temperatures or smoke-derived compounds may provide additional germination stimuli. *Sarracenia* are dispersed barochorously, and median dispersal distance in *S. purpurea* is only 5 cm (Ellison and Parker 2002). No information is available of the longevity of *Sarracenia* seeds in the soil seed bank.

22.4.10 Byblidaceae

The seeds of *Byblis* are small and angular to ovoid (Fig. 22.3l), with a heavily sculptured testa (Conran and Carolin 2004, Cross et al. 2013, A.T. Cross *unpublished data*). They have an axile, fully developed, linear embryo occupying roughly one-quarter of the seed volume (Table 22.1). *Byblis* seeds have PD, and dormancy is alleviated by warm stratification in species from the Mediterranean southwest of Western Australia (Cross et al. 2013), and by after-ripening in species from tropical northern Australia (A.T. Cross *unpublished data*). Germination in all species is stimulated by KAR_1, and ethylene gas also is a germination cue for the northern Australian species (Cross et al. 2013, A.T. Cross *unpublished data*). Seeds are dispersed barochorously, and all species have a persistent soil seed bank. Although recruitment occurs each season for annual *Byblis* species, seedling emergence appears to be significantly greater in seasons following dry-season fires (Cross et al. 2013, A.T. Cross *unpublished data*). Recruitment in *B. gigantea* occurs almost exclusively following fire (Cross et al. 2013). Ziemer (2012) reports that *B. gigantea* seeds retained viability and germinated after 22 years of storage in a refrigerator, and it is likely that the seeds of *Byblis*, particularly the perennial subshrubs *B. gigantea* and *B. lamellata*, are capable of persisting in the soil seed bank for decades (Cross et al. 2013).

22.4.11 Plantaginaceae

Seeds have not been observed for some species of *Philcoxia* (Taylor et al. 2000, Scatigna et al. 2015), but those of *P. goiasensis* and *P. minensis* are tiny (Table 22.1), ovoid to broadly ellipsoid (Fig. 22.3m), with a longitudinally ribbed, foveate-reticulate testa (Taylor et al. 2000, Fritsch et al. 2007, A.V. Scatigna *unpublished data*). No information is available on the embryos and seed dormancy of *Philcoxia*, although

PD is common in Plantaginaceae (Baskin and Baskin 2014). Based on their size and mass, seeds are likely to be dispersed anemochorously or hydrochorously. At least some species (i.e., the annual species *P. maranhensis* and *P. goiasensis*) are likely to have a persistent soil seed bank.

22.4.12 Lentibulariaceae

The seeds of *Genlisea* are tiny (Table 22.1), prismatic, pyramidal, globose, or obovoid (Fig. 22.3 n), and generally have a smooth, foveate, or reticulate testa (Fleischmann 2012a). They have virtually no endosperm, and the dwarf embryo appears to be undifferentiated (sometimes poorly differentiated) and occupies most of the seed volume (Fleischmann 2012a). *Genlisea* seeds are dispersed hydrochorously or anemochorously (Fleischmann 2012a). Seeds are photophilous, and those of perennial species germinate readily in one to three months after seeds are released onto wet soil (A. Fleischmann *unpublished data*). Annual species have either PD or MPD alleviated by after-ripening, and require a period of 6–12 months dry storage to alleviate dormancy prior to germination (Fleischmann 2012a). Perennial species do not produce a persistent soil seed bank, and the seeds of annual species are relatively short-lived.

Most *Pinguicula* species produce small, ellipsoidal or linear seeds (Fig. 22.3 o), with a foveate to reticulate testa (Dwyer 1983, Degtjareva et al. 2004). The embryo in all species for which information is available is large, fully developed, linear, and surrounded by only a very thin endosperm layer (Degtjareva et al. 2004). Seeds are most likely anemochorous and hydrochorous (Table 22.1). *Pinguicula* species have variable numbers of cotyledons, including monocotyly (Degtjareva et al. 2004). Seeds generally are dormant at maturity (Maas 1989, Studnička 1993, Baskin and Baskin 2014) and thus have PD. Dormancy at least in temperate *Pinguicula* is alleviated by cold stratification, and seeds germinate readily once dormancy has been alleviated with no light requirement (Maas 1989, Studnička 1993). Seeds of temperate species appear to be somewhat desiccation sensitive and probably do not form a persistent soil seed bank. Studnička (1993) reports that seeds of several European species lost viability over seven months of storage and were inviable by 14 months, but seeds remained viable after three and for up to five years of dark storage in water. Little information is available on the germination requirements of *Pinguicula* from non-temperate regions, but seed longevity in these species also appears to be less than six months (A. Fleischmann *unpublished data*).

Utricularia is a large genus with representatives from many different climatic regions, and seed morphology and biology vary markedly among species. Most have small to very small (Table 22.1), subglobose, ovoid, cylindrical, or prismatic (Fig. 22.3p) seeds with a reticulate, rugose, or striate testa sometimes with numerous papillae or hair-like extensions (Taylor 1989). The dwarf embryo is undifferentiated and occupies most of the seed volume (Kondo et al. 1978, Taylor 1989, Płachno and Świątek 2010), and seeds are dormant at release (Kondo 1971, A.T. Cross *unpublished data*). *Utricularia* seeds have MD or MPD, with dormancy apparently alleviated in species from temperate regions by cold stratification (Baskin and Baskin 2014), and in several species from northern Australia apparently by after-ripening (A.T. Cross *unpublished data*). Some species in *U.* sect. *Iperua* appear to be viviparous (Taylor 1989). Dispersal in most species occurs barochorously, with some displaying adaptations for hydrochory, anemochory, or zoochory (Taylor 1989). Germination in *U. vulgaris* and several species from northern Australia is stimulated by exposure to ethylene gas (A.T. Cross *unpublished data*), and Kondo (1971) reports that germination of *U. juncea* and *U. cornuta* seeds was stimulated to high levels only when seeds were plated on Moore's Solution (which contains 500 mg/L of NH_4NO_3). Many species of *Utricularia* produce a persistent soil seed bank, particularly annual taxa from seasonally wet habitats and tropical regions such as northern Australia (Brewer 1999a, A.T. Cross *unpublished data*).

22.5 Conservation seed banking

Ex situ seed banking is the storage of seeds for conservation, horticulture, and restoration (Smith et al. 2003). The time between collection and use depends on user requirements, but it also is governed by seed lifespan, which is specific to each species (and seed lot) and varies by at least four orders of magnitude. The longevity of a seed collection also is governed by the quality of the material collected, and by storage conditions. Temperature and relative humidity (RH) must be carefully manipulated to achieve specific conservation goals. Seed banks work most effectively in storing species with orthodox seeds (seeds that are desiccation tolerant and tolerant to storage at sub-zero temperatures). Most carnivorous species that have persistent soil seed banks or exhibit some form of seed dormancy are likely to produce orthodox seeds, and studies to date indicate or imply that orthodox seeds are produced by *Aldrovanda* (Cross et al. 2016), *Byblis* (A.T. Cross *unpublished data*), *Drosera* (A.T. Cross *unpublished data*), and *Drosophyllum* (Gonçalves and Romano 2009).

Seed moisture content influences all key processes associated with seed aging, dormancy, and germination. Seeds gain or lose moisture depending on the RH of the surrounding air, allowing for effective manipulation of seed moisture content by managing ambient RH (Gold 2008). Seeds age rapidly at high moisture levels, and for orthodox seeds, moisture levels need to be reduced for low-temperature (including sub-zero) storage that increases longevity. Seed moisture content equilibrates at approximately 12% of fresh weight at 50% RH, and approximately 5% at 10–15% RH. A "safe" moisture level to prevent rapid aging is at 50% RH or below, although most orthodox seeds tolerate desiccation to 10–15% RH and should be dried to this level for longer-term storage.

The temperature at which seeds are stored depends on the desired storage life of the seed collection and the facilities available. Large walk-in freezers operating at temperatures of −18 °C or lower, or even cryostorage in liquid nitrogen (−196 °C; Martyn et al. 2009), are preferred for long-term conservation collections. These are the standard conditions for active large-scale seed banks such as the Svalbard Global Seed Vault and the Millennium Seed Bank at Kew. However, other storage environments suitable for shorter-term storage include refrigerators (0–5 °C), and commercial chest freezers (−18 °C).

To store seeds with short lifespans, including those of *Brocchinia, Catopsis, Genlisea, Heliamphora, Nepenthes, Triphyophyllum*, and *Utricularia*, it is critical to place them under optimal conditions as soon as possible after collection. Suboptimal storage conditions can result in rapid viability loss in just a few weeks for some species (Martyn et al. 2009). Seeds should be collected when fully mature and on the cusp of dehiscing and then stored in cool, dry conditions (15 °C and 15% RH), ideally through the use of a controlled environment drying room, or, for small collections, over silica gel for several weeks. If subsequently sealed inside an airtight container (e.g., laminated foil bags) at 5 °C or −18 °C, they can be kept for several years without a significant risk of viability loss. Beyond this point, however, there may be some viability decline, and regular germination testing should be done to manage risk of loss of important conservation and horticultural collections. To store longer-lived seeds, especially those of *Byblis, Drosera, Drosophyllum*, and *Roridula*, air-drying (at ≤ 50% RH) upon collection should be sufficient if seeds are to be used within 1–2 years (indeed, this "moderate" level of drying may be beneficial if seeds are dormant and respond to after-ripening) (Martyn et al. 2009, Ziemer 2012, Cross et al. 2013). *Aldrovanda vesiculosa* seeds have been described as desiccation tolerant, but they also have been reported to lose viability rapidly when stored at −18 °C and were dead after storage for only three months under these conditions (Cross et al. 2016). Seeds held at 15 °C lost viability more gradually, but < 15% of seeds showed embryo growth after 12 months storage (Cross et al. 2016). Seeds of *Cephalotus follicularis* also lose viability rapidly when desiccated and stored at −18 °C (D.J. Merritt *unpublished data*). However, *Cephalotus* seeds stored for several years at −18 °C but without prior desiccation germinated readily (A. Shade *personal communication*).

22.6 Future research

Available data on pollinator–prey overlap suggest that prey and pollinators almost never overlap, the risk of pollinators being trapped is generally relatively minimal, and that PPC is rare or non-existent.

Contributions from disciplines such as plant reproductive biology, pollination biology, neuroethology, sensory ecology, and mathematical modeling should be integrated to determine fitness consequences of traits (and trait variation) such as phenology, morphology, and signaling cues that attract prey and pollinators and whether they are adaptions for avoiding PPC.

In the few cases where pollinators are attracted to and captured by traps, the hypothesis that trapping of pollinators is an evolutionarily stable strategy only if fitness is affected more by (macro-) nutrient deficiency than by pollen or pollinator limitation needs to be tested.

Detailed empirical studies into the seed dormancy and germination biology of the majority of carnivorous species are lacking. Future research should focus on rare or threatened taxa and elucidate the germination requirements of species from regions where ongoing environmental change (Chapter 28) is altering important ecological processes such as hydrology and fire regimes. There is still much to be learned about *ex situ* seed storage of carnivorous plant seeds. Given the increasing number of taxa recognized as threatened or endangered (Chapter 27), *ex situ* seed banking needs to be explored more seriously and receive additional resources.

Finally, a major goal of future studies should be the compilation of an accurate database including both inter- and intrapopulation variation of reproductive traits, seed dormancy characteristics, and germination requirements of all carnivorous plants. Synthetic analysis of such a dataset accounting for phytogeography and phylogeny will yield significant advances in understanding of evolutionary dynamics of carnivorous plants.

CHAPTER 23

Commensals of *Nepenthes* pitchers

Leonora S. Bittleston

23.1 Introduction

Carnivorous pitcher plants are elegant systems for biological study. The aquatic pool inside of a pitcher represents a small, relatively self-contained ecosystem. In these pools, pitchers host complex communities of arthropods, fungi, protozoa, and bacteria ("inquilines:" organisms commonly found living within another's space [e.g., nest, burrow, or pitcher]). Many inquilines prosper within pitcher pools, despite the fact that the pitchers trap, drown, and digest other creatures. But pitchers are not just passive housing—they are parts of living plants, and their internal conditions change over time. To thrive in *Nepenthes* pitchers, inquilines have to contend with acidic conditions, free radicals, viscous fluids, plant chemicals, and other hungry inhabitants. It is still unclear if and how adaptations are necessary for living in pitchers, and also if, how, and to what extent pitcher inquilines influence their host plants. Researchers are just beginning to uncover information about the full diversity of pitcher communities, and which environmental and host characteristics most strongly affect inquiline colonization and persistence.

With over one hundred species ranging from Madagascar to Australia, *Nepenthes* pitcher plants are widespread and have diverse strategies for acquiring nutrients (Chapters 5, 17, 26). Pitchers generally attract insect prey with extrafloral nectar, volatiles, and ultraviolet patterning (Moran et al. 1999, Merbach et al. 2001, Di Giusto et al. 2008, Kurup et al. 2013; Chapters 12, 15). The slippery peristome lip combined with a slippery, often waxy, region inside the top third of the pitcher cause prey to fall in and be unable to escape (Bohn and Federle 2004, Gorb et al. 2005; Chapter 15). The aquatic pool within a pitcher is a mix of rainwater and secretions from the plant. *Nepenthes* plants can produce their own digestive enzymes, including proteases and chitinases, to break down compounds in insect bodies (Tökés et al. 1974, Eilenberg et al. 2006, Hatano and Hamada 2008; Chapter 16). Autolytic enzymes from drowned prey and free radicals also contribute to the process, and it is likely that microbial enzymes play a role as well (Juniper et al. 1989, Kitching 2000, Chia et al. 2004; Chapter 16). *Nepenthes* pitchers contain some secreted liquid even before opening (Bauer et al. 2009). Upon opening, pitchers are colonized by insects, mites, bacteria, fungi, and protozoa (Kitching 2000, Bauer et al. 2009), and an aquatic food web rapidly forms. Many inquilines living in *Nepenthes* are found only in pitcher habitats, and do not complete their life cycles in other habitats (Beaver 1983).

In this chapter, I briefly review *Nepenthes* pitcher plants as habitats and their common inquilines. I then describe in more detail the recent literature on *Nepenthes* microbial communities, and patterns across pitcher systems. The last sections compare *Nepenthes* pitcher communities to those of surrounding habitats and other pitcher genera, and discuss future directions in the study of pitcher microecosystems. I illuminate what makes pitchers unique as habitats, and hypothesize how pitchers might influence the assembly of their contained communities.

23.2 History of *Nepenthes* inquiline studies

Rumphius (1750, in Beekmann 2004) may have been the first to report—in the 1600s—the presence

Bittleston, L. S., *Commensals of* Nepenthes *pitchers*. In: *Carnivorous Plants: Physiology, ecology, and evolution*. Edited by Aaron M. Ellison and Lubomír Adamec: Oxford University Press (2018).
 DOI: 10.1093/oso/9780198779841.003.0023

of living organisms in *Nepenthes* pitchers, but research on these inquilines began in earnest in the early 1900s. The Dutch entomologist Johannes de Meijere published extensive descriptions of his observations of *Nepenthes* inquilines in Singapore (in Part I of Jensen 1910), and proposed a new genus, *Nepenthosyrphus*, for three syrphid flies whose larvae live in *Nepenthes* pitchers. Günther (1913) wrote about the insects of Sri Lankan *Nepenthes*; Dover et al. (1928), working in Singapore, completed what was likely the first comprehensive study of *Nepenthes* pitchers as habitat; and Thienemann (1932) coined three terms: nepenthebiont—specialist inhabitants of *Nepenthes* pitchers; nepenthephils—common inhabitants that also live elsewhere; and nepenthexenes—accidental inhabitants that do not establish persistent populations.

Since Thienemann's time, researchers have studied *Nepenthes* food webs in multiple species and regions. For example, food webs tend to be more complex in regions closer to the center of *Nepenthes* diversity, where more *Nepenthes* species coexist on the same landmass (Beaver 1985, Clarke and Kitching 1993, Clarke 1998, Kitching 2000). Differences among *Nepenthes* species and their physical shapes appear to influence the inquilines that colonize them (Clarke and Kitching 1993, Clarke 1998), and even different forms of the same *Nepenthes* species can have different food-web structures (Ratsirarson and Silander 1996). There are observable successional patterns within pitchers: Sota et al. (1998) documented how abundances of different inquiline species changed with pitcher age, and how bacterial density increased as pitchers aged and then decreased as they senesced. Bacteria and other microbial organisms generally have been included in depictions of *Nepenthes* food webs as one combined node, because of difficulties in identifying microbial species and their functional roles.

Recent progress in understanding associations among *Nepenthes* pitchers and other organisms has gone in two directions: the very small and the very unusual. Detailed data on bacterial communities have only emerged in recent literature (Chou et al. 2014, Takeuchi et al. 2015, Sickel et al. 2016, Bittleston 2016, Chan et al. 2016, Kanokratana et al. 2016) but the (even smaller) viruses living within *Nepenthes* pitchers have not yet been characterized, and roles of viruses in assembly of inquiline communities are completely unknown. In unusual cases, *Nepenthes* species have been found to form mutualistic associations with ants, bats, and tree shrews (Clarke and Kitching 1995, Clarke et al. 2009, Grafe et al. 2011, Bazile et al. 2012, Scharmann et al. 2013; Chapters 15, 26). Inqulines recorded to date from *Nepenthes* pitchers are listed in Table 23.1.

Table 23.1 Organisms living within pitcher microcosms. Based on Table S2 from Adlassnig et al., (2011), with modifications and updates. For metabarcoding studies only the top reported organisms are included; bacteria are reported at the family level.

Taxon		Family or species	Host species	Ref.*
Bacteria	Actinobacteria	Brevibacteriaceae	*N. hemsleyana*	1
		Conexibacteraceae	*N. ampullaria, N. andamana, N. gracilis, N. mirabilis* var. *globosa, N. mirabilis* var. *mirabilis, N. smilesii, N. suratensis*	2
		Corynebacteriaceae	*N. hemsleyana, N. rafflesiana*	1
		Dermabacteraceae	*N. hemsleyana*	1
		Dietziaceae	*N. hemsleyana*	1
		Microbacteriaceae	*N. ampullaria, N. andamana, N. gracilis, N. mirabilis* var. *globosa, N. mirabilis* var. *mirabilis, N. smilesii, N. suratensis, N.* sp.	2, 3
		Mycobacteriaceae	*N. albomarginata, N. ampullaria, N. andamana, N. gracilis, N. hirsuta, N. hemsleyana, N. mirabilis* var. *echinostoma, N. mirabilis* var. *globosa, N. mirabilis* var. *mirabilis, N. rafflesiana, N. reinwardtiana, N. smilesii, N. stenophylla, N. suratensis, N. tentaculata, N. veitchii*	1, 2, 4, 5
		Propionibacteriaceae	*N. albomarginata, N. ampullaria, N. hirsuta, N. mirabilis* var. *echinostoma*	4

(*continued*)

Table 23.1 (*Continued*)

Taxon		Family or species	Host species	Ref.*
	Bacteroidetes	Bacteroidaceae	*N. ampullaria, N. andamana, N. gracilis, N. mirabilis* var. *globosa, N. mirabilis* var. *mirabilis, N. smilesii, N. suratensis*	2
		Chitinophagaceae	*N. ampullaria, N. andamana, N. gracilis, N. hemsleyana, N. hirsuta, N. mirabilis* var. *globosa, N. mirabilis* var. *mirabilis, N. rafflesiana, N. reinwardtiana, N. smilesii, N. stenophylla, N. suratensis, N. tentaculata, N. veitchii*	1, 2, 5
		Cryomorphaceae	*N. albomarginata, N. ampullaria, N. hemsleyana, N. hirsuta, N. mirabilis* var. *echinostoma*	1, 4
		Cytophagaceae	*N. hemsleyana*	1
		Flavobacteriaceae	*N. ampullaria, N. andamana, N. gracilis, N. mirabilis* var. *globosa, N. mirabilis* var. *mirabilis, N. smilesii, N. suratensis, N.* sp.	2, 3
		Porphyromonadaceae	*N. ampullaria, N. andamana, N. gracilis, N. mirabilis* var. *globosa, N. mirabilis* var. *mirabilis, N. smilesii, N. suratensis*	2
		Sphingobacteriaceae	*N. albomarginata, N. ampullaria, N. andamana, N. gracilis, N. hemsleyana, N. hirsuta, N. mirabilis* var. *echinostoma, N. mirabilis* var. *globosa, N. mirabilis* var. *mirabilis, N. reinwardtiana, N. rafflesiana, N. smilesii, N. stenophylla, N. suratensis, N. tentaculata, N. veitchii, N.* sp.	1, 2, 3, 4, 5
		Weeksellaceae	*N. hemsleyana*	1
	Firmicutes	Bacillaceae	*N. rafflesiana, N.* sp.	1, 3
		Clostridiaceae	*N. ampullaria, N. andamana, N. gracilis, N. mirabilis* var. *globosa, N. mirabilis* var. *mirabilis, N. smilesii, N. suratensis*	2
		Staphylococcaceae	*N. hemsleyana, N. rafflesiana*	1
		Streptococcaceae	*N. hemsleyana*	1
	Proteobacteria	Aeromonadaceae	*N. ampullaria, N. andamana, N. gracilis, N. mirabilis* var. *globosa, N. mirabilis* var. *mirabilis, N. smilesii, N. suratensis*	2
		Alcaligenaceae	*N. ampullaria, N. andamana, N. gracilis, N. mirabilis* var. *globosa, N. mirabilis* var. *mirabilis, N. smilesii, N. suratensis, N.* sp.	2, 3
		Acetobacteraceae	*N. ampullaria, N. andamana, N. gracilis, N. hemsleyana, N. hirsuta, N. mirabilis, N. rafflesiana, N. mirabilis* var. *globosa, N. mirabilis* var. *mirabilis, N. reinwardtiana, N. smilesii, N. stenophylla, N. suratensis, N. tentaculata, N. veitchii*	1, 2, 5, 6
		Bradyrhizobiaceae	*N. ampullaria, N. gracilis, N. hirsuta, N. hemsleyana, N. rafflesiana, N. reinwardtiana, N. stenophylla, N. tentaculata, N. veitchii*	1, 5
		Burkholderiaceae	*N. ampullaria, N. andamana, N. gracilis, N. hirsuta, N. mirabilis* var. *globosa, N. mirabilis* var. *mirabilis, N. mirabilis* var. *globosa, N. mirabilis* var. *mirabilis, N. rafflesiana, N. reinwardtiana, N. smilesii, N. stenophylla, N. suratensis, N. tentaculata, N. veitchii*	1, 2, 5
		Caulobacteraceae	*N. ampullaria, N. andamana, N. gracilis, N. mirabilis* var. *globosa, N. mirabilis* var. *mirabilis, N. smilesii, N. suratensis*	2
		Comamonadaceae	*N. ampullaria, N. andamana, N. gracilis, N. hirsuta, N. mirabilis* var. *globosa, N. mirabilis* var. *mirabilis, N. rafflesiana, N. reinwardtiana, N. smilesii, N. stenophylla, N. suratensis, N. tentaculata, N. veitchii*	2, 5
		Enterobacteriaceae	*N. ampullaria, N. andamana, N. gracilis, N. hemsleyana, N. hirsuta, N. mirabilis, N. mirabilis* var. *globosa, N. mirabilis* var. *mirabilis, N. rafflesiana, N. reinwardtiana, N. smilesii, N. stenophylla, N. suratensis, N. tentaculata, N. veitchii, N.* sp.	1, 2, 3, 5, 6

(*continued*)

Table 23.1 (*Continued*)

Taxon		Family or species	Host species	Ref.*
		Hyphomicrobiaceae	*N. ampullaria, N. andamana, N. gracilis, N. mirabilis* var. *globosa, N. mirabilis* var. *mirabilis, N. smilesii, N. suratensis*	2
		Methylocystaceae	*N. ampullaria, N. gracilis, N. mirabilis*	6
		Methylophilaceae	*N. albomarginata, N. ampullaria, N. mirabilis* var. *echinostoma, N. hirsuta*	4
		Moraxellaceae	*N. albomarginata, N. ampullaria, N. andamana, N. gracilis, N. hirsuta, N. mirabilis* var. *echinostoma, N. mirabilis* var. *globosa, N. mirabilis* var. *mirabilis, N. smilesii, N. suratensis*	2, 4
		Nannocystaceae	*N. rafflesiana*	1
		Neisseriaceae	*N. ampullaria, N. andamana, N. gracilis, N. mirabilis* var. *globosa, N. mirabilis* var. *mirabilis, N. smilesii, N. suratensis*	2
		Oxalobacteraceae	*N. ampullaria, N. andamana, N. gracilis, N. mirabilis* var. *globosa, N. mirabilis* var. *mirabilis, N. smilesii, N. suratensis*	2
		Planctomycetaceae	*N. ampullaria, N. andamana, N. gracilis, N. mirabilis* var. *globosa, N. mirabilis* var. *mirabilis, N. smilesii, N. suratensis*	2
		Pseudomonadaceae	*N. albomarginata, N. ampullaria, N. andamana, N. gracilis, N. hemsleyana, N. hirsuta, N. mirabilis* var. *echinostoma, N. mirabilis* var. *globosa, N. mirabilis* var. *mirabilis, N. rafflesiana, N. smilesii, N. suratensis, N.* sp.	1, 2, 3, 4
		Rhodocyclaceae	*N. hemsleyana*	1
		Rhizobiaceae	*N. ampullaria, N. gracilis, N. hirsuta, N. rafflesiana, N. reinwardtiana, N. stenophylla, N. tentaculata, N. veitchii*	1, 5
		Sphingomonadaceae	*N. ampullaria, N. andamana, N. gracilis, N. hirsuta, N. mirabilis, N. mirabilis* var. *globosa, N. mirabilis* var. *mirabilis, N. rafflesiana, N. reinwardtiana, N. smilesii, N. stenophylla, N. suratensis, N. tentaculata, N. veitchii*	1, 2, 5, 6
		Xanthobacteraceae	*N. ampullaria, N. andamana, N. gracilis, N. mirabilis, N. mirabilis* var. *globosa, N. mirabilis* var. *mirabilis, N. smilesii, N. suratensis*	2, 6
		Xanthomonadaceae	*N. ampullaria, N. andamana, N. gracilis, N. hemsleyana, N. hirsuta, N. mirabilis, N. mirabilis* var. *globosa, N. mirabilis* var. *mirabilis, N. rafflesiana, N. reinwardtiana, N. smilesii, N. stenophylla, N. suratensis, N. tentaculata, N. veitchii*	1, 2, 5, 6
	Undetermined phyla	*Bacterium colianindolicum, Bacterium diffusum, Bacterium gastricum*	*N. mirabilis*	7
Fungi	Ascomycota	*Aspergillus glaucus, Penicillium glaucum*	*N. mirabilis*	7
		Aureobasidium pullans, Candida diffluens	*N. gracilis, N. macfarlanei, N. sanguinea*	8
		Candida sp.	*N. ampullaria, N. gracilis, N. hirsuta, N. rafflesiana, N. reinwardtiana, N. stenophylla, N. tentaculata, N. veitchii*	5
		Undetermined Herpotrichiellaceae	*N. ampullaria, N. gracilis, N. hirsuta, N. rafflesiana, N. reintwardiana, N. stenophylla, N. tentaculata, N. veitchii*	5
		Undetermined yeasts	*N. madagascariensis*	9
	Basidiomycota	*Cryptococcus albidus, Tilletiopsis* sp.	*N. gracilis, N. mirabilis, N. macfarlanei, N. sanguinea*	8

(*continued*)

Table 23.1 (*Continued*)

Taxon		Family or species	Host species	Ref.*
		Bullera alba	*N. macfarlanei, N. sanguinea*	8
		Cryptococcus laurentii, Trichosporon pullulans	*N. macfarlanei*	8
		Rhodotorula rubra	*N. gracilis, N. macfarlanei, N. sanguinea*	8
		Sporobolomyces roseus	*N. mirabilis, N. macfarlanei, N. sanguinea*	8
		Undetermined Agaricomycetes	*N. ampullaria, N. gracilis, N. hirsuta, N. rafflesiana, N. reinwardtiana, N. stenophylla, N. tentaculata, N. veitchii*	5
	Chitridiomycota	Undetermined Monoblepharidales	*N. ampullaria, N. gracilis, N. hirsuta, N. rafflesiana, N. reinwardtiana, N. stenophylla, N. tentaculata, N. veitchii*	5
	Myxomycota	*Merismopedium glaucum*	*N. melamphora*	10
	Zygomycota	*Mucor mucido, Mucor racemosus, Rhizopus nigricans*	*N. mirabilis*	7
		Undetermined Mucoromycotina	*N. ampullaria, N. gracilis, N. hirsuta, N. rafflesiana, N. reinwardtiana, N. stenophylla, N. tentaculata, N. veitchii*	5
Algae	Bacilliarophyta	*Achnanthes lanceolata, Achnanthes minutissima, Cocconeis placentula* var. *lineata, Epithemia sorex, Navicula elliptica, Navicula viridis*	*N. melamphora*	10
	Chlorophyta	*Euastrum* sp.	*N. melamphora*	10
		Goniomonas sp.	*N. ampullaria, N. gracilis, N. rafflesiana*	11
		Microthamnion sp., *Pseudomuriella* sp.	*N. ampullaria, N. rafflesiana*	11
Protozoa	Amoebozoa	*Acanthamoeba* sp.	*N. gracilis*	11
	Apicomplexa	Undetermined Gregarinasina	*N. ampullaria, N. gracilis, N. rafflesiana*	11
	Cercozoa	Undetermined Cercomonas	*N. ampullaria, N. gracilis, N. rafflesiana*	11
		Undetermined Heteromita	*N. gracilis, N. rafflesiana*	11
		Undetermined Phaeodarea	*N. ampullaria, N. gracilis, N. rafflesiana*	11
	Choanozoa	*Lagenoeca* sp.	*N. ampullaria, N. rafflesiana*	11
	Ciliophora	Undetermined CONthreeP	*N. ampullaria, N. gracilis, N. rafflesiana*	11
	Euglenophyta	Undetermined Euglenida	*N. ampullaria, N. gracilis, N. rafflesiana*	11
	Euglenozoa	Undetermined Kinetoplastea	*N. ampullaria, N. gracilis, N. rafflesiana*	11
	Rhizopoda	*Amoeba guttula, Amoeba nepenthesi, Arcella vulgaris, Centropyxis aculeata, Cochliopodium bilimbosum, Difflugia constricta, Lesquereusia epistomium*	*N. melamphora*	10
Rotifera	Bdelloidea	Undetermined Philodinidae	*N. ampullaria, N. gracilis, N. hirsuta, N. rafflesiana, N. reinwardtiana, N. stenophylla, N. tentaculata, N. veitchii*	5

(*continued*)

Table 23.1 (*Continued*)

Taxon		Family or species	Host species	Ref.*
Vermiform	Nematoda	*Baujardia mirabilis*	*N. mirabilis*	12
		Dorylaimus sp., *Subanguina* sp.	*N.* sp.	13
		Undetermined nematodes	*N. alata*	14
	Oligochaeta	*Aeolosoma* sp.	*N. ampullaria, N. gracilis, N. rafflesiana*	11
		Undetermined Naididae	*N. ampullaria*	11
		Undetermined Oligochaeta	*N.* sp.	15
Crustacea	Copepoda	*Epactophanes richardi*	*N. ampullaria, N.* sp.	15, 16
		Parastenocaris incerta, Phyllognathopus viguieri	*N. ampullaria*	16
		Phyllognathopus viguieri ssp. *menzeli*	*N.* sp.	16
		Undetermined Cyclopoida	*N. ampullaria, N. gracilis, N. hirsuta, N. rafflesiana, N. reinwardtiana, N. stenophylla, N. tentaculata, N. veitchii*	5
		Undetermined Harpacticoida	*N. ampullaria, N. gracilis, N. hirsuta, N. rafflesiana, N. reinwardtiana, N. stenophylla, N. tentaculata, N. veitchii*	5, 17
	Decapoda	*Geosesarma malayanum*	*N. ampullaria, N. bicalcarata*	18
		Undetermined Decapoda Dutch "Garneeltjes"	*N.* cf. *mirabilis*	19, 20
	Ostracoda	*Cypridiopsis* sp.	*N. ampullaria, N. gracilis, N. hirsuta, N. rafflesiana, N. reinwardtiana, N. stenophylla, N. tentaculata, N. veitchii*	5
Arachnida	Acari	*Anoetus nepenthesiana*	*N. albomarginata, N. gracilis, N. ampullaria, N. gymnamphora, N. mirabilis, N. tobaica*	21
		Creutzeria seychellensis	*N. pervillei*	22
		Creutzeria sp.	*N. madagascariensis*	9
		Creutzeria tobaica	*N. albomarginata, N. madagascariensis, N. mirabilis, N. reinwardtiana, N. tobaica*	21, 22
		Hormosianoetus sp.	*N. ampullaria, N. gracilis, N. rafflesiana*	11
		Nepenthacarus warreni	*N. mirabilis*	23
		Rostrozetes sp.	*N. ampullaria, N. gracilis, N. rafflesiana*	11
		Zwickia guentheri	*N. albomarginata, N. ampullaria, N. distillatoria, N. gracilis, N. gymnamphora, N. mirabilis*	13, 21
		Zwickia nepenthesiana	*N. ampullaria, N. gracilis*	13
		Undetermined Anoetidae	*N. alata*	14
	Araneae	*Misumenops nepenthicola*	*N. albomarginata, N. gracilis, N. gymnamphora, N. rafflesiana, N. reinwardtiana, N. tobaica*	13, 21, 22
		Misumenops thienemanni	*N. tobaica*	21
		Peucetia sp., *Theridion decaryi, Thyena* sp., *Synaema obscuripes*	*N. madagascariensis*	9, 24
		Theridion sp.	*N. stenophylla*	21
		Thomisus callidus	*N. tobaica*	21

(*continued*)

Table 23.1 (*Continued*)

Taxon		Family or species	Host species	Ref.*
		Thomisus nepenthiphilus	*N. mirabilis, N. rafflesiana, N. reinwardtiana, N. tobaica, N.* sp.	13, 21
		Undetermined spiders	*N. mirabilis*	25
Insects	Collembola	*Podura aquatic*	*N. melamphora*	10
	Odonata	*Lyriothemis salva*	*N.* sp.	26
	Lepidoptera	*Eublemma radda*	*N. bicalcarata, N. mirabilis, N. rafflesiana*	21
		Nepenthophilus tigrinus	*N. distillatoria*	21
		Phyllocnistis nepenthae	*N. tobaica*	21
	Hymenoptera	*Allocata* sp.	*N. tobaica*	21
		Camponotus sp.	*N. bicalcarata*	27
		Camponotus schmitzi	*N. bicalcarata, N. gracilis, N. mirabilis*	21, 28
		Dolichoderus bituberculatus	*N. gracilis, N. mirabilis*	21
		Polyrachis nepenthicola	*N. stenophylla*	29
		Tachinaephagus sp.	*N. tobaica*	21
		Trichopria sp.	*N. ampullaria*	21
		Undetermined Elachertida	*N. tobaica*	21
		Undetermined Encyrtida	*N. albomarginata, N. gracilis*	21
	Diptera	*Aedes albopictus*	*N. ampullaria*	17
		Aedes brevitibia	*N. ampullaria, N. bicalcarata, N. rafflesiana*	17, 21
		Aedes dybasi, A. maehleria	*N. mirabilis*	30
		Aedes gani, A. medialis	*N.* sp.	21
		Aedes treubi	*N. gymnamphora*	21
		Armigeres conjugens, A. flavus, A. hybridus, A. malati	*N.* sp.	21
		Armigeres durami, A. giveni, A. kuchingensis	*N. ampullaria*	21
		Armigeres magnus	*N. mirabilis*	21
		Corethrella calathicola	*N. ampullaria*	17
		Corethrella spp.	*N. ampullaria, N. bicalcarata*	31, 32
		Culex acutipalus, C. hewitti	*N. ampullaria*	17, 21
		Culex coerulescens	*N. ampullaria, N. bicalcarata, N.* sp.	13, 17, 31, 32
		Culex curtipalpis	*N. alata, N. mirabilis, N. albomarginata, N. gracilis*	14, 21
		Culex eminentia	*N. ampullaria, N. gracilis, N. rafflesiana*	21
		Culex jenseni	*N. gymnamphora, N. reinwardtiana*	21
		Culex lucaris	*N. albomarginata*	21
		Culex navalis	*N. ampullaria, N. alata, N. bicalcarata, N. rafflesiana*	14, 17, 21
		Culex shebbearei, C. sumatranus	*N. mirabilis*	21

(*continued*)

Table 23.1 (*Continued*)

Taxon	Family or species	Host species	Ref.*
	Culicoides confinis	*N. mirabilis*	15
	Dasyhelea ampullariae	*N. ampullaria, N. gracilis*	21
	Dasyhelea biseriata	*N. ampullaria*	21
	Dasyhelea confinis, D. subgrata	*N. mirabilis*	21
	Dasyhelea nepenthicola	*N. ampullaria, N. albomarginata, N. gracilis*	21
	Dasyhelea sp.	*N. ampullaria, N. bicalcarata, N. alata, N. rafflesiana, N. mirabilis, N. albomarginata*	14, 15, 17, 31, 32
	Endonepenthia cambodiae	*N. ampullaria*	21
	Endonepenthia campylonympha	*N. mirabilis*	21
	Endonepenthia gregalis	*N. gymnamphora*	21
	Endonepenthia schuitemakeri	*N. albomarginata, N. gracilis, N. rafflesiana*	13, 21
	Endonepenthia spp.	*N. ampullaria, N.* sp.	17, 22
	Endonepenthia tobaica	*N. tobaica*	21
	Forcipomyia sp.	*N. ampullaria*	17
	Lestodiplosis sp.	*N. ampullaria, N. alata, N. gracilis, N. mirabilis, N. rafflesiana*	13, 14, 17, 33
	Lestodiplosis syringopais	*N. albomarginata, N. gracilis, N. tobaica*	21
	Megarhinus metallicus	*N. sanguinea*	22
	Megaselia cf. *bivesicata*	*N. ampullaria*	21
	Megaselia cambodiae	*N.* sp.	15
	Megaselia campylonympha	*N. ampullaria, N. mirabilis*	15
	Megaselia corkerae	*N. mirabilis*	22
	Megaselia decipiens	*N. gymnamphora*	21
	Megaselia deningsi, M. meningi	*N. distillatoria*	21, 22
	Megaselia gregalis	*N. distillatoria, N. gymnamphora*	15
	Megaselia nepenthina	*N. albomarginata, N. gracilis, N. mirabilis, N.* sp.	15, 21
	Megaselia schuitemakeri	*N. albomarginata, N. gracilis*	22
	Megaselia spp.	*N. ampullaria, N. alata*	14, 17
	Megaselia tobaica	*N. tobaica*	15
	Metriocnemus sp.	*N. ampullaria, N. tentaculata, N.* cf. *villosa, N.* sp.	15, 21
	Mimomyia jeansottei	*N. madagascariensis*	21
	Nepenthomyia malayana	*N. ampullaria*	22

(*continued*)

Table 23.1 (*Continued*)

Taxon	Family or species	Host species	Ref.*
	Nepenthosyrphus cf. *capitatus*	*N. albomarginata, N. reinwardtiana, N. tobaica*	21
	Nepenthosyrphus malayanus, N. venustus	*N.* sp.	21
	Nepenthosyrphus oudemansi	*N. ampullaria, N. rafflesiana*	21
	Nepenthosyrphus spp.	*N. ampullaria, N. bicalcarata*	17, 31, 32
	Phaonia nepenthincola	*N. gymnamphora*	21
	Phaonia sp.	*N. alata*	14
	Pierretia urceola	*N. albomarginata, N. gracilis*	21, 22
	Polypedilum convexum	*N. bicalcarata, N. ampullaria*	15
	Polypedilum sp.	*N. ampullaria, N. bicalcarata, N.* cf. *villosa, N.* sp.	13, 15, 31, 32
	Sarcophaga sp.	*N. sanguinea*	21
	Sarcosolomonia pauensis	*N. mirabilis*	15
	Sarcosolomonia carolinensis	*N.* sp.	15
	Succingulum fransseni	*N. mirabilis*	21
	Systenus spp.	*N. ampullaria, N. bicalcarata*	31, 32
	Toxorhynchites acaudatus	*N. ampullaria, N. rafflesiana*	21
	Toxorhynchites ater	*N. rafflesiana*	21
	Toxorhynchites aurifluus, T. coeruleus, T. nepenthicola, T. nepenthes, T. nigripes, T. pendleburyi, T. quasiferox, T. sumatranus,	*N.* sp.	21, 22
	Toxorhynchites indicus	*N. ampullaria*	17
	Toxorhynchites klossi	*N. albomarginata*	21
	Toxorhynchites metallicus	*N. sanguinea*	21
	Toxorhynchites sp.	*N. ampullaria, N. bicalcarata, N. rafflesiana, N. alata*	13,14, 31, 32
	Toxorhynchites splendens	*N. ampullaria, N. rafflesiana*	21
	Tripteroides adentata, T. apoensis, T. barraudi, T. belkini, T. bimaculipes, T. christophersi, T cuttsi, T. delpilari, T. digoelensis, T. dyari, T. elegans, T. flabelliger, T. intermediatus, T. longipalpatus, T. malvari, T. mathesoni, T. mendacis, T. microcala, T. microlepis, T. obscurus, T. pallidus, T. papua, T. pillosus, T. reiseni, T. roxasi, T. simplex, T. simulatus, T. werneri	*N.* sp.	21
	Tripteroides aranoides	*N. sanguinea*	22

(*continued*)

Table 23.1 (*Continued*)

Taxon	Family or species	Host species	Ref.*
	Tripteroides bambusa	*N. albomarginata, N. ampullaria, N. gracilis*	21
	Tripteroides bisquamatus, T. brevirhynchus, T. caledonicus, T. filipes, T. kingi, T. subobscurus	*N. mirabilis*	21, 22
	Tripteroides dofleini	*N. distillatoria*	21
	Tripteroides nepenthicola	*N. alata*	21
	Tripteroides nepenthis	*N. ampullaria, N. bicalcarata, N. rafflesiana*	17, 21, 31, 32
	Tripteroides nepenthisimilis	*N. ampullaria, N. rafflesiana*	21
	Tripteroides spp.	*N. ampullaria, N. alata*	14, 17
	Tripteroides tenax	*N. albomarginata, N. ampullaria, N. gracilis, N. gymnamphora, N. rafflesiana, N. reinwardtiana, N. sanguinea, N. tobaica*	17, 21
	Tripteroides vicinus	*N. gracilis, N. lowii, N. mirabilis, N. sanguinea*	21, 22
	Uranotaenia ascidiicola	*N. gymnamphora*	21
	Uranotaenia belkini, U. bosseri, U. brunhesi, U. damasci	*N. madagascariensis*	9, 21
	Uranotaenia gigantea	*N. alata, N. tobaica*	14, 21
	Uranotaenia moultoni	*N. ampullaria, N. bicalcarata, N. alata, N. gracilis, N. rafflesiana*	17, 21, 31, 32
	Uranotaenia nepenthes	*N. pervillei*	22
	Uranotaenia nivipleura	*N. distillatoria*	22
	Uranotaenia sp.	*N. ampullaria*	17
	Uranotaenia xanthomelaena	*N. gracilis*	21
	Wilhelmina nepenthicola	*N. alata, N. ampullaria, N. rafflesiana*	14, 21
	Xenoplatyura beaveri	*N. ampullaria*	13
	Xylota sp.	*N. ampullaria*	21
	Undetermined Anoetidae	*N. albomarginata, N. ampullaria*	17, 21
	Undetermined Chironomidae	*N. ampullaria*	17
	Undetermined Chloropidae	*N. madagascariensis*	9
	Undetermined Dolichopodidae	*N. ampullaria*	15
	Undetermined Lauxanidae	*N.* sp.	21
	Undetermined Muscomorphae	*N. alata*	14
	Undetermined Sarcophagidae	*N. maxima*	15
	Undetermined Sciarida	*N. ampullaria, N. alata*	14, 17
Vertebrates Amphibia	*Kalophrynus pleurostigma* ssp. *pleurostigma*	*N. ampullaria*	34, 35

(*continued*)

Table 23.1 (*Continued*)

Taxon	Family or species	Host species	Ref.*
	Heterixalus tricolor	*N. madagascariensis*	9
	Dendrobates spp.	*N.* spp.	15
	Microhyla borneensis	*N.* sp.	34
	Microhyla nepenthicola	*N. ampullaria*	36
	Philautus aurifasciatus (dubious description)	*N. sanguinea*	34
	Undetermined tadpoles	*N. ampullaria, N. bicalcarata*	31, 32

*References: 1. Sickel et al. (2016); 2. Kanokratana et al. (2016); 3. Chan et al. (2016); 4. Takeuchi et al. (2015); 5. Bittleston (2016); 6. Chou et al. (2014); 7. Okahara (1933); 8. Shivas and Brown (1989); 9.Ratsirarson and Silander (1996); 10. Van Oye (1921); 11. Bittleston et al. (2016a); 12. Bert et al. (2003); 13. Choo et al. (1997); 14. Sota et al. (1998); 15. Kitching (2000); 16. Reid (2002); 17. Mogi and Yong (1992); 18. Carrow et al. (1997); 19. Beekmann (2004); 20. Rumphius (1750); 21. Beaver (1983); 22. Juniper et al. (1989); 23. Fashing (2002); 24. Rembold et al. (2013); 25. Hua and Li (2005); 26. Corbet (1983); 27. Clarke and Kitching (1995); 28. Barthlott et al. (2004); 29. Grafe and Kohout (2013); 30. Mogi (2010); 31. Cresswell (1998); 32. Cresswell (2000); 33. Clarke and Kitching (1993); 34. Lim and Ng (1991); 35. Ming (1997); 36. Das and Haas (2010).

23.3 Physical properties of *Nepenthes* pitchers

A living, growing pitcher creates the habitat for its inquiline community, and its properties influence the colonization and establishment of the inquilines. Pitchers attract prey with both olfactory and visual cues, and these cues also likely influence initial colonization of inquilines (Moran et al. 1999, Di Giusto et al. 2008, Bazile et al. 2015). Extrafloral nectar and pitcher fluid both produce volatile scents that attract arthropod prey and perhaps adult inquilines looking to oviposit in an appropriate habitat (Di Giusto et al. 2008). Likewise, ultraviolet reflectance around the pitcher rim and pitcher coloration or structure serve as prey attractants and possibly as signals to inquilines (Moran et al. 1999, Kurup et al. 2013). Microbial organisms are unlikely to use olfactory or visual cues for pitcher colonization, but bacteria, fungi, and protozoa may enter pitchers on or in arthropod prey or inquilines, and thus might be indirectly affected by these pitcher properties.

Once organisms have reached pitchers, internal properties of the fluid affect their establishment. Pitcher-fluid pH varies with species, age, prey content, decomposition, and possibly rainfall (Kitching 2000, Adlassnig et al. 2011, Bazile et al. 2015). Oxygen levels in pitcher fluid are higher than in the same fluids transferred to inert tubes, most likely because of photosynthesis by the pitcher itself (Juniper et al. 1989, Adlassnig et al. 2011). Oxygen content may influence the viability of arthropod inquilines in *Nepenthes*, and should alter the relative abundances of aerobic and anaerobic bacteria. *Nepenthes* pitcher fluid contains free radicals generated from hydrogen peroxide that likely create a harsher environment for inquilines (Chia et al. 2004, Adlassnig et al. 2011). Certain *Nepenthes* species, for example *N. rafflesiana, N. fusca*, and *N. maxima*, can produce very viscous fluids (Bonhomme et al. 2011b; Chapter 15), but the effects of viscosity on pitcher inquiline communities are unstudied.

23.4 *Nepenthes* inquilines and their functional roles

Some inquilines colonize pitchers opportunistically, whereas many others are specialists that are limited to pitchers for at least one stage of their life cycles (Beaver 1983). Specialist inquilines likely have adapted to pitcher cues and conditions, and take advantage of a protected habitat that attracts external nourishment in the form of insect prey.

23.4.1 Arthropods, vermiform organisms, and rotifers

Taxonomic composition. Aquatic insects and mites frequently colonize *Nepenthes* pitchers and often specialize on these unusual habitats. Diptera are the most obvious inquilines; the most common are mosquitoes (Culicidae), midges (Chironomidae, Ceratopogonidae, Cecidomyiidae, Chaoboridae) and flies in four other families (Phoridae, Calliphoridae, Syrphidae, and Sarcophagidae). Common genera of *Nepenthes* mosquitoes include *Armigeres, Culex, Toxorhynchites, Tripteroides*, and *Uranotaenia* (Beaver 1983, Clarke and Kitching 1993). Occasional *Aedes* mosquitoes occur as *Nepenthes* inquilines, but these have different traits from non-pitcher *Aedes* spp. and do not feed on human blood (Mogi 2010). In particular, pitchers are not suitable habitats for *Aedes albopictus* (Chou et al. 2015) and are not attractive to gravid females of *A. albopictus* or *A. aegypti* (Chou et al. 2016), so *Nepenthes* are unlikely to serve as breeding grounds for vectors of human disease.

Mites (Histiostomatidae and Acaridae) also are very common. Crustacea, vermiforms (worms and worm-like animals), and rotifers are sporadically collected. From a metabarcoding study using Illumina sequencing, the most abundant crustaceans were harpacticoid and cyclopoid copepods and podocopid ostracods (Bittleston 2016). The most abundant vermiform operational taxonomic unit (OTU) matched to Aeolosomatidae, a family of micro-annelids that live in soil and decaying matter in stagnant water (Bittleston 2016). Other abundant annelid OTUs were assigned to the Naididae (detritus worms). A sequence classified as a Platyhelminthes (flatworm) was very abundant and might represent an inquiline parasite.

Nematode OTUs, present in about 25% of sampled pitchers, were taxonomically assigned principally to the Panagrolaimidae and Rabditidae. Bert et al. (2003) isolated a new species of nematode from this family and described a new genus based on individuals found in *Nepenthes mirabilis* in Thailand. Choo et al. (1997) also noted two species of nematodes living in the pitchers of Singaporean *Nepenthes*, and undetermined species of nematodes were found in 11% of the *N. alata* pitchers examined in West Sumatra (Sota et al. 1998). Thienemann (1932) noted rotifers and Beaver (1983) classified them as occasional nepenthexenes, but they actually may play a significant role in the pitcher ecosystem. For example, in *Sarracenia purpurea* pitchers, the bdelloid rotifer *Habrotrocha rosa* is a common inquiline and contributes significant amounts of nitrogen and phosphorus to the plant's nutrient budget (Błędzki and Ellison 1998). In the metabarcoding dataset, abundant OTUs close to unnamed bdelloid rotifers in the Philodinidae were present in ≈30% of the sampled pitchers (Bittleston 2016).

Ecological functions. The macroscopic organisms living in *Nepenthes* pitchers generally are classified as saprophages or predators. The saprophages include filter feeders that feed on living or dead matter suspended in the water column, and detritivores that feed on settled detritus (Kitching 2000). Direct observations of feeding are rare, and the functional roles of inquilines often are extrapolated from close relatives. Aedine and anopheline mosquitoes, rotifers, copepods, and histiostigmatid mites are filter feeders that likely consume protozoa, bacteria, and suspended particles (Kitching 2000). Nematodes, ostracods, culicine mosquitoes, chironomid and ceratopogonid midges, phorid flies, and astigmatid mites are thought to feed primarily on fine detritus and any attached microbes. Calliphorid and sarcophagid flies feed on larger detritus and exploit recent prey, but also may act as predators (Sota et al. 1998).

Toxorhynchites mosquitoes and chaoborid midges are voracious predators of other insect larvae. Cannibalism has been observed in inquiline *Toxorhynchites* spp., and generally only one individual is present per pitcher (Beaver 1983). Other genera of mosquitoes, including *Culex* and *Topomyia*, may be facultative predators. Cecidomyiid midges are specialist predators of phorid fly larvae and other small dipterans (Clarke and Kitching 1993).

23.4.2 Fungi, protozoa, algae, and bacteria

Taxonomic composition. Fungi, protozoa, algae, and bacteria are common and abundant inquilines of all *Nepenthes* species examined to date. Microbial inquilines initially were studied in the early-to-mid 1900s (Hepburn 1918, Van Oye 1921, Okahara 1933) when classification of microbes was difficult. In recent years, "next-generation" sequencing and metabarcoding has improved greatly the ability to characterize and classify microbes (Baker et al. 2016, Bittleston et al. 2016a). Metabarcoding is extremely useful for uncovering microscopic or cryptic organisms but species identification is difficult because it relies on existing databases. DNA sequences of most pitcher organisms are not present in the databases and many represent new species, so exact identification is impossible and there is still a strong need for good morphological classification based on descriptive zoology and microbiology.

Nepenthes species produce napthoquinones with anti-fungal activity (Cannon et al. 1980, Shin et al. 2007a, Eilenberg et al. 2010), but fungi—particularly yeasts—still thrive within pitchers. Pitcher-fluid cultures from four *Nepenthes* species in West Malaysia yielded several different yeast species (Shivas and Brown 1989), and metabarcoding of eight *Nepenthes* species from Singapore and Borneo found *Candida* yeasts (Saccharomycetales) to be the most abundant fungi (Bittleston 2016, Bittleston et al. 2016a). Metabarcoding studies also found numerous basal fungi, including chitrids (Monoblepharidales) and several Mucoromycotina. Other relatively abundant fungal OTUs included an ascomycete with strong matches to the Herpotrichiellae (Pezizomycotina) and a basidiomycete (Agaricomycetes) (Bittleston 2016). Fungi live within pitcher tissue as endophytes as well as inside the fluid. A recent study isolated 26 endophytic fungi from *N. ampullaria* and *N. mirabilis* pitchers and leaves, mainly from the *Colletotrichum* species complex (Lee et al. 2014).

Protists are common in *Nepenthes* pitchers. Metabarcoding has revealed that most abundant protozoan inquilines are ciliates (Ciliophora), flagellates (Euglenophyta and Cercozoa), gregarines (Apicomplexa), and amoebae (multiple lineages) (Bittleston 2016, Bittleston et al. 2016a). In addition to protozoa, *Nepenthes* pitcher fluids can have high abundances of green algae (Chlorophyceae) (Bittleston 2016, Bittleston et al. 2016a). Green algae photosynthesize, and thus act as primary producers, adding organic carbon to the pitcher fluid ecosystems to be consumed by other inquilines. Algae may be present in pitcher fluid even when it is not noticeably green, and so are not always obvious to the naked eye.

In terms of numbers, bacteria are probably the most abundant and diverse organisms living in pitcher fluid. In a metabarcoding study of the 16S ribosomal RNA gene, the bacterial orders with the highest relative abundances were Rhodospirillales, Rhizobiales, Actinomycetales, Xanthomonadales, Neisseriales, and the family Chitinophagaceae in an uncertain order (Bittleston 2016, Bittleston et al. 2016a). Sickel et al. (2016) also found Rhodospirallales, Actinomycetales, and Rhizobiales as the top three represented orders in their study of *N. rafflesiana* and *N. hemsleyana* pitcher fluid. Two other recent studies from Malaysia and Thailand also report similar taxa, with highly abundant genera from the same phylogenetic groups (Chou et al. 2014, Kanokratana et al. 2016). Takeuchi et al.'s (2015) study does not show similar most-abundant taxa, but they took many of their samples from cultivated plants in Zurich, and did not report data from Bornean field samples separately from European *in vitro* ones.

Pitcher fluid has long been considered to be sterile before the lids open (Hepburn 1918, Okahara 1933). Sterility of unopened pitchers has been well-demonstrated for *Sarracenia purpurea*, the convergently evolved North American purple pitcher plant (Peterson et al. 2008; Chapter 9), but Chou et al. (2014) and Takeuchi et al. (2015) sequenced bacteria from unopened *Nepenthes* pitcher fluid, and Kanokratana et al. (2016) detected bacterial DNA in three of 14 closed *Nepenthes* pitchers. Sota et al. (1998) saw bacteria in their microscopy counts from fluid of unopened pitchers, but noted they could not rule out the possibility of contamination. Buch et al. (2013) did not find bacteria in unopened *Nepenthes* pitchers, and deemed the secreted plant fluid unsuitable for microbial growth. It remains to be definitively determined if unopened *Nepenthes* pitchers are in fact sterile, or if they contain bacteria of internal plant origin.

Ecological functions. Most bacteria and fungi in pitchers act as decomposers, breaking down the proteins, fats, and carbohydrates of insect prey and occasional plant material that falls into pitchers. They can secrete extracellular digestive enzymes that likely act in concert with plant-produced enzymes. Bacterial species isolated from *Nepenthes* pitchers in Malaysia were able to degrade protein, starch, xylan, chitin, and cellulose (Chan et al. 2016).

The microbial inquilines cannot be classified into only one functional group, because different trophic levels are represented. Some protozoa are predatory—consuming bacteria, fungi, or other protozoa. Others are parasitic, and feed on arthropods or other inquilines. Algae, photosynthetic protozoa, and some bacterial taxa (e.g., photosynthetic Cyanobacteria and nitrogen-fixing Rhizobiales) act as primary producers within pitcher ecosystems. The extent to which they increase levels of organic carbon or nitrogen within the pitchers is currently unknown. Ideally, future *Nepenthes* food webs will incorporate different microbial functional groups, and energy budgets will account for primary production by pitcher inhabitants.

23.4.3 Other inquilines

Certain arthropods spend only part of their life history within pitcher fluid, and are clearly parasitic, such as crab spiders and noctuid moths in the genus *Eublemma*. Crab spiders, including *Misumenops nepenthicola* (= *Henriksenia labuanica*) and *Thomisus nepenthiphilus* in Southeast Asia and *Synaema obscuripes* in Madagascar, live inside the lip of pitchers and feed on trapped prey or inquilines (Chua and Lim 2012, Rembold et al. 2013). If threatened, they can dive into pitcher pools, hide in the detritus at the bottom, and later climb back out. *Eublemma* moth larvae are specialist herbivores of *Nepenthes* species, and will feed on pitcher tissue—sometimes while partially submerged beneath the fluid. Their feeding can damage pitchers, destroying their water-holding capacity (Dover et al. 1928, Clarke and Kitching 1993).

Vertebrates that live within *Nepenthes* pools are rare, but frogs occasionally colonize pitchers and some frogs even have long-term associations with *Nepenthes* species. For example, one of the world's tiniest frogs, *Microhyla nepenthicola*, breeds only in *N. ampullaria* pitchers in Sarawak, Borneo (Das and Haas 2010). A review by Malkmus and Dehling (2008) includes numerous other instances of frogs colonizing *Nepenthes* pitchers, particularly senescent *N. ampullaria* and, to a lesser extent, *N. bicalcarata*.

23.4.4 Inquiline effects on hosts

Inquilines likely both increase and decrease plant-accessible nutrient levels within pitchers, and their overall effects are unclear. Scharmann et al. (2013) hypothesized that phorid dipterans are kleptoparasites of the *Nepenthes bicalcarata–Camponotus schmitzi* mutualism (Chapters 15, 26) and remove nitrogen from the system. When the ants are present, pitchers have higher levels of prey-derived nitrogen and fewer phorid fly adults emerge from the pitchers. This may be because the ants feed on the relatively large phorid flies, decreasing export of nitrogen from pitchers (Scharmann et al. 2013). Conversely, in an *in situ* experiment using fluid and insects from *Nepenthes gracilis*, both culicid and phorid inquilines increased levels of ammonium and soluble protein (Lam et al. 2017). Microbial organisms, in the absence of insects, also increased levels of ammonium. The increases in plant-accessible nitrogen should be beneficial for the plant host. In general, inquilines may both increase and decrease nutrient availability. Insect inquilines increase nutrient availability by ripping apart prey carcasses and making compounds more accessible to extracellular enzymes present in pitcher fluid, and by processing complex proteins and carbohydrates into simpler compounds through feeding and excreting. They decrease nutrient availability by incorporating nutrients into their bodies and removing them from the system when they emerge as adults and leave the pitcher pool. The same dual functions occur for micro-inquilines: bacteria, protozoa, and fungi all consume some resources and excrete others. Their contribution to, and removal from, the total nutrient pool is complex and likely shifts depending on circumstances and environmental conditions. Future experiments should identify nutrient tradeoffs and quantify the extents to which different species (or functional groups) should be considered mutualists, parasites, or commensals.

23.5 Geographic patterns

23.5.1 Patterns within and among pitchers

Dipteran inquilines have some degree of specificity in the *Nepenthes* species they colonize. Clark and Kitching (1993) examined six different Bornean species of *Nepenthes* that co-occur within 1 km of one another. The inquiline genera found were very similar to those in *Nepenthes* pitchers in Penang, although few species were shared (Clarke and Kitching 1993). Dipteran inquilines showed preferences for certain species. For example, among the predators, *Lestodiplosis* sp. (Cecidomyiidae) was found only in *N. mirabilis* and *N. gracilis*, whereas *Corethrella* sp. (Chaoboridae) was found only in *N. bicalcarata* and *N. ampullaria* (Clarke and Kitching 1993). Similarly, Bittleston et al. (2016a) found a *Corethrella* sp. only in *N. ampullaria*, and not *N. gracilis* or *N. rafflesiana*. Certain mites also are species-specific: *Naiadacarus nepenthicola* is restricted to the pitchers of *N. bicalcarata*, even when other species grow in close proximity (Fashing and Chua 2002).

Gregarine protozoa are obligate parasites of arthropods. Because they are parasites, one might expect gregarines to echo arthropod community diversity patterns. However, gregarine community diversity patterns depend more on pitcher location, whereas arthropod communities are better explained by identity of the host species (Baker et al. 2016).

Bacteria have distinct patterns of association with different pitcher plants even when they are in the same habitat, suggesting that certain *Nepenthes* species provide different environmental conditions that create different bacterial niches. Compositional diversity appears to be driven by different acidity levels and other unmeasured aspects of the *Nepenthes* species, and is less affected by geographic location. *Nepenthes* species differ in their average fluid pH levels, and acidity strongly influences bacterial communities in other habitats (e.g., Fierer and Jackson 2006). In a metabarcoding study across eight *Nepenthes* species from Singapore and Borneo, average fluid pH levels ranged from ≈3 to 5.5 (Figure 23.1). *N. stenophylla*, *N. gracilis*, and *N. rafflesiana* all had some pitchers with very low pH values (Figure 23.1a). Among measured variables, pH was the best predictor of community composition with a Mantel test (Bittleston 2016; Figures 23.1b, 23.2). A study of the bacterial communities from six species of *Nepenthes* pitchers grown together in an open-air nursery in Thailand also found a correlation between community composition and pitcher pH (Kanokratana et al. 2016).

Kanokratana et al. (2016) found bacterial community composition in *N. ampullaria* differed from that of other species growing together in the same habitat. Across eight *Nepenthes* species from Singapore and Borneo, pitcher-plant species identity was a better predictor than location of community composition (metabarcoding data in Bittleston 2016; Figure 23.2). The volume of pitcher fluid was not

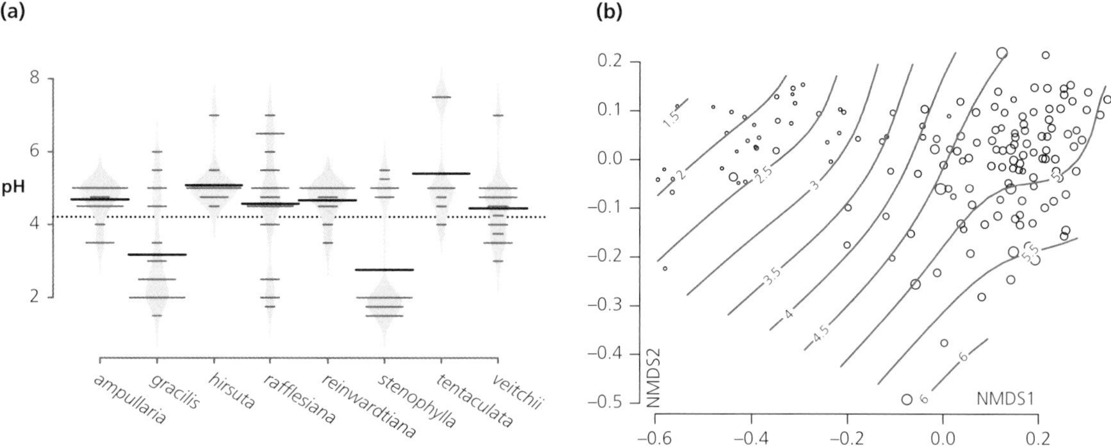

Figure 23.1 Acidity of *Nepenthes* fluid is different among species and correlates with bacterial community diversity. (**a**) Bean plots of pH in different *Nepenthes* species. The number of samples at each value is represented by the width of the thin lines, and thicker black lines represent mean values for each species. (**b**) Non-metric multidimensional scaling (NMDS) plot of *Nepenthes* bacterial communities using the phylogenetic unweighted UniFrac metric. The size of each circle corresponds to the pH value of each sample, and pH levels are mapped onto the plot as contour lines.

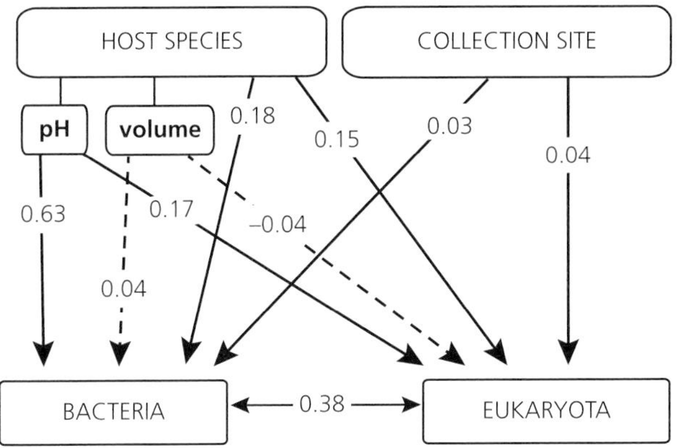

Figure 23.2 Correlations of measured factors with *Nepenthes* bacterial and eukaryotic communities. Solid lines indicate significant correlations ($P < 0.05$); dashed lines are not significant. The numbers are either Mantel *r* values (for continuous variables: pH, volume, and bacteria–eukaryote relationship), or r^2 values from permutational MANOVA tests (for discrete variables: host species and collection site).

significantly correlated with community composition, perhaps because volume changes frequently following rain and evaporation.

Patterns of co-occurrence (or co-presence) and mutual exclusion among inquilines across multiple pitcher samples can indicate species cooperation, competition, or preference for similar conditions (Faust and Raes 2012). It remains unclear whether inquiline community composition simply reflects inherent differences among host species, or if inquilines also contribute to these differences and are increasing niche separation among co-occurring *Nepenthes* species.

As a first step toward examining these alternative possibilities, I constructed a co-occurrence network of the most abundant inquiline species reported in Bittleston (2016). From OTU tables that were subsampled (rarefied) to the same numbers of sequences per sample (6778 for bacteria and 4852 for eukaryotes), I selected only OTUs with more than either 1000 (bacterial) or 600 (eukaryotic) sequences that were present in at least six pitchers across 175 samples, and combined the bacterial and eukaryotic tables. I used the Co-Net program (Faust et al. 2012) to measure four indices (Pearson and Spearman correlation coefficients, Bray–Curtis dissimilarity, and mutual information) and to retain the top and bottom 1000 edges. I then ran 100 permutations, bootstrapped them, and retained only significant edges (connections between two OTUs with alpha ≤ 0.05 after a Benjamini–Hochberg correction for multiple tests). The resulting network was visualized in Cytoscape, and shows nodes (OTUs) connected by edges representing either co-presence or mutual exclusion (Figure 23.3).

This co-occurrence network of *Nepenthes* inquilines reveals patterns related to pitcher fluid pH and inquiline identity. Modules, or tightly linked sets of organisms, are present (Figure 23.3). In general, bacteria cluster together most tightly and are surrounded closely by protists. Fungi, arthropods, and other metazoans are on the periphery of the network (Figure 23.3). There are two main clusters of bacteria separated by edges of mutual exclusion: the larger one contains species that are generally more abundant in pitchers with higher pH, and the smaller one contains species generally more abundant in pitchers with low pH. This result reflects other findings and provides further evidence of a strong influence of acidity on bacterial community composition. Insects also form a tight module (Figure 23.3). Insect inquilines may prefer similar conditions in pitchers, thus co-occurring because of shared niche space. That modules contain closely related organisms implies that competition may not be the primary force structuring inquiline communities.

Phylogenetic community compositions of bacteria and eukaryotes were correlated with each other (Figure 23.2). Other than the correlation of bacteria with pH, phylogeny was the strongest associational signal among the measured variables (Bittleston 2016; Figure 23.2). This result suggests that certain bacterial and eukaryotic lineages co-occur either because of real associations among them or because of preferences for similar environmental conditions.

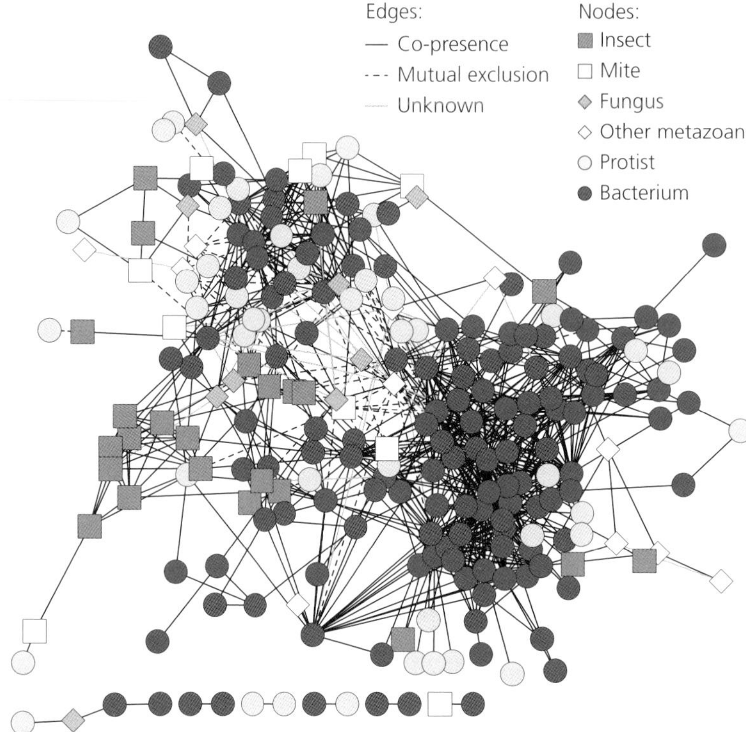

Figure 23.3 Co-occurrence network of abundant *Nepenthes* inquilines. Nodes represent operational taxonomic units (OTUs) connected either by solid black lines (edges) representing co-occurrence of those organisms, dashed lines representing mutual exclusion, or gray lines where there was not enough information about the relationship.

One possible mechanism for the correlation is introduction of bacteria to pitchers via eukaryotic hosts. However, this hypothesis was not supported: when eukaryotes were subset to arthropods, the correlation was weaker. The co-occurrence network also did not support a strong co-presence among bacteria and arthropods, as they were not attached by many edges (Figure 23.3); instead, I hypothesize that associations among bacteria and protists drive the pattern.

23.5.2 Comparisons with surrounding habitats

Pitchers are colonized from the surrounding environment, but the inquilines often are different from, or a subset of, species living in surrounding habitats (Bittleston 2016). These data suggest that environmental filtering plays a role in selecting the organisms that colonize and thrive in pitchers.

Inquiline community structure also differs from that of communities living in the surrounding leaf-litter and soil organic layer. Fewer eukaryotes and bacterial OTUs are found in pitcher plants than in the soil directly around the plants (Figure 23.4). Shannon diversity (H') also is significantly lower ($P < 0.001$) in pitchers than in soil communities, indicating that pitcher communities are less even and less predictable in composition (Figure 23.4). Because pitchers are temporary, ephemeral habitats, it is reasonable that they would contain fewer species and have more variable species composition than surrounding, more stable habitats. Chance also likely plays a role in colonization. Finally, prey capture occurs in pulses, and stochasticity in the availability of nutrients likely contributes further to the variability of inquiline community structure among pitchers.

23.5.3 Inquilines of *Nepenthes* and *Sarracenia*

The pitcher habit has evolved at least three times in different lineages of plants: pitchers of Nepenthaceae in Southeast Asia, Sarraceniaceae in the Americas, and Cephalotaceae in Australia all have pitchers that form from a single modified leaf

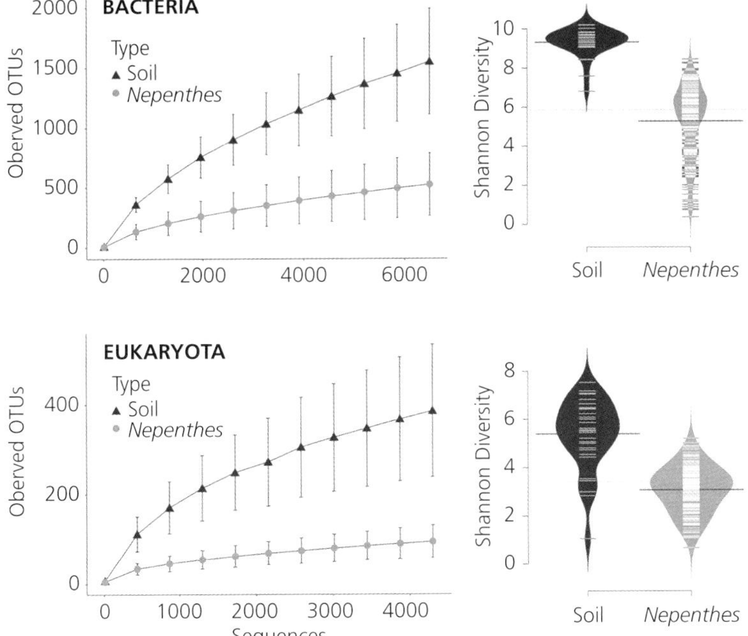

Figure 23.4 Species richness and Shannon diversity (H') of *Nepenthes* and soil organic layer bacterial (top) and eukaryotic (bottom) communities. *Nepenthes* communities are less species-rich and have lower H' than surrounding soil communities ($P < 0.001$). Error bars in the rarefaction curves represent standard deviations within each sequence-abundance category. The black lines in the bean plots are mean H' values within each group, and the dotted lines are the mean H' of the pooled data.

(Chapters 3, 9, 10). Certain bromeliads from South America are sometimes classified as pitcher plants, although the physical structure of the pitchers is different (Chapter 10). The convergent evolution of pitcher form and function provides an opportunity to examine how similarities in habitat can affect the evolution and maintenance of associated communities (Bittleston et al. 2016b).

I investigated the different families of insects and arachnids living in pitcher habitats, to test if the ones found in *Nepenthes* and *Sarracenia* systems (cf. Chapter 24) were more similar than expected by chance. Using data from Kitching (2000) and Adlassnig et al. (2011), I counted only insect and arachnid families that appeared to be regularly associated with *Nepenthes* or *Sarracenia* species (found more than once in the habitat, avoiding incidental organisms). As a control comparison, I did the same counts for tree-hole inquilines from North America and Southeast Asia. The numbers of inquiline families found in *Sarracenia* and *Nepenthes* were 10 and 21, respectively, with 9 shared between them, whereas for tree holes of North America and Southeast Asia the numbers were 13 and 21, respectively, with 8 shared (respectively, $P = 0.003$, and $P = 0.20$ based on permutation tests with 10,000 randomizations and keeping the number of shared families constant and using all families found in association with phytotelmata as the source population for both pitcher-plant and tree-hole samples). That is, certain families of insects, mites, and spiders appeared more likely to associate with pitcher plants, even on opposite sides of the planet.

Takeuchi et al. (2015) compared *Nepenthes* bacterial communities at the class level to those from other habitats with published data, including farm soil, coastal seawater, the Amazon River, termite gut, human gut, *Arabidopsis thaliana* phyllosphere, and *Sarracenia alata*. They found that *S. alata* bacterial community composition was most similar (Jaccard similarity) to the *Nepenthes* samples. A more extensive study comparing over 140 *Sarracenia* and 180 *Nepenthes* samples substantiated the result: community compositions (using the phylogenetic unweighted UniFrac metric) were more similar to each other than they were to the surrounding bog and soil communities from their respective habitats (Bittleston 2016). Furthermore, both *Nepenthes* and *Sarracenia* pitcher communities had similar measures of observed OTU richness and H' for bacteria and eukaryotes (Bittleston 2016). For both macro- and micro-inquilines, the data suggest that convergent evolution of a pitcher appears to lead to convergent interactions among the plants and their associated organisms (Bittleston et al. 2016b).

23.6 Future research

The pitchers of carnivorous pitcher plants are living habitats that host complex, diverse communities. Colonizing organisms contend with acidity, viscous fluid, digestive enzymes, and free radicals, but surviving inquilines have a nearly guaranteed source of host-trapped nutrients (prey), higher than normal oxygen levels, and a protective chamber. In general, inquilines are a highly variable subset of the available pool of organisms. Bacterial community structure is strongly correlated with fluid pH, and bacteria co-occur with many protists. Some insect and mite inquilines specialize on certain *Nepenthes* species, potentially because of differences in the physical structure of the pitchers. Because we see some convergence in organisms colonizing *Nepenthes* and *Sarracenia* pitchers, it is likely that pitcher form and function selects for related inquilines in particular functional groups.

Many aspects of *Nepenthes* inquiline communities are still unknown and are prime targets for future research. These include: viral dynamics; succession or unpredictable community assembly; which (if any) special adaptations are necessary for colonizing pitcher habitats; why some pitcher species acidify their fluid and whether low pH selects for mutualistic bacteria; the extent to which associations are parasitic or mutualistic; and if there are short-term host–inquiline feedback loops or evolutionary co-diversification of inquilines and *Nepenthes*.

Pitchers are ideal model systems for community ecology (Ellison et al. 2003, Srivastava et al. 2004). Because pitcher pools are isolated islands contained within a plant, defining the boundaries of the community is relatively straightforward. The abundance of pitchers and their diminutive size make them convenient ecosystems where one could measure replicate communities across all trophic levels. Future studies could use pitcher systems to measure the predictability of community assembly (e.g., Ellison et al. 2003), and to understand more fully processes that structure communities. New developments in network analyses and other methods for studying complex systems could be used to uncover patterns of species interactions within pitchers, and to examine tipping points in model ecosystems (Sirota et al. 2013). Perhaps studies of pitcher-plant micro-ecosystems will provide a better understanding of general ecosystem dynamics and contribute to conservation efforts in the face of global change (Ellison and Gotelli in press).

CHAPTER 24

Pitcher-plant communities as model systems for addressing fundamental questions in ecology and evolution

Thomas E. Miller, William E. Bradshaw, and Christina M. Holzapfel

24.1 Introduction

Many carnivorous plants have close associations with other species. The best studied of such plants is *Sarracenia purpurea*. The genus *Sarracenia* includes 11 species (Chapter 9), most of which have upright tubular leaves and occur across parts of the coastal plain of southeastern North America. Many of these species serve as a host for other obligate or facultative visitors (Juniper et al. 1989), referred to generally as "inquiline" species. *Sarracenia purpurea* (referring to the *S. purpurea* species complex, which includes *S. purpurea* ssp. *purpurea*, *S. purpurea* ssp. *venosa*, *S. purpurea* ssp. *venosa* var. *montana*, and *S. rosea*; Chapter 9) is unique in several regards. Its distribution is much greater than the other pitcher plants; it occurs in bogs and related habitats from Alabama and Florida to Canada, where it can be found growing from eastern British Columbia to the Maritime Provinces (Chapter 9, Figure 9.2). *Sarracenia purpurea* also has a more prostrate growth form and the pitcher mouth is open to precipitation, unlike any of the other species (*S. psittacina* is also prostrate, but has a well-covered leaf opening). Because of this morphology, *S. purpurea* is the only species in the genus that consistently holds liquid in its leaves and this liquid almost entirely is derived from precipitation.

Sarracenia purpurea also is unique in the genus in having a relatively large number of mostly invertebrate and microbial inquiline species that live in its water-filled leaves (Miller and Kneitel 2005). Juniper et al. (1989) describes many of these species and their associations with *S. purpurea*; since 1989, over 200 papers have been published on *S. purpurea* or its inhabitants, including studies of interactions among different inquilines; local and biogeographic patterns in inquiline community structure; and studies of their genetic and evolutionary processes. The *S. purpurea* inquiline community has become a model experimental system for studying contemporary questions in ecology and evolution (Srivastava et al. 2004).

In this chapter, we review how studies of *Sarracenia* inquilines have answered fundamental questions in ecology and evolution. Much of the work focuses on *Sarracenia purpurea*, but where appropriate we will discuss studies using other *Sarracenia* species.

24.2 Natural history of *Sarracenia* and its inquilines

Sarracenia purpurea s.l. is a clonal, long-lived species that occurs primarily in bogs and savannas and other wet areas with generally acidic, low-nutrient conditions. It can be locally abundant but regionally scarce, due to its highly specialized habitat needs. Gene-based studies show that *S. purpurea* forms two major clades: a southern clade ranging from the Gulf Coast to mainland Maryland (ssp. *venosa*) and a northern clade (ssp. *purpurea*), ranging from the Eastern Shore of Maryland northwards (Stephens et al. 2015b; Chapter 9). The large geographic

Miller, T. E., Bradshaw, W. E., and Holzapfel, C. M., *Pitcher-plant communities as model systems for addressing fundamental questions in ecology and evolution*. In: *Carnivorous Plants: Physiology, ecology, and evolution*. Edited by Aaron M. Ellison and Lubomír Adamec: Oxford University Press (2018). © Oxford University Press. DOI: 10.1093/oso/9780198779841.003.0024 distribution of *S. purpurea*, including both non- and postglacial populations, results in variation in evolutionary history and genetic architecture over identifiable time scales and environmental gradients. Although the plants can be long-lived, individual leaves seldom survive longer than a year; one study from a southern population found a median survival time for leaves of 40 weeks (Miller and terHorst 2012). The shape of the leaf minimizes desiccation and provides a stable habitat for aquatic species in the leaf (Kingsolver 1981).

The complex species interactions associated with *S. purpurea* have been studied for many years (Addicott 1974, Fish and Hall 1978, Bradshaw 1983). The species that occur within leaves are components of a detritivore-based or "brown" food web (Butler et al. 2008, Mouquet et al. 2008), with basal energy and resources derived from prey captured by the pitcher plant's leaves. Most of the energy and nutrients are captured by the pitcher early in the life of each leaf (Fish and Hall 1978, Bradshaw 1983, Miller and terHorst 2012) although leaves may continue to capture a significant number of prey if they persist into a second year (Heard 1998).

24.2.1 Prey capture

Some studies suggest that *S. purpurea* uses nectar to attract ants, which make up a majority of its prey (Miller and terHorst 2012), whereas others have suggested a more generalized prey spectrum (Folkerts 1982, Cresswell 1991). The discrepancy may result from differences in prey capture with leaf age; at least one study in a southern population shows that leaves attract only ants, presumably with nectar, during the first 4–6 weeks of life (Miller and terHorst 2012), after which leaves may act as a less selective pitfall trap (Fish and Hall 1978, Cresswell 1991, Heard 1998). Although algae have been reported in pitcher plants, they generally are rare and restricted to older leaves.

24.2.2 Microbes

Captured prey are primarily broken down by a diverse microbial community, with energy and nutrients transferred on to higher trophic levels. It is unclear how much the rest of the community contributes to the activity of the microbes. Consumers have the potential to affect microbial abundance and diversity (Heard 1994, Peterson et al. 2008, Hoekman et al. 2009), but Butler et al. (2008) found that the rest of the community had little effect on nutrients available to the plant itself. All *Sarracenia* species are known to have digestive glands that are embedded in the cell wall of the leaf and produce digestive enzymes (Juniper et al. 1989; Chapter 13). Digestive enzymes are produced by *S. purpurea* mostly in younger leaves or facultatively in older ones in response to prey (Gallie and Chang 1997), but the contribution of enzymes to prey digestion still needs exploration. Molecular assays of microbial communities in *S. purpurea* are revealing high densities and diversities of microbes, consisting largely of Proteobacteria and Bacteroidetes and a variety of yeasts (Gray et al. 2012, Paisie et al. 2014). The composition of the microbial community has been shown to change with different bacterivores (Paisie et al. 2014) and the presence of the pitcher-plant mosquito, *Wyeomyia smithii* (Peterson et al. 2008).

24.2.3 Bacterivores

The microbes are consumed by a suite of bacterivores, including protists and rotifers. The protists are thought to be mostly generalists, and include species from several different phyla (e.g., Ochrophyta, Euglenozoa, Ciliophora) and both flagellates and ciliates. Although several different rotifers have been also identified from pitcher plant leaves, the bdelloid *Habrotrocha rosa* is the most common rotifer throughout the range of *S. purpurea* (Buckley et al. 2010). This species may contribute to the host plant by excreting nitrogen and phosphorus into the leaf water (Błędzki and Ellison 1998). The specialist mite *Sarraceniopus gibsonii* is thought to be omnivorous, feeding on bacteria and protists, and can be quite common within leaves (Miller and terHorst 2012) and across populations (Buckley et al. 2010). Very little is known about its ecology, however.

24.2.4 *Wyeomyia smithii*

Many of the bacterivores are consumed in turn by filter-feeding larvae of the well-studied pitcher-plant mosquito, *Wyeomyia smithii*. Females oviposit directly into leaf-held water (Bradshaw 1983, Miller and terHorst 2012), where the eggs hatch, develop through four larval instars, and then pupate, eventually emerging as adults. *Wyeomyia smithii* are multivoltine in the southern range of *S. purpurea* but univoltine in northern Canada, where they overwinter in the frozen water in the leaf. *Wyeomyia* are omnivores, selectively feeding on larger protozoa and rotifers (Błędzki and Ellison 1998, Kneitel and Miller 2002, Kneitel 2012), while also directly consuming bacteria and detritus (Cochran-Stafira and von Ende 1998, Hoekman 2007).

Historically, *Wyeomyia smithii* was divided taxonomically into two species based on characteristics of the anal and ventral papillae: *W. smithii* ranging from New Jersey northwards, and *W. haynei* ranging from Maryland to the Carolina coastal plain (Bradshaw and Lounibos 1977). Bradshaw and Lounibos (1977) found that both northern and southern "species" of pitcher-plant mosquitoes were fully interfertile and that hybrids between a Massachusetts and an Alabama population were indistinguishable from *W. haynei*. They concluded that there is one single species of *Wyeomyia* in pitcher plants, *W. smithii*, and that *W. haynei* represents an intermediate morph along a geographic cline.

More recent work has shown that *Wyeomyia smithii* is divided into southern and northern clades (Figure 24.1, Merz et al. 2013), which matches the genetic variation among populations of the host plant (Stephens et al. 2015b). Morphological, physiological, developmental, and molecular characters consistently show that *W. smithii* has diverged from south to north, starting along the Gulf Coast, then along the Atlantic coastal plain and post-glacially to the southern Appalachians and northward into Canada (Bradshaw and Lounibos 1977, Merz et al. 2013). The southern clade of *W. smithii* splits, as does *S. purpurea*, between Gulf and Atlantic coasts with the Atlantic coastal populations being derived. Deeper taxon sampling of *W. smithii* shows that the northern clade splits into at least four monophyletic lineages (Merz et al. 2013).

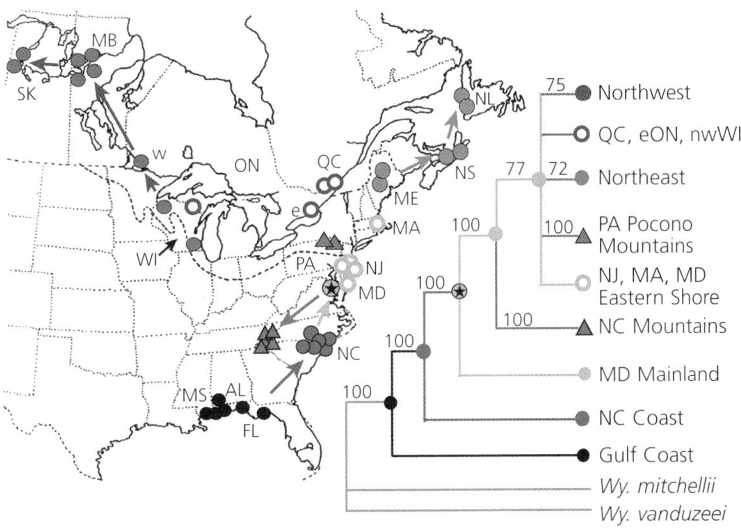

Figure 24.1 (Plate 16 on page P14) Phylogeography of *Wyeomyia smithii* based on the combined 46-population tree. Arrows indicate directions of range expansion. Maximum extent of the Laurentide Ice Sheet at the last glacial maximum is plotted as a dotted line. Two-letter abbreviations identify each state or province. USA states: AL—Alabama; FL—Florida; MA—Massachusetts; ME—Maine; MD—Maryland; MS—Mississippi; NC—North Carolina; NJ—New Jersey; PA—Pennsylvania; WI—Wisconsin. Canadian provinces: MB—Manitoba; NL—Newfoundland and Labrador; NS—Nova Scotia; ON—Ontario; QC—Quebec; SK—Saskatchewan. Triangles represent mountain populations in the southern Appalachians of North Carolina (southern triangles) or in the Pocono Mountains of Pennsylvania (northern triangles). The filled symbol with the star indicates the basal mainland Maryland population (MD2). Numbers associated with each node or branch tip represent maximum parsimony bootstrap support for that clade. There are no support numbers for the QC, eastern ON (eON) and northwest WI (nwWI), or the NJ, MA, and Eastern Shore MD populations because they do not constitute a monophyletic grouping. Adapted from Merz et al. (2013).

24.2.5 Other dipterans

The larvae of two other dipterans commonly occur in the leaves of *S. purpurea*. Larvae of the midge, *Metriocnemus knabi*, are shredders, and are thought to subsist on the prey carcasses that sink to the bottom of the leaves. In doing so, *M. knabi* larvae can increase resource availability for bacteria, which may have indirect positive effects on protozoa and mosquito larvae higher in the food chain in what Heard (1994) termed a "processing chain commensalism." However, the role of *M. knabi* in the inquiline community is not clear and may vary among sites. Bradshaw (1983) found that *Wyeomyia smithii* can inhibit pupation success of *M. knabi*, whereas *M. knabi* can facilitate the rate of *W. smithii* development, especially at high larval densities. In some cases, midges may have direct negative effects on some bacterivores through predation (Hoekman et al. 2009).

Larvae of the flesh fly (*Fletcherimyia fletcheri*) also are common in *S. purpurea* leaves. The genus *Fletcherimyia* is tightly associated with *Sarracenia*: there are up to eight other species of *Fletcherimyia* that occur in the leaves of various *Sarracenia* (Dahlem and Naczi 2006). They all share a similar life history, including feeding on dead prey on the water surface, ovolarvipary, and cannibalism (Forsyth and Robertson 1975), except that flesh flies found in *S. purpurea* are necessarily more aquatic than those found in other *Sarracenia*. However, species in this group often are misidentified and some species of *Fletcherimyia* appear to live in more than one species of *Sarracenia* (Stephens and Folkerts 2012). The larvae complete three instars while consuming prey captured by the leaf (Forsyth and Robertson 1975). The larvae have unusual cup-like posterior spiracles, which, when spread out, allow the larvae to float at the surface of the water within the leaf. Cannibalism often results in one survivor per leaf (Rango 1999); the surviving instar crawls from the leaf and pupates in the moist vegetation around the plant. Adults of *F. fletcheri* also may act as a pollinator for *S. purpurea* (Ne'eman et al. 2006).

24.2.6 Inquiline dispersal

Leaves are generally colonized within days of opening, so significant dispersal appears to occur among leaves. The mechanisms of dispersal of rotifers, protozoa, mites, and microbes such as bacteria and yeast are unknown, and their effects on the plant or the dipterans are rarely studied. Of the dipterans, *Wyeomyia smithii* are thought to be poor flyers, and generally lay 1–2 eggs per leaf in mostly young leaves. Oviposition by both *W. smithii* and *Metriocnemus knabi* is higher in larger leaves and higher local leaf densities (Trzcinski et al. 2003). Overall, dispersal remains an important gap in our understanding of the natural history of these communities.

24.2.7 Non-aquatic associates: moths

Several non-aquatic species are also associated with *Sarracenia*, including at least four noctuid moths that are obligate herbivores on *S. purpurea: Papaipema appassionata* and three species of *Exyra*. Little is known about the former species, except that it is a rhizome feeder that generally kills its host plant (Atwater et al. 2006). Species in the genus *Exyra* all use *Sarracenia* as hosts, including *E. fax* on *S. purpurea*, *E. ridingsii* on *S. flava*, and *E. semicrocea* on several different species of *Sarracenia* (Folkerts 1999). For all three *Exyra* species, adults lay eggs on the leaves, where they go through five instars, slowly filling leaves with frass. The larvae consume the inside of the leaf, leaving a thin membranous outer wall. They generally weave a fine web over the opening of the leaf, possibly to protect the larvae from predators. Larvae can move from leaf to leaf, but eventually move to a final leaf, often cut a hole in the bottom of the leaf to prevent drowning, skeletonize the leaf, and then develop into a pupa.

24.2.8 Pollinators

The common lore is that flowers of *Sarracenia* do not self-pollinate and require animal pollinators. For *Sarracenia* with larger flowers (including *S. purpurea*), it has often been assumed that natural pollinators include native bumblebees, whereas other species required smaller bees and flies. However, our current understanding of pollination in *Sarracenia* comes almost entirely from observations (e.g., Folkerts 1982, Ne'eman et al. 2006). Juniper et al. (1989: 271) noted that, "[a] critical comparative study of pollination processes in *Sarracenia* is needed to determine how they relate to reproductive success and hybridization." This need still remains.

24.2.9 Spiders

Various spiders also often are found in association with carnivorous plants, where they may be prey, competitors, or even mutualists (Juniper et al. 1989). Some ground-foraging spiders are known to be occasional prey of *S. purpurea* (e.g., Cresswell 1991, Heard 1998). The Green Lynx spider, *Peucetia viridans*, often co-occurs with *Sarracenia* in the southeastern United States; these and other spiders may use leaves and flower heads for foraging or laying and guarding egg masses (Jennings et al. 2008). Because *Sarracenia* and spiders may have common insect prey, it has been suggested that they can be competitors (Cresswell 1991, Folkerts 1999), but further studies are required to demonstrate the implications of such interactions for either species.

24.3 *Sarracenia purpurea* and its associates as a model ecological system

Several features of *S. purpurea* and its inquilines make for a near-ideal experimental system in which to study questions about species interactions, species diversity, and community structure. Individual leaves create small and well-defined phytotelmata that can be easily manipulated. The communities found inside the leaves are transient and occur at well-defined spatial scales (e.g., host leaves, plants, and populations) and thus are ideal for studying dispersal, invasion, and succession. *Sarracenia purpurea* is highly constrained to specific habitats (bogs, seeps, and savannas), yet occurs over a very large geographic range from North Florida to northern Canada, providing significant variation in climate. The leaves host several obligate species over the entire range of the host plant. Further, both the plant and its inquilines can be maintained in growth chambers or greenhouses for more intensive study, enabling the evaluation of inquiline physiology, ecology, and evolution in their natural microhabitat under controlled conditions. The result is that many studies have been conducted with *S. purpurea* and its inquilines on a variety of questions.

Other pitcher plants share some of these characteristics (Adlassnig et al. 2011; Chapter 23). *Sarracenia alata* (e.g., Satler et al. 2016) and *Darlingtonia californica* (Naeem 1988) also host inquilines, but other related species do not (e.g., *S. psittacina*). In many cases, detailed studies of possible inquilines in other New World pitcher plants are simply lacking (e.g., *Heliamphora*; Adlassnig et al. 2011). Finally, other types of carnivorous plants also host inquilines, including those found in the bladder traps of *Utricularia* (Chapter 25). Unlike *S. purpurea* and its inquilines, however, most of these systems neither have been well described nor generally have been used to address broader questions in ecology and evolution.

24.3.1 Mutualism between *Sarracenia purpurea* and its aquatic inquilines

Sarracenia purpurea and its inquilines comprise a mutualistic, but not necessarily co-adapted relationship (Bradshaw and Creelman 1984). Decomposing prey provide the nutrient base for the inquilines community that respires CO_2 and excretes ammonium ions. Both CO_2 and ammonia production increase at higher temperatures while higher temperature and brighter light accelerate uptake of both molecules by the host plant. Ammonium ions are taken up directly by the glutamate cycle without further reduction and the concomitant reduction of CO_2 generates oxygen that is directly transferred from the leaves to the contained water. In essence, this is a feed-forward system: inquilines are producing maximum CO_2 and ammonium ion under exactly the same conditions that the plant is taking up these molecules and generating excess oxygen. There are two consequences. First, unless the system is overloaded by the capture of large prey, digestion remains aerobic. Second, pitcher plants are located within the boundary layer of their habitats and often exposed to direct sunlight. Water temperatures in leaves routinely range 40–50 °C in the south and 30–40 °C in the north (Bradshaw et al. 2004). At low levels of convection and high temperatures, carbon can become limiting for photosynthesis so that carnivory can provide a source of carbon as well as nitrogen (Bradshaw and Creelman 1984).

24.3.2 Consumer versus resource control of communities

Ecology has long grappled with understanding the roles of resource availability (competition), consumer abundance (predation), and their interactions (e.g., Hairston et al. 1960, Paine 1966). Theoretical work has suggested that both top-down and bottom-up forces can be important in communities (Powers 1992), but has also suggested that interactions among competition and predation can lead to more complex patterns (e.g., Paine 1966, Holt 1977). Experiments to disentangle resource and consumer effects can be complex, requiring manipulating both resources and predators in factorial designs.

Pitcher plants are ideal for such studies (Kitching 2001), as both the resources (dead prey) and top consumers (e.g., mosquitoes) are easily manipulated and community responses occur over relatively short timescales. One of the earliest experimental studies with *Sarracenia* demonstrated that mosquitoes suppressed protozoa without any "keystone effect" in increasing protozoa diversity (Addicott 1974, Cochran-Stafira and von Ende 1998, Peterson et al. 2008). Kneitel and Miller (2002) showed that effects of adding resources translate up through the food web to increase bacterial, protozoan, rotifer, and mite abundances, but increased predation affected only rotifers, with a weak trophic cascade on bacteria. Hoekman (2007, 2011) found similar patterns, although top predators were found to have different effects in Michigan and Florida populations that may be related to temperature-associated differences in microbial productivity.

The food webs in *S. purpurea* also have been used to demonstrate more complex interactions between competition and predation. Heard (1994) demonstrated that *M. knabi* larvae increase bacterial abundances, which can result in increased growth of *W. smithii* larvae further up the food web. Hoekman et al. (2009) found similar positive effects of *M. knabi* on bacteria, but found that it also suppressed rotifers and some protozoa, apparently though direct consumption. Finally, Kneitel and Miller (2003) demonstrated that moderate levels of dispersal among pitcher plant leaves can increase species richness, but only in the absence of the top predator, *W. smithii*. Collectively, these manipulative studies of inquiline food webs show that entire communities are not dominated by only resource or only consumer effects and that omnivory can create more complex species associations in communities.

24.3.3 Testing theories of succession

Succession has remained a fundamental but frustrating concept in ecology, with longstanding questions about mechanisms and predictability (McIntosh 1999). Much of our understanding comes from plant succession and much of the evidence is indirect and obtained from chronosequences. Inquilines can be used to study succession in aquatic systems; entire successional sequences can be observed directly over a reasonable time frame.

The inquiline communities found within *S. purpurea* leaves are highly variable from leaf to leaf within and between plants in a single population. There have been several attempts to understand this variation; earlier studies identified specific leaf traits that are correlated with this variation, including leaf size (Cresswell 1993, Harvey and Miller 1996, Buckley et al. 2004) and density (Trzcinski et al. 2003). However, it appears that the primary factor determining variation among leaves is leaf age, and successional patterns of prey capture and *W. smithii* abundance. Fish and Hall (1978) documented dipteran abundances through succession in natural populations of *S. purpurea* in Massachusetts and prey capture rates in leaves of greenhouse-grown plants. They found that leaf age determines the strong resource and consumer drivers in inquiline communities. Most prey are captured in leaves < 5-wks old, when nectar is present, and most mosquitoes also occur in leaves < 7-wks old.

Miller and terHorst (2012) tested general theory associated with succession using a 28-month study of *S. purpurea* leaves in North Florida. Successional theory suggests that communities become more stable and converge with similarly aged communities during succession, and that species diversity peaks at some intermediate stage. Individual *S. purpurea* leaves survived for a median of 40 wks, with most of the successional changes occurring in the first 20 wks (Miller and terHorst 2012). Mosquitoes oviposit in young leaves (Figure 24.2a), except in the early spring before young leaves are available. At the bottom of the food web, most ant prey are captured in the first few weeks, with virtually no ants captured after week ten (Figure 24.2c). Bacteria appear to be limited mostly by resources, whereas protozoa are limited by mosquito predation (Figure 24.2c and 24.2b, respectively). Communities did become more similar through time and stability peaked at intermediate levels, but diversity showed no consistent pattern with leaf age (Miller and terHorst 2012). The study demonstrated that understanding system-specific drivers of succession may be more important for understanding succession than general principles.

The pattern of succession in Miller and terHorst (2012) is very similar to that described by Fish and Hall (1978), despite having been studied in very different locations and for a longer time. Both studies note that leaf age often is neglected in studies of *Sarracenia* inquilines and needs to be explicitly considered in future studies because of its strong effects on succession.

24.3.4 Dispersal and metacommunities

Dispersal among local habitats is known to affect species persistence and diversity in patchy habitats (Levin and Paine 1974) and islands (MacArthur and Wilson 1967). More recently, metacommunity theory has unified ideas about the role of dispersal, competition, and local habitat heterogeneity and stability (Leibold et al. 2004). This theory makes predictions about the maintenance of

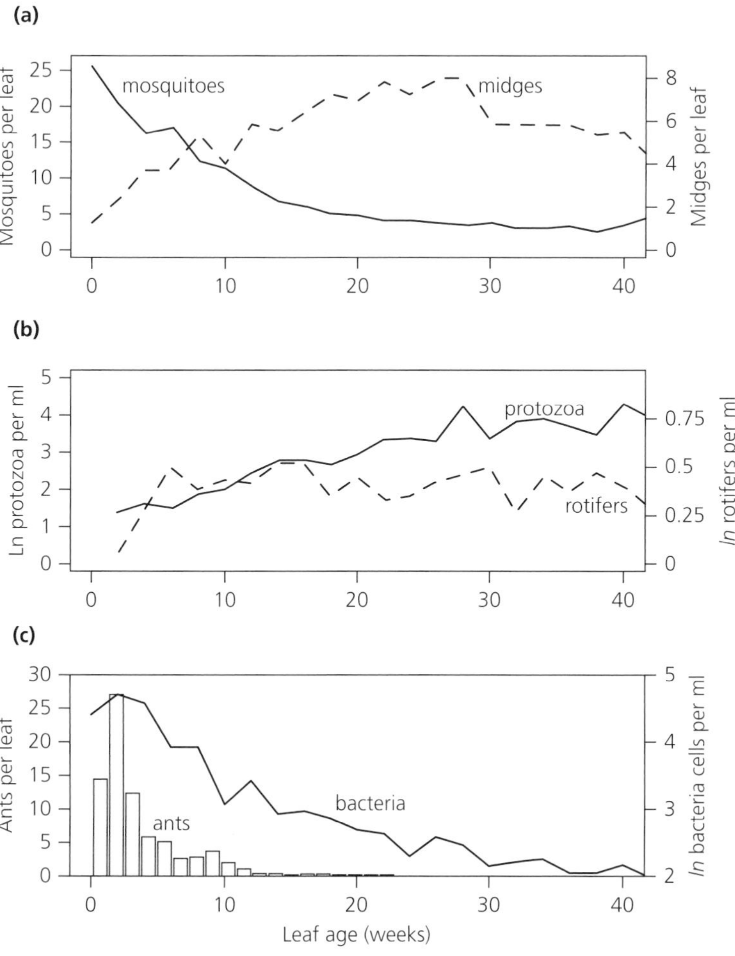

Figure 24.2 Successional patterns of selected inquilines from leaves followed from opening through the first 40 weeks. (**a**) Oviposition behavior by the Dipteran *W. smithii* (solid line) causes high abundances in younger leaves, with a gradual loss of individuals as they eclose. *Metriocnemus knabii* (dashed line) shows a much lower overall abundance and a more intermediate peak in abundance. (**b**) The bacterivores include various protozoa species (solid line) whose numbers increase as predation declines, and rotifers (dashed line) whose densities are less affected by leaf age. (**c**) Virtually all ant capture (bars) is by leaves <10 weeks of age, which generates a corresponding pattern in bacteria densities (dashed line). Adapted from Miller and terHorst (2012).

communities at different spatial scales, especially as a function of dispersal among local habitats. Testing such theory remains difficult, especially in natural communities.

Several studies have used patterns of inquiline species distributions among *S. purpurea* leaves to infer scales of dispersal (Harvey and Miller 1996, Trzcinski et al. 2003, Buckley et al. 2004). However, these studies generally explain only a small amount of the total variation in the composition of local communities, perhaps because they did not incorporate leaf age (§24.3.3). Baiser et al. (2013) analyzed inquiline data from a variety of spatial scales, and concluded that metacommunity models using either species-sorting or patch dynamics best predicted the variation in community structure. Studies also have used direct experiments to test if some inquiline species are limited by dispersal. Conditions for establishment success vary among protozoa species, with some species having higher success with higher resources, whereas others establish best when predators are absent (Miller et al. 2002). Finally, in the one study that varied rates of dispersal among leaves to test metacommunity theory, Kneitel and Miller (2003) demonstrated that dispersal can increase diversity of bacterivores when the mosquito predator is missing from the food web.

24.3.5 Biogeography at the scale of a community

An important goal of ecology is to understand the role of spatial and temporal scales in determining species abundances and community composition (Levin 1992). Reaching this goal could help to unify core concepts of population biology, evolutionary history, and ecosystem ecology. A unique characteristic of the *S. purpurea* inquiline community is its large range, which is ideal for biogeographic questions about diversity, stability, and coevolution. The host plant is similar across the entire range, although leaf morphology varies with temperature and precipitation (Ellison et al. 2004). Virtually all populations of *S. purpurea* have the same dipterans, rotifers, mites, and protozoa; although within-population variation in inquiline composition can be significant, among-population variation is relatively low (Buckley et al. 2003, 2010).

Rasic and Keyghobadi (2012) used microsatellite loci for the midge, *Metriocnemus knabi*, in *S. purpurea* to suggest that bog size and plant density influence oviposition behavior, which in turn determines patterns of genetic differentiation within and across local populations. *Sarracenia purpurea* inquilines also have been used to address several questions in biogeography. For example, diversity increases with latitude (Buckley et al. 2003), contrary to the general pattern found in other communities, and species composition is correlated with climatic variables, including temperature and precipitation (Buckley et al. 2010). Variation in food-web structure among leaves within sites decreased with latitude (Baiser et al. 2012), perhaps because shorter growing seasons constrain variation in leaf age and conditions. However, despite large sample sizes and scale of sampling, only 40% of the variation in community structure among leaves could be explained by among-leaf, within-site, or among-site variables.

It is difficult to conduct experiments across the large scales necessary to test the factors determining biogeographic patterns for entire communities. Natural microhabitats, such as rock pools, tree holes, bromeliads, and pitcher plants are well suited for such studies (Srivastava et al. 2004). Hoekman (2007) found that bacterivores in Michigan and Florida responded differently to resource and consumer control, perhaps due to differences in temperature. Gray et al. (2016) investigated the importance of temperature, trophic structure, and local adaptation on ecosystem functioning. They collected leaf contents from five sites along a natural temperature gradient from Florida to Quebec, and then conducted a "reciprocal temperature" experiment using growth chambers. Communities from each site were maintained at a summer temperature appropriate for each of the five sites, with and without the top predator (mosquito larvae); ecosystem functions such as bacterial respiration, and ammonium and phosphorus production were monitored. Gray et al. (2016) found that temperature has a greater effect on ecosystem functioning than site of origin, and that top-down trophic regulation increases with temperature. Similar experiments may allow us to separate the effects of climate and host plant, and the contributions of various trophic levels to large-scale community and ecosystem patterns (Parain et al. 2016).

24.3.6 Evolution in a community context

The realization that ecological and evolutionary changes can occur within similar time frames (Hairston et al. 2005) has led to questions about how evolution occurs in a multispecies context and how evolution can lead to feedback loops between ecological and evolutionary change (e.g., Kokko and López-Sepulcre 2007). Because pitcher-plant leaves develop from high prey capture and mosquito abundances to low prey and mosquito abundances (Figure 24.2), the conditions for bacterivores go from favoring predation tolerance (consumer control) to favoring competitive traits (resource control). Further, the relatively short generation times of pitcher-plant protozoa allow for "rapid" evolution of either predator tolerance (terHorst et al. 2010) or growth in competition (terHorst 2011). An investigation of the growth of protozoa from leaves of different ages does show that predator tolerance and interspecific competitive abilities evolve (Miller et al. 2014). However, interspecific competitive ability evolves differently in subordinate and dominant competitors (Figure 24.3), perhaps because subordinate competitors experience largely interspecific competition, whereas dominant competitors experience intraspecific competition.

There are also several very interesting cases of coevolution of pitcher plants with other species. Different species of *Nepenthes*, for example, have apparently coevolved with ants, bats, and small mammals (Chapter 26). As obligates, many of the *Sarracenia* inquilines appear to be adapted to conditions in pitcher-plant leaves, but the reciprocal evolution of the plant to its inquilines is less clear (§24.3.1). The long shared history of pitcher plants and their inquilines over a large biogeographic scale means that they likely are interesting systems for studies of coevolution (e.g., Stephens et al. 2015b). *Sarracenia alata* has been shown to have population genetic structure that is congruent with several associated species, including mites and fungi (Satler et al. 2016). However, *S. alata* has a very restricted range; it would be very interesting to see similar work with the diverse community associated with *S. purpurea*.

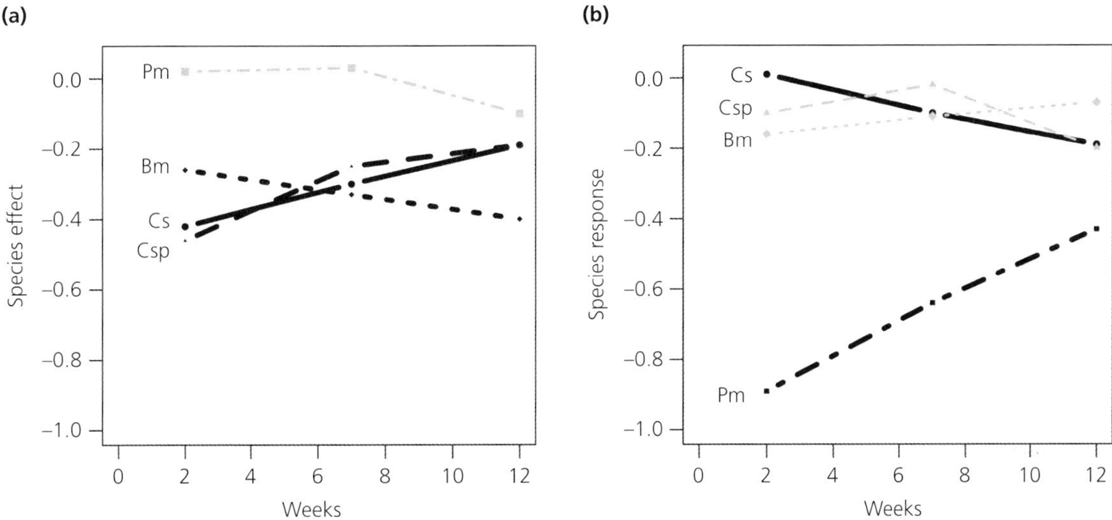

Figure 24.3 Average effect of protozoa species on other protozoans (**a**) and response to protozoa by other protozoans (**b**), using protozoa isolated from leaves of different ages (2, 7, and 12 wk old). Pm: *Poterioochromonas malhamensis*; Bm: *Bodo menges*; Cs: *Colpoda steinii*; Csp: an unidentified colpodid species. Values represent the growth of each species in mixture relative to its growth in monoculture; negative values indicate greater suppression of growth. Bold solid lines have significantly increasing effect or response with wk; bold dotted lines indicate significantly decreased effect or response with wk. Adapted from Miller et al. (2014).

24.4 *Wyeomyia* as a model system for inquiline species

A common characteristic of carnivory in plants is that they serve as habitats or hosts for unique and diverse assemblages of other species (Chapters 23–26), and many such associations remain to be described. Because inquilines are nearly ubiquitous, an understanding of these species may be necessary to fully understand the ecology and evolution of carnivorous plants themselves. Here we consider the biology of one inquiline in more detail, and illustrate that it can serve as a model for understanding the ecology of other inquilines worldwide.

The mosquito *Wyeomyia smithii* is the best studied inquiline among all carnivorous plants. The climates where different populations of *W. smithii* occur vary from relatively benign in the south where growth seasons are long and the dipterans are multivoltine, to harsh conditions in the north where there may be only a few months for the plants to grow and there is only one brood per year. These climatic differences have resulted in significant variation in the life history of *W. smithii* across its geographic range. For example, southern populations require a blood meal for their second and subsequent ovarian cycles, whereas northern populations do not bite. There are also latitudinal gradients in traits including number of egg batches per season, rates of egg production, and lifetime fecundity. This variation sets the stage for studies of the evolution of important life-history traits.

24.4.1 Density-dependent selection

The theory of *r*- and *K*-selection was developed by MacArthur (1962), who considered genetic models of how density-dependent selection should act. He concluded that in equilibrium populations, "the carrying capacity, *K*, replaces fitness (*r*) as the agent controlling the action of natural selection" (MacArthur 1962: 1897). This basic concept spawned a myriad of tests of MacArthur's theory based on life-history traits presumably associated with *r*- and *K*-"strategies." The ability to rear *W. smithii* in its natural microhabitats under controlled conditions provided the opportunity to test this theory in its most fundamental form: density-dependent selection imposed in nature directing the evolution of life histories, especially *r* and *K*.

Within pitchers, *W. smithii* larvae are limited primarily by intraspecific scramble competition for available resources, which in turn are determined by the prey captured by that leaf (Bradshaw 1983, Broberg and Bradshaw 1995). Mean crowding of *W. smithii* larvae per unit resource is negatively correlated with latitude and altitude (Bradshaw and Holzapfel 1986). This geographic gradient in mean crowding translates into a similar gradient of density-dependent development and pupation success. Southern populations experiencing a long growing season and mild winters exist near the carrying capacity of their environment much of the year (Miller and terHorst 2012). Density-dependent effects on growth and development then abate with increasing latitude and altitude and no density-dependent effects can be observed in some northern and mountain populations (Bradshaw and Holzapfel 1996, Bergland et al. 2005).

Bradshaw and Holzapfel (1989) sampled 12 populations from the Gulf of México to Canada that varied 25-fold in larval density and density-dependent development observed in the field. They varied larval density in the leaves of intact pitcher plants in computer-controlled climate rooms, with the daily light and temperature cycle programmed to that of a mid-latitude population during the summer. Larvae were fed freeze-dried *Drosophila* in a regimen mimicking the time-course of prey capture by leaves in nature (Bradshaw 1983). In these experiments fitness (capacity for increase, r_c) declined with increasing density. Regression of fitness on density then estimated r_{max} from the intercept on the *y*-axis at zero density and *K* on the *x*-axis when $r_c = 0$. Neither r_{max} nor *K* was correlated with density in nature, leading to the conclusion that *r*- and *K*-selection based on the effects of density alone are inadequate to explain the evolution of major demographic traits in *W. smithii* (Bradshaw and Holzapfel 1989) and casting doubt on the universality of MacArthur's theory in natural populations.

24.4.2 Evolution of protandry

The emergence of males before females (protandry) in plants and animals usually has been interpreted as a consequence of sexual selection (Darwin 1871) on both sexes. Early emerging males have access to more females and females have immediate access to mates. Selection on protandry should be especially strong in monogamous species such as mosquitoes that emerge synchronously in a seasonal environment and have non-overlapping generations. Selection on protandry should be enhanced even more when density-dependent effects result in larger early-emerging (more fecund) females. However, there is a tradeoff in the degree of protandry between competition for early-emerging females and survivorship of males from the time of male emergence to the time of female emergence. In addition, the optimal degree of protandry is dependent not only on female emergence time, but also on the degree of protandry of other males in the population (Holzapfel and Bradshaw 2002).

Protandry in *W. smithii* is a heritable trait that increases with larval density (Bradshaw et al. 1997). Larval densities and density-dependent development decrease with increasing latitude or altitude (Bradshaw and Holzapfel 1986, Buckley et al. 2003). Consequently, northern populations tend to have discrete summer generations whereas southern populations have overlapping generations. From the standpoint of sexual selection, protandry confers a fitness advantage only when female emergence is predictable in time, i.e., when generations are discrete. Sexual selection theory then predicts that protandry should be more pronounced in the north than in the south. Yet, over a geographic gradient from southern populations with multiple, overlapping generations to northern populations with a single synchronous generation, there is a non-significant trend toward greater protandry in southern than northern populations (Bradshaw et al. 1997). From the standpoint of natural, as compared to sexual, selection, female fitness represents a compromise between minimizing development time and maximizing size at pupation (fecundity). Male fitness is little affected by size (Benjamin and Bradshaw 1994) and males conserve development time while losing pupal mass with increasing larval density, thereby minimizing pre-adult development time with little cost on subsequent lifetime offspring sired. When confronted with variable larval resources, females and males then make different contributions to fitness: females through optimizing fecundity and development, and males through minimizing development, and therefore generation time. Hence, in southern populations with overlapping generations, protandry is maintained through natural, not sexual selection. In contrast, sexual selection should become more important at higher latitudes and altitudes with lower larval densities and increasingly discrete generations.

24.4.3 The evolution of diapause and photoperiodism in *Wyeomyia smithii*

In temperate seasonal environments, selection should favor exploitation of the growing season, mitigation of the unfavorable season, and a means to switch from summer to winter lifestyles and back again in a timely manner. The optimal timing of development and diapause depends upon four geographic patterns of light and temperature (Figure 24.4). First, the climatic gradient in eastern and central North America is primarily one of winter cold, not summer heat. Second, with increasing latitude, spring arrives later and fall arrives earlier (e.g., 15 °C isotherm). Third, the length of the growing season declines with increasing latitude, leading to a decrease in the number of generations per year at higher latitudes. Fourth, the earlier arrival of winter at progressively higher latitudes imposes selection in response to day length.

At any specific spot on Earth, the day length today is the same as it was 10,000 years ago and will be 10,000 years into the future, regardless of temperature. Apart from equatorial latitudes, day-length cycles repeat with time of year (Figure 24.4b) and provide the most reliable predictive cue of any variable for future seasonal variation in the environment. At any given latitude, day length does not change with elevation and is the same in a coastal pine savanna as in a mountain bog at the same latitude. However, duration of summer and winter do change with elevation: a southern mountain bog has a similar seasonal climate as a coastal bog farther north.

Consequently, the optimal time to enter diapause at any locality depends upon the correct response to day length at that latitude and altitude.

Physiological response to day length is under selection and the selective force is the timing of the onset of winter, not day length itself. Entering diapause too early loses opportunity for reproduction and consumes nutritional reserves at warm temperatures; entering diapause too late exposes

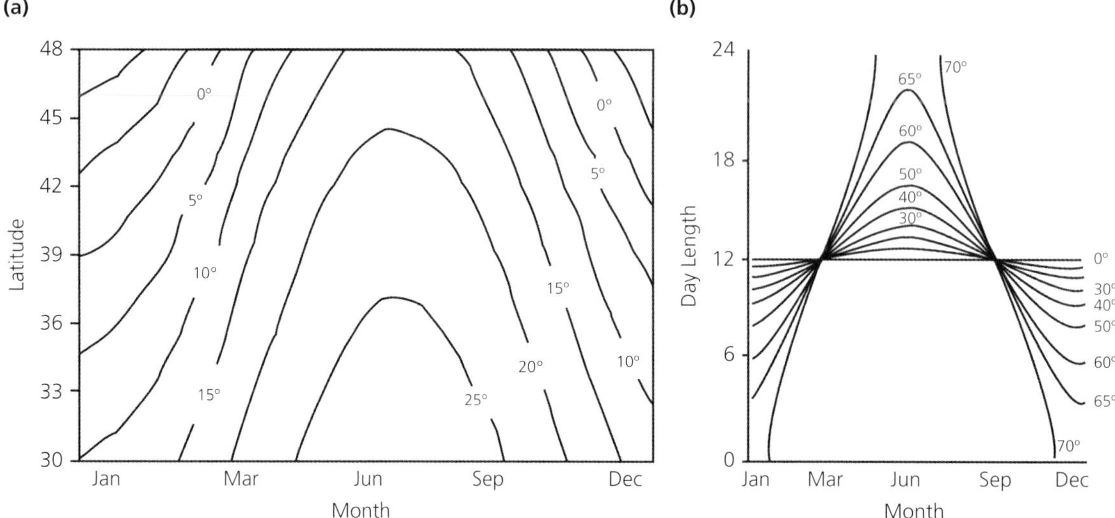

Figure 24.4 Geographic and seasonal variation in day length and temperature. (**a**) Isotherms for mean monthly temperature in central and eastern North America. The latitudinal variation in climate is less a matter of summer warmth (June isotherms are farther apart) than winter cold (January isotherms are closer together) and northern populations experience a shorter growing season than southern populations. Hence, changes in season length and the timing of spring and fall activities have a greater effect on animal populations than do the direct effects of temperature. (**b**) Seasonal patterns in day length (sunrise to sunset) at different latitudes (°N) in the northern hemisphere. Day length at temperate and polar latitudes predicts future seasons more reliably than any other environmental cue.

an individual to the lethal effects of winter. The optimal time to switch to diapause occurs at the point where there are insufficient thermal units remaining during the fall to support development through an entire generation (Taylor 1980). Therein lies a special relevance of day length as an environmental cue for anticipating future seasonal conditions: larvae of *Wyeomyia smithii* (Lounibos and Bradshaw 1975; Bergland et al. 2005), like many other insects, are able to enter diapause at the optimal time of year even though current conditions are otherwise favorable for development and reproduction.

Adaptation to different climates requires genetic flexibility in a population's ability to use day length in the timing of seasonal events (photoperiodism). In *W. smithii*, the critical photoperiod has evolved 10 standard deviations between the Gulf of México and Canada (Lair et al. 1997) and is tightly correlated (R^2 repeatedly ≥ 0.92) with altitude as well as latitude of origin (Bradshaw and Holzapfel 2001). At higher latitudes, winter arrives earlier and when day lengths are longer (Figure 24.5); hence, switching from active development to diapause earlier in the year at higher latitudes involves reliance on longer day lengths. With increasing altitude, the growing season also becomes shorter, winter arrives earlier, and *W. smithii* use progressively longer day lengths to cue the onset of diapause earlier in the year. Over altitudinal gradients at the same latitude, the seasonal environment changes while the annual fluctuation in day length is constant. Hence, the increase in critical photoperiod with increasing altitude illustrates the concept that photoperiodic response and critical photoperiod are adaptations to the timing of seasonal life-cycle events and not to day length itself.

The question still remains as to the relative importance of photoperiodic versus thermal adaptation over the climatic gradient of North America. To answer this question, Bradshaw et al. (2004) used processor-controlled climate rooms to replicate the year-long climates of the Gulf Coast (30 °N), mid-latitudes (40 °N), and northern latitudes (50 °N). Experiments were run in the leaves of intact pitcher plants where four replicate *W. smithii* populations from each of these latitudinal regions were exposed to each annual climatic regimen in a fully crossed experiment. Fitness was assessed by the year-long

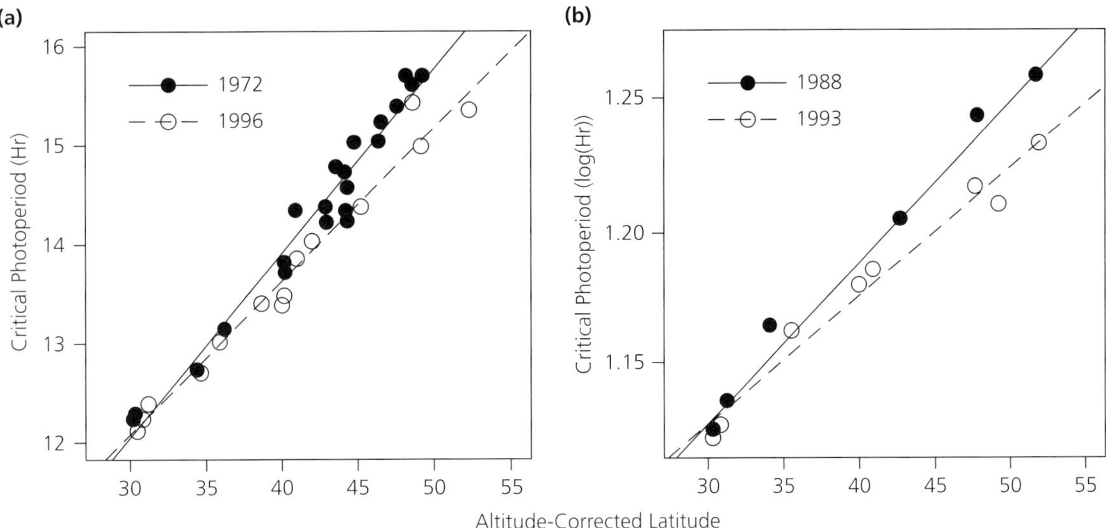

Figure 24.5 Critical photoperiods of *W. smithii* collected during the overwintering generation from 1972 to 1996 determined from (a) static (1972, 1996) or (b) changing (1988, 1993) photoperiods. Analysis of covariance indicated significantly steeper slopes for the earlier year in each comparison, meaning that shifts toward shorter critical photoperiods (more southern phenotypes) increased with latitude. Redrawn from Bradshaw and Holzapfel (2001).

cohort replacement rate integrated over all four seasons. Latitude of thermal year, region of mosquito origin, and their interaction all had significant effects on fitness. Along the climatic gradient of North America, there has been long-term thermal adaptation in *W. smithii*. However, fitness was highest in the moderate mid-latitude thermal year, indicating that extreme southern and northern climates imposed significant heat and cold stress, respectively, on mosquito populations living in pitcher plants at those latitudes.

This experiment allowed the effects of temperature and day-length to be disentangled (Bradshaw et al. 2004). When northern populations are "transplanted" to a mid-latitude thermal year, but kept at their native northern photic year, fitness increased 47% over the northern thermal-photic year. For northern populations, warmer is better. However, when northern populations were transplanted to the same benign mid-latitude thermal year, but exposed to the mid-latitude photic year, they lost 88% of fitness. Mid-latitude day lengths were simply too short to sustain non-diapause development and resulted in a negative rate of population growth, leading to rapid population decline and extinction in an otherwise favorable thermal climate. Thus, the immediate target of selection in a rapidly changing temperate climate is photoperiodic adaptation, rather than the expected direct effect of warmer temperatures on thermal adaptation. These experiments also illustrate the power of the pitcher-plant system to dissect the causes of geographic variation in fitness in a temperate climate. The primary determinant of fitness is the correct timing of seasonal activities, i.e., being in the right stage of development at the right time of year mediated through the optimal, genetically programmed response to day length.

24.4.4 Climatic change as a selective force driving evolution

During recent rapid climatic change, two patterns of biotic response commonly have been observed. Organisms are expanding their ranges poleward and are altering the timing of their seasonal activities. Part of these seasonal alterations represents phenotypic plasticity, especially among ectothermic plants and animals. Even after accounting for these phenotypic patterns, the question remains as to whether the observed effects of recent rapid climatic change also have a genetic basis. Long-term studies of photoperiodism in *W. smithii* over a 24-year period provide the first example of genetic shifts of any phenotype driven by recent rapid climatic change (Bradshaw and Holzapfel 2001). Larvae were collected from Florida to Canada and photoperiodic responses were quantified using rigorously controlled conditions in environmental chambers.

The results were conclusive (Figure 24.5). First, genetic change in response to selection imposed by climatic change occurred in a seasonal trait. Second, the genetic signature of climatic change can be detected in as little as five years (Figure 24.5b). Third, the genetic shift in critical photoperiod is greater at higher latitudes where the rate of climatic change is faster and genetic variation underlying photoperiodic response is higher than at lower latitudes. Fourth, there is always a concern when comparing phenotypes over time that identical environmental conditions were used over a period of years or decades. The southern populations (at 30 °N), where climatic change is proceeding more slowly and genetic variation is lower, provide an important between-year control. There is complete overlap in critical photoperiods over both a 24- and 5-yr interval in these southern latitudes (Figure 24.5a), demonstrating that exactly the same experimental conditions were imposed across the different ranges of years shown in Figure 24.5. Finally, documentation of genetic response to climatic change over a wide geographic range was possible because of the local nature of pitcher-plant populations and the long-lived nature of the plants themselves, enabling the establishment of a "library" of multiple, discrete, and readily resampled mosquito populations through decades of time.

In comparison with geographic variation in demographic traits, the five-year genetic shift in photoperiodic response of *W. smithii* in nature represents evolution at breakneck speed. Genetic shifts related to seasonality subsequently have been documented in a number of species, including plants, birds, and mammals (Bradshaw and Holzapfel 2006). The accumulating documentation of rapid genetic shifts in the timing of seasonal events in diverse organisms means that, when confronted with climatic change, the initial evolutionary response involves seasonal adaptation, usually mediated by photoperiodism, and that thermal adaptation is a secondary physiological response.

24.4.5 Genetic architecture of adaptive evolution

Response to selection depends upon underlying genetic variation (additive genetic variation) whose expression is independent of alleles at the same (dominance) and other (epistasis) loci. Cryptic genetic variation exists in populations due to masking by dominance and epistasis. This cryptic genetic variation can be unmasked by selection on, or genetic drift resulting from, inbreeding following population collapse or isolated colonizing events. At the same time, standing genetic variation can be enhanced by alleles that have opposing effects on fitness (antagonistic pleiotropy). Hence, understanding genetic architectures within and among natural populations can provide insights into the rates and modes of evolutionary change.

Genetic variation among populations of *W. smithii*, as represented by average heterozygosity at protein-coding loci, remains high from the Gulf of México to the Mid-Atlantic Region, and then declines with increasing latitude northwards. This decline in heterozygosity occurs frequently in many organisms and is interpreted as the consequence of sequential founder events, isolation, and drift following recession of the Laurentide Ice Sheet about 20,000 years ago. By marked contrast, genetic variability (additive genetic variation) underlying photoperiodic response increases exponentially over the same range (Armbruster et al. 1998). Additional crosses showed that directional epistasis (gene–gene interaction) contributes to the differences in photoperiod response over latitudinal and altitudinal gradients (Hard et al. 1993, Lair et al. 1997).

Hard et al. (1993) proposed that the increase in genetic variability with latitude in post-glacial populations resulted from the release of hidden additive variation from epistatic variation. However, for there to be a release of additive variation from epistatic variation, there must have been epistatic variation for photoperiodic response in the ancestral population. This proposition was tested by selecting for divergent critical photoperiod in three subpopulations collected from within a 200-m radius within the New Jersey Pine Barrens (40 °N) and then testing for differences due to epistatic effects in the diverged lines.

Hybrids between the long and short critical photoperiod lines revealed directional epistasis in all three cases (Bradshaw et al. 2005), demonstrating that there was epistatic variation within the three original subpopulations. Importantly, the "fingerprint" of digenic epistasis (additive × additive, dominance × dominance, and additive × dominance effects) differed among the three diverged lines. This result means either that there is fine-scale genetic structure even within a local habitat, or that response to selection involves separate, heritable genetic trajectories. In a separate study, quantitative trait loci (QTL) mapping revealed multiple regions of all three *W. smithii* chromosomes that contributed to the evolution of photoperiodic response (Bradshaw et al. 2012). The number and position of the QTL in the genome varied within a northern population, between populations within the northern clade, and between the northern and southern clades. Therefore, extensive genetic polymorphism and multiple alternative genetic pathways to climatic adaptation occur over the range of *W. smithii* and raise the question as to whether analogous genetic flexibility underlies geographic divergence in other inquilines or the pitcher plants themselves.

These results demonstrate that the genetic basis for adaptive evolution crucially depends on genetic background: the genetic context in which genes are expressed. Further, the genetic basis for adaptive evolution of photoperiodic response is highly variable within contemporary populations and between anciently diverged populations. Occupation of the pitcher-plant habitat and genetic flexibility underlying photoperiodic response has enabled *W. smithii*, a member of an otherwise tropical and subtropical genus, to invade and adapt to the climatic gradient of North America from the Gulf of México to northern Canada.

24.5 Future research

Communities living within the leaves of *Sarracenia*, especially *S. purpurea*, have provided a myriad of opportunities for addressing fundamental questions in ecology and evolution. The inquilines in *S. purpurea* have become a model system, in large part because of the widespread distribution of the host plant, the discrete patches in which it lives, the constancy of a suite of dominant species inhabiting it, and the tractability of the community and ecosystem for experiments.

At the scale of individual leaves, experiments have demonstrated that the community is structured by both consumer ("top-down") and resource-supply ("bottom-up") forces and that consumer and resource effects vary predictably with the age of the community. However, the ecology of many species, such as mites and midges, remains relatively unexplored, as are the interactions between the plant and its inquiline community.

Because of the different spatial scales created by leaves, plants, and plant populations, *Sarracenia* inquilines have been ideal for studying questions about dispersal and the structure of metacommunities. The variation in community composition among leaves within a site is much greater than the variation among sites, likely because of the large contribution of leaf age and successional stage to community structure. Biogeographic and evolutionary constraints are difficult to quantify, but experimental studies with pitcher plants and their inquilines will continue to exemplify how to document the importance of evolutionary history for understanding population and community patterns.

Wyeomyia smithii also has become a model system for studying the evolution of complex life histories. This species has migrated northwards and shows an evolutionary progression in stage and depth of winter dormancy (diapause) by using an evolutionarily flexible, genetically programmed response to day length that optimizes the timing of seasonal diapause and development. *Wyeomyia smithii* encounters severe density-dependent development in the southernmost portion of its range and declining densities with increasing latitude and altitude. Yet, *r*- and *K*-selection theory based on density is without predictive power over a 25-fold range of density in nature. Whether this applies to other pitcher-plant inquilines, or to other species with similarly broad geographic ranges remain open questions.

Studies of pitcher-plant mosquitoes were the very first to demonstrate a genetic (evolutionary) basis for phenotypic response driven by recent climatic change. Photoperiodic response in *W. smithii* shows a genetic shift in response to recent rapid climate change in as few as five years. Photoperiodic adaptation is a demonstrably more important evolutionary response to the climatic gradient of North America than has been thermal adaptation. This evolutionary flexibility is based on complex genetic architectures and QTL that vary in their position within the genome within as well as between populations. This variability highlights the importance of genetic background when evaluating the genetic foundations of evolutionary adaptation. Pitcher plants and their associates can provide the means to elucidate evolutionary patterns over multiple time scales, ranging from within-season evolution, over 5–25 years, post-glacial timescales of 20,000 years or less, and more ancient ancestral divergence over millennial time.

Scientists invested in a specific system often attempt to use that system for answering broader questions for which the system may be ill-suited. However, a large number of questions in modern biology are addressable with pitcher plants and their inquilines. The associated biota of pitcher plants range from microbes, protozoa, and rotifers to arthropods with complex life histories, and span multiple trophic levels, including herbivores of the host plant itself. Pitcher plants provide endless opportunities to address and resolve ever newer and developing broad questions in ecological, evolutionary, and genetic contexts at multiple levels of biological integration from molecules to ecosystems, and across a wide range of spatial and temporal scales. This microecosystem certainly will continue to be an important model for experimentally testing new concepts, theories, and questions in genetics, evolution, and ecology for the foreseeable future.

CHAPTER 25

The *Utricularia*-associated microbiome: composition, function, and ecology

Dagmara Sirová, Jiří Bárta, Jakub Borovec, and Jaroslav Vrba

25.1 Introduction

Leaves and other above-ground parts of plants represent an ecosystem highly colonized by microorganisms; carnivorous plants and their traps are no exception. In plant–microbe systems, the associated microbiota is crucial for host-plant survival, nutrient acquisition, protection from pathogens, or ability to cope with various abiotic stresses (Berg et al. 2014). Traps of many carnivorous plants, such as pitchers, are phytotelmic environments that are extreme in many ways. Temporarily erratic and inherently transient, they support large and diverse assemblages of microorganisms whose role in the ecophysiology of their hosts historically has been studied far less than the more conspicuous invertebrate prey.

Utricularia is one of the most species-rich and widely distributed genera of carnivorous plants (Chapter 8). Its unique bladder-like traps not only capture small organisms and other allochthonous material (Chapters 13, 14, 19) but also are a distinctive environment with characteristic microbial communities that colonize the trap lumen (Sirová et al. 2009, 2011, Caravieri et al. 2014). Many researchers have noticed the presence of living or reproducing algae and other eukaryotes in *Utricularia* traps, and lists of (mainly algal) species found there have been published (Juniper et al. 1989). However, the role of the host plant in structuring the associated microbial communities, the role of the commensal microorganisms in enhancing or reducing nutrient dynamics of their host, and other ecological interactions between *Utricularia* and its microbiome are just being addressed.

In many ways, the trap-associated microbial communities of *Utricularia* are similar to food webs described from the phytotelmata of other carnivorous plant species, especially those found in the pitchers of terrestrial genera such as *Nepenthes* and *Sarracenia* spp. (Chapters 23, 24). The traps of carnivorous *Utricularia* are perhaps the most miniaturized phytotelmata. They are essentially isolated from the surrounding environment and the near-zero concentration of oxygen in the traps is lethal to the types of metazoans that live in other phytotelmata (Adamec 2007b, 2011g). Researchers began to study the functional microbial ecology of the traps only after Richards (2001) questioned the importance of carnivory in some aquatic *Utricularia* species and argued that the presence of microbial commensals in the traps might be more beneficial to the plants than the capture of prey itself.

This chapter reviews recent research on plant–microbe interactions in aquatic *Utricularia*. We summarize data on the composition and function of trap commensals, compare the *Utricularia* microbiome with those associated with the traps of other carnivorous plant genera, evaluate the evidence for interactions between *Utricularia* and its microbiome, and highlight the importance of aquatic *Utricularia* traps as model environments for microbial ecology and studies of interrelationships between plants and microbes.

Sirová, D., Bárta, J., Borovec, J., and Vrba, J., *The* Utricularia*-associated microbiome: composition, function, and ecology.* In: *Carnivorous Plants: Physiology, ecology, and evolution.* Edited by Aaron M. Ellison and Lubomír Adamec: Oxford University Press (2018). © Oxford University Press. DOI: 10.1093/oso/9780198779841.003.0025

In this chapter, we consider those commensal microorganisms—bacteria and archaea, algae, microscopic fungi, and single-celled protozoa—that are able to survive in the traps for prolonged periods, are capable of reproducing there, and, in the case of motile, plant surface-crawling protozoa, can recolonize new traps following the senescence of the original traps. We consider larger metazoa, including rotifers, nematodes, mites, and some crustaceans to be prey (§25.4.4).

25.2 The environment of the trap lumen

The trap lumen ("lumen") is a miniature aquatic habitat, contained in few microliters of trap fluid. It is completely sealed from its surroundings, except for the very few milliseconds when the trap fires. The environmental conditions within the lumen are controlled much more directly and extensively than those within pitchers of terrestrial carnivorous plants (Chapters 23, 24). The generally low (acidic) pH inside the lumen varies among species (Sirová et al. 2003, 2009) and seems to be tightly regulated, especially in young and middle-aged traps. The extent of pigmentation of trap walls, which often contain anthocyanins, likely regulates the intensity and quality of light reaching the lumen, which in turn affects the photosynthesis of autotrophic commensals. Redox potential and oxygen concentrations in the lumen fluctuate markedly because of plant photosynthesis and intensive respiration of some trap structures (Adamec 2007b). Traps become deeply anoxic at night or during prey digestion and take in short bursts of oxygenated water each time the trap fires. Microbial commensals that can withstand such harsh conditions have at their disposal an abundant supply of both organic and mineral nutrients. These unique conditions allow for the presence of specialized organisms, such as a ciliate species recently described from the lumen of *U. reflexa* (§25.4.3).

The short (few weeks), species-specific trap life cycle (Friday 1989) has a profound influence on the development of the associated microbial community. All *Utricularia* traps appear to be sterile at the start of their life cycle. The plant exudes as much as 20% of organic carbon newly fixed by photosynthesis into the trap fluid of the youngest traps (Chapter 19), and this carbon provides a suitable growth medium for microbial colonists. The total biomass of commensals inside the traps increases steadily along with the amount of accumulated organic nutrients, and reaches its maximum in the oldest traps (Sirová et al. 2009). However, the amount of dissolved mineral nutrients, such as phosphate (Figure 25.1), does not follow this pattern; its concentration generally is lowest in middle-aged traps. Although direct evidence is still lacking, these findings, together with older results describing age-related ultrastructural changes in trap walls (Vintéjoux 1973, 1974), suggest that as they age, traps shift from predominantly secretion-based

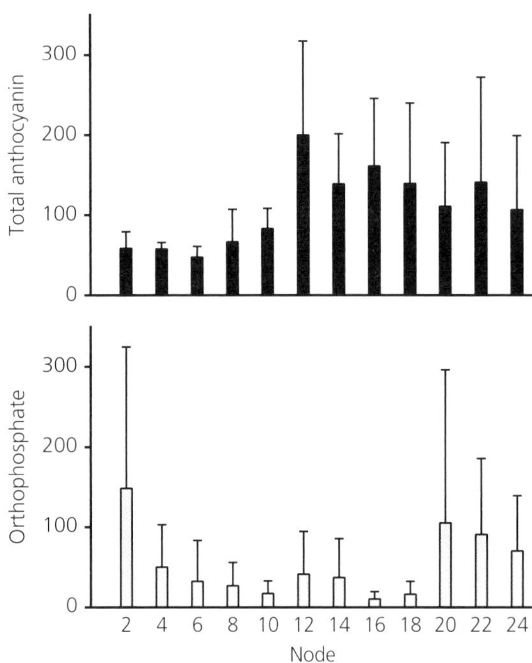

Figure 25.1 Total anthocyanin and orthophosphate content (as absorbance per mg dry weight and µg L^{-1}, respectively) in *U. reflexa* traps of increasing age. Each bar represents mean ± SD of 6–9 traps from a node of the same age on three replicate *U. reflexa* plants. Trap fluid was extracted according to Sirová et al. (2011), filtered on a miniature-size spin column (membrane filter, pore size 0.2 µm), and analyzed for orthophosphate ion concentrations using an ion chromatograph (Dionex ICS 5000, capillary column AS11-HC with a hydroxide gradient). The remaining empty trap was freeze-dried, weighed, and extracted using a mixture of acetone:water:hydrochloric acid (70:30:1) according to Barnes et al. (2009). Total anthocyanin content was measured at 520 nm by direct injection onto the Dionex DX320 with a PDA detector, and normalized to trap dry mass.

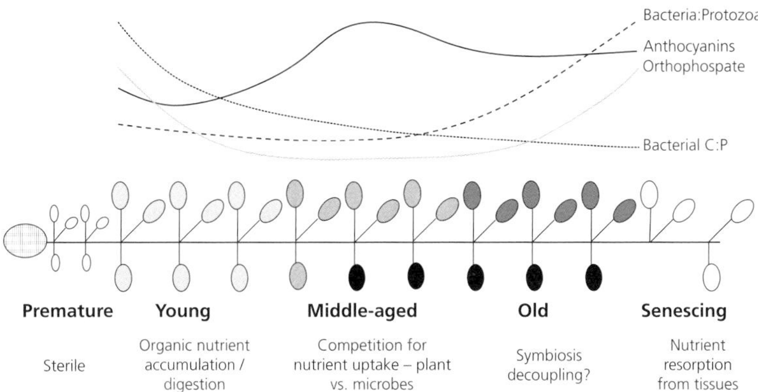

Figure 25.2 Schematic representation of an *Utricularia* shoot, with trap nodes of increasing age from the growth tip on the left and senescent traps on the right. Grayscale in the traps of the upper row indicates the level of their microbial colonization, whereas in the bottom row of traps it indicates the content of anthocyanin pigments in the trap walls. The curve approximating the development of bacterial C:P ratios and bacteria:protozoa abundances with increasing age of traps was drawn according to Sirová et al. (2009).

organs, through a stage when they rapidly decompose organic matter and absorb mineral nutrients from the lumen, and finally senesce and resorb nutrients from their own walls. These changes also appear to affect relationships between *Utricularia* and its microbiome. Based on bacterial C:P:N stoichiometry, these relationships appear to be primarily mutualistic in younger traps, competitive for mineral nutrients in middle-aged traps, and finally are decoupled in older, senescent traps (Sirová et al. 2009; Figure 25.2).

There is little data on variation in nutrients or their stoichiometry among similarly aged traps on a single *Utricularia* shoot. Rather, chemical properties and microbial composition of trap fluid so far has been done only on samples pooled across tens to hundreds of similarly aged traps sampled from multiple shoots. Such pooling has been done primarily because of technical limitations: most commonly available analytical methods require much larger sample volumes than the few μL of fluid contained in a single trap.

Using ion chromatography, we found that both phosphate concentrations in *U. reflexa* trap fluid and anthocyanin content in its trap walls can vary up to 20-fold among traps of the same age and six-fold among traps growing on the same leaf node (Figure 25.1). The observed variation in phosphate concentration most likely results from the stochastic nature of trap firing (Chapter 14) and consequent large differences in the amount and type of organic matter entering each trap. The reason for the high variation in total anthocyanin content in trap walls—a characteristic feature of many aquatic *Utricularia* species—is currently unknown. A "resorption protection hypothesis" explaining anthocyanin accumulation in senescing tree leaves (Hoch et al. 2001, 2003) posits that the shading of photosynthetic tissues by anthocyanins produced during senescence helps to protect the ability of the plant to resorb foliar nutrients by shielding leaves from potentially harmful light levels. Such resorption protection may occur in *Utricularia* as well: fluid phosphate concentration is inversely related to total anthocyanin content in *Utricularia* traps (Figures 25.1, 25.2) and traps are heavily involved in nutrient resorption (Adamec 2014). This plausible explanation merits further research.

It is also reasonable to assume that the concentration of chemical compounds other than phosphate in the trap fluid will vary in a similar manner from trap to trap and that this variation can affect the development, community composition, and activity of trap-associated microbes. We note that the data we summarize in the following sections were collected using sampling methods that did not account for inter-trap variation. Therefore, they represent either averages of traps sampled from entire shoots or from different age categories of shoots. There is also a strong bias toward data from aquatic *Utricularia* species because the traps of terrestrial species are generally smaller and their microbiomes are more challenging to describe.

25.3 Prokaryotes

Recent advances in cultivation-independent methods of analyzing microbial environmental samples, especially in "next-generation" sequencing, have allowed for more detailed insights about the species composition of Bacteria and Archaea inhabiting *Utricularia* traps. Because it is problematic to define prokaryotic "species," operational taxonomic units (OTUs) are used to refer to clusters of microorganisms grouped by DNA sequence similarity. Although *Utricularia* traps only contain a few μL of liquid, and despite the fact that the modern metagenome sequencing approaches may overestimate the number of OTUs found in environmental samples (Edgar 2013), the colonizing prokaryotic communities studied so far in *Utricularia* are surprisingly diverse (hundreds of distinct OTUs) and are comparable in diversity not only to samples from the sediment and water, but also with pitcher fluid of terrestrial carnivorous plants (Table 25.1; Chapters 23, 24).

The prokaryotic composition of trap microbial communities, like that of the pitchers, is dominated by Proteobacteria. This important and diverse group of gram-negative bacteria also includes genera responsible for biological nitrogen fixation. Two major sub-phyla of Proteobacteria dominate in similar proportion (most often 10–20% of total OTUs): Alpha- and Betaproteobacteria, followed by the less abundant Gamma- and Deltaproteobacteria (up to 10% of total OTUs each).

Table 25.1 Estimated species richness (Chao1) and diversity (Shannon *H* and Simpson's 1–*D*) of prokaryotes in three aquatic *Utricularia* species, *Sarracenia purpurea*, *Nepenthes* species, and freshwater/sediment habitats. Results shown were obtained with various next generation sequencing approaches using samples collected in the field or in semi-natural cultivation tanks (*U. vulgaris*).

Host	Observed OTUs	Chao 1	H′	1–*D*	Ref.*
U. australis	938	1715	7.74	0.98	1
U. gibba	1041	1168	4.27	0.94	2
U. vulgaris	720	1095	7.37	0.98	1
Nepenthes spp.	116–1683	291–2911	0.94–5.67	0.02–0.74	3
Sarracenia purpurea	238–875	344–1170	3.50–4.83	0.72–0.97	4
Freshwater and sediment	1087	1222	5.44	0.99	2

*References: 1. Sirová et al. (2018); 2. Alcaraz et al. (2016); 3. Kanokratana et al. (2016); 4. Takeuchi et al. (2015)

Actinobacteria and Bacteroidetes also are important phyla in *Utricularia* traps that have been reported from various pitcher plants (e.g., Takeuchi et al. 2015, Kanokratana et al. 2016). Actinobacteria appear to be more abundant in *Utricularia* species growing in dystrophic waters (Alcaraz et al. 2016, Sirová et al. 2018). This is not surprising because these gram-positive bacteria are, along with fungi, responsible for the degradation of complex organic matter of plant origin. The Bacteroidetes also are specialists in degrading high molecular weight organic matter—proteins and carbohydrates—and are an important part of the microflora in animal guts (Thomas et al. 2011). Other phyla, including Acidobacteria, Armatimonadetes, Chloroflexi, Cyanobacteria, Firmicutes, Fusobacteria, Gemmatimonadetes, Planctomycetes, Spirochaetes, and Verrucomicrobia have been found in the traps in varying abundance, generally well below 5% of total prokaryotes for most of the groups (Alcaraz et al. 2016, Sirová et al. 2018).

Archaea constitute <0.5% in most samples from *Utricularia* traps growing in oligotrophic conditions. Their abundance increases significantly and reaches ≈7% of the total prokaryotic community following the addition of mineral nitrogen and phosphorus into the cultivation medium (Sirová et al. 2011). This may be a form of "dysbiosis" caused by the high nutrient levels; if the lumen accumulates too much prey it becomes putrid (L. Adamec *unpublished data*).

There is a large variation in the microbial composition of the traps, both within and among *Utricularia* species (Sirová et al. 2018). Alcaraz et al. (2016) analyzed the metagenome associated with *U. gibba* plants and found that *Rubrivivax gelatinosus*, *Methylibium petroleiphilum*, and various *Pseudomonas* species (all Proteobacteria) dominated in the traps. Members of all of these genera also were found in traps of *U. vulgaris* and *U. australis* (Sirová et al. 2018) using 16S rRNA amplicon sequencing, but other bacteria had higher relative abundance in these latter species. These other bacteria included Gemmataceae (Planctomycetes), Hyphomicrobiaceae and Acetobacteraceae (Proteobacteria), and Paenibacillaceae (Firmicutes).

There is very little indication of a significant "core prokaryotic microbiome" common to all *Utricularia* species (Sirová et al. 2018), perhaps because of large differences in organic matter amount and type collected by the traps. These differences likely support distinct prokaryotic communities in different *Utricularia* species, at different times during the growing season, and in different traps on the same shoot (Figures 25.1, 25.2). High abundance of a particular bacterium need not necessarily correlate with its importance in the microbial community. Co-occurrence analyses of the bacterial communities in *U. vulgaris* and *U. australis* traps showed that the main keystone taxa, based on the number of potential interactions with other bacteria, showed low relative abundance (Sirová et al. 2018).

Prokaryotes also constitute a significant proportion of their total microbial biomass (Sirová et al. 2009, Šimek et al. 2018). Both single cells floating freely in the trap fluid and aggregates and biofilm-like structures associated with organic detritus in the traps have been observed using epifluorescence microscopy (Sirová et al. 2009). Biofilms covering the plant's major absorptive surfaces may represent a serious barrier to plant nutrient uptake, however, and significant growth of bacteria on the inner trap walls has been observed only in the oldest traps (Płachno et al. 2012a). A hypothesized mechanism explaining this observation is the exudation of xylitol—a sugar alcohol abundant in *Utricularia* trap fluid that prevents the attachment of bacteria to eukaryotic cells in younger active traps (Sirová et al. 2011).

25.4 Eukaryotes

Eukaryotic microorganisms are larger than prokaryotes, more conspicuous, and easier to identify and classify using traditional taxonomic methods. Various protozoa and particularly algae in *Utricularia* traps therefore have been the sole focus of past researchers (Goebel 1891a, Lemmermann 1914, Hegner 1926, Botta 1976, Mette et al. 2000).

25.4.1 Algae

Over a hundred different species from virtually all major freshwater algal groups can be found associated with a single *Utricularia* species at a particular location, with large differences occurring among plant species, sampling locations, or traps of different ages (Mette et al. 2000, Peroutka et al. 2008, Alkhalaf et al. 2009, Sirová et al. 2009, Koller-Peroutka et al. 2014, Płachno et al. 2014b). Most genera are osmotrophic, and because the lumen has a high concentration of dissolved organic nutrients, it is an ideal growth environment for algae. Few specific genera, however, seem to dominate and most are found in mature traps.

The most conspicuous algae are the highly motile *Euglena* spp. and *Phacus* spp. (Euglenophyta: Euglenophyceae). Several species have been described from traps (e.g., Alkhalaf et al. 2009, Płachno et al. 2012a, Šimek et al. 2017) and often more than one species can be found in a single trap (Figure 25.3).

Non-motile, colonial *Scenedesmus* spp. (Chlorophyceae; Figure 25.3) often exceed the biovolume of the euglenids (Płachno et al. 2012a, Šimek et al. 2017). More than ten different *Scenedesmus* species have been described from various *Utricularia* (Alkhalaf et al. 2009, Šimek et al. 2017). Although we can only speculate why *Scenedesmus* spp. so readily colonize *Utricularia* traps, these algae are capable of mixotrophy. Mixotrophic *Scenedesmus* cells show optimum ammonium uptake relative to solely autotrophic ones (Combres et al. 1994). High concentrations of both ammonium (Sirová

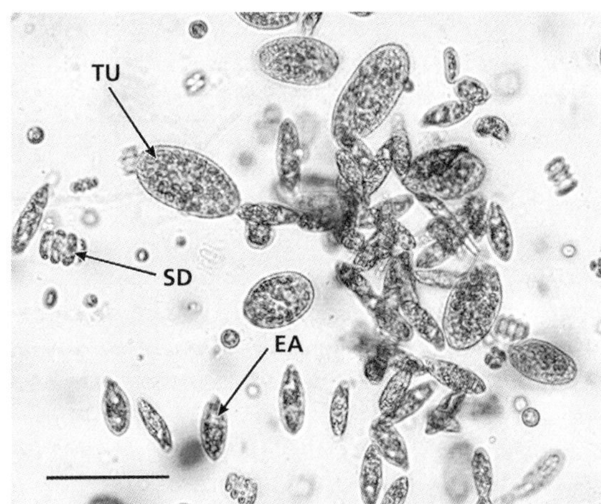

Figure 25.3 A typical example of a microbial community from *U. reflexa* traps as observed using light microscopy. The main eukaryotic commensals are clearly visible: TU—*Tetrahymena utriculariae*; EA—*Euglena agilis*; SD—*Scenedesmus* sp. bar = 400 μm. Photograph courtesy of T. Posch and G. Pitsch.

et al. 2014) and dissolved organic carbon may, as in the case of euglenids, stimulate the growth of various *Scenedesmus* spp. in the traps.

Last, as many as ten to 25 different desmid species (Charophyta: Desmidiales) were reported to be associated with the traps of single *Utricularia* species at one location (Schumacher 1960, Alkhalaf et al. 2009, Sirová et al. 2009). Mixotrophy is not known to occur in the desmids, but they grow best at low pH (Sirová et al. 2009). The specialized trap environment warrants more thorough phycological study focused on the Desmidiales, as species new to science are likely to be discovered associated with *Utricularia*, especially in less dystrophic locations outside of Europe (J. Komárek *unpublished data*).

25.4.2 Fungi

To our knowledge there are only two studies that have found evidence for the presence of viable microfungi—a paraphyletic group distinguished from macrofungi only by the absence of a large, multicellular fruiting body—in *Utricularia* traps. Sirová et al. (2009) used the concentration of phospholipid fatty acids as a proxy for viable microfungal biomass in two subtropical aquatic species and found it to be comparable to that of the algae, and some indication of an increase in microfungal abundance with increasing age. Sirová et al. (2018) found more than 100 different fungal OTUs in the traps of European *U. vulgaris* and *U. australis*, but it is unlikely that all of these are viable trap commensals; many may have been present as "accidentally" trapped spores.

Abundant OTUs, however, may represent *Utricularia*-associated fungi. These include species in the genera: *Fusarium* (Ascomycota: Sordariomycetes), a large genus of filamentous fungi, often associated with plants either as harmless saprobes or parasites; *Archaeorhizomyces* (Ascomycota: Archaeorhizomycetes), recorded from young traps of the rootless *U. vulgaris* (Sirová et al. 2018), that are globally distributed plant-associated saprotrophs that normally are found around roots, although they are not considered mycorrhizal; *Epicoccum* (Ascomycota: Dothideomycetes), globally distributed plant-associated, often endophytic, saprotrophs that also are known from the aquatic systems; and *Phialocephala* (Ascomycota: Leotiomycetes), which are known plant endophytes. The presence of taxa that have been reported to use plant exudates as a carbon source (e.g., *Archaeorhizomyces* spp.) is limited to the youngest traps that have highest rates of plant photosynthate exudation (Sirová et al. 2010).

Freshwater fungi in general are an understudied group, and little work has been done on interactions among them. For example, we barely know if species can inhibit one another's growth when competing for substrata. We also do not know the importance of microfungi relative to other microbes in freshwater nutrient cycling (Wong et al. 1998). The traps of aquatic *Utricularia* may be a useful experimental system with which to gain more information on the biodiversity, ecology, and role of microfungi in lentic environments.

25.4.3 Protozoa

Various protozoa, mainly larger ciliates, frequently have been reported from *Utricularia* traps, both as living commensals and as prey (Sorenson and Jackson 1968, Botta 1976, Płachno et al. 2012a, Šimek et al. 2017). The latter authors noted the conspicuous absence from traps of heterotrophic nanoflagellates—a group of protists that is common and abundant in most lentic systems—and observed that a single species of ciliate protozoa was dominant in *U. reflexa* traps. This recently described *Tetrahymena utriculariae* (Ciliophora, Oligohymenophorea; Figure 25.4) is the first known mixotrophic member of its genus—*Micractinium* sp. (Chlorophyta) is its endosymbiont (Pitsch et al. 2017)—and is a bacterial grazer that is the top predator in the microbial food web within *U. reflexa* traps. Although *T. utriculariae* has been observed only in the traps of culture-grown *U. reflexa* originally collected in the Okavango delta region of Botswana in 2005 (Płachno et al. 2012a misclassified as *Paramecium bursaria*; Pitsch et al. 2017, Šimek et al. 2017), Alcaraz et al. (2016) and Sirová et al. (2018) have recovered sequences of *Tetrahymena* sp. from *U. gibba* and *U. vulgaris*, respectively. *Tetrahymena utriculariae* has never been found outside of *Utricularia* traps and so far colonizes only two cultured

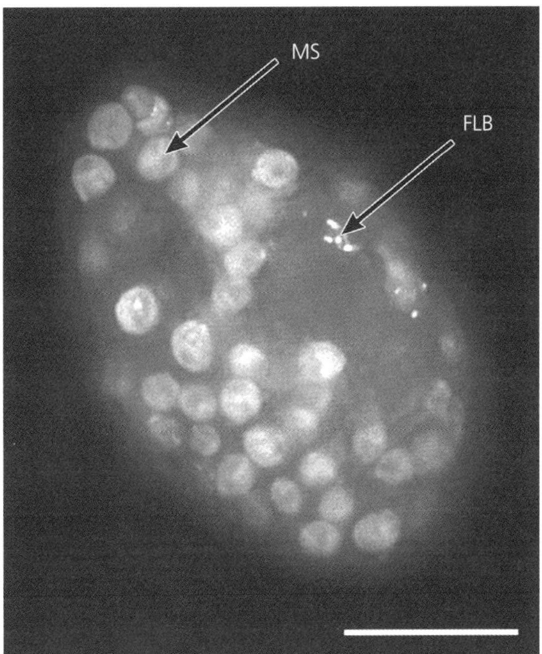

Figure 25.4 A *Tetrahymena utriculariae* individual from the traps of *U. reflexa*, imaged using epifluorescence microscope after fixation and DAPI staining (Šimek et al. 2017). The image was converted into a negative for clarity. MS: *Micractinium* symbiont; FLB: fluorescently labeled bacteria in a food vacuole as an indication of active grazing by the ciliate. bar = 150 μm. Photograph courtesy of K. Šimek.

populations of *U. reflexa*—from Botswana and Zambia—not other co-occurring *Utricularia* species (Šimek et al. 2017). The association between *T. utriculariae* and *U. reflexa* therefore seems to be the closest to a specialized, symbiotic plant–microbe relationship out of all of the known, and otherwise loose, associations.

25.4.4 Are metazoa capable of long-term survival in *Utricularia* traps?

There is no published evidence that metazoa, such as rotifers, nematodes, mites, and microcrustaceans, can survive inside traps for more than several hours. Although there are anecdotal reports that these animals have been observed in the traps alive, these most likely describe recent capture events that occurred while handling or manipulating the plants; crawling and surface-dwelling organisms (i.e., those normally present on the plant surface, often in close proximity to the traps) are captured more frequently than pelagic ones (Meyers and Strickler 1979). Eukaryotes tend to have a much larger demand for oxygen due to their size and complex metabolism (Stamati et al. 2011). Under natural growth conditions there occur long periods of anoxia inside of traps and these severely limit the lifespan of any multicellular animal within the lumen. The same also likely holds true for *Wolffia* sp., a small aquatic vascular plant that also has been reported to occur in traps (Roberts 1972).

25.5 Periphyton

Shoots of aquatic *Utricularia* serve as the substrate for species-rich periphytic communities and often are colonized significantly more frequently and abundantly than other aquatic macrophytes (Bosserman 1983, Friday 1989). The morphological complexity of *Utricularia* shoots may provide a large number of microhabitats for colonization, or the living substrate may also provide a nutrient advantage. For example, periphytic phosphorus content on tropical *U. foliosa* was unrelated to availability of phosphorus in the surrounding water, unlike on artificial substrata, whose P-availability was correlated significantly and positively with their surrounds (dos Santos et al. 2013). *Utricularia* also accumulated much more organic matter and algal cells on its shoot (dos Santos et al. 2013).

Microbial communities inhabiting *Utricularia* shoots are, like those in the traps, diverse and their composition is affected by plant location, shoot age, and time within the growing season (Díaz-Olarte et al. 2007, Peroutka et al. 2008, dos Santos et al. 2013, Sirová et al. 2018). Members of virtually all major freshwater algal groups, including Cyanobacteria, Bacillariophyceae, Zygnemophyceae, Dinophyceae, Cryptophyceae, and Chrysophyceae, have been found on shoots in varying proportions. Fungi also seem to be present and are likewise diverse. Most fungi appear to be associated predominantly with the youngest *Utricularia* shoots; exceptions are plant pathogens (e.g., *Davidiella* or *Pyrenophora* spp. [Ascomycota: Dothideomycetes]), which are more abundant on older, senescent shoot parts (Sirová et al. 2018).

Alpha- and Betaproteobacteria appear to be the most important groups of attached bacterial organisms (Sirová et al. 2011).

Published data suggest that the periphyton itself, not the surrounding water, is the main source of microbial inocula for traps. Sirová et al. (2009) found that ≈60–70% of algal species associated with the field-grown *U. foliosa* and *U. purpurea* plants were present in both the traps and the periphyton. Likewise, Sirová et al. (2017) showed that traps and periphyton share up to 70% of bacterial OTUs. Ulanowicz (1995) proposed a positive feedback model for *Utricularia* carnivory, in which there is a benefit to *Utricularia* from periphyton that comes through the ingestion of animal grazers. On the other hand, it also has been hypothesized that algae may serve as source of nutrients for *Utricularia* and that algal prey may supplement animal prey in oligotrophic waters (Peroutka et al. 2008, Koller-Peroutka et al. 2015). If this is indeed the case, periphyton could also provide a source of algae for the traps.

Despite the similarity in the microbial composition between *Utricularia* traps and periphyton, the two communities are clearly distinct. Co-occurrence analyses (Sirová et al. 2018) revealed that whereas trap microbial communities consist of several apparently non-interacting sub-communities, the periphytic microbes create one, highly interactive network. The diversity of sub-communities within traps may result from the high spatial and physico-chemical heterogeneity of trap microenvironments that result from different organic matter pieces of varying chemical composition, in varying stages of decomposition, or grazing pressure by ciliates. Similarly, bacterial keystone taxa also differ between traps and periphyton at the ordinal level. Actinomycetales, Rhodospirillales, and Cytophagales are keystone prokaryotes inside the traps, whereas Pseudomonadales, Sphingobacteriales, and Rhizobiales are the most interacting taxa in the periphyton.

Biological nitrogen fixation also has been reported to occur in the periphyton of aquatic *Utricularia* (Wagner and Mshigeni 1986, Sirová et al. 2014) and may be an important nitrogen source in the nutrient-limited littoral zone. Estimates are that the *Utricularia*–periphyton system can contribute hundreds of mg of N m^{-2} yr^{-1}.

25.6 Effects of microbial activity on *Utricularia* growth

All recent studies confirm that microbial assemblages, both inside *Utricularia* traps and in the periphyton, are highly diverse in composition. But the mere presence of an organism or its phylogenetic identity says very little about its function, especially in prokaryotes.

Sirová et al. (2003) proposed that microbes could contribute extracellular enzymes to aid in digestion of prey carcasses and organic matter captured by *Utricularia*. Sirová et al. (2009) found that both bacterial and algal cells displayed extracellular phosphatase activity and that the proportion of phosphatase activity bound to cell membranes of microorganisms increased with increasing trap age, and was highest in old traps. The production of extracellular enzymes is energetically costly for both the plant and the microbes. Bacteria apparently gain sufficient energy by the rapid respiration of the readily available carbon in the form of excess plant photosynthates (Sirová et al. 2011). In return, the plants receive considerable help in mineralizing N and P from organic detritus in prey-free traps.

Based on metagenomics data (Alcaraz et al. 2016) and by comparing 16S rRNA data with functional gene databases (Sirová et al. 2018), there appears to be the potential for microbes in the traps to contribute to nitrogen cycling (including nitrogen fixation, denitrification, and urea cycling), iron scavenging and metabolism, production of various hydrolytic enzymes, respiration, osmoprotection, and oxidative stress protection, or antibiotic production. The list is indicative of a metabolically diverse microbial community and many of these hypothesized potential functions have been supported by metatranscriptomic analysis of the *U. vulgaris* microbiome (Sirová et al. 2018). However, even though potential nitrogen-fixing microbes (mainly bacteria and some cyanobacteria) were detected in *U. vulgaris* traps, actual N-fixation measured by stable isotope labeling and by the extent of *nif* gene expression was negligible, and is not considered to be ecologically important for the plants (Sirová et al. 2014). N-fixation within traps may be inhibited there by high concentrations of ammonium in the trap fluid.

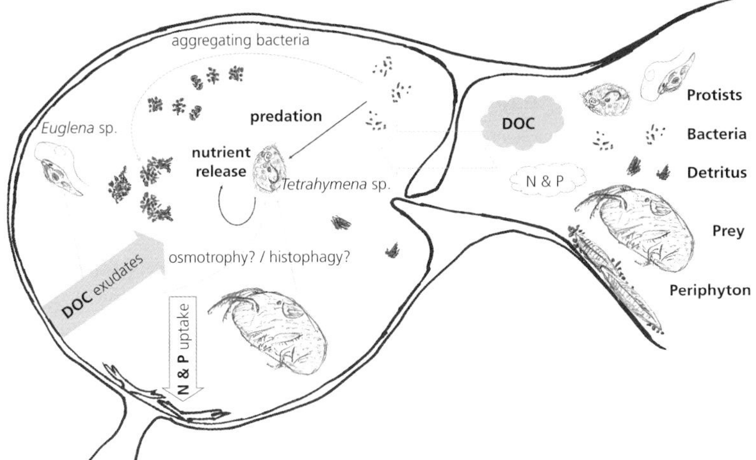

Figure 25.5 Schematic representation of important plant–microbe interactions within *Utricularia* traps, with main potential exogenous sources of nutrients shown at the trap opening to the right. DOC: dissolved organic carbon; N: nitrogen; P: phosphorus.

Algae could be a major source of nutrients for *Utricularia* if complex organic matter of algal origin is hydrolyzed by bacteria or fungi in the traps. Metatranscriptomic data provide some support for this hypothesis. Sirová et al. (2018) reported high transcriptional activity of α-galactosidase, which catalyzes the cleavage of the terminal α-1,6-linked D-galactosyl residues of plant and algal hemicelluloses, and of UDP-glucose 4-epimerase, which performs the final step in the Leloir pathway catalyzing the reversible conversion of UDP-galactose to UDP-glucose. These two highly expressed enzymes underscore the importance of microbial galactose metabolism in the traps. Propionyl-CoA synthetase also is highly expressed in the *Utricularia* traps. This enzyme is involved in the catabolism of propionate and its high expression also may indicate that complex substrates of plant origin are fermented in the trap to produce short-chain fatty acids such as propionate, which is then utilized by certain microbial groups (the presence of short-chain fatty acids in the trap fluid has been confirmed by ion chromatography by Sirová et al. [2011] and Borovec et al. [2012]).

The nutrients mineralized from the complex organic matter through decomposition by bacteria, however, are first used to increase their own biomass and are not accessible to the plant. Ciliate grazers such as *T. utriculariae* serve as mineral nutrient regenerating "machines," which, through their grazing activities (Figure 25.4), turn over the entire standing stock of bacteria in the traps as many as five times each day (Šimek et al. 2017, Sirová et al. 2018). The activity of grazers alone may be responsible for the high concentration of dissolved mineral nutrients observed in the trap fluid that are available for plant uptake (Figure 25.5).

Organisms that typically occur in the rhizosphere and in animal guts are present in traps, and meta-analysis supports the conclusion that trap-associated communities fall between those occurring in these two environments (Sirová et al. 2018). We hypothesize that these symbiotic plant–microbial interactions, involving a microbial food web in the trap and a supply of organic matter (whether alive or dead, algal or animal) from the periphyton, positively support the growth of *Utricularia* and may have contributed to the ecological and evolutionary success of this genus. Omnivory, rather than carnivory alone, appears to be the most probable mode of nutrition in *Utricularia*, at least for aquatic species.

25.7 Future research

Methodological advances and sequencing technologies have profoundly changed our understanding of the origin, evolution, biology, and ecology not only of microorganisms, but also of plants and animals (McFall-Ngai et al. 2013). It is increasingly evident that all higher eukaryotes live in a relationship with diverse microorganisms that live on and in their bodies. Biological and even evolutionary entities are ever more frequently defined not as single organisms, but rather as chimeras composed of many interacting organisms (Zilber-Rosenberg and Rosenberg 2008). Carnivorous plants are no exception.

Relatively simple symbioses involving only two species, such as mycorrhizae or plants and their rhizobia, are well known and well-studied, but we have very little understanding of the more loosely associated plant microbiomes with immense diversity and functional potential to enhance the growth, survival, and interaction of their hosts within the context of populations or ecosystems. There is considerable similarity between the microbiomes of colonizing plant surfaces such as roots, and those of the animal gut, despite the fact that one is considerably more influenced directly by the surrounding environment than the other (cf. Bittleston et al. 2016b). This often overlooked parallel offers a surprising number of ways to understand both plant and animal biology from a different perspective (Ramírez-Puebla et al. 2013). Animal digestive tracts and plant roots strongly influence the chemical and physical parameters in their immediate vicinity. Large surface areas enhanced by either microvilli or root hairs are highly structured, with steep gradients in pH, oxygen, or nutrient concentrations, and create microhabitats suitable for microbial growth and conditions allowing for high functional diversity and ecological dynamics characteristic for both systems.

Utricularia offers an ideal model system for studying host–microbe interactions and their ecological dynamics. Although *Utricularia* is rootless, its traps, with their very large inner absorption surfaces created by numerous glands, have characteristics of both the plant rhizosphere and the animal gut. But unlike these environments, traps are relatively easy to define and experimentally manipulate in both the laboratory and the field. Despite their relatively low trophic complexity, the miniaturized ecosystems within *Utricularia* traps are far from simple. Finally, genome size of *Utricularia* ranges from ultra-small to extremely large (Chapter 11), providing potential for new studies of plant–microbe evolutionary ecology at molecular and genetic levels (e.g., Bárta et al. 2015, Sirová et al. 2014, 2018).

There is a range of open topics in these areas that, if addressed, could add great insights to our understanding of plant–microbe relationships in the *Utricularia* system. Of particular interest would be determining whether patterns and occurrences of plant–microbe and microbe–microbe interactions discovered in aquatic *Utricularia* also occur in terrestrial species. It would be useful to completely describe and model the biogeochemical cycles of nutrients and energy within traps. Such a model would need measurements of the proportions of nutrients that both aquatic and terrestrial *Utricularia* obtain from plant or animal sources in the field. Follow-up experiments could examine the effects of eliminating top predators and keystone species (e.g., *Tetrahymena* spp.) on microbial community structure, within-trap ecosystem processes, and overall plant growth and fitness. Last, much more remains to be learned about effects of microbial community composition and activity on the ecophysiology of *Utricularia* as manifest at the metabolic and transcriptomic levels.

CHAPTER 26

Nutritional mutualisms of *Nepenthes* and *Roridula*

Jonathan A. Moran, Bruce Anderson, Lijin Chin, Melinda Greenwood, and Charles Clarke

26.1 Introduction

Traps of several carnivorous plant species display characteristics commonly associated with attraction of pollinating insects. These may include a combination of color, scent, and a nectar reward for visitation (Chapter 12). This apparent similarity between carnivorous organs and arthropod-pollinated flowers led some authors to suggest that pitchers were examples of mimicry (e.g., Wiens 1978, Pasteur 1982). Juniper et al. (1989) devoted a chapter to this question, and concluded that convergent evolution of attractive traits, rather than mimicry, was responsible for the observed "floral" characteristics of pitcher traps. Moran (1996) reached a similar conclusion based on field studies of *Nepenthes rafflesiana* in Borneo.

Juniper et al. (1989) developed their argument beyond merely refuting the idea of mimicry. Rather, they hypothesized that some interactions between invertebrates and carnivorous plants possessing extrafloral nectaries (EFNs in e.g., *Sarracenia, Nepenthes, Cephalotus*; Chapter 12) constituted not a predator–prey relationship, but a mutualism: an obligate or facultative interaction between species that is beneficial to both (Boucher et al. 1982). Within this broad definition, many mutualistic associations have been identified, including: resource/service (e.g., pollination of flowers by an animal for a nectar reward); service/service (e.g., ants protecting a host plant from herbivory in exchange for nesting space); and resource/resource or resource exchange (e.g., association between mycorrhizal fungi and green plants).

Subsequent studies have provided support for the mutualistic hypothesis of Juniper et al. (1989). In this chapter, we review and synthesize the evidence for mutualistic associations between several animal taxa and members of the Nepenthaceae and Roridulaceae that facilitate nutrient acquisition by the plants via their trapping structures.

26.2 *Nepenthes* and Formicidae

26.2.1 *Nepenthes rafflesiana*

Eusocial Hymenoptera are a conspicuous component of the tropical lowland biota. Ants (Formicidae) occur at high densities (e.g., >500 individuals m^{-2} for Bornean lowland forest; Collins 1980). The widely noted affinity of many ant species for sugars (e.g., Faegri and Van der Pijl 1979, Blüthgen et al. 2000) would be expected to render them open to exploitation by carnivorous plants such as *Nepenthes* spp. that produce EFNs to attract prey. This appears to be the case: Moran (*unpublished data*) found that both aerial and terrestrial pitchers of *N. rafflesiana* caught significantly more ants over a three-month period in northwestern Borneo than did adjacently placed aerial and terrestrial pitfall traps, demonstrating specialization in the capture of this taxon (N = 18 for each trap type; Figure 26.1). Chin et al. (2014) reported a similar pattern of ant-specialization in Bornean *Nepenthes gracilis, N. ampullaria, N. mirabilis, N. bicalcarata,* and *N. rafflesiana*.

Moran, J. A., Anderson, B., Chin, L., Greenwood, M., and Clarke, C., *Nutritional mutualisms of* Nepenthes *and* Roridula.
In: *Carnivorous Plants: Physiology, ecology, and evolution*. Edited by Aaron M. Ellison and Lubomír Adamec:
Oxford University Press (2018). © Oxford University Press. DOI: 10.1093/oso/9780198779841.003.0026

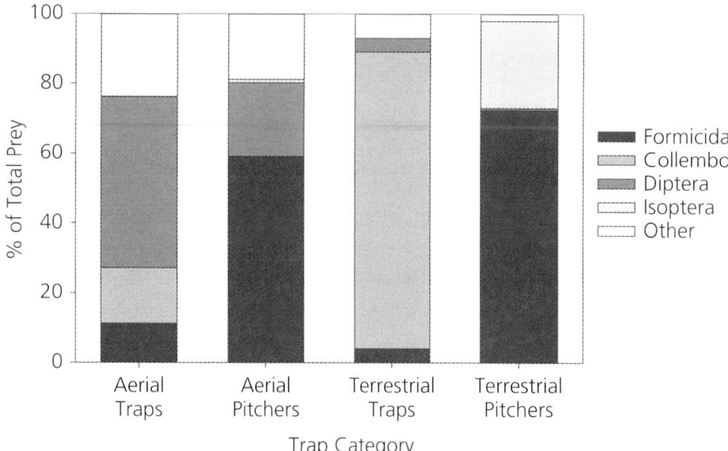

Figure 26.1 Comparison of invertebrate prey items caught by pitchers of *Nepenthes rafflesiana* and artificial pitfall traps, 19 May–10 August, 1989, in Brunei, northwest Borneo. $N = 18$ for each trap type. Data from Moran (1991).

Despite these data, *N. rafflesiana* pitchers appear to be inefficient ant traps. Ants visiting the peristome and feeding on extrafloral nectar produced there (Chapters 12, 13, 15), are seldom observed falling into the pitcher (Moran 1996). Although this observation seems to be at odds with the data presented in Figure 26.1, Bohn and Federle (2004) demonstrated that the peristome of another *Nepenthes* species, *N. bicalcarata*, provides a stable foothold for ants and other insects under dry conditions, but becomes slippery when wet. Subsequent work showed the same mechanism operating in *N. rafflesiana*, whose ridged peristome functions as a wettable surface (Bauer et al. 2008; Chapter 15). Bauer et al. (2008) also demonstrated a diurnal pattern of peristome efficiency in *N. rafflesiana* pitchers. During the heat of the day, the peristome is dry and ants can feed without being trapped. However, near dawn and dusk, elevated ambient moisture wets the peristome and pitchers become effective traps. This switch to an effective trap also occurs during rainy conditions and periods of high nectar secretion by EFNs located between the "teeth" of the peristome (Chapter 12).

How does the intermittent inefficiency of the peristome as a trapping mechanism facilitate mutualism? To explore this question, consider how ants exploit the landscape for food. Typically, a worker ant will scout out a patch of habitat until a food source is located, at which point it recruits others to the resource by laying down a pheromone trail back to the colony, or by leading them to the site of the resource (Carroll and Janzen 1973, Putyatina 2007). Given this pattern of foraging, it is possible to envisage three different types of interactions between *Nepenthes* and foraging ants: completely efficient prey capture; complete inability to capture ants; or somewhere in between—intermittent capture.

The first type of interaction is an exploitative predator–prey relationship in which the pitcher traps as many individuals as encounter it by chance, input of ant-derived nutrients to the plant is low, and those few scouting ants that do encounter the plant will not survive to relay the location of the nectar source to the colony. The second is a kleptoparasitic relationship in which scouts survive and relay information to the colony. The colony then returns to the pitcher to collect nectar, but because the plant captures no prey, it receives no nutritional benefit. In the third type of relationship, which has been observed in the field (Moran 1996), *N. rafflesiana* with intermittently wet peristomes catches more ants (via "batch capture"), than pitchers with peristomes kept artificially wet over the same time period (Bauer et al. 2015a).

Such batch capture of ants by *N. rafflesiana* appears to meet the requirements for a resource-exchange mutualism: each party receives a net nutritional benefit from the reciprocal exchange of a resource that the other requires. The colony gains access to a perennial source of carbohydrate in the form of sugars, and the plant receives periodic pulses of nitrogen (N) via the bodies of workers that slip into pitchers when the peristome is activated by moisture. The loss of workers to the pitcher can be viewed as the "cost of doing business" (Joel 1988), analogous to the losses encountered by *Pseudomyrmex* sp. ant colonies in the protection of bullhorn acacia trees (*Vachellia* [*Acacia*] *cornigera*), which in turn, provide them with nutritional resources and accommodation (Janzen 1985). Conversely, carbohydrates are a relatively inexpensive resource for the plant to provide, as they are readily produced via photosynthesis in an environment in which water and light are not usually limited (Givnish et al. 1984; Chapter 18).

The nutritional benefit that *N. rafflesiana* derives from its relationship with ants is well established. Relative to controls, *Nepenthes rafflesiana* plants denied access to ants (and other invertebrates) over an 18-week period showed a significant reduction in growth, and increased nutrient stress (as detected by foliar reflectance; Moran and Moran 1998). In a subsequent study, natural abundance stable isotope ($\delta^{15}N$) modeling provided an estimate of ant-derived N in *N. rafflesiana* tissues of 68.1 ± 2.4% (mean ± 1 SE of the mean; Moran et al. 2001). Ants should benefit too, because *N. rafflesiana* pitchers provide a year-round, predictable, and abundant source of carbohydrate in the form of nectar, in contrast to the flowers of many other plant species in the same habitat that provide it only seasonally (Moran 1991). There are as yet no quantitative data for this hypothesized net nutritional benefit to the ants.

26.2.2 *Nepenthes bicalcarata*

Nepenthes bicalcarata, discovered in the peat-swamp forests of Borneo in the late 18th century, is named for the presence of a pair of sharp, fanglike structures (derived from the upper part of the peristome), which overhang the pitcher mouth (Figure 26.2). These structures house some of the largest EFNs yet described (Merbach et al. 1999). *Nepenthes bicalcarata* also is the only known member of the genus that produces domatia (hollow spaces within the plant that houses, e.g., ant colonies), in this case within the pitcher tendril. *Camponotus schmitzi* ants obligately nest in these domatia (Figure 26.2). In contrast to the situation with *N. rafflesiana* and its ant partners, which appears to be a simple resource/resource mutualism, the relationship between *N. bicalcarata* and *C. schmitzi* is considerably more complex.

Both *N. bicalcarata* and *C. schmitzi* have evolved unique features that facilitate their association. The main adaptive features of *N. bicalcarata* are the domatia (Figure 26.2) and the giant EFNs in the "fangs" that provide a reliable source of carbohydrates for the colony (Merbach et al. 1999). *Camponotus schmitzi* can swim (Clarke and Kitching 1995, Bohn et al. 2012), and this unusual ability allows the ant to regulate nutrient inputs and losses to the pitcher. The ionic composition of the pitcher fluid is maintained at optimum levels for nutrient

Figure 26.2 Aerial pitcher of *Nepenthes bicalcarata*. An individual *Camponotus schmitzi* worker ant can be seen on one of the pitcher's paired fang-like structures, which house large extrafloral nectaries. The antennae of a second worker can be seen emerging from the opening to the domatium, situated in the looped section of the pitcher tendril.

uptake by active transport across the pitcher wall (Moran et al. 2010). This optimum can be upset by excessive nutrient loading when, for example, excessive biochemical oxygen demand and putrefaction occur following capture of an excessive number of prey or an oversized prey item. This can overwhelm the homeostatic capacity of the pitcher, leading to its loss of function as an organ of nutrient acquisition. Clarke and Kitching (1995) demonstrated that *C. schmitzi* workers act cooperatively to remove large prey items from the pitcher to prevent putrefaction, thereby ensuring that the ammonium ion concentration of the pitcher fluid does not exceed a level that the pitcher can tolerate.

Recent studies have quantified the physiological benefits for *N. bicalcarata* from its association with *C. schmitzi*. Bazile et al. (2012) demonstrated that *N. bicalcarata* plants without *C. schmitzi* showed a significant reduction in growth compared to occupied plants. The presence of the ants resulted in a 200% increase in foliar N content, and natural-abundance ^{15}N modeling estimated that 42% of the plant's N was derived from ant waste material (Bazile et al. 2012). The nutritional benefit of this association was also identified by hyperspectral analysis of leaf reflectance. Using the structure independent pigment index (SIPI), Bazile et al. (2012) showed that unoccupied plants experienced a significant degree of nutrient stress, compared to occupied plants. The SIPI provides a proxy estimate of the ratio of total carotenoids to chlorophyll *a* (Peñuelas et al. 1995), and is effective at detecting nutrient stress in plants growing in the variable ambient light conditions encountered under a forest canopy (Moran et al. 2000).

The *Nepenthes* pitcher typically contains an aquatic ecosystem (Chapter 23). Aquatic dipteran larvae in this phytotelm feed on captured prey, sequester some of its nutrients, and carry those nutrients away with them when they eclose. Scharmann et al. (2013) showed that swimming *C. schmitzi* reduce this airborne loss of nutrients from *N. bicalcarata* pitchers by preying on dipteran larvae in the pitcher fluid. Some of the nutrients that would have been lost are then returned to the plant via the ants' waste. The ants also appear to capture and kill insects that visit the pitcher, preventing their escape (Bonhomme et al. 2011a).

In addition to regulating nutrient inputs and outputs to the pitcher, *C. schmitzi* provides other services to its host. To function efficiently as a trapping surface, the peristome must be kept clear of surface contamination to allow for wetting by capillary action. Thornham et al. (2012) demonstrated that the ants clean contaminants such as fungi from the peristome, thereby maintaining its functionality. Scharmann et al. (2013) quantified the effects of this active maintenance of pitcher function by *C. schmitzi*: using ^{15}N as a tracer, they estimated that nearly all foliar N is prey-derived in ant-occupied plants, compared to 77% in plants lacking ants. The authors concluded that, since the ants feed only on nectar and prey items (and do not bring in N from outside of the pitcher itself), the net increase in foliar N must be due to increased pitcher efficiency when the ants are present. A final service rendered by *C. schmitzi* is to prevent *Alcidodes* sp. weevils (Curculionidae) from feeding on and destroying young pitcher buds. In carrying out this protective service, the ant partner ensures a reliable supply of new domatia for its colony and of trapping organs for its host plant (Merbach et al. 2007, Bazile et al. 2012).

The relationship between *N. bicalcarata* and *C. schmitzi* incorporates elements of both resource/service and resource-exchange mutualisms. The plant provides domatia and a food supply (in the form of nectar, prey, and aquatic dipteran larvae), and in return the ants prevent damage to young pitcher buds, maintain functionality of the pitcher as a trapping organ, ensure that the pitcher is not overloaded with large prey, and reduce nutrient loss by hunting aquatic dipteran larvae. Together, these actions combine to provide measurable nutritional benefits to their host.

26.3 *Nepenthes* and vertebrates

26.3.1 Types of interactions with vertebrates

Nepenthes are known to interact with a variety of vertebrate species, including small reptiles, birds, and mammals (Clarke 1997, Clarke et al. 2009, Bauer et al. 2016). Until recently, almost all documented observations of *Nepenthes*–vertebrate interactions involved occasional observations of vertebrates captured by pitchers. The first and perhaps most famous account of this type was by St. John (1862), who found a giant pitcher of *Nepenthes rajah* on Mount Kinabalu in Borneo that contained the remains of a drowned rat (Phillipps and Lamb 1996). Since then, vertebrate remains have been found occasionally in *N. rajah* pitchers (Phillipps and Lamb 1996) and it was assumed that the enormous pitchers of this species possess the capability to trap and digest small mammals as prey on a regular basis. Over the last 25 years, two of us (CC, JM) have observed occasional captures of geckos, skinks, and mice by several *Nepenthes* species, but have found no evidence that the frequency of these events is sufficient to suggest that any species have evolved traps with features that enable them to target vertebrate prey. These observations suggested two questions. First, under what circumstances do vertebrates encounter and interact with *Nepenthes* pitchers? And second, if species with outsized pitchers do not target vertebrate prey but capture mostly small invertebrates, is there any benefit to the production of such large pitchers?

The first question was addressed by Bauer et al. (2016), who monitored vertebrate visitors to pitchers of *N. gracilis* and *N. rafflesiana* in the lowlands of northwestern Borneo. Five species of vertebrates visited pitchers of both species to feed on nectar produced by glands on the pitcher lids, including four species of sunbird and a lesser tree shrew (*Tupaia minor*). Bauer et al. (2016) concluded that, in all cases, these *Nepenthes*–vertebrate interactions were of no apparent benefit to the plants: no vertebrates were trapped and nectar that might have served to attract invertebrate prey to the pitchers was lost to the vertebrate visitors. These interactions are therefore likely to be kleptoparasitic. The importance of pitcher nectar to the diets of these vertebrate species has yet to be determined, but these observations demonstrate that a range of vertebrates visit pitchers, providing the potential for more sophisticated interactions to evolve.

26.3.2 Highland *Nepenthes* and terrestrial mammals

Chin et al. (2010) addressed the second question in a study of *Nepenthes*–vertebrate interactions that revealed that three species of giant-pitchered, montane *Nepenthes* from northwestern Borneo engage in resource-exchange mutualisms with vertebrates, and that shared trap geometry facilitates this interaction. Mutualism between *Nepenthes* and vertebrates was first detected for *Nepenthes lowii*, a species that produces highly modified pitchers bearing a remarkable resemblance to a toilet bowl. Through a series of field-based experiments and observations, Clarke et al. (2009) demonstrated that these pitchers are visited by mountain tree shrews (*Tupaia montana*), which feed on exudates produced by glands on the underside of the pitcher lid. The structure and orientation of the pitcher lid is such that the only way the tree shrews can access the food source is to straddle the pitcher rim, with their hindquarters located over the orifice. Tree shrews mark the location of valuable resources with feces (Kawamichi and Kawamichi 1979), and when visiting *N. lowii* pitchers, scats are generally deposited within the pitchers. These inputs account for 57–100% of foliar N in *N. lowii*, and the pitcher modifications that facilitate the association with *T. montana* have rendered its aerial pitchers largely ineffective for attracting and trapping invertebrate prey (Chapter 17). In contrast, terrestrial pitchers produced by young *N. lowii* plants appear to function as invertebrate traps in the usual way (Clarke et al. 2009).

Chin et al. (2010) found that *N. rajah* and *Nepenthes macrophylla* have a similar mutualistic interaction with *T. montana*, but unlike *N. lowii*, these species retain the ability to trap arthropods throughout all stages of growth and development. *Nepenthes rajah* also engages in an additional resource-exchange mutualism with the largely nocturnal summit rat (*Rattus baluensis*) (Greenwood et al. 2011, Wells et al. 2011).

Nepenthes macrophylla pitchers are frequented by a bird, the mountain blackeye (*Chlorocharis emiliae*) (C. Clarke *unpublished data*) and by *T. montana* (Chin et al. 2010); both species defecate into the pitchers. However, no detailed investigations of the relationships between *N. macrophylla* and these vertebrates have yet been undertaken. Despite extensive observations of other montane *Nepenthes* species from northwestern Borneo, no other resource-exchange mutualistic interactions with vertebrates have been detected. *Nepenthes ephippiata* from the highlands of central Borneo has pitchers that are very similar in structure to those of *N. lowii*, and a resource-exchange mutualism with *T. montana* is suspected (Chin et al. 2010).

The oversize pitchers of *N. lowii, N. macrophylla*, and *N. rajah* also share a suite of modifications to trap geometry, and Chin et al. (2010) tested the hypothesis that these modifications facilitate mutualisms with mountain tree shrews. This hypothesis was suggested because two other large-pitchered species from northwestern Borneo—*Nepenthes burbidgeae* and *N. villosa*—have no apparent association with *T. montana* (Figure 26.3a, 26.3b). Chin et al. (2010) demonstrated that to receive vertebrate scats, the pitcher orifice must be large enough to accommodate the head + body ("*hb*") length of the mammal. The *hb* of *T. montana* is 157–227 mm (Payne et al. 1985), so a large pitcher size is fundamental (Figure 26.3a, 26.3c), but Chin et al. (2010) found that pitcher size is not the sole morphological facilitator of the association: the mean capacity of *N. lowii* pitchers is significantly less than that of *N. macrophylla* and *N. rajah*, but not significantly different from those of *N. burbidgeae, N. villosa, N. reinwardtiana*, and *N. stenophylla*, none of which "capture" vertebrate scats.

Chin et al. (2010) focused their study on the geometry of the pitcher orifice and lid, noting that the large, vaulted, and reflexed pitcher lids of all three species place the tree shrews' food source some distance *beyond* the rear of the pitcher orifice, making it inaccessible to them from all directions except when they directly face the underside of the lid by straddling the pitcher rim. A comparative analysis of 15 morphological pitcher traits, involving eight montane *Nepenthes* species from northwestern Borneo, including *N. lowii, N. macrophylla, N. rajah*, and five species that do not capture tree shrew scats, revealed that the distance from the front of the pitcher orifice to the tree shrews' food source on the pitcher lid ("*fmfs*": the distance from the front of the mouth to the food source) corresponds closely to the minimum *hb* of adult *T. montana*. This distance is greater than the depth of the pitcher orifice alone (Figure 26.3d): *fmfs* is determined by both orifice depth and the degree of reflexion of the pitcher lid. A strong allometric relationship among these variables was detected across all eight species examined: $\log_{10} fmfs = 1.18[\log_{10}(\textit{mouth length} + 2 \times \textit{lid angle})] - 0.212$ ($F_{1,174} = 2073.31$; $P < 0.001$; $r^2 = 0.923$). Of the pitchers that received *T. montana* scats, all but one had *fmfs* values >156 mm. Chin et al. (2010) argued that the combination of pitcher orifice size, degree of lid reflexion, and the vaulted shape of the lid forced visiting tree shrews to straddle the pitchers to feed on the lid gland exudates, and that only *Nepenthes* species that possessed this combination of characteristics could engage in resource-exchange mutualism with *T. montana*.

Nepenthes rajah has an additional resource-exchange mutualism with the slightly smaller *Rattus baluensis* (*hb* = 150–188 mm; Payne et al. 1985). Greenwood et al. (2011) and Wells et al. (2011) showed that *N. rajah* pitchers are visited by two members of the local mammal community, at all times of day and night. The pitcher geometry relationship detected by Chin et al. (2010) did not account for *R. baluensis*, which could conceivably take nectar from pitchers with $fmfs \approx 150$ mm; Chin et al. (2010) sampled few pitchers whose dimensions were less than or equal to the minimum *hb* of shrews (*T. montana*) and rats (*R. baluensis*). To address this deficiency, two of us (LC, MG) randomly sampled additional pitchers of *N. macrophylla* ($N = 22$) and *N. rajah* ($N = 16$). The range of *fmfs* values from this new sample exceeded the sample of Chin et al. (2010), but the allometric relationship of the combined data was virtually identical: $\log_{10} fmfs = 1.15[\log_{10}(\textit{mouth length} + 2 \times \textit{lid angle})] - 0.189$ ($F_{1,190} = 2407.94$; $P < 0.001$; $r^2 = 0.919$; Figure 26.3e).

Among *N. macrophylla* individuals, only pitchers with $fmfs > 156$ mm caught *T. montana* scats. By contrast, six *N. rajah* pitchers with $fmfs \leq 156$ mm received fecal inputs. In all cases, the angle of reflexion of the lid was constrained by adjacent vegetation and terrain, but even when this happens, vertebrates will feed at, and defecate into, *N. rajah* pitchers as long as they are physically capable of straddling the pitcher rim and reaching the nectar glands on the underside of the lid.

We also repeated the principal component analysis (PCA) of pitcher dimensions performed by Chin et al. (2010), using the additional measurements. The results (Figure 26.3f, Table 26.1) show that the scores for PC1 are > 0 for all the species that do not receive

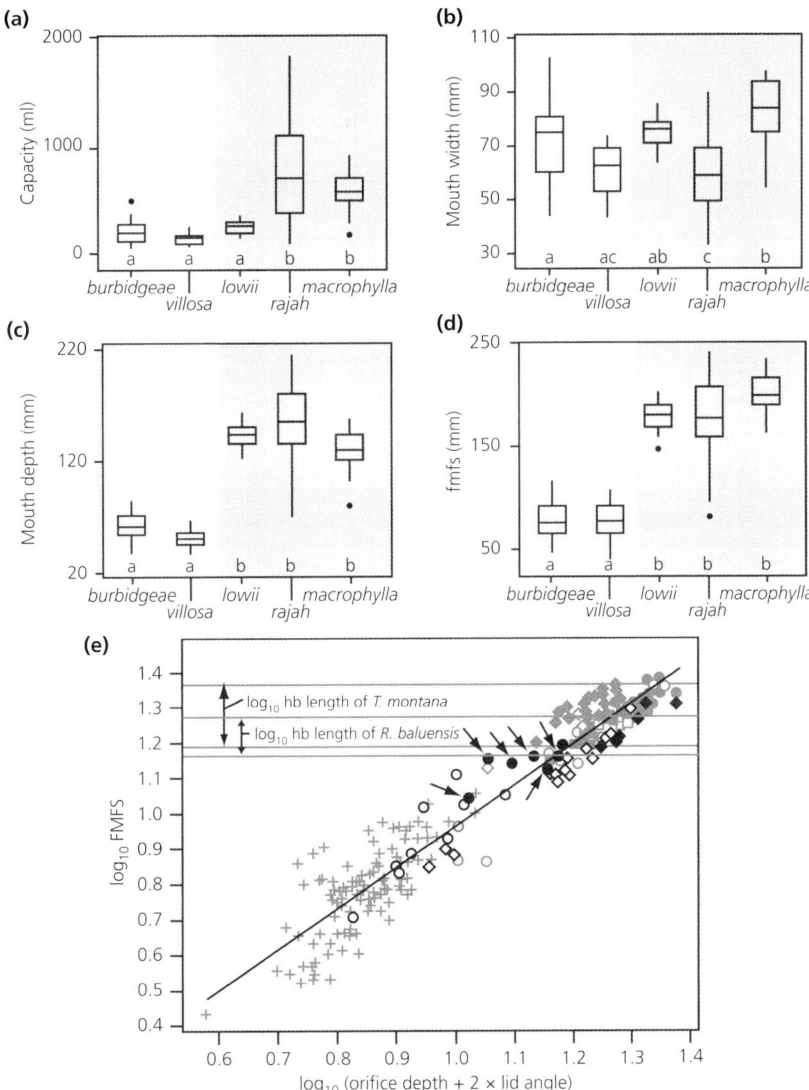

Figure 26.3 (**a–d**) Box-and-whisker plots for various pitcher characteristics for five montane *Nepenthes* species studied by Chin et al. (2010). The midline of each box denotes the median, boxes denote the inner quartiles, and the whiskers extend to 1.5× the interquartile range. Lower case letters above the *x*-axis denote species for which there is no significant difference for the relevant pitcher characteristic, based on Tukey's *post hoc* pairwise comparisons of means. (**e**) Regression of the distance from the front of the mouth to the food source (*fmfs*) vs. orifice depth and lid angle for the eight *Nepenthes* species studied by Chin et al. (2010). Arrows denote pitchers with *fmfs* ≤ 150 mm that trapped feces. (**f**) Principal component biplot (PC1 vs. PC2) for pitcher morphology. Key to symbols for (**e**) and (**f**): squares, *N. lowii*; diamonds, *N. macrophylla*; large circles, *N. rajah*; +, pitchers of *N. burbidgeae, N. reinwardtiana, N. stenophylla, N. tentaculata*, and *N. villosa*. Closed symbols for *N. lowii, N. macrophylla*, and *N. rajah* denote pitchers that trapped feces. Grey-toned symbols represent original data analyzed by Chin et al. (2010); black symbols denote additional data collected by L. Chin (*N. rajah*) and M. Greenwood (*N. macrophylla*).

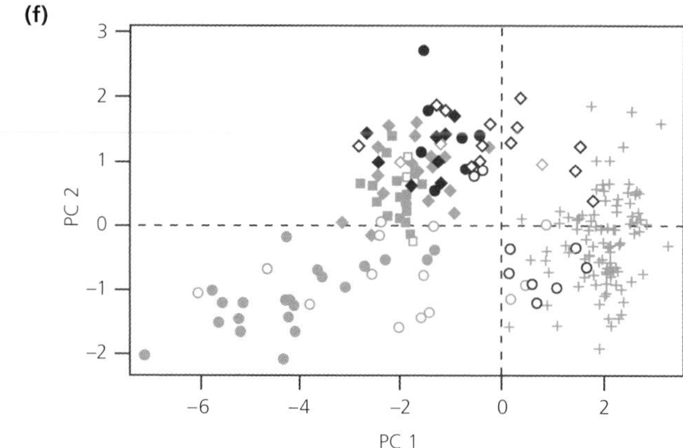

Figure 26.3 (*Continued*)

Table 26.1 Eigenvectors and contributions of the principal components and scores for variables included in the principal component analysis (PCA).

	PC1	PC2	PC3
Eigenvalue	5.39	0.91	0.67
Proportional contribution	0.68	0.11	0.08
Cumulative contribution	0.67	0.79	0.87
Variable			
Mouth length	−0.41	0.12	−0.06
Capacity	−0.37	−0.24	0.03
Front of the mouth to the food source *(fmfs)*	−0.41	0.19	−0.05
Lid angle	−0.25	0.79	−0.18
Mouth angle	0.28	0.48	0.47
Lid area	−0.40	−0.17	0.05
Lid concavity	−0.38	−0.05	−0.13
Midrib protrusion	−0.28	−0.06	0.85

fecal inputs, and for the small pitchers produced by immature *N. macrophylla* and *N. lowii*. However, PC1 < 0 for all pitchers that captured scats.

These results clearly illustrate ontogenic changes in morphology that occur in these three species as they reach maturity. Small, immature plants cannot produce pitchers large enough to accommodate vertebrates and rely entirely on arthropod capture for supplementary nutrients. Their pitchers are morphologically similar to those of typical, arthropod-trapping *Nepenthes*. As the plants reach maturity and become capable of producing pitchers with *fmfs* ≥ *hb* of *T. montana* (or *R. baluensis* in the case of *N. rajah*), their pitcher geometry changes, and their traps bear little resemblance to those of species that catch only arthropods. Chin et al. (2010) and Greenwood et al. (2011) argued that these plant–vertebrate facilitation interactions required only relatively minor modifications to trap geometry. Because arthropod diversity and abundance decrease as altitude increases (Collins 1980, Samson et al. 1997), there may be significant selective pressure for highland *Nepenthes* to exploit alternative sources of nutrition. Although tree shrew feces contain less N per unit dry weight than ants (Chin et al. 2010), the steady supply and amount of fecal material represents a substantial contribution to the plants' nutrient budgets (Clarke et al. 2009).

26.3.3 *Nepenthes hemsleyana* and bats

Low levels of arthropod diversity and abundance associated with highlands are not the only driver of resource-exchange mutualisms between *Nepenthes* and vertebrates. One species from the lowlands of northwestern Borneo, *N. hemsleyana*, engages in such a relationship with Hardwicke's Woolly Bat (*Kerivoula hardwickii*) (Grafe et al. 2011). The pitchers of this species are used as daytime roosts by *K. hardwickii*, which deposit substantial amounts of feces into them while roosting. *N. hemsleyana* is thought to be a close relative of *N. rafflesiana* (Chapter 5), a species that also occurs in the lowlands of northern Borneo and has a more sophisticated invertebrate trapping strategy (Moran 1996, Bauer et al. 2008, 2011, Chin et al. 2014). *Nepenthes rafflesiana* has funnel-shaped aerial pitchers that are rigid and leathery in texture, emit a strong fragrance, and have a high-contrast ultraviolet pattern (Moran et al. 1999). In contrast, *N. hemsleyana* pitchers are funnel-shaped only in their lower half; the upper half is cylindrical with a well-developed waxy zone. Their texture is papery and flexible, they do not emit any discernible fragrance or possess the high-contrast ultraviolet pattern found in *N. rafflesiana*, produce only small amounts of nectar, and trap few invertebrates (Moran 1996, Bauer et al. 2011, Clarke et al. 2011). These morphological differences are minor with respect to resource allocation by the plant (C.R. Schöner et al. 2013), but their consequences for nutrient acquisition are significant. The aerial pitchers of *N. rafflesiana* may be highly effective arthropod traps, but are not suitable roost sites for *K. hardwickii* because of their conical shape, whereas those of *N. hemsleyana* are less effective at attracting arthropods, but provide ideal roost sites (Grafe et al. 2011, C. R. Schöner et al. 2013, 2015). Grafe et al. (2011) demonstrated that 11–56% of foliar N in *N. hemsleyana* is derived from bat feces.

Studies of the interactions between *Nepenthes* and vertebrates, along with the relationships between *Nepenthes* pitcher structure and climate (Bauer et al. 2012a, Moran et al. 2013), have revealed a great deal about key evolutionary drivers of trap geometry and characteristics, suggesting that the pitcher is a highly plastic organ that is capable of rapid evolutionary responses to changing selective pressures (Chapter 5). Most *Nepenthes* species that produce pitchers with unique, atypical characteristics have been found to have specialized or modified prey capture or nutrient sequestration strategies (Clarke 2001, Moran et al. 2001, 2003, Merbach et al. 2002, Clarke et al. 2009, Chin et al. 2010, Grafe et al. 2011). However, *N. hemsleyana* represents an example of a species whose pitchers, at least in terms of geometry, appear relatively typical, emphasizing the point that minor modifications to trap structure can facilitate major shifts in nutrient sequestration strategy. To date, the prey capture and nutrient sequestration strategies of few *Nepenthes* species have been examined and there is a high likelihood of further discoveries of novel interactions between *Nepenthes* and vertebrates. Candidate species for such associations are most likely to be those that produce pitchers with enlarged orifices and copious amounts of nectar. Any additional modifications will, most likely, reflect the body plans of the vertebrates with which they interact.

26.3.4 The future

It is possible that multiple resource-exchange mutualisms between *Nepenthes* and vertebrates have come and gone as climates in the areas they inhabit have changed. How long the resource-exchange mutualisms we have documented recently from northern Borneo persist could depend on the plasticity of responses of both the plant and animal participants to regional changes in climate (Chapter 28). Clarke and Moran (2016) noted that the *N. rajah–T. montana* interaction broke down during an extended period (≈2 months) of dry weather: pitchers died or failed to produce nectar and rates of *T. montana* visitation and scat deposition fell substantially. Even subtle shifts in rainfall patterns have the potential to disrupt these interactions, indicating that over evolutionary time, they could be highly ephemeral. It is possible that anthropogenic climate change will disrupt rainfall patterns in north Borneo, leading to increased seasonality (Foster 2001), threatening the persistence of these unusual plant–vertebrate mutualisms.

26.4 Other potential mutualists with *Nepenthes*

26.4.1 *Nepenthes albomarginata*

Nepenthes albomarginata is named for the cream-colored, tomentose band of tissue beneath the peristome (Figure 26.4). Its pitchers produce very little nectar relative to other *Nepenthes* species (Merbach et al. 2001). Research in Borneo has found that it specializes in the capture of processionary termites in the genus *Hospitalitermes*, which feed on the tomentose tissue comprising the band (Moran et al. 2001, Merbach et al, 2002; Chapter 21). Because of the large numbers of termites present in processional columns (Jones and Gathorne-Hardy 1995, Miura and Matsumoto 1998), several hundred individuals may be caught in a pitcher over a very short period of time. The tomentose band is not

Figure 26.4 *Nepenthes albomarginata.* (**a**) Terrestrial pitcher with a pale, tomentose band visible beneath the peristome. (**b**) Scanning electron micrograph of the area outlined by the small black square in (**a**); note the dense mat of highly branched trichomes that make up the pale band. The lower edge of the peristome, identifiable by its vertically ridged surface, is visible at top of the panel. (**c**) Close-up of the tomentose band with a dense network of fungal hyphae covering the trichomes. Bars = (**a**) 1.5 cm, (**b, c**) 100 µm. Scanning electron microscopy by Brent Gowen.

regrown, so the *N. albomarginata* pitcher functions as a "single-shot" trap, unlike pitchers of *Nepenthes* species that rely on nectar to encourage insects to remain at the peristome.

Hospitalitermes spp. feed predominantly on epiphytic lichens (Jones and Gathorne-Hardy 1995), and the tomentose band appears to function as a fungus garden (J. Moran, C. Clarke, and B.E. Gowen *unpublished data*). The tomentose band is colonized by a dense network of fungal hyphae (Figure 26.4), which are not found anywhere else on the *N. albomarginata* pitcher, nor anywhere on pitchers of six other *Nepenthes* species—including some sympatric with *N. albomarginata*—examined using scanning electron microscopy (J. Moran *unpublished data*). It is conceivable that the tightly packed and complex hair-like processes maintain a layer of moist air at the surface of the band, and that this humid microenvironment facilitates fungal colonization and growth. Maintenance of such a boundary layer may be important for *N. albomarginata*, given that it inhabits drier microsites than sympatric species such as *N. rafflesiana* (Moran et al. 2001).

Fungal growth on the tomentose band is not an artifact of a pitcher having been grown at a particular location; the hyphal network has been observed on the tomentose band of *N. albomarginata* pitchers grown from tissue culture in British Columbia, Canada, and on pitchers collected from the wild in Peninsular Malaysia (J. Moran and C. Clarke *unpublished data*). It is still uncertain whether the relationship between *N. albomarginata* and the fungi is exploitative or mutualistic. On the one hand, the fungi are consumed by the prey, but on the other, the fungi on some pitchers may never be consumed, as not every pitcher will be visited by termites. It is also unclear what specific attractive cues may be involved in the targeting of termites. Fungi may produce volatiles that in some way mimic the scent of lichens; this hypothesis awaits further experimentation.

26.4.2 *Nepenthes ampullaria*

The lowland *N. ampullaria* has terrestrial pitchers with a suite of morphological features that facilitate capture of leaf litter from the forest canopy, an organic input that accounts for 35–55% of the plant's N budget; the remainder is derived from invertebrate prey and perhaps root uptake (Moran et al. 2003, Pavlovič et al. 2011b). *Nepenthes* pitchers produce an array of digestive enzymes to facilitate breakdown of organic inputs (Chapter 16), but leaf litter is not easy to decompose.

The *Nepenthes* phytotelm (§26.2.2) includes bacteria, fungi, protozoa, and arthropods (Beaver 1983, Clarke and Kitching 1993, Adlassnig et al. 2011, Bittleston et al. 2016a; Chapter 23). Moran et al. (2003) hypothesized that the phytotelm of *N. ampullaria* could break down the leaf material; bacteria form a biofilm that is grazed on by mosquito larvae, which in turn excrete N in the form of ammonium ions that are absorbed by the pitcher. Active transport of ammonium ions from the pitcher fluid into the plant tissues was demonstrated by Moran et al. (2010). The same study also showed that *N. ampullaria* maintains a higher pH in its pitcher fluid (by active transport of protons out of the fluid) than more typical species such as *N. rafflesiana*. This may account in part for the high diversity of inquilines found in *N. ampullaria* pitchers, which host more species than those of any other *Nepenthes* (Clarke and Kitching 1993, Adlassnig et al. 2011). The role of inquilines in the uptake of nutrients from leaf litter has yet to be investigated.

26.5 *Roridula* and Hemiptera

Using a sticky trap mechanism that is superficially similar to that of the more well-known Droseraceae, *Roridula* captures very large numbers of prey items (Bruce 1907, Barthlott et al. 2004; Chapter 21). Although much of the prey captured is small in size (3.6 ± 0.57 mm; Marloth 1903, 1910, Ellis and Midgley 1996), *Roridula* is capable of catching large and robust flying insects (wingspans > 50 mm), including Hymenoptera, Diptera, Coleoptera, Hemiptera, Lepidoptera, and Thysanoptera (B. Anderson *unpublished data*), and even small birds (Anderson 2003). The large number and size of insects captured by *Roridula* led to the belief that it produces the strongest glue of all insect-trapping plants (Marloth 1910, Barthlott et al. 2004). However, Voigt et al. (2009) suggested that it is not only the resinous secretion that is important for insect trapping, but also the hierarchical, cascade-like organization of the plant and the specialized functions played by different glandular trichomes, some of which are used for entangling insects, some for dampening insect movement, and others for their adherence capabilities (Chapter 15). Indeed, the adherence capabilities of some trichomes are stronger than those of any other insect-trapping plants and can be four times stronger than commercial Tanglefoot® flypaper (Voigt et al. 2009). The excellent adherence results from the use of triterpenoid resins to trap prey (Simoneit et al. 2008, Frenzke et al. 2016), and not the mucilaginous polysaccharides that are used by other carnivorous plants with sticky leaves (Adlassnig et al. 2010; Chapters 12, 13, 15). The resins have much greater adhesive power and do not desiccate, even when the plants dry out (Voigt and Gorb 2010a). In addition, the resins are not water soluble and their trapping ability is maintained under water (Voigt and Gorb 2015). They are such effective trappers of insects that early settlers to South Africa used to hang *Roridula* cuttings from the ceilings of their homes to capture bothersome flies (Marloth 1925), and it is not surprising that the first biologists to observe *Roridula* regarded them as carnivorous plants.

26.5.1 Digestive mutualism

This idea that *Roridula* is carnivorous was challenged by Marloth (1925), who noted the presence of large numbers of *Pameridea* bugs (Miridae: Heteroptera), crawling apparently unhindered over the leaves of the plants. The bugs are obligately associated with *Roridula* and each *Roridula* species is associated with a different *Pameridea* species (Chapter 10). These bugs attack the insects trapped by *Roridula* and suck their fluids using piercing proboscides (Figure 26.5). Marloth (1925) and Lloyd (1934) hypothesized that *Roridula* gains few nutrients from trapping insects, and investigations into the presence of digestive enzymes suggested that *Roridula* does not possess proteases (Marloth 1925). Consequently, Marloth (1925) and Lloyd (1934) both hypothesized that the sticky hairs of *Roridula* did not evolve to trap prey, but rather to deter herbivores.

The idea of stickiness for defense persisted until Ellis and Midgley (1996) showed that the hemipterans defecate on the leaves of *Roridula* plants, and using N isotopes they established that the presence of hemipterans facilitates the uptake of N via its leaves. Studies on N uptake demonstrated that up to 73% of N found in *Roridula* originates from insect prey (Anderson and Midgley 2003). Phosphatase enzymes also show high activity on leaf surfaces and may facilitate foliar uptake of phosphates, but their importance is unknown (Płachno et al. 2006).

Figure 26.5 (Plate 17 on page P15) *Pameridea* bugs attacking a fly trapped on the sticky surface of a *Roridula* leaf. The bug in the center of the picture is using its piercing mouthparts to suck fluids from the body of the fly.

Mirid bugs, like those found on *Roridula*, occur frequently on plants bearing sticky hairs, and the majority of them suck the sap of their hosts (Schuh 1995, Wheeler 2001, Anderson et al. 2012, Wheeler and Krimmel 2015). However, the omnivorous genus *Pameridea* appears to have evolved carnivory secondarily (Dolling and Palmer 1991), and it is likely that the relationship between *Roridula* and *Pameridea* may not always have been mutualistic. The stickiness of *Roridula* may have evolved first to exclude herbivorous insects like a proto-*Pameridea*. As *Roridula* evolved ever-more sticky leaves, *Pameridea* could have evolved counter-adaptations to combat the stickiness. *Pameridea* does not appear to avoid *Roridula*'s sticky trichomes, but instead holds its body very close to the plant, making frequent contact with the sticky surface (Voigt and Gorb 2010b). Its claws grasp the trichomes close to their thick bases (instead of near their apices), where the bugs are able to generate stronger traction forces. This, combined with heavy musculature, helps them to power through the sticky traps using quick movements, jumps, and short flights (Voigt and Gorb 2010b). But perhaps their most important adaptation is an unusually thick epicuticular secretion which is cohesively weak and grease-like. This layer sloughs off on contact with the resinous droplets and further prevents the bugs from being captured (Voigt and Gorb 2008).

Pameridea still shows traits from its herbivorous past and will suck sap from *Roridula* plants. Anderson and Midgley (2007) demonstrated experimentally that plants without hemipterans may not grow, whereas plants with hemipteran densities close to what one would normally find in the wild had positive growth rates. By increasing hemipteran densities beyond what is normally found in the field, they showed that the sucking of sap can result in negative growth rates. Almost the entire *Pameridea* life cycle is completed on the *Roridula* plants, making it difficult to imagine how isolated populations are recolonized after being burned by the frequent fires on the fynbos where the plants and bugs occur. However, Anderson (2003) found that although *Pameridea* cannot complete its life cycle on other plants, it can suck the sap of many noncarnivorous plants and survive for several weeks without insect food. It is likely that this ability allows it to traverse landscapes which do not have *Roridula* plants growing in them, so that it is able to recolonize *Roridula* populations after they have burned.

26.5.2 Other symbionts

Although *Pameridea* is the only symbiont frequently encountered on *R. gorgonias*, the density of captured insects on *R. dentata* attracts a diversity of other symbionts. Assassin bugs are commonly associated with sticky plants (Krimmel and Pearse 2013, 2014), and a particularly large species (*Rhynocoris disciventris*, ≈30-mm long) can be found occasionally on *Roridula dentata* plants in the extreme north of its range, where the bugs move about very slowly, consuming captured insects (B. Anderson 2006). *Rhynocoris* keeps its body away from *Roridula* traps by walking along the woody stems, rather than its leaves, which hinder the assassin bug quite significantly, despite its large size. *Rhynocoris* has also been observed on other plant species, suggesting that its relationship with *Roridula* is not specialized.

Several spiders have also been recorded on *Roridula*, and their ability to walk on sticky surfaces may help them avoid the traps. Although most observations of spiders on *Roridula* are incidentals, three species are common. Lynx spiders are commonly associated with sticky plants and one species (*Peucetia nicolae*) was frequently encountered on *R. dentata* in the northern parts of the range, and very occasionally found on *R. gorgonias* (B. Anderson 2006). This is a wide-ranging spider, and its affinity for sticky plants has probably allowed it to take advantage of the trapping ability of *Roridula*. The northern *R. dentata* populations are all associated closely with an unidentified, but specialized Araneidae spider (B. Anderson 2006, A. Dippenaar *unpublished data*). Unlike most other Araneidae, those on *R. dentata* do not spin orb webs. Instead, they build a lattice highway of threads over the leaves of *Roridula*, allowing them to move mostly unimpeded over the surface. They build and live in nests constructed from silk and old leaves in the axils of living *Roridula* leaves and branches. They dart out of these nests to grab trapped insects and *Pameridea* bugs that are not fast enough to get out of their way (Figure 26.6). Southern *R. dentata* populations have an ecologically equivalent species that

Figure 26.6 (Plate 18 on page P15) Unidentified orb-web spider (Araneidae) attacking a *Pameridea* bug on a leaf of *Roridula dentata*.

also appears to be specialized on *R. dentata* plants: the crab spider, *Synema marlothi* (named for Marloth, one of the most prolific of the early Cape researchers, who had a special interest in *Roridula*), which makes similar nests and preys upon captured insects and *Pameridea* (B. Anderson 2006).

The ecological role of spiders on *Roridula* has been little studied. It is possible that they perform a similar role to the hemipterans and facilitate N uptake, but most prey consumption takes place in their nests where feces are unlikely to make contact with the leaves of *Roridula*, which have cuticular gaps that facilitate N absorption. Anderson and Midgley (2002) reported a negative association between hemipteran and spider densities, suggesting that spiders may lower hemipteran densities either by consuming them directly or by competing with them for prey. They also showed that *Roridula* plants received less N from fecal sources in populations with high spider densities and low hemipteran densities, suggesting that the spiders contribute very little toward the plants' N budget. However, in *R. dentata* populations where prey capture rates are especially high in summer, spiders may play an important role in keeping the hemipteran populations under control and mitigating the negative effects of sap-sucking.

The restricted distributions of the specialist spiders suggest that their associations with *Roridula* evolved after the extreme geographic separation of the two species and after the geographic separation between the northern and southern *R. dentata* populations (B. Anderson 2006). In contrast, the ubiquitous association between *Roridula* and *Pameridea* suggests a much older relationship that precedes the split between the two *Roridula* species (B. Anderson 2006).

26.6 Future research

Several instances of nutritional mutualisms between carnivorous plants and other organisms have been documented, but there is considerable scope for further investigations of such relationships. Useful immediate avenues of investigation include quantifying nutritional benefits to ants associated with *N. rafflesiana* and identification of possible chemical cues produced by the fungi associated with the tomentose band of *N. albomarginata*. The potential role of inquilines in the nutrient fluxes and budgets of *N. ampullaria* and other *Nepenthes* species also needs further research. Finally, additional natural history observations and detailed experiments may reveal new mutualisms between vertebrates and unstudied giant-pitchered *Nepenthes*, and between animals and the other pitcher plant genera.

PART V

The Future of Carnivorous Plants

CHAPTER 27

Conservation of carnivorous plants

Charles Clarke, Adam Cross, and Barry Rice

27.1 Introduction

The ≈800 known species of carnivorous plants are taxonomically and ecologically diverse (Juniper et al. 1989; Chapters 3–10, Appendix), resulting in a corresponding diversity in conservation-related issues. The conservation status of a small number of species, particularly those from the USA and a few high-profile species from other regions, is reasonably well known (Jennings and Rohr 2011). However, the conservation status of the vast majority of species is so poorly documented that by early 2016 most had never been assessed for the International Union for the Conservation of Nature's (IUCN) Red List of threatened species. This knowledge deficit is being addressed. More than 300 species were assessed for the Red List in 2016, but nearly 200 species still await their first assessment.

As of this writing, in early 2017, we can divide the global carnivorous plant flora into three groups based on our knowledge of their conservation status: well-known species whose conservation status is also reasonably well understood (≈10% of species); species whose conservation status has been assessed, but for which detailed additional information on threats and conservation status is unavailable (≈60%); and species whose conservation status has yet to be assessed (≈30%).

The reasons for the uneven spread of knowledge and survey effort are relatively simple. Public awareness of the depletion of wild populations of the iconic *Dionaea*, *Sarracenia*, and *Nepenthes* arose in the USA because species within these genera were the most popular in cultivation, and plants collected from the wild accounted for a substantial portion of this trade (Simpson 1995, D'Amato 1998). This threat was compounded by land clearing for domestic and commercial development in the southeastern USA, which led to the loss of a number of key habitats for *Dionaea* and several *Sarracenia* species (Rice 2006). Similar circumstances applied to several spectacular highland *Nepenthes* from Borneo (Clarke 1997). The community of scientists and enthusiasts, and the governments of the countries concerned, have responded to these threats primarily through the restriction of domestic and international trade and the protection of some remnant habitat patches. The task of conserving these species is ongoing, but our collective focus has now broadened to encompass the remaining ≈90% of species that are less scientifically or horticulturally well known. Many of these species grow in developing countries where rates of habitat loss and degradation typically are high (Laurance 2010), and where many species are cryptic, occur only in remote or inaccessible terrain where fieldwork is logistically challenging, or are endemic to regions that are politically restive or in overt conflict.

In this chapter, we review the key conservation issues concerning carnivorous plants, focusing on two geographical regions—the USA and southwest Australia—and one widespread genus (*Nepenthes*), as these are the only groups currently amenable to analysis. Southwest Australia hosts the highest number of carnivorous plant species in the world, whereas *Nepenthes* is the largest genus of pitcher plants and includes the highest proportion of threatened taxa. Although the carnivorous flora of North America comprises relatively few species, it contains an unusually high number of iconic taxa. These three groups are

Clarke, C., Cross, A., and Rice, B., *Conservation of carnivorous plants.* In: *Carnivorous Plants: Physiology, ecology, and evolution.* Edited by Aaron M. Ellison and Lubomír Adamec:
Oxford University Press (2018). © Oxford University Press. DOI: 10.1093/oso/9780198779841.003.0027

taxonomically, geographically, and ecologically disparate and our approach to discussing them varies accordingly.

A lack of data from South America, Africa, and Europe precludes a more complete global analysis, and we cannot cover countries such as Brazil, South Africa, and Venezuela that harbor many carnivorous plant species. We do not assess all the conservation threats to individual species, nor do we consider potential impacts of climatic change (Chapter 28). Conservation is frequently a crisis discipline with action often required without all the desirable and relevant information. Significant and widespread data deficiency remains the greatest constraint to conservation of the global carnivorous flora, and we highlight the need for increased survey efforts and detailed assessments of the threats to individual species and their habitats.

27.2 The conservation status of carnivorous plants

The depletion of wild populations of *Dionaea muscipula*, and several *Nepenthes* and *Sarracenia* species, in the 1980s led to their listing on the Appendices of the Convention on International Trade in Endangered Species (CITES), a treaty that signatory states use to provide a legal framework to control the legal trade of plants across international boundaries. In the 1990s, increasing concern about other, non-trade related threats to wild populations led to the establishment of the Carnivorous Plant Specialist Group (CPSG) by the IUCN. The task of the CPSG was to generate conservation assessments of all CITES-listed carnivorous species for publication on the IUCN Red List. This task was completed by Arx et al. (2001), but the CPSG ceased activities shortly afterwards and, until its reactivation in 2012, there was no cohesive international strategy (or effort) for carnivorous plant conservation. This lack of action resulted from the fact that many plant taxa are unstudied and most of the limited resources directed toward conservation are spent on charismatic animal and habitat-related actions; only a small number of iconic plant species compete effectively for these scarce resources.

The IUCN and its affiliate, the Species Survival Commission (SSC) reactivated the CPSG in 2012 in response to an analysis of conservation status and key threatening processes to 48 species by Jennings and Rohr (2011). The scope of their study was severely constrained by a lack of published data and only included species whose conservation status had been assessed and for which supporting data regarding threats were available: 48 species, ≈50% from the USA, whose flora includes < 10% of the world's carnivorous taxa. Jennings and Rohr (2011) cited 87 references, only 13 of which were direct studies of conservation or threats to carnivorous taxa. An additional nine were citations of Red List assessments for individual species, and the remaining 65 were general papers on conservation biology or carnivorous plant biology.

Jennings and Rohr (2011) asserted that conservation biologists had no useful data for >90% of their estimated ≈600 species of carnivorous plants. This reflected a lack of recent, detailed field observations for many species (typically because of difficulties in accessing wild populations). Those data that did exist had been collected by a disparate group of professional and amateur enthusiasts and not published. Without a functioning CPSG, the IUCN and SSC lacked the ability to organize and coordinate efforts to assess and monitor the conservation status of carnivorous plants globally.

The goal of the reactivated CPSG is to assess the conservation status of all known carnivorous plant species, first at a global scale, and subsequently at regional or national scales. By early 2017, more than 570 of the ≈750 species have been assessed, many for the first time (Figure 27.1). For *Nepenthes*—the only genus to have been assessed twice (in 2001 and again in 2012–2016)—the proportions of species considered to be Data Deficient (DD), Endangered (EN), and Critically Endangered (CR) are similar (Figure 27.1). By contrast, the proportion of non-threatened, Least Concern (LC) species was significantly greater in 2016 and was offset by a corresponding reduction in the number of Vulnerable (VU) species. This shift did not result from a genuine change in the status of any species. Rather, it reflects improvements in our understanding of the ecology and distribution of many of the lesser-known species.

By the end of 2016, approximately 21% of species are threatened, whereas 74% are LC or Near Threatened (NT) (Figure 27.1c); proportions that are similar to those for all types of plants examined

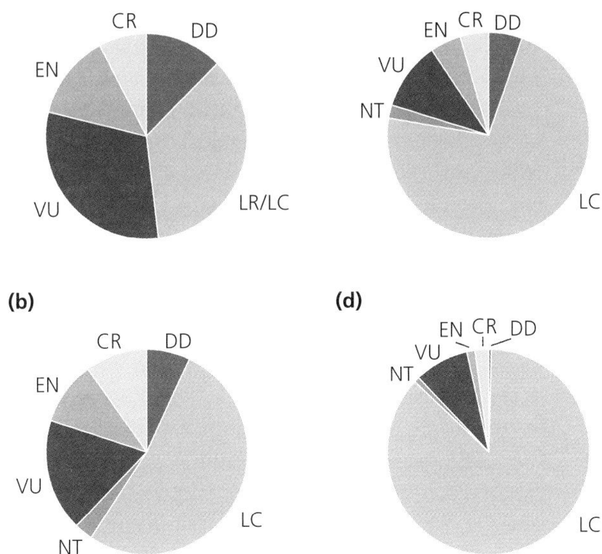

Figure 27.1 Pie charts showing the proportions of taxa in IUCN Red List categories. (**a**) *Dionaea, Sarracenia*, and *Nepenthes* species assessed by Arx et al. (2001); (**b**) *Dionaea, Sarracenia*, and *Nepenthes* species in 2017 (*Nepenthes* species re-assessed between 2012 and 2016); (**c**) all carnivorous taxa assessed for the IUCN Red List as of early 2017; (**d**) all Australian and New Zealand carnivorous taxa assessed for the IUCN Red List in 2016. Source of data for **b-d**: IUCN (2016).

by RBG Kew (2016). Although progress toward the goal of assessing the conservation status of all carnivorous plants has been rapid in recent years, the carnivorous floras of Africa and South America remain largely unassessed. Kalema et al. (2016) provide regional assessments for carnivorous plants of Uganda, but we still lack assessments for Brazil, Venezuela, and South Africa—hotspots of carnivorous plant diversity and regions with significant conservation threats.

27.3 Key threats

Our understanding of the processes threatening carnivorous plants is increasing slowly. Habitat loss and degradation are the most significant threats, followed by biological resource use (Jennings and Rohr 2011). The primary threats to habitats are agricultural activities, pollution, and altered fire regimes. Of the 28 threatened Australian carnivorous plant species assessed by the CPSG in 2016, 13 are threatened by natural system modifications (primarily changes to fire regimes and overuse of ground or surface water), whereas 10 are affected by agricultural activities. Five species each are threatened by development and invasive species. Other threats include pollution, collection of plants for trade, mining, and road or railroad construction. The overwhelming majority of threatened species occur in southwest Western Australia and the Northern Territory. Kalema et al. (2016) detected a similar range of threats in Uganda. Collectively, these three surveys demonstrate that habitat degradation and destruction for agriculture represent the most significant threats, followed by environmental variation caused by changes in water or fire management. Collection of plants for trade significantly threatens species that are prized for horticulture, especially *Dionaea, Nepenthes*, and *Sarracenia*. Although we have ranked the threats, we rarely know how they threaten individual species. For example, some *Nepenthes* species respond poorly to habitat disturbances, whereas others are unable to persist without them (§27.5). Negative effects of habitat destruction and degradation generally can be ameliorated only by habitat protection, supported by monitoring and law enforcement. However, one threatening process that is well understood is the collection of plants from the wild for trade (§27.4.1).

After the CPSG completes its Red List assessments (scheduled for 2020), we will identify those taxa that are in greatest need of protection to help direct funding for effective conservation actions. These actions will require triage, with resources distributed based on perceived threats and a host country's ability to implement the necessary actions (Bottrill et al. 2008, Gerber 2016). For example, many recently described *Nepenthes* taxa occur in parts of the Philippines where there is political and civil unrest. In other countries, ineffective law enforcement cannot protect threatened species, even when their habitat lies within an officially "protected" area (Gaveau et al. 2009, Clark et al. 2013). Without support of local governments and communities, effective conservation measures are unlikely to succeed (Porter-Bolland et al. 2012).

27.4 Carnivorous plant conservation in North America

Six genera of carnivorous plants are native to the USA and Canada, including the endemic *Dionaea muscipula* (Chapter 4), *Darlingtonia californica* and *Sarracenia* (Chapter 9), and representatives of *Drosera* (Chapter 4), *Pinguicula* (Chapter 6), and *Utricularia* (Chapter 8). Although relatively species poor, the three endemic genera account for a disproportionately large number of the world's iconic carnivorous plant species, and many of these face significant threats. In the USA, organizations responsible for conserving threatened species include: federal entities that operate nationally (e.g., US Forest Service, National Park Service, US Army Corps of Engineers, and US Fish and Wildlife Service; Stein et al. 2008); state agencies that oversee state-owned and managed lands; and private individuals and nongovernmental organizations who own and manage their own properties or oversee deeded conservation easements.

27.4.1 Threats

The primary threat facing carnivorous plant biodiversity in the USA is habitat loss. By 2009, wetland area in the USA had fallen to ≈50% of that present in the 1600s, irrespective of wetland quality or type (Dahl and Johnson 1991, Dahl 2011). Widespread destruction and subsequent development of wetland habitats have affected adversely the range and population sizes of many North American carnivorous plants (D'Amato 1998, Schnell 2002). Rates of ongoing degradation of remaining wetlands (e.g., by pollutants or draining) are substantial (Dahl and Johnson 1991, Osmond et al. 1995, Gutzwiller and Flather 2011). Despite widespread habitat loss and degradation, no carnivorous plant species have yet gone extinct in North America (Schnell 2002).

Coordinated actions are needed to slow the rate of loss of wild populations. First, land that is already protected must be managed appropriately to ensure that it retains its ability to support viable (sub)populations of the species it is intended to conserve. Unfortunately, management teams are almost always understaffed and underfunded and efforts tend to focus on the most pressing and immediate actions, rather than the full suite of measures needed. Furthermore, public policy often requires guidance from experts: government administrators rarely are experts in conservation biology and need to be educated about ecosystem services and intrinsic values that protected areas and species provide (Millennium Ecosystem Assessment 2005). Focused NGOs such as the CPSG and the International Carnivorous Plant Society (ICPS) play pivotal roles in raising public awareness of the value of conserving carnivorous plants.

Habitat protection is a fundamental conservation action. Unless there are strong economic demands for a given site, it may be relatively easy to garner public support for protecting it. However, simply protecting habitats does not constitute effective conservation, especially if the habitat or resident populations are very small. Many sites require ongoing management or remediation. For example, local hydrological processes must be preserved in (or restored to) a manner that is as close to their original state as possible. Sites must be burned at appropriate time intervals and intensities, lest species that are adapted to highly specific fire regimes disappear. Both woody and herbaceous non-native species are major threats at some sites and their control or removal is labor intensive, expensive to implement, and rarely successful in the long term (Lippincott 2000, Randall and Rice 2003). A notable exception is power-line rights of way that are maintained free of invading woody vegetation. If herbicide is not used to keep them clear, these corridors can become refugia for native species that colonize gaps or disturbed sites, including carnivorous plants in the USA, Australia, and Southeast Asia (Clarke 1997, Sheridan et al. 1997).

Legal and illegal collection of plants for the horticultural and cut-flower trade continue to impact many populations. For instance, cut pitchers, especially of *Sarracenia leucophylla*, are readily available for purchase online (B. Rice *unpublished data*). Poaching has resulted in the depletion and even complete destruction of some subpopulations of *Dionaea* and *Sarracenia* [USFWS 1994, N. Murdock (USFWS) *unpublished data*]. The public debate about the costs and benefits of collecting wild plants for trade has led to a deep philosophical divide between some carnivorous plant enthusiasts and conservation practitioners; this division can hinder implementation of conservation programs. Continued efforts at education within the horticultural community are directed at curbing poaching by hobbyists.

27.4.2 Species at risk

Dionaea muscipula is not only the most iconic carnivorous plant, it is one of the world's most spectacular and popular plants of any kind. It is endemic to the border region between North and South Carolina, where it can still be found in about 15 counties (Franklin and Finnegan 2004). Its range continues to decline due to ongoing habitat destruction and development (Luken 2012). Viable populations require a mosaic of early-successional patches that are maintained by regular fires (Luken 2005).

Three species of *Sarracenia* (*S. alabamensis, S. jonesii*, and *S. oreophila*) are listed as federally endangered and thus fall under the purview of the US Endangered Species Act (16 U.S.C. §1531 *et seq.* [1973]; USFWS 1994, Godt and Hamrick 1996, Brewer 2005); *Pinguicula ionantha* is listed as federally threatened. At this point, all known localities for these plants appear to be reasonably secure, although subpopulations on private lands could be threatened by ownership changes. All known locations are being monitored and, in most cases, are under some degree of active management. Of the remaining carnivorous plant species in North America, the majority appear to be secure. Species that could become threatened in the event of sudden, significant habitat degradation or loss include *Pinguicula primuliflora, Utricularia amethystina, U. simulans*, and perhaps some subpopulations in the *Drosera filiformis–Drosera tracyi* complex (Taylor 1989, Schnell 2002, Rice 2011b).

27.4.3 Expert assessments

Arx et al. (2001) assessed *Dionaea* and *Sarracenia* for the IUCN Red List, and Jennings and Rohr (2011) analyzed threats to carnivorous plants in the USA. For this chapter, one of the authors (B. Rice) surveyed conservation practitioners in 2016 to assess expert opinion about species of concern and key threats to carnivorous plants in the USA and Canada.

Between January and March 2016, Rice contacted 340 conservation workers. One hundred and seventy-five responded, including individuals from 45 of 50 states (none from Arizona, Missouri, Oklahoma, Pennsylvania, or West Virginia); and nine of the 13 Canadian provinces (none from Newfoundland and Labrador, Northwest Territories, Nunavut Territory, or Ontario). The survey asked respondents to rank the importance of "Habitat development," "Invasive Species," "Poaching," and "Changes in ecosystem processes (hydrology, fire, pollutants, etc.)." The respondents also could provide and rank any other threats they felt were important. Approximately 28% of surveys were completed by individuals from data-based groups such as state Natural Heritage Programs, NatureServe, or Canada Conservation Data Centres; 26% were from federal agencies (e.g., USFS, FWS, NPS); 11% were from state agencies; 14% were from The Nature Conservancy; 12% were from academic researchers; and 9% were from other sources (NGOs, botanical gardens, consultants). Overall, rankings of threats by these experts (Table 27.1) were comparable to those of Jennings and Rohr (2011) and the CPSG (§27.3).

Table 27.1 Expert assessment of threats to carnivorous plants. Values shown are percent of 175 respondents who, in 2016, ranked each threat as moderately to very harmful for widespread or rare carnivorous plant species.

Threat	Widespread species	Rare species
Habitat development	57	56
Invasive species	37	31
Poaching	23	30
Changes in ecosystem processes (fire, hydrology)	27	34

27.4.4 Conservation and management of threatened species

The USA appears to be the only country to have developed and implemented species-specific conservation plans for carnivorous plants. As required by US federal law, the recovery plan for *Sarracenia oreophila* was developed shortly after the species was listed as a federally endangered species (USFWS 1994). The plan is designed to engage all stakeholders in the effective management of all remaining subpopulations, with the ultimate goal of removing it from the endangered species list. Actions include surveys to improve understanding of the species' ecology, maintenance and restoration of suitable habitat patches, removal of invasive species, and *ex situ* preservation in the form of an indexed living collection maintained by the Atlanta Botanical Garden.

Studies of the genetic diversity of *S. oreophila* indicate that both overall genetic diversity and rates of gene flow among subpopulations are low. Long distance dispersal of pollen or seeds from one patch to another is rare and the loss of additional subpopulations likely will reduce its overall genetic diversity. Carter et al. (2006) characterized the floristic composition of plant communities and their relationships to soil types where *S. oreophila* grows. Such studies are used in gap analysis to detect potential additional localities for conservation or restoration of *S. oreophila*, including previously unrecorded subpopulations, degraded sites where the species had been extirpated, or patches that are unoccupied but potentially suitable for natural or assisted (re)colonization.

Ecological restoration of degraded habitat patches is a popular approach to the conservation of threatened plant species, but outcomes are difficult to predict. Brewer (2005) found that the relatively common *Sarracenia alata* responded to artificial and fire-mediated reduction in competition from other plants by increasing its investment in carnivory. However, the endangered *S. alabamensis* responded poorly to the removal of competing plants. The reasons for these divergent responses to restoration actions are not known, but they demonstrate that effective conservation actions rely upon a detailed knowledge of the ecology of the target species.

Implementation of recovery plans has arrested the decline of *S. oreophila*, but its long-term persistence depends on ongoing research and monitoring programs. There is no room for complacency. Reduced funding or resources, changes in government attitudes to conservation and even stochastic, environmental events can result in the rapid loss of additional subpopulations.

27.4.5 The role of horticulture

Growing carnivorous plants is a popular hobby, and horticulture represents a significant threatening process to species with desirable traits (Simpson 1995). Market demand is twofold: non-specialist growers of carnivorous plants are concerned only with cultivating species that are readily available at low prices, whereas dedicated enthusiasts seek to build collections of species, often investing considerable resources in the process. The large-scale production of affordable, commercially propagated plants of *Dionaea* and *Sarracenia* appeals to the former group.

Because legally sourced plants can be produced and sold for lower cost than plants collected from the wild, such cultivation should reduce or eliminate the market for wild-collected plants. But it has not. Wild populations of *D. muscipula* continue to decline because plants continue to be collected illegally from the wild (Platt 2015, Love 2016). Growers who believe that their right to collect and grow carnivorous plants outweighs the rights of the plants to persist unmolested in the wild continue to drive strong demands for wild-collected plants of desirable species.

Horticulture also can mitigate losses by legally assembling *ex situ* subpopulations to preserve the genetic diversity of threatened or endangered species. However, it is difficult to assemble such subpopulations, even for easily grown species. Many horticultural collections are biased toward eye-catching phenotypes and inadvertently or deliberately select for traits that would have low fitness in the wild. This point may be obvious to conservation biologists, but individuals who evaluate traits from a purely aesthetic standpoint rarely think about it (B. Rice and C. Clarke *unpublished data*).

The Atlanta Botanical Gardens (ABG) in Georgia (USA) is an exemplar of an institution that has formed a partnership with conservation agencies to develop and implement effective conservation programs in which horticulture plays a role. The gardens maintain indexed selections of endangered *Sarracenia*—including specimens from every known site of *S. alabamensis, S. jonesii*, and *S. oreophila*—and preserves nearly all of the extant genetic diversity of these species (R. Determann *unpublished data*). The success of this (and other) *ex situ* preservation schemes depends entirely upon the persistence (or restoration) of suitable habitat so that species can be re-established in the wild.

Havens et al. (2006) provide criteria to gauge the success of conservation-related programs involving horticulture, including: strict adherence to laws and relevant policies; promoting education and public awareness; social and ethical activism toward sustainability and conservation; and adequate, sustained, effective investment in conservation actions such as managing protected areas and cultivating genetically diverse stock for eventual reintroduction. One of the most important aspects of horticulture is education and public awareness. Botanic gardens and arboreta play vital roles in focusing the attention of their visitors on rare and threatened plant species, and effective education programs have the potential to change the way people perceive and treat threatened species and their habitats.

27.5 Conservation of *Nepenthes* in Southeast Asia

The tropical pitcher plant genus *Nepenthes* comprises 130–160 species, most of which occur in the Malay Archipelago (Chapter 5). Some of the largest and most famous carnivorous plants belong to this genus, including *Nepenthes rajah*, whose traps have evolved to capture the feces of tree shrews and may exceed 3 L in capacity (Chin et al. 2010). *Nepenthes clipeata* grows on a single mountain in western Borneo and has been considered critically endangered since the 1990s. More than any other *Nepenthes, N. clipeata* species symbolizes the effects of over-collection of plants from the wild for horticulture and, more recently, the degradation of habitat due to fire (Lee 2007). However, we know much less about species that lack the high public profiles of *N. rajah* and *N. clipeata* (Clarke 1997, 2001, Arx et al. 2001).

27.5.1 Poaching

Collection of plants from the wild is the principal threat to most *Nepenthes*. The great diversity in pitcher morphology generates strong market demand for those species that produce extraordinary pitchers, leading to over-collection and significant reductions in the sizes of natural populations (Simpson 1995, Phillipps and Lamb 1996). In contrast, widespread species lacking unique trap characteristics generally are not collected for trade except in some local markets in Thailand and Indonesia (C. Clarke and Hernawati *unpublished data*). Such practices can still result in significant declines in some subpopulations, but in most cases (e.g., *N. suratensis* and *N. mirabilis* var. *globosa* from southern Thailand), habitat degradation or loss remains the primary threat (Clarke and Sarunday 2013).

27.5.2 Habitat fragmentation

Individual species respond to habitat disturbances differently, and generalized, yet accurate, assessment of the conservation status of *Nepenthes* is difficult. Most rare species with specialized pitcher morphologies are narrow endemics (*sensu* Rabinowitz 1981), typically occurring on mountain summits and cliffs, or primary heath and peat-swamp forests (Clarke 1997, 2001, van der Ent et al. 2015). Generally, these narrow endemics respond poorly to habitat disturbances: within a few years after disturbance, adult plants die and seedlings fail to establish. In contrast, widespread species with unspecialized pitcher morphologies (e.g., *N. mirabilis*), typically colonize secondary, inundated habitats. Others (e.g., *N. rafflesiana, N. sanguinea*) naturally colonize ephemeral clearings in forests, such as tree-fall gaps and landslips. These latter two species have benefitted greatly from widespread habitat modification for agriculture and infrastructure; they were far less abundant during the colonial period in Malaysia (Wallace 1869), when suitable habitat patches were scarce.

The peat-swamp and heath forest communities near the Baram River in northwestern Borneo support seven lowland *Nepenthes* species, and these have been studied at several localities on a regular basis since 1987 (Figure 27.2; C. Clarke *unpublished data*). Three of these species—*Nepenthes bicalcarata*,

N. hemsleyana, and *N. ampullaria*—grow predominantly in intact heath and peat-swamp forest, whereas *N. albomarginata* grows in heath forests on sandstone ridges. *Nepenthes gracilis* and *N. rafflesiana* colonize open, highly disturbed sites and forest edges; *N. mirabilis* var. *echinostoma* is confined largely to permanent, deep swamps dominated by *Pandanus* spp. and sedges; and a giant-pitchered form of *N. rafflesiana* appears to be associated only with *Gymnostoma nobile* trees in heath forest.

The seeds of all of these taxa germinate readily in highly disturbed sites (where introgression also may be common), but *N. albomarginata, N. bicalcarata, N. hemsleyana*, and *N. mirabilis* var. *echinostoma* fail to persist for more than a few years unless the vegetation returns to something resembling its original state. *Nepenthes ampullaria* is somewhat more tolerant, whereas the relationship between *G. nobile* and the large-pitchered form of *N. rafflesiana* is unstudied. *Nepenthes gracilis* and *N. rafflesiana* occur in all habitat types, but are abundant only in open, highly disturbed sites. These contrasting strategies are representative of the entire genus and provide a clear message to conservationists: without an adequate understanding of the ecology of an individual species, it is impossible to determine whether the degradation (or even complete removal) of the vegetation at a given site represents a threat or an opportunity. Unfortunately, ecological data are rare in the taxonomic literature, and conservation assessments that rely on this literature likely will be inaccurate.

The habitats of *Nepenthes* typically are spatially and temporally patchy. Research is just beginning on the ecological effects of habitat fragmentation on, and its relevance for conservation of, widespread *Nepenthes* species. Chin et al. (2014) studied prey capture patterns in a cohort of five lowland *Nepenthes* species in Sarawak, and found that species confined to intact peat-swamp forest trapped combinations of prey that differed from those growing at the forest edge or in adjacent, open, and highly disturbed patches. Whether this reflects divergent prey-trapping strategies among *Nepenthes* species, or simply exposure to different combinations of arthropod prey in different habitat niches is not known, but it does demonstrate that the distribution of *Nepenthes* co-varies with the physical environment and the types of prey captured (Chin et al. 2014, van der Ent et al. 2015).

Species such as *N. rafflesiana* and *N. gracilis* that have evolved strategies to colonize ephemeral, disturbed sites are favored by current land-management practices, which typically involve rapid development of large-scale infrastructure that leads to extensive areas of bare ground or sparse shrublands. Consequently, these species have flourished in, for example, the Baram River area of northwestern Borneo in recent years, but at the same time, *N. bicalcarata, N. hemsleyana*, and *N. mirabilis* var. *echinostoma* have experienced catastrophic declines because of the almost complete destruction of the primary swamplands, heaths, and peat-swamp forests upon which they depend. With the exception of Brunei, where peat-swamp forests are relatively intact, this pattern has been repeated throughout the range of *N. bicalcarata*, which once extended from near Sintang in West Kalimantan to southern Sabah (Figure 27.2). Despite a historic extent of occurrence (EOO) of ≈95,000 km^2, much of this is now committed to plantation agriculture. The scale of loss of peat-swamp forests in western Borneo is reflected in the latest IUCN Red List assessment for this species, which lists it as globally endangered (Clarke et al. in press). More than 60% of mature *N. bicalcarata* individuals are thought to have been lost over the last 30 years and the population continues to decline.

27.5.3 Narrow endemics

Many lowland *Nepenthes* species have extensive geographical ranges, but most montane taxa are confined to the summits of one (or, at most, a few) mountains. An example is *Nepenthes muluensis*, a montane species with a small EOO and a small area of occupancy (AOO) that has been found growing at 1800–2350 m a.s.l. on three, or possibly four, mountains in northwestern Borneo (Figure 27.2; Clarke 2018). The seeds of all *Nepenthes* spp. are dispersed by wind, and the ability of *N. muluensis* to establish in new sites depends on the direction and strength of air currents and the proximity of sites at appropriate elevations. Specialized species such as *N. muluensis* have the greatest risk of extinction from random events (Kitching

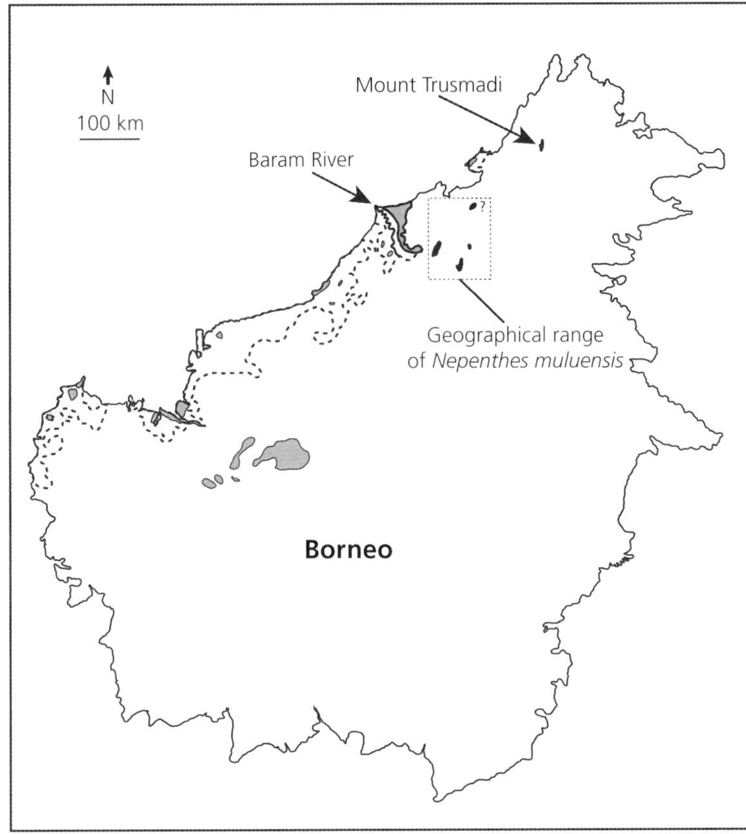

Figure 27.2 Outline of the island of Borneo, showing historical (light gray shading with dashed boundaries) and extant (dark gray shading with solid boundaries) peat-swamp forests located within the distributional range of *Nepenthes bicalcarata*. Also shown are the four montane regions that comprise the range of *Nepenthes muluensis*, the location of the Baram River mouth, and the summit area of Mount Trusmadi.

and Beaver 1990). Although development does not threaten the habitats of most montane *Nepenthes*, collection of plants for trade, which typically targets small, immature individuals that are more easily smuggled across borders, reduces the number of plants that can mature.

The entire population of *N. muluensis* lies within a protected area, so its future seems to be secure. Two other species, *N. lowii* and *N. macrophylla* are not so fortunate. These two montane species grow on Mount Trusmadi in northern Borneo (Figure 27.2) and have been collected by poachers since the 1980s. Seedlings used to be abundant along the trail to the summit of this mountain, but nearly all were removed in the 1990s and now only mature plants occur there (C. Clarke *unpublished data*). Once these die, the potential will be lost for these plants either to recolonize Mount Trusmadi or disperse to other patches. For the critically endangered *N. macrophylla*, this could lead to extinction in the wild, whereas for *N. lowii*, one of the largest subpopulations would be lost, increasing the degree of fragmentation within the surviving population.

27.5.4 Taxonomic fragmentation

Revisionary taxonomy also fragments *Nepenthes* populations. For example, Arx et al. (2001) did not consider *N. alata* to be of conservation concern (i.e., it was of "least concern"). Cheek and Jebb (2013b) subsequently fragmented (i.e., taxonomically revised) *N. alata*. The resulting 18 species in the new *N. alata* group include three species with extensive ranges, three that are known from at most 2–3 localities, and 12 that are known from only a single location or have very small EOOs (Figure 27.3). Six of these taxa are poorly known or have not been observed in the wild since they were first collected. These were assessed as Data Deficient by the CPSG,

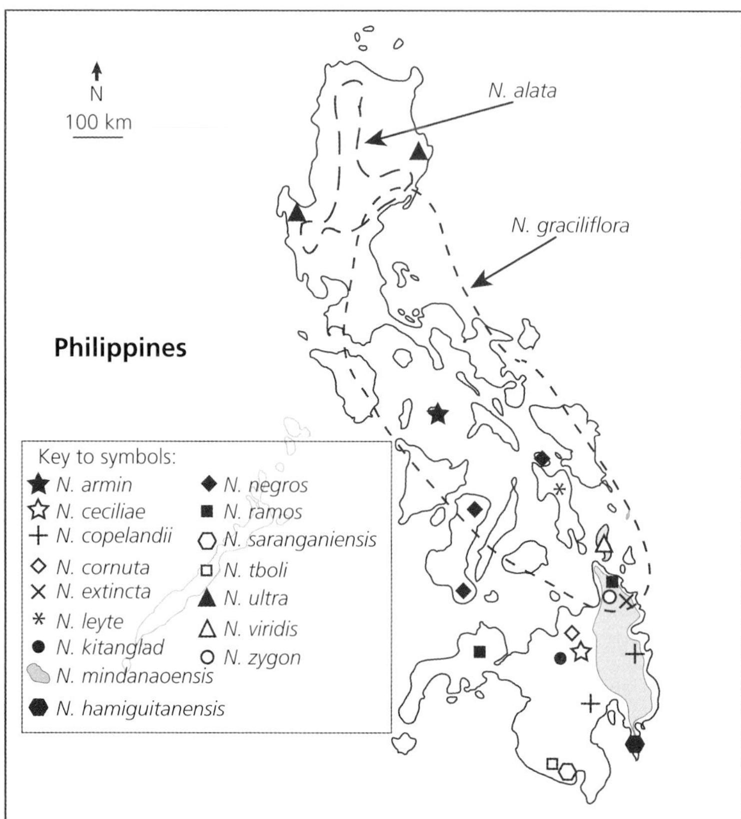

Figure 27.3 Outline of the Philippines, showing the distributions of taxa in the *Nepenthes alata* group. Three species with extensive geographical ranges are depicted using shading or dashed lines. All other taxa occur at point localities, which are denoted by symbols.

because no determination of threats was possible without field surveys to determine how they respond to habitat disturbance.

Taxonomic "splitting" is not confined to *Nepenthes*, and it presents challenges to conservation biologists (Zachos 2013). The IUCN uses estimates of EOO, AOO, and population size to define boundaries between Red List threat categories (IUCN 2000), and taxonomic fragmentation typically gives rise to one or more taxa with very small EOOs or AOOs, increasing the number of threatened species. This can be beneficial or problematic. Taxa whose status was unknown or unassessed because they had been lumped within widespread, non-threatened species are more likely to attract the attention of conservationists. On the other hand, if splitting is poorly supported or unjustified, costly conservation actions could become irrelevant if a protected taxon is subsequently synonymized with a non-threatened one. *Nepenthes kurata*, recently described by Cheek and Jebb (2013b) but later synonymized with *N. ramos* (Gronemeyer et al. 2016) provides a relevant example from the *N. alata* group.

27.6 Conservation of Australian carnivorous plants

Australia is one of the world's 17 "megadiverse" countries and both its flora and fauna exhibit very high levels of endemism (84% of plants, 83% of mammals). Australia is the only developed and industrialized megadiverse country, and it has a sufficient number of specialists and national wealth to adequately study and conserve its biodiversity (AGWA 2009). Many ecosystems within Australia's ancient, arid landscape are fragile, and, as in many other countries, the standard list of threats—habitat loss, land-use change, invasive species, diseases, and data deficiency—threaten much of its biodiversity (Brooks et al. 2006).

27.6.1 The Southwest Australian floristic region

Western Australia contains over half of Australia's recognized National Biodiversity Hotspots (AGWA 2009), and the Southwest Floristic Region (SWAFR) is a global biodiversity hotspot (Myers et al. 2000). The SWAFR encompasses > 300,000 km^2 and harbors ≈8000 native plant taxa. It represents one of the oldest landscapes on earth, and its Mediterranean climate, surrounded by ocean or desert, provides a spatial refuge for many species. Nearly 50% of the plant species that grow in the SWAFR are endemic, and over a third have been described since 1970 (Hopper and Gioia 2004).

More than 2500 species of plants in the SWAFR are of conservation concern (Hopper and Gioia 2004), and this number continues to increase at a much higher rate than the number of species being down-listed or delisted (AGWA 2009). Recovery plans are in place for less than half of the threatened plants in Western Australia, and implementation of plans rarely occurs because of limited funding and personnel constraints (AGWA, 2009). Although biodiversity is high throughout the SWAFR, there are several loci of very high species richness, such as the 297,000-ha Fitzgerald River National Park, which contains ≈1800 flowering plant species (AGWA, 2009).

27.6.2 Diversity

The highest diversity of carnivorous plant species—more than 125 taxa—of any single biogeographic area on Earth occurs in the SWAFR. This number includes > 50% of the Droseraceae (*Aldrovanda vesiculosa* and at least 110 *Drosera* spp.; Chapter 4), at least 15 species of *Utricularia* (Chapter 8), two of the eight *Byblis* spp. (Chapter 10), and the monotypic, endemic *Cephalotus follicularis* (Chapter 10). Carnivorous plants are 4.5× more diverse in Western Australia than would be expected in the regional flora (Brundrett 2009). This is a likely consequence of the extremely nutrient-impoverished sandy soils, long isolation, and great geological and climatic stability of the region (Lambers et al. 2014; Chapter 18).

Although many species are widespread in the SWAFR (e.g., *Drosera macrantha, D. glanduligera, D. drummondii, D. neesii, D. menziesii, U. tenella, U. menziesii*), a significant number have small, restricted ranges within the SWAFR because of close associations with specific soils or landforms (Lowrie 2013). As a result, natural fragmentation and disjunction of populations commonly occurs in the regional flora (Hopper and Gioia 2004, Phillips et al. 2011). Small geographic range size is associated strongly with a high extinction risk (Purvis et al. 2000), and ≈10% of carnivorous plant species in the SWAFR are threatened or endangered.

27.6.3 Threats

Conservation issues in the SWAFR include large-scale habitat loss, habitat fragmentation, weed invasion, impacts of *Phytophthora cinnamomi* dieback disease, and dryland salinity (Hopper and Gioia 2004).

Habitat loss. As in most cases around the world, habitat loss is the most significant driver of species declines. Since permanent European settlements were established in Australia in the late 1700s, ≈92.5 million ha, equal to ≈50% of the woody vegetation in non-arid and non-monsoonal regions, has been cleared (Barson et al. 2000); more than 50% of that between 1945 and 1982 (Saunders 1989). By 2000, ≈90% of the primary vegetation in the SWAFR had been cleared (Myers et al. 2000), and by 2009, ≈97% of the original vegetation in some areas of the Central Wheatbelt region of Western Australia had been lost (Bradshaw 2012). The remaining vegetation often is fragmented, unmanaged, and degraded (Table 27.2), and the biodiversity of Western Australia is in serious decline (Hopper and Gioia 2004, Bradshaw 2012).

The effects of habitat loss and fragmentation are exacerbated by the ecology and biogeography of southwestern Western Australia: ecosystems in the SWAFR are often island-like, characterized by high species turnover, prone to weed invasion, and harboring numerous range-restricted and ecologically specialized species. Many plant species in the SWAFR are engaged in very complex biotic relationships. Population regeneration and recruitment often is closely linked to sporadic environmental events (such as fire) or occurs more slowly than in most other floristic regions (Hopper and Gioia 2004,

Table 27.2 Number of species of each carnivorous plant genus in the SWAFR by IBRA bioregion, and conservation status of IBRA Bioregions as of 2000 (State of the Environment Committee 2011).

	Avon Wheatbelt	Esperance Plains	Geraldton Sandplains	Jarrah Forest	Mallee	Swan Coastal Plain	Warren
Carnivorous taxa							
Drosera	50	45	46	68	37	55	31
Byblis	0	0	1	1	0	1	0
Cephalotus	0	1	0	1	0	0	1
Aldrovanda	0	1	0	0	0	0	0
Utricularia	3	11	4	11	1	11	10
Total	**53**	**58**	**51**	**81**	**38**	**67**	**42**
Remaining woody vegetation (%)	10–30	10–50	10–50	50–70	10–30	30–50	70–90
Woody vegetation in reserves (%)	0–5	20–50	10–20	0–5	10–20	10–20	20–50
Habitat connectivity	Very low	Low	Low	High	Low	Low	Medium

Mucina et al. 2014). Hopper and colleagues (Hopper et al. 1996, Hopper 1997, 2000) hypothesized that selective pressures have favored ecological mechanisms that facilitate local persistence rather than wide dispersive capability. Thus, habitat loss reduces species ranges and further isolates already disjunct and poorly connected populations, while coincident fragmentation increases perimeter-to-area ratios, thereby facilitating colonization by exotic species that can deleteriously affect the reproductive success and genetic integrity of indigenous plant communities.

Detailed habitat modeling has not been done for carnivorous plants in the SWAFR. The legacy of past habitat loss and fragmentation is apparent in the very restricted modern distributions of species that were once more widespread: *Drosera leioblastus, D. bicolor, D. gibsonii, D. micra, D. orbiculata, D. nivea, D. leucostigma, D. pedicellaris, D. oreopodion, Utricularia helix*, and *Byblis gigantea* (Cross et al. 2013, Lowrie 2013). The effects of habitat deterioration and loss may also be particularly significant for species restricted to insular or specialized habitats such as granite outcrops, where connectivity is naturally low and disturbance rates high (e.g., *D. graniticola, D. browniana*), the non-saline margins of salt lakes (e.g., *D. zigzagia, D. bicolor, D. salina*), and the summits of mountains (e.g., *D. monticola*). The species confined to these habitats are often highly specialized and sensitive to habitat disturbance and degradation.

Altered hydrology. Land clearing and the replacement of woody vegetation with shallow-rooted, annual crop species progressively has altered the hydrology of southwestern Australia on a landscape scale, and altered hydrology now represents one of the most prevalent and serious threats to both terrestrial and freshwater habitats in the state (Department of Environment and Conservation 2012a). Waterlogged and moisture-retaining soils in the SWAFR provide habitat for numerous phylogenetically relictual taxa (Hopper and Gioia 2004). Cramer and Hobbs (2002) found that both the drying of wetlands and secondary salinization of soils is putting native vegetation increasingly at risk. Many carnivorous plants in the SWAFR are restricted to seasonally or permanently wet freshwater habitats, and hydrological change already has been implicated directly in the decline of species *Cephalotus follicularis, Utricularia helix*, and *Byblis gigantea*, and the localized extirpation of *Aldrovanda vesiculosa* (Cross 2012b, Cross et al. 2013, Lowrie 2013). Alteration to the hydrology of remnant vegetation, particularly small habitat fragments, represents a serious threat to the long-term persistence of many species. The hydrology of individual fragments can be influenced by disturbance and change throughout the watershed (often tens of km^2), and effective hydrological management must be undertaken at catchment level.

Fire is an integral component of the ecology of the SWAFR, and many ecosystems in Western Australia are fire-maintained (Hopper and Gioia 2004, Burrows 2008). However, different organisms and communities exhibit variable responses to different fire regimes, and it is evident that a diversity of fire regimes that accommodate ecological needs of all elements of the ecosystem are essential for maintaining biodiversity (Burrows 2008).

Fire management is one of the most ecologically challenging and socially contentious topics faced by land managers (Burrows and McCaw 2013). Nearly 300,000 ha of forested land in the SWAFR are burned each year by prescribed fires or wildfires, and individual fires > 20,000 ha are becoming increasingly common (Burrows and McCaw 2013). In some habitats, plant litter approaches 40–50 metric tons ha^{-1} in the prolonged absence of fire (Burrows and McCaw 2013). Although these are fire-resilient ecosystems able to cope with hot summer fires that periodically consume litter and restart ecological processes, a severe wildfire can be detrimental to areas already stressed by prolonged drought. For example, although *C. follicularis* recovers vigorously even after very hot summer wildfires (Lowrie 2013), significant population declines have been observed following extensive damage to peat soils supporting its *Callistemon glaucus/Homalospermum firmum* swamp thicket habitat after several intense wildfires that have occurred since 2000.

Many carnivorous taxa in the SWAFR need fire to establish or reproduce (Cross et al. 2013, Lowrie 2013). For example, for many species, fire induces flowering after fire, and smoke-derived chemicals stimulate seed germination. Current fire management policy sets an unconditional target of 200,000 ha per year to be burned in the SWAFR (Burrows and McCaw 2013), with most fires set in late autumn, winter, or spring, when geophytic or summer-dormant taxa are in active growth or flowering (Lowrie 2013). It is likely that species with highly restricted ranges—e.g., *D. leioblastus*, *D. gibsonii*, and *D. oreopodion*—may be threatened by aseasonal prescribed burns; future studies are needed to understand the fire ecology of carnivorous plants that grow in the SWAFR.

27.6.4 Conservation and management

Conservation and management of the SWAFR flora is complex and challenging and is compounded by constraints of scale, data deficiency, personnel, and funding. Effective conservation relies not only upon possessing a good knowledge of the ecology and distribution of a species or community, but also of the threats affecting individual species and entire communities. To date, management strategies for threatened species in the SWAFR have been driven by types of threats, geographic arrangement of species, and the amount of knowledge available for any given species (K. Atkins *unpublished data*). If management has occurred at all, species in Western Australia tend to be managed individually, and rarely are species delisted or even moved to a less vulnerable category (AGWA, 2009).

More recently, land managers in Western Australia increasingly are adopting holistic landscape or regional approaches to conservation. These strategies are an efficient way to conserve multiple species, particularly those that occur in close proximity to one another and face the same threats. Regional management plans also afford protection to other species and communities that lie within the boundaries of protected areas. For example, the Fitzgerald Biosphere Recovery Plan applies to an area ≈1.3 million ha and includes 2500 described vascular plant species, including many carnivorous ones (Department of Environment and Conservation 2012b). The management of hydrology and wildfire at the landscape scale also may be more efficient and economical than focusing independently on small habitat fragments or the threats to individual species.

27.7 Future research and conservation prospects

The conservation status of many carnivorous plants remains unknown, but recent research is resolving data deficiency; the CPSG aims to assess all carnivorous plant species for the IUCN Red List by

2020. Because knowledge of the basic ecology of many species, especially those in Africa and South America, is wanting, sustained research into their population structure and dynamics is required to achieve this target. Practical and methodological challenges limit our ability to assess accurately the conservation status of many carnivorous plant species. It is difficult to determine AOOs and population sizes of species in large, poorly accessible, and biodiverse regions such as the SWAFR or Malay Archipelago. Given the potential impacts to industry, land managers, and government conservation agencies of listing a species, it is critical that the integrity of the listing process be maintained (K. Atkins *unpublished data*).

Little yet has been achieved in designing and implementing effective conservation actions for carnivorous plants apart from identifying the habitats of some species as protected areas. In many parts of Southeast Asia, such gazetting affords little protection, because illegal logging, mining, land clearing, and collection of plants for the horticultural trade persist in protected areas where there is inadequate law enforcement.

The future of the world's carnivorous flora is uncertain. While many are widespread and abundant, or well protected on existing conservation lands, we cannot manage most others because resources are lacking. The scale and severity of threats faced by many species is increasing faster than our understanding of their basic ecology. Better data allow us to identify those species most at risk of extinction; the next, more difficult step is conservation triage: which of the many threatened or endangered species are most deserving of limited resources, and how can we best allocate those resources to alleviate threats and achieve successful conservation. For some species already poised on the brink of extinction, such as *Drosera oreopodion* and *Nepenthes suratensis*, it may already be too late.

CHAPTER 28

Estimating the exposure of carnivorous plants to rapid climatic change

Matthew C. Fitzpatrick and Aaron M. Ellison

28.1 Introduction

Forecasting how carnivorous plant species will respond to climatic change is a key issue in their conservation and management (Chapter 27) but presents a number of challenges. These challenges derive from interactions between the relatively simplistic statistical methods typically used to forecast species responses to climatic change, which to date have been limited mainly to species distribution models ("SDMs;" Elith and Leathwick 2009, Franklin 2009), and particular aspects of the ecology of carnivorous plants, including their rarity, habitat specialization (Chapter 2), and limited dispersal ability (Chapter 22).

The small ranges and oftentimes low local abundance of carnivorous plants provide few occurrence records, which increase the potential for poorly or over-fitted SDMs and misspecification of relationships with their "optimal" environments. The unique habitats in which carnivorous plants often grow (Chapter 2) also are difficult to characterize using the basic temperature and precipitation data that often undergird SDMs. Rather, habitats in which carnivorous plants are common often are decoupled from broader climatic patterns (e.g., many retain high soil moisture even during seasonal drought) and may be associated with frequent disturbance (e.g., fire; Chapter 2). Last, dispersal limitation also may constrain range shifts of carnivorous plants as the climate changes. These three issues raise two related questions that are critical for understanding and forecasting the future of carnivorous plants. First, to what extent are current carnivorous plants distributions constrained by climate; and second, how readily, if at all, might carnivorous plants disperse to colonize new habitat as it becomes climatically suitable?

In this chapter, we estimate the vulnerability of carnivorous plants to climatic change in light of challenges identified with SDMs in general and their particular application to these unique species. The modeling approaches we use partially overcome some of these challenges, and may be applicable to other sparse or rare species. We begin by reviewing the basics of SDMs. We then highlight specific ecological characteristics of carnivorous plants and their geographic distributions that limit the utility of classical SDMs for forecasting their future distributions. We combine two approaches: "ensembles of small models" (Breiner et al. 2015), which attempt to deal with the challenges of fitting SDMs for data-limited species; and "bioclimatic velocity" (Serra-Diaz et al. 2014), which is an estimate of how fast a species would have to migrate to track its climatic niche (as opposed to a prediction of the potential shift in distribution, the typical output from SDM projections), to provide initial assessments of the vulnerability of carnivorous plants to climatic change.

28.2 The basics of species distribution models

Efforts to quantify the vulnerability of species to climatic change typically rely on SDMs. These models (also called bioclimatic envelope models, habitat suitability models, or ecological niche models;

Fitzpatrick, M. C., and Ellison, A. M., *Estimating the exposure of carnivorous plants to rapid climatic change.* In: *Carnivorous Plants: Physiology, ecology, and evolution.* Edited by Aaron M. Ellison and Lubomír Adamec: Oxford University Press (2018).
 DOI: 10.1093/oso/9780198779841.003.0028

Guisan and Zimmermann 2000, Elith and Leathwick 2009, Franklin 2009) are relatively simple statistical models that predict habitat suitability across an area of interest using empirical relationships between the distribution of a species (expressed as a set of point locations at which the species is known to occur) and coincident environmental variables (typically derived from digital maps of interpolated climatic data). When applied to current climates or simulations of future climate, SDMs respectively infer current species distributions or forecast potential changes in species distributions from predictions of habitat suitability (Guisan and Thuiller 2005).

The field of species distribution modeling has grown rapidly over the last few decades (Guisan et al. 2013), fueled by three primary factors: increased availability of species occurrence and environmental datasets (Graham et al. 2004); the development of more powerful statistical techniques and user-friendly software packages (Phillips et al. 2006, Thuiller et al. 2009); and an overall need for comprehensive information on species distributions, including quantitative assessments of the vulnerability of species to climatic change (Chapter 27).

28.2.1 Challenging species distribution models with sparse or rare species

Despite advancements in data and algorithms, modeling the impacts of climatic change on sparse species (*sensu* Rabinowitz 1981), rare species, or habitat specialists like carnivorous plants using SDMs remains a major challenge. These challenges arise primarily from fitting models with insufficient or unreliable point-occurrence records and environmental predictors that inadequately characterize habitats. Whereas rare species may be under greatest threat from climatic change and thus might benefit most from SDM-based climate-impact assessments, they are often are most difficult to model using SDMs, a conflict Lomba et al. (2010) summarized as the "rare species modeling paradox."

28.2.2 Critiques of species distribution models

Even when data are adequate and statistical issues can be minimized, numerous critiques have argued that SDMs stand on weak theoretical footing, especially when used to forecast species distributions under scenarios of climatic change (Hampe 2004, Heikkinen et al. 2006, Dormann 2007). The primary critique centers on whether empirical species–climate relationships derived solely from observations of point-occurrence records and associated environmental conditions can reliably predict responses of species to climatic change. Detractors argue that multiple interacting abiotic and biotic factors determine range limits (Gaston 2003), yet SDMs typically use only a small number of climatic variables (most frequently temperature and precipitation) in model fitting and often ignore other causal abiotic drivers, biotic interactions (Wisz et al. 2012), dispersal processes (Fitzpatrick et al. 2008), or adaptation (Fitzpatrick and Keller 2015).

Although carnivorous plants are found worldwide and several species have large geographic ranges, all of them occupy patchy, restricted microhabitats within their geographic range (Juniper et al. 1989; Chapter 2). Local environmental conditions including soils and hydrology likely are at least as important, if not more so, than climate *per se* in determining habitat suitability and occurrence patterns. In terms of dispersal constraints, few species have any obvious mechanisms for long-distance seed dispersal. Moreover, the successful establishment of some carnivorous plants far outside their native ranges suggests that distributions of some species are limited by their ability to disperse and colonize suitable, but distant, habitats (e.g., Ellison and Parker 2002; Chapter 22). These dispersal constraints suggest carnivorous plants may not be able to track rapid climatic shifts by shifting their geographic ranges.

Even if all habitat factors could be included in perfectly calibrated SDMs, fitted species–climate relationships still may not reflect true distributional constraints. There also is no reason to expect that current species–climate relationships will remain constant in altered ecological contexts (e.g., Fitzpatrick et al. 2007, Veloz et al. 2012) or novel climates of the future (Williams and Jackson 2007, Fitzpatrick and Hargrove 2009).

Last, when applied to future scenarios of climatic change, SDMs forecast potential changes in species distributions, not actual changes in where populations occur on the landscape. The extent to which species will be able to follow these forecasted range shifts over the next several decades will depend on dispersal and population dynamics, both of which are uncertain and stochastic. Moreover, the increasing isolation and fragmentation of natural habits and the rapid rates of projected climatic change likely will make rapid dispersal unfeasible for all but the most vagile and widespread species (Hill et al. 1999, Malcolm et al. 2002, Loarie et al. 2009).

Pragmatic users of SDMs are quick to acknowledge these, and other, shortcomings of the models, but counter that SDMs are one of the only tools available that can be applied across multiple taxa, regions, times, and spatial scales (Guisan and Thuiller 2005). Proponents also contend that both the fossil record and contemporary observations provide evidence that, at broad spatial scales, species often predictably respond to climatic change by shifting their geographic ranges to track suitable habitat (Davis 1989, Parmesan et al. 1999, Chen et al. 2011) and SDMs have been shown to predict such responses under modest degrees of extrapolation in the distant (Maguire et al. 2016) and recent past (Araújo et al. 2005, Hijmans and Graham 2006, Kearney et al. 2010, Dobrowski et al. 2011). Although these suggestions may be true for certain taxa in certain regions (e.g., temperate trees), the same cannot necessarily be said of carnivorous plants or other sparse or rare species.

28.3 Characteristics of carnivorous plants that challenge SDMs

28.3.1 Rarity and sparse distributions

Although carnivorous plants are found worldwide and several species are abundant or have large geographic ranges, all occupy patchy, restricted microhabitats within their geographic range (Chapter 2). Typically, carnivorous plant species are rare *sensu* Rabinowitz (1981): they tend to have small ranges, high habitat specificity, and some either have small population sizes or are sparse within their range. Each of these components of rarity limits the ability of SDMs to forecast reliably future distributions of carnivorous plants (Sbragia et al. 2014). Small ranges reduce the number of locations (occurrence records) that can be used in fitting SDMs. Besides limiting information content, low numbers of point-occurrence records can bias fitted statistical relationships that are extrapolated across the study area to map habitat current and future suitability (Barry and Elith 2006). Loss of populations because of habitat conversion or over-collecting (Chapter 27) only adds to the challenges of reliably applying SDMs to carnivorous plants: not only are the number of occurrences reduced, but also relationships between their distributions and climate may be altered or obscured.

A small geographic distribution is not necessarily problematic; SDMs often perform best for narrowly distributed species with specialized climate niches. However, performance of SDMs under such circumstances is high only when species are well sampled (Guisan et al. 2007). Good predictive performance based on current distributions also may be indicative of over-fitting, which is most likely in models fit with fewer than approximately ten occurrence records per predictor variable (Franklin 2009). Over-fit models may perform poorly when extrapolated to new climatic conditions (Maguire et al. 2016) and could lead to errors of omission (failure to predict suitable habitat where the species in reality can survive).

28.3.2 Habitat specialization

Carnivorous plants are adapted to unique microhabitats that exist as distinct, sometimes minute, patches on the landscape (Chapter 2). Identifying the environmental or habitat factors that most constrain a species' distribution is central to modeling it accurately; failure to include true and direct drivers of habitat suitability may lead predictions of suitable habitat where the species in reality cannot survive (Guisan and Zimmermann 2000). For carnivorous plants, these limiting factors are relatively well understood (Chapters 2, 18), but environmental datasets describing microhabitats are rare and such habitats are poorly represented by the spatially and temporally homogenized climatic data commonly used as predictors in SDMs.

Studies often include non-climatic predictors such as land cover, substrate, geology, soils, topography, or remotely sensed data when fitting SDMs (reviewed in Franklin 2009), but these can be properly employed only if habitats containing carnivorous plants have been identified and such patches are as large as the resolution (pixel size) of gridded data, and if the spatial precision and accuracy of the occurrence data are sufficiently high to ensure that point-occurrence records are assigned to the correct grid cells. Because occurrence data for many species, including carnivorous plants, were collected before the advent of modern GPS technology, many occurrence records have low precision and accuracy. Thus, it is impossible to assign accurately occurrence records to precise spatial locations. A related problem for historic occurrence records is that land cover may have changed and populations no longer exist where they once did. Many available predictor (climatic, habitat) variables are collected at widely spaced geographic intervals. Subsequent interpolation to fill intervening spaces between collection locations at best only poorly represents suitable microhabitats. Although variables (e.g., slope, soils, vegetation) might be combined to improve characterization of particular habitats, the use of numerous predictor variables only exacerbates the problem of over-fitting SDMs from a small number of point-occurrence records. In general, SDMs should have a minimum of ten point-occurrence records per predictor variable, plus another ten for the model intercept (Franklin 2009, Feeley and Silman 2011). Based on our experience, the number of available occurrence records for most carnivorous plants seems to be well below this threshold (Figure 28.1).

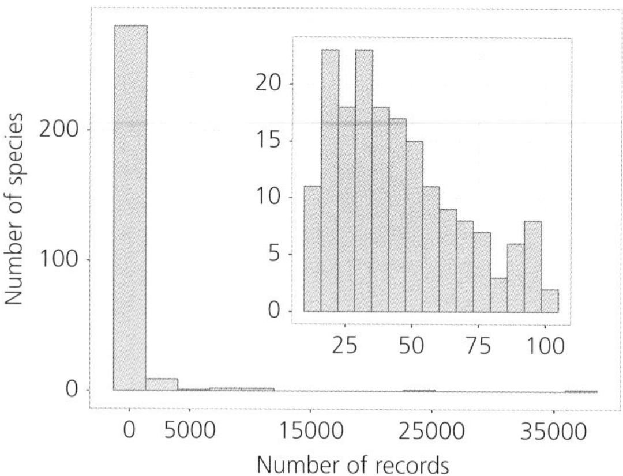

Figure 28.1 Frequency distribution of occurrence records obtained for carnivorous plant species from regional and national herbaria and global biodiversity databases. Inset expands the range of records between zero and 100.

28.3.3 Are carnivorous plant distributions constrained by climate?

Just because an SDM performs well at fitting the distribution of a carnivorous plant species is not evidence that its range is actually determined by climate. Besides strong habitat constraints, geographic distributions of carnivorous plants are constrained by biotic factors (e.g., Kesler et al. 2008; Paniw et al. 2015), including dispersal limitation (Ellison and Parker 2002; Chapter 22), competition (Chapter 2; but see Brewer 2003), pollination (Chapter 22), and trophic mutualisms (Chapters 23–26).

Of these, dispersal limitation may be the most important, as many carnivorous plants have been introduced successfully well beyond their native ranges. For example, *Dionaea muscipula* is now established in areas within the US states of Florida, New Jersey, Pennsylvania, and Washington (Giblin 2016). *Sarracenia* species have been introduced in parts of Europe and New Zealand (Walker 2014). Several species of *Utricularia*, as well as *Aldrovanda vesiculosa*, occur well outside of their native ranges (Compton et al. 2012, Lamont et al. 2013), and have become species of management concern in areas where they have been accidently or deliberately introduced. The extent to which these introductions are outside of the climatic limits implied by their native distributions is not known; dispersal limitation for these and other carnivorous plants might be determined by geography and history rather than climate.

28.4 Species distribution models for carnivorous plants and other rare species

The need to develop reliable methods for modeling the distributions of rare species and forecasting their response to ongoing and intensifying climatic change is gaining greater attention. We use two such approaches to assess the vulnerability of carnivorous plants to climatic change: ensembles of small models (Breiner et al. 2015); and estimation of bioclimatic velocity (Serra-Diaz et al. 2014).

28.4.1 Ensembles of small models

Lomba et al. (2010) described a strategy for avoiding over-fitting when modeling rare species that Breiner et al. (2015) termed "ensembles of small models" ("ESM"). Instead of fitting one, potentially over-fit, SDM with numerous predictor variables, numerous bivariate models are fit, evaluated, and then averaged to an ensemble weighted by model performance. Breiner et al. (2015) tested this method on 107 species and found that ESM performed significantly better than standard SDMs, and that the performance gains were greatest for rare species. They concluded that an average of many simple models avoids over-fitting while retaining explanatory power.

28.4.2 Controlling complexity and over-fitting

In addition to using ESM, it can be beneficial to control the complexity of the statistical model used to relate occurrence with climate. A statistical approach known as MaxEnt is one of the most popular techniques for fitting SDMs. It is an implementation of a statistical approach called maximum entropy that characterizes probability distributions from incomplete information (Phillips et al. 2006). Modeling distributions of species using MaxEnt assumes that occurrence data represent an incomplete sample of an empirical probability distribution; this unknown distribution can be estimated most appropriately as the distribution with "maximum entropy:" the probability distribution that is most uniform, subject to constraints imposed by environmental variables; and that this distribution of maximum entropy approximates the potential distribution of the species (Phillips et al. 2006, Elith et al. 2011).

Software for implementing MaxEnt models (Phillips et al. 2006, Hijmans et al. 2016) allows users to control model complexity, but most studies rely on the default settings instead of taking advantage of this feature. Warren and Seifert (2011) demonstrated that optimizing model complexity can help avoid models that are overly complex or simple and that exhibit poor inference regarding habitat suitability. Controlling model complexity also can help to identify correctly the relative importance of variables in constraining species' distributions and improve their transferability to other time periods (Warren and Seifert 2011).

28.4.3 Estimating bioclimatic velocity

The vulnerability of species to climatic change can be assessed with three primary components: exposure, sensitivity, and adaptive capacity (Dawson et al. 2011). Exposure refers to the rate and magnitude of climatic change where a species currently occurs; exposure is positively related to vulnerability. In contrast, sensitivity considers the extent to which a species is dependent on prevailing climatic conditions; sensitivity also is positively related to vulnerability. Last, adaptive capacity refers to how well a species may persist *in situ* as the climate changes; adaptive capacity is negatively related to vulnerability. SDMs such as MaxEnt, whether used individually or in ensembles, assess only exposure to climatic change.

Because exposure includes the rate and magnitude of climatic change where the species now occurs, one method of estimating exposure is to ignore species entirely and instead simply estimate the magnitude of climatic change in that particular region. Loarie et al. (2009) estimated exposure as being related to the velocity of climatic change (measured in km/year): the local rate of displacement of climate expressed as the ratio of the rate of climatic change through time and the local rate of climatic change across space. For temperature, climatic change velocity is:

$$\frac{{}^{\circ}C/_{year}}{{}^{\circ}C/_{km}} = {}^{km}/_{year}$$

Biologically, the velocity of climatic change estimates the speed at which species must migrate to track constant climatic conditions (in the context of a single dimension of climate such as temperature). All else being equal, larger velocities of climatic change represent greater exposure because the magnitude of a species' response (in terms of rate and/or distance of migration) must be greater so that it can track its niche. Indeed, some studies have suggested that the magnitudes or velocities of past climatic changes are related positively to rates of species extinction and evolution (Dynesius and Jansson 2000, Nogués-Bravo et al. 2010, Sandel et al. 2011), including the diversification of *Drosera* in southwestern Australia (Yesson and Culham 2006).

Although estimation of the velocity of climatic change can be considered a "species-free" method of estimating exposure because it does not require any data on species occurrences, several studies have estimated species-specific velocities of climatic change. For example, Serra-Diaz et al. (2014) estimated the species-specific exposure of oak species to climatic change in California by calculating the velocity of change in habitat suitability from an SDM instead of the velocity of change of a single climatic variable (see also Ordonez and Williams 2013, Carroll et al. 2015). Substituting habitat suitability for climate allows the calculation of exposure to be scaled specifically to the change in overall habitat suitability for each species rather than to a single dimension of climate (e.g., temperature alone). As such, the species-specific bioclimatic velocity accounts for the multidimensional nature of climatic change (i.e., multiple variables are changing at once and at different rates) and appropriately weights these changes given the strength of their relationship to the current distribution of a particular species.

28.5 Modeling exposure of carnivorous plants to climatic change

We estimate the exposure of carnivorous plants to climatic change by fitting ESM for a large set of carnivorous plants. We then project these models using numerous scenarios of climatic change for 2050 and calculate bioclimatic velocity for each species to estimate their potential exposure to climatic change.

28.5.1 Species occurrence data

We assembled a comprehensive database of occurrence records from multiple sources, including the Global Biodiversity Information Facility (GBIF), the Australia Virtual Herbarium, and numerous regional herbaria (see Acknowledgments for herbaria that contributed data). We sought records for all currently recognized carnivorous plant genera (Appendix).

We applied numerous quality control measures to the occurrence records before using them for modeling. First, we used literature searches, books, and online databases to update and standardize species names to the latest taxonomy (Chapters 4–10) and to identify hybrids, which were removed from further consideration. Second, species records provided without geographic (e.g., latitude–longitude) coordinates but with collection location information were georeferenced using GEOLocate 2.0 (Rios and Bart 2010). Third, we took four additional steps to ensure that only data of high spatial integrity were used in modeling fitting: we removed occurrence records with low spatial precision, whose coordinates were far outside the known range, whose coordinates matched the centroids of political boundaries, or whose coordinates fell within botanical gardens or museums. Last, we removed all spatial replicates (i.e., records for the same species with identical coordinates). The remaining points were those that fell within the known native distribution of each species and which had sufficient spatial precision given the resolution of the climate data.

28.5.2 Climate data

We obtained gridded data layers for both current and potential future climatic conditions. To describe current climate (average for the period 1971–2000), we worked with the 19 bioclimatic variables from WorldClim (http://www.worldclim.org; Hijmans et al. 2005), a database of globally contiguous, gridded representations of climate developed from interpolations of observed data. Bioclimatic variables include estimates of minima, maxima, and seasonality in temperature (°C) and precipitation (mm). We used data at 2.5 arc-minute spatial resolution. We reduced the full set of 19 variables by retaining those that had low multicollinearity ($r < 0.7$), but retained uncorrelated pairs of variables that were, in our opinion, most biologically informative. This selection process reduced the 19 covariates to four variables that we used in model fitting and prediction: maximum temperature of the warmest month, mean temperature of the coldest quarter, annual precipitation, and precipitation seasonality.

For scenarios of future climates, we used decadal averages of the same four bioclimatic variables for 2050 from 32 future climate simulations statistically downscaled to 2.5 arc-minute spatial resolution.

These climate simulations allowed us to project our models for a range of possible future conditions. The simulations included output from numerous general circulation models and one representative concentration pathway (RCP 8.5) developed as part of the IPCC AR5 and available from the Research Program on Climate Change, Agriculture and Food Security.[1]

28.5.3 Species distribution modeling

We modeled habitat suitability for carnivorous plants using MaxEnt version 3.3.3E as implemented in the dismo package (Hijmans et al. 2016) in R (R Core Team 2016). Except for optimizing model complexity (§28.4.2; Warren and Seifert 2011), we fit models using the default values for all settings. We fit MaxEnt models for any species with at least 10 spatially unique occurrence records and constrained selection of background data (points selected at random from the study region and also used to inform model fitting) to those areas within any terrestrial ecoregion in which the species has been observed and immediately adjacent ecoregions, as defined using the World Wildlife Fund's definitions of terrestrial ecoregions (Olson et al. 2001). We partitioned the occurrence data for each species randomly 10 times into calibration (70%) and evaluation (30%) datasets, and models were run on each of the 10 resulting datasets. The multiple models for each species resulting from different random splits of the occurrence data into training and test partitions were combined into a single ensemble by weighted averaging (§28.5.4). To evaluate model performance, we used the continuous Boyce Index (Hirzel et al. 2006), calculated using the ecospat package (Broennimann et al. 2016) in R (R Core Team 2016). The Boyce Index varies from −1 to 1. Values greater than zero indicate that predictions from the model are consistent with the distribution of the species (i.e., high habitat suitability predicted at known presences), values near zero indicate a model no better than random, and negative values indicate a model that predicts low habitat suitability at presences and higher suitability in locations where the species is not known to occur (Hirzel et al. 2006).

[1] http://www.ccafs-climate.org

28.5.4 Ensembles of small models (ESM)

All possible combinations of bivariate MaxEnt models (i.e., models that consider only two predictors at a time out of the set of all possible predictors) were fit for each species. Our set of four total climatic variables resulted in six possible bivariate models, which were then combined into a weighted ensemble based on predictive performance (Breiner et al. 2015).

28.5.5 Model projections, bioclimatic velocity, and exposure metrics

For MaxEnt models with a Boyce Index of at least 0.4 (model appreciably better than random), we projected the models for each of the 32 global scenarios of future climate for 2050. Using the ensemble maps of current and future habitat suitability, we calculated the species-specific bioclimatic velocity (Serra-Diaz et al. 2014) as:

$$\frac{\Delta HS/_{year}}{\Delta HS/_{km}} = km/_{year}$$

where ΔHS is the projected change in habitat suitability between current and future climate from MaxEnt (i.e., $\Delta HS/50$, the number of years between 2000 and 2050). If the spatial gradient in climatic suitability, $\Delta HS/\text{km} = 0$, we removed that point to avoid an infinite bioclimatic velocity.

We quantified the exposure of carnivorous plants to climatic change using two metrics: bioclimatic velocity at each occurrence point for each modeled species between current climate and 2050 (averaged across the 32 future climate scenarios); and the percent change in range area.

Bioclimatic velocities can be either positive or negative, indicating the rate of gain or loss of climatically suitable habitat through time respectively. Negative bioclimatic velocity indicates a decline in habitat suitability where the species is known to occur: i.e., a loss of climatically suitable habitat from which populations must migrate or possibly face local extinction. Positive bioclimatic velocity requires more cautious interpretation. Positive values at locations where the species has been observed represent habitats where populations may persist in the future, but such increases are not necessarily an improvement from a vulnerability perspective. Projected increases in habitat suitability can offset habitat losses only if individuals are able to disperse to and establish in newly suitable habitats beyond the current distribution.

To estimate the extent to which habitat losses might be offset through increases in habitat suitability, we calculated the percent change in range area as:

$$\Delta A_{range} = \frac{A_{gained} - A_{lost}}{A_{present}} \times 100,$$

where A_{gained} is the area of suitable habitat in 2050 but not at present, A_{lost} is the area of suitable habitat at present but absent in 2050, and $A_{present}$ and is the predicted range area based on current climate. To calculate ΔA_{range}, we converted the continuous habitat suitability values (0–1) from MaxEnt to suitable/unsuitable (0 or 1) using the threshold that maximized model performance based on the true skill statistic (Allouche et al. 2006). Positive values of ΔA_{range} indicate spatial expansion of climatically suitable habitat that offer opportunities for establishment outside the current range should those locations provide other aspects of habitat that carnivorous plants require. Negative values indicate a net loss of suitable range area. Based on these metrics, species with greatest exposure to climate change are those with both bioclimatic velocities and ΔA_{range} less than zero.

28.6 Results

28.6.1 Occurrence data for carnivorous plants

We obtained more than 160,000 occurrence records for 296 carnivorous plants (Table 28.1; Figure 28.1). Of these, 223 carnivorous plants had sufficient data to fit models (number of spatially unique occurrence records ≥ 10). The majority of carnivorous plants had fewer than 70 occurrence records (Figure 28.1) but a few had many thousands of records (e.g., *Drosera rotundifolia*, *D. intermedia*, and *Pinguicula vulgaris*). The highest densities of occurrence records are in Western Europe, followed by eastern North America and Australia (Figure 28.2). North Africa, the Arabian Peninsula, Russia, and China were lacking in occurrence records for carnivorous plants, likely due to both a lack of suitable habitat (e.g., north Africa, Arabian Peninsula) and a lack of reporting (e.g., Russia, China).

28.6.2 Performance of species distribution models for carnivorous plants

In general, SDMs were able to predict high suitability habitat at known presences (as quantified using the weighted average Boyce Index across ESMs for each species) for most carnivorous plants. Of the 223 species for which we could fit models, 180 (80.7%) had a Boyce Index greater than 0.4 (Table 28.1)—our selected cutoff value for a model to be considered of high enough quality to be projected to future scenarios. Seventy-five carnivorous plants (33.6%) had a Boyce Index of at least 0.75, whereas 18 (8.1%) carnivorous plants had a Boyce Index less than 0, indicating a model worse than random.

28.6.3 Vulnerability of carnivorous plants to climatic change

The majority of modeled carnivorous plants (68.3%) had a negative median bioclimatic velocity (Figure 28.3), indicating that most current locations

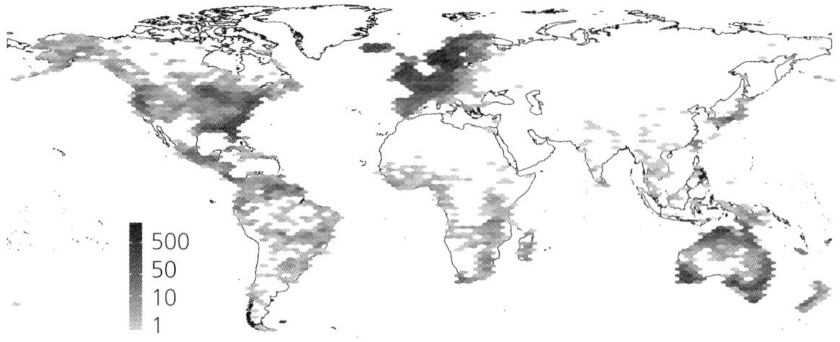

Figure 28.2 Spatial distribution and density of occurrence records obtained for carnivorous plant species from regional and national herbaria and global biodiversity databases.

Table 28.1 Summary of occurrence data for carnivorous plant species obtained from regional and national herbaria and global biodiversity databases. The number of records is the total number of spatially unique records for each genus at 2.5 arc-minute resolution remaining after quality control (§28.5.1). The number of modeled species indicates the number of species with at least ten occurrence records and the number of projected species is the number of modeled species with an ensemble weighted average Boyce Index of at least 0.4.

Family	Genus	Number of Species with Data	Number of Records	Number of Modeled/ Projected Species
Bromeliaceae				
	Brocchinia	2	104	2/0
	Catopsis	1	143	1/0
Byblidaceae				
	Byblis	3	791	3/3
Cephalotaceae				
	Cephalotus	1	97	1/1
Dioncophyllaceae				
	Triphyophyllum	1	131	1/1
Droseraceae				
	Aldrovanda	1	370	1/1
	Dionaea	1	503	1/1
	Drosera	104	72,064	73/63
Drosophyllaceae				
	Drosophyllum	1	256	1/1
Lentibulariaceae				
	Genlisea	6	170	4/2
	Pinguicula	33	34,663	22/19
	Utricularia	124	52,099	96/74
Martyniaceae				
	Ibicella	1	139	1/1
Nepenthaceae				
	Nepenthes	5	388	5/2
Sarraceniaceae				
	Darlingtonia	1	758	1/1
	Heliamphora	3	130	3/2
	Sarracenia	8	4,989	8/8
TOTAL		**296**	**167,795**	**223/180**

where carnivorous plants have been observed are projected to decrease in climatic suitability through time; declines exceed projected increases in habit suitability. Median bioclimatic velocities ranged from a minimum of –4.62 km year^{-1} for *Utricularia olivacea* to a maximum of 4.80 km year^{-1} for *U. floridana*, with bioclimatic velocity depending on the spatial distribution, magnitude, and nature of climatic change relative to the current distribution of the species (Figure 28.4).

Bioclimatic velocities also varied by genus and latitude. When averaged within genera, only *Byblis*, *Nepenthes*, and *Sarracenia* were projected to have more increases than losses in habitat suitability in

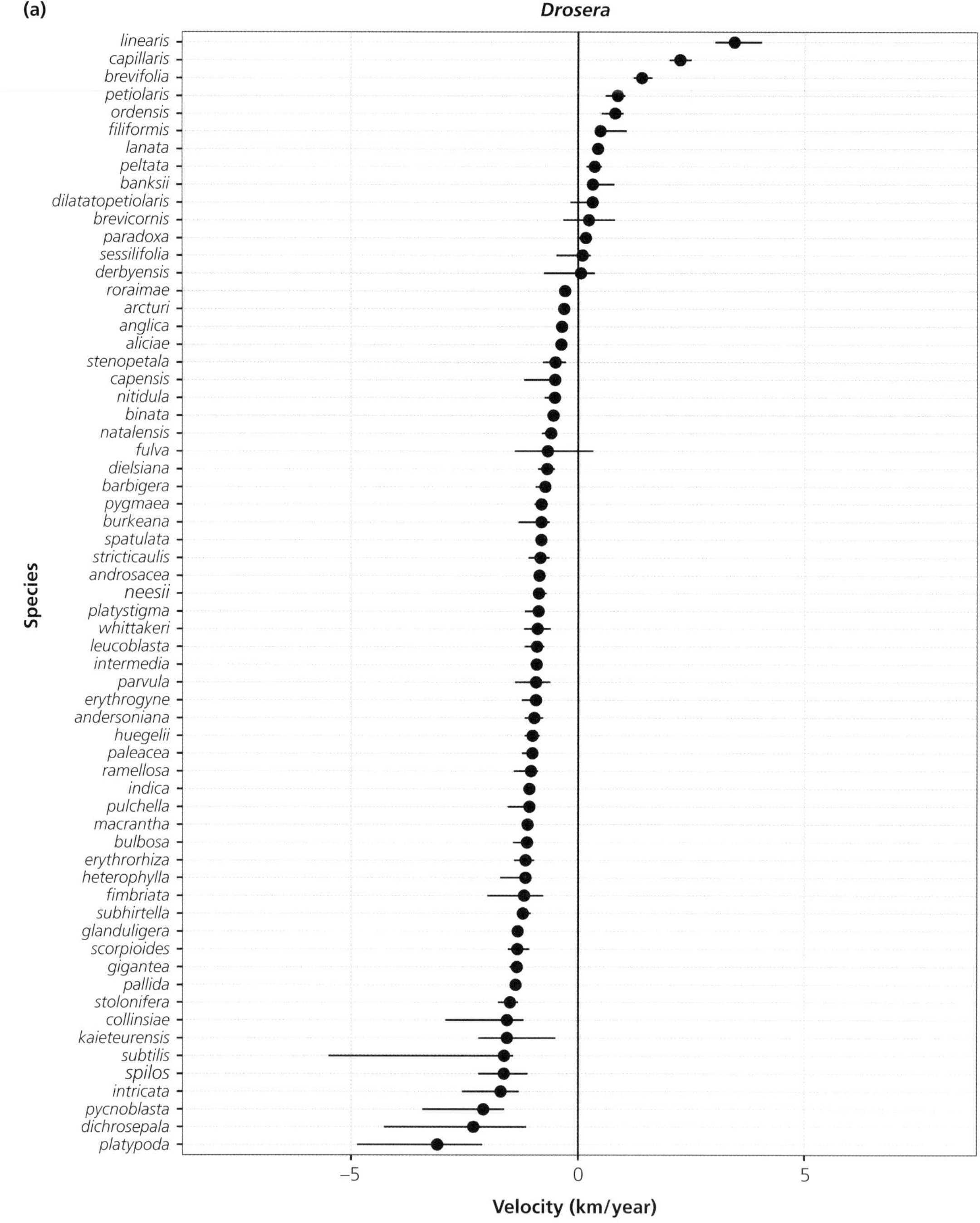

Figure 28.3 The bioclimatic velocity between current climate and year 2050 for (**a**) *Drosera* species, (**b**) *Utricularia* species, and (**c**) all other projected carnivorous plant species. Points indicate the median velocity across all occurrence locations and 32 future climate scenarios and lines indicate the 95% confidence interval.

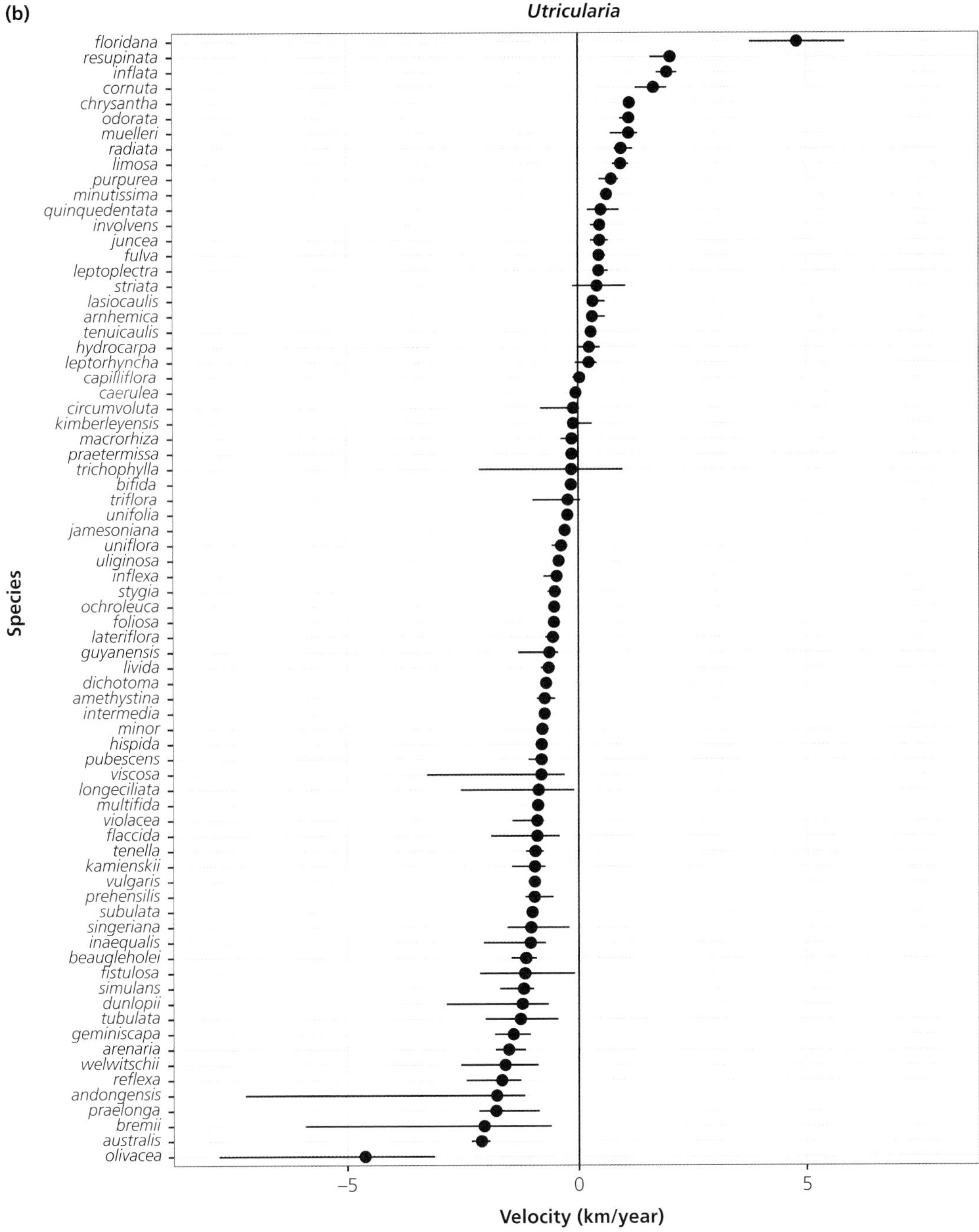

Figure 28.3 *(continued)*

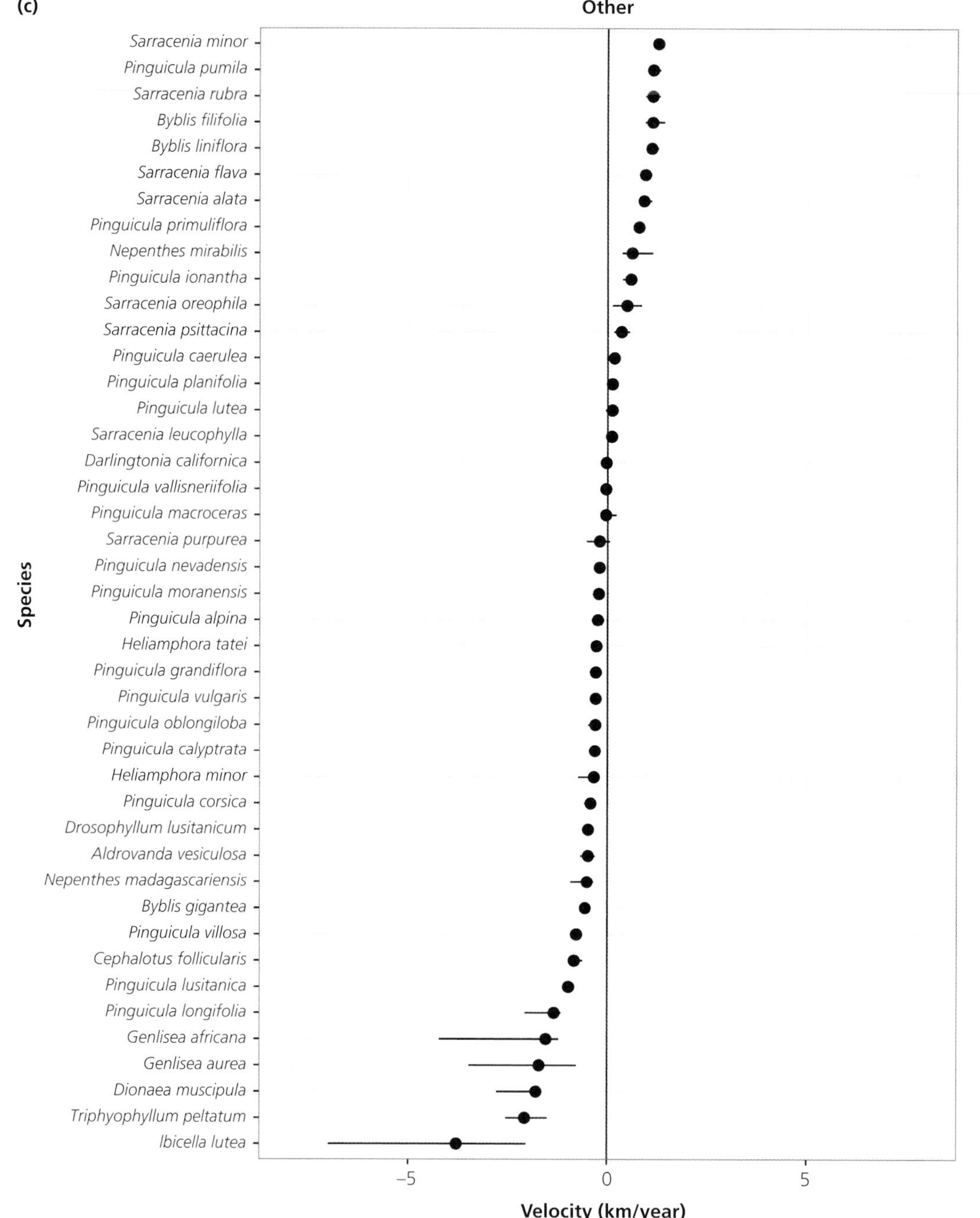

Figure 28.3 *(continued)*

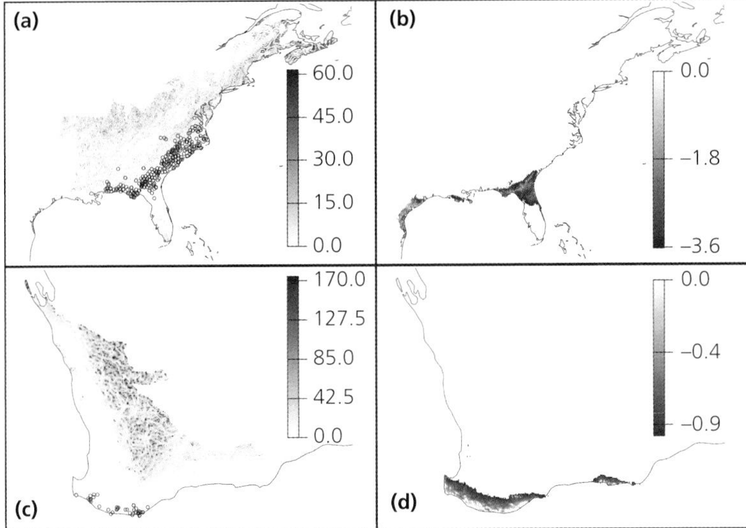

Figure 28.4 (Plate 19 on page P16) Maps of (**a, c**) positive and (**b, d**) negative bioclimatic velocity (km yr^{-1}) between current climate and year 2050 for (top row) *Sarracenia flava* and (bottom row) *Cephalotus follicularis* averaged across 32 future climate scenarios. Black circles represent occurrence locations used in model fitting. For *S. flava*, most locations where populations have been observed are projected to experience increases in habitat suitability and low velocity declines only at southernmost locations. In contrast, *C. follicularis* is projected to experience relatively low velocity decreases in suitability across all locations where populations have been observed.

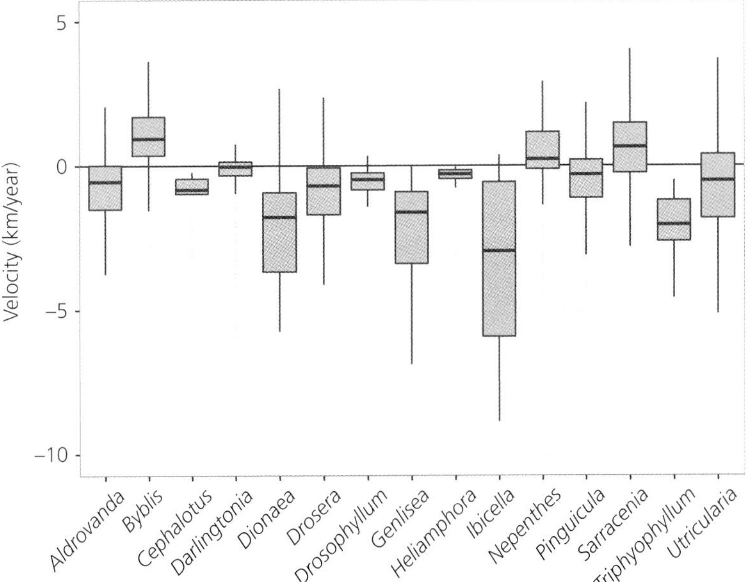

Figure 28.5 Genus-level boxplots of the bioclimatic velocity (km yr^{-1}) between current climate and year 2050 across 32 future climate scenarios. Values less than zero indicate declines in habitat suitability, while values greater than zero indicate increases in habitat suitability at locations where populations of each genus have been observed. For clarity, outliers are not shown.

locations where carnivorous plants in these genera have been observed, implying opportunities for persistence in current habitats (Figure 28.5). When viewed as a function of latitude (Figure 28.6), bioclimatic velocities tended to be mainly negative in the southern hemisphere, with the exception of between ≈10° and 25° south latitude, a range that includes northern South America, south central Africa, and northern Australia, where there was a near even mix of positive and negative bioclimatic velocities. The northern hemisphere exhibited a near equal distribution of positive and negative bioclimatic velocities, likely reflecting temperature-driven declines in habitat suitability in the southern portion of carnivorous plants' ranges and increases in climatic suitability in the northern portion of their ranges. All latitudinal bands between 50° south and 70° north contained at least one occurrence record of

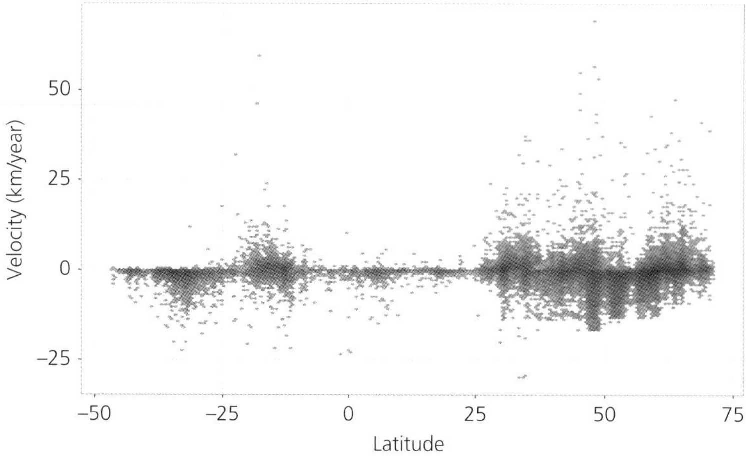

Figure 28.6 The bioclimatic velocity (km yr^{-1}) between current climate and year 2050 by latitude. Each point indicates the average bioclimatic velocity across 32 future climate scenarios at that latitude. Gray shading indicates the density of occurrence records.

a carnivorous species, but the extent to which these results reflect spatial sampling bias (Figure 28.2) is unclear.

Because bioclimatic velocity tends to be negative for most species in locations where carnivorous plants occur at present, a majority of species (57.8%) also were projected to experience overall declines in range area, even after factoring in any potential gains (Figure 28.7). Thirty-nine carnivorous plants were projected to undergo > 50% reductions in range area and 13 were projected to experience ≥ 80% declines in range area. The model projections also suggested potential opportunities for range expansion under future climate, should carnivorous plants be able to disperse to, and establish in, those regions that are projected to increase in climatic suitability.

28.7 Discussion

Our goal was to provide a first-pass estimate of the vulnerability of carnivorous plants to climatic change, given their unique ecology, and the statistical and other limitations of SDMs when fitting models for small-ranged, often rare, habitat specialists. We used ESM (Breiner et al. 2015) to overcome the statistical challenges of fitting SDMs for species with few occurrence records and bioclimatic velocity (Serra-Diaz et al. 2014) as a means to provide an integrated, species-specific metric of exposure to climatic change that simultaneously considers the magnitude, rate, and spatial distribution of change in habitat suitability. Because we quantified vulnerability mainly in the context of bioclimatic velocity at locations where carnivorous plants have been observed, we consciously discounted the potential for establishment in areas outside the current distribution. This approach seems reasonable given most carnivorous plants are unlikely to track rapid changes in climate over the next few decades. Nonetheless, expansion of climatically suitable habitat does represent potential opportunities for range expansion, which we quantified using percent increase in range area (Figure 28.7).

For > 65% of the species we considered, climatic change would lead to substantial declines in habitat suitability at locations where carnivorous plants have been observed. These projected declines suggest that many populations could be under threat of local extinction because of climatic change. If these projections are correct, then the large percent reductions in range area for many carnivorous plants might increase the number of at-risk species and perhaps increase the rank of climatic change as a threat to their conservation (Jennings and Rohr 2011).

Because our models considered only climate-related aspects of habitat, the extent to which the SDM-based projected reductions in habitat suitability may translate into population declines depends on the amount of coupling between microhabitats and the regional climate. The extent to which habitat losses may be offset by the emergence of

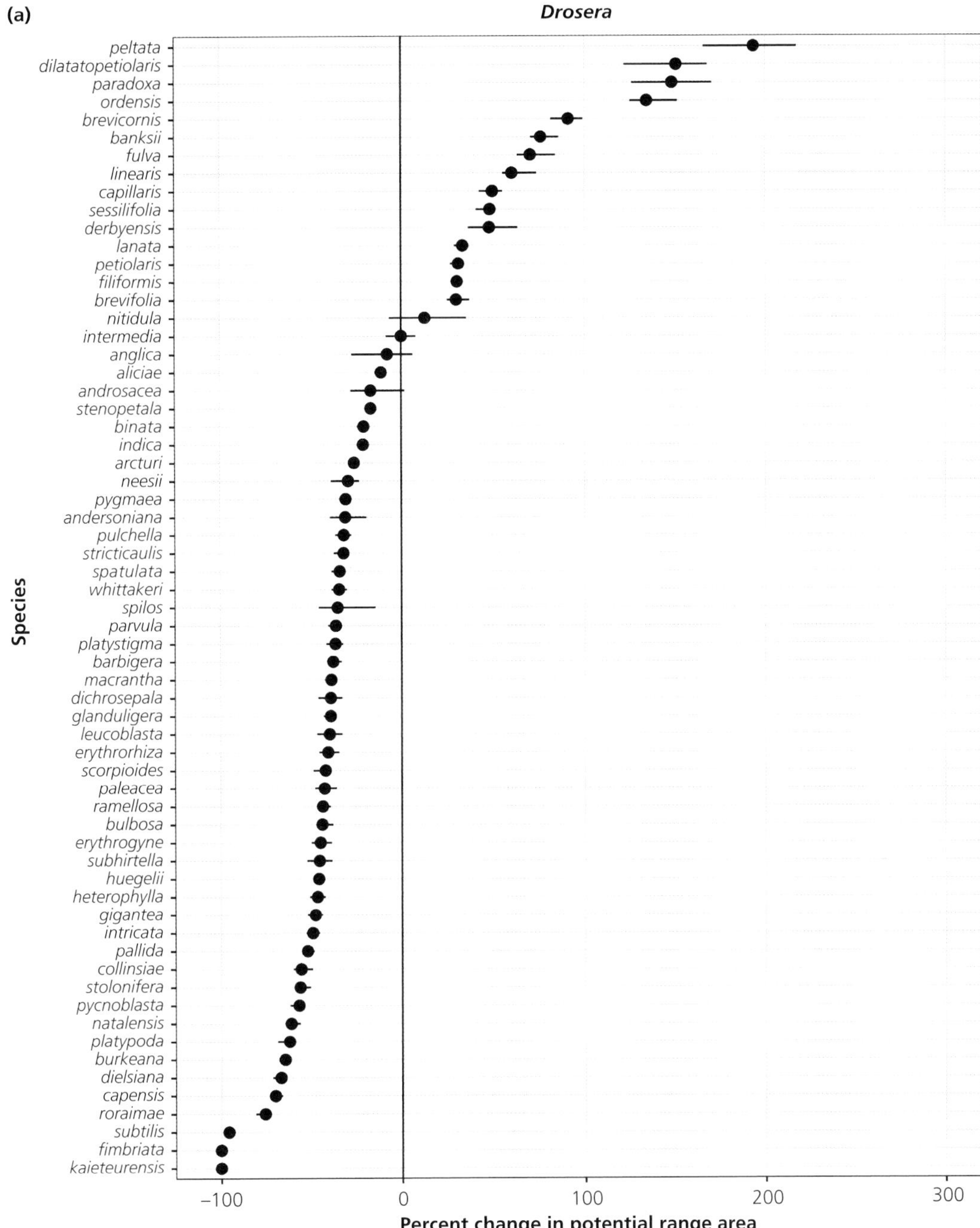

Figure 28.7 Percent change in range area between current climate and year 2050 for (**a**) *Drosera* species, (**b**) *Utricularia* species, and (**c**) all other projected carnivorous plant species. Points indicate the median projected percent change in range area across all occurrence locations and 32 future climate scenarios and lines indicate 95% confidence intervals.

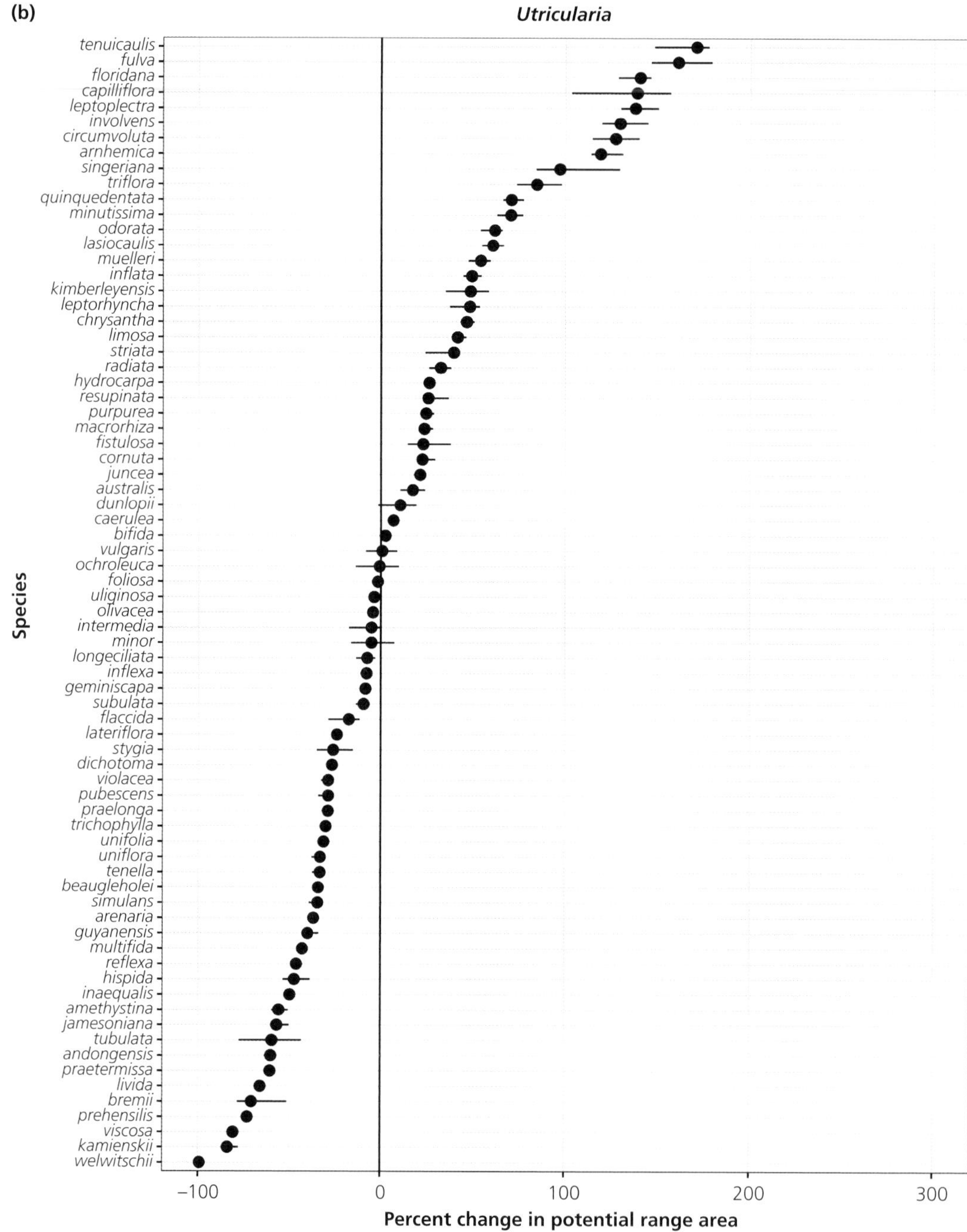

Figure 28.7 *(continued)*

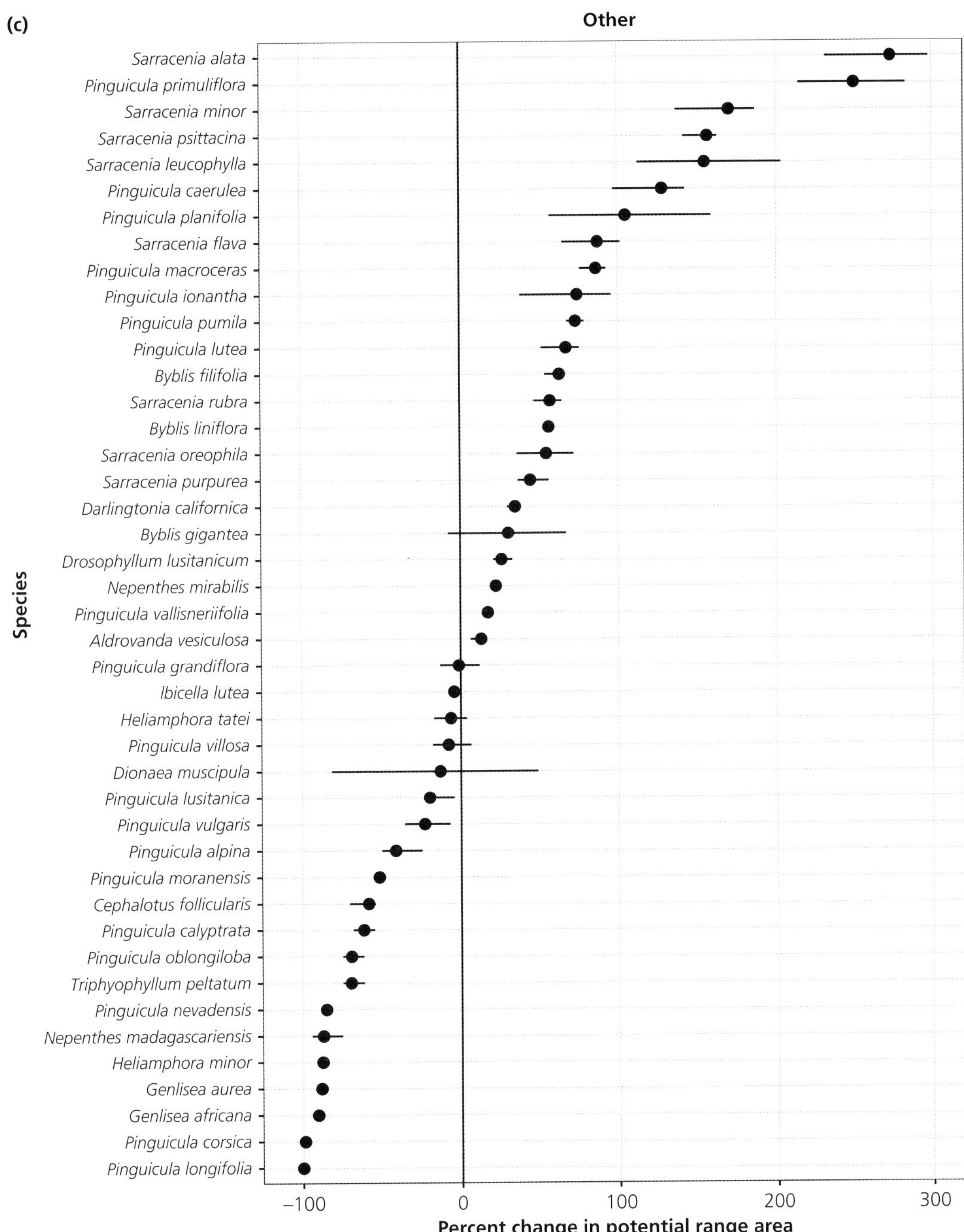

Figure 28.7 *(continued)*

climatically suitable habitat beyond the current distribution will depend strongly not only on the ability of individuals to reach these habitats, but also on whether unoccupied areas also provide unique habitat properties, including soil characteristics, hydrology, and disturbance regimes, none of which were included in our modeling framework. Finally, we used only a single algorithm to fit SDMs. While MaxEnt is regarded as one of the more robust algorithms when dealing with small sample sizes (Hernandez et al. 2006; Wisz et al. 2008), it is good practice to consider multiple algorithms. Algorithm type represents one of the primary sources of uncertainty in future projections of species distributions (Diniz-Filho et al. 2009).

Fifty-seven of the carnivorous plants modeled in this study had projected increases in habitat suitability at occurrence locations, mainly poleward of 25° north latitude; what do these increases actually mean in the context of vulnerability to climatic change? They could suggest either that current climatically marginal habitats will improve in the future or that current high-quality microhabitats are not identified as such by broad-brush climatic variables. Alternatively, they could be indicative of populations adapted to local conditions beyond the mean climatic conditions experienced by the species as a whole. We are unable to distinguish among these alternatives because of challenges with modeling habitat specialists such as carnivorous plants and the shortcomings of SDMs more generally for assessing the exposure of species to climatic change.

Beyond direct effects, climatic change also could have indirect effects on carnivorous plants. For example, changes in habitat suitability could alter interactions with co-occurring species, leading to declines in population growth rates (Nordbakken et al. 2004). In the southeastern United States, some of the most diverse carnivorous plant habitats occur in low-lying coastal areas (Chapter 2). Carnivorous plants endemic to such regions may be particularly susceptible to sea level rise or storm surges (Abbott and Battaglia 2015) associated with climatic change.

Overall, our results suggest that the climatic conditions where many carnivorous plants grow now may become less suitable for their persistence in the future. Given other, arguably more immediate, conservation and management challenges facing carnivorous plants (Jennings and Rohr 2011; Chapter 27), climatic change may be best viewed as an additional stressor. Managing climate itself is not a near-term solution, so a good strategy for conserving carnivorous plants as the climate changes would be to decrease other stressors on populations to the greatest extent possible, thereby potentially increasing resilience to climatic change. Actions should focus on immediate threats: loss of populations due to habitat conversion; poaching; native and non-native invasive species that compete with carnivorous plants; hydrological regimes; and natural disturbance such as fire that no longer occur with some degree of regularity. A more controversial approach would be to identify rear-edge populations that may contain unique genotypes pre-adapted to future climate (Hampe and Petit 2005) and consider these populations for transplant to new areas (Vitt et al. 2010) projected to become suitable in the future. On the other hand, because the habitats in which carnivorous plants occur are naturally fragmented, there may be little to be gained by increasing connectivity.

Different carnivorous plants often co-occur in the same habitat or have similar habitat requirements, so one of the most robust strategies for conservation would be to protect habitats that currently support carnivorous plants. Such a strategy might allow for the establishment of future range-shifting species with similar habitat requirements. For example, in eastern North America, many of the habitats that currently support cold-tolerant species such as *Sarracenia purpurea* and *Drosera rotundifolia* likely could support other carnivorous plants as winter temperatures continue to warm rapidly. Some habitats already may be suitable for more southerly species. For example, *Dionaea muscipula* already is established in the New Jersey Pine Barrens, well north of its native distribution (Chapters 4, 27). That numerous carnivorous plants are established well outside their native ranges suggests that these species may be more dispersal limited than constrained by narrow climatic tolerances.

28.8 Future research

How might we improve our ability to estimate the vulnerability of carnivorous plants and other rare plant species to climatic change? One strategy, which would require little in the way of additional data or modeling techniques, would be to use common plants as indicators for carnivorous plants (Smart et al. 2015) in a community-level modeling framework (Maguire et al. 2016). Although environmental data that characterize unique microhabitats could help improve our ability to map current distributions of carnivorous plants, such data would be useful only for vulnerability assessments if we also understand relationships between climate, hydrology, disturbance regimes, and soil characteristics in carnivorous plant habitats. Last, a better understanding of the climatic tolerances of different carnivorous plants and their ability to adapt to new climatic regimes, perhaps through the coupling of emerging genomic techniques (Chapter 11) and spatial modeling (Fitzpatrick and Keller 2015), would help assess whether we can expect these species to tolerate or adapt to climatic change *in situ*.

CHAPTER 29

The future of research with carnivorous plants

Aaron M. Ellison and Lubomír Adamec

The reviews and syntheses in the preceding chapters illustrate how much we know about the basic biology of carnivorous plants, and how much is still left to learn. Each chapter concludes with a section that outlines directions and opportunities for future research in specific areas; addressing all of them would lead to dozens of theses and dissertations and provide decades of interesting work for researchers ranging from undergraduates to senior scientists. Across all the concluding sections, there is a suite of common themes that could provide a framework for increasing progress in understanding carnivorous plants.

29.1 Phylogeny, evolution, and convergence

Carnivorous plants have evolved in multiple angiosperm lineages and, as a group, is an exemplar of how physiological and ecological constraints drive convergent evolution (Chapter 3). The placement of the various carnivorous plant families and genera within the overall angiosperm phylogeny is reasonably well settled (Chapter 3: Figure 3.1), but phylogenetically based infrageneric classifications need much more work (Chapters 4–10). Most carnivorous plant lineages have diversified relatively recently, hybrids occur in many genera, and molecular methods are just beginning to be used to clarify interspecific relationships in the more speciose carnivorous plant genera (*Drosera* [Chapter 4], *Nepenthes* [Chapter 5], *Pinguicula* [Chapter 6], and *Utricularia* [Chapter 8]). It would be especially useful if carnivorous plant taxonomists and systematists could coalesce around a common species concept that could be applied to each genus, or better yet, to all of them.

Genomic analysis of carnivorous plants also is in its infancy (Chapter 11). As the quality and quantity of knowledge about carnivorous plant genomes increases, we can expect to obtain new insights into patterns of convergent evolution and the underlying molecular and physiological processes that determine the observed patterns. The genomes of some *Utricularia* and *Genlisea* species are, as far as is known to date, the smallest of all plant genomes (Chapter 11), and for this trait they have become suitable models for studying the general patterns of the structure, organization, and functioning of plant genomes. Genomic data also promise to inform studies of biochemical functioning of carnivorous plant traps and the exogenous and endogenous regulation of their development or investment in carnivory (Chapter 16); to reveal aspects of the multiple interrelationships between traps and their commensals (Chapter 25); and to identify pathways with potential technological applications (Chapter 20). Because intraspecific variability affects patterns of distribution and abundance (Chapter 2); successful reproduction (Chapter 22); and interactions with prey (Chapters 12, 21), pollinators (Chapter 22), and inquilines (Chapter 23–25), sequencing technology also should be applied within species to help understand the population-level variation that is the primary subject of natural selection.

Ellison, A. M., and Adamec, L., *The future of research with carnivorous plants*. In: *Carnivorous Plants: Physiology, ecology, and evolution*. Edited by Aaron M. Ellison and Lubomír Adamec: Oxford University Press (2018).
 DOI: 10.1093/oso/9780198779841.003.0029

29.2 Field observations and experiments

In the last several decades, it has become technologically possible to do more complex experiments with carnivorous plants in the field; the results have been informative and, at times, quite surprising (Chapters 14, 15, 26). Environmental variation likely affects most elements of the carnivorous syndrome, including how plants attract and kill prey (Chapters 12, 13); rates of digestion (Chapters 13, 16); uptake of nutrients from prey carcasses (Chapters 16, 17, 19); and the relative contributions of traps and roots to the nutrient budget of each individual plant, including the physiological linkages between trap- and root-based nutrient uptake (Chapters 17, 19). Similarly, taxonomy heretofore based solely on herbarium specimens would be improved with additional information on intraspecific variability, habitat requirements, and plant–animal interactions that can be obtained much more easily in the field (Bateman 2011; Chapter 5).

Field experiments also could address a central conundrum: if carnivory is so beneficial in bright, wet, and nutrient-poor habitats (Givnish et al. 1984, Benzing 2000; Chapter 18), why aren't all plants in those habitats carnivorous? The multiple alternative hypotheses for the evolution of carnivory and mechanisms of coexistence between carnivorous and noncarnivorous plants can be tested only in the field (Chapters 2, 18). Conservation of carnivorous plants also must be field-based, so more detailed understanding of their population dynamics in the field will improve conservation strategies for all carnivorous plants (Chapter 27). It will be especially important to determine which factors will limit or facilitate geographic range shifts of carnivorous plants as the climate changes (Chapter 28). Such studies will lead naturally to discussions of whether and how we should "assist" in the migration of carnivorous plants into more favorable habitats (McLachlan et al. 2007).

29.3 Plant–animal and plant–microbe interactions

Carnivorous plants interact with animals and microbes in a variety of ways. Animals are prey, pollinators, inquilines, mutualists, and herbivores of carnivorous plants (Chapters 21–24, 26). Microbes are commensals of many carnivorous plants (Chapters 23–25), and their interactions with the plants are just beginning to be explored. Researchers have studied the degree to which attracting and releasing a pollinator may conflict with attracting and killing prey (Chapter 22), and how inquilines and mutualists—both microbial and macrobial—increase the nutrient supply of the plants (Chapters 23–26). Much less work has been done on the basic biology of the animals or microbes themselves or their population dynamics (cf. Chapter 24). This undoubtedly reflects the botanical emphasis in the education of most carnivorous plant researchers, but learning as much about the associated fauna as is already known about the plants will provide a much richer understanding of the key interactions between carnivorous plants and all of their associates.

29.4 Comparisons with noncarnivorous plants

Of the more than 350,000 species of flowering plants, fewer than 1000 are carnivorous. On the one hand, there are far more researchers working on noncarnivorous plants; their physiology, ecology, and evolution; interactions between them and animals; and their conservation and management. Application of general theories of plant biogeography, nutrient physiology, intra- and interspecific interactions, animal–plant interactions, and conservation and management to carnivorous plants will provide additional tests of these theories and could also yield informative counter-examples. On the other hand, particularities of carnivorous plant biology, including the widespread prevalence of inquilines and mutualisms (Chapters 23–26), repeated examples of exaptations such as the co-opting of defense mechanisms for trapping or digestion (Chapters 11, 16), physiological adaptations to extreme environments, especially nutrient-poor or barren ones (Chapters 2, 17, 19), and the physical boundaries of container habitats that make pitcher plants unique microecosystems (Ellison et al. 2003, Srivastava et al. 2004; Chapters 23, 24) should make carnivorous plants useful model systems with which to approach a wide range of mainstream questions in the physiology, ecology, and evolutionary biology of all plants.

APPENDIX

Species of carnivorous plants

Compiled by Andreas Fleischmann and Paulo M. Gonella (*Drosera*); Charles Clarke, Jan Schlauer, and Alastair Robinson (*Nepenthes*); Aymeric Roccia and Andreas Fleischmann (*Pinguicula*); Andreas Fleischmann (*Genlisea*); Paulo C. Baleeiro and Richard W. Jobson (*Utricularia*); Robert F. C. Naczi (Sarraceniaceae); and Adam Cross and André Vito Scatigna (*Aldrovanda, Dionaea, Philcoxia, Brocchinia, Catopsis*, and *Paepalanthus*).

Taxonomy of carnivorous plants, like those of all species, reflects current knowledge, is in a state of continual revision and flux, and should be treated as an hypothesis to be tested. This list was compiled by the authors of Chapters 4–10, and includes names and taxonomic authors that are currently used and validly published (as of 1 January 2018). We note that some of the names of *Nepenthes* species are not accepted by at least some of the authors of Chapter 5. Most synonyms are excluded.

Family	Genus	Subgenus	Section	Species
Droseraceae Salisb.	*Drosera* L.	*Regiae* Seine & Barthlott		*regia* Stephens
		Arcturia (Planch.) Schlauer		*arcturi* Hook.
				murfetii Lowrie & Conran
		Ergaleium DC.	*Coelophylla* Planch.	*glanduligera* Lehm.
			Bryastrum Planch.	*allantostigma* (N.G. Marchant & Lowrie) Lowrie & Conran
				androsacea Diels
				barbigera Planch.
				bindoon Lowrie
				callistos N.G. Marchant & Lowrie
				citrina Lowrie & Carlquist
				closterostigma N.G. Marchant & Lowrie
				coalara Lowrie
				coomallo Lowrie
				dichrosepala Turcz.
				echinoblastus N.G. Marchant & Lowrie
				eneabba N.G. Marchant & Lowrie
				enodes N.G. Marchant & Lowrie

(*Continued*)

Family	Genus	Subgenus	Section	Species
				gibsonii P. Mann
				grievei Lowrie & N.G. Marchant
				helodes N.G. Marchant & Lowrie
				hyperostigma N.G. Marchant & Lowrie
				lasiantha Lowrie & Carlquist
				leioblastus N.G. Marchant & Lowrie
				leucoblasta Benth.
				leucostigma (N.G. Marchant & Lowrie) Lowrie & Conran
				mannii Cheek
				meristocaulis Maguire & Wurdack
				micrantha Lehm.
				microscapa Debbert
				miniata Diels
				minutiflora Planch.
				nitidula Planch.
				nivea Lowrie & Carlquist
				occidentalis Morrison
				omissa Diels
				oreopodion N.G. Marchant & Lowrie
				paleacea DC.
				patens Lowrie & Conran
				pedicellaris Lowrie
				platystigma Lehm.
				pulchella Lehm.
				pycnoblasta Diels
				pygmaea DC.
				rechingeri Strid
				roseana N.G. Marchant & Lowrie
				sargentii Lowrie & N.G. Marchant
				scorpioides Planch.
				sewelliae Diels
				silvicola Lowrie & Carlquist
				spilos N.G. Marchant & Lowrie
				stelliflora Lowrie & Carlquist
				trichocaulis (Diels) Lowrie
				verrucata Lowrie & Conran
				walyunga N.G. Marchant & Lowrie
			Lasiocephala Planch.	*banksii* R.Br. ex DC.
				brevicornis Lowrie
				broomensis Lowrie

(*Continued*)

Family	Genus	Subgenus	Section	Species
				caduca Lowrie
				darwinensis Lowrie
				derbyensis Lowrie
				dilitatopetiolaris K. Kondo
				falconeri K. Kondo & Tsang
				fulva Planch.
				kenneallyi Lowrie
				lanata K. Kondo
				ordensis Lowrie
				paradoxa Lowrie
				petiolaris R. Br. ex DC
				subtilis N.G. Marchant
			Phycopsis Planch.	*binata* Labill.
			Ergaleium (DC.) Planch.	*aberrans* (Lowrie & Carlquist) Lowrie & Conran
				andersoniana W. Fitzg. ex Ewart & Jean White
				auriculata Backh. ex Planch.
				basifolia (N.G. Marchant & Lowrie) Lowrie
				bicolor Lowrie & Carlquist
				browniana Lowrie & N.G. Marchant
				bulbigena Morrison
				bulbosa Hook.
				calycina Planch.
				collina (N.G. Marchant & Lowrie) Lowrie
				drummondii Planch.
				eremaea (N.G. Marchant & Lowrie) Lowrie & Conran
				erythrogyne N.G. Marchant & Lowrie
				erythrorhiza Lindl.
				esperensis Lowrie
				fimbriata DeBuhr
				gigantea Lindl.
				gracilis Hook. f. ex Planch.
				graniticola N.G. Marchant
				heterophylla Lindl.
				hirsuta Lowrie & Conran
				hookeri R.P. Gibson, B.J. Conn & Conran
				huegelii Endl.
				humilis Planch.
				indumenta Lowrie & Conran
				intricata Planch.
				lowriei N.G. Marchant

(*Continued*)

Family	Genus	Subgenus	Section	Species
				lunata Buch.-Ham. ex DC.
				macrantha Endl.
				macrophylla Lindl.
				magna (N.G. Marchant & Lowrie) Lowrie
				major (Diels) Lowrie
				marchantii DeBuhr
				menziesii R.Br. ex DC.
				microphylla Endl.
				modesta Diels
				monantha (Lowrie & Carlquist) Lowrie
				monticola (Lowrie & N.G. Marchant) Lowrie
				moorei (Diels) Lowrie
				myriantha Planch.
				neesii Lehm.
				orbiculata N.G. Marchant & Lowrie
				pallida Lindl.
				peltata Thunb.
				planchonii Hook. f. ex Planch.
				platypoda Turcz.
				porrecta Lehm.
				praefolia Tepper
				prophylla (Marchant & Lowrie) Lowrie
				prostrata (N.G. Marchant & Lowrie) Lowrie
				prostratascaposa Lowrie & Carlquist
				purpurascens Schlotth.
				radicans N.G. Marchant
				ramellosa Lehm.
				rosulata Lehm.
				rupicola (N.G. Marchant) Lowrie
				salina N.G. Marchant & Lowrie
				schmutzii Lowrie & Conran
				squamosa Benth.
				stolonifera Endl.
				stricticaulis (Diels) O. H. Sargent
				subhirtella Planch.
				sulphurea Lehm.
				thysanosepala Diels
				tubaestylus N.G. Marchant & Lowrie
				whittakeri Planch.
				yilgarnensis R.P. Gibson & B.J. Conn

(*Continued*)

Family	Genus	Subgenus	Section	Species
				zigzagia Lowrie
				zonaria Planch.
		Drosera L.	*Thelocalyx* Planch.	*burmannii* Vahl
				sessilifolia A. St.-Hil.
			Prolifera C.T. White	*adelae* F. Muell.
				prolifera C.T. White
				schizandra Diels
			Stelogyne Diels	*hamiltonii* C.R.P. Andrews
			Arachnopus Planch.	*aquatica* Lowrie
				aurantiaca Lowrie
				barrettiorum Lowrie
				cucullata Lowrie
				finlaysoniana Wall. ex Arn.
				fragrans Lowrie
				glabriscapa Lowrie
				hartmeyerorum Schlauer
				indica L.
				nana Lowrie
				serpens Planch.
			Psychophila Planch.	*stenopetala* Hook. f.
				uniflora Willd.
			Drosera L.	*amazonica* Rivadavia, A. Fleischm. & Vicent.
				anglica Huds.
				arenicola Steyerm.
				biflora Willd. ex Roem. & Schult.
				brevifolia Pursh
				capillaris Poir.
				cayennensis Sagot ex Diels
				communis A. St.-Hil.
				esmeraldae (Steyerm.) Maguire & Wurdack
				felix Steyerm. & L.B. Sm.
				filiformis Raf.
				hirtella A. St.-Hil.
				hirticalyx Duno & Culham
				intermedia Hayne
				kaieteurensis Brumm.-Ding.
				lutescens (A. St.-Hil.) Gonella
				linearis Goldie
				neocaledonica Raym.-Hamet

(*Continued*)

Family	Genus	Subgenus	Section	Species
				oblanceolata Y.Z. Ruan
				roraimae (Klotzsch ex Diels) Maguire & Laundon
				rotundifolia L.
				solaris A.Fleischm., Wistuba & S. McPherson
				spatulata Labill.
				tokaiensis (Komiya & Shibata) T. Nakamura & Ueda
				tracyi Macfarl.
				ultramafica A. Fleischm., A.S. Rob. & S. McPherson
				viridis Rivadavia
				yutajensis Duno & Culham
			Brasiliae Rivadavia, Gonella & A. Fleischm.	*ascendens* A. St.-Hil.
				camporupestris Rivadavia
				cendeensis Tamayo & Croizat
				chimaera Gonella & Rivadavia
				chrysolepis Taub.
				condor Gonella, A. Fleischm. & Rivadavia
				graminifolia A. St.-Hil.
				grantsaui Rivadavia
				graomogolensis T.R.S. Silva
				latifolia (Eichler) Gonella & Rivadavia
				magnifica Rivadavia & Gonella
				montana A. St.-Hil.
				peruensis T.R.S. Silva & M.D. Correa
				quartzicola Rivadavia & Gonella
				riparia Gonella & Rivadavia
				schwackei (Diels) Rivadavia
				spiralis A. St.-Hil.
				spirocalyx Rivadavia & Gonella
				tentaculata Rivadavia & Gonella
				tomentosa A. St.-Hil.
				villosa A. St.-Hil.
			Ptycnostigma Planch.	*acaulis* L. f.
				admirabilis Debbert
				affinis Welw. ex Oliv.
				alba E. Phillips
				aliciae Raym.-Hamet
				bequaertii Taton
				burkeana Planch.
				capensis L.
				cistiflora L.

(*Continued*)

Family	Genus	Subgenus	Section	Species
				collinsiae N.E. Br. in Burtt Davy
				cuneifolia L. f.
				curvipes Planch.
				dielsiana Exell & Laundon
				elongata Exell & Laundon
				ericgreenii A. Fleischm., R.P. Gibson & Rivadavia
				esterhuyseniae (T.M. Salter) Debbert
				flexicaulis Welw. ex Oliv.
				glabripes (Harv. ex Planch.) Stein
				grandiflora Bartl.
				helianthemum Planch.
				hilaris Cham. & Schltdl.
				humbertii Exell & Laundon
				katangensis Taton
				liniflora Debbert
				madagascariensis DC.
				natalensis Diels
				nidiformis Debbert
				pauciflora Banks ex DC.
				pilosa Exell & Laundon
				ramentacea Burch. ex DC.
				rubrifolia Debbert
				slackii Cheek
				speciosa C. Presl
				trinervia Spreng.
				variegata Debbert
				venusta Debbert
				violacea Willd.
				zeyheri T.M. Salter
	Dionaea J. Ellis			*muscipula* J. Ellis
	Aldrovanda L.			*vesiculosa* L.
Nepenthaceae Dum.	*Nepenthes* L.		*Insignes* Danser	*aenigma* Nuytemans, W. Suarez & Calaramo
				alzapan Jebb & Cheek
				barcelonae Tadang & Cheek
				bellii Kondo
				burkei Hort. Veitch ex Mast.
				campanulata Kurata
				insignis Danser
				merrilliana Macf.
				samar Jebb & Cheek

(*Continued*)

Family	Genus	Subgenus	Section	Species
				sibuyanensis Nerz
				surigaoensis Elm.
				ventricosa Blanco
			Montanae Danser	*adnata* Tamin & Hotta ex Schlauer
				angasanensis R. Maulder, D.Schubert, B. Salmon & B.Quinn
				aristolochioides Jebb & Cheek
				bongso Korth.
				densiflora Danser
				diatas Jebb & Cheek
				dubia Danser
				eustachya Miq.
				flava Wistuba, Nerz & A. Fleischm.
				gymnamphora Reinw. ex Nees.
				inermis Danser
				izumiae T. Davis, C. Clarke & Tamin
				jacquelineae C. Clarke, T. Davis & Tamin
				jamban C.C. Lee, Hernawati & P. Akriadi
				lavicola Wistuba & Rischer
				lingulata C.C. Lee, Hernawati & P. Akriadi
				longifolia Nerz & Wistuba
				mikei Salmon & Maulder
				naga Akriadi, Hernawati, Primhaldi & Hambali
				ovata Nerz & Wistuba
				rhombicaulis Kurata
				rigidifolia Akriadi, Hernawati & Tamin
				singalana Becc.
				spathulata Danser
				spectabilis Danser
				sumatrana (Miq.) G.Beck
				talangensis Nerz & Wistuba
				tenuis Nerz & Wistuba
				tobaica Danser
			Nepenthes L.	*danseri* Jebb & Cheek
				distillatoria L.
				halmahera Jebb & Cheek
				khasiana Hook.f.
				lamii Jebb & Cheek
				madagascariensis Poir.
				masoalensis Schmid-Hollinger

(*Continued*)

Family	Genus	Subgenus	Section	Species
				monticola A.S. Robinson, Wistuba, Nerz, M. Mansur & S. McPherson
				neoguineensis Macf.
				paniculata Danser
				pervillei Bl.
				tomoriana Danser
				treubiana Warb.
				vieillardii Hook.f.
				weda Jebb & Cheek
			Pyrophytae Cheek & Jebb	*alba* Ridl.
				albomarginata T. Lobb ex Lindl.
				andamana M. Catal.
				benstonei C. Clarke
				bokorensis Mey
				chang M. Catal.
				gracillima Ridl.
				hemsleyana Macf.
				holdenii Mey
				kampotiana Lecomte
				kerrii M. Catal. & T. Kreutr.
				kongkandana Cheek ex M. Catal.
				krabiensis Nuanlaong, Onsanit, Chusangrach & Suraninpong
				macfarlanei Hemsl.
				rafflesiana Jack
				ramispina Ridl.
				reinwardtiana Miq.
				rosea M. Catal. & T. Kruetr.
				sanguinea Lindl.
				smilesii Hemls.
				suratensis M. Catal.
				thai Jebb & Cheek
				thorelii Lecomte
			Regiae Danser	*abalata* Jebb & Cheek
				abgracilis Jebb & Cheek
				alata Blanco
				appendiculata C.C. Lee, G. Bourke, Rembold, W. Taylor & S.T. Yeo
				argentii Jebb & Cheek
				armin Jebb & Cheek

(*Continued*)

Family	Genus	Subgenus	Section	Species
				attenboroughii A.S. Robinson, S. McPherson & V.R. Heinrich
				boschiana Korth.
				burbidgeae Hook. f. ex Burb.
				chaniana C. Clarke, C.C. Lee & S. McPherson
				cid Jebb & Cheek
				clipeata Danser
				deaniana Macf.
				ephippiata Danser
				epiphytica A.S. Robinson, Nerz & Wistuba
				eymae Kurata
				faizaliana Adam & Wilcock
				fallax G. Beck
				fusca Danser
				gantungensis S. McPherson, J. Cervancia, C.C. Lee, M. Jaunzems, Mey & A.S. Robinson
				glandulifera C.C. Lee
				graciliflora Elm.
				hurrelliana Cheek & A. Lamb
				kitanglad Jebb & Cheek
				klossii Danser
				leonardoi S. McPherson, G. Bourke, J. Cervancia, M. Jaunzems & A.S. Robinson
				leyte Jebb & Cheek
				lowii Hook. f.
				mantalingajanensis Nerz & Wistuba
				maxima Reinw. ex Nees.
				minima Jebb & Cheek
				mira Jebb & Cheek
				mollis Danser
				negros Jebb & Cheek
				palawanensis S. McPherson, J. Cervancia, C.C. Lee, M. Jaunzems, Mey & A.S. Robinson
				pantaronensis Gieray, Gronem., Wistuba, Marwinski, Micheler, Coritico & V.B. Amoroso
				peltata Kurata
				philippinensis Macf.
				pilosa Danser
				platychila C.C. Lee
				ramos Jebb & Cheek
				saranganiensis Kurata
				stenophylla Mast.

(*Continued*)

Family	Genus	Subgenus	Section	Species
				tboli Jebb & Cheek
				veitchii Hook. f.
				viridis Micheler et al.
				vogelii Schuit. & de Vogel
				ultra Jebb & Cheek
			Tentaculatae Jebb & Cheek	*glabrata* Turnbull & Middleton
				hamata Turnbull & Middleton
				mariae Jebb & Cheek
				muluensis Hotta
				murudensis Culham ex Jebb & Cheek
				nigra Nerz, Wistuba, C.C. Lee, G. Bourke, U. Zimm. & S. McPherson
				pitopangii C.C. Lee, S. McPherson, G. Bourke, & M. Mansur
				tentaculata Hook. f.
				undulatifolia Nerz, Wistuba, U. Zimm., C.C. Lee, Pirade & Pitopang
			Urceolatae Danser	*ampullaria* Jack
				bicalcarata Hook.f.
				gracilis Korth.
				mirabilis (Lour.) Rafarin
				papuana Danser
			Villosae Jebb & Cheek	*ceciliae* Gronem., Coritico, Micheler, Marwinski, Acil & V.B. Amoroso
				copelandii Merr. ex Macf.
				cornuta Marwinski, Coritico, Wistuba, Micheler, Gronem., Gieray & V.B. Amoroso
				edwardsiana Low ex Hook. f.
				hamiguitanensis Gronem., Wistuba, V.B. Heinrich, S. McPherson, Mey & V.B. Amoroso
				justinae Gronem., Wistuba, Mey & V.B. Amoroso
				macrophylla (Marabini) Jebb & Cheek
				micramphora V.B. Heinrich, S.R. McPherson, T. Gronemeyer & V.B. Amoroso
				mindanaoensis Kurata
				nebularum G. Mansell
				petiolata Danser
				pulchra Gronem., S. McPherson, Coritico, Micheler, Marwinski & V.B. Amoroso
				rajah Hook.f.
				robcantleyi Cheek
				sumagaya Cheek
				talaandig Gronem., Coritico, Wistuba, Micheler, Marwinski, Gieray & V.B. Amoroso

(*Continued*)

Family	Genus	Subgenus	Section	Species
				truncata Macf.
				villosa Hook.f.
				zygon Jebb & Cheek
Drosophyllaceae Chrtek, Slavíková & Studnička	*Drosophyllum* Link			*lusitanicum* (L.) Link
Dioncophyllaceae Airy Shaw	*Triphyophyllum* Airy Shaw			*peltatum* (Hutch. & Dalziel) Airy Shaw
Sarraceniaceae Dumort.	*Darlingtonia* Torr.			*californica* Torr.
	Heliamphora Benth.			*arenicola* Wistuba, A.Fleischm., Nerz & S.McPherson
				ceracea Nerz, Wistuba, Grantsau, Rivadavia, A.Fleischm. & S.McPherson
				chimantensis Wistuba, Carow & Harbarth
				ciliata Wistuba, Nerz & A.Fleischm.
				collina Wistuba, Nerz, S.McPherson & A.Fleischm.
				elongata Nerz
				exappendiculata (Maguire & Steyerm.) Nerz & Wistuba
				folliculata Wistuba, Harbarth & Carow
				glabra (Maguire) Nerz, Wistuba & Hoogenstrijd
				heterodoxa Steyerm.
				hispida Wistuba & Nerz
				huberi A.Fleischm., Wistuba & Nerz
				ionasi Maguire
				macdonaldae Gleason
				minor Gleason
				neblinae Maguire
				nutans Benth.
				parva (Maguire) S. McPherson, A. Fleischm., Wistuba & Nerz
				pulchella Wistuba, Carow, Harbarth & Nerz
				purpurascens Wistuba, A. Fleischm., Nerz & S. McPherson
				sarracenioides Carow, Wistuba & Harbarth
				tatei Gleason
				uncinata Nerz, Wistuba & A. Fleischm.
	Sarracenia L.			*alabamensis* Case & R.B. Case
				alata (Alph. Wood) Alph. Wood
				flava L.
				jonesii Wherry
				leucophylla Raf.
				minor Walter
				oreophila Wherry
				psittacina Michx.
				purpurea L.

(*Continued*)

Family	Genus	Subgenus	Section	Species
				rosea Naczi, Case, & R.B. Case
				rubra Walter
Roridulaceae	*Roridula* L.			*dentata* L.
				gorgonias Planch.
Cephalotaceae Dumort.	*Cephalotus* Labill.			*follicularis* Labill.
Byblidaceae Domin	*Byblis* Salisb.			*aquatica* Lowrie & Conran
				filifolia Planch.
				gigantea Lindl.
				guehoi Lowrie & Conran
				lamellata Conran & Lowrie
				liniflora Salisb.
				pilbarana Lowrie & Conran
				rorida Lowrie & Conran
Lentibulariaceae Rich.	*Pinguicula* L.	*Isoloba* Barnhart	*Isoloba* Casper	*caerulea* Walter
				ionantha R.K.Godfrey
				lutea Walter
				planifolia Chapm.
				primuliflora C.E.Wood & R.K.Godfrey
				pumila Michx.
			Cardiophyllum Casper	*crystallina* Sm.
				hirtiflora Ten.
				megaspilaea Boiss. & Heldr. ex Boiss.
			Pumiliformis (Casper) Roccia & A. Fleischm.	*lusitanica* L.
			Ampullipalatum Casper	*antarctica* Vahl
				calyptrata Kunth
				chilensis Clos
				involuta Ruiz & Pav.
				jarmilae Halda & Malina
		Pinguicula L.	*Pinguicula* L.	*apuana* Casper & Ansaldi
				arvetii Genty
				balcanica Casper
				bohemica Krajina
				caussensis (Casper) Roccia
				christinae Peruzzi & Gestri
				corsica Bernard & Gren. ex Gren. & Godr.
				dertosensis (Cañig.) Mateo & M.B. Crespo
				fiorii Tammaro & L. Pace
				fontiqueriana Romo, Peris & Stübing

(*Continued*)

Family	Genus	Subgenus	Section	Species
				grandiflora Lam.
				leptoceras Rchb.
				longifolia Ramond ex DC.
				macroceras Link
				mariae Casper
				mundi Blanca, Jamilena, Ruíz Rejón & Reg. Zamora
				nevadensis H. Lindb.
				poldinii J. Steiger & Casper
				reichenbachiana Schindler
				vallisneriifolia Webb
				vallis-regiae F. Conti & Peruzzi
				vulgaris L.
		Temnoceras Barnhart	*Temnoceras* Casper	*acuminata* Benth.
				agnata Casper
				albida C. Wright ex Griseb.
				benedicta Barnhart
				bissei Casper
				calderoniae Zamudio
				casabitoana J. Jiménez Alm.
				clivorum Standl. & Steyerm.
				colimensis McVaugh & Mickel
				conzattii Zamudio & van Marm
				crassifolia Zamudio
				crenatiloba DC.
				cubensis Urquiola & Casper
				cyclosecta Casper
				debbertiana Speta & F. Fuchs
				ehlersiae Speta & F. Fuchs
				elizabethiae Zamudio
				emarginata Zamudio & Rzed.
				esseriana B. Kirchn.
				filifolia C. Wright ex Griseb.
				gigantea Luhrs
				gracilis Zamudio
				greenwoodii Cheek
				gypsicola Brandegee
				hemiepiphytica Zamudio & Rzed.
				heterophylla Benth.
				ibarrae Zamudio
				imitatrix Casper

(*Continued*)

Family	Genus	Subgenus	Section	Species
				immaculata Zamudio & Lux
				infundibuliformis Casper
				jackii Barnhart
				jaraguana Casper
				kondoi Casper
				laueana Speta & F. Fuchs
				laxifolia Luhrs
				lignicola Barnhart
				lilacina Schltdl. & Cham.
				macrophylla Kunth
				martinezii Zamudio
				mesophytica Zamudio
				mirandae Zamudio & Salinas
				moctezumae Zamudio & R.Z. Ortega
				moranensis Kunth
				nivalis Luhrs & Lampard
				oblongiloba A.DC.
				orchidioides A.DC.
				parvifolia B.L. Rob.
				pilosa Luhrs, Studnička & Gluch
				pygmaea Rivadavia, E.L. Read & A. Fleischm.
				rotundiflora Studnička
				takakii Zamudio & Rzed.
				utricularioides Zamudio & Rzed.
			Micranthus Casper	*alpina* L.
			Nana Casper	*algida* Malyschev
				ramosa Miyoshi
				spathulata Ledeb.
				villosa L.
			Heterophylliformis (Casper) A. Fleischm. & Roccia	*elongata* Benj.
	Genlisea A. St.-Hil.	*Tayloria* (Fromm) Eb. Fisch., S. Porembski & Barthlott		*exhibitionista* Rivadavia & A. Fleischm.
				flexuosa Rivadavia, A. Fleischm. & Gonella
				lobata Fromm
				metallica Rivadavia & A. Fleischm.
				nebulicola Rivadavia, Gonella & A. Fleischm.
				oligophylla Rivadavia & A. Fleischm.
				uncinata P. Taylor & Fromm
				violacea A. St.-Hil.

(*Continued*)

Family	Genus	Subgenus	Section	Species
		Genlisea A. St.-Hil.	*Africanae* A. Fleischm., Kai Müll., Barthlott & Eb. Fisch.	*africana* Oliv.
				angolensis R.D. Good
				barthlottii S. Porembski, Eb. Fisch. & Gemmel
				hispidula Stapf
				stapfii A. Chev.
				subglabra Stapf
			Recurvatae A. Fleischm., Kai Müll., Barthlott & Eb. Fisch.	*glandulosissima* R.E. Fr.
				margaretae Hutch.
				pallida Fromm & P. Taylor
			Genlisea A. St.-Hil.	*aurea* A. St.-Hil.
				filiformis A. St.-Hil.
				glabra P. Taylor
				guianensis N.E. Br.
				multiflora A. Fleischm. & S.M. Costa
				nigrocaulis Steyerm.
				oxycentron P. Taylor
				pulchella Tutin
				pygmaea A. St.-Hil.
				repens Benj.
				roraimensis N.E. Br.
				sanariapoana Steyerm.
				tuberosa Rivadavia, Gonella & A. Fleischm.
	Utricularia L.	*Polypompholyx* (Lehm.) P. Taylor	*Polypompholyx* (Lehm.) P. Taylor	*multifida* R. Br.
				tenella R. Br.
			Tridentaria P. Taylor	*westonii* P. Taylor
			Pleiochasia Kamieński	*albiflora* R. Br.
				ameliae R.W. Jobson
				barkeri R.W. Jobson
				beaugleholei Gassin
				benthamii P. Taylor
				blackmanii R.W. Jobson
				byrneana R.W. Jobson & Baleeiro
				dichotoma Labill.
				fistulosa P. Taylor
				grampiana R.W. Jobson
				hamiltonii F.E. Lloyd
				helix P. Taylor
				inaequalis A. DC.
				linearis H. Wakab.

(*Continued*)

Family	Genus	Subgenus	Section	Species
				menziesii R. Br.
				monanthos Hook. f.
				novae-zelandiae Hook. f.
				paulineae Lowrie
				petertaylorii Lowrie
				singeriana F. Muell.
				terrae-reginae P. Taylor
				triflora P. Taylor
				tubulata F. Muell.
				violacea R. Br.
				volubilis R. Br.
			Lasiocaules R.W. Jobson, Reut & Baleeiro	*antennifera* P. Taylor
				arnhemica P. Taylor
				capilliflora F. Muell.
				cheiranthos P. Taylor
				dunlopii P. Taylor
				dunstaniae F.E. Lloyd
				georgei P. Taylor
				holtzei F. Muell.
				kamienskii F. Muell.
				kenneallyi P. Taylor
				kimberleyensis C.A. Gardner
				lasiocaulis F. Muell.
				leptorhyncha O. Schwarz
				lowriei R.W. Jobson
				quinquedentata F. Muell. ex P. Taylor
				rhododactylos P. Taylor
				tridactyla P. Taylor
				uniflora R. Br.
				wannanii R.W. Jobson & Baleeiro
		Bivalvaria Kurz	*Aranella* (Barnhart) P. Taylor	*blanchetii* A. DC.
				costata P. Taylor
				fimbriata Kunth
				laciniata A. St.-Hil. & Girard
				longeceliata A. DC.
				parthenopipes P. Taylor
				purpureocaerulea A. St.-Hil. & Girard
				rostrata A. Fleischm. & Rivadavia
				sandwithii P. Taylor
				simulans Pilg.

(*Continued*)

Family	Genus	Subgenus	Section	Species
			Australes P. Taylor	*delicatula* Cheeseman
				lateriflora R. Br.
				simplex R. Br.
			Avesicarioides Komiya	*rigida* Benj.
				tetraloba P. Taylor
			Benjaminia P. Taylor	*nana* A. St.-Hil. & Girard
			Calpidisca (Barnh.) Komiya	*arenaria* A. DC.
				bisquamata Schrank
				firmula Welw. ex Oliv.
				livida E. Mey.
				microcalyx (P. Taylor) P. Taylor
				pentadactyla P. Taylor
				sandersonii Oliv.
				troupinii P. Taylor
				welwitschii Oliv.
			Enskide (Raf.) P. Taylor	*chrysantha* R. Br.
				fulva F. Muell.
				jobsonii Lowrie
			Lloydia P. Taylor	*pubescens* Sm.
			Meionula (Raf.) P. Taylor	*geoffrayi* Pellegr.
				hirta J.G. Klein ex Link
				minutissima Vahl
				ramosissima Wakabayashi
				subramanyamii Janarth. & A.N. Henry
			Minutae Lowrie, Cowie & Conran	*simmonsii* Lowrie, Cowie & Conran
			Nigrescentes Kamieński	*bracteata* R.D. Good
				caerulea L.
				warburgii K.I. Goebel
			Oligocista A. DC.	*adpressa* Salzm. ex A. St.-Hil. & Girard
				albocaerulea Dalzell
				andongensis Hiern.
				arcuata Wight
				babui S.R. Yadav, Sardesai & S.P. Gaikwad
				bifida L.
				bosminifera Ostenf.
				cecilii P. Taylor
				chiribiquitensis A. Fernández
				circumvoluta P. Taylor

(*Continued*)

Family	Genus	Subgenus	Section	Species
				delphinioides Thorel ex Pellegr.
				densiflora Baleeiro & Bove
				erectiflora A. St.-Hil. & Girard
				foveolata Edgew.
				graminifolia Vahl
				heterosepala Benj.
				involvens Ridl.
				jackii J. Parn.
				janarthanamii S.R. Yadav, Sardesai & S.P. Gaikwad
				laxa A. St.-Hil. & Girard
				lazulina P. Taylor
				letestui P. Taylor
				lloydii Merl ex F.E. Lloyd
				macrocheilos (P. Taylor) P. Taylor
				malabarica Janarth. & A.N. Henry
				meyeri Pilg.
				micropetala Sm.
				naikii S.R. Yadav, Sardesai & S.P. Gaikwad
				odorata Pellegr.
				pierrei Pellegr.
				pobeguinii Pellegr.
				polygaloides Edgew.
				praeterita P. Taylor
				prehensilis E. Mey.
				recta P. Taylor
				reticulata Sm.
				scandens Benj.
				scandens Benj. subsp. *firmula* (Oliv.) Z.Y. Li
				smithiana Wight
				spiralis Sm.
				tortilis Welw. ex Oliv.
				uliginosa Vahl
				vitellina Ridl.
				wightiana P. Taylor
			Phyllaria (Kurtz) Kamieński	*brachiata* Oliv.
				christopheri P. Taylor
				corynephora P. Taylor
				forrestii P. Taylor
				furcellata Oliv.
				garrettii P. Taylor

(*Continued*)

Family	Genus	Subgenus	Section	Species
				inthanonensis Suksathan & J. Parn.
				kumaonensis Oliv.
				moniliformis P. Taylor
				multicaulis Oliv.
				phusoidaoensis Suksathan & J. Parn.
				pulchra P. Taylor
				spinomarginata Suksathan & J. Parn.
				salwinensis Hand.-Mazz.
				steenisii P. Taylor
				striatula Sm.
			Stomoisia (Raf.) Kuntze	*cornuta* Michx.
				juncea Vahl
		Utricularia L.	*Avesicaria* Kamiński	*neottioides* A. St.-Hil. & Girard
				oliveriana Steyerm.
			Candollea P. Taylor	*podadena* P. Taylor
			Chelidon P. Taylor	*mannii* Oliv.
			Choristothecae P. Taylor	*choristotheca* P. Taylor
				determannii P. Taylor
			Foliosa Kamieński	*amethystina* Salzm. ex A. St.-Hil. & Girard
				calycifida Benj.
				hintonii P. Taylor
				hispida Lam.
				huntii P. Taylor
				longifolia Gardner
				panamensis Steyerm. ex P. Taylor
				petersoniae P. Taylor
				praelonga A. St.-Hil. & Girard
				regia Zamudio & Olvera
				schultesii A. Fernández
				tricolor A. St.-Hil.
				tridentata Sylvén
			Kamienskia P. Taylor	*mangshanensis* G.W. Hu
				peranomala P. Taylor
			Lecticula (Barnh.) Komiya	*resupinata* B.D. Greene ex Bigelow
				spruceana Benth. ex Oliv.
			Martinia P. Taylor	*tenuissima* Tutin
			Mirabiles P. Taylor	*heterochroma* Steyerm.
				mirabilis P. Taylor

(*Continued*)

Family	Genus	Subgenus	Section	Species
			Nelipus (Raf.) P. Taylor	*biloba* R. Br.
				leptoplectra F. Muell.
				limosa R. Br.
			Oliveria P. Taylor	*appendiculata* E.A. Bruce
			Orchidioides A. DC.	*alpina* Jacq.
				asplundii P. Taylor
				buntingiana P. Taylor
				campbelliana Oliv.
				cornigera Studnička
				endresii Rchb. f.
				geminiloba Benj.
				humboldtii R.H. Schomb.
				humboldtii R.W. Chomb f. *albiflora* Komiya & C. Shibata
				jamesoniana Oliv.
				nelumbifolia Gardner
				nephrophylla Benj.
				praetermissa P. Taylor
				reniformis A. St.-Hil.
				quelchii N.E. Br.
				uniflora R. Br.
				unifolia Ruiz & Pavon
				uxoris Gómez-Laur.
			Setiscapela (Barnh.) P. Taylor	*flaccida* A.DC.
				nervosa G. Weber ex Benj.
				nigrescens Sylvén
				physoceras P. Taylor
				pusilla Vahl
				stanfieldii P. Taylor
				subulata L.
				trichophylla Spruce ex Oliv.
				triloba Benj.
			Sprucea P. Taylor	*viscosa* Spruce ex Oliv.
			Steyermarkia P. Taylor	*aureomaculata* Steyerm.
				cochleata C.P. Bove
				steyermarkii P. Taylor
			Stylotheca A. DC.	*guyanensis* A.DC.
			Utricularia L.	*aurea* Lour.
				aurea Lour. var. *gracilis* (Oliv.) Phuong

(*Continued*)

Family	Genus	Subgenus	Section	Species
				australis R. Br.
				australis R. Br. var. *tenuicaulis* (Miki) Hatus.
				benjaminiana Oliv.
				bentensis Komyia
				biovularioides (Kuhlm.) P. Taylor
				bremii Heer ex Koelliker
				breviscapa C. Wright ex Griseb.
				chiakiana Komiya & C. Shibata
				corneliana R.W. Jobson
				cymbantha Oliv.
				dimorphantha Makino
				foliosa L.
				floridana Nash
				geminiscapa Benj.
				gibba L.
				hydrocarpa Vahl.
				incisa (A. Rich.) Alain
				inflata Walter
				inflexa Forssk.
				intermedia Hayne
				macrorhiza Leconte
				minor L.
				muelleri Kamieński
				naviculata P. Taylor
				ochroleuca R.W. Hartm.
				olivacea C. Wright ex Griseb.
				perversa P. Taylor
				platensis Speg.
				poconensis Fromm
				punctata Wall. ex A. DC.
				radiata Small
				raynalii P. Taylor
				reflexa Oliv.
				stellaris L. f.
				striata Leconte ex Torr.
				stygia G. Thor
				vulgaris L.
				warmingii Kamieński

(*Continued*)

Family	Genus	Subgenus	Section	Species
			Vesiculina (Raf.) P. Taylor	*cucullata* A. St.-Hil. & Girard
				myriocista A. St.-Hil. & Girard
				purpurea Walter
Plantaginaceae Juss.	*Philcoxia* P. Taylor & V.C. Souza			*bahiensis* V.C. Souza & Harley
				courensis Scatigna
				goiasensis P. Taylor
				maranhensis Scatigna
				minensis V.C. Souza & Giul.
				tuberosa M.L.S. Carvalho & L.P. Queiroz
				rhizomatosa Scatigna & V.C. Souza
Bromeliaceae Juss.	*Brocchinia* Schult. & Schult. f.			*hechtioides* Mez
				reducta Baker
	Catopsis Griseb.			*berteroniana* (Schult. & Schult. f.) Mez
Eriocaulaceae Martinov	*Paepalanthus* Mart.	*Platycaulon* Mart. ex Körn.		*bromelioides* Silveira

References

Abbott, M.J. and Battaglia, L.L. (2015). Purple pitcher plant (*Sarracenia rosea*) dieback and partial community disassembly following experimental storm surge in a coastal pitcher plant bog. *PLoS ONE*, 10, e0125475.

Abbott, M.J. and Brewer, J.S. (2016). Competition does not explain the absence of a carnivorous pitcher plant from a nutrient-rich marsh. *Plant and Soil*, 409, 495–504.

Adam, J.H. (1997). Prey spectra of Bornean *Nepenthes* species (Nepenthaceae) in relation to their habitat. *Pertanika Journal of Tropical Agricultural Science*, 20, 121–34.

Adam, J.H. (1998). Reproductive biology of Bornean *Nepenthes* (Nepenthaceae) species. *Journal of Tropical Forest Science*, 10, 456–71.

Adamec, L. (1995). Ecological requirements and recent European distribution of the aquatic carnivorous plant *Aldrovanda vesiculosa* L.—a review. *Folia Geobotanica et Phytotaxonomica*, 30, 53–61.

Adamec, L. (1997a). Mineral nutrition of carnivorous plants: a review. *Botanical Review*, 63, 273–99.

Adamec, L. (1997b). Photosynthetic characteristics of the aquatic carnivorous plant *Aldrovanda vesiculosa*. *Aquatic Botany*, 59, 297–306.

Adamec, L. (1999). Seasonal growth dynamics and overwintering of the aquatic carnivorous plant *Aldrovanda vesiculosa* at experimental field sites. *Folia Geobotanica*, 34, 287–97.

Adamec, L. (2000). Rootless aquatic plant *Aldrovanda vesiculosa*: physiological polarity, mineral nutrition, and importance of carnivory. *Biologia Plantarum*, 43, 113–19.

Adamec, L. (2002). Leaf absorption of mineral nutrients in carnivorous plants stimulates root nutrient uptake. *New Phytologist*, 155, 89–100.

Adamec, L. (2003a). Ecophysiological characterization of dormancy states in turions of the aquatic carnivorous plant *Aldrovanda vesiculosa*. *Biologia Plantarum*, 47, 395–402.

Adamec, L. (2003b). Zero water flows in the carnivorous genus *Genlisea*. *Carnivorous Plant Newsletter*, 32, 46–48.

Adamec, L. (2005a). Ecophysiological characterization of carnivorous plant roots: oxygen fluxes, respiration, and water exudation. *Biologia Plantarum*, 49, 247–55.

Adamec, L. (2005b). Ten years after the introduction of *Aldrovanda vesiculosa* to the Czech Republic. *Acta Botanica Gallica*, 152, 239–45.

Adamec, L. (2006). Respiration and photosynthesis of bladders and leaves of aquatic *Utricularia* species. *Plant Biology*, 8, 765–69.

Adamec, L. (2007a). Investment in carnivory in *Utricularia stygia* and *U. intermedia* with dimorphic shoots. *Preslia*, 79, 127–39.

Adamec, L. (2007b). Oxygen concentrations inside the traps of the carnivorous plants *Utricularia* and *Genlisea* (Lentibulariaceae). *Annals of Botany*, 100, 849–56.

Adamec, L. (2008a). Mineral nutrient relations in the aquatic carnivorous plant *Utricularia australis* and its investment in carnivory. *Fundamental and Applied Limnology*, 171, 175–83.

Adamec, L. (2008b). Respiration of turions and winter apices in aquatic carnivorous plants. *Biologia*, 63, 515–20.

Adamec, L. (2008c). Soil fertilization enhances growth of the carnivorous plant *Genlisea violacea*. *Biologia*, 63, 201–03.

Adamec, L. (2008d). The influence of prey capture on photosynthetic rate in two aquatic carnivorous plant species. *Aquatic Botany*, 89, 66–70.

Adamec, L. (2009a). Ecophysiological investigation on *Drosophyllum lusitanicum*: Why doesn't the plant dry out? *Carnivorous Plant Newsletter*, 38, 71–74.

Adamec, L. (2009b). Photosynthetic CO_2 affinity of the aquatic carnivorous plant *Utricularia australis* (Lentibulariaceae) and its investment in carnivory. *Ecological Research*, 24, 327–33.

Adamec, L. (2010a). Dark respiration of leaves and traps of terrestrial carnivorous plants: are there greater energetic costs in traps? *Central European Journal of Biology*, 5, 121–24.

Adamec, L. (2010b). Field growth analysis of *Utricularia stygia* and *U. intermedia*—two aquatic carnivorous plants with dimorphic shoots. *Phyton*, 49, 241–51.

Adamec, L. (2010c). Mineral cost of carnivory in aquatic carnivorous plants. *Flora*, 205, 618–21.

Adamec, L. (2010d). Tissue mineral nutrient content in turions of aquatic plants: does it represent a storage function? *Fundamental and Applied Limnology*, 176, 145–51.

Adamec, L. (2011a). By which mechanism does prey capture enhance plant growth in aquatic carnivorous plants: Stimulation of shoot apex? *Fundamental and Applied Limnology*, 178, 171–76.

Adamec, L. (2011b). Dark respiration and photosynthesis of dormant and sprouting turions of aquatic plants. *Fundamental and Applied Limnology*, 179, 151–58.

Adamec, L. (2011c). Ecophysiological look at plant carnivory: why are plants carnivorous? In J. Seckbach and Z. Dubinski, eds. *All Flesh is Grass. Plant–Animal Interrelationships*, pp. 455–89. Cellular Origin, Life in Extreme Habitats and Astrobiology Vol. 16. Springer Science + Business Media B. V., Dordrecht.

Adamec, L. (2011d). Functional characteristics of traps of aquatic carnivorous *Utricularia* species. *Aquatic Botany*, 95, 226–33.

Adamec, L. (2011e). Shoot branching of the aquatic carnivorous plant *Utricularia australis* as the key process of plant growth. *Phyton*, 51, 133–48.

Adamec, L. (2011f). The comparison of mechanically stimulated and spontaneous firings in traps of aquatic carnivorous *Utricularia* species. *Aquatic Botany*, 94, 44–49.

Adamec, L. (2011g). The smallest but fastest: ecophysiological characteristics of traps of aquatic carnivorous *Utricularia*. *Plant Signaling & Behavior*, 6, 640–46.

Adamec, L. (2012a). Firing and resetting characteristics of carnivorous *Utricularia reflexa* traps: physiological or only physical regulation of trap triggering? *Phyton*, 52, 281–90.

Adamec, L. (2012b). Why do aquatic carnivorous plants prefer growing in dystrophic waters? *Acta Biologica Slovenica*, 55, 3–8.

Adamec, L. (2013). A comparison of photosynthetic and respiration rates in six aquatic carnivorous *Utricularia* species differing in morphology. *Aquatic Botany*, 111, 89–94.

Adamec, L. (2014). Different reutilization of mineral nutrients in senescent leaves of aquatic and terrestrial carnivorous *Utricularia* species. *Aquatic Botany*, 119, 1–6.

Adamec, L. (2015a). Regulation of the investment in carnivory in three aquatic *Utricularia* species: CO_2 or prey availability? *Phyton*, 55, 131–48.

Adamec, L. (2015b). Soil pH values at sites of terrestrial carnivorous plants in south-west Europe. *Carnivorous Plant Newsletter*, 44, 185–88.

Adamec, L. (2016). Mineral nutrition in aquatic carnivorous plants: effect of carnivory, nutrient reutilization and K^+ uptake. *Fundamental and Applied Limnology*, 188, 41–49.

Adamec, L., Fleischmann, A., and Pásek, K. (2015). Biology of the trapless rheophytic *Utricularia neottioides*: Is it possible to grow this specialized species in cultivation? *Carnivorous Plant Newsletter*, 44, 104–14.

Adamec, L., Gastinel, L., and Schlauer, J. (2006). Plumbagin content in *Aldrovanda vesiculosa* shoots. *Carnivorous Plant Newsletter*, 35, 52–55.

Adamec, L. and Kondo, K. (2002). Optimization of medium for growing the aquatic carnivorous plant *Aldrovanda vesiculosa* in vitro. *Plant Biotechnology*, 19, 283–86.

Adamec, L. and Kovářová, M. (2006). Field growth characteristics of two aquatic carnivorous plants, *Aldrovanda vesiculosa* and *Utricularia australis*. *Folia Geobotanica*, 41, 395–406.

Adamec, L. and Kučerová, A. (2013). Overwintering temperatures affect freezing temperatures of turions of aquatic plants. *Flora*, 208, 497–501.

Adamec, L. and Lev, J. (1999). The introduction of the aquatic carnivorous plant *Aldrovanda vesiculosa* to new potential sites in the Czech Republic: a five-year investigation. *Folia Geobotanica*, 34, 299–305.

Adamec, L. and Poppinga, S. (2016). Measurement of the critical negative pressure inside traps of aquatic carnivorous *Utricularia* species. *Aquatic Botany*, 133, 10–16.

Adamec, L. and Pásek, K. (2009). Photosynthetic CO_2 affinity of aquatic carnivorous plants growing under nearly natural conditions and in vitro. *Carnivorous Plant Newsletter*, 38, 107–13.

Adamec, L., Sirová, D., and Vrba, J. (2010). Contrasting growth effects of prey capture in two aquatic carnivorous plant species. *Fundamental and Applied Limnology*, 176, 153–60.

Adamec, L. and Tichý, M. (1997). Flowering of *Aldrovanda vesiculosa* in outdoor culture in the Czech Republic and isozyme variability of its European populations. *Carnivorous Plant Newsletter*, 26, 99–103.

Adamec, L., Vrba, J., and Sirová, D. (2011). Fluorescence tagging of phosphatase and chitinase activity on different structures of *Utricularia* traps. *Carnivorous Plant Newsletter*, 40, 68–73.

Addicott, J.F. (1974). Predation and prey community structure: an experimental study of the effect of mosquito larvae on the protozoan communities of pitcher plants. *Ecology*, 55, 475–92.

Adlassnig, W., Koller–Peroutka, M., Bauer, S., Koshkin, E., Lendl, T., and Lichtscheidl, I.K. (2012). Endocytotic uptake of nutrients in carnivorous plants. *Plant Journal*, 71, 303–13.

Adlassnig, W., Lendl, T., Peroutka, M., and Lang, I. (2010). Deadly glue—adhesive traps of carnivorous plants. In J.

Von Byern and I. Grunwald, eds. *Biological Adhesive Systems: From Nature to Technical and Medical Applications*, pp. 15–28. Springer-Verlag, London.

Adlassnig, W., Peroutka, M., Eder, G., Pois, W., and Lichtscheidl, I.K. (2006). Ecophysiological observations on *Drosophyllum lusitanicum*. *Ecological Research*, 21, 255–62.

Adlassnig, W., Peroutka, M., Lambers, H., and Lichtscheidl, I.K. (2005b). The roots of carnivorous plants. *Plant and Soil*, 274, 127–40.

Adlassnig, W., Peroutka, M., Lang, I., and Lichtscheidl, I.K. (2005a). Glands of carnivorous plants as a model system in cell biological research. *Acta Botanica Gallica*, 152, 111–24.

Adlassnig, W., Peroutka, M., and Lendl, T. (2011). Traps of carnivorous pitcher plants as a habitat: composition of the fluid, biodiversity and mutualistic activities. *Annals of Botany*, 107, 181–94.

Adlassnig, W., Steinhauser, G., Peroutka, M., Musilek, A., Sterba, J.H., Lichtscheidl, I.K., and Bichler, M. (2009). Expanding the menu for carnivorous plants: uptake of potassium, iron and manganese by carnivorous pitcher plants. *Applied Radiation and Isotopes*, 67, 2117–22.

Adler, P.B., Ellner, S.P., and Levine, J.M. (2010). Coexistence of perennial plants: an embarrassment of niches. *Ecology Letters*, 13, 1019–29.

Aerts, R., Verhoeven, J.T.A., and Whigham, D.F. (1999). Plant-mediated controls on nutrient cycling in temperate fens and bogs. *Ecology*, 80, 2170–81.

AGWA (Auditor General Western Australia) (2009). *Rich and Rare: Conservation of Threatened Species. Report 5*. Office of the Auditor General, Perth.

Ah-Lan, L. and Prakash, N. (1973). Life history of *Nepenthes gracilis*. *Malaysian Journal of Science*, 2, 45–53.

Airy Shaw, H.K. (1951). On the Dioncophyllaceae, a remarkable new family of flowering plants. *Kew Bulletin*, 51, 327–47.

ALA (2016). Atlas of Living Australia. Available online: http://www.ala.org.au.

Alamsyah, F. and Ito, M. (2013). Phylogenetic analysis of Nepenthaceae, based on internal transcribed spacer nuclear ribosomal DNA sequences. *Acta Phytotaxa et Geobotanica*, 64, 113–26.

Albach, D.C., Soltis, P.S., Soltis, D.E., and Olmstead, R.G. (2001). Phylogenetic analysis of asterids based on sequences of four genes. *Annals of the Missouri Botanical Garden*, 88, 163–212.

Alberch, P., Gould, S.J., Oster, G.F., and Wake, D.B. (1979). Size and shape in ontogeny and phylogeny. *Paleobiology*, 5, 296–317.

Albert, V.A., Jobson, R.W., Michael, T.P., and Taylor, D.J. (2010). The carnivorous bladderwort (*Utricularia*, Lentibulariaceae): a system inflates. *Journal of Experimental Botany*, 61, 5–9.

Albert, V.A., Williams, S.E., and Chase, M.W. (1992). Carnivorous plants: phylogeny and structural evolution. *Science*, 257, 1491–95.

Alcalá, R.E. and Domínguez, C.A. (2003). Patterns of prey capture and prey availability among populations of the carnivorous plant *Pinguicula moranensis* (Lentibulariaceae) along an environmental gradient. *American Journal of Botany*, 90, 1341–48.

Alcalá, R.E. and Domínguez, C.A. (2005). Differential selection for carnivory traits along an environmental gradient in *Pinguicula moranensis*. *Ecology*, 86, 2652–60.

Alcalá, R.E., Mariano, N.A., Osuna, F., and Abarca, C.A. (2010). An experimental test of the defensive role of sticky traps in the carnivorous plant *Pinguicula moranensis* (Lentibulariaceae). *Oikos*, 119, 891–95.

Alcaraz, L.D., Martínez-Sánchez, S., Torres, I., Ibarra-Laclette, E., and Herrera-Estrella, L. (2016). The metagenome of *Utricularia gibba*'s traps: Into the microbial input to a carnivorous plant. *PLoS ONE*, 11, e0148979.

Alkhalaf, I.A., Hübener, T., and Porembski, S. (2009). Prey spectra of aquatic *Utricularia* species (Lentibulariaceae) in northeastern Germany: the role of planktonic algae. *Flora*, 204, 700–08.

Alkhalaf, I.A., Hübener, T., and Porembski, S. (2011). Microalgae trapped by carnivorous bladderworts (*Utricularia*, Lentibulariaceae): analysis, attributes and structure of the microalgae trapped. *Plant Diversity and Evolution*, 1129, 125–38.

Allouche, O., Tsoar, A., and Kadmon, R. (2006). Assessing the accuracy of species distribution models: prevalence, kappa and the true skill statistic (TSS). *Journal of Applied Ecology*, 43, 1223–32.

An, C.I., Fukusaki, E.I., and Kobayashi A. (2001). Plasma-membrane H^+-ATPases are expressed in pitchers of the carnivorous plant *Nepenthes alata* Blanco. *Planta*, 212, 547–55.

Anderberg, A.A. (1992). The circumscription of the Ericales, and their cladistic relationships to other families of "higher" dicotyledons. *Systematic Botany*, 17, 660–75.

Anderberg, A.A., Rydin, C., and Källersjö, M. (2002). Phylogenetic relationships in the order Ericales s.l.: analyses of molecular data from five genes from the plastid and mitochondrial genomes. *American Journal of Botany*, 89, 677–87.

Anderson, B. (2003). *The Ecology, Evolution and Persistence of an Obligate, One-on-One Mutualism*. Ph.D. dissertation, University of Cape Town, South Africa.

Anderson, B. (2005). Adaptations to foliar absorption of faeces: a pathway in plant carnivory. *Annals of Botany*, 95, 757–61.

Anderson, B. (2006). Inferring evolutionary patterns from the biogeographical distributions of mutualists and exploiters. *Biological Journal of the Linnean Society*, 88, 593–602.

Anderson, B. (2010). Did *Drosera* evolve long scapes to stop their pollinators from being eaten? *Annals of Botany*, 106, 653–57.

Anderson, B., Kawakita, A., and Tayasu, I. (2012). Sticky plant captures prey for symbiotic bug: is this digestive mutualism? *Plant Biology*, 14, 888–93.

Anderson, B. and Midgley, J.J. (2001). Food or sex; pollinator–prey conflict in carnivorous plants. *Ecology Letters*, 4, 511–13.

Anderson, B. and Midgley, J.J. (2002). It takes two to tango but three is a tangle: mutualists and cheaters on the carnivorous plant *Roridula*. *Oecologia*, 132, 369–73.

Anderson, B. and Midgley, J.J. (2003). Digestive mutualism, an alternate pathway in plant carnivory. *Oikos*, 102, 221–24.

Anderson, B. and Midgley, J.J. (2007). Density-dependent outcomes in a digestive mutualism between carnivorous *Roridula* plants and their associated hemipterans. *Oecologia*, 152, 115–20.

Anderson, B., Midgley, J.J., and Stewart, B. (2003). Facilitated selfing offers reproductive assurance: a mutualism between a hemipteran and carnivorous plant. *American Journal of Botany*, 90, 1009–15.

Anderson, B., Olivieri, I., Lourmas, M., and Stewart B.A. (2004). Comparative genetic structures of *Roridula* and local adaptation of two mutualists. *Evolution*, 58, 1730–47.

Anderson, J.A.R. and Müller, J. (1975). Palynological study of a Holocene peat and a Miocene coal deposit from NW Borneo. *Review of Palaeobotany and Palynology*, 19, 291–351.

Anderson, M.J. (2006). Distance-based tests for homogeneity of multivariate dispersions. *Biometry*, 62, 245–53.

Anderson M.J., Crist T.O., Chase J.M., et al. (2011). Navigating the multiple meanings of β diversity: A roadmap for the practicing ecologist. *Ecology Letters*, 14, 19–28.

Antor, R.J. and Garcia, M.B. (1994). Prey capture by a carnivorous plant with hanging adhesive traps: *Pinguicula longifolia*. *American Midland Naturalist*, 131, 128–35.

Antor, R.J. and García, M.B. (1995). A new mite–plant association: mites living amidst the adhesive traps of a carnivorous plant. *Oecologia*, 101, 51–54.

APG (2009). An update of the Angiosperm Phylogeny Group classification for the orders and families of flowering plants: APG III. *Botanical Journal of the Linnean Society*, 161, 105–21.

APG (2016). An update of the Angiosperm Phylogeny Group classification for the orders and families of flowering plants: APG IV. *Botanical Journal of the Linnean Society*, 181, 1–20.

Araki, S. and Kadono, Y. (2003). Restricted seed contribution and clonal dominance in a free-floating aquatic plant *Utricularia australis* R. Br. in southwestern Japan. *Ecological Research*, 18, 599–609.

Aranguren Díaz, Y.C. (2016). *Genética Molecular de* Genlisea violacea *A. St.-Hil. e* Genlisea aurea *A. St.-Hil. (Lentibulariaceae)*. Ph.D. dissertation, Universidade Estadual Paulista, Sao Paulo.

Araújo, M.B., Pearson, R.G., Thuiller, W., and Erhard, M. (2005). Validation of species–climate impact models under climate change. *Global Change Biology*, 11, 1504–13.

Arber, A. (1941). On the morphology of the pitcher-leaves in *Heliamphora, Sarracenia, Darlingtonia, Cephalotus*, and *Nepenthes*. *Annals of Botany*, 5, 563–78.

Armbruster, P., Bradshaw, W.E., and Holzapfel, C.M. (1998). Effects of postglacial range expansion on allozyme and quantitative genetic variation of the pitcher-plant mosquito, *Wyeomyia smithii*. *Evolution*, 52, 1697–704.

Armitage, D.W. (2016a). Bacteria facilitate prey retention by the pitcher plant *Darlingtonia californica*. *Biology Letters*, 12, 20160577.

Armitage, D.W. (2016b). The cobra's tongue: rethinking the function of the "fishtail appendage" on the pitcher plant *Darlingtonia californica*. *American Journal of Botany*, 103: 780–85.

Arx, B., Schlauer, J., and Groves, M. (2001). *CITES Carnivorous Plant Checklist*. Royal Botanic Gardens, Kew, U.K.

Ashida, J. (1934). Studies on the leaf movement of *Aldrovanda vesiculosa* L. I. Process and mechanism of the movement. *Memoirs of the College of Science, Kyoto Imperial University, Series B*, 9, 141–244.

Ashida, J. (1935). Studies on the leaf movement of *Aldrovanda vesiculosa* L. II. Effects of mechanical, thermal, electrical, osmotic and chemical influence. *Memoirs of the College* of *Science, Kyoto Imperial University, Series B*, 11, 55–113.

Asim, M., Guerrero-Analco, J.A., Martineau, L.C., et al. (2012). Antidiabetic compounds from *Sarracenia purpurea* used traditionally by the Eeyou Istchee Cree First Nation. *Journal of Natural Products*, 75, 1284–88.

Asirvatham, R. and Christina, A.J.M. (2013). Anticancer activity of *Drosera indica* L., on Dalton's Lymphoma Ascites (DLA) bearing mice. *Journal of Intercultural Ethnopharmacology*, 2, 9–14.

Athauda, S.B.P., Matsumoto, K., Rajapakshe, S., et al. (2004). Enzymic and structural characterization of nepenthesin, a unique member of a novel subfamily of aspartic proteinases. *Biochemical Journal*, 381, 295–306.

Atwater, D.Z., Butler, J.L., and Ellison, A.M. (2006). Spatial distribution and impacts of moth herbivory on northern pitcher plants. *Northeastern Naturalist*, 13, 43–56.

Aung, H.H., Chia, L.S., Goh, N.K., et al. (2002). Phenolic constituents from the leaves of the carnivorous plant *Nepenthes gracilis*. *Fitoterapia*, 73, 445–47.

BOJA no. 107. (1994). Decreto 104/1994, de 10 de Mayo, por el que se establece el Catálogo Andaluz de la Flora Silvestre Amenazada. Junta de Andalucía, Sevilla.

Babula, P., Adam, V., Havel, L., and Kizek, R. (2009). Noteworthy secondary metabolites naphthoquinones—their occurrence, pharmacological properties and analysis. *Current Pharmaceutical Analysis*, 5, 47–68.

Bailey, T.S. (2008). *Miraculum Naturae, Venus's Flytrap*. Trafford Publishing, Victoria.

Bailey, T. and McPherson, S. (2012). Dionaea. *The Venus's Flytrap*. Redfern Natural History Productions, Poole, Dorset, U.K.

Baillon, H. (1870). Sur le developpement des feuilles des *Sarracenia*. *Comptes Rendus de l'Académie des Sciences*, 71, 331–33.

Baird, A.M. (1984). Observations on regeneration after fire in the Yule Brook Reserve near Perth, Western Australia. *Journal of the Royal Society of Western Australia*, 67, 1–13.

Baiser, B., Buckley, H.L., Gotelli, N.J., and Ellison, A.M. (2013). Predicting food-web structure with metacommunity models. *Oikos*, 122, 492–506.

Baiser, B., Gotelli, N.J., Buckley, H.L., Miller, T.E., and Ellison, A.M. (2012). Geographic variation in network structure of a Nearctic aquatic food web: network structure in an aquatic food web. *Global Ecology and Biogeography*, 21, 579–91.

Bak, S., Beisson, F., Bishop, G., et al. (2011). Cytochrome P450. *The Arabidopsis Book*, e0144.

Baker, H.G. and Baker, I. (1973). Amino-acids in nectar and their evolutionary significance. *Nature*, 241, 543–45.

Baker, C.C.M., Bittleston, L.S., Sanders, J.G., et al. (2016). Dissecting host-associated communities with DNA barcodes. *Philosophical Transactions of the Royal Society B*, 371, 20150328.

Baker, H.G., Opler, P.A., and Baker, I. (1978). A comparison of amino acid complements of floral and extrafloral nectars. *Botanical Gazette*, 139, 322–32.

Balajee, A.S., May, A., and Bohr, V.A. (1999). DNA repair of pyrimidine dimers and 6–4 photoproducts in the ribosomal DNA. *Nucleic Acids Research*, 27, 2511–20.

Balcerowicz, D., Schoenaers, S., and Vissenberg, K. (2015). Cell fate determination and the switch from diffuse growth to planar polarity in *Arabidopsis* root epidermal cells. *Frontiers in Plant Science*, 6, 1163.

Baleeiro, P.C., Jobson, R.W., and Sano, P.T. (2016). Morphometric approach to address taxonomic problems: The case of *Utricularia* sect. *Foliosa* Kamieński (Lentibulariaceae Rich.). *Journal of Systematics and Evolution*, 54, 175–86.

Banasiuk, R., Kawiak, A., and Krolicka, A. (2012). In vitro cultures of carnivorous plants from the *Drosera* and *Dionaea* genus for the production of biologically active secondary metabolites. *BioTechnologia*, 93, 87–96.

Barker, N.G. and Williamson, G.B. (1988). Effects of a winter fire on *Sarracenia alata* and *S. psittacina*. *American Journal of Botany*, 75, 138–43.

Barnes, J.S., Nguyen, H.P., Shen, S., and Schug, K.A. (2009). General method for extraction of blueberry anthocyanins and identification using high performance liquid chromatography–electrospray ionization–ion trap–time of flight–mass spectrometry. *Journal of Chromatography A*, 1216, 4728–35.

Barnhart, J.H. (1916). Segregation of genera in Lentibulariaceae. *Memoirs of the New York Botanical Garden*, 6, 39–64.

Barrett, S.C.H. (2015). Influences of clonality on plant sexual reproduction. *Proceedings of the National Academy of Sciences, USA*, 112, 8859–66.

Barry, S. and Elith, J. (2006). Error and uncertainty in habitat models. *Journal of Applied Ecology*, 43, 413–23.

Barson, M.M., Randal, L.A., and Bordas, V. (2000). *Land Cover Change in Australia. Results of The Collaborative Bureau of Rural Sciences—State Agencies' Project on Remote Sensing of Cover Change*. Bureau of Rural Sciences, Canberra.

Bárta, J., Stone, J.D., Pech, J., et al. (2015). The transcriptome of *Utricularia vulgaris*, a rootless plant with minimalist genome, reveals extreme alternative splicing and only moderate sequence similarity with *Utricularia gibba*. *BMC Plant Biology*, 15, 78.

Barthlott W (1989). Cuticular surfaces in plants. *Progress in Botany*, 51, 48–53.

Barthlott, W., Mail, M., Bushan, B., and Koch, K. (2017). Plant surfaces: structures and functions for biomimetic innovations. *Nano–Micro Letters*, 9, 23.

Barthlott, W. and Neinhuis, C. (1997). Purity of the sacred lotus, or escape from contamination in biological surfaces. *Planta*, 202, 1–8.

Barthlott, W., Porembski, S., Fischer, E., and Gemmel, B. (1998). First protozoa-trapping plant found. *Nature*, 392, 447.

Barthlott, W., Porembski, S., Seine, R., and Theisen, I. (2004). *Karnivoren. Biologie und Kultur fleischfressender Pflanzen*. Ulmer Verlag, Stuttgart.

Bartley, M.R. and Spence, D.H.N. (1987). Dormancy and propagation in helophytes and hydrophytes. *Archiv für Hydrobiologie, (Beiheft)*, 27, 139–55.

Baskin, C.C. and Baskin, J.M. (2014). *Seeds: Ecology, Biogeography, and Evolution of Dormancy and Germination*, 2nd edition. Academic Press, San Diego.

Baskin, C.C., Milberg, P., Andersson, L., and Baskin, J.M. (2001). Seed dormancy-breaking and germination requirements of *Drosera anglica*, an insectivorous species of the Northern Hemisphere. *Acta Oecologica*, 22, 1–8.

Basso, G. (2009). The Mexican *Pinguicula*. *AIPC Magazine*, Special Issue 3, 1–65.

Bateman, R.M. (2011). The perils of addressing long-term challenges in a short-term world: making descriptive taxonomy predictive. In T.R. Hodkinson, M.B. Jones, S. Waldren, and J.A.N. Parnell, eds. *Climate Change, Ecology and Systematics*, pp. 67–98. Cambridge University Press, Cambridge.

Bauer, U. (2010). *Mechanisms, Ecology and Evolution of Prey Capture by* Nepenthes *Pitcher Plants*. Ph.D. dissertation, University of Cambridge, Cambridge.

Bauer, U., Bohn, H.F., and Federle, W. (2008). Harmless nectar source or deadly trap: *Nepenthes* pitchers are activated by rain, condensation and nectar. *Proceedings of the Royal Society B*, 275, 259–65.

Bauer, U., Clemente, C.J., Renner, T., and Federle, W. (2012a). Form follows function: morphological diversification and alternative trapping strategies in carnivorous *Nepenthes* pitcher plants. *Journal of Evolutionary Biology*, 25, 90–102.

Bauer, U., Di Giusto, B., Skepper, J., Grafe, T.U., and Federle, W. (2012b). With a flick of the lid: A novel trapping mechanism in *Nepenthes gracilis* pitcher plants. *PLoS ONE*, 7, e38951.

Bauer, U. and Federle, W. (2009). The insect-trapping rim of *Nepenthes* pitchers: surface structure and function. *Plant Signaling & Behavior*, 4, 1019–23.

Bauer, U., Federle, W., Seidel, H., Grafe, T.U., and Ioannou, C.C. (2015a). How to catch more prey with less effective traps: explaining the evolution of temporarily inactive traps in carnivorous pitcher plants. *Proceedings of the Royal Society B*, 282, 20142675.

Bauer, U., Grafe, U., and Federle, W. (2011). Evidence for alternative trapping strategies in two forms of the pitcher plant, *Nepenthes rafflesiana*. *Journal of Experimental Botany*, 62, 383–92.

Bauer, U., Paulin, M., Robert, D., and Sutton, G.P. (2015b). Mechanism for rapid passive-dynamic prey capture in a pitcher plant. *Proceedings of the National Academy of Sciences, USA*, 112, 13384–89.

Bauer, U., Rembold, K., and Grafe, T.U. (2016). Carnivorous *Nepenthes* pitcher plants are a rich food source for a diverse vertebrate community. *Journal of Natural History*, 50, 483–95.

Bauer, U., Scharmann, M., Skepper, J., and Federle, W. (2013). "Insect aquaplaning" on a superhydrophilic hairy surface: how *Heliamphora nutans* Benth. pitcher plants capture prey. *Proceedings of the Royal Society B*, 280, 20122569.

Bauer, U., Willmes, C., and Federle, W. (2009). Effect of pitcher age on trapping efficiency and natural prey capture in carnivorous *Nepenthes rafflesiana* plants. *Annals of Botany*, 103, 1219–26.

Baumgartl, W. (1993). The genus *Heliamphora*. *Carnivorous Plant Newsletter*, 22, 86–92.

Bayer, R.J., Hufford, L., and Soltis, D.E. (1996). Phylogenetic relationships in Sarraceniaceae based on *rbcL* and ITS sequences. *Systematic Botany*, 21, 121–34.

Bazile, V., Le Moguédec, G., Marshall, D.J., and Gaume, L. (2015). Fluid physico-chemical properties influence capture and diet in *Nepenthes* pitcher plants. *Annals of Botany*, 115, 705–16.

Bazile, V., Moran, J.A., Le Moguédec, G., Marshall, D.J., and Gaume, L. (2012). A carnivorous plant fed by its symbiont: a unique multi-faceted nutritional mutualism. *PLoS ONE*, 7, e36179.

Beal, W.J., 1875. Carnivorous plants. *Proceedings of the American Association for the Advancement of Science*, 1875B, 251–53.

Beaver, R.A. (1983). The communities living in *Nepenthes* pitcher plants: fauna and food webs. In J.H. Frank and L.P. Lounibos, eds. *Phytotelmata: Terrestrial Plants as Hosts for Aquatic Insect Communities*, pp. 129–59. Plexus Publishing, Medford.

Beaver, R.A. (1985). Geographical variation in food web structure in *Nepenthes* pitcher plants. *Ecological Entomology*, 10, 241–48.

Beccari, O. (1904). *Wanderings in The Great Forests of Borneo*. Archibold and Constable, London.

Beck, S.G., Fleischmann, A., Huaylla, H., Müller, K.F., and Borsch, T. (2008). *Pinguicula chuquisacensis* (Lentibulariaceae), a new species from the Bolivian Andes, and first insights on phylogenetic relationships among South American *Pinguicula*. *Willdenowia*, 38, 201–12.

Beekmann, E.M. (2004). A note on the priority of Rumphius' observation of decapod crustacea living in *Nepenthes*. *Crustaceana*, 77, 1019–21.

Bell, C.R. (1949). A cytotaxonomic study of the Sarraceniaceae of North America. *Journal of the Elisha Mitchell Scientific Society*, 65, 137–66.

Bell, C.R. (1952). Natural hybrids in the genus *Sarracenia*: I. History, distribution, and taxonomy. *Journal of the Elisha Mitchell Scientific Society*, 68, 55–80.

Bell, C.R. and Case, F.W. (1956). Natural hybrids in the genus *Sarracenia*: II. Current notes on distribution. *Journal of the Elisha Mitchell Scientific Society*, 72, 142–52.

Bell, C.D., Soltis, D.E., and Soltis, P.S. (2010). The age and diversification of the angiosperms re-revisited. *American Journal of Botany*, 97, 1296–303.

Bemm, F., Becker, D., Larisch, C., et al. (2016). Venus flytrap carnivorous lifestyle builds on herbivore defense strategies. *Genome Research*, 26, 812–25.

Benjamin, S.N. and Bradshaw, W.E. (1994). Body-size and flight activity effects on male reproductive success in the pitcher plant mosquito (Diptera, Culicidae). *Annals of the Entomological Society of America*, 87, 331–36.

Bennett, M.D., Cox, A.V., and Leitch, I.J. (1998). Angiosperm DNA C-values database (release 2.0). Available online: http://www.rbgkew.org.uk/cval/databasel.ht.

Bennett, K.F. and Ellison, A.M. (2009). Nectar, not colour, may lure insects to their death. *Biology Letters*, 5, 469–72.

Bennetzen, J.L. and Wang, H. (2014). The contributions of transposable elements to the structure, function, and evolution of plant genomes. *Annual Review of Plant Biology*, 65, 505–30.

Bentham, G. and Hooker, J.D. (1865). *Genera Plantarum I.* Reeve, London.

Benz, M.J., Gorb, E.V., and Gorb, S.N. (2012). Diversity of the slippery zone microstructure in pitchers of nine carnivorous *Nepenthes* taxa. *Arthropod–Plant Interactions*, 6, 147–58.

Benzing, D.H. (1987). The origin and rarity of botanical carnivory. *Trends in Ecology and Evolution*, 2, 364–69.

Benzing, D.H. (1990). *Vascular Epiphytes*. Cambridge University Press, Cambridge.

Benzing, D.H. (2000). *Bromeliaceae: Profile of an Adaptive Radiation*. Cambridge University Press, Cambridge.

Berendse, F. (1982). Competition between plant populations with different rooting depths: III. Field experiments. *Oecologia*, 53, 50–55.

Berg, G., Grube, M., Schloter, M., and Smalla, K. (2014). Unraveling the plant microbiome: looking back and future perspectives. *Frontiers in Microbiology*, 5, 148.

Bergland, A.O., Agotsch, M., Mathias, D., Bradshaw, W.E., and Holzapfel, C.M. (2005). Factors influencing the seasonal life history of the pitcher-plant mosquito, *Wyeomyia smithii*. *Ecological Entomology*, 30, 129–37.

Bern, A.L. (1997). *Studies on Nitrogen and Phosphorus Uptake by The Carnivorous Bladderwort* Utricularia foliosa *L. in South Florida Wetlands*. M.Sc. thesis, Florida International University, Miami, Florida.

Berry, P.E., Riina, R., and Steyermark, J.A. (2005). Sarraceniaceae. In P.E. Berry, K. Yatskievych, and B.K. Holst, eds. *Flora of the Venezuelan Guayana*, Vol. 9, pp. 138–44. Missouri Botanical Garden Press, St. Louis.

Bert, W., De Ley, I.T., Van Driessche R., Segers, H., and De Ley, P. (2003). *Baujardia mirabilis* gen. n., sp. n. from pitcher plants and its phylogenetic position within Panagrolaimidae (Nematoda: Rhabditida). *Nematology*, 5, 405–20.

Bertol, N., Paniw, M., and Ojeda, F. (2015a). Dynamics of prey capture for a flypaper-trap carnivorous plant: luring insects versus passive trap. *American Journal of Botany*, 102, 1–6.

Bertol, N., Paniw, M., and Ojeda, F. (2015b). Effective prey attraction in the rare *Drosophyllum lusitanicum*, a flypaper-trap carnivorous plant *American Journal of Botany*, 102, 689–94.

Bessey, C.E. (1915). The phylogenetic taxonomy of flowering plants. *Annals of the Missouri Botanical Garden*, 2, 109–64.

Bewley, J.D., Bradford, K., and Hilhorst, H. (2013). *Seeds: Physiology of Development, Germination and Dormancy*. Springer, New York.

Bezanger-Beauquesne, L. (1954). Sur le pigment jaune du *Drosera*. *Comptes rendus d'Académie des Sciences de Paris*, 239, 618–20.

Bhattarai, G.P. and Horner, J.D. (2009). The importance of pitcher size in prey capture in the carnivorous plant, *Sarracenia alata* Wood (Sarraceniaceae). *American Midland Naturalist*, 161, 264–72.

Bilz, M., Kell, S.P., Maxted, N., and Lansdown, R.V. (2011). *European Red List of Vascular Plants*. Publications Office of the European Union, Luxembourg.

Biteau, F., Bourgaud, F., Gontier, E., and Fevre, J.P. (2012). Process for the production of recombinant proteins using carnivorous plants. *U.S. Patent and Trademark Office*, No. 8,178,749, Washington, DC.

Biteau, F., Nisse, E., Miguel, S., et al. (2013). A simple SDS–PAGE protein pattern from pitcher secretions as a new tool to distinguish *Nepenthes* species (Nepenthaceae). *American Journal of Botany*, 100, 2478–84.

Bittleston, L.S. (2016). *Convergent Interactions among Pitcher Plant Microcosms in North America and Southeast Asia*. Ph.D. dissertation, Harvard University, Cambridge, Massachusetts.

Bittleston, L.S., Baker, C.C.M., Strominger, L.B., Pringle, A., and Pierce, N.E. (2016a). Metabarcoding as a tool for investigating arthropod diversity in *Nepenthes* pitcher plants. *Austral Ecology*, 41, 120–32.

Bittleston, L.S., Pierce, N.E., Ellison, A.M., and Pringle, A. (2016b). Convergence in multispecies interactions. *Trends in Ecology and Evolution*, 31, 269–80.

Błędzki, L.A. and Ellison, A.M. (1998). Population growth and production of *Habrotrocha rosa* Donner (Rotifera: Bdelloidea) and its contribution to the nutrient supply of its host, the northern pitcher plant, *Sarracenia purpurea* L. (Sarraceniaceae). *Hydrobiologia*, 385, 193–200.

Blehová, A., Švubová, R., Lukačová, Z., Moravčíková, J., and Matušíková, I. (2015). Transformation of sundew: pitfalls and promises. *Plant Cell, Tissue and Organ Culture*, 120, 681–87.

Bleuyard, J.Y., Gallego, M.E., and White, C.I. (2006). Recent advances in understanding of the DNA double-strand break repair machinery of plants. *DNA Repair*, 5, 1–12.

Blüthgen, N., Verhaagh, M., Goitía, W., Jaffé, K., Morawetz, W., and Barthlott, W. (2000). How plants shape the ant community in the Amazonian rainforest canopy: The key role of extrafloral nectaries and hemipteran honeydew. *Oecologia*, 125, 229–40.

Bobák, M., Blehová, A., Krištín, J., Ovečka, M., and Šamaj, J. (1995). Direct plant regeneration from leaf explants of

Drosera rotundifolia cultured in vitro. *Plant Cell, Tissue and Organ Culture*, 43, 43–49.

Böhm, J., Scherzer, S., Król, E., et al. (2016). The Venus flytrap *Dionaea muscipula* counts prey-induced action potentials to induce sodium uptake. *Current Biology*, 26, 286–95.

Bohn, H.F. and Federle, W. (2004). Insect aquaplaning: *Nepenthes* pitcher plants capture prey with the peristome, a fully wettable water-lubricated anisotropic surface. *Proceedings of the National Academy of Sciences, USA*, 101, 14138–43.

Bohn, H.F., Thornham, D.G., and Federle, W. (2012). Ants swimming in pitcher plants: kinematics of aquatic and terrestrial locomotion in *Camponotus schmitzi*. *Journal of Comparative Physiology A*, 198, 465–76.

Boldt, J.K., Meyer, M.H., and Erwin, J.E. (2014). Foliar anthocyanins: a horticultural review. In: J. Janick ed. *Horticultural Reviews*, Volume 42. doi: 10.1002/9781118916827.ch04.

Bonhomme, V., Gounand, I., Alaux, C., et al. (2011a). The plant-ant *Camponotus schmitzi* helps its carnivorous host-plant *Nepenthes bicalcarata* to catch its prey. *Journal of Tropical Ecology*, 27, 15–24.

Bonhomme, V., Pelloux-Prayer, H., Jousselin, E., et al. (2011b). Slippery or sticky? Functional diversity in the trapping strategy of *Nepenthes* carnivorous plants. *New Phytologist*, 191, 545–54.

Bopp, M. and Weber, I. (1981). Hormonal regulation of the leaf blade movement of *Drosera capensis*. *Physiologia Plantarum*, 53, 491–96.

Bopp, M. and Weiler, E.W. (1985). Leaf blade movement of *Drosera* and auxin distribution. *Naturwissenschaften*, 72, 434.

Borovec, J., Sirová, D., and Adamec, L. (2012). Light as a factor affecting the concentration of simple organics in the traps of aquatic carnivorous *Utricularia* species. *Fundamental and Applied Limnology*, 181, 159–66.

Bosserman, R.W. (1983). Elemental composition of *Utricularia*–periphyton ecosystems from Okefenokee swamp. *Ecology*, 64, 1637–45.

Bott, T., Meyer, G.A., and Young, E.B. (2008). Nutrient limitation and morphological plasticity of the carnivorous pitcher plant *Sarracenia purpurea* in contrasting wetland environments. *New Phytologist*, 180, 631–41.

Botta, S.M., (1976). Sobre las trampas y las víctimas o presas de algunas especies argentinas del género *Utricularia*. *Darwiniana*, 20, 127–54.

Bottrill, M.C., Joseph, L.N., Carwadine, J., et al. (2008). Is conservation triage just smart decision making? *Trends in Ecology and Evolution*, 23, 649–54.

Boucher, D.H., James, S., and Keeler, K.H. (1982). The ecology of mutualism. *Annual Review of Ecology and Systematics*, 13, 315–47.

Bourke, G. (2009). The three sisters—ancient survivors of the tropical rainforest. *Carniflora Australis*, 7, 9–18.

Bowman, J. L. (2000). The YABBY gene family and abaxial cell fate. *Current Opinion in Plant Biology*, 3, 17–22.

Boyer, M.L.H. and Wheeler, B.D. (1989). Vegetation patterns in spring-fed calcareous fens—calcite precipitation and constraints on fertility. *Journal of Ecology*, 77, 597–609.

Bradford, J.C. and Barnes, R.W. (2001). Phylogenetics and classification of Cunoniaceae (Oxalidales) using chloroplast DNA sequences and morphology. *Systematic Botany*, 26, 354–85.

Bradshaw, C.J.A. (2012). Little left to lose: deforestation and forest degradation in Australia since European colonization. *Journal of Plant Ecology*, 5, 109–20.

Bradshaw, W.E. (1983). Interaction between the mosquito *Wyeomyia smithii*, the midge *Metriocnemus knabi*, and their carnivorous host, *Sarracenia purpurea*. In J.H. Frank and L.P. Lounibos, eds. *Phytotelmata: Terrestrial Plants as Hosts of Aquatic Insect Communities*, pp. 161–89. Plexus, Medford.

Bradshaw, W.E. and Creelman, R.A. (1984). Mutualism between the carnivorous purple pitcher plant and its inhabitants. *American Midland Naturalist*, 112, 294–304.

Bradshaw, W.E., Emerson, K.J., Catchen, J.M., Cresko, W.A., and Holzapfel, C.M. (2012). Footprints in time: comparative quantitative trait loci mapping of the pitcher-plant mosquito, *Wyeomyia smithii*. *Proceedings of the Royal Society B*, 279, 4551–58.

Bradshaw, W.E., Haggerty, B.P., and Holzapfel, C.M. (2005). Epistasis underlying a fitness trait within a natural population of the pitcher-plant mosquito, *Wyeomyia smithii*. *Genetics*, 169, 485–88.

Bradshaw, W.E. and Holzapfel, C.M. (1986). Geography of density-dependent selection in pitcher-plant mosquitoes. In F. Taylor and R. Karban, eds. *The Evolution of Insect Life Cycles*, pp. 48–65. Springer-Verlag, New York.

Bradshaw, W.E. and Holzapfel, C.M. (1989). Life-historical consequences of density-dependent selection in the pitcher-plant mosquito, *Wyeomyia smithii*. *American Naturalist*, 133, 869–87.

Bradshaw, W.E. and Holzapfel, C.M. (1996). Genetic constraints to life-history evolution in the pitcher-plant mosquito, *Wyeomyia smithii*. *Evolution*, 50, 1176.

Bradshaw, W.E. and Holzapfel, C.M. (2001). Genetic shift in photoperiodic response correlated with global warming. *Proceedings of the National Academy of Sciences, USA*, 98, 14509–11.

Bradshaw, W.E. and Holzapfel, C.M. (2006). Evolutionary response to rapid climate change. *Science*, 312, 1477–78.

Bradshaw, W.E., Holzapfel, C.M., Kleckner, C.A., and Hard, J.J. (1997). Heritability of development time and protandry in the pitcher-plant mosquito, *Wyeomyia smithii*. *Ecology*, 78, 969–76.

Bradshaw, W.E. and Lounibos, L.P. (1977). Evolution of dormancy and its photoperiodic control in pitcher-plant mosquitos. *Evolution*, 31, 546–67.

Bradshaw, W.E., Zani, P.A., and Holzapfel, C.M. (2004). Adaptation to temperate climates. *Evolution*, 58, 1748–62.

Braunberger, C., Zehl, M., Conrad, J., et al. (2015). Flavonoids as chemotaxonomic markers in the genus *Drosera*. *Phytochemistry*, 118, 74–82.

Breiner, F.T., Guisan, A., Bergamini, A., and Nobis, M.P. (2015). Overcoming limitations of modelling rare species by using ensembles of small models. *Methods in Ecology and Evolution*, 6, 1210–18.

Bremer, B., Bremer, K., Heidari, N., et al. (2002). Phylogenetics of asterids based on 3 coding and 3 non-coding chloroplast DNA markers and the utility of non-coding DNA at higher taxonomic levels. *Molecular Phylogenetics and Evolution*, 24, 274–301.

Brewer, J.S. (1998a). Effects of competition and litter on a carnivorous plant, *Drosera capillaris* (Droseraceae). *American Journal of Botany*, 85, 1592–96.

Brewer, J.S. (1998b). Patterns of plant species richness in a wet slash pine (*Pinus elliottii*) savanna. *Journal of the Torrey Botanical Society*, 125, 216–24.

Brewer, J.S. (1999a). Effects of competition, litter, and disturbance on an annual carnivorous plant (*Utricularia juncea*). *Plant Ecology*, 140, 159–65.

Brewer, J.S. (1999b). Effects of fire, competition, and soil disturbances on regeneration of a carnivorous plant, *Drosera capillaris*. *American Midland Naturalist*, 141, 28–42.

Brewer, J.S. (1999c). Short-term effects of fire and competition on growth and plasticity of the yellow pitcher plant, *Sarracenia alata* (Sarraceniaceae). *American Journal of Botany*, 86, 1264–71.

Brewer, J.S. (2003). Why don't carnivorous pitcher plants compete with non-carnivorous plants for nutrients? *Ecology*, 84, 451–62.

Brewer, J.S. (2005). The lack of favorable responses of an endangered pitcher plant to habitat restoration. *Restoration Ecology*, 13, 710–17.

Brewer, J.S. (2006). Resource competition and fire-regulated nutrient demand in carnivorous plants of longleaf pine savannas. *Applied Vegetation Science*, 9, 11–16.

Brewer, J.S. (2017). Stochastic losses of fire-dependent endemic herbs revealed by a 65-year chronosequence of dispersal-limited woody plant encroachment. *Ecology and Evolution*, 7, 4377–4389.

Brewer, J.S., Baker D.J., Nero, A.S., et al. (2011). Carnivory in plants as a beneficial trait in wetlands. *Aquatic Botany*, 94, 62–70.

Bringmann, G., Rischer, H., Schlauer, J., et al. (2002). The tropical liana *Triphyophyllum peltatum* (Dioncophyllaceae): formation of carnivorous organs is only a facultative prerequisite for shoot elongation. *Carnivorous Plant Newsletter*, 31, 44–52.

Bringmann, G., Rischer, H., Wohlfarth, M., Schlauer, J. and Assi, L.A. (2000). Droserone from cell cultures of *Triphyophyllum peltatum* (*Dioncophyllaceae*) and its biosynthetic origin. *Phytochemistry*, 53, 339–43.

Bringmann, G., Schlauer, J., Wolf, K., et al. (1998). Cultivation of *Triphyophyllum peltatum* (Dioncophyllaceae), the part-time carnivorous plant. *Carnivorous Plant Newsletter*, 28, 7–13.

Bringmann, G., Wenzel, M., Bringmann, H.P., et al. (2001). Uptake of the amino acid alanine by digestive leaves: proof of carnivory of the tropical liana *Triphyophyllum peltatum* (Dioncophyllaceae). *Carnivorous Plant Newsletter*, 30, 15–21.

Briscoe, A.D. and Chittka, L. (2001). The evolution of color vision in insects. *Annual Review of Entomology*, 46, 471–510.

Broberg, L. and Bradshaw, W. (1995). Density-dependent development in *Wyeomyia smithii* (Diptera, Culicidae)—intraspecific competition is not the result of interference. *Annals of the Entomological Society of America*, 88, 465–70.

Brockington, S.F., Alexandre, R., Ramdial, J., et al. (2009). Phylogeny of the Caryophyllales *sensu lato*: revisiting hypotheses on pollination biology and perianth differentiation in the core Caryophyllales. *International Journal of Plant Sciences*, 170, 627–43.

Brockington, S.F., Walker, R.H., Glover, B.J., Soltis, P.S., and Soltis, D.E. (2011). Complex pigment evolution in the Caryophyllales. *New Phytologist*, 190, 854–64.

Broennimann, O., Di Cola, V., and Guisan, A. (2016). Ecospat: Spatial ecology miscellaneous methods. R package version 2.1.1. Available online: https:// cran.r–project.org/ package=ecospat.

Brook, B.W., Sodhi, N.S., and Bradshaw, C.J.A. (2008). Synergies among extinction drivers under global change. *Trends in Ecology & Evolution*, 23, 453–60.

Brooks, T.M., Mittermeier, R.A., Fonseca, G.A.B., et al. (2006). Global biodiversity conservation priorities. *Science*, 313, 58–61.

Brownfield, L., Yi, J., Jiang, H., et al. (2015). Organelles maintain spindle position in plant meiosis. *Nature Communications*, 6, 6492.

Bruce, A.N. (1905). On the activity of the glands of *Byblis gigantea*. *Notes of the Royal Botanic Garden of Edinburgh*, 16, 9–14.

Bruce, A.N. (1907). On the distribution, structure, and function of the tentacles of *Roridula*. *Notes from the Royal Botanic Garden Edinburgh*, 17, 83–98.

Brugger, J. and Rutishauser, R. (1989). Bau und Entwicklung landbewohnender *Utricularia*–Arten. *Botanica Helvetica*, 99, 91–146.

Brundrett, M.C. (2009). Mycorrhizal associations and other means of nutrition of vascular plants: understanding the global diversity of host plants by resolving conflicting information and developing reliable means of diagnosis. *Plant and Soil*, 320, 37–77.

Bryant, J.P., Chapin, F.S., III, and Klein, D.R. (1983). Carbon/nutrient balance of boreal plant in relation to vertebrate history. *Oikos*, 40, 357–86.

Buch, F., Kaman, W.E., Bikker, F.J., Yilamujiang, A., and Mithöfer, A. (2015). Nepenthesin protease activity indicates digestive fluid dynamics in carnivorous *Nepenthes* plants. *PLoS ONE*, 10, e0118853.

Buch, F., Pauchet, Y., Rott, M., and Mithöfer, A. (2014). Characterization and heterologous expression of a PR–1 protein from traps of the carnivorous plant *Nepenthes mirabilis*. *Phytochemistry*, 100, 43–50.

Buch, F., Rott, M., Rottloff, S., et al. (2013). Secreted pitfall-trap fluid of carnivorous *Nepenthes* plants is unsuitable for microbial growth. *Annals of Botany*, 111, 375–83.

Buchen, B., Hensel, D., and Sievers, A. (1983). Polarity in mechanoreceptor cells of trigger hairs of *Dionaea muscipula* Ellis. *Planta*, 158, 458–68.

Buckley, H.L., Burns, J.H., Kneitel, J.M., et al. (2004). Small-scale patterns in community structure of *Sarracenia purpurea* inquilines. *Community Ecology*, 5, 181–88.

Buckley, H.L., Miller, T.E., Ellison, A.M., and Gotelli, N.J. (2003). Reverse latitudinal trends in species richness of pitcher-plant food webs. *Ecology Letters*, 6, 825–29.

Buckley, H.L., Miller, T.E., Ellison, A.M., and Gotelli, N.J. (2010). Local to continental-scale variation in the richness and composition of an aquatic food web: pitcher-plant food webs from local to continental scales. *Global Ecology and Biogeography*, 19, 711–23.

Budzianowski, J., Skrzypczak, L., and Kukulczanka, K. (1995). Phenolic compounds of *Drosera intermedia* and *D. spathulata* from in vitro cultures. *WOCMAP—Medicinal and Aromatic Plants Conference*, Part 4.

Bunawan, H., Yen, C.C., Yaakop, S., and Noor, N.M. (2017). Phylogenetic inferences of *Nepenthes* species in Peninsular Malaysia revealed by chloroplast (*trnL* intron) and nuclear (ITS) DNA sequences. *BMC Research Notes*, 10, 67.

Burbidge, F.W. (1880). *The Gardens of the Sun: A Naturalist's Journal of Borneo and The Sulu Archipelago*. John Murray, London.

Burdic, D.M. (1985). Propagation and culture of some western North American carnivorous plants. *Combined Proceedings of the International Plant Propagation Society*, 35, 285–93.

Burdon-Sanderson, J. and Page, F.J.M. (1876). On the mechanical effects and on the electrical disturbance consequent on excitation of the leaf of *Dionaea muscipula*. *Proceedings of the Royal Society of London*, 25, 411–34.

Burrows, N.D. (2008). Linking fire ecology and fire management in south-west Australian forest landscapes. *Forest Ecology and Management*, 255, 2394–406.

Burrows, N.D. and McCaw, L. (2013). Prescribed burning in southwestern Australian forests. *Frontiers in Ecology and the Environment*, e25–e34.

Bustamante, C.D., Townsend, J.P., and Hartl, D.L. (2000). Solvent accessibility and purifying selection within proteins of *Escherichia coli* and *Salmonella enterica*. *Molecular Biology and Evolution*, 17, 301–08.

Butler, J.L. and Ellison, A.M. (2007). Nitrogen cycling dynamics in the carnivorous northern pitcher plant, *Sarracenia purpurea*. *Functional Ecology*, 21, 835–43.

Butler, J.L., Gotelli, N.J., and Ellison, A.M. (2008). Linking the brown and the green: transformation and fate of allochthonous nutrients in the *Sarracenia* microecosystem. *Ecology*, 89, 898–904.

Butts, C.T., Bierma, J.C., and Martin, R.W. (2016a). Novel proteases from the genome of the carnivorous plant *Drosera capensis*: Structural prediction and comparative analysis. *Proteins: Structure, Function and Bioinformatics*, 84, 1517–33.

Butts, C.T., Zhang, X., Kelly, J.E., et al. (2016b). Sequence comparison, molecular modeling, and network analysis predict structural diversity in cysteine proteases from the Cape sundew, *Drosera capensis*. *Computational and Structural Biotechnology Journal*, 14, 271–82.

Byrne, M., Steane, D., Joseph, L., et al. (2011). Decline of a biome: evolution, contraction, fragmentation, extraction and invasion of the Australian mesic zone biota. *Journal of Biogeography*, 38, 1635–56.

CHAH (2006). *AVH: Australia's Virtual Herbarium*. Available online: http://www.chah.gov.au/avh/.

Cameron, K.M., Chase, M.W., and Swensen, S.M. (1995). Molecular evidence for the relationships of *Triphyophyllum* and *Ancistrocladus*. *American Journal of Botany*, 82, 117–18.

Cameron, K.M., Wurdack, K.J., and Jobson, R.W. (2002). Molecular evidence for the common origin of snap-traps among carnivorous plants. *American Journal of Botany*, 89, 1503–09.

Cannon, J., Lojanapiwatna, V., Raston, C.L., Sinchai W., and White A.H. (1980). The quinones of *Nepenthes rafflesiana*. The crystal structure of 2,5-dihydroxy-3,8-dimethoxy-7-methylnaphtho-1,4-quinone (Nepenthone-E) and a synthesis of 2,5-dihydroxy-3-methoxy-7-methylnaphtho-1,4-quinone (Nepenthone-C). *Australian Journal of Chemistry*, 33, 1073–93.

Cao, H.X., Schmutzer, T., Scholz, U., et al. (2015). Metatranscriptome analysis reveals host–microbiome interactions in traps of carnivorous *Genlisea* species. *Frontiers in Microbiology*, 6, e526.

Caravieri, F.A, Ferreira, A.J., Ferreira, A., et al. (2014). Bacterial community associated with traps of the carnivorous plants *Utricularia hydrocarpa* and *Genlisea filiformis*. *Aquatic Botany*, 116, 8–12.

Carlquist, S. (1976). Wood anatomy of Roridulaceae: ecological and phylogenetic implications. *American Journal of Botany*, 63, 1003–08.

Carlquist, S. and Wilson, E.J. (1995). Wood anatomy of *Drosophyllum* (Droseraceae): ecological and phylogenetic considerations. *Bulletin of the Torrey Botanical Club*, 122, 185–89.

Carretero-Paulet, L., Chang, T.-H., Librado, P., et al. (2015a). Genome-wide analysis of adaptive molecular evolution in the carnivorous plant *Utricularia gibba*. *Genome Biology and Evolution*, 7, 444–56.

Carretero-Paulet, L., Librado, P., Chang, T.-H., et al. (2015b). High gene family turnover rates and gene space adaptation in the compact genome of the carnivorous plant *Utricularia gibba*. *Molecular Biology and Evolution*, 32, 1284–95.

Carow, T., Wistuba, A., and Harbarth, P. (2005). *Heliamphora sarracenioides*, a new species of *Heliamphora* (Sarraceniaceae) from Venezuela. *Carnivorous Plant Newsletter*, 34, 4–6.

Carroll, C., Lawler, J.J., Roberts, D.R., and Hamann, A. (2015). Biotic and climatic velocity identify contrasting areas of vulnerability to climate change. *PLoS ONE*, 10, e0140486.

Carroll, C.R. and Janzen, D.H. (1973). Ecology of foraging by ants. *Annual Review of Ecology and Systematics*, 4, 231–57.

Carrow, T., Hirschel, K., and Radke, R. (1997). *Fleischfressende Pflanzen: Tödliche Fallen*. Film, ZDF.

Carstens, B.C. and Satler, J.D. (2013). The carnivorous plant described as *Sarracenia alata* contains two cryptic species. *Biological Journal of the Linnean Society*, 109, 737–46.

Carter, R., Boyer, T., McCoy, H., and Londo, A. (2006). Classification of green pitcher plant (*Sarracenia oreophila* (Kearney) Wherry) communities in the Little River National Preserve, Alabama. *Natural Areas Journal*, 26, 84–93.

Carvalho, M.L.S. and Queiroz, L.P. (2014). *Philcoxia tuberosa* (Plantaginaceae), a new species from Bahia, Brazil. *Neodiversity*, 7, 14–20.

Case, F.W. and Case, R.B. (1976). The *Sarracenia rubra* complex. *Rhodora*, 78: 270–325.

Casper, S.J. (1962). Revision der Gattung *Pinguicula* in Eurasien. *Feddes Repertorium*, 66: 1–148.

Casper, S.J. (1963). Gedanken zur Gliederung der Gattung *Pinguicula* L. *Jahrbücher für Botanische Systematik*, 82: 321–35.

Casper, S.J. (1966). Monographie der Gattung *Pinguicula* L. *Bibliotheca Botanica*, 127/128, 1–225.

Casper, S.J. and Manitz, H. (1975). Beiträge zur Taxonomie und Chorologie der mitteleuropäischen *Utricularia*–Arten. 2. Androsporogenese, Chromosomenzahlen und Pollenmorphologie. *Feddes Repertorium*, 86, 211–32.

Casper, S.J. and Stimper, R. (2006). New and revised chromosome numbers in *Pinguicula* (Lentibulariaceae). *Haussknechtia*, 11, 3–8.

Casper, S.J. and Stimper, R. (2009). Chromosome numbers in *Pinguicula* (Lentibulariaceae): survey, atlas, and taxonomic conclusions. *Plant Systematics and Evolution*, 277, 21–60.

Catalano, M. (2010). *Nepenthes della Thailandia: Diario di Viaggio*. Prague.

Chadwick, Z.D. and Darnowski, D.W. (2002). Observations on prey preference and other associations of *Aldrovanda vesiculosa* in a new culture system. *Proceedings of the 4th International Carnivorous Plant Conference, Tokyo, Japan*. pp. 39–47. Higashi-Hiroshima University, Higashihiroshima.

Chadwick, O.A., Derry, L.A., Vitousek, P.M., Huebert, B.J. and Hedin, L.O. (1999). Changing sources of nutrients during four million years of ecosystem development. *Nature*, 397, 491–97.

Chae, L., Kim, T., Nilo-Poyanco, R. and Rhee, S.Y. (2014). Genomic signatures of specialized metabolism in plants. *Science*, 344, 510–13.

Chan, X.Y., Hong, K.W., Yin, W.F., and Chan, K.G. (2016). Microbiome and biocatalytic bacteria in monkey cup (*Nepenthes* pitcher) digestive fluid. *Scientific Reports*, 6, 20016.

Chandler, G.E. and Anderson, J.W. (1976). Uptake and metabolism of insect metabolites by leaves and tentacles of *Drosera* species. *New Phytologist*, 77, 625–34.

Chapin, C.T. and Pastor, J. (1995). Nutrient limitations in the northern pitcher plant *Sarracenia purpurea*. *Canadian Journal of Botany*, 73, 728–34.

Chase, M.W., Christenhusz, M.J.M., Sanders, D., and Fay M.F. (2009). Murderous plants: Victorian Gothic, Darwin and modern insight into vegetable carnivory. *Botanical Journal of the Linnean Society*, 161, 329–56.

Chase, J.M., Kraft, N.J.B., Smith, K.G., Vellund, M., and Inouye, B.D. (2011). A null models approach to disentangle community similarity from α-diversity. *Ecosphere*, 2, 24.

Chase, M.W., Soltis, D.E., Olmstead, R.G., et al. (1993). Phylogenetics of seed plants: an analysis of nucleotide sequences from the plastid gene *rbcL*. *Annals of the Missouri Botanical Garden*, 80, 528–80.

Cheek, M. (1988). *Sarracenia psittacina*: Sarraceniaceae. *Curtis's Botanical Magazine*, 5, 60–65.

Cheek, M. and Jebb, M. (2001). *Nepenthaceae. Flora Malesiana Series 1, Vol. 15*, pp. 1–164. National Herbarium of the Netherlands, Leiden.

Cheek, M. and Jebb, M. (2013a). *Nepenthes ramos* (Nepenthaceae), a new species from Mindanao, Philippines. *Wildenowia*, 43, 107–11.

Cheek, M. and Jebb, M. (2013b). Recircumscription of the *Nepenthes alata* group (Caryophyllales: Nepenthaceae), in the Philippines, with four new species. *European Journal of Taxonomy*, 69, 1–23.

Cheek, M. and Jebb, M. (2013c). Typification and redelimitation of *Nepenthes alata* with notes on the *N. alata* group, and *N. negros* sp. nov. from the Philippines. *Nordic Journal of Botany*, 31, 616–22.

Cheek, M. and Jebb, M. 2014. Expansion of the *Nepenthes alata* group (Nepenthaceae), Philippines, and descriptions of three new species. *Blumea*, 59, 144–54.

Cheek, M. and Jebb, M. (2015). *Nepenthes*, three new infrageneric names and leptotypifications. *Planta Carnivora*, 37, 34–42.

Cheek, M. and Jebb, M. (2016a). A new section in *Nepenthes* (Nepenthaceae) and a new species from Sulawesi. *Blumea*, 61, 59–62.

Cheek, M. and Jebb, M. (2016b). *Nepenthes minima* (Nepenthaceae), a new pyrophytic grassland species from Sulawesi, Indonesia. *Blumea*, 61, 181–85.

Cheek, M. and Jebb, M. (2016c). *Nepenthes* section *Pyrophytae*. *Planta Carnivora*, 38, 44–45.

Chen, I.C., Hill, J.K., Ohlemüller, R., Roy, D.B. and Thomas, C.D. (2011). Rapid range shifts of species associated with high levels of climate warming. *Science*, 333, 1024–26.

Chen, L., James, S.H., and Stace, H.M. (1997). Self-incompatibility, seed abortion and clonality in the breeding systems of several western Australian *Drosera* species (Droseraceae). *Australian Journal of Botany*, 45, 191–201.

Chen, H., Zhang, P., Zhang, L., et al. (2016). Continuous directional water transport on the peristome surface of *Nepenthes alata*. *Nature*, 532, 85–91.

Cheng, Y., Dai, X., and Zhao, Y. (2007). Auxin synthesized by the YUCCA flavin monooxygenases is essential for embryogenesis and leaf formation in *Arabidopsis*. *The Plant Cell*, 19, 2430–39.

Cheng, F., Sun, C., Wu, J., et al. (2016). Epigenetic regulation of subgenome dominance following whole genome triplication in *Brassica rapa*. *New Phytologist*, 211, 288–99.

Cheng, F., Wu, J., Fang, L., et al. (2012). Biased gene fractionation and dominant gene expression among the subgenomes of *Brassica rapa*. *PLoS ONE*, 7, e36442.

Chesson, P. (2000). Mechanisms of maintenance of species diversity. *Annual Review of Ecology and Systematics*, 31, 343–66.

Chia, T.F., Aung, H.H., Osipov, A.N., Goh, N.K., and Chia, L.S. (2004). Carnivorous pitcher plant uses free radicals in the digestion of prey. *Redox Report*, 9, 255–61.

Chin, L., Chung, A.Y.C., and Clarke, C. (2014). Interspecific variation in prey capture behavior by co-occurring *Nepenthes* pitcher plants—evidence for resource partitioning or sampling scheme artifacts? *Plant Signaling & Behavior*, 9, e27930.

Chin, L., Moran, J.A., and Clarke, C. (2010). Trap geometry in three giant montane pitcher plant species from Borneo is a function of tree shrew body size. *New Phytologist*, 186, 461–70.

Chinese Herbal Medicine Compilation Group (1975). *Compilation of Chinese Medicine Encyclopaedia*. Chinese Herbal Medicine Group, Beijing.

Choi, H.W. and Hwang, B.K. (2012). The pepper extracellular peroxidase CaPO_2 is required for salt, drought and oxidative stress tolerance as well as resistance to fungal pathogens. *Planta*, 235, 1369–82.

Choi, S.S., Vallender, E.J., and Lahn, B.T. (2006). Systematically assessing the influence of 3-dimensional structural context on the molecular evolution of mammalian proteomes. *Molecular Biology and Evolution*, 23, 2131–33.

Choo, J.P.S., Koh, T.L., and Ng, P.K.L. (1997). Nematodes and arthropods. In H.T.W. Tan, ed. *A Guide to the Carnivorous Plants of Singapore*, pp. 141–54. Singapore Science Centre, Singapore.

Chou, L.Y., Clarke, C.M., and Dykes, G.A. (2014). Bacterial communities associated with the pitcher fluids of three *Nepenthes* (Nepenthaceae) pitcher plant species growing in the wild. *Archives of Microbiology*, 196, 709–17.

Chou, L.Y., Dykes, G.A., Wilson, R.F., and Clarke, C.M. (2016). *Nepenthes ampullaria* (Nepenthaceae) pitchers are unattractive to gravid *Aëdes aegypti* and *Aëdes albopictus* (Diptera: Culicidae). *Environmental Entomology*, 45, 201–06.

Chou, L.Y., Wilson, R.F., Dykes, G.A., and Clarke, C.M. (2015). Why are *Aëdes* mosquitoes rare colonisers of *Nepenthes* pitcher plants? *Ecological Entomology*, 40, 603–11.

Christiansen, N.L. (1976). The role of carnivory in *Sarracenia flava* L. with regard to specific nutrient deficiencies. *Journal of the Elisha Mitchell Society*, 92, 144–47.

Chrtek, J., Slavíková, Z., Studnička, M. (1989). Beitrag zur Leitbündelanordnung in den Kronblättern von ausgewählten Arten der fleischfressenden Pflanzen. *Preslia*, 61, 107–24.

Chrtek, J., Slavíková, Z., Studnička, M. (1992). Beitrag zur Morphologie und Taxonomie der Familie Sarraceniaceae. *Preslia*, 64, 1–10.

Chua, L.S.L. (2000). The pollination biology and breeding system of *Nepenthes macfarlanei* (Nepenthaceae). *Journal of Tropical Forest Research*, 12, 635–42.

Chua, T. and Lim, M. (2012). Cross-habitat predation in *Nepenthes gracilis*: the red crab spider *Misumenops nepenthicola* influences abundance of pitcher dipteran larvae. *Journal of Tropical Ecology*, 28, 97–104.

Cieniak, C., Walshe-Roussel, B., Liu, R., et al. (2015). Phytochemical comparison of the water and ethanol leaf extracts of the Cree medicinal plant, *Sarracenia purpurea* L. (*Sarraceniaceae*). *Journal of Pharmacy and Pharmaceutical Sciences*, 18, 484–93.

Cieslak, T., Polepalli, J.S., White, A., et al. (2005). Phylogenetic analysis of *Pinguicula* (Lentibulariaceae): chloroplast DNA sequences and morphology reveal several geographically distinct radiations. *American Journal of Botany*, 92, 1723–36.

Cipollini, D.F., Jr, Newell, S.J. and Nastase, A.J. (1994). Total carbohydrates in nectar of *Sarracenia purpurea* L. (northern pitcher plant). *American Midland Naturalist*, 131, 374–77.

Clancy, F.G. and Coffey, M.D. (1977). Acid phosphatase and protease release by the insectivorous plant *Drosera rotundifolia*. *Canadian Journal of Botany*, 55, 480–88.

Clark, N.E., Boakes, E.H., McGowan, P.J.K., Mace, G.M., and Fuller, R.A. (2013). Protected areas in South Asia have not prevented habitat loss: a study using historical models of land use. *PLoS ONE*, 8, e65298.

Clarke, C. (1997). *Nepenthes of Borneo*. Natural History Publications, Kota Kinabalu, Sabah, Malaysia.

Clarke, C.M. (1998). A re-examination of geographical variation in *Nepenthes* food webs. *Ecography*, 21, 430–36.

Clarke, C. (1999). *Nepenthes benstonei* (Nepenthaceae)—a new pitcher plant from Peninsular Malaysia. *Sandakania*, 13, 79–87.

Clarke, C. (2001). *Nepenthes of Sumatra and Peninsular Malaysia*. Natural History Publications, Kota Kinabalu, Sabah, Malaysia.

Clarke, C.M., Bauer, U., Lee, C.C., et al. (2009). Tree shrew lavatories: a novel sequestration strategy in a tropical pitcher plant. *Biology Letters*, 5, 632–35.

Clarke, C. and Kitching, R.L. (1993). The metazoan food webs from six Bornean *Nepenthes* species. *Ecological Entomology*, 18, 7–16.

Clarke, C.M. and Kitching, R.L. (1995). Swimming ants and pitcher plants: a unique ant–plant interaction from Borneo. *Journal of Tropical Ecology*, 11, 589–602.

Clarke, C., Lee, C.C., and Enar, V. (2014). Observations on the natural history and ecology of *Nepenthes campanulata*. *Carnivorous Plant Newsletter*, 43, 7–13.

Clarke, C.M., Lee, C.C., and Sarunday, C. (in press). *Nepenthes bicalcarata*. The IUCN Red List of Threatened Species 2017.

Clarke, C. and Moran J.A. (2011). Incorporating ecological context: a revised protocol for the preservation of *Nepenthes* pitcher plant specimens (Nepenthaceae). *Blumea*, 56, 225–28.

Clarke, C. and Moran, J.A. (2016). Climate, soils and vicariance—their roles in shaping the diversity and distribution of *Nepenthes* in Southeast Asia. *Plant and Soil*, 403, 37–51.

Clarke, C., Moran, J.A., and Lee, C.C. (2011). *Nepenthes baramensis* (Nepenthaceae)—a new species from north-western Borneo. *Blumea*, 56, 229–33.

Clarke, C.M. and Sarunday, C. (2013). *Nepenthes suratensis*. The IUCN Red List of Threatened Species 2013, e.T21848236A21848252.

Clarke, C.M. (2018). *Nepenthes muluensis*. The IUCN Red List of Threatened Species 2018, e.T39680A21844351.

Clivati, D., Cordeiro, G.D., Płachno, B.J., and de Miranda, V.F. (2014). Reproductive biology and pollination of *Utricularia reniformis* A.St.–Hil. (Lentibulariaceae). *Plant Biology*, 16, 677–82.

Clivati, D., Gitzendanner, M.A., Hilsdorf, A.W.S., Araujo, W.L., and de Miranda V.F. (2012). Microsatellite markers developed for *Utricularia reniformis* (Lentibulariaceae). *American Journal of Botany*, e375–e378.

Close, D.C. and Beadle, C.L. (2003). The ecophysiology of foliar anthocyanin. *Botanical Review*, 69, 149–61.

Cochran-Stafira, D.L. and von Ende, C.N. (1998). Integrating bacteria into food webs: studies with *Sarracenia purpurea* inquilines. *Ecology*, 79, 880–98.

Collett, C., Ardron, A., Bauer, U., et al. (2015). A portable extensional rheometer for measuring the viscoelasticity of pitcher plant and other sticky liquids in the field. *Plant Methods*, 11, 16.

Collingsworth, D. (2015). Do bears disperse *Darlingtonia*? *Carnivorous Plant Newsletter*, 44, 44–47.

Collins, N.M. (1980). The distribution of soil macrofauna on the west ridge of Gunung (Mount) Mulu, Sarawak. *Oecologia*, 44, 263–75.

Colmer, T.D. and Pedersen, O. (2008). Underwater photosynthesis and respiration in leaves of submerged wetland plants: gas films improve CO_2 and O_2 exchange. *New Phytologist*, 177, 918–26.

Colombani, M. and Forterre, Y. (2011). Biomechanics of rapid movements in plants: poroelastic measurements at the cell scale. *Computer Methods in Biomechanics and Biomedical Engineering*, 14, 115–17.

Combres, C., Laliberté, G., Sevrin Reyssac, J., and de la Noüe, J. (1994). Effect of acetate on growth and ammonium uptake in the microalga *Scenedesmus obliquus*. *Physiologia Plantarum*, 91, 729–34.

Compton, T.J., De Winton, M., Leathwick, J.R., and Wadhwa, S. (2012). Predicting spread of invasive macrophytes in New Zealand lakes using indirect measures of human accessibility. *Freshwater Biology*, 57, 938–48.

Conran, J.G. (1996). The embryology and relationships of the Byblidaceae. *Australian Systematic Botany*, 9, 243–54.

Conran, J.G. (2004a). Cephalotaceae. In K. Kubitzki, ed. *The Families and Genera of Vascular Plants*, volume 6, pp. 65–68. Springer, Berlin.

Conran, J.G. (2004b). Roridulaceae. In K. Kubitzki, ed. *The Families and Genera of Vascular Plants*, volume 6, pp. 339–42. Springer, Berlin.

Conran, J.G. and Carolin, R. (2004). Byblidaceae. In J.W. Kadereit, ed. *The Families and Genera of Vascular Plants*, volume 7, pp. 45–49. Springer, Berlin.

Conran, J. and Denton, M. (1996). Germination in the Western Australian pitcher plant *Cephalotus follicularis* and its unusual early seedling development. *Western Australian Naturalist*, 21, 37–42.

Conran, J.G. and Dowd, J.M. (1993). The phylogenetic relationships of *Byblis* and *Roridula* (Byblidaceae–Roridulaceae) inferred from partial 18S RNA ribosomal sequences. *Plant Systematics and Evolution*, 188, 73–86.

Conran, J.G., Lowrie, A., and Moyle-Croft, J. (2002). A revision of *Byblis* (Byblidaceae) in south-western Australia. *Nuytsia*, 15, 11–19.

Conran, J.G., Jaudzems, G., and Hallam, N.D. (2007). Droseraceae gland and germination patterns revisited: support for recent molecular phylogenetic studies. *Carnivorous Plant Newsletter*, 36, 14–20.

Cooper, G.P. and Record, S.J. (1931). The evergreen forests of Liberia. *Yale University School of Forestry Bulletin*, 31, 27–142.

Corbet, P.S. (1983). Odonata in phytotelmata. In J.H. Frank and L.P. Lounibos, eds. *Phytotelmata: Terrestrial Plants as Hosts for Aquatic Insect Communities*, pp. 29–54. Plexus Publishing, Medford.

Coritico, F.P. and Fleischmann, A. (2016). The first record of the boreal bog species *Drosera rotundifolia* L. (Droseraceae) from the Philippines, and a key to the Philippine sundews. *Blumea*, 61, 24–28.

Corker, B. (1986). Germination and viability of seeds of the pitcher plant, *Nepenthes mirabilis* Druce. *Malaysian Naturalist Journal*, 39, 259–64.

Cornelissen, J.H.C., Lavorel, S., Garnier, E., et al. (2003). A handbook of protocols for standardized and easy measurement of plant functional traits worldwide. *Australian Journal of Botany*, 51, 335–80.

Corner, E.J.H. (1976). *The Seeds of Dicotyledons. Volume 1.* Cambridge University Press, London.

Correia, E. and Freitas, H. (2002). *Drosophyllum lusitanicum*, an endangered West Mediterranean endemic carnivorous plant: threats and its ability to control available resources. *Botanical Journal of the Linnean Society*, 140, 383–90.

Cowan, I.R. and Farquhar, G.D. (1977). Stomatal function in relation to leaf metabolism and environment. In D.H. Jennings, ed. *Integration of Activity in the Higher Plant*, pp. 471–505. Cambridge University Press, Cambridge.

Cowling, R. and Richardson, D. (1996). *Fynbos: South Africa's Unique Floral Kingdom*. Fernwood Press, Vlaeberg.

Cramer, V.A. and Hobbs, R.J. (2002). Ecological consequences of altered hydrological regimes in fragmented ecosystems in southern Australia: impacts and possible management responses. *Austral Ecology*, 27, 546–64.

Crawford, R.M.M. (1989). *Studies in Plant Survival: An Ecophysiological Investigation of Plant Distribution*. Blackwell Scientific Publications, Oxford.

Cresswell, J.E. (1991). Capture rates and composition of insect prey of the pitcher plant *Sarracenia purpurea*. *The American Midland Naturalist*, 125, 1–9.

Cresswell, J.E. (1993). The morphological correlates of prey capture and resource parasitism in pitchers of the carnivorous plant *Sarracenia purpurea*. *American Midland Naturalist*, 129, 35–41.

Cresswell, J.E. (1998). Morphological correlates of necromass accumulation in the traps of an Eastern tropical pitcher plant, *Nepenthes ampullaria* Jack, and observations on the pitcher infauna and its reconstitution following experimental removal. *Oecologia*, 113, 383–90.

Cresswell, J.E. (2000). Resource input and the community structure of larval infaunas of an eastern tropical pitcher plant *Nepenthes bicalcarata*. *Ecological Entomology*, 25, 362–66.

Croizat, L. (1960). *Principia Botanica, or Beginnings of Botany*. Published by the author, Caracas.

Cronquist, A. (1981). *An Integrated System of Classification of Flowering Plants*. Columbia University Press, New York.

Cronquist, A. (1988). *The Evolution and Classification of Flowering Plants*. New York Botanic Gardens, New York.

Cross, A (2012a). Aldrovanda. *The Waterwheel Plant*. Redfern Natural History Production, Poole, Dorset.

Cross, A. (2012b). *Aldrovanda vesiculosa*. The IUCN Red List of Threatened Species 2012, e.T162346A901031.

Cross, A.T. (2013). Turion development is an ecological trait in all populations of the aquatic carnivorous plant *Aldrovanda vesiculosa*. *Carnivorous Plant Newsletter*, 42, 1–3.

Cross, A.T., Adamec, L., Turner, S.R., Dixon, K.W., and Merritt, D.J. (2016). Seed reproductive biology of the rare aquatic carnivorous plant *Aldrovanda vesiculosa* (Droseraceae). *Botanical Journal of the Linnean Society*, 180, 515–29.

Cross, A.T., Cawthray, G.R., Merritt, D.J., et al. (2014). Biogenic ethylene promotes seedling emergence from the sediment seed bank in an ephemeral tropical rock pool habitat. *Plant and Soil*, 380, 73–87.

Cross, A.T., Merritt, D.J., Turner, S.R., and Dixon, K.W. (2013). Seed germination of the carnivorous plant *Byblis gigantea* (Byblidaceae) is cued by warm stratification and karrikinolide. *Botanical Journal of the Linnean Society*, 173, 143–52.

Cross, A.T., Skates, L.M., Adamec, L., et al. (2015). Population dynamics of the endangered aquatic carnivorous macrophyte *Aldrovanda vesiculosa* at a naturalised site in North America. *Freshwater Biology*, 60, 1772–83.

Crouch, I.J., Finnie, J.F., and van Staden, J. (1990). Studies on the isolation of plumbagin from in vitro and in vivo grown *Drosera* species. *Tissue and Organ Culture*, 21, 79–82.

Crowder, A.A, Pearson, M.C., Grubb, P.J., and Langlois, P.H. (1990). Biological flora of the British Isles, 167. *Drosera* L. *Journal of Ecology*, 78, 233–67.

Crowley, P.H., Hopper, K.R., and Krupa, J.J. (2013). An insect-feeding guild of carnivorous plants and spiders: does optimal foraging lead to competition or facilitation? *American Naturalist*, 182, 801–19.

Cruden, R.W. (1977). Pollen-ovule ratios: a conservative indicator of breeding systems in flowering plants. *Evolution*, 31, 32–46.

Cui, H., Gobbato, E., Kracher, B., et al. (2016). A core function of EDS1 with PAD4 is to protect the salicylic acid defense sector in *Arabidopsis* immunity. *New Phytologist*, 213, 1802–17.

Culham, A. and Gornall, R. J. (1994). The taxonomic significance of naphthoquinones in the Droseraceae. *Biochemical Systematics and Ecology*, 22, 507–15.

Cuénoud, P., Savolainen, V., Chatrou, L.W., et al. (2002). Molecular phylogenetics of Caryophyllales based on nuclear 18S rDNA and plastid *rbcL*, *atpB*, and *matK* DNA sequences. *American Journal of Botany*, 89, 132–44.

Czaja, A.T. (1924). Reizphysiologische Untersuchungen an *Aldrovanda vesiculosa* L. *Pflüger's Archiv für die gesamte Physiologie des Menschen und der Tiere*, 206, 635–58.

D'Amato, P. (1998). *The Savage Garden*. Ten Speed Press, Berkeley.

D'Amato, P. (2013). *The Savage Garden, Revised*. Ten Speed Press, Berkeley.

Dahl, T.E. (2011). *Status and Trends of Wetlands in The Conterminous United States 2004–2009*. U.S. Department of the Interior. U.S. Fish and Wildlife Service, Washington, DC.

Dahl, T.E. and Johnson, C.E. (1991). *Status and Trends of Wetlands in The Conterminous United States, mid-1970s to mid-1980s*. U.S. Department of the Interior. U.S. Fish and Wildlife Service, Washington, DC.

Dahlem, G.A. and Naczi, R.F.C. (2006). Flesh flies (Diptera: Sarcophagidae) associated with North American pitcher plants (Sarraceniaceae), with descriptions of three new species. *Annals of the Entomological Society of America*, 99, 218–40.

Dahlgren, R.M.T. (1980). A revised system of classification of the angiosperms. *Botanical Journal of the Linnean Society*, 80, 91–124.

Danser, B.H. (1928). Nepenthaceae of the Netherlands' Indies. *Bulletin du Jardin de Botanique, Buitenzorg. Série III*, 9, 249–438.

Darnowski, D.W. (2016). Attraction of preferred prey by carnivorous plants. In J. Seckbach and R. Gordon, eds. *Biocommunication: Sign-Mediated Interactions between Cells and Organisms*, pp. 309–25. World Scientific Publishing, London.

Darnowski, D.W., Carroll, D.M., Płachno, B.J., Kabanoff, E., and Cinnamon, E. (2006). Evidence of protocarnivory in triggerplants (*Stylidium* spp.; Stylidiaceae). *Plant Biology*, 8, 805–12.

Darnowski, D.W. and Fritz, S. (2010). Prey preference in *Genlisea*. Small crustaceans, not protozoa. *Carnivorous Plant Newsletter*, 39, 114–16.

Darwin, C. (1871). *The Descent of Man, and Selection in Relation to Sex*. John Murray, London.

Darwin, C. (1875). *Insectivorous Plants*. John Murray, London.

Darwin, F. (1878). Experiments on the nutrition of *Drosera rotundifolia*. *Journal of the Linnean Society—Botany*, 17, 17–32.

Das, I. and Haas, A. (2010). New species of *Microhyla* from Sarawak: Old World's smallest frogs crawl out of miniature pitcher plants on Borneo (Amphibia: Anura: Microhylidae). *Zootaxa*, 2571, 37–52.

Davis, M.B. (1989). Insights from paleoecology on global change. *Bulletin of the Ecological Society of America*, 70, 222–28.

Dawson, T.P., Jackson, S.T., House, J.I., Prentice, I.C., and Mace, G.M. (2011). Beyond predictions: biodiversity conservation in a changing climate. *Science*, 332, 53–58.

de Almeida, A.M.R., Yockteng, R., Schnable, J., et al. (2014). Co-option of the polarity gene network shapes filament morphology in angiosperms. *Scientific Reports*, 4, 6194.

De Castro, O., Innangi, M., Di Maio, A., et al. (2016). Disentangling phylogenetic relationships in a hotspot of diversity: The butterworts (*Pinguicula* L., Lentibulariaceae) endemic to Italy. *PLoS ONE*, 11, e0167610.

DeBuhr, L. (1975a). Observations on *Byblis gigantea* in southwestern Australia. *Carnivorous Plant Newsletter*, 4, 60–63.

DeBuhr, L.E. (1975b). Phylogenetic relationships of the Sarraceniaceae. *Taxon*, 24, 297–306.

de Candolle, A.P. (1844). *Prodromus Systematis Naturalis Regni Vegetabilis*, 8. Fortin, Masson et Sociorum, Paris.

Degreef, J.D. (1997). Fossil *Aldrovanda*. *Carnivorous Plant Newsletter*, 26, 93–97.

Degtjareva, G.V., Casper, S.J., Hellwig, F.H., et al. (2006). Morphology and nrITS phylogeny of the genus *Pinguicula* L. (Lentibulariaceae), with special attention to embryo evolution. *Plant Biology*, 8, 778–90.

Degtjareva, G.V., Casper, S.J., Hellwig, F.H., and Sokoloff, D.D. (2004). Seed morphology in the genus *Pinguicula* (Lentibulariaceae) and its relation to taxonomy and phylogeny. *Botanische Jahrbücher für Systematik*, 125, 431–52.

Degtjareva, G.V. and Sokoloff, D.D. (2012). Inflorescence morphology and flower development in *Pinguicula alpina* and *P. vulgaris* (Lentibulariaceae: Lamiales): monosymmetric flowers are always lateral and occurrence of early sympetaly. *Organisms Diversity & Evolution*, 12, 99–111.

Department of Environment and Conservation (2012a). *A Guide to Managing and Restoring Wetlands in Western Australia*. Department of Environment and Conservation, Perth.

Department of Environment and Conservation, (2012b). *Fitzgerald Biosphere Recovery Plan: A Landscape Approach to Threatened Species and Ecological Communities Recovery and Biodiversity Conservation*. Western Australian Department of Environment and Conservation, Albany.

Deppe, J.L., Dress, W.J., Nastase, A.J., Newell, S.J., and Luciano, C.S. (2000). Diel variation of sugar amount in nectar from pitchers of *Sarracenia purpurea* L. with and without insect visitors. *American Midland Naturalist*, 144, 123–32.

Depuy, N.C. and Dreyfus, B.L. (1992). *Bradyrhizobium* populations occur in deep soil under the leguminous tree *Acacia albida*. *Applied Environmental Microbiology*, 58, 2415–19.

Devi, S.P., Kumaria, S., Rao, S.R., and Tandon, P. (2013). In vitro propagation and assessment of clonal fidelity of *Nepenthes khasiana* Hook. f.: a medicinal insectivorous plant of India. *Acta Physiologiae Plantarum*, 35, 2813–20.

Devi, S.P., Kumaria, S., Rao, S.R., and Tandon, P. (2015). Genetic fidelity assessment in micropropagated plants using cytogenetical analysis and heterochromatin distribution: a case study with *Nepenthes khasiana* Hook f. *Protoplasma*, 252, 1305–12.

Dhamecha, D., Jalalpure, S., and Jadhav, K. (2016). *Nepenthes khasiana* mediated synthesis of stabilized gold nanoparticles: characterization and biocompatibility studies. *Journal of Photochemistry and Photobiology B*, 154, 108–17.

Díaz-Olarte, J., Valoyes-Valois, V., Guisande, C., et al. (2007). Periphyton and phytoplankton associated with the tropical carnivorous plant *Utricularia foliosa*. *Aquatic Botany*, 87, 285–91.

Di Giusto, B., Bessiere, J.M., Gueroult, M., et al. (2010). Flower-scent mimicry masks a deadly trap in the carnivorous plant *Nepenthes rafflesiana*. *Journal of Ecology*, 98, 845–56.

Di Giusto, B., Grosbois, V., Fargeas, E., Marshall, D.J., and Gaume, L. (2008). Contribution of pitcher fragrance and fluid viscosity to high prey diversity in a *Nepenthes* carnivorous plant from Borneo. *Journal of Biosciences*, 33, 121–36.

Didry, N., Dubreuil, L., Trotin, F., and Pinkas, M. (1998). Antimicrobial activity of aerial parts of *Drosera peltata* Smith on oral bacteria. *Journal of Ethnopharmacology*, 60, 91–96.

Diels, L. (1906). Droseraceae. In A. Engler, ed. *Das Pflanzenreich 26, 4*, pp. 11–36. Wilhelm Engelmann, Leipzig.

Diels, L. (1928). Cephalotaceae. In A. Engler and K. Prantl, eds. *Die natürlichen Pflanzenfamilien*, 2nd edition, *18a*, pp. 71–74. Engelmann, Leipzig.

Ding, Z., Wang, B., Moreno, I., et al. (2012). ER-localized auxin transporter PIN8 regulates auxin homeostasis and male gametophyte development in *Arabidopsis*. *Nature Communications*, 3, 941.

Diniz-Filho, J.A., Bini, L.M., Rangel, T.F., et al. (2009). Partitioning and mapping uncertainties in ensembles of forecasts of species turnover under climate change. *Ecography*, 32, 897–906.

DiPalma, J.R., McMichael, R., and DiPalma, M. (1966). Touch receptor of Venus Flytrap, *Dionaea muscipula*. *Science*, 152, 539–40.

Dixon, P.M., Ellison, A.M., and Gotelli, N.J. (2005). Improving the precision of estimates of the frequency of rare events. *Ecology*, 86, 1114–23.

Dixon, K.W. and Pate, J.S. (1978). Phenology, morphology, and reproductive biology of the tuberous sundew, *Drosera erythrorhiza* Lindl. *Australian Journal of Botany*, 26, 441–54.

Dixon, K.W., Pate, J.S., and Bailey, W.J. (1980). Nitrogen nutrition of the tuberous sundew *Drosera erythrorhiza* Lindl. with special reference to catch of arthropod fauna by its glandular leaves. *Australian Journal of Botany*, 28, 283–97.

Dobrowski, S.Z., Thorne, J.H., Greenberg, J.A., et al. (2011). Modeling plant ranges over 75 years of climate change in California, USA: temporal transferability and species traits. *Ecological Monographs*, 81, 241–57.

Dolling, W.R. and Palmer, J.M. (1991). *Pameridea* (Hemiptera: Miridae): predaceous bugs specific to the highly viscid plant genus *Roridula*. *Systematic Entomology*, 16, 319–28.

Domin, K. (1922). Byblidaceae: A new archichlamydeous family. *Acta Botanica Bohemica*, 1, 3–4.

Domínguez, Y., da Silva, S.R., Panfet Valdés, C.M., and de Miranda, V.F.O. (2014). Inter- and intra-specific diversity of Cuban *Pinguicula* (Lentibulariaceae) based on morphometric analyses and its relation with geographical distribution. *Plant Ecology & Diversity*, 7, 519–31.

Dormann, C.F. (2007). Promising the future? Global change projections of species distributions. *Basic and Applied Ecology*, 8, 387–97.

dos Santos, T.R., Ferragut, C., and de Mattos Bicudo, C.E. (2013). Does macrophyte architecture influence periphyton? Relationships among *Utricularia foliosa*, periphyton assemblage structure and its nutrient (C, N, P) status. *Hydrobiologia*, 714, 71–83.

Dover, C., Fage, L., Hirst, S., et al. (1928). Notes on the fauna of pitcher-plants from Singapore Island. *Journal of the Malaysian Branch of the Royal Asiatic Society*, 6, 1–27.

Doxey, A.C., Yaish, M.W.F., Moffatt, B.A., Griffith, M., and McConkey, B.J. (2007). Functional divergence in the *Arabidopsis* β-1,3-glucanase gene family inferred by phylogenetic reconstruction of expression states. *Molecular Biology and Evolution*, 24, 1045–55.

Drenovsky, R.E., Khasanova, A., and James, J.J. (2012). Trait convergence and plasticity among native and invasive species in resource–poor environments. *American Journal of Botany*, 99, 629–39.

Dress, W.J., Newell, S.J., and Ford, J.C. (1997). Analysis of amino acids in nectar from pitchers of *Sarracenia purpurea* (Sarraceniaceae). *American Journal of Botany*, 84, 1701–06.

Dressler, S. and Bayer, C. (2004). Actinidiaceae. In K. Kubitzki, ed. *The Families and Genera of Vascular Plants*, volume 6, pp. 14–19. Springer, Berlin.

Durand, R. and Zenk, M.H. (1974). The homogenisate ring-cleavage pathway in the biosynthesis of acetate-derived naphthoquinones of the *Droseraceae*. *Phytochemistry*, 13, 1483–92.

Dwyer, T.P. (1983). Seed structure of carnivorous plants. *Carnivorous Plant Newsletter*, 12, 8–23.

Dybzinski, R. and Tilman, D. (2007). Resource use patterns predict long-term outcomes of plant competition for nutrients and light. *American Naturalist*, 170, 305–18.

Dynesius, M. and Jansson, R. (2000). Evolutionary consequences of changes in species' geographical distributions driven by Milankovitch climate oscillations. *Proceedings of the National Academy of Sciences, USA*, 97, 9115–20.

Eckardt, N.A. (2010). YABBY genes and the development and origin of seed plant leaves. *The Plant Cell*, 22, 2103.

Eckstein, R.L. and Karlsson, P.S. (2001). The effect of reproduction on nitrogen use–efficiency of three species of the carnivorous genus *Pinguicula*. *Journal of Ecology*, 89, 798–806.

Edgar, R.C. (2013). UPARSE: highly accurate OTU sequences from microbial amplicon reads. *Nature Methods*, 10, 996–98.

Edwards, E.J., de Vos, J.M., and Donoghue, M.J. (2015). Doubtful pathways to cold tolerance in plants. *Nature*, 521, E5–E6.

Egan, P.A. and van der Kooy, F. (2012). Coproduction and ecological significance of naphthoquinones in carnivorous sundews (*Drosera*). *Chemistry and Biodiversity*, 9, 1033–44.

Egan, P.A. and van der Kooy, F. (2013). Phytochemistry of the carnivorous sundew genus *Drosera* (*Droseraceae*)—future perspectives and ethnopharmacological relevance. *Chemistry and Biodiversity*, 10, 1774–90.

Ehrlich, P.R. and Raven, P.H. (1964). Butterflies and plants: a study in coevolution. *Evolution*, 18, 586–608.

Eilenberg, H., Pnini-Cohen, S., Rahamim, Y., et al. (2010). Induced production of antifungal naphthoquinones in the pitchers of the carnivorous plant *Nepenthes khasiana*. *Journal of Experimental Botany*, 61, 911–22.

Eilenberg, H., Pnini-Cohen, S., Schuster, S., Movtchan, A., and Zilberstein, A. (2006). Isolation and characterization of chitinase genes from pitchers of the carnivorous plant *Nepenthes khasiana*. *Journal of Experimental Botany*, 57, 2775–84.

El-Sayed, A.M., Byers, J.A. and Suckling, D M. (2016). Pollinator–prey conflicts in carnivorous plants: when flower and trap properties mean life or death. *Scientific Reports*, 6, e21065.

Elansary, H., Adamec, L., and Štorchová, H. (2010). Uniformity of organellar DNA in *Aldrovanda vesiculosa*, an endangered aquatic carnivorous species, distributed across four continents. *Aquatic Botany*, 92, 214–20.

Elith, J. and Leathwick, J.R. (2009). Species distribution models: ecological explanation and prediction across space and time. *Annual Review of Ecology, Evolution, and Systematics*, 40, 677–97.

Elith, J., Phillips, S.J., Hastie, T., et al. (2011). A statistical explanation of MaxEnt for ecologists. *Diversity and Distributions*, 17, 43–57.

Ellis, J. (1770). *Directions for Bringing over Seeds and Plants, from the East Indies and Other Distant Countries, in a State of Vegetation: Together with a Catalogue of Such Foreign Plants as Are Worthy of Being Encouraged in Our American Colonies, for the Purposes of Medicine, Agriculture, and Commerce. To Which is Added, the Figure and Botanical Description of a New Sensitive Plant, Called Dionaea muscipula: or, Venus's Fly-trap*. L. Davis, London.

Ellis, A.G. and Midgley, J.J. (1996). A new plant–animal mutualism involving a plant with sticky leaves and a resident hemipteran. *Oecologia*, 106, 478–81.

Ellison, A.M. (2001). Interspecific and intraspecific variation in seed size and germination requirements of *Sarracenia* (Sarraceniaceae). *American Journal of Botany*, 88, 429–37.

Ellison, A.M. (2006). Nutrient limitation and stoichiometry of carnivorous plants. *Plant Biology*, 8, 740–47.

Ellison, A.M. (2014). They really do eat insects: learning from Charles Darwin's experiments with carnivorous plants. In M.J. Reiss, C.J. Boulter, and D.L Sanders, eds. *Darwin-Inspired Learning*, pp. 243–56. Sense Publishers, Rotterdam.

Ellison, A.M. (2016). It's time to get real about conservation. *Nature*, 538, 141.

Ellison, A.M. and Adamec, L. (2011). Ecophysiological traits of terrestrial and aquatic carnivorous plants: are the costs and benefits the same? *Oikos*, 120, 1721–31.

Ellison, A.M., Buckley, H.L., Miller, T.E., and Gotelli, N.J. (2004). Morphological variation in *Sarracenia purpurea* (Sarraceniaceae): geographic, environmental, and taxonomic correlates. *American Journal of Botany*, 91, 1930–35.

Ellison, A.M., Butler, E.D., Hicks, E.J., et al. (2012). Phylogeny and biogeography of the carnivorous plant family Sarraceniaceae. *PLoS ONE*, 7, e39291.

Ellison, A.M., Davis, C.C., Calie, P.J. and Naczi, R.F.C. (2014). Pitcher plants (*Sarracenia*) provide a 21st-century perspective on infraspecific ranks and interspecific hybrids: a modest proposal for appropriate recognition and usage. *Systematic Botany*, 39, 939–49.

Ellison, A.M. and Farnsworth, E.J. (2005). The cost of carnivory for *Darlingtonia californica* (Sarraceniaceae): evidence from relationships among leaf traits. *American Journal of Botany*, 92, 1085–93.

Ellison, A.M. and Gotelli, N.J. (2001). Evolutionary ecology of carnivorous plants. *Trends in Ecology and Evolution*, 16, 623–29.

Ellison, A.M. and Gotelli, N.J. (2002). Nitrogen availability alters the expression of carnivory in the northern pitcher plant, *Sarracenia purpurea*. *Proceedings of the National Academy of Sciences, USA*, 99, 4409–12.

Ellison, A.M. and Gotelli, N.J. (2009). Energetics and the evolution of carnivorous plants—Darwin's "most wonderful plants in the world." *Journal of Experimental Botany*, 60, 19–42.

Ellison, A.M., Gotelli, N.J., Brewer, J.S., et al. (2003). The evolutionary ecology of carnivorous plants. *Advances in Ecological Research*, 33, 1–74.

Ellison, A.M. and Gotelli, N.J. (in press). *Scaling Sarracenia*. Princeton University Press, Princeton.

Ellison, A.M. and Parker, J.N. (2002). Seed dispersal and seedling establishment of *Sarracenia purpurea* (Sarraceniaceae). *American Journal of Botany*, 89, 1024–26.

Endress, P.K. and Stumpf, S. (1991). The diversity of stamen structures in "lower" Rosidae (Rosales, Fabales, Proteales, Sapindales). *Botanical Journal of the Linnean Society*, 107, 217–93.

Engler, A. (1897). Brunelliaceae. In A. Engler and K. Prantl, eds. *Die natürlichen Pflanzenfamilien, Nachtrag zu Teil III*, p. 184. Engelmann, Leipzig.

Englund, G. and Harms, S. (2003). Effects of light and microcrustacean prey on growth and investment in carnivory in *Utricularia vulgaris*. *Freshwater Biology*, 48, 786–94.

Erber, D. (1979). Untersuchungen zur Biozönose und Nekrozönose in Kannenpflanzen auf Sumatra. *Archiv für Hydrobiologie*, 87, 37–48.

Erickson, R. (1978). *Plants of Prey in Australia*. University of Western Australia Press, Nedlands.

Erni, P., Varagnat, M., Clasen, C., Crest, J. and McKinley, G.H. (2011). Microrheometry of sub-nanolitre biopolymer samples: non-Newtonian flow phenomena of carnivorous plant mucilage. *Soft Matter*, 7, 10889–98.

Escalante-Pérez, M., Scherzer, S., Al-Rasheid, K.A., et al. (2014). Mechano-stimulation triggers turgor changes associated with trap closure in the Darwin plant *Dionaea muscipula*. *Molecular Plant*, 7, 744–46.

Escalante-Pérez, M., Król E., Stange, A., et al. (2011). A special pair of phytohormones controls excitability, slow closure, and external stomach formation in the Venus flytrap. *Proceedings of the National Academy of Sciences, USA*, 108, 15492–97.

Evans, J.R. (1989). Photosynthesis and nitrogen relationships in leaves of C3 plants. *Oecologia*, 78, 9–19.

Exell, A.W. and Laundon, J.R. (1956). New and noteworthy species of *Drosera* from Africa and Madagascar. *Boletim da Sociedade Broteriana*, 30, 213–20.

Fabian-Galan, G. and Salageanu, N. (1968). Considerations on the nutrition of certain carnivorous plants (*Drosera capensis* and *Aldrovanda vesiculosa*). *Review Roumain de Biologie, Botany*, 13, 275–80.

Faegri, K. and van der Pijl, L. (1979). *The Principles of Pollination Ecology*, 3rd edition. Pergamon Press, Oxford.

Fagerberg, W.R. and Allain, D. (1991). A quantitative study of tissue dynamics during closure in the traps of Venus's flytrap *Dionaea muscipula* Ellis. *American Journal of Botany*, 78, 647–57.

Fagerberg, W.R. and Howe, D.G. (1996). A quantitative study of tissue dynamics in Venus's flytrap *Dionaea muscipula* (Droseraceae). II. Trap reopening. *American Journal of Botany*, 83, 836–42.

Fagerstrom, T. (1989). Anti-herbivory chemical defense in plants—a note on the concept of cost. *American Naturalist*, 133, 281–87.

Farnsworth, E. (2000). The ecology and physiology of viviparous and recalcitrant seeds. *Annual Review of Ecology*, 31, 107–38.

Farnsworth, E.J. and Ellison, A.M. (2008). Prey availability directly affects physiology, growth, nutrient allocation and scaling relationships among leaf traits in ten carnivorous plant species. *Journal of Ecology*, 96, 213–21.

Fasbender, L., Maurer, D., Kreuzwieser, J., et al. (2017). The carnivorous Venus flytrap uses prey-derived amino acid carbon to fuel respiration. *New Phytologist*, 214, 597–606.

Fashing, N.J. (2002). *Nepenthacarus*, a new genus of Histiostomatidae (Acari: Astigmata) inhabiting the pitchers of *Nepenthes mirabilis* (Lour.) Druce in far north Queensland, Australia. *Australian Journal of Entomology*, 41, 7–17.

Fashing, N.J. and Chua, T.H. (2002). Systematics and ecology of *Naiadacarus nepenthicola*, a new species of Acaridae (Acari: Astigmata) inhabiting the pitchers of *Nepenthes bicalcarata* Hook. f. in Brunei Darussalam. *International Journal of Acarology*, 28, 157–67.

Fashing, N.J. and OConnor, B.M. (1984). *Sarraceniopus*—a new genus for histiostomatid mites inhabiting the pitchers of the Sarraceniaceae (Astigmata: Histiostomatidae). *International Journal of Acarology*, 10, 217–27.

Faust, K. and Raes, J. (2012). Microbial interactions: from networks to models. *Nature Reviews Microbiology*, 10, 538–50.

Faust, K., Sathirapongsasuti, J.F., Izard, J., et al. (2012). Microbial co-occurrence relationships in the human microbiome. *PLoS Computational Biology*, 8, e1002606.

Federle, W., Maschwitz, U., Fiala, B., Riederer, M., and Hölldobler, B. (1997). Slippery ant-plants and skillful climbers: selection and protection of specific ant partners by epicuticular wax blooms in *Macaranga* (Euphorbiaceae). *Oecologia*, 112, 217–24.

Fedotov, V.V. (1982). [The genus *Dioncophyllites* in the Eocene flora of Raitschikha from the Amur district.] *Botanicheskii Zhurnal*, 67, 985–87. (In Russian).

Feeley, K.J. and Silman, M.R. (2011). Keep collecting: accurate species distribution modelling requires more collections than previously thought. *Diversity and Distributions*, 17, 1132–40.

Feild, T.S., Sage, T.L., Czerniak, C., and Iles, W.J.D. (2005). Hydathodal leaf teeth of *Chloranthus japonicas* (Chloranthaceae) prevent guttation-induced flooding of the mesophyll. *Plant, Cell and Environment*, 28, 1179–90.

Fejfarová, K., Kádek, A., Mrázek, H., et al. (2016). Crystallization of nepenthesin I using a low-pH crystallization screen. *Acta Crystallographica Section F: Structural Biology Communications*, 72, 24–28.

Feller, U., Anders, I., and Mae, T. (2008). Rubiscolytics: fate of Rubisco after its enzymatic function in a cell is terminated. *Journal of Experimental Botany*, 59, 1615–24.

Fennane, M. and Ibn Tattou, M. (1998). Catalogue des plantes vasculaires rares, menacées ou endémiques du Maroc. *Bocconea*, 8, 1–243.

Fertig, B. (2001). *Importance of Prey Derived and Absorbed Nitrogen to New Growth; Preferential Uptake of Ammonia or Nitrate for Three Species of* Utricularia. M.Sc. thesis, Brandeis University, Waltham.

Fierer, N. and Jackson, R.B. (2006). The diversity and biogeography of soil bacterial communities. *Proceedings of the National Academy of Sciences, USA*, 103, 626–31.

Figueira, J.E.C., Vasconcellos-Neto, J., and Jolivet, P. (1994). Une nouvelle plante protocarnivore *Paepalanthus bromelioides* Silv. (Eriocaulaceae) du Brésil. *Revue d'Écologie*, 49, 3–9.

Fineran, B.A. (1985). Glandular trichomes in *Utricularia*: a review of their structure and function. *Israel Journal of Botany*, 34, 295–330.

Fineran, B.A. and Gilbertson, J.M. (1980). Application of lanthanum and uranyl salts as tracers to demonstrate apoplastic pathways for transport in glands of the carnivorous plant *Utricularia monanthos*. *European Journal of Cell Biology*, 23, 66–72.

Fisahn, J., Herde, O., Willmitzer, L., and Peña-Cortés, H. (2004). Analysis of the transient increase in cytosolic Ca^{2+} during the action potential of higher plants with high temporal resolution: requirement of Ca^{2+} transients for induction of jasmonic acid biosynthesis and PINII gene expression. *Plant and Cell Physiology*, 45, 456–59.

Fischer, E., Porembski, S., and Barthlott, W. (2000). Revision of the genus *Genlisea* (Lentibulariaceae) in Africa and Madagascar with notes on ecology and phytogeography. *Nordic Journal of Botany*, 20, 291–318.

Fish, D. (1976). *Structure and Composition of the Aquatic Invertebrate Community Inhabiting Epiphytic Bromeliads in South Florida and the Discovery of an Insectivorous Bromeliad*. Ph.D. dissertation, University of Florida, Gainesville.

Fish, D. and Hall, D.W. (1978). Succession and stratification of aquatic insects inhabiting the leaves of the insectivorous pitcher plant, *Sarracenia purpurea*. *American Midland Naturalist*, 99, 172–83.

Fitzpatrick, M.C., Gove, A.D., Sanders, N.J., and Dunn, R.R. (2008). Climate change, plant migration, and range collapse in a global biodiversity hotspot: the *Banksia* (Proteaceae) of Western Australia. *Global Change Biology*, 14, 1337–52.

Fitzpatrick, M. and Hargrove, W. (2009). The projection of species distribution models and the problem of no-analog climate. *Biodiversity and Conservation*, 18, 2255–61.

Fitzpatrick, M.C. and Keller, S.R. (2015). Ecological genomics meets community-level modelling of biodiversity: mapping the genomic landscape of current and future environmental adaptation. *Ecology Letters*, 18, 1–16.

Fitzpatrick, M.C., Weltzin, J.F., Sanders, N.J., and Dunn, R.R. (2007). The biogeography of prediction error: why does the introduced range of the fire ant over-predict its native range? *Global Ecology and Biogeography*, 16, 24–33.

Fleischmann, A. (2010). Evolution of carnivorous plants. In S. McPherson, ed. *Carnivorous plants and their habitats*, pp. 68–123. Redfern Natural History Productions, Poole.

Fleischmann, A. (2011a). Do we have any evidence that any plants have given up carnivory? *Carnivorous Plant Newsletter*, 40, 37.

Fleischmann, A. (2011b). *Phylogenetic Relationships, Systematics, and Biology of Carnivorous Lamiales, with Special Focus on The Genus* Genlisea *(Lentibulariaceae)*. Ph.D. dissertation, University of Munich, Munich.

Fleischmann, A. (2012a). *Monograph of the Genus* Genlisea. Redfern Natural History Productions, Poole.

Fleischmann, A. (2012b). *Philcoxia*—a new genus of carnivorous plant. *Carnivorous Plant Newsletter*, 41, 77–81.

Fleischmann, A. (2015a). Taxonomic *Utricularia* news. *Carnivorous Plant Newsletter*, 44, 13–16.

Fleischmann, A. (2015b). The intricate *Pinguicula crystallina/hirtiflora* complex. *Carnivorous Plant Newsletter*, 44, 48–61.

Fleischmann, A. (2016a). Olfactory prey attraction in *Drosera*? *Carnivorous Plant Newsletter*, 45, 19–25.

Fleischmann, A. (2016b). *Pinguicula* flowers with pollen imitations close at night—some observations on butterwort flower biology. *Carnivorous Plant Newsletter*, 45, 84–92.

Fleischmann, A., Costa, S.M., Bittrich, V., Amaral, M.C.E., and Hopkins, M. (2017). A new species of corkscrew plant (*Genlisea*, Lentibulariaceae) from the Amazon lowlands of Brazil, including a key to all species occurring north of the Amazon River. *Phytotaxa*, 319, 289–287.

Fleischmann, A., Gibson, R., and Rivadavia, F. (2008). Notes of African plants: Droseraceae: *Drosera ericgreenii*, a new species from the fynbos of South Africa. *Bothalia*, 38, 141–44.

Fleischmann, A. and McPherson, S. (2010). Some ecological notes on *Heliamphora* (Sarraceniaceae) from Ptari-tepui. *Carniflora Australis*, 7, 19–31.

Fleischmann, A., Michael, T.P., Rivadavia, F., et al. (2014). Evolution of genome size and chromosome number in the carnivorous plant genus *Genlisea* (Lentibulariaceae), with a new estimate of the minimum genome size in angiosperms. *Annals of Botany*, 114, 1651–63.

Fleischmann, A., Rivadavia, F., Gonella, P.M., et al. (2016). Where is my food? Brazilian flower fly steals prey from carnivorous sundews in a newly discovered plant–animal interaction. *PLoS ONE*, 11, e0153900.

Fleischmann, A., Rivadavia, F., Gonella, P.M., and Heubl, G. (2011a). A revision of *Genlisea* subgenus *Tayloria* (Lentibulariaceae). *Phytotaxa*, 33, 1–40.

Fleischmann, A., Robinson, A.S., McPherson, S., et al. (2011b). *Drosera ultramafica* (Droseraceae), a new sundew species of the ultramafic flora of the Malesian highlands. *Blumea*, 56, 10–15.

Fleischmann, A., Schäferhoff, B., Heubl, G., et al. (2010). Phylogenetics and character evolution in the carnivorous plant genus *Genlisea* A. St.-Hil. (Lentibulariaceae). *Molecular Phylogenetics and Evolution*, 56, 768–83.

Fleischmann, A., Wistuba, A., and McPherson, S. (2007). *Drosera solaris* (Droseraceae), a new sundew from the Guayana Highlands. *Willdenowia*, 37, 551–55.

Fleischmann, A., Wistuba, A., and Nerz, J. (2009). Three new species of *Heliamphora* (Sarraceniaceae) from the Guayana Highlands of Venezuela. *Willdenowia*, 39, 273–83.

Folkerts, G.W. (1977). Endangered and threatened carnivorous plants of North America. In G.T. Prance and T.S. Elias, eds. *Extinction is Forever: Threatened and Endangered Species of Plants in the Americas and their Significance in Ecosystems Today and in the Future*, pp. 301–13. The New York Botanical Garden, Bronx.

Folkerts, G.W. (1982). The Gulf Coast pitcher plant bogs: one of the continent's most unusual assemblages of organisms depends on an increasingly rare combination of saturated soil and frequent fires. *American Scientist*, 70, 260–67.

Folkerts, D.R. (1992). *Interactions of Pitcher Plants* (Sarracenia: *Sarraceniaceae*) *with their Arthropod Prey in the Southeastern United States*. Ph.D. dissertation, University of Georgia, Athens.

Folkerts, D. (1999). Pitcher plant wetlands of the southeastern United States: arthropod associates. In D. Batzer, R. Rader, and S. Wissinger, eds. *Invertebrates in Freshwater Wetlands of North America: Ecology and Management*, pp. 247–75. John Wiley & Sons, New York.

Foot, G., Rice, S.P., and Millett, J. (2014). Red trap colour of the carnivorous plant *Drosera rotundifolia* does not serve a prey attraction or camouflage function. *Biology Letters*, 10, 20131024.

Forsyth, A.B. and Robertson, R.J. (1975). *K* reproductive strategy and larval behavior of the pitcher plant sarcophagid fly, *Blaesoxipha fletcheri*. *Canadian Journal of Zoology*, 53, 174–79.

Forterre, Y. (2013). Slow, fast and furious: understanding the physics of plant movements. *Journal of Experimental Botany*, 64, 4745–60.

Forterre, Y., Skotheim, J.M., Dumais, J., and Mahadevan, L. (2005). How the Venus flytrap snaps. *Nature*, 433, 421–25.

Foster, P. (2001). The potential negative impacts of global climate change on tropical montane cloud forests. *Earth Science Reviews*, 55, 73–106.

Franck, D.H. (1976) The morphological interpretation of epiascidiate leaves—a historical perspective. *Botanical Review*, 42, 345–88.

Frank, J.H. and O'Meara, G.F. (1984). The bromeliad *Catopsis berteroniana* traps terrestrial arthropods but harbors *Wyeomyia* larvae (Diptera: Culicidae). *Florida Entomologist*, 67, 418–24.

Franklin, J. (2009). *Mapping Species Distributions: Spatial Inference and Prediction*. Cambridge University Press, Cambridge.

Franklin, M.A. and Finnegan, J.T. (2004). *Natural Heritage Program List of Rare Plant Species of North Carolina*. North Carolina Department of Environment and Natural Resources, Raleigh.

Freeling, M. (2009). Bias in plant gene content following different sorts of duplication: tandem, whole-genome, segmental, or by transposition. *Annual Review of Plant Biology*, 60, 433–53.

Frenzke, L., Lederer, A., Malanin, M., et al. (2016). Plant pressure sensitive adhesives: similar chemical properties in distantly related plant lineages. *Planta*, 244, 145–54.

Friday, L.E. (1989). Rapid turnover of traps in *Utricularia vulgaris* L. *Oecologia*, 80, 272–77.

Friday, L.E. (1991). The size and shape of the traps of *Utricularia vulgaris* L. *Functional Ecology*, 5, 602–07.

Friday, L.E. (1992). Measuring investment in carnivory: seasonal and individual variation in trap number and biomass in *Utricularia vulgaris* L. *New Phytologist*, 121, 439–45.

Friday, L.E. and Quarmby, C. (1994). Uptake and translocation of prey-derived ^{15}N and ^{32}P in *Utricularia vulgaris* L. *New Phytologist*, 126, 273–81.

Fritsch, P.W., Almeda, F., Martins, A.B., Cruz, B.C., and Estes, D. (2007). Rediscovery and phylogenetic placement of *Philcoxia minensis* (Plantaginaceae), with a test of carnivory. *Proceedings of the California Academy of Sciences*, 58, 447–67.

Fromm-Trinta, E. (1977). *Tayloria* Fromm-Trinta—nova seção do gênero *Genlisea* St.-Hil. (Lentibulariaceae). *Boletim do Museu Nacional Rio de Janeiro, Botanica*, 44, 1–4.

Fromm-Trinta, E. (1981). Revisão do gênero *Genlisea* St.-Hil. (Lentibulariaceae) no Brasil. *Boletim do Museu Nacional Rio de Janeiro, Botanica*, 61, 1–21.

Froebe, H.A. and Baur, N. (1988). Die Morphogenese der Kannenblätter von *Cephalotus follicularis* Labill. *Abhandlungen der Akademie der Wissenschaften und der Literatur, Mainz, Mathematisch–Naturwissenschaftliche Klasse*, 3, 3–19.

Fukushima K., Fang X., Alvarez-Ponce D., et al. (2017). Genome of the pitcher plant *Cephalotus* reveals genetic changes associated with carnivory. *Nature Ecology & Evolution*, 1, e0059.

Fukushima, K., Fujita, H., Yamaguchi, T., et al. (2015). Oriented cell division shapes carnivorous pitcher leaves of *Sarracenia purpurea*. *Nature Communications*, 6, 6450.

Fukushima, K. and Hasebe, M. (2014). Adaxial–abaxial polarity: the developmental basis of leaf shape diversity. *Genesis*, 52, 1–18.

Fukushima, K., Imamura, K., Nagano, K., and Hoshi, Y. (2011). Contrasting patterns of the 5S and 45S rDNA evolutions in the *Byblis liniflora* complex (Byblidaceae). *Journal of Plant Research*, 124, 231–44.

Fukushima, K., Nagai, K., Hoshi, Y., et al. (2009). *Drosera rotundifolia* and *Drosera tokaiensis* suppress the activation of HMC-1 human mast cells. *Journal of Ethnopharmacology*, 125, 90–96.

Furches, M.S., Small, R.L. and Furches, A. (2013a). Genetic diversity in three endangered pitcher plants species (*Sarracenia*; Sarraceniaceae) is lower than widespread congeners. *American Journal of Botany*, 100, 2092–101.

Furches, M.S., Small, R.L. and Furches, A. (2013b). Hybridization leads to interspecific gene flow in *Sarracenia* (Sarraceniaceae). *American Journal of Botany*, 100, 2085–91.

Gaascht, F., Dicato, M., and Diederich, M. (2013). Venus flytrap (*Dionaea muscipula* Solander ex Ellis) contains powerful compounds that prevent and cure cancer. *Frontiers in Oncology*, 3, 1–18.

Galek, H., Osswald, W.F., and Elstner, E.F. (1990). Oxidative protein modification as predigestive mechanism of the carnivorous plant *Dionaea muscipula*: an hypothesis based on in vitro experiments. *Free Radical Biology and Medicine*, 9, 427–34.

Gallie, D.R. and Chang, S.-C. (1997). Signal transduction in the carnivorous plant *Sarracenia purpurea*. *Plant Physiology*, 115, 1461–71.

Gao, X., Hou, Y., Ebina, H., Levin, H.L., and Voytas, D.F. (2008). Chromodomains direct integration of retrotransposons to heterochromatin. *Genome Research*, 18, 359–69.

Gao, P., Loeffler, T.S., Honsel, A., et al. (2015). Integration of trap- and root-derived nitrogen nutrition of carnivorous *Dionaea muscipula*. *New Phytologist*, 205, 1320–29.

Garrido, B., Hampe, A., Marañón, T., and Arroyo, J. (2003). Regional differences in land use affect population performance of the threatened insectivorous plant *Drosophyllum lusitanicum* (Droseraceae). *Diversity and Distributions*, 9, 335–50.

Gassin, R. (1993). *Utricularia beaugleholei* (Lentibulariaceae: subgenus *Utricularia*: section *Pleiochasia*), a new species from south-eastern Australia. *Muelleria*, 8, 37–42.

Gaston, K.J. (1994). *Rarity*. Chapman & Hall, London.

Gaston, K.J. (2003). *The Structure and Dynamics of Geographic Ranges*. Oxford University Press, Oxford.

Gaston, K.J. and Mound, L.A. (1993). Taxonomy, hypothesis testing and the biodiversity crisis. *Proceedings of the Royal Society B*, 251, 139–42.

Gaume, L., Bazile, V., Huguin, M., and Bonhomme, V. (2016). Different pitcher shapes and trapping syndromes explain resource partitioning in *Nepenthes* species. *Ecology and Evolution*, 6, 1378–92.

Gaume, L. and Di Giusto, B. (2009). Adaptive significance and ontogenetic variability of the waxy zone in *Nepenthes rafflesiana*. *Annals of Botany*, 104, 1281–91.

Gaume, L. and Forterre, Y. (2007). A viscoelastic deadly fluid in carnivorous pitcher plants. *PLoS ONE*, 11, e1185.

Gaume, L., Gorb, S., and Rowe, N. (2002). Function of epidermal surfaces in the trapping efficiency of *Nepenthes alata* pitchers. *New Phytologist*, 156, 479–89.

Gaume, L., Perret, P., Gorb, E., et al. (2004). How do plant waxes cause flies to slide? Experimental tests of wax-based trapping mechanisms in three pitfall carnivorous plants. *Arthropod Structure & Development*, 33, 103–11.

Gaveau, D.L.A., Epting, J., Lyne, O., et al. (2009). Evaluating whether protected areas reduce tropical deforestation in Sumatra. *Journal of Biogeography*, 36, 2165–75.

GBIF (2016). GBIF Occurrence Download. Available online: https://www.gbif.org.

Gerber, L.R. (2016). Conservation triage or injurious neglect in endangered species recovery. *Proceedings of the National Academy of Sciences, USA*, 113, 3563–66.

Giblin, D. (2016). *Dionaea muscipula*. WTU Image Collection, Burke Museum of Natural History and Culture. Available online: http://biology.burke.washington.edu/herbarium/imagecollection.php?Genus=Dionaea.

Gibson, T.C. (1983). *Competition, Disturbance, and the Carnivorous Plant Community in the Southeastern United States*. Ph.D. dissertation, University of Utah, Salt Lake City, UT.

Gibson, T.C. (1991a). Competition among threadleaf sundews for limited insect resources. *American Naturalist*, 138, 785–89.

Gibson, T.C. (1991b). Differential escape of insects from carnivorous plant traps. *American Midland Naturalist*, 125, 55–62.

Gibson, R. (1999). *Drosera arcturi* in Tasmania and a comparison with *Drosera regia*. *Carnivorous Plant Newsletter*, 28, 76–80.

Gibson, R. (2013). Variation in floral fragrance of tuberous *Drosera*. *Carnivorous Plant Newsletter*, 42, 117–21.

Gibson, N., Brown, M.J., Williams, K., and Brown, A.V. (1992). Flora and vegetation of ultramafic areas in Tasmania. *Australian Journal of Ecology*, 17, 297–303.

Gibson, R., Conn, B.J., and Bruhl, J.J. (2012). Morphological evaluation of the *Drosera peltata* complex (Droseraceae). *Australian Systematic Botany*, 25, 49–80.

Gibson, T.C. and Waller, D.M. (2009). Evolving Darwin's "most wonderful" plant: ecological steps to a snap-trap. *New Phytologist*, 183, 575–87.

Gilbert, B.M. and Wolpert, T.J. (2013). Characterization of the LOV1-mediated, victorin-induced, cell-death response with virus-induced gene silencing. *Molecular Plant–Microbe Interactions*, 26, 903–17.

Gillissen, B., Bürkle, L., André, B., et al. (2000). A new family of high-affinity transporters for adenine, cytosine, and purine derivatives in *Arabidopsis*. *The Plant Cell*, 12, 291–300.

Gillmor, C.S., Park, M.Y., Smith, M.R., et al. (2010). The MED12-MED13 module of mediator regulates the timing of embryo patterning in *Arabidopsis*. *Development*, 137, 113–22.

Givnish, T.J. (1979). On the adaptive significance of leaf form. In O.T. Solbrig, S. Jain, G.B. Johnson, and P.H. Raven, eds. *Topics in Plant Population Biology*, pp. 375–407. Columbia University Press, New York.

Givnish, T.J. (1989). Ecology and evolution of carnivorous plants. In W.G. Abrahamson, ed. *Plant–Animal Interactions*, pp. 243–90. McGraw-Hill Book Co., New York.

Givnish, T.J. (2003). How a better understanding of adaptations can yield better use of morphology in plant systematics: toward eco-evo-devo. *Regnum Vegetabile*, 141, 273–95.

Givnish, T.J. (2015). New evidence on the origin of carnivorous plants. *Proceedings of the National Academy of Sciences, USA*, 112, 10–11.

Givnish, T.J. (2017). One hundred million years of bromeliad evolution. *Journal of the Bromeliad Society*, 66, 199–221.

Givnish, T.J., Barfuss, M.H.J., Van Ee, B., et al. (2011). Phylogeny, adaptive radiation, and historical biogeography in Bromeliaceae: insights from an 8-locus plastid phylogeny. *American Journal of Botany*, 98, 872–95.

Givnish, T.J., Barfuss, M.H.J., Van Ee, B., et al. (2014a). Adaptive radiation, correlated and contingent evolution, and net species diversification in Bromeliaceae. *Molecular Phylogenetics and Evolution*, 71, 55–78.

Givnish, T.J., Burkhardt, E.L., Happel, R.E., and Weintraub, J.D. (1984). Carnivory in the bromeliad *Brocchinia reducta*, with a cost/benefit model for the general restriction of carnivorous plants to sunny, moist, nutrient-poor habitats. *American Naturalist*, 124, 479–97.

Givnish, T.J., McDiarmid, R.W., and Buck, W.R. (1986). Fire adaptation in *Neblinaria celiae* (Theaceae), a high-elevation rosette shrubs endemic to a wet equatorial tepui. *Oecologia*, 70, 481–85.

Givnish, T.J., Spalink, D., Ames, M., et al. (2015). Orchid phylogenomics and multiple drivers of their extraordinary diversification. *Proceedings of the Royal Society of London, Series B*, 282, 171–80.

Givnish, T.J., Sytsma, K.J., Smith, J.F., et al. (1997). Molecular evolution and adaptive radiation in *Brocchinia* (Bromeliaceae: Pitcairnioideae) atop tepuis of the Guayana Shield. In T.J. Givnish and K.J. Sytsma, eds. *Molecular Evolution and Adaptive Radiation*, pp. 259–311. Cambridge University Press, New York.

Givnish, T. J. and Vermeij, G. J. (1976). Sizes and shapes of liane leaves. *American Naturalist*, 110, 743–78.

Givnish, T.J., Wong, S.C, Stuart-Williams, H., Holloway-Phillips, M., and Farquhar, G.D. (2014b). Determinants of maximum tree height of *Eucalyptus* along a rainfall gradient in Victoria, Australia. *Ecology*, 95, 2991–3007.

Glen, T.C. (2012). Field guide to next-generation DNA sequencers. *Molecular Ecology Resources*, 11, 759–69.

Glitzenstein, J.S., Streng, D.R., Masters, R.E., Robertson, K.M., and Hermann, S.M. (2012). Fire frequency effects on vegetation in north Florida pinelands: another look at the long-term Stoddard Fire Research Plots at Tall Timbers Research Station. *Forest Ecology and Management*, 264, 197–209.

Glitzenstein, J.S., Streng, D.R., and Wade, D.D. (2003). Fire frequency effects on longleaf pine (*Pinus palustris*

P. Miller) vegetation in southeast Florida and northeast Florida, USA. *Natural Areas Journal*, 23, 22–37.

Gloßner, F. (1992). Ultraviolet patterns in the traps and flowers of some carnivorous plants. *Botanische Jahrbücher für Systematik*, 113, 577–87.

Godt, M.J.W. and Hamrick, J.L. (1996). Genetic structure of two endangered pitcher plants, *Sarracenia jonesii* and *Sarracenia oreophila* (Sarraceniaceae). *American Journal of Botany*, 83, 1016–23.

Godt, M.J.W. and Hamrick, J.L. (1998). Allozyme diversity in the endangered pitcher plants *Sarracenia rubra* ssp. *alabamensis* (Sarraceniaceae) and its close relative *S. rubra* ssp. *rubra*. *American Journal of Botany*, 85, 802–10.

Goebel, K. (1891a). Morphologische und biologische Studien. V. *Annales du Jardin botanique de Buitenzorg*, 9, 41–119.

Goebel, K. (1891b). *Pflanzenbiologische Schilderungen II*. N.G. Elwert, Marburg.

Goebel, K. (1893). Zur Biologie von *Genlisea*. *Flora*, 77, 208–12.

Goebel, K. (1908). Morphologische und biologische Bemerkungen. 18. Brutknospenbildung bei *Drosera pygmaea* und einigen Monokotylen. *Flora*, 98, 324–35.

Gold, K. (2008). *Post-harvest handling of seed collections*. Technical Information Sheet, 4, Millennium Seed Bank Project, Royal Botanical Garden, Kew.

Goldblatt, P. and Manning, J.C. (2000). *Cape Plants: A Conspectus of the Cape Flora of South Africa*. National Botanical Institute of South Africa, Cape Town.

Goldman, N., Thorne, J.L., and Jones, D.T. (1998). Assessing the impact of secondary structure and solvent accessibility on protein evolution. *Genetics*, 149, 445–58.

Gomes Rodrigues, F., Franco Marulanda, N., Silva, S.R., et al. (2017). Phylogeny of the 'orchid-like' bladderworts (gen. *Utricularia* sect. *Orchidioides* and *Iperua*: Lentibulariaceae) with remarks on the stolon-tuber system. *Annals of Botany* doi: 10.1093/aob/mcx056.

Gonçalves, A.Z., Mercier, H., Mazzafera, P., and Romero, G.Q. (2011). Spider-fed bromeliads: seasonal and interspecific variation in plant performance. *Annals of Botany*, 107, 1047–55.

Gonçalves, S. and Romano, A. (2005). Micropropagation of *Drosophyllum lusitanicum* (Dewy pine), an endangered West Mediterranean endemic insectivorous plant. *Biodiversity and Conservation*, 14, 1071–81.

Gonçalves, S. and Romano, A. (2009). Cryopreservation of seeds from the endangered insectivorous plant *Drosophyllum lusitanicum*. *Seed Science and Technology*, 37, 485–90.

Gonella, P.M., Fleischmann, A., Rivadavia, F., Neill, D.A., and Sano, P.T. (2016). A revision of *Drosera* (Droseraceae) from the central and northern Andes, including a new species from the Cordillera del Cóndor (Peru and Ecuador). *Plant Systematics and Evolution*, 302, 1419–32.

Gonella, P.M., Rivadavia, F., and Fleischmann, A. (2015). *Drosera magnifica* (Droseraceae): the largest New World sundew, discovered on Facebook. *Phytotaxa*, 220, 257–67.

Gonella, P.M., Rivadavia, F., and Sano, P.T. (2012). Re-establishment of *Drosera spiralis* (Droseraceae), and a new circumscription of *D. graminifolia*. *Phytotaxa*, 75, 43–57.

Gonsiska, P.A. (2010). *Aspects of The Evolutionary Ecology of The Genus* Catopsis *(Bromeliaceae)*. Ph.D. dissertation, University of Wisconsin, Madison.

González, J.M., Jaffe, K., and Michelangeli, F. (1991). Competition for prey between the carnivorous Bromeliaceae *Brocchinia reducta* and Sarraceniaceae *Heliamphora nutans*. *Biotropica*, 23, 602–04.

Gorb, E., Haas, K., Henrich, A., et al. (2005). Composite structure of the crystalline epicuticular wax layer of the slippery zone in the pitcher of the carnivorous plant *Nepenthes alata* and its effect on insect attachment. *Journal of Experimental Biology*, 208, 4651–62.

Gorb, E., Kastner, V., Peressadko, A., et al. (2004). Structure and properties of the glandular surface in the digestive zone of the pitcher in the carnivorous plant *Nepenthes ventrata* and its role in insect trapping and retention. *Journal of Experimental Biology*, 207, 2947–63.

Gordon, E. and Pacheco, S. (2007). Prey composition in the carnivorous plants *Utricularia inflata* and *U. gibba* (Lentibulariaceae) from Paria Peninsula, Venezuela. *Revista de Biologia Tropical*, 55, 795–803.

Gotelli, N.J. and Ellison, A.M. (2002). Nitrogen deposition and extinction risk in the northern pitcher plant, *Sarracenia purpurea*. *Ecology*, 83, 2758–65.

Gowda, D.C., Reuter, G., and Schauer, R. (1982). Structural features of an acidic polysaccharide from the mucin of *Drosera binata*. *Phytochemistry*, 21, 2297–300.

Gowda, D.C., Reuter, G., and Schauer, R. (1983). Structural studies of an acidic polysaccharide from the mucin secreted by *Drosera capensis*. *Carbohydrate Research*, 113, 113–24.

Grafe, T.U. and Kohout, R.J. (2013). A new case of ants nesting in *Nepenthes* pitcher plants. *Ecotropica*, 19, 77–80.

Grafe, T.U., Schöner, C.R., Kerth, G., Junaidi, A., and Schöner, M.G. (2011). A novel resource-service mutualism between bats and pitcher plants. *Biology Letters*, 7, 436–39.

Graham, C.H., Ferrier, S., Huettman, F., Moritz, C., and Peterson, A.T. (2004). New developments in museum-based informatics and applications in biodiversity analysis. *Trends in Ecology and Evolution*, 19, 497–503.

Gray, S.M., Akob, D.M., Green, S.J., and Kostka, J.E. (2012). The bacterial composition within the *Sarracenia purpurea* model system: local scale differences and the relationship with the other members of the food web. *PLoS ONE*, 7, e50969.

Gray, S.M., Poisot, T., Harvey, E., et al. (2016). Temperature and trophic structure are driving microbial productivity along a biogeographical gradient. *Ecography*, 39, 981–89.

Green, S. (1967). Notes on the distribution of *Nepenthes* species in Singapore. *Garden Bulletin* (Singapore), 22, 53–65.

Green, S., Green, T.L., and Heslop-Harrison, Y. (1979). Seasonal heterophylly and leaf gland features in *Triphyophyllum* (Dioncophyllaceae), a new carnivorous plant genus. *Botanical Journal of the Linnean Society*, 78, 99–116.

Green, M.L. and Horner, J.D. (2007). The relationship between prey capture and characteristics of the carnivorous pitcher plant, *Sarracenia alata* Wood. *American Midland Naturalist*, 158, 424–31.

Greenwood, M., Clarke, C., Lee, C.C., Gunsalam, A., and Clarke, R.H. (2011). A unique resource mutualism between the giant Bornean pitcher plant, *Nepenthes rajah*, and members of a small mammal community. *PLoS ONE*, 6, e21114.

Greilhuber, J., Borsch, T., Müller, K., et al. (2006). Smallest angiosperm genomes found in Lentibulariaceae, with chromosomes of bacterial size. *Plant Biology*, 8, 770–77.

Grevenstuk, T., Gonçalves, S., Almeida, S., et al. (2009). Evaluation of the antioxidant and antimicrobial properties of in vitro cultured *Drosera intermedia* extracts. *Natural Product Communications*, 4, 1063–68.

Grevenstuk, T., Gonçalves, S., Domingos, T., et al. (2012a). Inhibitory activity of plumbagin produced by *Drosera intermedia* on food spoilage fungi. *Journal of the Science of Food and Agriculture*, 92, 1638–42.

Grevenstuk, T., Gonçalves, S., Nogueira, J.M.F., Bernardo-Gil, M.G., and Romano, A. (2012b). Recovery of high purity plumbagin from *Drosera intermedia*. *Industrial Crops and Products*, 35, 257–60.

Grevenstuk, T., Gonçalves, S., Nogueira, J.M.F., and Romano, A. (2008). Plumbagin recovery from field specimens of *Drosophyllum lusitanicum* (L.) Link. *Phytochemical Analysis*, 19, 229–35.

Grevenstuk, T. and Romano, A. (2012). In vitro plantlet production of the endangered *Pinguicula vulgaris*. *Open Life Sciences*, 7, 48–53.

Grime, J.P. (1979). *Plant Strategies and Vegetation Processes*. John Wiley & Sons, Chichester.

Grime, J.P., Mason, G., Curtis, A.V., et al. (1981). A comparative study of germination characteristics in a local flora. *Journal of Ecology*, 69, 1017–59.

Gronemeyer, T., Coritico, F., Micheler, M., et al. (2011). *Nepenthes ceciliae*, a new pitcher plant species from Mount Kiamo, Mindanao. In S.R. McPherson, ed. *New Nepenthes: Volume One*, pp. 412–23. Redfern Natural History Productions, Poole.

Gronemeyer, T., Coritico, F., Wistuba, A., et al. (2014). Four new species of *Nepenthes* L. (Nepenthaceae) from the central mountains of Mindanao, Philippines. *Plants*, 3, 284–303.

Gronemeyer, T., Suarez, W., Nuytemans, H., et al. (2016). Two new *Nepenthes* species from the Philippines and an emended description of *Nepenthes ramos*. *Plants*, 5, 23.

Gronemeyer, T., Wistuba, A., Heinrich, V., et al. (2010). *Nepenthes hamiguitanensis* (Nepenthaceae), a new pitcher plant species from Mindanao Island, Philippines. In S.R. McPherson, ed. *Carnivorous Plants in their Habitats, Volume Two*, pp. 1296–305. Redfern Natural History Productions, Poole, Dorset, UK.

Groom, P. (1897). On the leaves *of Lathraea squamaria* and of some allied Scrophulariaceae. *Annals of Botany*, 11, 385–98.

Grossenbacher, D., Runquist, R.B., Goldberg, E.E., and Brandvain, Y. (2015). Geographic range size is predicted by plant mating system. *Ecology Letters*, 18, 706–13.

Grover, A. (2012). Plant chitinases: genetic diversity and physiological roles. *Critical Reviews in Plant Sciences*, 31, 57–73.

Guimarães, E.F., Queiroz, G.A., Negrão R., Santos Filho, L., and Serrano, T. (2014). Plantaginaceae. In G. Martinelli, T. Messina, and L. Santos Filho, eds. *Livro Vermelho da Flora do Brasil—Plantas Raras do Cerrado*, pp. 218–19. Centro Nacional de Conservação da Flora, Rio de Janeiro.

Guiral, D. and Rougier, C. (2007). Trap size and prey selection of two coexisting bladderwort (*Utricularia*) species in a pristine tropical pond (French Guiana) at different trophic levels. *International Journal of Limnology*, 43, 147–59.

Guisan, A. and Thuiller, W. (2005). Predicting species distribution: offering more than simple habitat models. *Ecology Letters*, 8, 993–1009.

Guisan, A., Tingley, R., Baumgartner, J.B., et al. (2013). Predicting species distributions for conservation decisions. *Ecology Letters*, 16, 1424–35.

Guisan, A. and Zimmermann, N.E. (2000). Predictive habitat distribution models in ecology. *Ecological Modelling*, 135, 147–86.

Guisan, A., Zimmermann, N.E., Elith, J., et al. (2007). What matters for predicting the occurrences of trees: techniques, data, or species' characteristics? *Ecological Monographs*, 77, 615–30.

Guisande, C., Andrade, C., Granado-Lorencio, C., Duque, S.R., and Núñez-Avellaneda, M. (2000). Effects of zooplankton and conductivity on tropical *Utricularia foliosa* investment in carnivory. *Aquatic Ecology*, 34, 137–42.

Guisande, C., Aranguren, N., Andrade-Sossa, C., et al. (2004). Relative balance of the cost and benefit associated with carnivory in the tropical *Utricularia foliosa*. *Aquatic Botany*, 80, 271–82.

Guisande, C., Granado-Lorencio, C., Andrade-Sossa, C., and Duque, S.R. (2007). Bladderworts. *Functional Plant Science and Biotechnology*, 1, 58–68.

Guitton, Y., Legendre, L., and Adamec, L. (2012). Mineral nutrition in hydroponically grown *Pinguicula*. *Carnivorous Plant Newsletter*, 41, 8–15.

Gulmon, S.L. and Chu, C.C. (1981). The effect of light and nitrogen on photosynthesis, leaf characteristics, and dry matter allocation in the chaparral shrub, *Diplacus aurantiacus*. *Oecologia*, 49, 207–12.

Günther, K. (1913). Die lebenden Bewohner der Kannen der Insektfressenden Pflanze *Nepenthes destillatoria* auf Ceylon. *Zeitschrift für Wissenschaftliche Insektenbiologie*, 9, 90–270.

Guo, Q., Dai, E., Han, X., et al. (2015). Fast nastic motion of plants and bioinspired structures. *Journal of the Royal Society Interface*, 12, 20150598.

Gutzwiller, K.J. and Flather, C.H. (2011). Wetland features and landscape context predict the risk of wetland habitat loss. *Ecological Applications*, 21, 968–82.

Gwee, P.S., Khoo, K.S., Ong, H.C., and Sit, N.W. (2014). Bioactivity-guided isolation and structural characterization of the antifungal compound, plumbagin, from *Nepenthes gracilis*. *Pharmaceutical Biology*, 52, 1526–31.

Haccius, B. and Hartle-Baude, E. (1957). Embryologische und histogenetische Studien an "monokotylen Dikotylen" II. *Pinguicula vulgaris* L. und *Pinguicula alpina* L. *Österreichische Botanische Zeitung*, 103, 567–87.

Haas, K., Brune, T., and Rücker, E. (2001). Epicuticular wax crystalloids in rice and sugar cane leaves are reinforced by polymeric aldehydes. *Journal of Applied Botany*, 75, 178–87.

Hackl, T. (2016). *A Draft Genome for The Venus Flytrap,* Dionaea muscipula: *Evaluation of Assembly Strategies for a Complex Genome—Development of Novel Approaches and Bioinformatics Solutions*. Ph.D. dissertation, Julius–Maximilians–Universität, Würzburg.

Hagan, D.V., Grogan, W.L. Jr., Murza, G.L., and Davis, A.R. (2008). Biting midges (Diptera: Ceratopogonidae) from the English sundew, *Drosera anglica* Hudson (Droseraceae), at two fens in Saskatchewan, Canada. *Proceedings of the Entomological Society of Washington*, 110, 397–401.

Hairston, N.G., Ellner, S.P., Geber, M.A., Yoshida, T., and Fox, J.A. (2005). Rapid evolution and the convergence of ecological and evolutionary time. *Ecology Letters*, 8, 1114–27.

Hairston, N.G., Smith, F.E., and Slobodkin, L.B. (1960). Community structure, population control, and competition. *American Naturalist*, 94, 421–25.

Hájek, T. and Adamec, L. (2010). Photosynthesis and dark respiration of leaves of terrestrial carnivorous plants. *Biologia*, 65, 69–74.

Hamilton, A.G. (1903). Notes on *Byblis gigantea*. *Proceedings of the Linnean Society of New South Wales*, 28, 680–84.

Hamilton, A.G. (1904). Notes on the West Australian pitcher plant (*Cephalotus follicularis* Labill.). *Proceedings of the Linnean Society of New South Wales*, 29, 36–53.

Hampe, A. (2004). Bioclimate envelope models: what they detect and what they hide. *Global Ecology and Biogeography*, 13, 469–71.

Hampe, A. and Petit, R.J. (2005). Conserving biodiversity under climate change: the rear edge matters. *Ecology Letters*, 8, 461–67.

Hanslin, H.M. and Karlsson, P.S. (1996). Nitrogen uptake from prey and substrate as affected by prey capture level and plant reproductive status in four carnivorous plant species. *Oecologia*, 106, 370–75.

Hanson, P.C., Carpenter, S.R, Armstrong, D.E, Stanley, E.H., and Kratz, T.K. (2006). Lake dissolved inorganic carbon and dissolved oxygen: changing drivers from days to decades. *Ecological Monographs*, 76, 343–63.

Hanson, L., McMahon, K.A., Johnson, M.A.T., and Bennett, M.D. (2001). First nuclear DNA C-values for another 25 angiosperm families. *Annals of Botany*, 5, 851–58.

Hard, J., Bradshaw, W.E., and Holzapfel, C.M. (1993). The genetic basis of photoperiodism and its evolutionary divergence among populations of the pitcher-plant mosquito, *Wyeomyia smithii*. *American Naturalist*, 142, 457–73.

Harder, R. (1970). *Utricularia* als Objekt für Heterotrophieuntersuchungen bei Blütenpflanzen (Wechselwirkung von Saccharose und Acetat). *Zeitschrift für Pflanzenphysiologie*, 63, 181–84.

Harms, S. (1999). Prey selection in three species of the carnivorous aquatic plant *Utricularia* (bladderwort). *Archiv für Hydrobiologie*, 146, 449–70.

Harms, S. and Johansson, F. (2000). The influence of prey behaviour on prey selection of carnivorous plant *Utricularia vulgaris*. *Hydrobiologia*, 247, 113–20.

Harold, F. M. (2016). *To Make the World Intelligible: A Scientist's Journey*. FriesenPress, Victoria.

Harper, R.M. (1903). Botanical explorations in Georgia during the summer of 1901. II. Noteworthy species. *Bulletin of the Torrey Botanical Club*, 30, 319–42.

Harris, C.S., Asim, M., Saleem, A., et al. (2012). Characterizing the cytoprotective activity of *Sarracenia purpurea* L., a medicinal plant that inhibits glucotoxicity in PC12 cells. *BMC Complementary and Alternative Medicine*, 12, 245–55.

Hartmeyer, I. and Hartmeyer, S.R.H. (2010). Snap-tentacles and runway lights. *Carnivorous Plant Newsletter*, 39, 101–13.

Hartmeyer, S.R.H. and Hartmeyer, I. (2015). Several pygmy sundew species possess catapult-flypaper traps with repetitive function, indicating a possible evolutionary change into aquatic snap traps similar to *Aldrovanda*. *Carnivorous Plant Newsletter*, 44, 172–84.

Hartmeyer, I., Hartmeyer, S.R.H., Masselter, T., et al. (2013). Catapults into a deadly trap: the unique prey-capture mechanism of *Drosera glanduligera*. *Carnivorous Plant Newsletter*, 42, 4–14.

Harvey, E. and Miller, T.E. (1996). Variance in composition of inquiline communities in leaves of *Sarracenia purpurea* L. on multiple spatial scales. *Oecologia*, 108, 562–66.

Hatano, N. and Hamada, T. (2008). Proteome analysis of pitcher fluid of the carnivorous plant *Nepenthes alata*. *Journal of Proteome Research*, 7, 809–16.

Hatano, N. and Hamada, T. (2012). Proteomic analysis of secreted protein induced by a component of prey in pitcher fluid of the carnivorous plant *Nepenthes alata*. *Journal of Proteomics*, 75, 4844–52.

Hatcher, C.R. and Hart, A.G. (2014). Venus flytrap seedlings show growth-related prey size specificity. *International Journal of Ecology*, 2014, e135207.

Havens, K., Vitt, P., Mauner, M., Guerrant, E.O., and Dixon, K. (2006). *Ex situ* plant conservation and beyond. *BioScience*, 56, 525–31.

He, J. and Zain, A. (2012). Photosynthesis and nitrogen metabolism of *Nepenthes alata* in response to inorganic NO_3^- and organic prey N in the greenhouse. *International Scholarly Research Network Botany*, 2012, 263270.

Heacock, M.L., Idol, R.A., Friesner, J.D., Britt, A.B., and Shippen, D.E. (2007). Telomere dynamics and fusion of critically shortened telomeres in plants lacking DNA ligase IV. *Nucleic Acids Research*, 35, 6490–500.

Heard, S.B. (1994). Pitcher-plant midges and mosquitoes: a processing chain commensalism. *Ecology*, 75, 1647–60.

Heard, S.B. (1998). Capture rates of invertebrate prey by the pitcher plant, *Sarracenia purpurea* L. *American Midland Naturalist*, 139, 79–89.

Hecht, A. (1949). The somatic chromosomes of *Sarracenia*. *Bulletin of the Torrey Botanical Club*, 76, 7–9.

Hegnauer, R. (1990). Nepenthaceae. In R. Hegnauer, ed. *Chemotaxonomie der Pflanzen 9*, pp. 132–33. Birkhäuser, Basel.

Hegner, R.W. (1926). The interrelations of protozoa and the utricles of *Utricularia*. *Biological Bulletin*, 50, 239–70.

Heibl, C. and Renner, S.S. (2012). Distribution models and a dated phylogeny for Chilean *Oxalis* species reveal occupation of new habitats by different lineages, not rapid adaptive radiation. *Systematic Biology*, 61, 823–34.

Heikkinen, R.K., Luoto, M., Araujo, M.B., et al. (2006). Methods and uncertainties in bioclimatic envelope modelling under climate change. *Progress in Physical Geography*, 30, 751–77.

Heil, M. (2011). Nectar: generation, regulation, and ecological functions. *Trends in Plant Science*, 16, 191–200.

Heil, M. (2015). Extrafloral nectar at the plant–insect interface: a spotlight on chemical ecology, phenotypic plasticity, and food webs. *Annual Review of Entomology*, 60, 213–32.

Hepburn, J.S. (1918). Biochemical studies of the pitcher liquor of *Nepenthes*. *Proceedings of the American Philosophical Society*, 57, 112–29.

Hernandez, P.A., Graham, C.H., Master, L.L., and Albert, D.L. (2006). The effect of sample size and species characteristics on performance of different species distribution modeling methods. *Ecography*, 29, 773–85.

Heslop-Harrison, Y. (1962). Winter dormancy and vegetative propagation in Irish *Pinguicula grandiflora* Link. *Proceedings of the Royal Irish Academy, Section B*, 62: 23–30.

Heslop-Harrison, Y. (1975). Enzyme release in carnivorous plants. In J.T. Dingle and R.T. Dean, eds. *Lysozymes in Biology and Pathology*, pp. 525–78. North Holland Publishing Company, Amsterdam.

Heslop-Harrison, Y. (1976). Carnivorous plants a century after Darwin. *Endeavor*, 35, 114–22.

Heslop-Harrison, Y. (1978). Carnivorous plants. *Scientific American*, 238, 104–15.

Heslop-Harrison, Y. (2004). *Pinguicula* L. *Journal of Ecology*, 92, 1071–118.

Heslop-Harrison, Y. and Knox, R.B. (1971). A cytochemical study of the leaf-gland enzymes of insectivorous plants of the genus *Pinguicula*. *Planta*, 96, 183–211.

Heubl, G., Bringmann, G., and Meimberg, H. (2006). Molecular phylogeny and character evolution of carnivorous plant families in Caryophyllales—revisited. *Plant Biology*, 8, 821–30.

Heubl, G. and Wistuba, A. (1995). A cytological study of the genus *Nepenthes* L. (Nepenthaceae). *Sendtnera*, 4, 169–74.

Heřmanová, Z. and Kvaček, J. (2010). Late Cretaceous *Palaeoaldrovanda*, not seeds of a carnivorous plant, but eggs of an insect. *Journal of the National Museum (Prague), Natural History Series*, 179, 105–18.

Hijmans, R.J., Cameron, S.E., Parra, J.L., Jones, P.G., and Jarvis, A. (2005). Very high resolution interpolated climate surfaces for global land areas. *International Journal of Climatology*, 25, 1965–78.

Hijmans, R.J. and Graham, C.H. (2006). The ability of climate envelope models to predict the effect of climate change on species distributions. *Global Change Biology*, 12, 2272–81.

Hijmans, R.J., Phillips, S.J., Leathwick, J.R., and Elith, J. (2016). Dismo: species distribution modeling. Available online: https://cran.r-project.org/package=dismo.

Hill, J.K., Thomas, C.D., and Huntley, B. (1999). Climate and habitat availability determine 20th century changes in a butterfly's range margin. *Proceedings of the Royal Society of London Series B*, 266, 1197–206.

Hinman, S.E. and Brewer, J.S. (2007). Responses of two frequently burned wet pine savannas to an extended period without fire. *Journal of Torrey Botanical Society*, 134, 512–26.

Hinman, S.E., Brewer, J.S., and Ashley, S.W. (2008). Shrub establishment is limited by dispersal, slow growth, and fire in two wet pine savannahs in Mississippi. *Natural Areas Journal*, 28, 37–43.

Hirsikorpi, M., Kämäräinen, T., Teeri, T., and Hohtola, A. (2002). *Agrobacterium*-mediated transformation of round leaved sundew (*Drosera rotundifolia* L.). *Plant Science*, 162, 537–42.

Hirzel, A.H., Le Lay, G., Helfer, V., Randin, C., and Guisan, A. (2006). Evaluating the ability of habitat suitability models to predict species presences. *Ecological Modelling*, 199, 142–52.

Ho, W., Kutz, N., Ng, J., and Riffell, J. (2016). Variable rates of scent evolution in functionally distinct organs of the NA Sarraceniaceae. *bioRxiv*, e079947. Available online: https://doi.org/10.1101/079947.

Hobbhahn, N., Küchmeister, H., and Porembski, S. (2006). Pollination biology of mass flowering terrestrial *Utricularia* species (Lentibulariaceae) in the Indian Western Ghats. *Plant Biology*, 8, 791–804.

Hoch, W.A., Singsaas, E.L., and McCown, B.H. (2003). Resorption protection. Anthocyanins facilitate nutrient recovery in autumn by shielding leaves from potentially damaging light levels. *Plant Physiology*, 133, 1296–305.

Hoch, W.A., Zeldin, E.L., and McCown, B.H. (2001). Physiological significance of anthocyanins during autumnal leaf senescence. *Tree Physiology*, 21, 1–8.

Hodick, D. and Sievers, A. (1986). The influence of Ca^{2+} on the action potential in mesophyll cells of *Dionaea muscipula* Ellis. *Protoplasma*, 133, 83–84.

Hodick, D. and Sievers, A. (1988). The action-potential of *Dionaea muscipula* Ellis. *Planta*, 174, 8–18.

Hodick, D. and Sievers, A. (1989). On the mechanism of trap closure of Venus flytrap (*Dionaea muscipula* Ellis). *Planta*, 179, 32–42.

Hoekman, D. (2007). Top-down and bottom-up regulation in a detritus-based aquatic food web: a repeated field experiment using the pitcher plant (*Sarracenia purpurea*) inquiline community. *American Midland Naturalist*, 157, 52–62.

Hoekman, D. (2011). Relative importance of top-down and bottom-up forces in food webs of *Sarracenia* pitcher communities at a northern and a southern site. *Oecologia*, 165, 1073–82.

Hoekman, D., Winston, R., and Mitchell, N. (2009). Top-down and bottom-up effects of a processing detritivore. *Freshwater Science*, 28, 552–59.

Holt, R.D. (1977). Predation, apparent competition, and the structure of prey communities. *Theoretical Population Biology*, 12, 197–229.

Holzapfel, C.M. and Bradshaw, W.E. (2002). Protandry: the relationship between emergence time and male fitness in the pitcher-plant mosquito. *Ecology*, 83, 607–11.

Hooker, J.D. (1859). On the origin and development of the pitchers of *Nepenthes*, with an account of some new Bornean plants of that genus. *Transactions of the Linnean Society of London*, 22, 415–24.

Hopper, S.D. (1997). An Australian perspective on plant conservation biology in practice. In P.L. Fiedler and P.M. Kareiva, eds. *Conservation Biology for the Coming Decade*, pp. 255–78. Chapman & Hall, New York.

Hopper, S.D. (2000). Climate change, dispersal mechanisms and revegetation with WA plants. *Western Wildlife*, 4, 4–5.

Hopper, S. and Gioia, P. (2004). The Southwest Australian floristic region: evolution and conservation of a global hot spot of biodiversity. *Annual Review of Ecology, Evolution, and Systematics*, 35, 623–50.

Hopper, S.D., Harvey, M.S., Chappill, J.A., Main, A.R., and Main, B.Y. (1996). The Western Australian biota as Gondwanan heritage—a review. In S.D. Hopper, J.A. Chappill, M.S. Harvey, and A.S. George, eds. *Gondwanan heritage*, pp. 1–46. Surrey Beatty & Sons, Chipping Norton.

Horne, I., Sutherland, T.D., Oakeshott, J.G., and Russell, R.J. (2002). Cloning and expression of the phosphotriesterase gene hocA from *Pseudomonas monteilii* C11b. *Microbiology*, 148, 2687–95.

Horner, J.D. (2014). Phenology and pollinator–prey conflict in the carnivorous plant, *Sarracenia alata*. *American Midland Naturalist*, 171, 153–56.

Horner, J.D. and Schatz, B.A. (2016). Resorption of trap nitrogen during senescence and the benefit of prey capture in the carnivorous plant, *Sarracenia alata*. *Plant Ecology*, 217, 985–91.

Horner, J.D., Steele, J.C., Underwood, C.A., and Lingamfelter, D. (2012). Age-related changes in characteristics and prey capture of seasonal cohorts of *Sarracenia alata* pitchers. *American Midland Naturalist*, 167, 13–27.

Hoshi, Y., Shirakawa, J., and Hasebe, M. (2006). Nucleotide sequence variation was unexpectedly low in an endangered species, *Aldrovanda vesiculosa* L. (Droseraceae). *Chromosome Botany*, 1, 27–32.

Hoyo, Y. and Tsuyuzaki, S. (2013). Characteristics of leaf shapes among two parental *Drosera* species and a hybrid examined by canonical discriminant analysis and a hierarchical Bayesian model. *American Journal of Botany*, 100, 817–23.

Hu, T.T., Pattyn, P., Bakker, E.G., et al. (2011). The *Arabidopsis lyrata* genome sequence and the basis of rapid genome size change. *Nature Genetics*, 43, 476–81.

Hu, X. and Xu, L. (2016). Transcription factors WOX11/12 directly activate WOX5/7 to promote root primordia initiation and organogenesis. *Plant Physiology*, 172, 2363–73.

Hua, Y.J. and Li, H. (2005). Food web and fluid in pitchers of *Nepenthes mirabilis* in Zhuhai, China. *Acta Botanica Gallica*, 152, 165–75.

Huang, T.-C. (1978). Miocene palynomorphs of Taiwan II. Tetrad Grains. *Botanical Bulletin of Academia Sinica*, 19, 77–81.

Huang, Y., Wang, Y., Sun, L., Agrawal, R., and Zhang, M. (2015). Sundew adhesive: a naturally occurring hydrogel. *Journal of the Royal Society Interface*, 12, 20150226.

Hufford, L. (1992). Rosidae and their relationships to other nonmagnoliid dicotyledons: A phylogenetic analysis using morphological and chemical data. *Annals of the Missouri Botanical Garden*, 79, 218–48.

Hutchens, J.J. and Luken, J.O. (2009). Prey capture in the Venus flytrap: collection or selection? *Botany*, 87, 1007–10.

Huynh, K. (1968). Étude de la morphologie du pollen du genre *Utricularia* L. *Pollen et Spores*, 5, 11–55.

Ibarra-Laclette, E., Albert, V.A., Pérez-Torres, C.A., et al. (2011). Transcriptomics and molecular evolutionary rate analysis of the bladderwort (*Utricularia*), a carnivorous plant with a minimal genome. *BMC Plant Biology*, 11, 101.

Ibarra-Laclette, E., Lyons, E., Hernández-Guzmán G., et al. (2013). Architecture and evolution of a minute plant genome. *Nature*, 498, 94–98.

Ichiishi, S., Nagamitsu, T., Kondo, Y., et al. (1999). Effects of macro-components and sucrose in the medium on in vitro red-color pigmentation in *Dionaea muscipula* Ellis and *Drosera spathulata* Labill. *Plant Biotechnology*, 16, 235–38.

Idei, S. and Kondo, K. (1998). Effects of NO_3^- and BAP on organogenesis in tissue-cultured shoot primordia induced from shoot apices of *Utricularia praelonga* St. Hil. *Plant Cell Reports*, 17, 451–56.

Iosilevskii, G. and Joel, D. M. (2013). Aerodynamic trapping effect and its implications for capture of flying insects by carnivorous pitcher plants. *European Journal of Mechanics B/Fluids*, 38, 65–72.

Ishii, H. (2011). How do changes in leaf/shoot morphology and crown architecture affect growth and physiological function of tall trees? In F.C. Meinzer, B. Lachenbruch and T.E. Dawson, eds. *Size- and Age-Related Changes in Tree Structure and Function*, pp. 215–32. Springer, New York.

Ishisaki, K., Arai, S., Hamada, T., and Honda, Y. (2012a). Biochemical characterization of a recombinant plant class III chitinase from the pitcher of the carnivorous plant *Nepenthes alata*. *Carbohydrate Research*, 361, 170–74.

Ishisaki, K., Honda, Y., Taniguchi, H., Hatano, N., and Hamada, T. (2012b). Heterogonous expression and characterization of a plant class IV chitinase from the pitcher of the carnivorous plant *Nepenthes alata*. *Glycobiology*, 22, 345–51.

IUCN (2000). *IUCN Red List Categories and Criteria: Version 3.1*. Second Edition. IUCN, Gland.

IUCN (2016). The IUCN Red List of Threatened Species, version 2016–12. Available online: http://www.iucnredlist.org.

Jaffe, M.J. (1973). The role of ATP in mechanically stimulated rapid closure of Venus's flytrap. *Plant Physiology*, 51, 17–18.

Jaffe, K., Blum, M.S., Fales, H.M., Mason, R.T., and Cabrera, A. (1995). On insect attractants from pitcher plants of the genus *Heliamphora* (Sarraceniaceae). *Journal of Chemical Ecology*, 21, 379–84.

Jaffe, K., Michelangeli, F., Gonzalez, J.M., Miras, B., and Ruiz, M.C. (1992). Carnivory in pitcher plants of the genus *Heliamphora* (Sarraceniaceae). *New Phytologist*, 122, 733–44.

Janzen, D.H. (1985). The natural history of mutualism. In D.H. Boucher, ed. *The Biology of Mutualism: Ecology and Evolution*, pp. 40–99. Oxford University Press, New York.

Jayaram, K. and Prasad, M.N.V. (2006). *Drosera indica* L. and *D. burmannii* Vahl., medicinally important insectivorous plants in Andhra Pradesh—regional threats and conservation. *Current Science*, 91, 943–47.

Jayaram, K. and Prasad, M.N.V. (2007). Rapid in vitro multiplication of *Drosera indica* L.: a vulnerable, medicinally important insectivorous plant. *Plant Biotechnology Reports*, 1, 79–84.

Jebb, M.H.P. and Cheek, M. (1997). A skeletal revision of *Nepenthes* (Nepenthaceae). *Blumea*, 42, 1–106.

Jeffree, C.E. (1986). The cuticle, epicuticular waxes and trichomes of plants, with reference to their structure, functions and evolution. In B. Juniper and R. Southwood, eds. *Insects and the Plant Surface*, pp. 23–64. Edward Arnold, London.

Jennings, D.T., Cutler, B., and Connery, B. (2008). Spiders (Arachnida: Araneae) associated with seed heads of *Sarracenia purpurea* (Sarraceniaceae) at Acadia National Park, Maine. *Northeastern Naturalist*, 15, 523–40.

Jennings, D.E., Krupa, J.J., Raffel, T.R., and Rohr, J.R. (2010). Evidence for competition between carnivorous plants and spiders. *Proceedings of the Royal Society B*, 277, 3001–08.

Jennings, D.E. and Rohr, J.R. (2011). A review of the conservation threats to carnivorous plants. *Biological Conservation*, 144, 1356–63.

Jensen, H. (1910). Nepenthes—Tiere II. *Biologische Notizen. Annales du Jardin botanique de Buitenzorg*, Suppl. 3, 941–46.

Jensen, M.K., Vogt, J.K., Bressendorff, S., et al. (2015). Transcriptome and genome size analysis of the Venus flytrap. *PLoS ONE*, 10, e0123887.

Jérémie, J. and Jeune, B. (1985). Un cas probable de speciation sympatrique chez *Utricularia alpina* Jacq. (Lentibulariaceae) aux Petites Antilles. *Bulletin du Museum D'Histoire Naturelle, B. Andansonia*, 7, 213–37.

Jobson, R.W. (2013). Five new species of *Utricularia* (Lentibulariaceae) from Australia. *Telopea*, 15, 127–42.

Jobson, R.W. and Albert, V.A. (2002). Molecular rates parallel diversification contrasts between carnivorous plant sister lineages. *Cladistics*, 18, 127–36.

Jobson, R.W. and Baleeiro, P.C. (2015). Two new species of *Utricularia* (Lentibulariaceae) from the northwest region of Western Australia. *Telopea*, 18, 201–08.

Jobson, R.W., Baleeiro, P.C., and Reut, M. (2017). Molecular phylogeny of subgenus *Polypompholyx* (*Utricularia*; Lentibulariaceae) based on three plastid markers: diversification and proposal for new section. *Australian Systematic Botany*, 30, 259–278.

Jobson, R.W. and Morris, E.C. (2001). Feeding ecology of a carnivorous bladderwort (*Utricularia uliginosa*, Lentibulariaceae). *Austral Ecology*, 26, 680–91.

Jobson, R.W., Morris, E.C., and Burgin, S. (2000). Carnivory and nitrogen supply affect the growth of the bladderwort *Utricularia uliginosa*. *Australian Journal of Botany*, 48, 549–60.

Jobson, R.W., Nielsen, R., Laakkonen, L., Wikström, M., and Albert, V.A. (2004). Adaptive evolution of cytochrome *c* oxidase: infrastructure for a carnivorous plant radiation. *Proceedings of the National Academy of Sciences, USA*, 101, 18064–68.

Jobson, R.W., Playford, J., Cameron, K.M., and Albert, V.A. (2003). Molecular phylogenetics of Lentibulariaceae inferred from plastid *rps16* intron and *trnL-F* DNA sequences: implications for character evolution and biogeography. *Systematic Botany*, 28, 157–71.

Jolivet, P. and Vasconcellos-Neto, J. (1993). Convergence chez les plantes carnivores. *La Recherche*, 24, 456–58.

Joel, D.M. (1988). Mimicry and mutualism in carnivorous pitcher plants (Sarraceniaceae, Nepenthaceae, Cephalotaceae, Bromeliaceae). *Biological Journal of the Linnean Society*, 35, 185–97.

Joel, D.M. (2002). Carnivory and parasitism in plants. In K. Kondo, ed. *Proceedings of the 4th International Carnivorous Plant Conference, Tokyo, Japan*, pp. 55–60. Higashi-Hiroshima University, Higashihiroshima.

Joel, D.M. and Juniper, B.E. (1982). Cuticular gaps in carnivorous plant glands. In D.F. Cutler, K.L. Alvin, and C.E. Price, eds. *The Plant Cuticle*, pp. 121–30. Academic Press, London.

Joel, D.M., Juniper, B.E., and Dafni, A. (1985). Ultraviolet patterns in the traps of carnivorous plants. *New Phytologist*, 101, 585–93.

Joel, D.M., Rea, P.A., and Juniper, B.E. (1983). The cuticle of *Dionaea muscipula* Ellis. (Venus's Flytrap) in relation to stimulation, secretion and absorption. *Protoplasma*, 114, 44–51.

Jonathan (1992). A letter from Sierra Leone. *Carnivorous Plant Newsletter*, 21, 51–53.

Jones, D.T. and Gathorne-Hardy F. (1995). Foraging activity of the processional termite *Hospitalitermes hospitalis* (Termitidae: Nasutitermitinae) in the rain forest of Brunei, north-west Borneo. *Insectes Sociaux*, 42, 359–69.

Joppa, L., Visconti, P., Jenkins, C.N., and Pimm, S.L. (2013). Achieving the convention on biological diversity's goals for plant conservation. *Science*, 341, 1100.

Joyeux, M. (2013). Elastic models of the fast traps of carnivorous *Dionaea* and *Aldrovanda*. *Physical Review E*, 88, 034701.

Juang, T.C.C., Juang, S.D.C., and Liu, Z.H. (2011). Direct evidence of the symplastic pathway in the trap of the bladderwort *Utricularia gibba* L. *Botanical Studies*, 42, 47–54.

Junichi, S., Nagano, K., and Hoshi, Y. (2011). A chromosome study of two centromere differentiating *Drosera* species, *D. arcturi* and *D. regia*. *Caryologia*, 64, 453–63.

Juniper, B.E. and Burras, J.K. (1962). How pitcher plants trap insects. *New Scientist*, 269, 75–77.

Juniper, B.E., Robins, R.J., and Joel, D.M. (1989). *The Carnivorous Plants*, Academic Press Ltd, London.

Jürgens, A., El-Sayed, A.M., and Suckling, D.M. (2009). Do carnivorous plants use volatiles for attracting prey insects? *Functional Ecology*, 23, 875–87.

Jürgens, A., Sciligo, A., Witt, T., El-Sayed, A.M., and Suckling, D.M. (2012). Pollinator–prey conflict in carnivorous plants. *Biological Reviews*, 87, 602–15.

Jürgens, A., Witt, T., Sciligo, A., and El-Sayed, A.M. (2015). The effect of trap colour and trap–flower distance on prey and pollinator capture in carnivorous *Drosera* species. *Functional Ecology*, 29, 1026–37.

Kačániová, M., Ďurechová, D., Vuković, N. et al. (2014). Antimicrobial activity of *Drosera rotundifolia* L. *Animal Science and Biotechnologies*, 47, 366–69.

Kadek, A., Mrázek, H., Halada, P., et al. (2014a). Aspartic protease Nepenthesin–1 as a tool for digestion in hydrogen/deuterium exchange mass spectrometry. *Analytical Chemistry*, 86, 4287–94.

Kadek, A., Tretyachenko V., Mrázek H., et al. (2014b). Expression and characterization of plant aspartic protease nepenthesin-1 from *Nepenthes gracilis*. *Protein Expression and Purification*, 95, 121–28.

Kalema, J., Namganda, M., Bbosa, G., and Oqwal-Okeng, J. (2016). Diversity and status of carnivorous plants in

Uganda: towards identification of sites most critical for their conservation. *Biodiversity and Conservation*, 25, 2035–53.

Kämäräinen, T., Uusitalo, J., Jalonen, J., Laine, K., and Hohtola, A. (2003). Regional and habitat differences in 7-methyljuglone content of Finnish *Drosera rotundifolia*. *Phytochemistry*, 63, 309–14.

Kameyama, Y. and Ohara, M. (2006). Genetic structure in aquatic bladderworts: clonal propagation and hybrid perpetuation. *Annals of Botany*, 98, 1017–24.

Kamieński, F. (1895). Lentibulariaceae. In A. Engler and K.A.E. Prantl, eds. *Die naturalichen Pflanzenfamilien IV, 3b*. Leipzig.

Kamiński, R. (1987a). Studies on the ecology of *Aldrovanda vesiculosa* L. I. Ecological differentiation of *A. vesiculosa* population under the influence of chemical factors in the habitat. *Ekologia Polska*, 35, 559–90.

Kamiński, R. (1987b). Studies on the ecology of *Aldrovanda vesiculosa* L. II. Organic substances, physical and biotic factors and the growth and development of *A. vesiculosa*. *Ekologia Polska*, 35, 591–609.

Kanokratana, P., Mhuanthong, W., Laothanachareon, T., et al. (2016). Comparative study of bacterial communities in *Nepenthes* pitchers and their correlation to species and fluid acidity. *Microbial Ecology*, 72, 381–93.

Karagatzides, J.D., Butler, J.L., and Ellison, A.M. (2009). The pitcher plant *Sarracenia purpurea* can directly acquire organic nitrogen and short-circuit the inorganic nitrogen cycle. *PLoS ONE*, 4, e6164.

Karagatzides, J.D. and Ellison, A.M. (2009). Construction costs, payback times, and the leaf economics of carnivorous plants. *American Journal of Botany*, 96, 1612–19.

Karlsson, P.S. (1988). Seasonal patterns of nitrogen, phosphorus and potassium utilization by three *Pinguicula* species. *Functional Ecology*, 11, 203–209.

Karlsson, P.S. and Carlsson, B. (1984). Why does *Pinguicula vulgaris* L. trap insects? *New Phytologist*, 97, 25–30.

Karlsson, P.S., Nordell, K.O., Eirefelt, S., and Svensson, A. (1987). Trapping efficiency of three carnivorous *Pinguicula* species. *Oecologia*, 73, 518–21.

Karlsson, P.S. and Pate, J.S. (1992). Contrasting effects of supplementary feeding of insects or mineral nutrients on the growth and nitrogen and phosphorus economy of pygmy species of *Drosera*. *Oecologia*, 92, 8–13.

Karlsson, P.S., Thorén, L.M., and Hanslin, H.M. (1994). Prey capture by three *Pinguicula* species in a subarctic environment. *Oecologia*, 99, 188–93.

Katagiri, F., Thilmony, R., and He, S.Y. (2002). The *Arabidopsis thaliana–Pseudomonas syringae* interaction. *The Arabidopsis Book/American Society of Plant Biologists*, 1, e0039.

Kato, M. (1993). Floral biology of *Nepenthes gracilis* (Nepenthaceae) in Sumatra. *American Journal of Botany*, 80, 924–27.

Kato, M., Hotta, M., Tamin, R., and Itino, T. (1993). Inter- and intra-specific variation in prey assemblages and inhabitant communities in *Nepenthes* pitchers in Sumatra. *Tropical Zoology*, 6, 11–25.

Kato, M. and Kawakita, A. (2004). Plant–pollinator interactions in New Caledonia influenced by introduced honey bees. *American Journal of Botany*, 91, 1814–27.

Kaul, R.B. (1982). Floral and fruit morphology of *Nepenthes lowii* and *N. villosa*, montane carnivores of Borneo. *American Journal of Botany*, 69, 793–803.

Kawamichi, T. and Kawamichi, M. (1979). Spatial organization and territory of tree shrews (*Tupaia glis*). *Animal Behaviour*, 27, 381–93.

Kawiak, A., Królicka, A., and Łojkowska, E. (2003). Direct regeneration of *Drosera* from leaf explants and shoot tips. *Plant Cell, Tissue and Organ Culture*, 75, 175–78.

Kawiak, A., Królicka, A., and Łojkowska, E. (2011). In vitro cultures of *Drosera aliciae* as a source of a cytotoxic naphthoquinone: ramentaceone. *Biotechnology Letters*, 33, 2309–16.

Kearney, M.R., Wintle, B.A., and Porter, W.P. (2010). Correlative and mechanistic models of species distribution provide congruent forecasts under climate change. *Conservation Letters*, 3, 203–13.

Keighery, G. (1979). Chromosome counts in *Cephalotus* (Cephalotaceae). *Plant Systematics and Evolution*, 133, 103–04.

Keller, J.A., Herendeen, P.S., and Crane, P.R. (1996). Fossil flowers and fruits of the Actinidiaceae from the Campanian (late Cretaceous) of Georgia. *American Journal of Botany*, 83, 528–41.

Kellermann, C. and von Raumer, E. (1878). Vegetationsversuche an *Drosera rotundifolia*, mit und ohne Fleischfütterung. *Botanische Zeitung*, 36, 209–18; 225–29.

Kerner von Marilaun, A. (1878). *Flowers and their Unbidden Guests*. C. Kegan Paul & Co., London.

Kerner von Marilaun, A. and Wettstein, R. (1886). Die rhizopoiden Verdauungsorgane tierfangender Pflanzen. *Sitzungsberichte der kaiserlichen Akademie der Wissenschaften in Wien*, 93, 4–15.

Kesler, H.C., Trusty, J.L., Hermann, S.M., and Guyer, C. (2008). Demographic responses of *Pinguicula ionantha* to prescribed fire: a regression-design LTRE approach. *Oecologia*, 156, 545–57.

Khosla, C., Shivanna, K.R., and Mohan Ram H.Y. (1998). Pollination in the aquatic insectivore *Utricularia inflexa* var. *stellaris*. *Phytomorphology*, 48, 417–25.

Kibriya, S. and Jones, J.I. (2007). Nutrient availability and the carnivorous habit in *Utricularia vulgaris*. *Freshwater Biology*, 52, 500–09.

Kilian, C. (1951). Germination et développement post-embryonnaire de *Genlisea africana*. *Bulletin de l'Institut français d'Afrique noire*, 13, 1029–36.

Kim, J.K., Kim, J.J., and Lee, C.H. (2006). Effect of media components on in vitro propagation of *Darlingtonia californica* and *Heliamphora minor. 27th International Horticultural Congress and Exhibition*, 2006.8, 355 (Abstract).

Kingsolver, J. (1981). The effect of environmental uncertainty on morphological design and fluid balance in *Sarracenia purpurea* L. *Oecologia*, 48, 364–70.

Kirscht, A., Kaptan, S.S., Bienert, G.P., et al. (2016). Crystal structure of an ammonia-permeable aquaporin. *PLoS Biology*, 14, e1002411.

Kitching, R.L. (2000). *Food Webs and Container Habitats: The Natural History and Ecology of Phytotelmata*. Cambridge University Press, Cambridge.

Kitching, R.L. (2001). Food webs in phytotelmata: "Bottom–up" and "top–down" explanations for community structure. *Annual Review of Entomology*, 46, 729–60.

Kitching, R.L. and Beaver, R. (1990). Patchiness and community structure. In B. Shorrocks and I.R. Swingland, eds. *Living in a Patchy Environment*, pp. 147–76.Oxford University Press, Oxford.

Kneitel, J.M. (2012). Are trade-offs among species' ecological interactions scale dependent? A test using pitcher-plant inquiline species. *PLoS ONE*, 7, e41809.

Kneitel, J.M. and Miller, T.E. (2002). Resource and top-predator regulation in the pitcher plant (*Sarracenia purpurea*) inquiline community. *Ecology*, 83, 680–88.

Kneitel, J.M. and Miller, T.E. (2003). Dispersal rates affect species composition in metacommunities of *Sarracenia purpurea* inquilines. *American Naturalist*, 162, 165–71.

Knight, S.E. (1988). *The Ecophysiological Significance of Carnivory in* Utricularia vulgaris. Ph.D. dissertation, University of Wisconsin, Madison.

Knight, S.E. (1992). Costs of carnivory in the common bladderwort, *Utricularia macrorhiza. Oecologia*, 89, 348–55.

Knight, S.E. and Frost, T.M. (1991). Bladder control in *Utricularia macrorhiza*—lake-specific variation in investment in carnivory. *Ecology*, 72, 728–34.

Knobloch, E. and Mai, D.H. (1984). Neue Gattungen nach Früchten und Samen aus dem Cenoman bis Maastricht (Kreide) von Mitteleuropa. *Feddes Repertorium*, 95, 341.

Knoll, F. (1914). Über die Ursache des Ausgleitens der Insektenbeine an wachsbedeckten Pflanzenteilen. *Jahrbücher für Wissenschaftliche Botanik*, 54, 448–97.

Knápek, O. (2012). [*Evolution of The Genome Size and DNA Base Content in The* Nepenthes *Genus*] Ph.D. dissertation, Masaryk University, Brno. (In Czech.)

Koch, K. and Barthlott, W. (2009). Superhydrophobic and superhydrophilic plant surfaces: an inspiration for biomimetic materials. *Philosophical Transactions of the Royal Society A*, 367, 1487–509.

Koch, K. and Ensikat, H.-J. (2008). The hydrophobic coatings of plant surfaces: epicuticular wax crystals and their morphologies, crystallinity and molecular self-assembly. *Micron*, 39, 759–72.

Koch, G.W., Sillett, S.C., Jennings, G.M., and Davis, S.V. (2004). The limits to tree height. *Nature*, 428, 851–54.

Kokko, H. and López-Sepulcre, A. (2007). The ecogenetic link between demography and evolution: can we bridge the gap between theory and data? *Ecology Letters*, 10, 773–82.

Kok, P.J.R., Ratz, S., Tegelaar, M., Aubret, F., and Means, D.B. (2015). Out of taxonomic limbo: a name for the species of *Tepuihyla* (Anura: Hylidae) from the Chimantá Massif, Pantepui region, northern South America. *Salamandra*, 51, 283–314.

Koller-Peroutka, M., Lendl, T., Watzka, M., and Adlassnig, W. (2015). Capture of algae promotes growth and propagation in aquatic *Utricularia*. *Annals of Botany*, 115, 227–36.

Kolodziej, H., Pertz, H.H., and Humke, A. (2002). Main constituents of a commercial *Drosera* fluid extract and their antagonist activity at muscarinic M3 receptors in guinea-pig ileum. *Pharmazie*, 57, 201–03.

Kondo, K. (1971). Germination and developmental morphology of seed in *Utricularia cornuta* Michx. and *Utricularia juncea* Vahl. *Rhodora*, 73, 541–47.

Kondo, K. (1972a). A comparison of variability in *Utricularia cornuta* and *Utricularia juncea*. *American Journal of Botany*, 59, 23–37.

Kondo, K. (1972b). The chromosome number of *Heliamphora heterodoxa*. *Journal of Japanese Botany*, 47: 238.

Kondo, K. (1973). The chromosome numbers of four species of carnivorous plants. *Phyton*, 31, 93–94.

Kondo, K., Segawa, M., and Nehira, K. (1978). Anatomical studies on seeds and seedlings of some *Utricularia* (Lentibulariaceae). *Brittonia*, 30, 89–95.

Kondo, K. and Shimai, H. (2006). Phylogenetic analysis of the northern *Pinguicula* (Lentibulariaceae) based on internal transcribed spacer (ITS) sequence. *Acta Phytotaxonomica et Geobotanica*, 57, 155–64.

Konno, K., Hirayama, C., Nakamura, M., et al. (2004). Papain protects papaya trees from herbivorous insects: role of cysteine proteases in latex. *The Plant Journal for Cell and Molecular Biology*, 37, 370–78.

Konopka, A.S., Herendeen, P.S., and Crane, P.R. 1998. Sporophytes and gametophytes of Dicranaceae from the Santonian (late Cretaceous) of Georgia, USA. *American Journal of Botany*, 85, 714–23.

Kopp, B., Wawrosch, C., Buol, I., and Dorfer, T. (2006). Efficient production of Sundew (*Drosera rotundifolia* L.) in vitro using a temporary immersion system. *Planta Medica*, 72, 336.

Kosiba, P. (1992). Studies on the ecology of *Utricularia vulgaris* L. I. Ecological differentiation of *Utricularia vulgaris* L. population affected by chemical factors of the habitat. *Ekologia Polska*, 40, 147–92.

Kottek, M., Grieser, J., Beck, C., Rudolf, B., and Rubel, F. (2006). World map of Köppen–Geiger climate classification updated. *Meteorologische Zeitschrift*, 15, 259–63.

Kováčik, J., Klejdus, B., Štork, F., and Hedbavny, J. (2012). Prey-induced changes in the accumulation of amino acids and phenolic metabolites in the leaves of *Drosera capensis* L. *Amino Acids*, 42, 1277–85.

Krausko, M., Perutka, Z., Šebela, M., et al. (2017). The role of electrical and jasmonate signalling in the recognition of captured prey in the carnivorous sundew plant *Drosera capensis*. *New Phytologist*, 213, 1818–35.

Krenn, L., Beyer, G., Pertz, H.H., et al. (2004). In vitro antispasmodic and anti-inflammatory effects of *Drosera rotundifolia*. *Arzneimittelforschung*, 54, 402–05.

Kress, A. (1970). Zytotaxonomische Untersuchungen an einigen Insektenfängern (Droseraceae, Byblidaceae, Cephalotaceae, Roridulaceae, Sarraceniaceae). *Berichte der Deutschen Botanischen Gesellschaft*, 83, 55–62.

Kreuzwieser, J., Scheerer, U., Kruse, J., et al. (2014). The Venus flytrap attracts insects by the release of volatile organic compounds. *Journal of Experimental Botany*, 65, 755–66.

Krimmel, B.A. and Pearse, I.S. (2013). Sticky plant traps insects to enhance indirect defence. *Ecology Letters*, 16, 219–24.

Krimmel, B.A. and Pearse, I.S. (2014). Generalist and sticky plant specialist predators suppress herbivores on a sticky plant. *Arthropod–Plant Interactions*, 8, 403–10.

Krogan, N.T. and Berleth, T. (2007). From genes to patterns: auxin distribution and auxin-dependent gene regulation in plant pattern formation. *Botany*, 85, 353–68.

Krolicka, A., Szpitter, A., Gilgenast, E., et al. (2008). Stimulation of antibacterial naphthoquinones and flavonoids accumulation in carnivorous plants grown in vitro by addition of elicitors. *Enzyme and Microbial Technology*, 42, 216–21.

Krolicka, A., Szpitter, A., Stawujak, K., et al. (2010). Teratomas of *Drosera capensis* var. *alba* as a source of naphthoquinone: ramentaceone. *Plant Cell, Tissue and Organ Culture*, 103, 285–92.

Kruse, J., Gao, P., Honsel, A., et al. (2014). Strategy of nitrogen acquisition and utilization by carnivorous *Dionaea muscipula*. *Oecologia*, 174, 839–51.

Krutzsch, W. (1970). Zur Kenntnis fossiler disperser Tetradenpollen. *Paläontologische Abhandlungen, Abteilung B*, 3, 399–433.

Krutzsch, W. (1985). Über *Nepenthes*-Pollen (alias "*Droseridites*" p. p.) im europäischen Tertiär. *Gleditschia*, 13, 89–93.

Krychowiak, M., Grinholc, M., Banasiuk, R., et al. (2014). Combination of silver nanoparticles and *Drosera binata* extract as a possible alternative for antibiotic treatment of burn wound infections caused by resistant *Staphylococcus aureus*. *PLoS ONE*, 9, e115727.

Król, E., Dziubinska, H., Stolarz, M., and Trebacz, K. (2006). Effects of ion channel inhibitors on cold- and electrically-induced action potentials in *Dionaea muscipula*. *Biologia Plantarum*, 50, 411–16.

Król, E., Płachno, B.J., Adamec, L., et al. (2012). Quite a few reasons for calling carnivores "the most wonderful plants in the world." *Annals of Botany*, 109, 47–64.

Kubitzki, K. (2003a). Drosophyllaceae. In K. Kubitzki and C. Bayer, eds. *The Families and Genera of Vascular Plants*, volume 5, pp. 203–05. Springer, Berlin.

Kubitzki, K. (2003b). Nepenthaceae. In K. Kubitzki and C. Bayer, eds. *The Families and Genera of Vascular Plants*, volume 5, pp. 320–24. Springer, Berlin.

Kubitzki, K. (2004). Sarraceniaceae. In K. Kubitzki, ed. *The Families and Genera of Vascular Plants*, volume 6, pp. 422–25. Springer, Berlin.

Kuhlmann, J.G. (1938). Notas biológicas sobre Lentibulariáceas. *Anais da Primeira Reunião Sul–Americana de Botânica*, 3, 311–22.

Kumar, M. (1995). Pollen tetrads from Palaeocene sediments of Meghalaya, India: comments on their morphology, botanical affinity and geological records. *Palaeobotanist*, 43, 68–81.

Kumazawa, M. (1967). An experimental study on the seedling of *Utricularia pilosa* Makino. *Phytomorphology*, 17, 524–28.

Kurata, S. (2001). Two new *Nepenthes* species from Sumatra (Indonesia) and Mindanao (Philippines). *Journal of Insectivorous Plant Society*, 52, 30–34.

Kurata, S. (2003). A new Philippine pitcher plant, the third species having a saddle-shaped stem. *Journal of Insectivorous Plant Society*, 54, 41–44.

Kurata, K., Jaffré, T. and Setoguchi, H. (2008). Genetic diversity and geographical structure of the pitcher plant *Nepenthes vieillardii* in New Caledonia: a chloroplast DNA haplotype analysis. *American Journal of Botany*, 95, 1632–44.

Kurup, R., Johnson, A.J., Sankar, S., et al. (2013). Fluorescent prey traps in carnivorous plants. *Plant Biology*, 15, 611–15.

Laakkonen, L., Jobson, R.W., and Albert, V.A. (2006). A new model for the evolution of carnivory in the bladderwort plant (*Utricularia*): adaptive changes in cytochrome *c* oxidase (COX) provide respiratory power. *Plant Biology*, 8, 758–64.

Lair, K.P., Bradshaw, W.E., and Holzapfel, C.M. (1997). Evolutionary divergence of the genetic architecture underlying photoperiodism in the pitcher-plant mosquito, *Wyeomyia smithii*. *Genetics*, 147, 1873–83.

Laisk, A., Oja, V., and Eichelmann, H. (2007). Kinetics of leaf oxygen uptake represent in planta activities of respiratory electron transport and terminal oxidases. *Physiologia Plantarum*, 131, 1–9.

Lam, W.N., Chong, K.Y., Anand, G.S., and Tan, H.T.W. (2017). Dipteran larvae and microbes facilitate nutrient sequestration in the *Nepenthes gracilis* pitcher plant host. *Biology Letters*, 13, 20160928.

Lambers, H. (2014). *Plant Life on The Sandplains in Southwest Australia: A Global Biodiversity Hotspot*. University of Western Australia Press, Nedlands.

Lambers, H., Hayes, P., Laliberté, E., and Zemunik, G. (2014). The role of phosphorus in explaining plant biodiversity patterns and processes in a global biodiversity hotspot. In L. Mucina, J.N. Price, and J.M. Kalwij, eds. *Biodiversity and Vegetation: Patterns, Processes, Conservation*, pp. 41–42. Kwongan Foundation, Perth.

Lamont, E.E., Sivertsen, R., Doyle, C., and Adamec, L. (2013). Extant populations of *Aldrovanda vesiculosa* (Droseraceae) in the New World. *Journal of the Torrey Botanical Society*, 140, 517–22.

Lampard, S., Gluch, O., Robinson, A., et al. (2016). Pinguicula *of Latin Americ*a. Redfern Natural History Production, Poole, Dorset.

Lampert, K. and Schartl, M. (2008). The origin and evolution of a unisexual hybrid: *Poecilia formosa*. *Philosophical Transactions of the Royal Society, B*, 363, 2901–09.

Lan, T., Renner, T., Ibarra-Laclette, E., et al. (2017). Long-read sequencing uncovers the adaptive topography of a carnivorous plant genome. *Proceedings of the National Academy of Sciences, USA*, **114**, E4435–E4441.

Lang, F.X. (1901). Untersuchungen über Morphologie, Anatomie und Samenentwicklung von *Polypompholyx* und *Byblis*. *Flora*, 88, 149–206.

Länger, R., Pein, I., and Kopp, B. (1995). Glandular hairs in the genus *Drosera* (Droseraceae). *Plant Systematics and Evolution*, 194, 163–72.

Laurance, W.F. (2010). Habitat destruction: death by a thousand cuts. In N.S. Sodhi and P.R. Ehrlich, eds. *Conservation Biology for All*, pp. 73–87. Oxford University Press, Oxford.

Leavitt, R.G. (1903). Reversionary stages experimentally induced in *Drosera intermedia*. *Rhodora*, 5, 265–72.

Leduc, C., Coonishish, J., Haddad, P., and Cuerrier, A. (2006). Plants used by the Cree Nation of Eeyou Istchee (Quebec, Canada) for the treatment of diabetes: a novel approach in quantitative ethnobotany. *Journal of Ethnopharmacology*, 105, 55–63.

Lee, C.C. (2007). A preliminary conservation assessment of *Nepenthes clipeata* (Nepenthaceae). In C.C. Lee and C. Clarke, eds. *Proceedings of the 2007 Sarawak Nepenthes Summit*, pp. 96–100. Sarawak Forestry Corporation, Kuching.

Lee, J.M., Tan, W.S., and Ting, A. (2014). Revealing the antimicrobial and enzymatic potentials of culturable fungal endophytes from tropical pitcher plants (*Nepenthes* spp.). *Mycosphere*, 5, 364–77.

Lee, L., Zhang, Y., Ozar, B., Sensen, C.W., and Schriemer, D.C. (2016). Carnivorous nutrition in pitcher plants (*Nepenthes* spp.) via an unusual complement of endogenous enzymes. *Journal of Proteome Research*, 15, 3108–17.

Legendre, L. (2000). The genus *Pinguicula* L. (*Lentibulariaceae*): An overview. *Acta Botanica Gallica*, 147, 77–95.

Legendre, L. (2012). An improved mineral nutrient solution for the in vitro propagation of *Pinguicula* species. *Carnivorous Plant Newsletter*, 41, 16–19.

Leibold, M.A., Holyoak, M., Mouquet, N., et al. (2004). The metacommunity concept: a framework for multi-scale community ecology. *Ecology Letters*, 7, 601–13.

Lemmermann, E. (1914). Algologische Beitrage 23. Über das Vorkommen von Algen in den Schlauchen von *Utricularia*. *Abhandlungen des Naturwissenschaftlichen Vereins zu Bremen*, 23, 261–67.

Lenihan, W. and Schultz, R. (2014). Carnivorous pitcher plant species (*Sarracenia purpurea*) increases root growth in response to nitrogen addition. *Botany*, 92, 917–21.

Leushkin, E.V., Sutormin, R.A., Nabieva, E.R., et al. (2013). The miniature genome of a carnivorous plant *Genlisea aurea* contains a low number of genes and short non-coding sequences. *BMC Genomics*, 14, 476.

Levin, S.A. (1992). The problem of pattern and scale in ecology. *Ecology*, 73, 1943–67.

Levin, S.A. and Paine, R.T. (1974). Disturbance, patch formation, and community structure. *Proceedings of the National Academy of Sciences, USA*, 71, 2744–47.

Li, H. (2005). Early Cretaceous sarraceniacean-like pitcher plants from China. *Acta Botanica Gallica*, 152, 227–34.

Li, P. and Johnston, M.O. (2000). Heterochrony in plant evolutionary studies through the twentieth century. *Botanical Review*, 66, 57–88.

Li, J., Liu, Z., Tan, C., et al. (2010). Dynamics and mechanism of repair of ultraviolet-induced (6–4) photoproduct by photolyase. *Nature*, 466, 887–90.

Li, S. and Wang, K.W. (2015). Fluidic origami with embedded pressure dependent multi-stability: a plant inspired innovation. *Journal of the Royal Society Interface*, 12, 20150639.

Li, F., Wu, X., Lam, P., et al. (2008). Identification of the wax ester synthase/acyl-coenzyme A: diacylglycerol acyltransferase WSD1 required for stem wax ester biosynthesis in *Arabidopsis*. *Plant Physiology*, 148, 97–107.

Liao, Y.K. and Ji, Y.Y. (2014). Mass propagation of *Drosera burmannii* Vahl via induction of shoot fasciation and recovery of morphologically normal plantlets. *Propagation of Ornamental Plants*, 14, 158–70.

Libantová, J., Kämäräinen, T., Moravčíková, J., Matušíková, I., and Salaj, J. (2009). Detection of chitinolytic enzymes with different substrate specificity in tissues of intact sundew (*Drosera rotundifolia* L.): chitinases in sundew tissues. *Molecular Biology Reports*, 36, 851–56.

Libiaková, M., Floková, K., Novák, O., Slováková, L., and Pavlovič, A. (2014). Abundance of cysteine endopeptidase dionain in digestive fluid of Venus flytrap (*Dionaea muscipula* Ellis) is regulated by different stimuli from prey through jasmonates. *PLoS ONE*, 9, e104424.

Lichtner, F.T. and Williams, S.E. (1977). Prey capture and factors controlling trap narrowing in *Dionaea* (Droseraceae). *American Journal of Botany*, 64, 881–86.

Likhitwitayawuid, K., Kaewamatawong, R., Ruangrungsi, N., and Krungkrai, J. (1998). Antimalarial naphthoquinones from *Nepenthes thorelii*. *Planta Medica*, 64, 237–41.

Lim, K.K.P. and Ng, P.K.L. (1991). Nepenthophilous larvae and breeding habitats of the sticky frog, *Kalophrynus pleurostigma* Tschudi, Amphibia: Microhylidae. *Raffles Bulletin of Zoology*, 39, 209–14.

Lim, Y.S., Schöner, C.R., Schöner, M.G., et al. (2014). How a pitcher plant facilitates roosting of mutualistic woolly bats. *Evolutionary Ecology Research*, 16, 581–91.

Linder, H.P., Meadows, M.E., and Cowling, R.M. (1992). History of the Cape flora. In R.M. Cowling, ed. *The Ecology of Fynbos*, pp. 113–34. Oxford University Press, Cape Town.

Linnaeus, C. (1753). *Species Plantarum 1*. Impensis G.C. Nauk, Stockholm.

Lippincott, C.L. (2000). Effects of *Imperata cylindrica* (L.) Beauv. (cogon grass) invasion on fire regime in Florida Sandhill (USA). *Natural Areas Journal*, 20, 140–49.

Lloyd, F.E. (1934). Is *Roridula* a carnivorous plant? *Canadian Journal of Research*, 10, 780–86.

Lloyd, F.E. (1942). *The Carnivorous Plants*. Chronica Botanica, Waltham, MA, USA.

Lobreau-Callen, D., Jérémie, J., and Suarez-Cervera, M. (1999). Morphologie et ultrastructure du pollen dans le genre *Utricularia* L. (Lentibulariaceae). *Canadian Journal of Botany*, 77, 744–67.

Loarie, S.R., Duffy, P.B., Hamilton, H., et al. (2009). The velocity of climate change. *Nature*, 462, 1052–55.

Lockhart, J. (2014). Never let a good crisis go to waste: the kinesin ARK1 promotes microtubule catastrophe during root hair development. *The Plant Cell*, 26, 3221.

Löfstrand, S.D. and Schönenberger, J. (2015). Molecular phylogenetics and floral evolution in the sarracenioid clade (Actinidiaceae, Roridulaceae, and Sarraceniaceae) of Ericales. *Taxon*, 64, 1209–24.

Lomba, A., Pellissier, L., Randin, C., et al. (2010). Overcoming the rare species modelling paradox: a novel hierarchical framework applied to an Iberian endemic plant. *Biological Conservation*, 143, 2647–57.

Lorang, J., Kidarsa, T., Bradford, C., et al. (2012). Tricking the guard: exploiting plant defense for disease susceptibility. *Science*, 338, 659–62.

Lounibos, L. and Bradshaw, W. (1975). Second diapause in *Wyeomyia smithii*—seasonal incidence and maintenance. *Canadian Journal of Zoology*, 53, 215–21.

Love, S. (2016). A poacher who stole 970 Venus flytraps in N.C. is sentenced to prison. *The Washington Post*, July 28, 2016.

Lowrey, T.K. (1991). Chromosome and isozyme number in the Nepenthaceae. *American Journal of Botany*, 78 Suppl., 200–201.

Lowrie, A. (1996). An easy method to smoke treat carnivorous plant seed. *Bulletin of the Australian Carnivorous Plant Society*, 15, 3–5.

Lowrie, A. (1998). *Carnivorous Plants of Australia*, volume 3. University of Western Australia Press, Perth.

Lowrie, A. (2001). Floral mimicry and pollinator observations in carnivorous plants. *Bulletin of the Australian Carnivorous Plant Society*, 20, 10–15.

Lowrie, A. (2013). *Carnivorous Plants of Australia—Magnum Opus*. Redfern Natural History Productions, Poole.

Lowrie, A. and Conran, J.G. (1998). A taxonomic revision of the genus *Byblis* (Byblidaceae) in northern Australia. *Nuytsia*, 12, 59–74.

Lowrie, A. and Conran, J.G. (2007). *Byblis guehoi* (Byblidaceae), a new species from the Kimberley, Western Australia. *Telopea*, 12, 23–29.

Lowrie, A., Cowie, I., and Coran, J. (2008). A new species and section of *Utricularia* (Lentibulariaceae) from Northern Australia. *Telopea*, 12, 31–46.

Luetzelburg, P. (1910). Beitrage zur Kenntniss der Utricularien. *Flora*, 100, 145–212.

Luken, J.O. (2005). Habitats of *Dionaea muscipula* (Venus' Fly Trap), Droseraceae, associated with Carolina Bays. *Southeastern Naturalist*, 4, 573–84.

Luken, J.O. (2007). Performance of *Dionaea muscipula* as influenced by developing vegetation. *Journal of the Torrey Botanical Society*, 134, 45–52.

Luken, J.O. (2012). Long term outcomes of Venus Flytrap (*Dionaea muscipula*) establishment. *Restoration Ecology*, 20, 669–70.

Luther, H.E. (2014). *An Alphabetical List of Bromeliad Binomials*, 14th edition. Marie Selby Botanical Gardens and Bromeliad Society International, Sarasota.

Lüttge, U. (1983). Ecophysiology of carnivorous plants. In O.L. Lange, P.S. Nobel, C.B. Osmond, and H. Ziegler, eds. *Encyclopedia of Plant Physiology*, volume 12C, pp. 489–517. Springer-Verlag, Heidelberg.

Lynch, M. (2007). *The Origins of Genome Architecture*. Sinauer Associates, Sunderland.

Lyons, E., Pederson, B., Kane, J., et al. (2008). Finding and comparing syntenic regions among *Arabidopsis* and the outgroups papaya, poplar and grape: CoGe with rosids. *Plant Physiology*, 148, 1772–81.

Maas, D. (1989). Germination characteristics of some plant species from calcareous fens in southern Germany and their implications for the seed bank. *Holarctic Ecology*, 12, 337–44.

Maberly, S.C. and Spence, D.H.N. (1983). Photosynthetic inorganic carbon use by freshwater plants. *Journal of Ecology*, 71, 705–24.

MacArthur, R.H. (1962). Some generalized theorems of natural selection. *Proceedings of the National Academy of Sciences, USA*, 48, 1893–97.

MacArthur, R.H. and Wilson, E.O. (1967). *The Theory of Island Biogeography*. Princeton University Press, Princeton, NJ.

Macaya-Sanz, D., Heuertz, M., Lindtke, D., et al. (2016). Causes and consequences of large clonal assemblies in a poplar hybrid zone. *Molecular Ecology*, 25, 5330–44.

Macfarlane, J.M. (1908). Nepenthaceae. In A. Engler, ed. *Das Pflanzenreich, Heft 36, 4, 3*, 1–92.

Macphail, M.K. and Truswell, E.M. (2004). Palynology of Site 1166, Prydz Bay, East Antarctica. *Proceedings of the Ocean Drilling Program, Scientific Results*, 188, 1–43.

Magallón, S., Gómez-Acevedo, S., Sánchez-Reyes, L.L., and Hernández-Hernández, T. (2015). A metacalibrated time-tree documents the early rise of flowering plant phylogenetic diversity. *New Phytologist*, 207, 437–53.

Maffei, M.E., Mithöfer, A., and Boland, W. (2007). Before gene expression: early events in plant–insect interaction. *Trends in Plant Science*, 12, 310–16.

Maguire, B. (1978). Sarraceniaceae. *Memoirs of the New York Botanical Garden*, 29, 36–62.

Maguire, B. and Wurdack, J.J. (1957). The botany of the Guayana Highland, part II. *Memoirs of the New York Botanical Garden*, 9, 331–36.

Maguire, K.C., Nieto-Lugilde, D., Blois, J.L., et al. (2016). Controlled comparison of species- and community-level models across novel climates and communities. *Proceedings of the Royal Society of London B*, 283, 20152817.

Mäkela, A., Givnish, T.J., Berninger, F., et al. (2002). Challenges and opportunities of the optimality approach in plant ecology. *Silva Fennica*, 36, 605–14.

Malcolm, J.R., Markham, A., Neilson, R.P., and Garaci, M. (2002). Estimated migration rates under scenarios of global climate change. *Journal of Biogeography*, 29, 835–49.

Maldonado San Martin, A., Adamec, L., Suda, J., Mes, T., and Štorchová, H. (2003). Genetic variation within the endangered species *Aldrovanda vesiculosa* (Droseraceae) as revealed by RAPD analysis. *Aquatic Botany*, 75, 159–72.

Malkmus, R. and Dehling, J.M. (2008). Anuran amphibians of Borneo as phytotelma-breeders—a synopsis. *Herpetozoa*, 20, 164–72.

Mameli, E. (1916). Ricerche anatomiche, fisiologiche e biologiche sulla *Martynia lutea* Lindl. *Atti dell'Istituto Botanico e del Laboratorio Crittogamico dell'Universita di Pavia, Serie II*, 16, 137–88.

Manda, A.J. (1892). Insect-eating plants. *Journal of the Royal Horticultural Society*, 15, 135–43.

Manjarres-Hernandez, A., Guisande, C., Torres, N.N., et al. (2006). Temporal and spatial change of the investment in carnivory of the tropical *Utricularia foliosa*. *Aquatic Botany*, 85, 212–18.

Marabini, J. (1987). Eine neue Unterart von *Nepenthes edwardsiana* Hook.fil. sowie Anmerkungen zur Taxonomie der Gattung *Nepenthes* L. *Mitteilungen der Botanischen Staatssammlung München*, 23, 423–29.

Marburger, J.E. (1979). Glandular leaf structure of *Triphyophyllum peltatum* (Dioncophyllaceae): a "fly-paper" insect trapper. *American Journal of Botany*, 66, 404–11.

Markin, V.S., Volkov, A.G., and Jovanov, E. (2008). Active movements in plants. Mechanism of trap closure by *Dionaea muscipula* Ellis. *Plant Signaling & Behavior*, 3, 778–83.

Markstädter, C., Federle, W., Jetter, R., Riederer, M., and Hölldobler, B. (2000). Chemical composition of the slippery epicuticular wax blooms on *Macaranga* (Euphorbiaceae) ant-plants. *Chemoecology*, 10, 33–40.

Marloth, R. (1903). Some recent observations on the biology of *Roridula*. *Annals of Botany*, 17, 151–58.

Marloth, R. (1910). Further observations on the biology of *Roridula*. *Transactions of the Royal Society of South Africa*, 2, 59–62.

Marloth, R. (1925). Roridulaceae. *The Flora of South Africa Vol. 2, Part I*, pp. 26–30. Darter Bros., Cape Town.

Martyn, A.J., Merritt, D.J., and Turner, S.R. (2009). Seed banking. In C.A. Offord and P.F. Meagher, eds. *Plant Germplasm Conservation in Australia*, pp. 63–86. Australian Network for Plant Conservation Inc., Canberra.

Masi, E., Ciszak, M., Colzi, I., Adamec, L., and Mancuso, S. (2016). Resting electrical network activity in traps of the aquatic carnivorous plants of the genera *Aldrovanda* and *Utricularia*. *Scientific Reports*, 6, e24989.

Masters, M. (1890). *Nepenthes stenophylla* spec. nova. *The Gardener's Chronicle*, 3, 240.

Mastretta-Yanes, A., Moreno-Letelier, A., Piñero, D., Jorgensen, T.H., and Emerson, B.C. (2015). Biodiversity in the Mexican highlands and the interaction of geology, geography and climate within the Trans-Mexican Volcanic Belt. *Journal of Biogeography*, 42, 1586–600.

Matsuzaki, Y., Ogawa-Ohnishi, M., Mori, A., and Matsubayashi, Y. (2010). Secreted peptide signals required for maintenance of root stem cell niche in *Arabidopsis*. *Science*, 329, 1065–67.

Matthews, J.V. and Ovenden, L.E. (1990). Late Tertiary macrofossils from localities in arctic/subarctic North America: a review of the data. *Arctic*, 43, 364–92.

Matušíková, I., Salaj, J., Moravčíková, J., et al. (2005). Tentacles of in vitro-grown round-leaf sundew (*Drosera rotundifolia* L.) show induction of chitinase activity upon mimicking the presence of prey. *Planta*, 222, 1020–27.

McAlpine, D.K. (1990). A new apterous micropezid fly (Diptera: Schizophora) from Western Australia. *Systematic Entomology*, 15, 81–86.

McCormick, P.V., Harvey, J.W., and Crawford, E.S. (2011). Influence of changing water sources and mineral chemistry on the Everglades ecosystem. *Critical Reviews in Environmental Science and Technology*, 41, S28–S63.

McDaniel, S. (1971). The genus *Sarracenia* (Sarraceniaceae). *Bulletin of the Tall Timbers Research Station*, 9, 1–36.

McFall-Ngai M., Hadfield, M.G., Bosch, T.C.G., et al. (2013). Animals in a bacterial world, a new imperative for the life sciences. *Proceedings of the National Academy of Sciences, USA*, 110, 3229–36.

McGregor, J.P., Moon, D.C., and Rossi, A.M. (2016). Role of areoles on prey abundance and diversity in the hooded pitcher plant (*Sarracenia minor*, Sarraceniaceae). *Arthropod–Plant Interactions*, 10, 133–41.

McIntosh, R.P. (1999). The succession of succession: a lexical chronology. *Bulletin of the Ecological Society of America*, 80, 256–65.

McLachlan, J.S., Hellmann, J.J., and Schwartz, M.W. (2007). A framework for debate of assisted migration in an era of climate change. *Conservation Biology*, 21, 297–302.

McPherson, S. (2007). *Pitcher Plants of the Americas*. The McDonald & Woodward Publishing Company, Blacksburg.

McPherson, S. (2008). *Glistening Carnivores—The Sticky-Leaved Insect-Eating Plants*. Redfern Natural History Productions, Poole.

McPherson, S. (2009). *Pitcher Plants of the Old World*. Redfern Natural History Productions, Poole.

McPherson, S. (2010). *Carnivorous Plants and Their Habitats*. Redfern Natural History Productions, Poole.

McPherson, S. (2011). *New Nepenthes: Volume One*. Redfern Natural History Productions, Poole.

McPherson, S. and Schnell, D. (2011). *Sarraceniaceae of North America*. Redfern Natural History Productions, Poole.

McPherson, S., Wistuba, A., Fleischmann, A., and Nerz, J. (2011). *Sarraceniaceae of South America*. Redfern Natural History Productions, Poole.

Meierhofer, H. (1902). Beiträge zur Kenntnis der Anatomie und Entwicklungsgeschichte der *Utricularia*–Blasen. *Flora*, 90, 84–113.

Meimberg, H., Dittrich, P., Bringmann, G., Schlauer, J., and Heubl, G. (2000). Molecular phylogeny of Caryophyllidae s.l. based on *matK* sequences with special emphasis on carnivorous taxa. *Plant Biology*, 2, 218–28.

Meimberg, H. and Heubl, G. (2006). Introduction of a nuclear marker for phylogenetic analysis of Nepenthaceae. *Plant Biology*, 8, 831–40.

Meimberg, H., Wistuba, A., Dittrich, P., and Heubl, G. (2001). Molecular phylogeny of Nepenthaceae based on cladistic analysis of plastid *trnK* intron sequence data. *Plant Biology*, 3, 164–75.

Meindl, G.A. (2009). *Pollination Biology of* Darlingtonia californica, *The California Pitcher plant*. M.Sc. thesis, Humboldt State University, Arcata.

Meindl, G.A. and Mesler, M.R. (2011). Pollination biology of *Darlingtonia californica* (Sarraceniaceae), the California pitcher plant. *Madroño*, 58, 22–31.

Mellichamp, T.L. (2009). *Darlingtonia*. In Flora of North America Editorial Committee, eds. *Flora of North America North of Mexico*, volume 8, pp. 349–50. Oxford University Press, New York.

Mellichamp, T.L. and Case, F.W. (2009). *Sarracenia*. In Flora of North America Editorial Committee, eds. *Flora of North America North of Mexico*, volume 8, pp. 350–63. Oxford University Press, New York.

Melzig, M.F., Pertz, H.H., and Krenn, L. (2001). Anti-inflammatory and spasmolytic activity of extracts from Droserae herba. *Phytomedicine*, 8, 225–29.

Mencuccini, M. (2003). The ecological significance of long distance water transport, short-term regulation, and long-term acclimation across plant growth forms. *Plant, Cell and Environment*, 26,163–82.

Méndez, M. and Karlsson, P.S. (1999). Costs and benefits of carnivory in plants: insights from the photosynthetic performance of four carnivorous plants in a subarctic environment. *Oikos*, 86, 105–12.

Menninger, E.A. (1965). An African vine with three kinds of leaves for three different jobs. *The Garden Journal*, 15, 29–31.

Menz, J., Li, Z., Schulze, W.X., and Ludewig, U. (2016). Early nitrogen-deprivation responses in *Arabidopsis* roots reveal distinct differences on transcriptome and (phospho-) proteome levels between nitrate and ammonium nutrition. *Plant Journal*, 88, 717–34.

Merbach, M.A., Merbach, D.J., Maschwitz, U., et al. (2002). Mass march of termites into the deadly trap. *Nature*, 415, 36–37.

Merbach, M.A., Zizka, G., Fiala, B., et al. (2007). Why a carnivorous plant cooperates with an ant: selective defense against pitcher-destroying weevils in the myrmecophytic pitcher plant *Nepenthes bicalcarata* Hook. f. *Ecotropica*, 13, 45–56.

Merbach, M.A., Zizka, G., Fiala, B., Maschwitz, U., and Booth, W.E. (2001). Patterns of nectar secretion in five *Nepenthes* species from Brunei Darussalam, Northwest Borneo, and implications for ant–plant relationships. *Flora*, 196, 153–60.

Merbach, M.A., Zizka, G., Fiala, B., Merbach, D., and Maschwitz, U. (1999). Giant nectaries in the peristome thorns of the pitcher plant *Nepenthes bicalcarata* Hooker f. (Nepenthaceae): anatomy and functional aspects. *Ecotropica*, 5, 45–50.

Merckx, V.S.F.T., Hendriks, K.P., and Beentjes, K.K. (2015). Evolution of endemism on a young tropical mountain. *Nature*, 424, 347–50.

Merz, C., Catchen, J.M., Hanson-Smith, V., et al. (2013). Replicate phylogenies and post-glacial range expansion of the pitcher-plant mosquito, *Wyeomyia smithii*, in North America. *PLoS ONE*, 8, e72262.

Metcalfe, C.R. (1952). The anatomical structure of the Dioncophyllaceae in relation to the taxonomic affinities of the family. *Kew Bulletin*, 6, 351–68.

Mette, N., Wilbert, N., and Barthlott, W. (2000). Food composition of aquatic bladderworts (*Utricularia*, Lentibulariaceae) in various habitats. *Beiträge zur Biologie der Pflanzen*, 72, 1–13.

Meudt, H.M., Rojas-Andrés, B.M., Prebble, J.M, et al. (2015). Is genome downsizing associated with diversification in polyploid lineages of *Veronica*? *Botanical Journal of the Linnean Society*, 178, 243–66.

Meusel, H., Jäger, E.J., and Weinert, E. (1965). *Vergleichende Chorologie der zentraleuropäischen Flora. Text u. Karten* 1. VEB Fischer, Jena.

Meyers, D.G. and Strickler, J.R. (1979). Capture enhancement in a carnivorous aquatic plant: function of antennae and bristles in *Utricularia vulgaris*. *Science*, 203, 1022–25.

Meyers-Rice, B.A. (1994). Are *Genlisea* traps active? A crude calculation. *Carnivorous Plant Newsletter*, 23, 40–42.

Michalko J., Mészáros, P., Renner, T., et al. (2017). Molecular characterization and evolution of carnivorous sundew (*Drosera rotundifolia* L.) class V β-1,3-glucanase. *Planta*, 245, 77–91.

Michalko, J., Socha, P., Mészáros, P., et al. (2013). Glucan-rich diet is digested and taken up by the carnivorous sundew (*Drosera rotundifolia* L.): implication for a novel role of plant β-1,3-glucanases. *Planta*, 238, 715–25.

Midgley, J.J. and Stock, W.D. (1998). Natural abundance of δ^{15}N confirms insectivorous habit of *Roridula gorgonias*, despite its having no proteolytic enzymes. *Annals of Botany*, 82, 387–88.

Mildenhall, D.C. (1980). New Zealand Late Cretaceous and Cenozoic plant biogeography: a contribution. *Palaeogeography, Palaeoclimatology, Palaeoecology*, 31, 197–233.

Miles, D.H., Kokpol, U., Mody, N.V., and Hedin, P.A. (1975). Volatiles in *Sarracenia flava*. *Phytochemistry*, 14, 845–46.

Millennium Ecosystem Assessment (2005). *Ecosystems and Human Well-being: Synthesis*. Island Press, Washington, DC.

Miller, T.E. and Kneitel, J.M. (2005). Inquiline communities in pitcher plants as a prototypical metacommunity. In M. Holyoak, M.A. Leibold and R.D. Holt, eds. *Metacommunities: Spatial Dynamics and Ecological Communities*, pp. 122–45. University of Chicago Press, Chicago.

Miller, T.E., Kneitel, J.M., and Burns, J.H. (2002). Effect of community structure on invasion success and rate. *Ecology*, 83, 898–905.

Miller, T.E., Moran, E.R., and terHorst, Casey. (2014). Rethinking niche evolution: experiments with natural communities of protozoa in pitcher plants. *American Naturalist*, 184, 277–83.

Miller, T.E. and terHorst, C.P. (2012). Testing successional hypotheses of stability, heterogeneity, and diversity in pitcher-plant inquiline communities. *Oecologia*, 170, 243–51.

Millett, J., Foot, G.W., and Svensson, B.M. (2015). Nitrogen deposition and prey nitrogen uptake control the nutrition of the carnivorous plant *Drosera rotundifolia*. *Science of the Total Environment*, 512–13, 631–36.

Millett, J., Jones, R.I., and Waldron, S. (2003). The contribution of insect prey to the total nitrogen content of sundews (*Drosera* spp.) determined in situ by stable isotope analysis. *New Phytologist*, 158, 527–34.

Ming, L.T. (1997). Chordata. In H.T.W. Tan, ed. *A Guide to the Carnivorous Plants of Singapore*, pp. 155–57. Singapore Science Centre, Singapore.

Mithöfer, A. (2011). Carnivorous pitcher plants: insights in an old topic. *Phytochemistry*, 72, 1678–82.

Mithöfer, A., Reichelt, M., and Nakamura, Y. (2014). Wound and insect-induced jasmonate accumulation in carnivorous *Drosera capensis*: two sides of the same coin. *Plant Biology*, 16, 982–87.

Miura, T. and Matsumoto, T. (1998). Foraging organization of the open-air processional lichen-feeding termite *Hospitalitermes* (Isoptera, Termitidae) in Borneo. *Insectes Sociaux*, 45, 17–32.

Moeller, R.E. (1980). The temperature-determined growing season of a submerged hydrophyte: tissue chemistry and biomass turnover of *Utricularia purpurea*. *Freshwater Biology*, 10, 391–400.

Mogi, M. (2010). Unusual life history traits of *Aëdes* (*Stegomyia*) mosquitoes (Diptera: Culicidae) inhabiting *Nepenthes* pitchers. *Annals of the Entomological Society of America*, 103, 618–24.

Mogi, M. and Yong, H.S. (1992). Aquatic arthropod communities in *Nepenthes* pitchers: the role of niche differentiation, aggregation, predation and competition in community organization. *Oecologia*, 90, 172–84.

Moles, A.T., Ackerly, D.D., Webb, C.O., et al. (2005). A brief history of seed size. *Science*, 307, 576–80.

Moon, D.C., Rossi, A.M., Depaz, J., et al. (2010). Ants provide nutritional and defensive benefits to the carnivorous plant *Sarracenia minor*. *Oecologia*, 164, 185–92.

Moore, D. (1874). On a hybrid *Sarracenia*, with observations on other rare plants exhibited from Ireland. *The Gardener's Chronicle*, 1, 702–03.

Moran, J.A. (1991). *The Role and Mechanism of* Nepenthes rafflesiana *Pitchers as Insect Traps in Brunei*. Ph.D. dissertation, University of Aberdeen, Aberdeen.

Moran, J.A. (1996). Pitcher dimorphism, prey composition and the mechanisms of prey attraction in the pitcher plant *Nepenthes rafflesiana* in Borneo. *Journal of Ecology*, 84, 515–25.

Moran, J.A., Booth, W.E., and Charles, J.K. (1999). Aspects of pitcher morphology and spectral characteristics of six Bornean *Nepenthes* pitcher plant species: implications for prey capture. *Annals of Botany*, 83, 521–28.

Moran, J.A. and Clarke, C.M. (2010). The carnivorous syndrome in *Nepenthes* pitcher plants: current state of knowledge and potential future directions. *Plant Signaling & Behavior*, 5, 644–48.

Moran, J.A., Clarke, C., and Gowen, B.E. (2012). The use of light in prey capture by the tropical pitcher plant *Nepenthes aristolochioides*. *Plant Signaling & Behavior*, 7, 957–60.

Moran, J.A., Clarke, C.M., and Hawkins, B.J. (2003). From carnivore to detritivore? Isotopic evidence for leaf litter utilization by the tropical pitcher plant *Nepenthes ampullaria*. *International Journal of Plant Sciences*, 164, 635–39.

Moran, J.A., Gray, L.K., Clarke, C., and Chin, L. (2013). Capture mechanism in Palaeotropical pitcher plants (Nepenthaceae) is constrained by climate. *Annals of Botany*, 112, 1279–91.

Moran, J.A., Hawkins B.J., Gowen B.E., and Robbins S.L. (2010). Ion fluxes across the pitcher walls of three Bornean *Nepenthes* pitcher plant species: flux rates and gland distribution patterns reflect nitrogen sequestration strategies. *Journal of Experimental Botany*, 61, 1365–74.

Moran, J.A., Merbach, M.A., Livingston, N.J., Clarke, C.M., and Booth, W.E. (2001). Termite prey specialization in the pitcher plant *Nepenthes albomarginata*—evidence from stable isotope analysis. *Annals of Botany*, 88, 307–11.

Moran, J.A., Mitchell, A., Goodmanson, G., and Stockburger, K. (2000). Differentiation among effects of nitrogen fertilization treatments of conifer seedlings by foliar reflectance: a comparison of methods. *Tree Physiology*, 20, 1113–20.

Moran, J.A. and Moran, A.J. (1998). Foliar reflectance and vector analysis reveal nutrient stress in prey-deprived pitcher plants (*Nepenthes rafflesiana*). *International Journal of Plant Sciences*, 159, 996–1001.

Mouquet, N., Daufresne, T., Gray, S.M., and Miller, T.E. (2008). Modelling the relationship between a pitcher plant (*Sarracenia purpurea*) and its phytotelma community: mutualism or parasitism? *Functional Ecology*, 22, 728–37.

Mucina L., Laliberté E., Thiele K.R., Dodson J.R., and Harvey J. (2014). Biogeography of Kwongan: origins, diversity, endemism, and vegetation patterns. In H. Lambers, ed. *Plant Life on the Sand Plains in Southwest Australia*, pp. 35–80. University of Western Australia Publishing, Perth.

Müller, K. and Borsch, T. (2005). Phylogenetics of *Utricularia* (Lentibulariaceae) and molecular evolution of the *trnK* intron in a lineage with high substitutional rates. *Plant Systematics and Evolution*, 250, 39–67.

Müller, K., Borsch, T., Legendre, L., et al. (2004). Evolution of carnivory in Lentibulariaceae and the Lamiales. *Plant Biology*, 6, 477–90.

Müller, K., Borsch, T., Legendre, L., Porembski, S., and Barthlott, W. (2006). Recent progress in understanding the evolution of carnivorous Lentibulariaceae (Lamiales). *Plant Biology*, 8, 748–57.

Müller, J. and Deil, U. (2001). Ecology and structure of *Drosophyllum lusitanicum* (L.) Link populations in the south-west of the Iberian peninsula. *Acta Botanica Malacitana*, 26, 47–68.

Mullins, J.T. (2000). *Molecular Systematics of Nepenthaceae*. Ph.D. dissertation, University of Reading, Reading.

Munro, P.G. (2009). Deforestation: constructing problems and solutions on Sierra Leone's Freetown Peninsula. *Journal of Political Ecology*, 16, 104–22.

Muravnik, L.E. (1988). [The slime gland ultrastructure in *Pinguicula vulgaris* (Lentibulariaceae) in the course of their development and function]. *Botanicheskii Zhurnal*, 73, 1523–35. (In Russian.)

Muravnik, L.E. (1996). [Morphometrical approach to the secretory activity determination in digestive glands of *Aldrovanda vesiculosa* (Droseraceae)]. *Botanicheskii Zhurnal*, 81, 1–8. (In Russian.)

Muravnik, L.E. (2000). The effect of chemical stimulation on the ultrastructure of secretory cells of glandular hairs in the two species of *Drosera*. *Russian Journal of Plant Physiology*, 47, 614–23.

Muravnik, L.E. (2005). [Significance of the tentacle morphological and ultrastructural features in *Drosera* (Droseraceae) taxonomy]. *Botanicheskii Zhurnal*, 90, 14–24. (In Russian.)

Muravnik, L.E. (2008). [Ultrastructure on *Dionaea muscipula* and *Aldrovanda vesiculosa* (Droseraceae) digestive glands]. In Russian. *Botanicheskii Zhurnal*, 93, 289–99. (In Russian.)

Muravnik, L.E., Vassilyev, A.E., and Potapova, Y.Y. (1995). Ultrastructural aspects of digestive gland functioning in *Aldrovanda vesiculosa*. *Russian Journal of Plant Physiology*, 42, 5–13.

Murza, G.L. (2002). *Plant–Arthropod Interactions of The English Sundew* (Drosera anglica *Huds.) at The Macdowall Bog Protected Region, Saskatchewan*. M.Sc. thesis, University of Saskatchewan, Saskatoon.

Murza, G.L. and Davis, A.R. (2003). Comparative flower structure of three species of sundew (Droseraceae: *Drosera anglica, D. linearis* and *D. rotundifolia*) in relation to breeding system. *Canadian Journal of Botany*, 81, 1129–42.

Murza, G.L. and Davis, A.R. (2005). Flowering phenology and reproductive biology of *Drosera anglica* (Droseraceae). *Botanical Journal of the Linnean Society*, 147, 417–26.

Murza, G.L., Heaver, J.R., and Davis, A.R. (2006). Minor pollinator–prey conflict in the carnivorous plant, *Drosera anglica*. *Plant Ecology*, 184, 43–52.

Myers, J.A. and Harms, K.E. (2009). Local immigration, competition from dominant guilds, and the ecological assembly of high–diversity pine savannas. *Ecology*, 90, 2745–54.

Myers, N., Mittermeier, R.A., Mittermeier, C.G., da Fonseca, G.A.B., and Kent, J. (2000). Biodiversity hotspots for conservation priorities. *Nature*, 403, 853–58.

Naczi, R.F.C., Soper, E.M., Case, F.W., and Case, R.B. (1999). *Sarracenia rosea* (Sarraceniaceae), a new species of pitcher plant from the southeastern United States. *Sida*, 18, 1183–206.

Naeem, S. (1988). Resource heterogeneity fosters coexistence of a mite and a midge in pitcher plants. *Ecological Monographs*, 58, 215–27.

Nahálka, J., Nahálková, J., Gemeiner, P., and Blanárik, P. (1998). Elicitation of plumbagin by chitin and its release into the medium in *Drosophyllum lusitanicum* Link. suspension cultures. *Biotechnology Letters*, 20, 841–45.

Nakamura, Y., Reichelt, M., Mayer, V.E., and Mithöfer, A. (2013). Jasmonates trigger prey-induced formation of "outer stomach" in carnivorous sundew plants. *Proceedings of the Royal Society B*, 280, 20130228.

Nelson, E.C. (1986). *Sarracenia* hybrids raised at Glasnevin Botanic Gardens, Ireland: nomenclature and typification. *Taxon*, 35, 574–78.

Netolitzky, F. (1926). Anatomie der Angiospermen–Samen. Gebrüder Borntraeger, Berlin.

Neuhaus, J.M., Sticher, L., Meins, F., and Boller, T. (1991). A short C-terminal sequence is necessary and sufficient for the targeting of chitinases to the plant vacuole. *Proceedings of the National Academy of Sciences, USA*, 88, 10362–66.

Neumann, P., Navrátilová, A., Koblížková, A., et al. (2011). Plant centromeric retrotransposons: a structural and cytogenetic perspective. *Mobile DNA*, 2, 4.

Newell, S.J. and Nastase, A.J. (1998). Efficiency of insect capture by *Sarracenia purpurea* (Sarraceniaceae), the northern pitcher plant. *American Journal of Botany*, 85, 88–91.

Neyland, R., Bushnell, J., and Tangkham, W. (2015). An updated taxonomic treatment of the natural hybrids of *Sarracenia* L. (Sarraceniaceae). *Carnivorous Plant Newsletter*, 44, 4–12.

Neyland, R. and Merchant, M. (2006). Systematic relationships of Sarraceniaceae inferred from nuclear ribosomal DNA sequences. *Madroño*, 53, 223–32.

Ne'eman, G., Ne'eman, R., and Ellison, A.M. (2006). Limits to reproductive success of *Sarracenia purpurea* (Sarraceniaceae). *American Journal of Botany*, 93, 1660–66.

Nielsen, S.L. and Sand-Jensen, K. (1991). Variation in growth rates of submerged rooted macrophytes. *Aquatic Botany*, 39, 109–20.

Nishi, A.H., Vasconcellos-Neto, J., and Romero, G.Q. (2013). The role of multiple partners in a digestive mutualism with a protocarnivorous plant. *Annals of Botany*, 111, 143–50.

Nishimura, E., Jumyo, S., Arai, N., et al. (2014). Structural and functional characteristics of S-like ribonucleases from carnivorous plants. *Planta*, 240, 147–59.

Nishimura, E., Kawahara, M., Kodaira, R., et al. (2013). S-like ribonuclease gene expression in carnivorous plants. *Planta*, 238, 955–67.

Nogales, M., Heleno, R., Traveset, A., and Vargas, P. (2012). Evidence for overlooked mechanisms of long-distance seed dispersal to and between oceanic islands. *New Phytologist*, 194, 313–17.

Nogués-Bravo, D., Ohlemüller, R., Batra, P., and Araújo, M.B. (2010). Climate predictors of Late Quaternary extinctions. *Evolution*, 64, 2442–49.

Nordbakken, J.-F., Rydgren, K., and Økland, R.H. (2004). Demography and population dynamics of *Drosera anglica* and *D. rotundifolia*. *Journal of Ecology*, 92, 110–21.

Norment, C.J. (1987). A comparison of three methods for measuring arthropod abundance in tundra habitats and its implications in avian ecology. *Northwest Science*, 61,191–98.

Northcutt, C., Davies, D., Gagliardo, R., et al. (2012). Germination in vitro, micropropagation, and cryogenic storage for three rare pitcher plants: *Sarracenia oreophila* (Kearney) Wherry (federally endangered), *S. leucophylla* Raf. and *S. purpurea* spp. *venosa* (Raf.) Wherry. *HortScience*, 47, 74–80.

Noss, R.F., Platt, W.J, Sorrie, B.A., et al. (2015). How global biodiversity hotspots may go unrecognized: lessons from the North American coastal plain. *Diversity and Distributions*, 21, 236–44.

Nyoka, S.E. and Ferguson, C. (1999). Pollinators of *Darlingtonia californica* Torr., the California pitcher plant. *Natural Areas Journal*, 19, 386–91.

Obermeyer, A.A. (1970a). Droseraceae. In L.E. Codd, B. DeWinter, D.J.B. Killick and H. Rycroft, eds. *Flora of Southern Africa 13*, pp. 187–201. Kirstenbosch Botanic Garden Press, Cape Town.

Obermeyer, A.A. (1970b). Roridulaceae. In L.E. Codd, B. DeWinter, D.J.B. Killick, and H. Rycroft, eds. *Flora of Southern Africa 13*, pp. 201–4. Kirstenbosch Botanic Garden Press, Cape Town.

OECD (2016). http://www.oecd.org/sti/biotech/statisticaldefinitionofbiotechnology.htm.

Ogg, J.G. and Ogg, G. (2008). Late Cretaceous (65–100 Ma time-slice). https://engineering.purdue.edu/Stratigraphy/charts/Timeslices/3_Late_Cret.pdf.

Okabe, T., Futatsuya, C., Tanaka, O., and Ohyama, T. (2005b). Structural analysis of the gene encoding *Drosera adelae* S-like ribonuclease DA–I. *Journal of Advanced Science*, 17, 218–24.

Okabe, T., Iwakiri, Y., Mori, H., Ogawa, T., and Ohyama, T. (2005a). An S-like ribonuclease gene is used to generate a trap-leaf enzyme in the carnivorous plant *Drosera adelae*. *FEBS Letters*, 579, 5729–33.

Okabe, T., Mori, H., and Ohyama, T. (1997). Deoxyribonuclease secreted from an insectivorous plant *Drosera adelae*. *Nucleic Acids Symposium Series*, 37, 127–28.

Okahara, K. (1933). Physiological studies on *Drosera* IV. On the function of micro–organisms in the digestion of insect bodies by insectivorous plants. *Scientific Reports of Tohoku Imperial University*, 8, 151–68.

Okumoto, S., Schmidt, R., Tegeder, M., et al. (2002). High affinity amino acid transporters specifically expressed in xylem parenchyma and developing seeds of *Arabidopsis*. *Journal of Biological Chemistry*, 277, 45338–46.

Olde Venterink, H., Wassen, M.J., Verkroost, A.W.M., and de Ruiter, P.C. (2003). Species richness–productivity patterns differ between N-, P-, and K-limited wetlands. *Ecology*, 84, 2191–99.

Olff, H., van Andel, J., and Bakker, J.P. (1990). Biomass and shoot/root ratio allocation of five species from a grassland succession series at different combinations of light and nutrient supply. *Functional Ecology*, 4, 193–200.

Olivencia, A.O., Claver, J.P.C., and Alcaraz, J.A.D. (1995). Floral and reproductive biology of *Drosophyllum lusitanicum* (L.) Link (Droseraceae). *Botanical Journal of the Linnean Society*, 118, 331–51.

Ollerton, J., Winfree, R., and Tarrant, S. (2011). How many flowering plants are pollinated by animals? *Oikos*, 120, 321–26.

Olson, D.M., Dinerstein, E., Wikramanayake, E.D., et al. (2001). Terrestrial ecoregions of the world: a new map of life on earth. *BioScience*, 51, 933–38.

Ordonez, A. and Williams, J.W. (2013). Climatic and biotic velocities for woody taxa distributions over the last 16,000 years in eastern North America. *Ecology Letters*, 16, 773–81.

Orivel, J. and Leroy, C. (2011). The diversity and ecology of ant gardens (Hymenoptera: Formicidae; Spermatophyta: Angiospermae). *Myrmecological News*, 14, 73–85.

Ortega-Olivencia, A., Carrasco Claver, J.P., and Devesa Alcaraz, J.A. (1995). Floral and reproductive biology of *Drosophyllum lusitanicum* (L.) Link (Droseraceae). *Botanical Journal of the Linnean Society*, 118, 331–51.

Ortega-Olivencia, A., Paredes, J. A., Rodriguez-Riano, T., and Devesa, J. A. (1998). Modes of self-pollination and absence of cryptic self-incompatibility in *Drosophyllum lusitanicum* (Droseraceae). *Botanica Acta*, 111, 474–80.

Osmond, D.L., Line, D.E., Gale, J.A., et al. (1995). WATERSHEDSS: Water, Soil and Hydro-Environmental Decision Support System. Available online: http://h2osparc.wq.ncsu.edu.

Osunkoya, O.O., Bujang, D., Moksin, H., Wimmer, F.L., and Holige, T.M. (2004). Leaf properties and the construction cost of common, co-occurring plant species of disturbed heath forest in Borneo. *Australian Journal of Botany*, 52, 499–507.

Osunkoya, O.O., Daud, S.D., Di Giusto, B., Wimmer, F.L., and Holige, T.M. (2007). Construction costs and physico-chemical properties of the assimilatory organs of *Nepenthes* species in northern Borneo. *Annals of Botany*, 99, 895–906.

Owen, T.P., Benzing, D.H., and Thomson, W.W. (1988). Apoplastic and ultrastructural characterizations of the trichomes from the carnivorous bromeliad *Brocchinia reducta*. *Canadian Journal of Botany*, 66, 941–48.

Owen, T.P. and Lennon, K.A. (1999). Structure and development of the pitchers from the carnivorous plant *Nepenthes alata* (Nepenthaceae). *American Journal of Botany*, 86, 1382–90.

Owen, T.P., Jr., Lennon, K.A., Santo, M.J., and Anderson, A.N. (1999). Pathways for nutrient transport in the pitchers of the carnivorous plant *Nepenthes alata*. *Annals of Botany*, 84, 459–66.

Owen, T.P. and Thomson, W.W. (1991). Structure and function of a specialized cell wall in the trichomes of the carnivorous bromeliad *Brocchinia reducta*. *Canadian Journal of Botany*, 69, 1700–06.

Padhye, S., Dandawate, P., Yusufi, M., Ahmad, A., and Sarkar, F.H. (2012). Perspectives on medicinal properties of plumbagin and its analogs. *Medicinal Research Reviews*, 32, 1131–58.

Pagano, A.M. and Titus, J.E. (2004). Submersed macrophyte growth at low pH, contrasting responses of three species to dissolved inorganic carbon enrichment and sediment type. *Aquatic Botany*, 79, 65–74.

Pagano, A.M. and Titus, J.E. (2007). Submersed macrophyte growth at low pH, carbon source influences response to dissolved inorganic carbon enrichment. *Freshwater Biology*, 52, 2412–20.

Paine, R.T. (1966). Food web complexity and species diversity. *American Naturalist*, 100, 65–75.

Paisie, T.K., Miller, T.E., and Mason, O.U. (2014). Effects of a ciliate protozoa predator on microbial communities in pitcher plant (*Sarracenia purpurea*) leaves. *PLoS ONE*, 9, e113384.

Palací, C.A., Brown, G.K., and Tuthill, D.E. (2004). The seeds of *Catopsis* (Bromelliaceae: Tillandsioideae). *Systematic Botany*, 29, 518–27.

Palmquist, K.A., Peet, R.K., and Weakley, A.S. (2014). Changes in plant species richness following reduced fire frequency and drought in one of the most species rich savannas in North America. *Journal of Vegetation Science*, 25, 1426–37.

Panchy, N., Lehti-Shiu, M., and Shiu, S.H. (2016). Evolution of gene duplication in plants. *Plant Physiology*, 171, 2294–316.

Paniw, M., Gil-Cabeza, E. and Ojeda, F. (2017a). Plant carnivory beyond bogs: reliance on prey feeding in *Drosophyllum lusitanicum* (Drosophyllaceae) in dry Mediterranean heathland habitats. *Annals of Botany*, 119, 1035–41.

Paniw, M., Gil-López, M.J., and Segarra-Moragues, J.G. (2014) Isolation and characterization of microsatellite loci in the carnivorous subshrub *Drosophyllum lusitanicum*. *Biochemical Systematics and Ecology*, 57, 416–19.

Paniw, M., Quintana-Ascencio, P.F., Ojeda, F., and Salguero-Gómez, R. (2017b). Accounting for uncertainty in dormant life stages in stochastic demographic models. *Oikos*, 126, 900–909.

Paniw, M., Quintana-Ascencio, P.F., Ojeda, F., and Salguero-Gómez, R. (2017c). Interacting livestock and fire may both threaten and increase viability of a fire-adapted Mediterranean carnivorous plant. *Journal of Applied Ecology*, 54, 1884–94.

Paniw, M., Salguero-Gómez, R., and Ojeda, F. (2015). Local-scale disturbances can benefit an endangered, fire-adapted plant species in Western Mediterranean heathlands in the absence of fire. *Biological Conservation*, 187, 74–81.

Pant, B.D., Pant, P., Erban, A., et al. (2015). Identification of primary and secondary metabolites with phosphorus status-dependent abundance in *Arabidopsis*, and of the transcription factor PHR1 as a major regulator of metabolic changes during phosphorus limitation. *Plant, Cell and Environment*, 38, 172–87.

Paper, D.H., Karall, E., Kremser, M., and Krenn, L. (2005). Comparison of the anti-inflammatory effects of *Drosera rotundifolia* and *Drosera madagascariensis* in the HET–CAM assay. *Phytotherapy Research*, 19, 323–26.

Parain, E.C., Gravel, D., Rohr, R.P., Bersier, L.-F., and Gray, S.M. (2016). Mismatch in microbial food webs: predators but not prey perform better in their local biotic and abiotic conditions. *Ecology and Evolution*, 6, 4885–97.

Paris, R.R. and Delaveau, P. (1959). *Drosera*. Isolation of plumbagone from *Drosera auriculata* and ramentaceone from *D. ramentacea*. *Annales Pharmacologies Français*, 17, 585–92.

Paris, R. R. and Quevauvillier, A. (1947). Action de quelques drogues végétales sur les bronchospasms histaminiques et acetycholiniques. *Thérapie*, 2, 69–72.

Parisod, C., Trippi, C., and Galland, N. (2005). Genetic variability and founder effect in the pitcher plant *Sarracenia purpurea* (Sarraceniaceae) in populations introduced into Switzerland: from inbreeding to invasion. *Annals of Botany*, 95, 277–86.

Parkes, D.M. (1980). *Adaptive Mechanisms of Surfaces and Glands in Some Carnivorous Plants*. M.Sc. thesis, Monash University, Clayton.

Parkes, D. and Hallam, N. (1984). Adaptation for carnivory in the West Australian pitcher plant *Cephalotus follicularis* Labill. *Australian Journal of Botany*, 32, 595–604.

Parmesan, C., Ryrholm, N., Stefanescu, C., et al. (1999). Poleward shifts in geographical ranges of butterfly species associated with regional warming. *Nature*, 399, 579–83.

Pasteur, G. (1982). A classificatory review of mimicry systems. *Annual Review of Ecology and Systematics*, 13, 169–99.

Paszota, P., Escalante-Perez, M., Thomsen, L.R., et al. (2014). Secreted major Venus flytrap chitinase enables digestion of arthropod prey. *Biochimica et Biophysica Acta—Proteins and Proteomics*, 1844, 374–83.

Pate, J.S. (1986). Economy of symbiotic nitrogen fixation. In T.J. Givnish, ed. *On the Economy of Plant Form and Function*, pp. 299–35. Cambridge University Press, Cambridge, U.K.

Pate, J.S. and Dixon, K.W. (1978). Mineral nutrition of *Drosera erythrorhiza* Lindl. with special reference to its tuberous habit. *Australian Journal of Botany*, 26, 455–64.

Pavlovič, A. (2011). Photosynthetic characterization of Australian pitcher plant *Cephalotus follicularis*. *Photosynthetica*, 49, 253–58.

Pavlovič, A., Demko, V., and Hudák, J. (2010a). Trap closure and prey retention in Venus flytrap (*Dionaea muscipula*) temporarily reduces photosynthesis and stimulates respiration. *Annals of Botany*, 105, 37–44.

Pavlovič, A., Krausko, M., and Adamec, L. (2016). A carnivorous sundew plant prefers protein over chitin as a source of nitrogen from their traps. *Plant Physiology and Biochemistry*, 104, 11–16.

Pavlovič, A., Krausko, M., Libiaková, M., and Adamec, L. (2014). Feeding on prey increases photosynthetic efficiency in the carnivorous sundew *Drosera capensis*. *Annals of Botany*, 113, 69–78.

Pavlovič, A., Masarovičová, E., and Hudák, J. (2007). Carnivorous syndrome in Asian pitcher plants of the genus *Nepenthes*. *Annals of Botany*, 100, 527–36.

Pavlovič, A. and Saganová, M. (2015). A novel insight into the cost–benefit model for the evolution of botanical carnivory. *Annals of Botany*, 115, 1075–92.

Pavlovič, A., Singerová, L., Demko, V., and Hudák, J. (2009). Feeding enhances photosynthetic efficiency in the carnivorous pitcher plant *Nepenthes talangensis*. *Annals of Botany*, 104, 307–14.

Pavlovič, A., Singerová, L., Demko, V., Šantrůček, J., and Hudák, J. (2010b). Root nutrient uptake enhances photosynthetic assimilation in prey-deprived carnivorous pitcher plant *Nepenthes talangensis*. *Photosynthetica*, 48, 227–33.

Pavlovič, A., Slováková, L, Pandolfi, C., and Mancuso, S. (2011a). On the mechanism underlying photosynthetic limitation upon trigger hair irritation in the carnivorous plant Venus flytrap (*Dionaea muscipula* Ellis). *Journal of Experimental Botany*, 62, 1991–2000.

Pavlovič, A., Slováková, Ľ., and Šantrůček, J. (2011b). Nutritional benefit from leaf litter utilization in the pitcher plant *Nepenthes ampullaria*. *Plant, Cell and Environment*, 34, 1865–73.

Payne, J., Francis, C.M., and Phillipps, K. (1985). *A Field Guide to The Mammals of Borneo*. Sabah Society, Kota Kinabalu.

Peel, M.C., Finlayson, B.L., and McMahon, T.A. (2007). Updated world map of the Köppen–Geiger climate classification. *Hydrology and Earth System Sciences Discussions, European Geosciences Union*, 4, 439–73.

Peet, R.K. and Allard, D.J. (1993). Longleaf pine vegetation of the southern Atlantic and eastern Gulf Coast regions: a preliminary classification. *Proceedings of the Tall Timbers Fire Ecology Conference*, 18, 45–81.

Peet, R.K., Palmquist, K.A., and Tessel, S.M. (2014). Herbaceous layer species richness in southeastern forests and woodlands: patterns and causes. In F. Gilliam, ed. *The Herbaceous Layer in Forests in Eastern North America*, 2nd edition, pp. 255–76. Oxford University Press, Oxford.

Pelayo-Villamil, P., Guisande, C., Vari, R.P., et al. (2015). Global diversity patterns of freshwater fishes—potential victims of their own success. *Diversity and Distribution*, 21, 345–56.

Pereira, C.G., Almenara, D.P., Winter, C.E., et al. (2012). Underground leaves of *Philcoxia* trap and digest nematodes. *Proceedings of the National Academy of Sciences, USA*, 109, 1154–58.

Peroutka, M., Adlassnig, W., Volgger, M., et al. (2008). *Utricularia*, a vegetarian carnivorous plant? Algae as prey of bladderwort in oligotrophic bogs. *Plant Ecology*, 199, 153–62.

Peterson, C.N., Day, S., Wolfe, B.E., et al. (2008). A keystone predator controls bacterial diversity in the pitcher-plant (*Sarracenia purpurea*) microecosystem. *Environmental Microbiology*, 10, 2257–66.

Peterson, B.K., Weber, J.N., Kay, E.H., Fisher, H.S., and Hoekstra, H.E. (2012). Double digest RADseq; an inexpensive method for de novo SNP discovery and genotyping in model and non-model species. *PLoS ONE* 7, e37135.

Petricka, J.J., Clay, N.K., and Nelson, T.M. (2008). Vein patterning screens and the *defectively organized tributaries* mutants in *Arabidopsis thaliana*. *Plant Journal*, 56, 251–63.

Peñuelas, J., Baret, F., and Filella, I. (1995). Semi-empirical indices to assess carotenoids/chlorophyll *a* ratio from leaf spectral reflectance. *Photosynthetica*, 31, 221–30.

Phillipps, A. and Lamb, A. (1996). *Pitcher Plants of Borneo*. Natural History Publications, Kota Kinabalu.

Phillips, R.D., Brown, A.P., Dixon, K.W., and Hopper, S.D. (2011). Orchid biogeography and factors associated with rarity in a biodiversity hotspot, the Southwest Australian floristic region. *Journal of Biogeography*, 38, 487–501.

Phillips, S.J. Anderson, R.P., and Schapire, R.E. (2006). Maximum entropy modeling of species geographic distributions. *Ecological Modelling*, 190, 231–59.

Pietropaulo, J. and Pietropaulo, P. (1986). *Carnivorous Plants of The World*. Timber Press, Portland.

Pitman, A. J., Narisma, G. T., Pielke, R., and Holbrook, N. (2004). Impact of land cover change on the climate of southwest Western Australia. *Journal of Geophysical Research: Atmospheres*, 109, D18109.

Pitsch, G., Adamec, L., Dirren, S., et al. (2017). The green *Tetrahymena utriculariae* n. sp. (Ciliophora, Oligohymenophorea) with its endosymbiotic algae (*Micractinium* sp.), living in traps of a carnivorous aquatic plant. *Journal of Eukaryotic Microbiology*, 64, 322–35.

Płachno, B.J., Kozieradzka-Kiszkurno, M., and Świątek, P. (2007a). Functional ultrastructure of *Genlisea* (Lentibulariaceae) digestive hairs. *Annals of Botany*, 100, 195–203.

Płachno, B.J., Adamec, L., Kozieradzka-Kiszkurno, M., Świątek, P., and Kamińska, I. (2014a). Cytochemical and ultrastructural aspects of aquatic carnivorous plant turions. *Protoplasma*, 251, 1449–54.

Płachno, B.J., Świątek P., Sas-Nowosielska, H., and Kozieradzka-Kiszkurno, M. (2012b). Organisation of the endosperm and endosperm–placenta syncytia in bladderworts (*Utricularia*, Lentibulariaceae) with emphasis on the microtubule arrangement. *Protoplasma*, 250, 863–73.

Płachno, B.J., Kozieradzka-Kiszkurno, M., Świątek, P., and Darnowski, D.W. (2008). Prey attraction in carnivorous *Genlisea* (Lentibulariaceae). *Acta Biologica Cracoviensia Series Botanica*, 50, 87–94.

Płachno, B.J. and Adamec, L. (2007). Differentiation of *Utricularia ochroleuca* and *U. stygia* populations in

Třeboň Basin, Czech Republic, on the basis of quadrifid glands. *Carnivorous Plant Newsletter*, 36, 87–95.

Płachno, B.J., Adamec, L., and Huet, H. (2009a). Mineral nutrient uptake from prey and glandular phosphatase activity as a dual test of carnivory in semi-desert plants with glandular leaves suspected of carnivory. *Annals of Botany*, 104, 649–54.

Płachno, B.J., Adamec, L., and Kamińska, I. (2015a). Relationship between trap anatomy and function in Australian carnivorous bladderworts (*Utricularia*) of the subgenus *Polypompholyx*. *Aquatic Botany*, 120, 290–96.

Płachno, B.J., Adamec, L., Lichtscheidl, I.K., et al. (2006). Fluorescence labelling of phosphatase activity in digestive glands of carnivorous plants. *Plant Biology*, 8, 813–20.

Płachno, B.J., Adamus, K., Faber, J., and Kozlowski, J. (2005a). Feeding behaviour of carnivorous *Genlisea* plants in the laboratory. *Acta Botanica Gallica*, 152, 159–64.

Płachno, B.J., Faber, J., and Jankun, A. (2005b). Cuticular discontinuities in glandular hairs of *Genlisea* St.-Hil. in relation to their functions. *Acta Botanica Gallica*, 152, 125–30.

Płachno, B.J., Stpiczyńska, M., Świątek, P., and Davies, K.L. (2015b). Floral micromorphology of the Australian carnivorous bladderwort *Utricularia dunlopii*, a putative pseudocopulatory species. *Protoplasma*, 253, 1463–73.

Płachno, B.J. and Wołowski, K. (2008). Algae commensal community in *Genlisea* traps. *Acta Societatis Botanicorum Poloniae*, 77, 77–86.

Płachno, B.J., Wołowski, K., Fleischmann, A., Lowrie, A., and Lukaszek, M. (2014b). Algae and prey associated with traps of the Australian carnivorous plant *Utricularia volubilis* (Lentibulariaceae: *Utricularia* subgenus *Polypompholyx*) in natural habitat and in cultivation. *Australian Journal of Botany*, 62, 528–36.

Płachno, B.J., Łukaszek, M., Wołowski, K., Adamec, L., and Stolarczyk, P. (2012a). Aging of *Utricularia* traps and variability of microorganisms associated with that microhabitat. *Aquatic Botany*, 97, 44–48.

Płachno, B.J., Świátek, P., and Wistuba, A. (2007b). The giant extra-floral nectaries of carnivorous *Heliamphora folliculata*: Architecture and ultrastructure. *Acta Biologica Cracoviensia Series Botanica*, 49, 91–104.

Płachno, B.J. and Świątek, P. (2010). Unusual embryo structure in viviparous *Utricularia nelumbifolia*, with remarks on embryo evolution in genus *Utricularia*. *Protoplasma*, 239, 69–80.

Płachno, B.J., Świątek, P., Fleischmann, A., et al. (2009b). Structure and evolution of the carnivorous plant genus *Heliamphora*. *4th Conference of Polish Society of Experimental Plant Biology*. Cracow, Poland.

Planchon, J.É. (1848). Sur la familie des Droséracées. *Annales des Sciences Naturelles, Botanique, Série 3*, 9, 79–99.

Platt, J.R. (2015). Venus Flytraps at risk of extinction in the wild at the hands of poachers. Available online: https://blogs.scientificamerican.com/extinction-countdown/venus-flytraps-risk-extinction-in-the-wild-at-the-hands-of-poachers/.

Plummer, G.L. and Kethley, J.B. (1964). Foliar absorption of amino-acids, peptides and other nutrients by the pitcher-plant *Sarracenia flava*. *Botanical Gazette*, 125, 245–60.

Poppinga, S., Hartmeyer, S.R.H., Masselter, T., Hartmeyer, I., and Speck, T. (2013a). Trap diversity and evolution in the family Droseraceae. *Plant Signaling & Behavior*, 8: e24685.

Poppinga, S., Hartmeyer, S.R.H., Seidel, R., et al. (2012). Catapulting tentacles in a sticky carnivorous plant. *PLoS ONE*, 7, e45735.

Poppinga, S. and, Joyeux, M. (2011). Different mechanics of snap-trapping in the two closely related carnivorous plants *Dionaea muscipula* and *Aldrovanda vesiculosa*. *Physical Review E*, 84, 041928.

Poppinga, S., Kampowski, T., Metzger, A., Speck, O., and Speck T. (2016a). Comparative kinematical analyses of Venus flytrap (*Dionaea muscipula*) snap-traps. *Beilstein Journal of Nanotechnology*, 7, 664–74.

Poppinga S., Koch, K., Bohn, H.F., and Barthlott, W. (2010). Comparative and functional morphology of hierarchically structured anti-adhesive surfaces in carnivorous plants and kettle-trap flowers. *Functional Plant Biology*, 37, 952–61.

Poppinga, S., Masselter, T., and Speck, T. (2013b). Faster than their prey: New insights into the rapid movements of active carnivorous plants traps. *Bioessays*, 35, 649–57.

Poppinga, S., Weisskopf, C., Westermeier, A.S., Masselter, T., and Speck, T. (2016b). Fastest predators in the plant kingdom: Functional morphology and biomechanics of suction traps found in the largest genus of carnivorous plants. *AoB PLANTS*, 8, plv140.

Porembski, S., Theisen, I., and Barthlott, W. (2006). Biomass allocation patterns in terrestrial, epiphytic and aquatic species of *Utricularia* (Lentibulariaceae). *Flora*, 201, 477–82.

Porter-Bolland, L., Ellis, E.A., Guariguata, M.R., et al. (2012). Community managed forests and forest protected areas: an assessment of their conservation effectiveness across the tropics. *Forest Ecology and Management*, 268, 6–17.

Potts, L. and Krupa, J.J. (2016). Does the dwarf sundew (*Drosera brevifolia*) attract prey? *American Midland Naturalist*, 175, 233–41.

Powers, M.E. (1992). Top-down and bottom-up forces in food webs: Do plants have primacy? *Ecology*, 73, 733–46.

Prankevicius, A.B. and Cameron, D.M. (1991). Bacterial dinitrogen fixation in the leaf of the northern pitcher plant (*Sarracenia purpurea*). *Canadian Journal of Botany*, 69, 2296–98.

Prüm, B., Seidel, R., Bohn, H.F., and Speck, T. (2012). Plant surfaces with cuticular folds are slippery for beetles. *Journal of the Royal Society Interface*, 9, 127–35.

Purvis, A., Gittleman, J.L., Cowlishaw, G., and Mace, G.M. (2000). Predicting extinction risk in declining species. *Proceedings of the Royal Society B*, 267, 1947–52.

Putalun, W., Udomsin, O., Yusakul, G., et al. (2010). Enhanced plumbagin production from in vitro cultures of *Drosera burmannii* using elicitation. *Biotechnology Letters*, 32, 721–24.

Putyatina, T.S. (2007). The choice of foraging strategy as a mechanism for the coexistence of *Myrmica* species (Hymenoptera, Formicidae) in a multispecific ant association. *Entomological Review*, 87, 650–57.

R Core Team (2016). *R: A Language and Environment for Statistical Computing*. Vienna, Austria. Available online: https://www.r-project.org/.

Rabinowitz, D. (1981). Seven forms of rarity. In J. Synge, ed. *The Biological Aspects of Rare Plant Conservation*, pp. 205–17. John Wiley & Sons, New York.

Rafinesque, C.S. (1836). *Flora Telluriana* 4. H. Probasco, Philadelphia.

Raj, G., Kurup, R., Hussain, A.A., and Baby, S. (2011). Distribution of naphthoquinones, plumbagin, droserone, and 5-O-methyl droserone in chitin-induced and uninduced *Nepenthes khasiana*: molecular events in prey capture. *Journal of Experimental Botany*, 62, 5429–36.

Ramanamanjary, W. and Botteau, P. (1968). Active protectrice du Matahanando, *Drosera ramentacea* Burch. vis-à-vis du bronchospasme. *Comptes Rendues Hebdomadaires des Séances Académie des Sciences Naturelles Series D*, 266, 1787–89.

Ramírez-Puebla, S.T., Servín-Garcidueñas, L.E., Jiménez-Marín, B., et al. (2013). Gut and root microbiota commonalities. *Applied and Environmental Microbiology*, 79, 2–9.

Randall, J.M. and Rice, B.A. (2003). 1998–99 survey of invasive species on lands managed by the nature conservancy. Available online: http://tncinvasives.ucdavis.edu/survey.html.

Rango, J.J. (1999). Resource dependent larviposition behavior of a pitcher plant flesh fly, *Fletcherimyia fletcheri* (Aldrich) (Diptera: Sarcophagidae). *Journal of the New York Entomological Society*, 107, 82–86.

Rasic, G. and Keyghobadi, N. (2012). From broadscale patterns to fine-scale processes: habitat structure influences genetic differentiation in the pitcher plant midge across multiple spatial scales. *Molecular Ecology*, 21, 223–36.

Ratsirarson, J. and Silander, J.A., Jr. (1996). Structure and dynamics in *Nepenthes madagascariensis* pitcher plant micro-communities. *Biotropica*, 28, 218–27.

Ravikumar, K., Ved, D.K., Vijaya Sankar, R., and Udayan, P.S. (2000). *100 Red Listed Medicinal Plants of Conservation Concern in Southern India*. Foundation for Revitalisation of Local Health Traditions, Bangalore.

Raynal-Roques, A. and Jérémie, J. (2005). Biologie diversity in the genus *Utricularia* (Lentibulariaceae). *Acta Botanica Gallica*, 152, 177–86.

RBG Kew (2016). *State of the World's Plants Report—2016*. Royal Botanic Gardens, Kew.

Redbo-Torstensson, P. (1994). The demographic consequences of nitrogen fertilization of a population of sundew, *Drosera rotundifolia*. *Acta Botanica Neerlandica*, 43, 175–88.

Reddy, C.S., Reddy, K.N., and Jadhav, S.N. (2001). *Threatened (Medicinal) Plants of Andhra Pradesh*. Medicinal Plants Conservation Center, Hyderabad.

Reed, J.W. (2001). Roles and activities of Aux/IAA proteins in *Arabidopsis*. *Trends in Plant Science*, 6, 420–25.

Refulio-Rodriguez, N.F. and Olmstead, R.G. (2014). Phylogeny of Lamiidae. *American Journal of Botany*, 101, 287–99.

Reich, P.B., Wright, I.J., Cavender-Bares, J., et al. (2003). The evolution of plant functional variation: traits, spectra, and strategies. *International Journal of Plant Sciences*, 164 (Supplement 3), S143–S164.

Reid, J.W. (2002). A human challenge: discovering and understanding continental copepod habitats. *Hydrobiologia*, 453, 201–26.

Reifenrath, K., Theisen, I., Schnitzler, J., Porembski, S., and Barthlott, W. (2006). Trap architecture in carnivorous *Utricularia* (Lentibulariaceae). *Flora*, 201, 597–605.

Rejthar, J., Viehmannová, I., Cepková, P.H., Fernández, E., and Milella, L. (2014). In vitro propagation of *Drosera intermedia* as influenced by cytokinins, pH, sucrose, and nutrient concentration. *Emirates Journal of Food and Agriculture*, 26, 558–64.

Rembold, K., Fischer, E., Striffler, B.F., and Barthlott, W. (2013). Crab spider association with the Malagasy pitcher plant *Nepenthes madagascariensis*. *African Journal of Ecology*, 51, 188–91.

Rembold, K., Fischer, E., Wetzel, M.A., and Barthlott, W. (2010a). Prey composition of the pitcher plant *Nepenthes madagascariensis*. *Journal of Tropical Ecology*, 26, 365–72.

Rembold, K., Irmer, A., Poppinga, S., Rischer, H., and Bringmann, G. (2010b). Propagation of *Triphyophyllum peltatum* (Dioncophyllaceae) and observations on its carnivory. *Carnivorous Plant Newsletter*, 39, 71–77.

Renner, S.S. (1989). Floral biological observations on *Heliamphora tatei* Sarraceniaceae and other plants from Cerro de la Neblina in Venezuela. *Plant Systematics and Evolution*, 163, 21–30.

Renner, T. and Specht, C.D. (2011). A sticky situation: assessing adaptations for plant carnivory in the Caryophyllales by means of stochastic character mapping. *International Journal of Plant Sciences*, 172, 889–901.

Renner, T. and Specht, C.D. (2012). Molecular and functional evolution of class I chitinases for plant carnivory in the Caryophyllales. *Molecular Biology and Evolution*, 29, 2971–85.

Renner, T. and Specht, C.D. (2013). Inside the trap: gland morphologies, digestive enzymes, and the evolution of plant carnivory in the Caryophyllales. *Current Opinion in Plant Biology*, 16, 436–42.

Reut, M.S. (1993). Trap structure of the carnivorous plant *Genlisea* (Lentibulariaceae). *Botanica Helvetica*, 103, 101–11.

Reut, M.S. and Fineran, B.A. (2000). Ecology and vegetative morphology of the carnivorous plant *Utricularia dichotoma* (Lentibulariaceae) in New Zealand. *New Zealand Journal of Botany*, 38, 433–50.

Reut, M.S. and Jobson, R.W. (2010). A phylogenetic study of subgenus *Polypompholyx*: a parallel radiation of *Utricularia* (Lentibulariaceae) throughout Australasia. *Australian Systematic Botany*, 23, 152–61.

Rey, M., Yang, M., Burns, K.M., et al. (2013). Nepenthesin from monkey cups for hydrogen/deuterium exchange mass spectrometry. *Molecular & Cellular Proteomics*, 12, 464–72.

Rice, B.A. (1999). Testing the appetites of *Ibicella lutea* and *Drosophyllum*. *Carnivorous Plant Newsletter*, 28, 40–43.

Rice, B.A. (2006). *Growing Carnivorous Plants*. Timber Press, Portland.

Rice, B. (2007). Carnivorous plants with hybrid trapping strategies. *Carnivorous Plant Newsletter*, 36, 23–27.

Rice, B.A. (2011a). Reversing the roles of predator and prey—a review of carnivory in the botanical world. In: J. Seckbach and Z. Dubinsky, eds. *All Flesh is Grass. Plant-Animal Interrelationships*, pp. 493–518. Springer Science + Business Media B. V., Dordrecht.

Rice, B.A. (2011b). The thread-leaf sundews *Drosera filiformis* and *Drosera tracyi*. *Carnivorous Plant Newsletter*, 40, 4–16.

Rice, B.A. (2011c). What exactly is a carnivorous plant? *Carnivorous Plant Newsletter*, 40, 19–23.

Richards, J.H. (2001). Bladder function in *Utricularia purpurea* (Lentibulariaceae), is carnivory important? *American Journal of Botany*, 88, 170–76.

Ridley, H.N. (1905). On the dispersal of seeds by wind. *Annals of Botany*, 19, 351–63.

Riedel, M., Eichner, A., and Jetter, R. (2003). Slippery surfaces of carnivorous plants: composition of epicuticular wax crystals in *Nepenthes alata* Blanco pitchers. *Planta*, 218, 87–97.

Riedel, M., Eichner, A., Meimberg, H., and Jetter, R. (2007). Chemical composition of epicuticular wax crystals on the slippery zone in pitchers of five *Nepenthes* species and hybrids. *Planta*, 225, 1517–34.

Rios, N.E. and Bart, H.L. (2010). GEOLocate (Version 2.0). Tulane University Museum of Natural History, Belle Chasse, LA, USA.

Rischer, H., Hamm, A., and Bringmann, G. (2002). *Nepenthes insignis* uses a C2-portion of the carbon skeleton of L-alanine acquired via its carnivorous organs, to build up the allelochemical plumbagin. *Phytochemistry*, 59, 603–09.

Rissel, D., Losch, J., and Peiter, E. (2014). The nuclear protein Poly (ADP-ribose) polymerase 3 (AtPARP3) is required for seed storability in *Arabidopsis thaliana*. *Plant Biology*, 16, 1058–64.

Risør, M.W., Thomsen, L.R., Sanggaard, K.W., et al. (2016). Enzymatic and structural characterization of the major endopeptidase in the Venus flytrap digestion fluid. *Journal of Biological Chemistry*, 291, 2271–87.

Rivadavia, F., Gonella, P.M., and Fleischmann, A. (2013). A new and tuberous species of *Genlisea* (Lentibulariaceae) from the campos rupestres of Brazil. *Systematic Botany*, 38, 464–70.

Rivadavia, F., Kondo, K., Kato, M., and Hasebe, M. (2003). Phylogeny of the sundews, *Drosera* (Droseraceae), based on chloroplast *rbcL* and nuclear 18S ribosomal DNA sequences. *American Journal of Botany*, 90, 123–30.

Rivadavia, F., Read, E.L., and Fleischmann, A. (2017). *Pinguicula pygmaea* (Lentibulariaceae), a new annual gypsicolous species from Oaxaca state, Mexico. *Phytotaxa*, 292, 279–86.

Rivadavia, F., Vincentini, A. and Fleischmann, A. (2009) A new species of sundew (*Drosera*, Droseraceae), with water-dispersed seed, from the floodplains of the northern Amazon basin, Brazil. *Ecotropica*, 15, 13–21.

Rivadavia, F., de Miranda, V.F.O., Hoogenstrijd, G., et al. (2012). Is *Drosera meristocaulis* a pygmy sundew? Evidence of a long-distance dispersal between Western Australia and northern South America. *Annals of Botany*, 110, 11–21.

Roberts, M.L. (1972). *Wolffia* in the bladders of *Utricularia*: a "herbivorous" plant? *Michigan Botanist*, 11, 67–69.

Roberts, P.R. and Oosting, H.J. (1958). Responses of Venus fly trap (*Dionaea muscipula*) to factors involved in its endemism. *Ecological Monographs*, 28, 193–218.

Robins, R.J. and Juniper, B.E. (1980). The secretory cycle of *Dionaea muscipula* Ellis IV. The enzymology of the secretion. *New Phytologist*, 86, 401–12.

Robins, R.J. and Subramanyam, K. (1980). Scanning electron microscope study of the seed surface morphology of some *Utricularia* (Lentibulariaceae) species from India. *Proceedings of the Indian National Science Academy*, 46B, 310–24.

Roccia, A., Gluch, O., Lampard, S., et al. (2016). *Pinguicula of The Temperate North*. Redfern Natural History Production, Poole.

Roldán-Arjona, T. and Ariza, R.R. (2009). Repair and tolerance of oxidative DNA damage in plants. *Mutation Research*, 681, 169–79.

Roelfsema, M.R., Levchenko, V., and Hedrich, R. (2004). ABA depolarizes guard cells in intact plants, through a transient activation of R- and S-type anion channels. *The Plant Journal*, 37, 578–88.

Romero, G.Q., Mazzafera, P., Vasconcellos-Neto, J. and Trivelin, P.C.O. (2006). Bromeliad-living spiders improve host plant nutrition and growth. *Ecology*, 87, 803–08.

Romero, G.Q, Nomura, F., Goncalves, A.Z., et al. (2010). Nitrogen fluxes from treefrogs to tank epiphytic bromeliads: an isotopic and physiological approach. *Oecologia*, 162, 941–49.

Rosa, A.B., Malek, L., and Qin, W. (2009). The development of the pitcher plant *Sarracenia purpurea* into a potentially valuable recombinant protein production system. *Biotechnology and Molecular Biology Reviews*, 3, 105–10.

Rosin, F.M. and Kramer, E.M. (2009). Old dogs, new tricks: regulatory evolution in conserved genetic modules leads to novel morphologies in plants. *Developmental Biology*, 332, 25–35.

Rost, K. and Schauer, R. (1977). Physical and chemical properties of the mucin sectreted by *Drosera capensis*. *Phytochemistry*, 16, 365–68.

Rottloff, S., Miguel, S., Biteau, F., et al. (2016). Proteome analysis of digestive fluids in *Nepenthes* pitchers. *Annals of Botany*, 117, 479–95.

Rottloff, S., Müller, U., Kilper, R. and Mithöfer, A. (2009). Micropreparation of single secretory glands from the carnivorous plant *Nepenthes*. *Analytical Biochemistry*, 394, 135–37.

Rottloff, S., Stieber, R., Maischak, H., Turini, F.G., Heubl, G. and Mithöfer, A. (2011). Functional characterization of a class III acid endochitinase from the traps of the carnivorous pitcher plant genus, *Nepenthes*. *Journal of Experimental Botany*, 62, 4639–47.

Royal Botanic Gardens Kew (2017). Seed Information Database (SID). Version 7.1. Available online: http://data.kew.org/sid/.

Rumphius, G.E. (1750). *Het Amboinsche Kruitboek*. Joannes Burmannus, Amsterdam.

Rutishauser, R. (2016). Evolution of unusual morphologies in Lentibulariaceae (bladderworts and allies) and Podostemaceae (river-weeds): a pictorial report at the interface of developmental biology and morphological diversification. *Annals of Botany*, 117, 811–32.

Rutishauser, R. and Isler, B. (2001). Developmental genetics and morphological evolution of flowering plants, especially bladderworts (*Utricularia*): fuzzy Arberian morphology complements classical morphology. *Annals of Botany*, 88, 1173–202.

Sadowski, E-M., Seyfullah, L.J., Sadowski, F., et al. (2015). Carnivorous leaves from Baltic Amber. *Proceedings of the National Academy of Sciences, USA*, 112, 190–95.

Saetiew, K., Sang-in, V., and Arunyanart, S. (2011). The effects of BA and NAA on multiplication of Butterwort (*Pinguicula gigantea*) *in vitro*. *Journal of Agricultural Technology*, 7, 1349–54.

Sah, S.C.D. and Dutta, K. (1974). Palynostratigraphy of the sedimentary formations of Assam. 3. Biostratigraphic zonation of the Cherra Formation of South Shillong Plateau. *Palaeobotanist*, 21, 42–47.

Sakamoto, K., Sayama, H., Nakamoto, H., Matsushima, H., and Kaneko, Y. (2006). Ultrastructural changes during digestion and absorption in the aquatic plant *Aldrovanda vesiculosa*. 16th International Microscopy Congress, Sapporo, Japan, p. 460.

Salces-Castellano, A., Paniw, M., Casimiro-Soriguer, R., and Ojeda, F. (2016). Attract them anyway—benefits of large, showy flowers in a highly autogamous, carnivorous plant species. *AoB Plants*, 8, plw017.

Salmon, B. (1993). Some observations on the trapping mechanisms of *Nepenthes inermis* and *N. rhombicaulis*. *Carnivorous Plant Newsletter*, 21, 11–12.

Salmon, B. (2001). *Carnivorous Plants of New Zealand*. Ecosphere Publications, Auckland.

Samson, D.A., Rickart, E.A. and Gonzales, P.C. (1997). Ant diversity and abundance along an elevational gradient in the Philippines. *Biotropica*, 29, 349–63.

Samuels, L., Kunst, L., and Jetter, R. (2008). Sealing plant surfaces: cuticular wax formation by epidermal cells. *Annual Review of Plant Biology*, 59, 683–707.

Sanabria-Aranda, L., Gonzalez-Bermudez, A., Torres, N.N., et al. (2006). Predation by the tropical plant *Utricularia foliosa*. *Freshwater Biology*, 51, 1999–2008.

Sandel, B., Arge, L., Dalsgaard, B., et al. (2011). The influence of Late Quaternary climate-change velocity on species endemism. *Science*, 334, 660–64.

Sankoff, D. and Zheng, C. (2012). Fractionation, rearrangement and subgenome dominance. *Bioinformatics*, 28, i402–i408.

Sasago, A. and Sibaoka, T. (1985a). Water extrusion in the trap bladders of *Utricularia vulgaris* I. A possible pathway of water across the bladder wall. *Botanical Magazine*, 98, 55–66.

Sasago, A. and Sibaoka, T. (1985b). Water extrusion in the trap bladders of *Utricularia vulgaris*. II. A possible mechanism of water outflow. *Botanical Magazine*, 98, 113–24.

Satler, J.D., Zellmer, A.J., and Carstens, B.C. (2016). Biogeographic barriers drive co-diversification within associated eukaryotes of the *Sarracenia alata* pitcher plant system. *PeerJ*, 4, e1576.

Sattler, R. and Rutishauser, R. (1990). Structural and dynamic descriptions of the development of *Utricularia foliosa* and *U. australis*. *Canadian Journal of Botany*, 68, 1989–2003.

Saunders, D.A. (1989). Changes in the avifauna of a region, district and remnant as a result of fragmentation of native vegetation: the wheatbelt of Western Australia. A case study. *Biological Conservation*, 50, 99–135.

Saur, E., Sandrine, E., Carcelle, S., Guezennec, S., and Rousteau, A. (2000). Nodulation of legume species in wetlands of Guadeloupe (Lesser Antilles). *Wetlands*, 20, 730–34.

Savatin, D.V., Gramegna, G., Modesti, V., and Cervone, F. (2014). Wounding in the plant tissue: the defense of a dangerous passage. *Frontiers in Plant Science*, 5, 470.

Savolainen, V., Fay, M.F., Albach, D.C., et al. (2000). Phylogeny of the Eudicots: a nearly complete familial analysis based on *rbcL* gene sequences. *Kew Bulletin*, 55, 257–309.

Sbragia, P., Sibilio, G., Innangi, M., et al. (2014). Rare species habitat suitability assessment and reliability evaluation of an expert-based model: a case study of the insectivorous plant *Pinguicula crystallina* Sibth. et Smith subsp. *hirtiflora* (Ten.) Strid (Lentibulariaceae). *Plant Biosystems*, 150, 1–11.

Scatena, V.L. and Bouman, F. (2001). Embryology and seed development of *Paepalanthus* sect. *Actinocephalus* (Koem.) Ruhland (Eriocaulaceae). *Plant Biology*, 3, 341–50.

Scatigna, A.V., Amaral, A.G., Munhoz, C.B.R., Souza, V.C., and Simões, A.O. (2016a). The rediscovery of *Philcoxia goiasensis* (Plantaginaceae): lectotypification and notes on morphology, distribution and conservation of a threatened carnivorous species from the Serra Geral de Goiás, Brazil. *Kew Bulletin*, 71, 41.

Scatigna, A.V., Carmo, J.A.M., and Simões, A.O. (2016b). New records of *Philcoxia minensis* (Plantaginaceae) and *Mitracarpus pusillus* (Rubiaceae): conservation status assessment and notes on type specimens of two threatened species from the Espinhaço Range, Minas Gerais, Brazil. *Phytotaxa*, 243, 297–300.

Scatigna, A.V., Silva, N.G., Alves, R.J.V., Souza, V.C., and Simões, A.O. (2017). Two new species of the carnivorous genus *Philcoxia* (Plantaginaceae) from the Brazilian Cerrado. *Systematic Botany*, 42, 351–357.

Scatigna, A.V., Souza, V.C., Pereira, C.G., Sartori, M.A., and Simões, A.O. (2015). *Philcoxia rhizomatosa* (Gratioleae, Plantaginaceae): a new carnivorous species from Minas Gerais, Brazil. *Phytotaxa*, 226, 275–80.

Schäferhoff, B., Fleischmann, A., Fischer, E., et al. (2010). Towards resolving Lamiales relationships: insights from rapidly evolving chloroplast sequences. *BMC Evolutionary Biology*, 10, 352–74.

Schäferhoff, B., Müller, K.F., and Borsch, T. (2009). Caryophyllales phylogenetics: Disentangling Phytolaccaceae and Molluginaceae and description of Microteaceae as a new isolated family. *Willdenowia*, 39, 209–28.

Schaefer, H.M. and Ruxton, G.D. (2008). Fatal attraction: carnivorous plants roll out the red carpet to lure insects. *Biology Letters*, 4, 153–55.

Schaefer, H.M. and Ruxton, G.D. (2014). Fenestration: a window of opportunity for carnivorous plants. *Biology Letters*, 10, 20140134.

Scharmann, M. and Grafe, T.U. (2013). Reinstatement of *Nepenthes hemsleyana* (Nepenthaceae), an endemic pitcher plant from Borneo, with a discussion of associated *Nepenthes* taxa. *Blumea*, 58, 8–12.

Scharmann, M., Thornham, D.G., Grafe, T.U., and Federle, W. (2013). A novel type of nutritional ant–plant interaction: ant partners of carnivorous pitcher plants prevent nutrient export by dipteran pitcher infauna. *PLoS ONE*, 8, e63556.

Schell, C. (2003). A note on possible prey selectivity in the waterwheel plant (*Aldrovanda vesiculosa*) and a possible method of prey attraction. *Carniflora Australis*, 1, 18–19.

Scherzer, S., Böhm, J., Król, E., et al. (2015). Calcium sensor kinase activates potassium uptake systems in gland cells of Venus flytraps. *Proceedings of the National Academy of Sciences, USA*, 112, 7309–14.

Scherzer, S., Król, E., Kreuzer, I., et al. (2013). The *Dionaea muscipula* ammonium channel *DmAMT1* provides NH_4^+ uptake associated with Venus flytrap's prey digestion. *Current Biology*, 23, 1–9.

Schlauer, J. (1997a). Fossil *Aldrovanda*—additions. *Carnivorous Plant Newsletter*, 26, 98.

Schlauer, J. (1997b). "New" data relating to the evolution and phylogeny of some carnivorous plant families. *Carnivorous Plant Newsletter*, 26, 34–38.

Schlauer, J. (2000). Global carnivorous plant diversity. *Carnivorous Plant Newsletter*, 29, 75–82.

Schlauer, J. (2010). Carnivorous plant systematics. *Carnivorous Plant Newsletter*, 39, 8–24.

Schlauer, J. and Fleischmann, A. (2016). Chemical evidence for hybridity in *Drosera* (*Droseraceae*). *Biochemical Systematics and Ecology*, 66, 33–36.

Schlauer, J., Nerz, J., and Rischer, H. (2005). Carnivorous plant chemistry. *Acta Botanica Gallica*, 152, 187–95.

Schleicher, S., Lienhard, J., Poppinga, S., Speck, T., and Knippers, J. (2015). A methodology for transferring principles in plant movements to elastic systems in architecture. *Computer-Aided Design*, 60, 105–17.

Schmid, R. (1964). Die systematische Stellung der Dioncophyllaceen. *Botanische Jahrbücher*, 83, 1–56.

Schmid-Hollinger, R. (1970). *Nepenthes*–Studien I: Homologien von Deckel (operculum, lid) und Spitzchen (calcar, spur). *Botanische Jahrbücher*, 90, 275–96.

Schmidt, A. and Weber, H.C. (1983). Untersuchungen zur Mikroflora und Mikrofauna in den Schuppenblatthöhlen von *Tozzia alpina* L. (Scrophulariaceae). *Angewandte Botanik*, 57, 245–56.

Schneider, K., Mathur, J., Boudonck, K., et al. (1998). The ROOT HAIRLESS 1 gene encodes a nuclear protein required for root hair initiation in *Arabidopsis*. *Genes and Development*, 12, 2013–21.

Schnell, D.E. (1976). *Carnivorous Plants of The United States and Canada*. John F. Blair, Winston-Salem.

Schnell, D.E. (2002). *Carnivorous Plants of the United States and Canada*, 2nd edition. Timber Press, Portland.

Schoenly, K.G., Haskell, N.H., Hall, R.D., and Gbur, J.R. (2007). Comparative performance and complementarity of four sampling methods and arthropod preference tests from human and porcine remains at the forensic anthropology center in Knoxville, Tennessee. *Journal of Medical Entomology*, 44, 881–94.

Scholz, I., Bückins, M., Dolge, L., et al. (2010). Slippery surfaces of pitcher plants: *Nepenthes* wax crystals minimize insect attachment via microscopic surface roughness. *Journal of Experimental Biology*, 213, 1115–25.

Schönenberger, J., Anderberg, A.A., and Sytsma, K.J. (2005). Molecular phylogenetics and patterns of floral evolution in the Ericales. *International Journal of Plant Science*, 166, 265–88.

Schöner, C.R., Schöner, M.G., Kerth, G., and Grafe, T.U. (2013). Supply determines demand: influence of partner quality and quantity on the interactions between bats and pitcher plants. *Oecologia*, 173, 191–202.

Schöner, C.R., Schöner, M.G., Kerth, G., Suhaini, S.N.B.P., and Grafe, T.U. (2015). Low costs reinforce the mutualism between bats and pitcher plants. *Zoologischer Anzeiger*, 258, 1–5.

Schöner, M.G., Schöner, C.R., Simon, R., et al. (2015). Bats are acoustically attracted to mutualistic carnivorous plants. *Current Biology*, 25, 1911–16.

Schuh, R.T. (1995). *Plant Bugs of The World (Insecta: Heteroptera: Miridae): Systematic Catalog, Distributions, Host List, and Bibliography*. The New York Entomological Society, New York.

Schulze, E.D., Gebauer, G., Schulze, W., and Pate, J.S. (1991). The utilization of nitrogen from insect capture by different growth forms of *Drosera* from Southwest Australia. *Oecologia*, 87, 240–46.

Schulze, W., Frommer, W.B., and Ward J.M. (1999). Transporters for ammonium, amino acids and peptides are expressed in pitchers of the carnivorous plant *Nepenthes*. *Plant Journal*, 17, 637–46.

Schulze, W. and Schulze, E.D. 1990. Insect capture and growth of the insectivorous *Drosera rotundifolia* L. *Oecologia*, 82, 427–29.

Schulze, W., Schulze, E.D., Pate, J.S., and Gillison, A.N. (1997). The nitrogen supply from soils and insects during growth of the pitcher plants *Nepenthes mirabilis, Cephalotus follicularis* and *Darlingtonia californica*. *Oecologia*, 112, 464–71.

Schulze, W., Schulze, E.D., Schulze, I., and Oren, R. (2001). Quantification of insect nitrogen utilization by the venus fly trap *Dionaea muscipula* catching prey with highly variable isotope signatures. *Journal of Experimental Botany*, 52, 1041–49.

Schulze, W.X., Sanggaard, K.W., Kreuzer, I., et al. (2012). The protein composition of the digestive fluid from the venus flytrap sheds light on prey digestion mechanisms. *Molecular and Cellular Proteomics*, 11, 1306–19.

Schumacher, G.J. (1960). Further notes on the occurrence of desmids in *Utricularia* bladders. *Castanea*, 25, 62–65.

Schwaegerle, K.E. (1983). Population growth of the pitcher plant, *Sarracenia purpurea* L., at Cranberry Bog, Licking County, Ohio. *Ohio Journal of Science*, 83, 19–22.

Schwaegerle, K.E. and Schaal, B.A. (1979). Genetic variability and founder effect in the pitcher plant *Sarracenia purpurea* L. *Evolution*, 33, 1210–18.

Schwallier, R., Raes, N., de Boer, H.J., et al. (2016). Phylogenetic analysis of niche divergence reveals distinct evolutionary histories and climate change implications for tropical carnivorous pitcher plants. *Diversity and Distributions*, 22, 97–110.

Sciligo, A.R. (2009). *Food or Sex: Which Would You Choose? Pollinator–Prey Conflict and Reproductive Assurance in New Zealand* Drosera. Ph.D. dissertation, Lincoln University, Christchurch.

Seago J.L., Marsh, L.C., Stevens, K.J., et al. (2005). A reexamination of the root cortex in wetland flowering plants with respect to aerenchyma. *Annals of Botany*, 96, 565–79.

Secco, D., Wang, C., Arpat, B.A., et al. (2012). The emerging importance of the SPX domain-containing proteins in phosphate homeostasis. *New Phytologist*, 193, 842–51.

Segonzac, C., Boyer, J.-C., Ipotesi, E., et al. (2007). Nitrate efflux at the root plasma membrane: identification of an *Arabidopsis* excretion transporter. *Plant Cell*, 19, 3760–77.

Serra-Diaz, J.M., Franklin, J., Ninyerola, M., et al. (2014). Bioclimatic velocity: the pace of species exposure to climate change. *Diversity and Distributions*, 20, 169–80.

Seine, R. and Barthlott, W. (1993). On the morphology of trichomes and tentacles of Droseraceae Salisb. *Beiträge zur Biologie der Pflanzen*, 67, 345–66.

Seine, R., Porembski, S., Balduin, M., et al. (2002). Different prey strategies of terrestrial and aquatic species in the carnivorous genus *Utricularia* (Lentibulariaceae). *Botanische Jahrbücher für Systematik*, 124, 71–76.

Serraj, R. and Sinclair, T.R. (1996). Inhibition of nitrogenase activity and nodule oxygen permeability by water deficit. *Journal of Experimental Biology*, 47, 1067–73.

Sheridan, P.M. and Karowe, D.N. (2000). Inbreeding, outbreeding, and heterosis in the yellow pitcher plant, *Sarracenia flava* (Sarraceniaceae), in Virginia. *American Journal of Botany*, 87, 1628–33.

Sheridan, P.M. and Mills, R.R. (1998). Genetics of anthocyanin deficiency in *Sarracenia* L. *Hortscience*, 33, 1042–45.

Sheridan, P.M., Orzell, S.L., and Bridges, E.L. (1997). Powerline easements as refugia for state rare seepage and pineland plant taxa. In J.R. Williams, J.W. Goodrich-Mahoney, J.R. Wisniewski, and J. Wisniewski, eds. *The Sixth International Symposium on Environmental Concerns in Rights-of-Way Management*, pp. 451–60. Elsevier Science, Oxford.

Shimai, H. and Kondo, K. (2007). Phylogenetic analysis of Mexican and Central American *Pinguicula* (Lentibulariaceae) based on internal transcribed spacer (ITS) sequence. *Chromosome Botany*, 2, 67–77.

Shimai, H., Masuda, Y., Panfet Valdes, C.M., and Kondo, K. (2007). Phylogenetic analysis of Cuban *Pinguicula* (Lentibulariaceae) based on internal transcribed spacer (ITS) region. *Chromosome Botany*, 2, 151–58.

Shimizu, H., Torii, K., Araki, T., and Endo, M. (2016). Importance of epidermal clocks for regulation of hypocotyl elongation through PIF4 and IAA29. *Plant Signaling & Behavior*, 11, e1143999.

Shin, K.S., Lee, S.K., and Cha, B.J. (2007a). Antifungal activity of plumbagin purified from leaves of *Nepenthes ventricosa* × *maxima* against phytopathogenic fungi. *Plant Pathology Journal*, 23, 113–15.

Shin, K.S., Lee, S.K., and Cha, B.J. (2007b). Suppression of phytopathogenic fungi by hexane extract of *Nepenthes ventricosa* × *maxima* leaf. *Fitoterapia*, 78, 585–86.

Shirakawa, J., Hoshi, Y., and Kondo, K. (2011). Chromosome differentiation and genome organization in carnivorous plant family Droseraceae. *Chromosome Botany*, 6, 111–19.

Shivas, R.G. and Brown, J.F. (1989). Yeasts associated with fluid in pitchers of *Nepenthes*. *Mycological Research*, 93, 96–100.

Sibaoka, T. (1991). Rapid plant movements triggered by action potentials. *Botanical Magazine*, 104, 73–95.

Sickel, W., Grafe, T.U., Meuche, I., Steffan-Dewenter, I., Keller, A. (2016). Bacterial diversity and community structure in two Bornean *Nepenthes* species with differences in nitrogen acquisition strategies. *Microbial Ecology*, 71, 938–53.

Silva, S.R., Diaz, Y.C.A., Penha, H.A., et al. (2016). The chloroplast genome of *Utricularia reniformis* sheds light on the evolution of the *ndh* gene complex of terrestrial carnivorous plants from the Lentibulariaceae family. *PLoS ONE*, 11, e0165176.

Silva, S.R., Pinheiro, D.G., Meer, E.J., et al. (2017). The complete chloroplast genome sequence of the leafy bladderwort, *Utricularia foliosa* L. (Lentibulariaceae). *Conservation Genetics Resources*, 9, 213–216.

Šimek, K., Pitsch, G., Salcher, M.M., et al. (2017) Ecological traits of the algae-bearing *Tetrahymena utriculariae* (Ciliophora) from traps of the aquatic carnivorous plant *Utricularia reflexa*. *Journal of Eukaryotic Microbiology*, 64, 336–48.

Simões, M., Breitkreuz, L., Alvarado, M., et al. (2016). The evolving theory of evolutionary radiations. *Trends in Ecology & Evolution*, 31, 27–34.

Simon, S., Skůpa, P., Viaene, T., et al. (2016). PIN6 auxin transporter at endoplasmic reticulum and plasma membrane mediates auxin homeostasis and organogenesis in *Arabidopsis*. *New Phytologist*, 211, 65–74.

Simon, M. K., Williams, L.A., Brady-Passerini, K., Brown, R.H., and Gasser, C.S. (2012). Positive-and negative-acting regulatory elements contribute to the tissue-specific expression of INNER NO OUTER, a YABBY-type transcription factor gene in *Arabidopsis*. *BMC Plant Biology*, 12, 214.

Simoneit, B.R.T., Medeiros, P.M., and Wollenweber, E. (2008). Triterpenoids as major components of the insect-trapping glue of *Roridula* species. *Zeitschrift für Naturforschung*, 63, 625–30.

Simpson, R. 1995. *Nepenthes* and conservation. *Curtis's Botanical Magazine*, 12, 111–18.

Šimura, J., Spíchal, L., Adamec, L., et al. (2016). Cytokinin, auxin and physiological polarity in the aquatic carnivorous plants *Aldrovanda vesiculosa* and *Utricularia australis*. *Annals of Botany*, 117, 1037–44.

Singh, A.K., Prabhakar, S.P., and Sane, S.P. (2011). The biomechanics of fast prey capture in aquatic bladderworts. *Biology Letters*, 7, 547–50.

Sirota, J., Baiser, B., Gotelli, N.J., and Ellison, A.M. (2013). Organic-matter loading determines regime shifts and alternative states in an aquatic ecosystem. *Proceedings of the National Academy of Sciences, USA*, 110, 7742–47.

Sirová, D., Adamec, L., and Vrba, J. (2003). Enzymatic activities in traps of four aquatic species of the carnivorous genus *Utricularia*. *New Phytologist*, 159, 669–75.

Sirová, D., Bárta, J., Šimek, K., et al. (2018). Hunters or farmers? Microbiome characteristics help elucidate the diet composition in an aquatic carnivorous plant. *Microbiome* 6, 225.

Sirová, D., Borovec, J., Picek, T., et al. (2011). Ecological implications of organic carbon dynamics in the traps of aquatic carnivorous *Utricularia* plants. *Functional Plant Biology*, 38, 583–93.

Sirová, D., Borovec, J., Černá, B., et al. (2009). Microbial community development in the traps of aquatic *Utricularia* species. *Aquatic Botany*, 90, 129–36.

Sirová, D., Borovec, J., Šantrůčková, H., et al. (2010). *Utricularia* carnivory revisited: plants supply photosynthetic carbon to traps. *Journal of Experimental Botany*, 61, 99–103.

Sirová, D., Šantrůček, J., Adamec, L., et al. (2014). Dinitrogen fixation associated with shoots of aquatic carnivorous plants: is it ecologically important? *Annals of Botany*, 114, 125–33.

Skotheim, J.M. and Mahadevan, L. (2005). Physical limits and design principles for plant and fungal movements. *Science*, 308, 1308–10.

Slack, A. (1979). *Carnivorous Plants*. Ebury Press, London.

Slotkin, R.K. and Martienssen, R. (2007). Transposable elements and the epigenetic regulation of the genome. *Nature Reviews Genetics*, 8, 272–85.

Small, E. (1972). Photosynthetic rates in relation to nitrogen recycling as an adaptation to nutrient deficiency in peat bog plants. *Canadian Journal of Botany*, 50, 2227–33.

Smart, S.M., Jarvis, S., Walker, K.J., et al. (2015). Common plants as indicators of habitat suitability for rare plants: quantifying the strength of the association between threatened plants and their neighbours. *New Journal of Botany*, 5, 72–88.

Smith, C.M. (1931). Development of *Dionaea muscipula*. II. Germination of seed and development of seedling to maturity. *Botanical Gazette*, 91, 377–94.

Smith, R.D., Dickie, J.B., Linington, S.H., Pritchard, H.W., and Probert, R.J., eds. (2003). *Seed Conservation: Turning Science into Practice*. Royal Botanic Gardens, Kew.

Smith, L.B. and Downs, R.J. (1974). *Pitcairnioideae (Bromeliaceae). Flora Neotropica Monograph* 14, part 1. Hafner Press, New York.

Sohma, K. (1975). Pollen morphology of the Japanese species of *Utricularia* L. and *Pinguicula* L. with notes on fossil pollen of *Utricularia* from Japan (1). *Journal of Japanese Botany*, 50, 164–79.

Soltis, D.E., Albert, V.A., Leebens-Mack, J., et al. (2009). Polyploidy and angiosperm diversification. *American Journal of Botany*, 96, 336–48.

Soltis, D.E., Smith, S.A., Cellinese, N., et al. (2011). Angiosperm phylogeny: 17 genes, 640 taxa. *American Journal of Botany*, 98, 704–30.

Song, J., Keppler, B.D., Wise, R.R., and Bent, A.F. (2015). PARP2 is the predominant poly (ADP-ribose) polymerase in *Arabidopsis* DNA damage and immune responses. *PLoS Genetics*, 11, e1005200.

Sonoda, Y., Yao, S.G., Sako, K., et al. (2007). SHA1, a novel RING finger protein, functions in shoot apical meristem maintenance in *Arabidopsis*. *Plant Journal*, 50, 586–96.

Sorenson, D.R. and Jackson, W.T. (1968). The utilization of paramecia by the carnivorous plant *Utricularia gibba*. *Planta*, 83, 166–70.

Sota, T., Mogi, M., and Kato, K. (1998). Local and regional-scale food web structure in *Nepenthes alata* pitchers. *Biotropica*, 30, 82–91.

Speck, N.H. and Baird, A.M. (1984). Vegetation of the Yule Brook Reserve near Perth, Western Australia. *Journal of the Royal Society of Western Australia*, 66, 147–62.

Spomer, G.G. (1999). Evidence of protocarnivorous capabilities in *Geranium viscosissimum* and *Potentilla arguta* and other sticky plants. *International Journal of Plant Science*, 160, 98–101.

Spoor, D.C.A., Martineau, L.C., Leduc, C., et al. (2006). Selected plant species from the Cree pharmacopoeia of northern Quebec possess anti-diabetic potential. *Canadian Journal of Physiology and Pharmacology*, 84, 847–58.

Srivastava, J.K., Chandra, H., Kalra, S.J., Mishra, P., Khan, H. and Yadav, P. (2016). Plant–microbe interaction in aquatic system and their role in the management of water quality: a review. *Applied Water Science*, doi:10.1007/s13201-0415-2.

Srivastava, D.S., Kolasa, J., Bengtsson, J., et al. (2004). Are natural microcosms useful model systems for ecology? *Trends in Ecology and Evolution*, 19, 379–84.

Srivastava, A., Rogers, W.L., Breton, C.M., Cai, L., and Malmberg, R.L. (2011). Transcriptome analysis of *Sarracenia*, an insectivorous plant. *DNA Research*, 18, 253–61.

St. John, S. (1862). *Life in the Forests of the Far East*. Smith Elder & Co., London.

Stace, C. (1982). *Plant Taxonomy and Biosystematics*. Edward Arnold, London.

Stamati, K., Mudera, V., and Cheema, U. (2011). Evolution of oxygen utilization in multicellular organisms and implications for cell signalling in tissue engineering. *Journal of Tissue Engineering*, 2, 2041731411432365.

State of the Environment Committee (2011). *Australia State of the Environment 2011*. Department of Sustainability, Environment, Water, Population, and Communities, Canberra.

Stebbins, G.L.J. and Major, J. (1965). Endemism and speciation in the Californian flora. *Ecological Monographs*, 35, 1–35.

Stein, B.A., Scott, C., and Benton, N. (2008). Federal lands and endangered species: the role of military and other federal lands in sustaining biodiversity. *BioScience*, 58, 339–47.

Stephens, E.L (1926). A new sundew, *Drosera regia* (Stephens) from the Cape Province. *Transactions of the Royal Society of South Africa*, 13, 309–12.

Stephens, J.D. and Folkerts, D.R. (2012). Life history aspects of *Exyra semicrocea* (pitcher plant moth) (Lepidoptera: Noctuidae). *Southeastern Naturalist*, 11, 111–26.

Stephens, J.D., Godwin R.L. and Folkerts, D.R. (2015a). Distinctions in pitcher morphology and prey capture of the Okefenokee variety within the carnivorous plant species *Sarracenia minor*. *Southeastern Naturalist*, 14, 254–66.

Stephens, J.D., Rogers, W.L., Heyduk, K., et al. (2015b). Resolving phylogenetic relationships of the recently radiated carnivorous plant genus *Sarracenia* using target enrichment. *Molecular Phylogenetics and Evolution*, 85, 76–87.

Stephenson, P. and Hogan, J. (2006). Cloning and characterization of a ribonuclease, a cysteine proteinase, and an aspartic proteinase from pitchers of the carnivorous plant *Nepenthes ventricosa* Blanco. *International Journal of Plant Sciences*, 167, 239–48.

Stewart, C.-B., Schilling, J.W., and Wilson, A.C. (1987). Adaptive evolution in the stomach lysozymes of foregut fermenters. *Nature*, 330, 401–04.

Steyermark, J.A. (1984). Flora of the Venezuelan Guayana. I. *Annals of the Missouri Botanical Garden*, 71, 297–340.

Studnička, M. (1993). Solution to the problem of short vitality in seeds of European butterworts. *Carnivorous Plant Newsletter*, 22, 34.

Studnička, M. (1996). Several ecophysiological observations in *Genlisea*. *Carnivorous Plant Newsletter*, 25, 14–16.

Studnička, M. (2003a). *Genlisea* traps—a new piece of knowledge. *Carnivorous Plant Newsletter*, 32, 36–39.

Studnička, M. (2003b). Observations on life strategies of *Genlisea, Heliamphora* and *Utricularia* in natural habitats. *Carnivorous Plant Newsletter*, 32, 57–61.

Stützel, T. (1998). Eriocaulaceae. In K. Kubitzki, ed. *The Families and Genera of Vascular Plants*, volume 4, pp. 197–207. Springer, Berlin.

Sugiura, S. and Yamazaki, K. (2006). Consequences of scavenging behaviour in a plant bug associated with a glandular plant. *Botanical Journal of the Linnean Society*, 88, 593–602.

Sumsakul, W., Plengsuriyakarn, T., Chaijaroenkul, W., et al. (2014). Antimalarial activity of plumbagin in vitro and in animal models. *BMC Complementary and Alternative Medicine*, 14, 1–6.

Sun, M., Naaem, R., Su, J.–X., et al. (2016). Phylogeny of the Rosidae: A dense taxon sampling analysis. *Journal of Systematics and Evolution*, 54, 363–91.

Svensson, B.M. (1995). Competition between *Sphagnum fuscum* and *Drosera rotundifolia*: a case of ecosystem engineering. *Oikos*, 74, 205–12.

Sweat, T. and Bodri, M.S. (2014). Isolation of protoplasts from *Nepenthes*—a plant carnivore. *Plant Tissue Culture and Biotechnology*, 24, 93–100.

Sydenham, P.H. and Findlay, G.P. (1973). The rapid movement of the bladder of *Utricularia* sp. *Australian Journal of Biological Sciences*, 26, 1115–26.

Szczerba, M.W., Britto, D.T., Ali, S.A., Balkos, K.D., and Kronzucker H.J. (2008). NH_4^+-stimulated and -inhibited components of K^+ transport in rice (*Oryza sativa* L.). *Journal of Experimental Botany*, 59, 3415–23.

Szpitter, A., Narajczyk, M., Maciag-Dorszynska, M., et al. (2014). Effect of *Dionaea muscipula* extract and plumbagin on maceration of potato tissue by *Pectobacterium atrosepticum*. *Annals of Applied Biology*, 164, 404–14.

Takahashi, H. (1988). Ontogenetic development of pollen tetrads of *Drosera capensis* L. *Botanical Gazette*, 149, 275–82.

Takahashi, H. and Sohma, K. (1982). Pollen morphology of the Droseraceae and its related taxa. *Science Reports of the Research Institutes Tohoku University* (*Biology*), 38, 81–156.

Takahashi, K., Matsumoto, K., Nishii, W., Muramatsu, M., and Kubota, K. (2009). Comparative studies on the acid proteinase activities in the digestive fluid of *Nepenthes, Cephalotus, Dionaea*, and *Drosera*. *Carnivorous Plant Newsletter*, 38, 75–82.

Takahashi, K., Nishii, W., and Shibata, C. (2012). The digestive fluid of *Drosera indica* contains a cysteine endopeptidase ("droserain") similar to dionain from *Dionaea muscipula*. *Carnivorous Plant Newsletter*, 41, 132–34.

Takahashi, K., Suzuki, T., Nishii, W., et al. (2011). A cysteine endopeptidase ("dionain") is involved in the digestive fluid of *Dionaea muscipula* (Venus's fly-trap). *Bioscience, Biotechnology and Biochemistry*, 75, 346–48.

Takeuchi, Z., Chaffron, S., Salcher, M.M., et al. (2015). Bacterial diversity and composition in the fluid of pitcher plants of the genus *Nepenthes*. *Systematic and Applied Microbiology*, 38, 330–39.

Takhtajan, A.L. (1980). Outline of the classification of flowering plants (Magnoliophyta). *Botanical Review*, 46, 225–359.

Takhtajan, A.L. (1986). *Floristic Regions of the World*. University of California Press, Berkeley.

Takhtajan, A. (1992). *Anatomia Seminum Comparativa. Tomus 4. Dicotyledones. Rosidae I.* Nauka, Leningrad.

Takhtajan, A. (1996). *Anatomia Seminum Comparativa. Tomus 5. Dicotyledones. Rosidae I.* Nauka, Leningrad.

Takhtajan, A. (1997). *Diversity and Classification of Flowering Plants*. Columbia University Press, New York.

Tan, H.T.W. (1997). Prey. In H.T.W. Tan, ed. *A Guide to the Carnivorous Plants of Singapore*, pp. 125–31. Singapore Science Centre, Singapore.

Tank, D.C., Eastman, J.M., Pennell, M.W., et al. (2015). Nested radiations and the pulse of angiosperm diversification: increased diversification rates often follow whole genome duplications. *New Phytologist*, 207, 454–67.

Tarabini, C.T., Zanenga, S.K., and De Souza, P.G. (2001). Population ecology of *Paepalanthus polyanthus* (Bong.) Kunth: demography and life history of a sand dune monocarpic plant. *Brazilian Journal of Botany*, 24, 122–34.

Tatematsu, K., Kumagai, S., Muto, H., et al. (2004). MASSUGU2 encodes Aux/IAA19, an auxin-regulated protein that functions together with the transcriptional activator NPH4/ARF7 to regulate differential growth responses of hypocotyl and formation of lateral roots in *Arabidopsis thaliana*. *Plant Cell*, 16, 379–93.

Taylor, F. (1980). Optimal switching to diapause in relation to the onset of winter. *Theoretical Population Biology*, 18, 125–33.

Taylor, P. (1989). *The Genus* Utricularia: *A Taxonomic Monograph*. Kew Bulletin, Additional Series XIV, HMSO, London.

Taylor, P., Souza, V.C., Giulietti, A.M., and Harley, R.M. (2000). *Philcoxia*: A new genus of Scrophulariaceae with three new species from eastern Brazil. *Kew Bulletin*, 55, 155–63.

Teng, W.-L. (1999). Source, etiolation and orientation of explants affect in vitro regeneration of Venus fly-trap (*Dionaea muscipula*). *Plant Cell Reports*, 18, 363–68.

terHorst, C.P. (2011). Experimental evolution of protozoan traits in response to interspecific competition: Evolution in response to competition. *Journal of Evolutionary Biology*, 24, 36–46.

terHorst, C.P., Miller, T.E., and Levitan, D.R. (2010). Evolution of prey in ecological time reduces the effect size of predators in experimental microcosms. *Ecology*, 91, 629–36.

Thanh, N.V., Thao, N.P., Dat, L.D., et al. (2015). Two new naphthalene glucosides and other bioactive compounds from the carnivorous plant *Nepenthes mirabilis*. *Archives of Pharmacal Research*, 38, 1774–82.

Thanikaimoni, G. (1966). Pollen morphology of the genus *Utricularia*. *Pollen et Spores*, 8, 265–84.

Thanikaimoni, G. and Vasanthy, G. (1972). Sarraceniaceae: palynology and systematics. *Pollen et Spores*, 14, 143–55.

Thienemann, A. (1932). Die Tierwelt der *Nepenthes*-Kannen. *Archiv für Hydrobiologie Suppl.*, 13, 1–91.

Thomas, K. and Cameron, D.M. (1986). Pollination and fertilization in the pitcher plant (*Sarracenia purpurea* L.). *American Journal of Botany*, 73, 678–78.

Thomas, F., Hehemann, J.H., Rebuffet, E., Czjzek, M., and Michel, G. (2011). Environmental and gut Bacteroidetes: the food connection. *Frontiers in Microbiology*, 2, 93.

Thompson, J.N. (1981). Reversed animal–plant interactions: the evolution of insectivorous and ant-fed plants. *Biological Journal of the Linnean Society*, 16, 147–55.

Thor, G. (1988). The genus *Utricularia* in the Nordic countries, with special emphasis on *U. stygia* and *U. ochroleuca*. *Nordic Journal of Botany*, 8, 213–25.

Thorén, L.M. and Karlsson, P.S. (1998). Effects of supplementary feeding on growth and reproduction of three carnivorous plant species in a subarctic environment. *Journal of Ecology*, 86, 501–10.

Thorén, L.M., Tuomi, J., Kämäräinen, T., and Laine, K. (2003). Resource availability affects investment in carnivory in *Drosera rotundifolia*. *New Phytologist*, 159, 507–11.

Thorne, R.F. (1983). Proposed new realignments in the angiosperms. *Nordic Journal of Botany*, 3, 85–118.

Thorne, R.F. (1992). An updated phylogenetic classification of the flowering plants. *Aliso*, 13, 365–89.

Thornham, D.G., Smith, J.M., Grafe, T.U., and Federle, W. (2012). Setting the trap: cleaning behaviour of *Camponotus schmitzi* ants increases long-term capture efficiency of their pitcher plant host, *Nepenthes bicalcarata*. *Functional Ecology*, 26, 11–19.

Thuiller, W., Lafourcade, B., Engler, R., and Araújo, M.B. (2009). BIOMOD—a platform for ensemble forecasting of species distributions. *Ecography*, 32, 369–73.

Thum, M. (1986). Segregation of habitat and prey in two sympatric carnivorous plant species, *Drosera rotundifolia* and *Drosera intermedia*. *Oecologia*, 70, 601–05.

Thum, M. (1988). The significance of carnivory for the fitness of *Drosera* in its natural habitat. 1. The reactions of *Drosera intermedia* and *D. rotundifolia* to supplementary feeding. *Oecologia*, 75, 472–80.

Thum, M. (1989a). The significance of carnivory for the fitness of *Drosera* in its natural habitat. 2. The amount of captured prey and its effect on *Drosera intermedia* and *Drosera rotundifolia*. *Oecologia*, 81, 401–11.

Thum, M. (1989b). The significance of opportunistic predators for the sympatric carnivorous plant species *Drosera intermedia* and *Drosera rotundifolia*. *Oecologia*, 81, 397–400.

Tian, J., Chen, Y., Ma, B., et al. (2014). *Drosera peltata* Smith var. *lunata* (Buch.-Ham.) C.B. Clarke as a feasible source of plumbagin: phytochemical analysis and antifungal activity assay. *World Journal of Microbiology and Biotechnology*, 30, 737–45.

Tibetan People's Publishing House (1971). *The Common Chinese Herbal Medicines in Tibet*. Tibet People's Publishing House, Lhasa.

Tilman, D. (1982). *Resource Competition and Community Structure*. Princeton University Press, Princeton.

Tilman, D. (1988). *Plant Strategies and the Dynamics and Structure of Plant Communities*. Princeton University Press, Princeton.

Tjio, J.H. (1948). The somatic chromosomes of some tropical plants. *Hereditas*, 34, 135–46.

Tökés, Z.A., Woon, W.C., and Chambers, S.M. (1974). Digestive enzymes secreted by the carnivorous plant *Nepenthes macferlanei* L. *Planta*, 119, 39–46.

Tokunaga, T., Takada, N., and Ueda, M. (2004). Mechanism of antifeedant activity of plumbagin, a compound concerning the chemical defense in carnivorous plant. *Tetrahedron Letters*, 45, 7115–19.

Topp, C.N., Zhong, C.X., and Dawe, R.K. (2004). Centromere-encoded RNAs are integral components of the maize kinetochore. *Proceedings of the National Academy of Sciences, USA*, 101, 15986–91.

Towne, E.G. and Knapp, A.K. (1996). Biomass and density responses in tallgrass prairie legumes to annual fire and topographic position. *American Journal of Botany*, 83, 175–79.

Tran, T.D., Cao, H.X., Jovtchev, G., et al. (2015a). Centromere and telomere sequence alterations reflect the rapid genome evolution within the carnivorous plant genus *Genlisea*. *Plant Journal*, 84, 1087–99.

Tran, T.D., Cao, H.X., Jovtchev, G., et al. (2015b). Chromatin organization and cytological features of carnivorous

Genlisea species with large genome size differences. *Frontiers in Plant Science*, 6, e613.

Tran, T.D., Šimková, H., Schmidt, R., et al. (2016). Chromosome identification for the carnivorous plant *Genlisea margaretae*. *Chromosoma*, doi:10.1007/s00412-016-0599-0.

Trebacz, K. and Sievers, A. (1998). Action potentials evoked by light in traps of *Dionaea muscipula* Ellis. *Plant and Cell Physiology*, 39, 369–72.

Troll, W. and Dietz, H. (1954). Morphologische und histogenetische Untersuchungen an *Utricularia*-Arten. *Österreichische Botanische Zeitung*, 101, 165–207.

Troll, C. and Paffen, K.H. (1964). Karte der Jahreszeiten–Klimate der Erde. *Erdkunde*, 18, 5–28.

Trovó, M., de Andrade, M.J.G., Sano, P.T., Ribeiro, P.L., and van den Berg, C. (2013). Molecular phylogenetics and biogeography of Neotropical Paepalanthoideae with emphasis on Brazilian *Paepalanthus* (Eriocaulaceae). *Botanical Journal of the Linnean Society*, 171, 225–43.

Truswell, E.M. and Marchant, N. (1986). Early Tertiary pollen of probable droseracean affinity from Central Australia. *Special Papers in Palaeontology*, 35, 163–78.

Trzcinski, M.K., Walde, S.J., and Taylor, P.D. (2003). Colonisation of pitcher plant leaves at several spatial scales. *Ecological Entomology*, 28, 482–89.

Tuleja, M., Chmielowska, A., and Płachno, B.J. (2014). The preliminary attempts of in vitro regeneration of recalcitrant species of *Cephalotus follicularis* Labill. *Modern Phytomorphology*, 6, 37–38.

Tvrda, E. (2015). The *Drosera* extract as an alternative in vitro supplement to animal semen: Effects on bovine spermatozoa activity and oxidative balance. *Animal Science and Biotechnologies*, 48, 68–75.

Ulanowicz, R.E. (1995). *Utricularia*'s secret: the advantages of positive feedback in oligotrophic environments. *Ecological Modelling*, 79, 49–57.

Uphof, J.C.T. (1936). Sarraceniaceae. *Die Natürlichen Pflanzenfamilien*,17b, 704–27.

USFWS (1994). *Green Pitcher Plant Recovery Plan*, United States Fish and Wildlife Service, Jackson.

van Achterberg, C. (1973). A study of the Arthropoda caught by *Drosera* species. *Entomologische Berichten*, 33, 137–40.

van der Ent, A., Sumail, S., and Clarke, C. (2015). Habitat differentiation of obligate ultramafic *Nepenthes* endemic to Mount Kinabalu and Mount Tambuyukon (Sabah, Malaysia). *Plant Ecology*, 216, 789–807.

Van de Peer, Y., Maere, S., and Meyer, A. (2009). The evolutionary significance of ancient genome duplications. *Nature Reviews Genetics*, 10, 725–32.

Van Oye, P. (1921). Zur Biologie der Kanne von *Nepenthes melamphora* Reinw. *Biologisches Zentralblatt*, 41, 529–34.

Vanneste, K., Baele, G., Maere, S., and Van de Peer, Y., 2014. Analysis of 41 plant genomes supports a wave of successful genome duplications in association with the Cretaceous–Paleogene boundary. *Genome Research*, 24, 1334–47.

Varadarajan, G.S. and Brown, G.K. (1988). Morphological variation of some floral features of the subfamily Pitcairnioideae (Bromeliaceae) and their significance in pollination biology. *Botanical Gazette*, 149, 82–91.

Varadarajan, G.S. and Gilmartin, A.J. (1988). Seed morphology of the subfamily Pitcairnioideae (Bromeliaceae) and its systematic implications. *American Journal of Botany*, 75, 808–18.

Vassilyev, A.E. (1977). [*Functional morphology of plant secretory cells*.] Nauka Publishing House, Leningrad. (In Russian.)

Vassilyev, A.E. (2005). Dynamics of ultrastructural characters of *Drosophyllum lusitanicum* Link (Droseraceae) digestive glands during maturation and after stimulation. *Taiwania*, 50, 167–82.

Vassilyev, A.E. (2007). [The nectaries of the peristome in the closed pitchers of *Nepenthes khasiana* (Nepenthaceae) secrete polysaccharide slime.] *Botanicheskii Zhurnal*, 92, 1544–49. (In Russian.)

Vassilyev, A.E. and Muravnik, L.E. (1988). The ultrastructure of the digestive glands in *Pinguicula vulgaris* L. (Lentibulariaceae) relative to their function. II. The changes on stimulation. *Annals of Botany*, 62, 343–51.

Vassilyev, A.E. and Muravnik, L.E. (2007). [The nectaries of the lid in closed pitchers of *Nepenthes khasiana* (Nepenthaceae) secrete a digestive fluid.] *Botanicheskii Zhurnal*, 92, 1141–44. (In Russian.)

Veleba, A., Bureš, P., Adamec, L., et al. (2014). Genome size and genomic GC content evolution in the miniature genome-sized family Lentibulariaceae. *New Phytologist*, 203, 22–28.

Veleba, A., Šmarda, P., Zedek, F., et al. (2017). Evolution of genome size and genomic GC content in carnivorous holokinetics (Droseraceae). *Annals of Botany*, 119, 409–16.

Veloz, S.D., Williams, J.W., Blois, J.L., et al. (2012). No-analog climates and shifting realized niches during the late quaternary: implications for 21st-century predictions by species distribution models. *Global Change Biology*, 18, 1698–713.

Verbeek, N.A.M. and Boasson, R. (1993). Relationship between types of prey captured and growth form in *Drosera* in southwestern Australia. *Australian Journal of Ecology*, 18, 203–07.

Verburg, J.G., Rangwala, S.H., Samac, D.A., Luckow, V.A., and Huynh, Q.K. (1993). Examination of the role of tyrosine-174 in the catalytic mechanism of the *Arabidopsis thaliana* chitinase: comparison of variant chitinases generated by site-directed mutagenesis and expressed in insect cells using baculovirus vectors. *Archives of Biochemistry and Biophysics*, 300, 223–30.

Vincent, O., Roditchev, I., and Marmottant, P. (2011a). Spontaneous firings of carnivorous aquatic *Utricularia* traps: temporal patterns and mechanical oscillations. *PloS ONE*, 6, e20205.

Vincent, O., Weisskopf, C., Poppinga, S., et al. (2011b). Ultra-fast underwater suction traps. *Proceedings of the Royal Society B*, 278, 2909–14.

Vintéjoux, C. (1973). Études des aspects ultrastructuraux de certaines cellules glandulaires en rapport avec leur activité sécrétive chez l'*Utricularia neglecta* L. (Lentibulariaceae). *Comptes Rendues*, 277D, 2345–48.

Vintéjoux, C. (1974). Ultrastructural and cytochemical observations on the digestive glands of *Utricularia neglecta* L. (Lentibulariaceae). Distribution of protease and acid phosphatase activities. *Portugaliae Acta Biologica*, 14, 463–71.

Vintéjoux, C. and Shoar-Ghafari, A. (2000). Mucigenic cells in carnivorous plants. *Acta Botanica Gallica*, 147, 5–20.

Vintéjoux, C. and Shoar-Ghafari, A. (2005). Glandes digestives de l'Utriculaire: ultrastructures et fonctions. *Acta Botanica Gallica*, 152, 131–45.

Viscosi, V. and Cardini, A. (2011). Leaf morphology, taxonomy and geometric morphometrics: a simplified protocol for beginners. *PLoS ONE*, 6, e25630.

Vitousek, P.M. and Howarth, R.W. (1991). Nitrogen limitation on land and in the sea: how can it occur? *Biogeochemistry*, 13, 87–115.

Vitt, P., Havens, K., Kramer, A.T., Sollenberger, D., and Yates, E. (2010). Assisted migration of plants: changes in latitudes, changes in attitudes. *Biological Conservation*, 143, 18–27.

Vleeshouwers, L.M., Bouwmeester, H.J., and Karssen, C.M. (1995). Redefining seed dormancy: an attempt to integrate physiology and ecology. *Journal of Ecology*, 83, 1031–37.

Vogel, S. (1998). Remarkable nectaries: structure, ecology, organophyletic perspectives. II. Nectarioles. *Flora*, 193, 1–29.

Voigt, D. and Gorb, S. (2008). An insect trap as habitat: cohesion-failure mechanism prevents adhesion of *Pameridea roridulae* bugs to the sticky surface of the plant *Roridula gorgonias*. *Journal of Experimental Biology*, 211, 2647–57.

Voigt, D. and Gorb, S. (2010a). Desiccation resistance of adhesive secretion in the protocarnivorous plant *Roridula gorgonias* as an adaptation to periodically dry environment. *Planta*, 232, 1511–15.

Voigt, D. and Gorb, S. (2010b). Locomotion in a sticky terrain. *Arthropod–Plant Interactions*, 4, 69–79.

Voigt, D., Gorb, E., and Gorb, S. (2009). Hierarchical organisation of the trap in the protocarnivorous plant *Roridula gorgonias* (Roridulaceae). *Journal of Experimental Biology*, 212, 3184–91.

Voigt, D., Konrad, W., and Gorb, S. (2015). A universal glue: underwater adhesion of the secretion of the carnivorous flypaper plant *Roridula gorgonias*. *Interface Focus*, 5, 20140053.

Volkov, A.G., Adesina, T., and Jovanov, E. (2007). Closing of Venus flytrap by electrical stimulation of motor cells. *Plant Signaling & Behavior*, 2, 139–45.

Volkov, A.G., Adesina, T., and Jovanov, E. (2008a). Charge induced closing of *Dionaea muscipula* Ellis trap. *Bioelectrochemistry*, 74, 16–21.

Volkov, A.G., Adesina, T., Markin, V.S., and Jovanov, E. (2008b). Kinetics and mechanism of *Dionaea muscipula* trap closing. *Plant Physiology*, 146, 694–702.

Volkov, A.G., Carrell, H., Adesina, T., Markin, V.S., and Jovanov, E. (2008c). Plant electrical memory. *Plant Signaling & Behavior*, 3, 490–92.

Volkov, A.G., Coopwood, K.J., and Markin, V.S. (2008d). Inhibition of the *Dionaea muscipula* Ellis trap closure by ion and water channels blockers and uncouplers. *Plant Science*, 175, 642–49.

Volkov, A.G., Forde-Tuckett, V., Volkova, M.I., and Markin, V.S. (2014). Morphing structures of the *Dionaea muscipula* Ellis during the trap opening and closing. *Plant Signaling & Behavior*, 9, e27793.

Volkov, A.G., Harris II, S.L., Vilfranc, C.L., et al. (2013). Venus flytrap biomechanics: forces in the *Dionaea muscipula* trap. *Journal of Plant Physiology*, 170, 25–32.

Volkov, A.G., Murphy, V.A., Clemmons, J.I., Curley, M.J., and Markin, V.S. (2012). Energetics and forces of *Dionaea muscipula* trap closing. *Journal of Plant Physiology*, 169, 55–64.

Volkov, A.G., Pinnock, M.R., Lowe, D.C., Gay, M.S., and Markin, V.S. (2011). Complete hunting cycle of *Dionaea muscipula*: Consecutive steps and their electrical properties. *Journal of Plant Physiology*, 168, 109–20.

Volkova, P.A. and Shipunov, A.B. (2009). The natural behaviour of *Drosera*: sundews do not catch insects on purpose. *Carnivorous Plant Newsletter*, 38, 114–20.

Vu, G.T.H., Schmutzer, T., Bull, F., et al. (2015). Comparative genome analysis reveals divergent genome size evolution in a carnivorous plant genus. *Plant Genome*, 8, e3.

Wagner, G.M. and Mshigeni, K.E. (1986). The *Utricularia*-Cyanophyta association and its nitrogen-fixing capacity. *Hydrobiologia*, 141, 255–61.

Wakefield, A.E., Gotelli, N.J., Wittman, S.E., and Ellison, A.M. (2005). Prey addition alters nutrient stoichiometry of the carnivorous plant *Sarracenia purpurea*. *Ecology*, 86, 1737–43.

Walker, K.J. (2014). *Sarracenia purpurea* subsp. *purpurea* (Sarraceniaceae) naturalised in Britain and Ireland: distribution, ecology, impacts and control. *New Journal of Botany*, 4, 33–41.

Walker, A.P., Beckerman, A.P., Gu, L., et al. (2014). The relationship of leaf photosynthetic traits—V_{cmax} and

J_{max}—to leaf nitrogen, leaf phosphorus, and specific leaf area: a meta-analysis and modeling study. *Ecology and Evolution*, 4, 3218–35.

Wallace, A.R. (1869). *The Malay Archipelago*. Macmillan, London.

Wang, Z., Hamrick, J.L., and Godt, M.J.W. (2004). High genetic diversity in *Sarracenia leucophylla* (Sarraceniaceae), a carnivorous wetland herb. *Journal of Heredity*, 95, 234–43.

Wang, F., Muto, A., Van de Velde, J., et al. (2015). Functional analysis of *Arabidopsis* TETRASPANIN gene family in plant growth and development. *Plant Physiology*, 169, 2200–14.

Wang, L. and Zhou, Q. (2016). Surface hydrophobicity of slippery zones in the pitchers of two *Nepenthes* species and a hybrid. *Scientific Reports*, 6, 1–11.

Warming, E. (1874). Bidrag til Kundskaben om Lentibulariaceae. *Videnskabelige Meddelelser fra den naturhistoriske Forening i Kjöbenhavn*, 3–7, 33–45.

Warren, B.H. and Hawkins, J.A. (2006). The distribution of species diversity across a flora's component lineages: Dating the Cape's "relicts." *Proceedings of the Royal Society B*, 273, 2149–58.

Warren, D.L. and Seifert, S.N. (2011). Ecological niche modeling in Maxent: the importance of model complexity and the performance of model selection criteria. *Ecological Applications*, 21, 335–42.

Wasternack, C. and Hause, B. (2013). Jasmonates: biosynthesis, perception, signal transduction and action in plant stress response, growth and development. An update to the 2007 review in Annals of Botany. *Annals of Botany*, 111, 1024–58.

Waterworth, W.M., Masnavi, G., Bhardwaj, R.M., et al. (2010). A plant DNA ligase is an important determinant of seed longevity. *Plant Journal*, 63, 848–60.

Watson, A.P., Matthiessen, J.N., and Springett, B.P. (1982). Arthropod associates and macronutrient status of the red-ink sundew (*Drosera erythrorhiza* Lindl.). *Australian Journal of Ecology*, 7, 13–22.

Wee, Y.C. (1978). Concerning *Nepenthes* seedlings. *Malaysian Naturalist Journal*, 32, 105–06.

Weiss, T.E., Jr. (1980). *The Effects of Fire and Nutrient Availability on The Pitcher Plant* Sarracenia flava *L.* Ph.D. dissertation, University of Georgia, Athens.

Wells, K., Lakim, M.B., Schulz, S., and Ayasse, M. (2011). Pitchers of *Nepenthes rajah* collect faecal droppings from both diurnal and nocturnal small mammals and emit fruity odour. *Journal of Tropical Ecology*, 27, 347–53.

Westermeier, A.S., Fleischmann, A., Müller, K., et al. (2017). Trap diversity and character evolution in carnivorous bladderworts (*Utricularia*, Lentibulariaceae). *Scientific Reports*, 7, 12052.

Wheeler, A.G. (2001). *Biology of The Plant Bugs (Hemiptera: Miridae): Pests, Predators, Opportunists*. Cornell University Press, London.

Wheeler, A.G. and Krimmel, B.A. (2015). Mirid (Hemiptera: Heteroptera) specialists of sticky plants: adaptations, interactions, and ecological implications. *Annual Review of Entomology*, 60, 393–414.

Wiart, C., Mogana, S., Khalifah, S., et al. (2004). Antimicrobial screening of plants used for traditional medicine in the state of Perak, Peninsular Malaysia. *Fitoterapia*, 75, 68–73.

Wicke, S., Schäferhoff, B., dePamphilis, C.W., and Müller, K.F. (2014). Disproportional plastome-wide increase of substitution rates and relaxed purifying selection in genes of carnivorous Lentibulariaceae. *Molecular Biology and Evolution*, 31, 529–45.

Wickett, N.J., Mirarab, S., Nguyen, N., et al. (2014). Phylotranscriptomic analysis of the origin and early diversification of land plants. *Proceedings of the National Academy of Sciences, USA*, 111, 4859–68.

Wiens, D. (1978). Mimicry in plants. *Evolutionary Biology*, 11, 365–403.

Wikström, N., Savolainen, V., and Chase, M.W. (2001). Evolution of the angiosperms: calibrating the family tree. *Proceedings of the Royal Society B*, 268, 2211–20.

Willemsen, V., Bauch, M., Bennett, T., et al. (2008). The NAC domain transcription factors FEZ and SOMBRERO control the orientation of cell division plane in *Arabidopsis* root stem cells. *Developmental Cell*, 15, 913–22.

Williams, J. and Jackson, S.T. (2007). Novel climates, no-analog communities, and ecological surprises. *Frontiers in Ecology and the Environment*, 5, 475–82.

Williams, S.E. (1976). Comparative sensory physiology of the Droseraceae—the evolution of a plant sensory system. *Proceedings of the American Philosophical Society*, 120, 187–204.

Williams, S.E. (1992). Mechanisms of trap movement II: Does *Aldrovanda* close by a turgor mechanism? A question of how much, where and when. *Carnivorous Plant Newsletter*, 21, 46–51.

Williams, S.E., Albert, V.A., and Chase, M.W. (1994). Relationships of Droseraceae: a cladistic analysis of *rbcL* sequence and morphological data. *American Journal of Botany*, 81, 1027–37.

Williams, S.E. and Bennett, A.B. (1982). Leaf closure in the Venus Flytrap: an acid growth response. *Science*, 218, 1120–22.

Williams, S.E. and Pickard, B.G. (1972). Receptor potentials and action potentials in *Drosera* tentacles. *Planta*, 103, 193–221.

Williams, S.E. and Pickard, B.G. (1979). The role of action potentials in the control of capture movements of *Drosera* and *Dionaea*. *Plant Growth Substances*, 1979, 470–80.

Williams, S.E. and Spanswick, R.M. (1972). Intracellular recordings of the action potentials which mediate the thigmonastic movements of *Drosera*. *Plant Physiology*, 49, 64.

Wilson, J. (1890). The mucilage- and other glands of the Plumbagineae. *Annals of Botany*, 4, 231–58.

Wilson, S.D. (1985). The growth of *Drosera intermedia* in nutrient-rich habitats: the role of insectivory and interspecific competition. *Canadian Journal of Botany*, 63, 2468–69.

Wilson, P. (1995). Pollination in *Drosera tracyi*: selection is strongest when resources are intermediate. *Evolutionary Ecology*, 9, 382–96.

Winkler, M., Hulber, K., and Hietz, P. (2005). Effect of canopy position on germination and seedling survival of epiphytic bromeliads in a Mexican humid montane forest. *Annals of Botany*, 95, 1039–47.

Winston, R.D. and Gorham, P.R. (1979a). Roles of endogenous and exogenous growth regulators in dormancy of *Utricularia vulgaris*. *Canadian Journal of Botany*, 57, 2750–59.

Winston, R.D. and Gorham, P.R. (1979b). Turions and dormancy states in *Utricularia vulgaris*. *Canadian Journal of Botany*, 57, 2740–49.

Wistuba, A., Carow, T., and Harbarth, P. (2002). *Heliamphora chimantensis*, a new species of *Heliamphora* (Sarraceniaceae) from the "Macizo de Chimanta" in the south of Venezuela. *Carnivorous Plants Newsletter*, 31, 78–82.

Wistuba, A., Carow, T., Harbarth, P., and Nerz, J. (2005). *Heliamphora pulchella*, eine neue mit *Heliamphora minor* (Sarraceniaceae) verwadte Art aus der Chimanta Region in Venezuela. *Taublatt*, 53, 42–49.

Wisz, M.S., Hijmans, R.J., Li, J., et al. (2008). Effects of sample size on the performance of species distribution models. *Diversity and Distributions*, 14, 763–73.

Wisz, M.S., Pottier, J., Kissling, W.D., et al. (2012). The role of biotic interactions in shaping distributions and realised assemblages of species: implications for species distribution modelling. *Biological Reviews*, 88, 15–30.

Wong, S.C., Cowan, I.R., and Farquhar, G.D. (1979). Stomatal conductance correlates with photosynthetic capacity. *Nature*, 282, 424–26.

Wong, W.O., Dilcher, D.L., Labandeira, C.C., Sun, G., and Fleischmann, A. (2015). Early Cretaceous *Archaeamphora* is not a carnivorous angiosperm. *Frontiers in Plant Science*, 6, e326.

Wong, K.M.M., Goh, T.K., Hodgkiss, I.J., et al. (1998). Role of fungi in freshwater ecosystems. *Biodiversity and Conservation*, 7, 1187–206.

Wong, T.-S., Kang, S.H., Tang, S.K.Y., et al. (2011). Bioinspired self-repairing slippery surfaces with pressure-stable omniphobicity. *Nature*, 477, 443–47.

Wright, I.J., Reich, P.B., Cornelissen, J.H.C., et al. (2005). Modulation of leaf economic traits and trait relationships by climate. *Global Ecology and Biogeography*, 14, 411–21.

Yamamoto, I. and Kadono, Y. (1990). A study on the reproductive biology of aquatic *Utricularia* species in southwestern Japan. *Acta Phytotaxonomica et Geobotanica*, 41, 189–200.

Yang, Z. (2007). PAML 4: phylogenetic analysis by maximum likelihood. *Molecular Biology and Evolution*, 24, 1586–91.

Yang, Y.P., Liu, H.Y., and Chao, Y.S. (2009). Trap gland morphology and its systematic implications in Taiwan *Utricularia* (Lentibulariaceae). *Flora*, 204, 692–99.

Yeates, D. (1992). Immature stages of the apterous fly *Badisis ambulans* McAlpine (Diptera, Micropezidae). *Journal of Natural History*, 26, 417–24.

Yesson, C. and Culham, A. (2006). Phyloclimatic modeling: combining phylogenetics and bioclimatic modeling. *Systematic Biology*, 55, 785–802.

Yilamujiang, A., Reichelt, M., and Mithöfer, A. (2016). Slow food: insect prey and chitin induce phytohormone accumulation and gene expression in carnivorous *Nepenthes* plants. *Annals of Botany*, 118, 369–75.

Zachos, F.E. (2013). Taxonomy: species splitting puts conservation at risk. *Nature*, 494, 35.

Zakaria, W.-N.-A.W., Loke, K.-K., Goh, H.-H. and Noor, N.M. (2016). RNA-seq analysis for plant carnivory gene discovery in *Nepenthes* × *ventrata*. *Genomics Data*, 7, 18–19.

Zakaria, W.-N.-A.W., Loke, K.-K., Zulkapli M.-M., et al. (2015). RNA-seq Analysis of *Nepenthes ampullaria*. *Frontiers in Plant Science*, 6, 1229.

Zamora, R. (1990a). Observational and experimental study of a carnivorous plant–ant kleptobiotic interaction. *Oikos*, 58, 368–72.

Zamora, R. (1990b). The feeding ecology of a carnivorous plant (*Pinguicula nevadense*): prey analysis and capture constraints. *Oecologia*, 84, 376–79.

Zamora, R. (1995). The trapping success of a carnivorous plant, *Pinguicula vallisneriifolia*: the cumulative effects of availability, attraction, retention and robbery of prey. *Oikos*, 73, 309–22.

Zamora, R. (1999). Conditional outcomes of interactions: the pollinator–prey conflict of an insectivorous plant. *Ecology*, 80, 786–95.

Zamora, R. and Gómez, J.M. (1996). Carnivorous plant–slug interaction: a trip from herbivory to kleptoparasitism. *Journal of Animal Ecology*, 65, 154–60.

Zamora, R., Gómez, J.M., and Hódar, J.A. (1997). Responses of a carnivorous plant to prey and inorganic nutrients in a Mediterranean environment. *Oecologia*, 111, 443–51.

Zamora, R., Gómez, J.M., and Hódar, J.A. (1998). Fitness responses of a carnivorous plant in contrasting ecological scenarios. *Ecology*, 79,1630–44.

Zamudio, S. (2001). *Revisión de la sección* Orcheosanthus *del género* Pinguicula *(Lentibulariaceae)*. Ph.D. dissertation, Universidad Nacional Autónoma de México, Mexico City.

Zamudio, S. (2005). Dos especies nuevas de *Pinguicula* (Lentibulariaceae) de la Sierra Madre Oriental, México. *Acta Botanica Mexicana*, 70, 69–83.

Zanne, A.E., Tank, D.C., Cornwell, W.K., et al. (2014). Three keys to the radiation of angiosperms into freezing environments. *Nature*, 506, 89–92.

Zehl, M., Braunberger, C., Conrad, J., et al. (2011). Identification and quantification of flavonoids and ellagic acid derivatives in therapeutically important *Drosera* species by LC-DAD, LC-NMR, NMR, and LC-MS. *Analytical and Bioanalytical Chemistry*, 400, 2565–76.

Zeng, L., Zhang, Q., Sun, R., et al. (2014). Resolution of deep angiosperm phylogeny using conserved nuclear genes and estimates of early divergence times. *Nature Communications*, 5, e4956.

Zenk, M.H., Fürbringer, M., and Steglich, W. (1969). Occurrence and distribution of 7-methyljuglone and plumbagin in the Droseraceae. *Phytochemistry*, 8, 2199–200.

Zhang, J. (2006). Parallel adaptive origins of digestive RNases in Asian and African leaf monkeys. *Nature Genetics*, 38, 819–23.

Zhang, M., Lenaghan, S.C., Xia, L., et al. (2010). Nanofibers and nanoparticles from the insect-capturing adhesive of the sundew (*Drosera*) for cell attachment. *Journal of Nanobiotechnology*, 8, 20–30.

Zhao, D., Ferguson, A.A., and Jiang, N. (2015). What makes up plant genomes: the vanishing line between transposable elements and genes. *Biochimica et Biophysica Acta*, 1859, 366–80.

Ziaratnia, S.M., Kunert, K.J., and Lall, N. (2009). Elicitation of 7-methyljuglone in *Drosera capensis*. *South African Journal of Botany*, 75, 97–103.

Ziemer, B. (2012). Germination of 22-year-old *Drosophyllum lusitanicum* and *Byblis gigantea* seeds. *Carnivorous Plant Newsletter*, 41, 154.

Zilber-Rosenberg, I. and Rosenberg, E. (2008). Role of microorganisms in the evolution of animals and plants: the hologenome theory of evolution. *FEMS Microbiology Reviews*, 32, 723–35.

Zotz, G. (2013). The systematic distribution of vascular epiphytes—a critical update. *Botanical Journal of the Linnean Society*, 171, 453–81.

Zrenner, R., Riegler, H., Marquard, C.R., et al. (2009). A functional analysis of the pyrimidine catabolic pathway in *Arabidopsis*. *New Phytologist*, 183, 117–32.

Acknowledgments

This book would not exist without the dedication of the authors of the individual chapters. Every one of them thoughtfully synthesized broad areas of current research, worked cooperatively on related chapters, met deadlines, and responded to seemingly endless queries from the editors about details large and small. Each chapter was reviewed thoroughly and constructively by another chapter author and by at least one of the following external reviewers: Dave Armitage, František Baluška, Carol Baskin, Rebekah Bergkoetter, Leszek Błędzki, Pat Calie, John Conran, Charles Davis, Vitor de Miranda, Nicholas Gotelli, Angela Haas, Stephen Heard, Demetrios (Jim) Karagatzides, Steve Keller, Jamie Kneitel, Irene Lichtscheidl, Jiří Macas, Larry Mellichamp, Axel Mithöfer, Gidi Ne'eman, Zdeněk Opatrný, Bob Peet, Lorenzo Peruzzi, Sydne Record, Sergey Shabala, Robin Shin, Miloslav Studnička, Brita Svensson, John Titus, Dennis Whigham, and Ian Yates. In writing the Foreword, Danny Joel built a welcome bridge from Juniper et al. to this volume. The editors at Oxford University Press—Ian Sherman and Bethany Kershaw—kept the editors and contributors on track and, with their production team, saw this book through to publication. To all, the editors are most grateful.

Finally, like carnivorous plants, doing good science has both benefits and costs. The efforts of the editors on this book were supported in part by grants from the US National Science Foundation, awards DEB 11-44056, DEB 12-37491, and DBI 14-59519 to AME, by the Harvard Forest, and by Long-Term Research Development Project No. 67985939 from the Czech Academy of Sciences to LA.

Chapter 2

Steve Brewer thanks the Desoto National Forest for permitting access to field sites in Mississippi, and Sarah Hinman and Matthew Abbott for allowing Brewer to re-analyze some of their thesis and dissertation data. Both authors appreciate the helpful comments and encouragement from two external reviewers.

Chapter 8

This work was partly funded by a grant provided to Jobson through the Australian Biological Resources Study (ABRS) National Taxonomy Research Grant Programme (RFL212-45), and to Baleeiro through CAPES (Ciências sem Fronteiras 0242-13-6) and FAPESP (2013/02729-8) scholarships.

Chapter 9

Rob Naczi thanks Aaron M. Ellison and Lubomír Adamec for their invitation to contribute to this volume and for their improvements to the submission. Andreas Fleischmann helped with botanical nomenclature. Charles C. Davis and T. Lawrence Mellichamp reviewed the manuscript and suggested helpful revisions. Elizabeth Kiernan produced the map, using the resources of the Geographic Information Systems Laboratory at New York Botanical Garden.

Chapter 11

Tanya Renner acknowledges support from US National Science Foundation (NSF), award DEB 15-56931. Victor A. Albert acknowledges support from NSF awards IOS 09-22742 and DEB 14-42190. Mitsuyasu Hasebe acknowledges support from KAKENHI grants from MEXT and JSPS, Japan. Kenji Fukushima acknowledges support from JSPS Postdoctoral Fellowships for Research Abroad.

Chapter 12

The authors thank W.W. Ho, J.N Kutz, J. Ng, and J. Riffell, B. Molano-Flores, J.M. Annis, S.B. Primer, J. Coons, and M.A. Feist for allowing them to include some of their unpublished results in this chapter. J. Moran and A. Jürgens made helpful comments on an earlier draft of the manuscript. John Horner thanks his students for their enthusiastic interest and work on *Sarracenia*.

Chapter 13

Bartosz Płachno thanks The Foundation for Polish Sciences and the Ministry of Science and Higher Education (Poland) for their long-term support of his studies of carnivorous plants. Sincere thanks are due to Lubomír Adamec, Ulrike Bauer, Aaron Ellison, Irene Lichtscheidl, and Simon Poppinga for comments and corrections.

Chapter 14

Simon Poppinga would like to thank Siegfried Hartmeyer, Elżbieta Król, Anna Westermeier, and Stephen Williams for critical comments on an early draft of this chapter. The contributions of both Poppinga and Speck to this chapter were supported by the German Research Foundation (DFG: Deutsche Forschungsgemeinschaft) as part of the Transregional Collaborative Research Centre (SFB/Transregio) 141 "Biological Design and Integrative Structures"/project A04.

Chapter 15

The authors thank Anja and Holger Hennern for their kind permission to include Figures 15.1g and 15.1h in this chapter, Walter Federle for providing the images used for Figures 15.2e and 15.2f, and Holger Bohn for providing the image used for Figure 15.3a.

Chapter 16

This work was supported by the Grant Agency of the Czech Republic [project 16-07366Y], by the Slovak Research and Development Agency under the contract No. APVV-15-0051, and grant LO1204 from the National Program of Sustainability I.

Chapter 17

This study was supported by the long-term research development project RVO 67985939, project 16-07366Y from Czech Science Foundation, and by grant LO1204 from the National Program of Sustainability I. The authors extend their sincere thanks to all who provided literature or comments that helped to improve the manuscript.

Chapter 18

We thank Lubomír Adamec, Nick Gotelli, and Jim Karagatzides for their many thoughtful comments on the manuscript; Aaron Ellison and Lubomír Adamec for their kind invitation to participate in this important volume; and Sarah Friedrich for assembling the figures. Grant IOS 15-57906 from the US National Science Foundation to Givnish and grant LO1204 from the National Program of Sustainability to Pavlovič helped support this work.

Chapter 19

This study was partly funded by a long-term research development project (RVO 67985939) of the Grant Agency of the Czech Republic. The author extends his sincere thanks to all who reviewed and improved the manuscript.

Chapter 21

The authors wish to thank Drs. Lubomír Adamec, Dave Armitage, and Bruno Di Giusto for their helpful and considered comments on the draft chapter.

Chapter 23

Leonora Bittleston thanks the editors Aaron Ellison and Lubomír Adamec for their support and assistance, Seth Donoughe, Andrej Pavlovič, and Charles Clarke for helpful comments on the manuscript, the many people who helped with collecting and processing pitcher plant samples, and Naomi Pierce and Anne Pringle for supportive advice and insight during her years studying pitcher-plant communities. Her research was supported by a US National Science Foundation (NSF) Graduate Research Pre-doctoral Fellowship, NSF Doctoral Dissertation Improvement award DEB 14-00982, and a Putnam Expedition Grant from the Museum of Comparative Zoology, Harvard University.

Chapter 24

This chapter includes work supported by the US National Science Foundation (NSF) awards DEB 00-83617, 00-91776, 05-19170, 07-16891, and 14-56425 to Miller and NSF awards IOS 08-39998, 12-55628, and DEB 09-17827 and 14-55506 (OPUS) to Bradshaw and Holzapfel. The authors thank J. Kneitel, S. Heard, and B. Baiser for comments on earlier versions of the chapter.

Chapter 25
This study was partly funded by the Czech Science Foundation grant no. P504/11/0783, the programme Projects of Large Infrastructure for Research, Development, and Innovations (no. LM2010005), and RVO 67985939. The authors are grateful to all reviewers of the manuscript for their critical comments and insights. Special thanks are due to Dr. Lubomír Adamec: most of the research described in this chapter could not have been done without his knowledge of *Utricularia*, his dedication, and his help.

Chapter 26
Jonathan Moran thanks Barrie Juniper for his friendship and support. Bruce Anderson thanks Dagmar Voigt for editorial help. In memory of Tillie Moran (September 1, 2006–August 4, 2016).

Chapter 27
Adam Cross would like to thank Professors Kingsley Dixon and Grant Wardell-Johnson (Curtin University), Professor Hans Lambers (University of Western Australia), and Dr. Ken Atkins (Department of Parks and Wildlife) for discussion of conservation issues in Western Australia and critical review of the chapter.

Chapter 28
The authors thank the following herbaria for providing data for this project: ALA, BKL, BUPL, FSU, KSC, MIN, MISS, NCU, NMC, OSC, UNM, USF, and WTU. Matthew Lisk provided computational support and David Moon helped with data assembly. Fitzpatrick's work was supported by the University of Maryland's Center for Environmental Science, and Ellison's was supported by the US National Science Foundation (NSF), award DEB 11-44056.

Taxonomic Index

Only FAMILIES, *genera*, *infrageneric taxa*, and *species* are indexed. Names ending in -ACEAE or -YCEAE refer to families of bacteria, algae, fungi, and plants, whereas names ending in -IDAE refer to families of animals; current junior synonyms and other invalid taxa are set in roman type. Families and genera set in **bold** type include, at least in part, carnivorous plants. All chapters (**c**), figures (**f**), plates (**p**), and tables (**t**) are indexed, but the complete list of currently recognized carnivorous plants (Appendix) is not indexed. Note that if an entire chapter (figure, plate, or table) covers a given subject (*e.g.*, Chapter 5 covers *Nepenthes*), the successively lower categories of entries in the larger entry are not also indexed (*i.e.*, figure, plate, table, subject within chapter; or subject within figure, plate, or table).

C

D

E

F

G

H

O

P

R

S

V

W

X

Z

Subject Index

All chapters (**c**), figures (**f**), plates (**p**), and tables (**t**) are indexed. Note that if an entire chapter (figure, plate, or table) covers a given subject (*e.g.*, Chapter 12 covers attraction of prey), the successively lower categories of entries are not also indexed (*i.e.*, figure, plate, table, subject within chapter; or subject within figure, plate, or table).